20.	$\sin kt$	$\dfrac{k}{s^2+k^2}$
21.	$\cos kt$	$\dfrac{s}{s^2+k^2}$
22.	$\sinh kt$	$\dfrac{k}{s^2-k^2}$
23.	$\cosh kt$	$\dfrac{s}{s^2-k^2}$
24.	$t^n e^{at}$	$\dfrac{n!}{(s-a)^{n+1}}$
25.	$e^{at}\sin kt$	$\dfrac{k}{(s-a)^2+k^2}$
26.	$e^{at}\cos kt$	$\dfrac{s-a}{(s-a)^2+k^2}$
27.	$t\sin kt$	$\dfrac{2ks}{(s^2+k^2)^2}$
28.	$t\cos kt$	$\dfrac{s^2-k^2}{(s^2+k^2)^2}$

Some additional formulas useful for inverting Laplace transforms:

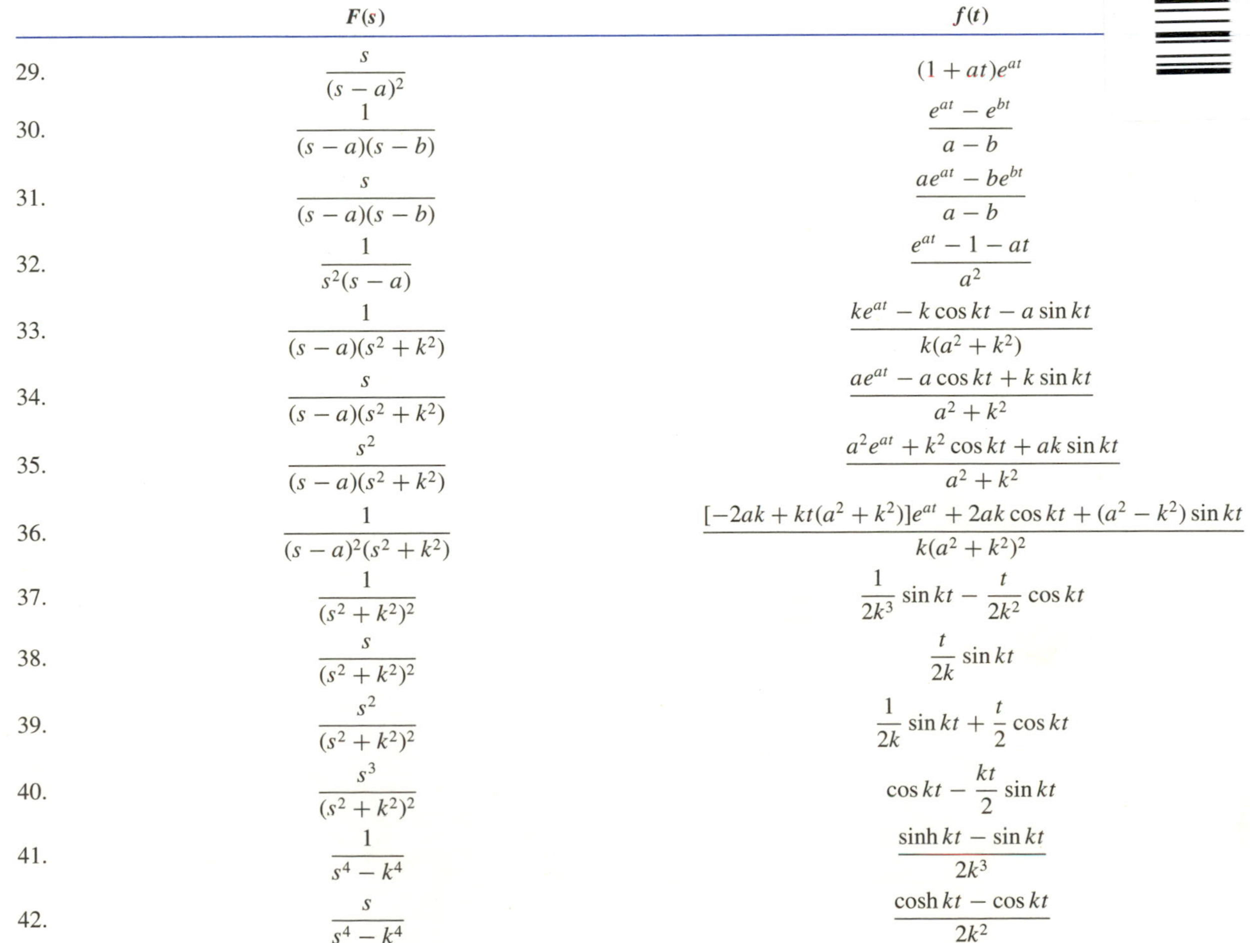

	$F(s)$	$f(t)$
29.	$\dfrac{s}{(s-a)^2}$	$(1+at)e^{at}$
30.	$\dfrac{1}{(s-a)(s-b)}$	$\dfrac{e^{at}-e^{bt}}{a-b}$
31.	$\dfrac{s}{(s-a)(s-b)}$	$\dfrac{ae^{at}-be^{bt}}{a-b}$
32.	$\dfrac{1}{s^2(s-a)}$	$\dfrac{e^{at}-1-at}{a^2}$
33.	$\dfrac{1}{(s-a)(s^2+k^2)}$	$\dfrac{ke^{at}-k\cos kt-a\sin kt}{k(a^2+k^2)}$
34.	$\dfrac{s}{(s-a)(s^2+k^2)}$	$\dfrac{ae^{at}-a\cos kt+k\sin kt}{a^2+k^2}$
35.	$\dfrac{s^2}{(s-a)(s^2+k^2)}$	$\dfrac{a^2e^{at}+k^2\cos kt+ak\sin kt}{a^2+k^2}$
36.	$\dfrac{1}{(s-a)^2(s^2+k^2)}$	$\dfrac{[-2ak+kt(a^2+k^2)]e^{at}+2ak\cos kt+(a^2-k^2)\sin kt}{k(a^2+k^2)^2}$
37.	$\dfrac{1}{(s^2+k^2)^2}$	$\dfrac{1}{2k^3}\sin kt-\dfrac{t}{2k^2}\cos kt$
38.	$\dfrac{s}{(s^2+k^2)^2}$	$\dfrac{t}{2k}\sin kt$
39.	$\dfrac{s^2}{(s^2+k^2)^2}$	$\dfrac{1}{2k}\sin kt+\dfrac{t}{2}\cos kt$
40.	$\dfrac{s^3}{(s^2+k^2)^2}$	$\cos kt-\dfrac{kt}{2}\sin kt$
41.	$\dfrac{1}{s^4-k^4}$	$\dfrac{\sinh kt-\sin kt}{2k^3}$
42.	$\dfrac{s}{s^4-k^4}$	$\dfrac{\cosh kt-\cos kt}{2k^2}$

DIFFERENTIAL EQUATIONS:
A MODELING APPROACH

DIFFERENTIAL EQUATIONS: A MODELING APPROACH

Glenn Ledder

Department of Mathematics

University of Nebraska–Lincoln

Boston Burr Ridge, IL Dubuque, IA Madison, WI New York San Francisco St. Louis
Bangkok Bogotá Caracas Kuala Lumpur Lisbon London Madrid Mexico City
Milan Montreal New Delhi Santiago Seoul Singapore Sydney Taipei Toronto

The McGraw·Hill Companies

DIFFERENTIAL EQUATIONS: A MODELING APPROACH

This book is printed on acid-free paper.

1 2 3 4 5 6 7 8 9 0 DOC/DOC 0 9 8 7 6 5 4

ISBN 0–07–242229–7

Publisher, Mathematics and Statistics: *William K. Barter*
Executive Editor: *Robert E. Ross*
Director of Development: *David Dietz*
Developmental Editor: *Amy R. Gembala*
Senior Marketing Manager: *Nancy Anselment*
Senior Project Manager: *Vicki Krug*
Senior Production Supervisor: *Sherry L. Kane*
Senior Media Project Manager: *Sandra M. Schnee*
Senior Designer: *David W. Hash*
Cover/Interior Designer: *Rokusek Design*
Supplement Producer: *Brenda A. Ernzen*
Compositor: *The GTS Companies/York, PA Campus*
Typeface: *10/12 Times Roman*
Printer: *R. R. Donnelley Crawfordsville, IN*

Library of Congress Cataloging-in-Publication Data

Ledder, Glenn.
Differential equations : a modeling approach / Glenn Ledder.—1st ed.
p. cm.
Includes index.
ISBN 0–07–242229–7 (acid-free paper)
1. Differential equations. 2. Mathematical models. I. Title.

QA371.L353 2005 2004049354
515′.35—dc22 CIP

www.mhhe.com

About the Author

Glenn Ledder

Glenn Ledder is Associate Professor of Mathematics at the University of Nebraska–Lincoln. His research encompasses mathematical modeling in a variety of settings, including combustion theory, groundwater flow, population dynamics, and physiological ecology, and has appeared in numerous journals in mechanical engineering, engineering science, geology, and biology, as well as applied mathematics. He has also published articles on undergraduate mathematics in the CODEE newsletter, *Mathematics Magazine,* and the *UMAP Journal.*

Glenn lives in Lincoln, Nebraska with his wife, Susan. They have two children, Louis and Becky. In his free time, Glenn enjoys running, canoeing, and folk music.

Contents

About the Author v
Preface xi
Acknowledgments xv
To the Student xvii

Chapter 1 Introduction 1

1.1 Natural Decay and Natural Growth 2
1.2 Differential Equations and Solutions 15
1.3 Mathematical Models and Mathematical Modeling 31
CASE STUDY 1 SCIENTIFIC DETECTION OF ART FORGERY 40

Chapter 2 Basic Concepts and Techniques 47

2.1 A Collection of Mathematical Models 47
2.2 Separable First-Order Equations 64
2.3 Slope Fields 76
2.4 Existence of Unique Solutions 85
2.5 Euler's Method 97
2.6 Runge–Kutta Methods 109
CASE STUDY 2 A SUCCESSFUL VOLLEYBALL SERVE 118

Chapter 3 Homogeneous Linear Equations 125

3.1 Linear Oscillators 126
3.2 Systems of Linear Algebraic Equations 138
3.3 Theory of Homogeneous Linear Equations 149
3.4 Homogeneous Equations with Constant Coefficients 163
3.5 Real Solutions from Complex Characteristic Values 170
3.6 Multiple Solutions for Repeated Characteristic Values 178
3.7 Some Other Homogeneous Linear Equations 188
CASE STUDY 3 HOW LONG SHOULD JELLYFISH HOLD THEIR FOOD? 200

Chapter 4 Nonhomogeneous Linear Equations 207

4.1 More on Linear Oscillator Models 208
4.2 General Solutions for Nonhomogeneous Equations 218

4.3 The Method of Undetermined Coefficients 227
4.4 Forced Linear Oscillators 240
4.5 Solving First-Order Linear Equations 253
4.6 Particular Solutions for Second-Order Equations by Variation of Parameters 265
CASE STUDY 4 A TUNING CIRCUIT FOR A RADIO 274

Chapter 5 Autonomous Equations and Systems 279

5.1 Population Models 279
5.2 The Phase Line 290
5.3 The Phase Plane 299
5.4 The Direction Field and Critical Points 313
5.5 Qualitative Analysis 319
CASE STUDY 5 A SELF-LIMITING POPULATION 329

Chapter 6 Analytical Methods for Systems 335

6.1 Compartment Models 336
6.2 Eigenvalues and Eigenspaces 345
6.3 Linear Trajectories 357
6.4 Homogeneous Systems with Real Eigenvalues 366
6.5 Homogeneous Systems with Complex Eigenvalues 378
6.6 Additional Solutions for Deficient Matrices 389
6.7 Qualitative Behavior of Nonlinear Systems 398
CASE STUDY 6 INVASION BY DISEASE 410

Chapter 7 The Laplace Transform 421

7.1 Piecewise-Continuous Functions 422
7.2 Definition and Properties of the Laplace Transform 430
7.3 Solution of Initial-Value Problems with the Laplace Transform 439
7.4 Piecewise-Continuous and Impulsive Forcing 448
7.5 Convolution and the Impulse Response Function 460
CASE STUDY 7 GROWTH OF A STRUCTURED POPULATION 469

Chapter 8 Vibrating Strings: A Focused Introduction to Partial Differential Equations 477

8.1 Transverse Vibration of a String 478
8.2 The General Solution of the Wave Equation 490
8.3 Vibration Modes of a Finite String 500

8.4 Motion of a Plucked String 508
8.5 Fourier Series 520
CASE STUDY 8 STRINGED INSTRUMENTS AND PERCUSSION 530

Chapter A Some Additional Topics 541

A.1 Using Integrating Factors to Solve First-Order Linear Equations 542
A.2 Proof of the Existence and Uniqueness Theorem for First-Order Equations 548
A.3 Error in Numerical Methods 559
A.4 Power Series Solutions 570
A.5 Matrix Functions 583
A.6 Nonhomogeneous Linear Systems 592
A.7 The One-Dimensional Heat Equation 603
A.8 Laplace's Equation 613

Answers to Odd-Numbered Problems 625
Index 661

Preface

Why Modeling?

Differential equations occupy a prominent place in the mathematics curriculum because they are central to many topics in science and engineering. Indeed, many core scientific principles can be efficiently stated in terms of differential equations. Theoretical work in science and design work in engineering are often done by *mathematical modeling,* which we can think of as the art of using mathematical models to discover scientific principles or to predict the behavior of a real-world system. The intimate connection between differential equations and mathematical models makes the differential equations course a natural place to study mathematical modeling. A modeling approach is of value not only to students who will pursue advanced studies in science and engineering, but also to anyone who merely wants to better understand the world around us.

In *Differential Equations: A Modeling Approach,* my goal is to teach the reader how to derive models, test models, find useful questions to ask about models, and identify limitations of models within the context of the study of differential equations. The modeling theme is emphasized in some optional parts of the text, and it is also integrated to some extent into the presentation of the mathematical theory and techniques. The text gives the reader a sense of the wide variety of topics that can be explored with mathematical models, including topics from physics, biology, and even some disciplines, such as history, that are not normally associated with mathematics. Many models are drawn from everyday experience, such as the cooling of a cup of coffee. Through this exploration of modeling, I hope to give readers a working knowledge of how to use the powerful problem-solving capabilities of differential equations in a variety of settings.

A Student-Centered Approach

In writing this text, I have made use of recent research on how people learn. In particular, I have incorporated the principle that learning is best facilitated by an approach that connects new and familiar concepts and that engages students in a dialogue with the material. Every chapter of this text opens with a section that treats a simple example of a new topic in a familiar way. Each section contains a *Model Problem* that examines a specific concept in depth prior to the discussion of the general case. Students are encouraged to test their reading comprehension at key points in most sections by working *Instant Exercises* before they continue. Examples that draw sharp distinctions between concepts are employed throughout the text to clarify new ideas. Exercises that involve comparing alternative methods for working a problem, or isolating the effect of a particular parameter on the behavior of a model, are included to encourage higher-order thinking.

Features

Differential Equations: A Modeling Approach employs a carefully developed set of pedagogical aids to engage the reader's interest and enhance his or her understanding of differential equations.

Case Studies The Case Studies are interesting in-depth explorations of mathematical modeling drawn from such disparate disciplines as art history, ecology, and physics. Some of them also focus on specific aspects of mathematical modeling, such as the derivation of differential equations from physical models and the effect of a parameter on the model behavior. The Case Study exercise sets provide an opportunity for additional practice in mathematical modeling. Solutions to these exercises are located in the *Instructor's Solutions Manual.*

Model Problems Each section of the text is organized around an interesting Model Problem that focuses attention on a specific example. These problems are presented early in the section so that the theory or method of the section arises from the study of the problem, encouraging the reader to learn by discovery.

Examples Worked examples from a wide variety of disciplines are employed throughout the text to illustrate key mathematical concepts. Each example is examined in sufficient depth to build connections with previous and subsequent material.

Instant Exercises These exercises appear at key points in most sections and ask the reader to check that he or she understands an important point before proceeding. Many Instant Exercises ask the reader either to do a quick calculation similar to one done in a previous example or to fill in details omitted from previous calculations. Complete solutions to the Instant Exercises appear after the exercise set for that section.

Exercise Sets Each section's exercise set has a balance of routine and challenging problems. Many of the exercise sets contain problems that are suitably challenging to be used as laboratory investigations or group projects.

Technology Usage This text supports the judicious use of graphing calculators and computer algebra systems, where appropriate, as helpful problem-solving tools. Exercises that can most easily be solved using a technological aid are marked with a T icon. For multipart exercises, the icon indicates that at least one part of the exercise is facilitated by a technological tool.

A Contemporary Organization

This text introduces a number of innovations in topical coverage and sequencing.

First-Order Linear Equations I have found that even the best students have difficulty retaining the integrating factor method, which is generally the only method presented for the solution of first-order linear equations. This text's solution is to bundle first-order linear equations with higher-order equations, in Chapters 3 and 4, rather than with basic introductory methods. First-order linear equations are then solved by the method of undetermined coefficients or variation of parameters. Instructors who prefer the integrating factor method will find it in Section A.1 and can incorporate it into Chapter 2.

Methods for Homogeneous Systems Utilizing computer algebra systems enables the study of systems via the phase plane either before or after the analytical treatment of homogeneous linear systems. I have chosen to present phase plane methods first. It is then possible to use the appearance of linear trajectories in a phase portrait for certain linear systems to motivate the analytical treatment of homogeneous linear systems. In Chapter 5, the concept of the phase plane is explored first through equations for which the trajectories can be determined analytically, and then for equations or systems for which trajectories have to be determined numerically.

I have also chosen to present qualitative analysis via nullclines in the same chapter as the introduction to phase portraits. My experience teaching mathematical biology has shown me that nullclines are not difficult for average students to understand. Nullclines have the virtue of providing insight into the behavior of linear and nonlinear systems without requiring any calculations.

Linear Algebra Because linear algebra is a prerequisite for the differential equations course at some colleges but not at others, I introduce linear algebra primarily through self-contained sections. These appear just before the material is needed. Section 3.2 introduces the theory and methods for the equation $\mathbf{Ax} = \mathbf{b}$, where $\mathbf{A}$ is a nonsingular square matrix. This material is subsequently used to determine general solutions for higher-order equations. Section 6.2 contains the theory and methods for the eigenvalue problem $\mathbf{Ax} = \lambda\mathbf{x}$, beginning with an introduction to vector spaces and a study of the nullspace of a singular square matrix. This section is followed by the theory and methods for linear systems with constant coefficients. Section 6.6 presents generalized eigenvectors in connection with the solution of systems of the form $\mathbf{x}' = \mathbf{Ax}$, where $\mathbf{A}$ is a deficient matrix.

The Laplace Transform Chapter 7, on Laplace transform methods, emphasizes problems with discontinuous or impulsive forcing, as the Laplace transform is clearly the method of choice for these problems. Section 7.5 develops the convolution theorem and its use as an alternative to the method of variation of parameters.

Partial Differential Equations The presentation in Chapter 8 focuses on the wave equation to make it easier for instructors to introduce partial differential equations in a limited amount of time. The wave equation is treated first in the context of wave motion on an infinite string and then as a setting in which to develop the method of separation of variables. Instructors may cover the optional sections (A.7 and A.8) on the heat and Laplace's equations for a more complete treatment of partial differential equations.

Options for Course Coverage

This text focuses on the essential topics for an elementary course in differential equations. However, to give instructors maximum flexibility in structuring their courses, Chapter A comprises a collection of optional topics. These topics may be used to customize the course and/or to create a more advanced course.

The heart of the differential equations course consists of the material in Chapters 1 through 6, exclusive of Sections 1.3, 2.6, 3.7, 4.4, 5.5, and 6.7. There are a number of ways to structure a course by adding material to this core.

- A course for an audience composed primarily of engineering students could cover Chapters 1 through 7 and add one or two Case Studies or partial differential equations (Chapter 8, Section A.7, and Section A.8), as time permits.
- A course for an audience composed primarily of math students could include Chapters 1 through 6, exclusive of Sections 1.3 and 3.7, along with as many of the topics from Chapter A as time permits.
- A course for math and science students that emphasizes mathematical modeling could use Chapters 1 through 6, including the Case Studies.

The accompanying schematic diagram shows the relationships among the chapters, with chapters at the tail of each arrow prerequisite for those at the head. It is not necessary to study the chapters in linear order, as long as the prerequisites indicated below are observed. Chapter 3 material is referenced only slightly in Chapter 5, and only a small amount of material from Chapter 4 is needed for Chapter 7.

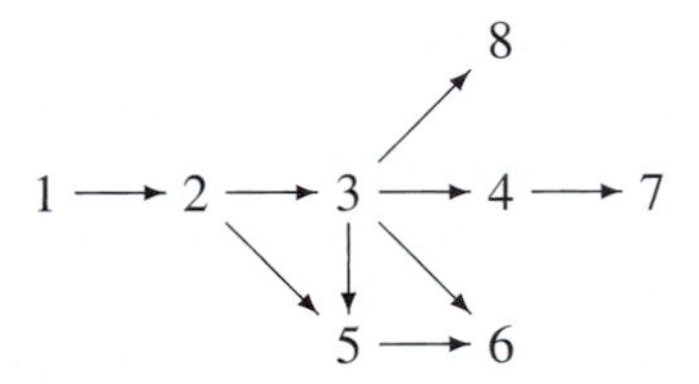

Supplements

Instructor's Solutions Manual This invaluable, time-saving resource contains complete worked-out solutions to all the exercises in the text, including the Case Study exercises, as well as tips for teaching from the text.

Student Solutions Manual Available on the Online Learning Center, this manual provides students with complete worked-out solutions to all the odd-numbered problems in the text.

Online Learning Center Online resources for the text may be accessed at *www.mhhe.com/ledder*.

Technology Guides for Maple and Mathematica These guides, available on the Online Learning Center, are available free to adopters and contain laboratory projects and worksheets for further practice in working with these computational tools.

Acknowledgments

No text is strictly the creation of its author. I have been dependent from start to finish on the professional staff at McGraw-Hill. My sincere thanks are due to Robert Ross, Executive Editor, for supporting the text through the entire process, and to Vicki Krug, Senior Project Manager, for skillfully managing the production.

The first draft of my manuscript bore only a cursory resemblance to the current version. The process of turning a rough draft into a finished text was ably directed by Amy Gembala, my Developmental Editor, who patiently endured endless revisions and wholesale rewrites, helped with content decisions but also gave me free rein when I wanted it, pushed me just hard enough to get the job done, and assembled an excellent panel of reviewers.

The following individuals provided invaluable feedback and guidance on the manuscript:

Deborah Brandon, *Carnegie Mellon University*
Jeff Dodd, *Jacksonville State University*
Paul DuChateau, *Colorado State University*
Richard Elderkin, *Pomona College*
David Ellis, *San Francisco State University*
John Fornaess, *University of Michigan–Ann Arbor*
Moses Glasner, *Pennsylvania State University*
Arek Goetz, *San Francisco State University*
David Gurarie, *Case Western Reserve University*
Tracy Dawn Hamilton, *California State University–Sacramento*
Donald Hartig, *California Polytechnic State University–San Luis Obispo*
James Heitsch, *University of Illinois–Chicago*
Michael Kirby, *Colorado State University*
Yang Kuang, *Arizona State University*
Alexandra Kurepa, *North Carolina A&T State University*
Irena Lasiecka, *University of Virginia*
Steven Leon, *University of Massachusetts–Dartmouth*
George Majda, *Ohio State University*
Douglas B. Meade, *University of South Carolina*
Piotr Mikusinski, *University of Central Florida*
John Neuberger, *Northern Arizona University*
V. W. Noonburg, *University of Hartford*
Jacek Polewczak, *California State University–Northridge*
James A. Reneke, *Clemson University*

Stephen Schecter, *North Carolina State University*
Brian Seymour, *University of British Columbia*
Christopher D. Sogge, *Johns Hopkins University*
Michael Stecher, *Texas A&M University*
Mark Sussman, *Florida State University*
Craig Tracy, *University of California–Davis*
Karl A. Voss, *Bucknell University*
Rick Ye, *University of California–Santa Barbara*
Dave Zachmann, *Colorado State University*
Steve Zelditch, *Johns Hopkins University*

The organization of the material on population models is largely based on a lecture I heard Rick Elderkin give. I am also grateful to Rick for his thoughtful comments on the derivation and presentation of various mathematical models and for his insights on the Wronskian and linear independence. The text also bears the influence of Donald Hartig, who convincingly pressed for greater use of linear algebra and more theory. The idea for Section 1.1 came from John Neuberger, who also deserves a special commendation for helping me improve the clarity of definitions, theorems, and overall wording. Bill Wolesensky helped prepare the answer key.

Finally, I want to thank my wife, Susan, and my children, Louis and Becky, for their support and for putting up with the stress that my project added to their lives. Susan made an important tangible contribution to the text as well: the cover design is based on her artwork.

Glenn Ledder

To the Student

Many textbooks are written as though they were meant for professional mathematicians rather than students. Mathematicians' preferred style of "definition, theorem, proof, example" is ideal for reference books, but it does not necessarily serve students' learning needs. This book was written with your needs in mind. As I wrote, I maintained a mental image of myself telling a story in which you are the main character. The action of the story is driven by *Model Problems* in each section that challenge your understanding or technical skill. The plot of the story unfolds with the gradual development of the concepts and skills you need to understand and solve these problems. The story's resolution consists of discussions that generalize the results of Model Problem investigations and solidify your understanding of key concepts. You will get the greatest benefit from your time if you play your role as actively as possible.

I recommend keeping a pencil and scratch paper handy as you read, so that you can work through details that were omitted or are hard to grasp. In particular, note the *Instant Exercises* that appear at key points in most sections. Some Instant Exercises ask you to fill in details omitted from worked examples. Others ask you to tackle problems virtually identical to preceding examples. You should solve these exercises to make sure you are ready to read the rest of the section. Solutions to Instant Exercises appear after the section exercise sets. Use these to check your solutions only after you think you have done the exercises correctly.

You will notice that this book places a great deal of emphasis on the *uses* of differential equations, particularly for modeling a variety of natural and physical processes. *Mathematical modeling,* as presented in this book, is much more than just "word problems": it is the process of developing, testing, and refining models. The models you will encounter here represent many different subject areas. My hope is that after reading this book, you will understand not only how to solve problems in differential equations, but also how to apply your knowledge of differential equations to areas outside of mathematics.

DIFFERENTIAL EQUATIONS:
A MODELING APPROACH

chapter

1

Introduction

This book is about two interrelated subjects: differential equations and mathematical modeling. You may have some experience with differential equations from your calculus course. You probably have some experience with mathematical modeling from courses leading all the way back to pre-algebra. The term *differential equation* is unambiguous: if you already know what it means, then you have a head start. However, the term *mathematical modeling* does not have a precise definition that is agreed upon by all mathematicians. You may have a clear idea of what mathematical modeling is, but your idea may not be the same as that used in this book. It is important that we have a shared understanding of the term right from the beginning, and this is part of the purpose of Chapter 1.

The chapter begins in Section 1.1 with a treatment of the phenomena of natural decay, broadly construed to include decay to a nonzero value. Examples include the natural cooling of a warm object in a cool environment, radioactive decay, and depreciation of the value of equipment owned by a business.

Section 1.2 presents the basic vocabulary of differential equations and solutions of differential equations. A mathematical model for the evaporation of a raindrop provides a context in which to discuss the relationship between mathematical models and differential equations and the key concept of uniqueness of solutions.

Section 1.3 deals with mathematical modeling as a subject in its own right, using as an example a simple investigation of a model to predict how high a person can throw a ball. The *model* is flawed, although the *differential equation* in the model is correct; one of the exercises guides the student through the development and analysis of a better model.

The chapter concludes with a careful examination in Case Study 1 of the celebrated Van Meegeren art forgery case, an example of the use of mathematical modeling to date a painting by dating the paint used in it. The treatment of the model differs from previous treatments by emphasizing the practical difficulty of trying to draw useful conclusions in the face of limited data.

Throughout the chapter, emphasis is given to the notion that mathematical modeling is not an exact science. Rare is the real-world situation that is simple enough to be "translated" into a mathematical model. It is far more common for mathematical models to be derived from an idealized oversimplification of a real-world problem. Results that run counter to one's intuition indicate that careful observation and experimentation are needed to verify the results of the model. Occasionally, one's intuition needs to be changed, but it is more common that the mathematical model is what needs to be changed. Surprising conclusions drawn from mathematical models should not be taken too seriously without first looking for flaws in the models.

1.1 Natural Decay and Natural Growth

Many processes in nature are characterized by the gradual approach of a quantity toward some final value. Perhaps the most common example is that of a cooling process. Pour a cup of hot coffee and leave it alone in a cooler environment. The coffee cools because it is at a temperature hotter than that of its surroundings. The excess heat disappears naturally into the environment, without help from any agent or machine. The cooling process can be studied mathematically. We consider a specific cooling problem.

MODEL PROBLEM 1.1

A cup of coffee has a temperature of 190°F when it is poured into a metal cup. The coffee is continually stirred with a plastic spoon and reaches a temperature of 150°F after 3 min. How much time will it take for the coffee to reach a temperature of 110°F?

We are going to have to postpone consideration of this question until we have developed some relevant ideas. In particular, we need a good understanding of rates of change.

Relative Rates of Change

Suppose you are watching the evening news reports and you hear that the people of a particular country eat about 4 million hamburgers each day. Is that a lot or a little? The question can only be answered in context. It all depends on how many people live in the country. If the country has a population of 1 million people, then 4 million hamburgers each day means 4 hamburgers per person. But if the population of the country is 100 million, then 4 million hamburgers in a day means only 0.04 hamburgers per person. Over the course of a year, this would amount to about 15 per person. In the context of the hamburger consumption data, 4 million hamburgers per day is an **absolute** quantity, because the figure is for a whole group. The rates of 4 hamburgers per day *per person* and 15 hamburgers per year *per person* are **relative** quantities.

Similarly, suppose the population of a certain region is growing at a rate of 1000 people per year. If there are currently only 1000 people, then a growth rate of 1000 people per year represents an increase of 100% in just 1 year. However, if the current population is 1 million people, then a growth rate of 1000 people per year represents a yearly increase of only 0.1%. In general, a relative quantity is obtained as a ratio of the corresponding absolute quantity to the current amount.

The **absolute rate of change** of a quantity y is the derivative dy/dt. The **relative rate of change** of a quantity y is the ratio of the absolute rate of change to the quantity y:

$$\text{Relative rate of change} = \frac{\text{absolute rate of change}}{\text{amount}} = \frac{1}{y}\frac{dy}{dt}.$$

The population of 1000 people has a relative rate of change of 1 yr^{-1} (100% per year), while the population of 1 million people with the same absolute rate of change has a relative rate of change of only 0.001 yr^{-1}. Note that the units of dy/dt are units of y per time; hence, the units of the relative rate of change are in inverse time. The relative rate of change of a given quantity does not depend on the unit of measurement chosen for the quantity, but it does depend on the unit of measurement chosen for the time.

Suppose x and y are decreasing quantities with unit initial value. The quantity x decreases with a constant negative absolute rate of change of 0.15 units of x per minute, while y decreases with a constant negative relative rate of change of 0.2 per minute. While x decreases by a fixed amount in each time unit, y loses a fixed percentage of its value (20%) during each time unit. Table 1.1.1 compares numerical values for the two functions; the graphs of x and y are compared in Figure 1.1.1.

t	0	1	2	3	4	5
x	1.000	0.850	0.700	0.555	0.400	0.250
y	1.000	0.800	0.640	0.512	0.410	0.328

Table 1.1.1 A quantity (x) with constant absolute rate of change and a quantity (y) with constant relative rate of change.

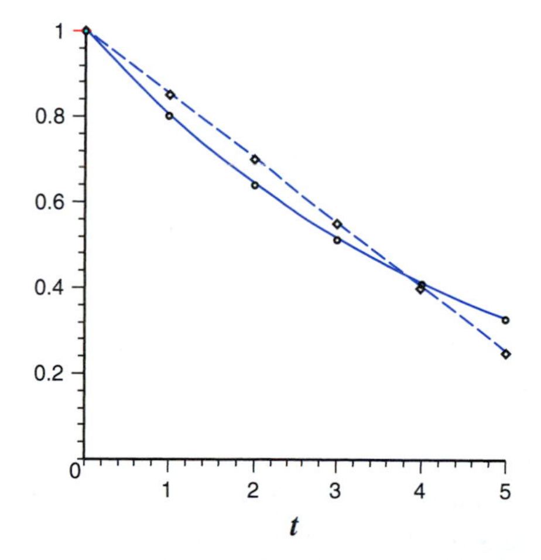

Figure 1.1.1
A quantity (x) with constant absolute rate of change (dashed) and a quantity (y) with constant relative rate of change (solid).

EXAMPLE 1

The height of a thrown ball is given in feet by the function $y = 4 + 20t - 16t^2$, where t is time in seconds. The derivative is $dy/dt = 20 - 32t$. Thus, the absolute rate of change of height at time 0 is 20 ft/s. The relative rate of change of height at time 0 is 5 s^{-1}. After 1 s, the absolute rate of change is -12 ft/s and the relative rate of change is -1.5 s^{-1}.

✦ INSTANT EXERCISE 1

Compute the relative rate of change of the function $y = 10 - \sin 2t$ at time $t = 0$.

The Natural Decay Equation

The word *decay* in standard English means "gradual disappearance." It is therefore natural to use the term in mathematics to describe any process whereby a quantity gradually increases or decreases to zero. In particular, a **natural decay** process is one that causes a quantity to have a constant negative relative rate of growth. It must be emphasized that what we are defining as natural decay is in fact an idealized concept. In practice, we should think of the resulting equation as a mathematical model, and we should always be prepared to question whether or not the model is a good one for the problem at hand.

By definition, a quantity y that undergoes natural decay satisfies the equation

$$\frac{1}{y}\frac{dy}{dt} = -k,$$

for some $k > 0$.[1] Algebraic manipulation of this equation yields the *differential equation*[2] of natural decay

$$\frac{dy}{dt} = -ky, \qquad k > 0. \tag{1}$$

Our task now is to determine what function or functions have the property of a constant negative relative rate of change. As an example, consider the differential equation

$$\frac{dy}{dt} = -y.$$

We can find solutions simply by interpreting the equation. We are looking for a function having the property that its derivative is the negative of the function. The function defined by $y(t) = e^{-t}$ has the right property, and so does any multiple of e^{-t}. The formula

$$y(t) = Ae^{-t}$$

defines a **solution** of the differential equation, for any value of A, because y so defined makes the equation true.

The differential equation can also be solved formally, with the same result.

EXAMPLE 2

Assume for the moment that the differential equation $y' = -y$ (using the prime symbol $'$ to indicate the derivative with respect to time) defines a one-to-one function $y(t)$. Such functions have an inverse, which

---

[1]We could omit the minus sign and think of the constant k as being negative; however, it is generally helpful to construct mathematical models in a way that makes the physical parameters positive rather than negative.

[2]The term *differential equation* is formally defined in Section 1.2.

means that we can interpret the differential equation as an equation for t in terms of y:

$$\frac{dt}{dy} = -\frac{1}{y}.$$

This differential equation is a simple calculus problem. Integrating with respect to y yields

$$t = -\ln|y| + C,$$

or

$$\ln|y| = C - t.$$

Raising both sides of the equation as exponents with base e yields

$$|y| = e^{C-t} = e^C e^{-t},$$

or

$$y = \pm e^C e^{-t}.$$

Note that the quantity $\pm e^C$ is an arbitrary constant. For convenience, we introduce a new arbitrary constant A by $A = \pm e^C$. Thus,

$$y = Ae^{-t}.$$

We emphasize that a *solution* is a function that makes the equation true, not a function that is found from the equation by some calculations. Our calculations have found the formula $y = Ae^{-t}$, suggesting that Ae^{-t} is a solution of the differential equation; the real test is to see if it makes the equation true. We have

$$y' = -Ae^{-t} = -y;$$

thus, $y = Ae^{-t}$ is a solution of $y' = -y$ for any value of A.

The Half-Life

Solutions of differential equations can be used to solve problems that include the differential equation and additional information. In particular, an **initial-value problem** for the decay equation consists of the differential equation and a condition indicating the value of y at some particular time (usually, but not necessarily, $t = 0$).

EXAMPLE 3

A quantity undergoes natural decay with relative growth rate -1. Initially there are 2 units of the quantity. At what time will there remain 1 unit of the quantity?

The decay process described here is an initial-value problem:

$$\frac{dy}{dt} = -y, \qquad y(0) = 2.$$

The differential equation has solutions

$$y = Ae^{-t};$$

of these, only

$$y = 2e^{-t}$$

satisfies the initial condition $y(0) = 2$. We want to know when $y = 1$, so we substitute this quantity and obtain

$$2e^{-t} = 1.$$

This can be rewritten as

$$e^t = 2.$$

Applying the natural logarithm to both sides gives

$$t = \ln 2,$$

or

$$t = \ln 2 \approx 0.693.$$

✦ INSTANT EXERCISE 2

Suppose initially we have 1 unit of the quantity in Example 3 and we want to know when there will be 1/2 unit. Show that the method of Example 3 results in the same equation $e^t = 2$; hence, the amount of time is again $t = \ln 2$.

Example 3 and Instant Exercise 2 illustrate a key property of natural decay: when the relative decay rate of a quantity is constant, so is the time required for one-half of the quantity to disappear. (Recall that constant relative rate of decay means that the function loses the same fraction of its value in each equal time increment.) The **half-life** t_h of a naturally decaying quantity is defined to be the amount of time required for the quantity to be reduced to one-half of its initial value.

Consider an initial quantity Q of some substance that decays with relative rate constant k. The quantity satisfies the initial-value problem

$$\frac{dy}{dt} = -ky, \qquad y(0) = Q.$$

Following the same plan as in Example 3, we begin with the solution

$$y = Ae^{-kt}$$

for the differential equation, with A arbitrary. Of these solutions, only

$$y = Qe^{-kt}$$

also satisfies the condition $y(0) = Q$. The half-life is the time t_h that corresponds to the quantity $y = Q/2$, so

$$e^{-kt_h} = \tfrac{1}{2},$$

or

$$e^{kt_h} = 2.$$

Thus,

$$kt_h = \ln 2. \tag{2}$$

Given either the relative decay rate constant or the half-life, we can easily calculate the other from the fact that their product is ln 2.

Newton's Law of Cooling

Among Isaac Newton's many contributions to science is the observation that a body at uniform temperature cools at a rate proportional to the difference between its temperature and that of its surroundings. Let $T(t)$ be the temperature of the object, and let S be the temperature of the surroundings. A direct translation of Newton's law of cooling into a differential equation follows from the interpretation of $-dT/dt$ as the rate at which the body cools. We obtain

$$-\frac{dT}{dt} = k(T - S), \tag{3}$$

where the proportionality constant k is positive.[3]

We can also interpret Newton's law of cooling as a decay process. Physically, it is clear that if the temperatures of the object and of the environment are the same, there will be no further change in the temperature of the object. In mathematical terms, the constant function defined by $T(t) \equiv S$ is a solution. A constant solution is called an *equilibrium solution.*

✦ INSTANT EXERCISE 3

Verify that $T(t) \equiv S$ is an equilibrium solution of Equation (3).

The Celsius temperature scale was devised to have 0°C be the freezing point of water and 100°C be the boiling point of water. A Celsius degree represents 1% of the temperature difference between the freezing and boiling points of water. Suppose we retain the use of Celsius degrees without retaining the definition of zero temperature as the freezing point of water. In the context of Newton's law of cooling, there is an advantage in measuring temperature as the number of Celsius degrees away from the equilibrium temperature S.[4] Mathematically, all we have to do is to define an adjusted temperature to be the amount by which the actual temperature of the object exceeds that of its environment:

$$y(t) = T(t) - S. \tag{4}$$

Note that the adjusted temperature y is positive for a cooling process (T is decreasing toward S) and negative for a heating process (T is increasing toward S), but the distinction does not affect the solution technique. Since the temperature S is constant, the rate of change of y is the same as the rate of change of T. That is,

$$\frac{dy}{dt} = \frac{dT}{dt} - \frac{dS}{dt} = \frac{dT}{dt}.$$

Substituting dy/dt for dT/dt and y for $T - S$ in Equation (3) yields the natural decay equation (1).

---

[3]Newton's law of cooling is based on observation, and observation is never 100% accurate. Equation (3) is not a description of the cooling process, but rather a mathematical model of the cooling process. It may be quite accurate in some circumstances and less accurate in others. This issue is explored in Exercises 1 through 3.

[4]The use of Celsius degrees here is as an example. Of course a similar argument can be made if temperature is to be measured using degrees Fahrenheit.

It makes sense to generalize the term *decay* to allow for a nonzero final value. In the case of Newton's law of cooling, the excess temperature y decays to zero, and the actual temperature T decays to S. It is helpful to think of Newton's law of cooling using the language of natural decay:

> Suppose a body at uniform temperature is placed in a uniform unchanging environment. The temperature of the body decays naturally toward that of the environment.

Solving the Model Problem

Conceptualizing the cooling of the coffee as a process by which the temperature decays naturally toward the environment temperature allows us to solve Model Problem 1.1, using what we already know about decay processes. Let T be the actual temperature of the coffee. We were not given the environment temperature in Model Problem 1.1,[5] so we assume it to be 70°F. Let y be the excess temperature, defined by

$$y = T - 70.$$

The excess temperature satisfies the natural decay equation, so we have the differential equation

$$\frac{dy}{dt} = -ky.$$

As before, the solution is $y = Ae^{-kt}$, where A is any constant. To complete the solution, we have to determine the value of A, and we also have to find the correct value of k, since this value was not given.

The excess temperature is initially $190 - 70 = 120$; we must choose $A = 120$ so that $y(0) = 120$. We do not yet know the relative rate of decay; nor do we know the half-life of the decay. The solution of the problem, in terms of the unknown rate constant k, is

$$y = 120e^{-kt}.$$

Figure 1.1.2 shows several possible solutions, each with a different value of k.

✦ INSTANT EXERCISE 4

Explain how we know that the top curve in Figure 1.1.2 is the one for $k = 0.1$.

There is additional information in Model Problem 1.1 that we have not yet used. The temperature must be 150°F at time $t = 3$. Thus, the excess temperature must satisfy $y(3) = 80$ as well as the initial condition $y(0) = 120$ and the differential equation. These two points appear along with the solution curves in Figure 1.1.2. The correct value of k is clearly between $k = 0.1$ and $k = 0.15$ because the point $(3, 80)$ lies between the curves corresponding to these values of k. We can find the exact value of k by substituting the point $(t, y) = (3, 80)$ into the solution formula:

$$80 = 120e^{-3k}.$$

[5] Real mathematical modeling investigations seldom begin with all the necessary data. The complete statement of the problem often develops during, rather than before, the investigation.

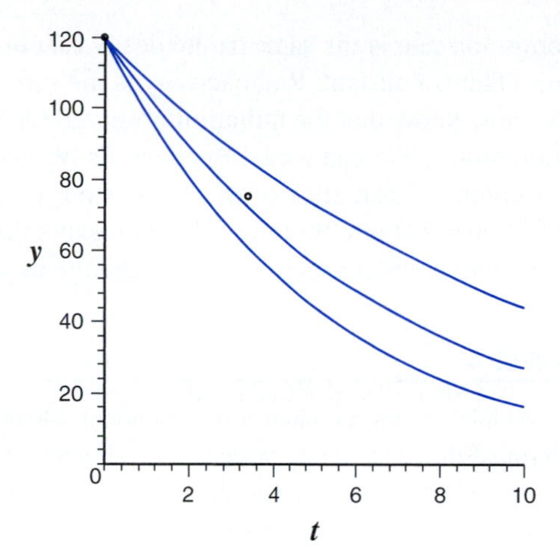

Figure 1.1.2
The functions $y = 120e^{-kt}$ with $k = 0.1, 0.15, 0.2$, and the data points (0, 120) and (3, 80).

This can be rewritten as

$$e^{-3k} = \tfrac{2}{3},$$

or

$$e^{3k} = 1.5.$$

Thus,

$$k = \frac{\ln 1.5}{3} \approx 0.135.$$

To find out when the temperature is 110°F, we must find the time corresponding to $y = 40$. Assume $y(t^*) = 40$. Substituting these data into the solution formula gives us

$$40 = 120e^{-kt^*}.$$

Thus, $kt^* = \ln 3$, and so

$$t^* = \frac{3 \ln 3}{\ln 1.5} \approx 8.1\,.$$

This is a reasonable answer. In exponential decay, equal changes take longer to be achieved as the quantity approaches the equilibrium value. The first drop of 40° takes only 3 minutes, while the second 40° drop takes about 8 minutes.

Radioactive Decay

Radioactivity was first observed by the French physicist Henri Becquerel in 1896. Becquerel observed that uranium salts emitted radiation that penetrated the black paper coverings used to protect photographic plates, thereby exposing the photographic emulsion. The rate at which this radiation was emitted seemed to depend in a simple way on the amount of uranium; specifically, it was proportional to the amount of uranium present. Another way to view Becquerel's observation is to note that the rate of radiation emission relative to the amount of uranium present is constant.

The emission rate is the same as the decay rate of the uranium, so it is the relative decay rate of uranium that is constant. Radioactivity is the standard example of natural decay.

We now know that the radiation observed by Becquerel is produced by radioactive decay of uranium atoms. We can recast Becquerel's observation in terms of atom decays measured by a Geiger counter. Each click of the counter corresponds to some fixed number of uranium atoms that have changed into thorium. If y represents the amount of uranium at time t, then each click of the Geiger counter represents some change in y.

EXAMPLE 4

A 1-g sample of pure uranium produces about 740,000 decays each minute. We can use these data to determine the relative rate of decay for uranium. Nearly a million decays in a minute may sound like a lot, but the units are misleading. The decay rate is given in atoms per minute, but the amount is given in grams. There are about 2.53×10^{21} atoms in 1 g of uranium. If we divide the decay rate in atoms per minute by the number of atoms, we get $k = 2.9 \times 10^{-16}\,\text{min}^{-1}$. We would get the same result if we had instead converted the decay rate to grams per minute. Relative rates of decay are always reported with units of inverse time; it doesn't matter whether quantities are reported as atoms or grams. The decay rate of uranium is usually given in inverse years rather than inverse minutes, and the result is $k = 1.537 \times 10^{-10}\,\text{yr}^{-1}$. This is a very slow rate of decay, but it is typical for naturally occurring radioactive materials. Any radioactive materials with a much faster decay rate (and present at the time of the earth's formation) have all disappeared.

✦ INSTANT EXERCISE 5

Determine the half-life of uranium, given the relative decay rate $k = 1.537 \times 10^{-10}\,\text{yr}^{-1}$ of uranium.

Investment and Depreciation

In economics, the term *capital asset* is used to denote the buildings, equipment, and other long-term assets of a company; the value of such assets is the company's *capital.* The capital decreases with time as the capital assets age and wear out. The term *depreciation* denotes methods of estimating the current value of an aging capital asset. Let K be the total capital of a company or other administrative unit. Assume that depreciation lowers the estimated capital at a constant relative rate D. Then in the absence of other factors, the capital satisfies the differential equation

$$\frac{1}{K}\frac{dK}{dt} = -D.$$

This is the decay equation (with different variables), and we know from our earlier analysis that solutions decay to zero. To prevent the loss of capital, money must be invested in capital improvements. Suppose investment is made at a constant absolute rate I. The capital is changed by the combination of depreciation and investment, leading to the differential equation

$$\frac{dK}{dt} = I - DK. \tag{5}$$

This equation is almost the same as the equation for Newton's law of cooling. It is solved in Exercise 12 by converting it to the decay equation.

Natural Growth

Natural growth is the name used to denote the equation

$$\frac{1}{y}\frac{dy}{dt} = k, \tag{6}$$

which follows from the mathematical assumption that a quantity has a constant positive relative rate of change. The absence of a minus sign in this equation, compared with the natural decay equation, does not affect the calculations; thus, the solution of the natural growth equation is

$$y = Ae^{kt}.$$

Of course the absence of the minus sign affects the properties of the solution considerably.

A Cautionary Note

Whenever a mathematical model is used to represent a physical problem, it is important to verify that the assumptions made in the model are at least approximately satisfied in the physical problem. Newton's law of cooling is based in part on the assumption that the body's temperature is uniform. What might we expect if the temperature is not uniform? As an example, consider the use of Newton's law of cooling to determine time of death. In principle, one could measure the temperature of a corpse at two different times and measure the temperature of the room. These data would be sufficient to uniquely determine a formula for $T(t)$. One could then determine the time at which the body was at normal body temperature, which could be taken as a close estimate of the time of death. There are several practical difficulties with this scheme. The primary difficulty from a modeling perspective is the uniform-temperature assumption. Endothermic (warm-blooded) creatures generate heat internally to maintain a body temperature higher than the environment temperature. With an internal temperature of about 98°F and an environment temperature of 70°F, the skin temperature is somewhere between the internal and environment temperatures. The heat transfer to the cooler environment takes place at the skin, so the temperature difference that drives the cooling process is less than the difference between the internal and environmental temperatures. In using Newton's law of cooling, we are in effect assuming that the skin temperature of a live person is actually 98°F. Nevertheless, Newton's law of cooling can give a first approximation for the time of death. To determine whether such an approximation is actually useful, we would have to test it either with experiments or with a more sophisticated mathematical model.

1.1 Exercises

1. Reread the statement of Model Problem 1.1. Newton's law of cooling would be less well suited for the problem if we did not stir the coffee, if the cup were made out of a material that holds in heat, or if the spoon were made of metal.

   a. How would you expect the real behavior to change if we did not stir the coffee?

b. How would you expect the real behavior to change if the cup were made of Styrofoam?
c. How would you expect the real behavior to change if the spoon were made of metal?

2. A bowl of water has a temperature of 125°F. It is put into a refrigerator where the temperature is 40°F. After 0.5 h, the water is stirred and its temperature is measured to be 70°F. It is then left to cool further.

   a. Use Newton's law of cooling to predict when the temperature will be 50°F.
   b. Do you think the actual time required will be larger or smaller than that predicted by your calculation? Explain your answer.
   c. Does the shape of the bowl make a difference?

3. Suppose there is a power failure in your house at 1:00 P.M. on a winter afternoon and your heating system stops working. The temperature in your house is 68°F when the power goes out. At 10:00 P.M., the temperature in the house is down to 57°F. Assume that the outside temperature is 10°F.

   a. Write down an initial value for the temperature in your house, assuming that Newton's law of cooling is valid.
   b. Solve the initial-value problem to estimate the temperature of the house when you get up at 7:00 A.M. the next morning. Should you worry about your water pipes freezing?
   c. What assumption did you have to make about the outside temperature? Given that this assumption is probably not correct, would you revise your estimate up or down? Why?

4. A scientist prepares a sample of a radioactive substance. One year later, the sample contains 3 g of the substance; 2 years later there is only 1 g. Determine how much of the radioactive substance was present initially.

5. The relative decay rate of naturally occurring uranium (U-238) was given in Example 4. Scientists estimate the age of the earth to be about 4.55 billion yr. Determine the fraction of the U-238 present when the earth was formed that is still present as U-238.

6. Strontium 90 (Sr-90) is a radioactive isotope produced in hydrogen bomb explosions. The aboveground nuclear test ban treaty of 1963 was based on evidence of Sr-90 contamination of milk and human bones. The half-life of Sr-90 is 29 yr. Suppose no new sources of Sr-90 have contaminated the atmosphere since 1963. Determine what fraction of the 1963 level of Sr-90 remained in the environment in the year 2003. Determine the approximate date when the Sr-90 level will be only 1% of what it was in 1963.

7. Data regarding decay rates are generally reported as the half-life, but other measures of decay rates could also be used.

   a. Suppose $k = 1$ in the natural decay equation. Determine the percentage of the initial amount that remains after 1 unit of time.
   b. Let t_t be the time required for a quantity to decay to 10% of its original value. Determine the relationship between t_t and k (taking k to be unspecified).

8. *a.* Let $y(t)$ be the amount of a radioactive material with relative decay rate k. Let $Q(t)$ be the decay rate. Use the differential equation for y (not the solution formula) to show that the quantity Q also undergoes exponential decay with rate constant k.

b. It is sometimes easier to measure the rate of radioactive decay than the amount of material. In Example 4, the composition of the sample is taken as given. The same date can be obtained without knowing the amount of material, by making use of the result of part *a*. To see this, first calculate the decay rate $Q(t) = -y'(t)$, using the solution of Example 4. Then determine the values of Q corresponding to time 0 and to the current time $t = 1.14 \times 10^9$ yr. Now assume that all you know are (1) the fact that Q decays exponentially to 0 with rate constant $k = 1.537 \times 10^{-10}\ \text{yr}^{-1}$, (2) the initial value of Q, and (3) the current value of Q. Use this information to determine the age of the sample.

9. Hydrogen peroxide is an unstable chemical that decomposes into a water molecule and an oxygen atom. Assume that the decomposition can be modeled by a decay process with relative decay rate k. Let y be the mass of hydrogen peroxide, and let w be the mass of water produced by the decay. Let z be the fraction of the total mass that is hydrogen peroxide. Determine $z(t)$ in terms of k. Note that 34 g of hydrogen peroxide produces 18 g of water and 16 g of oxygen (which bubbles out of the mixture).

T 10. When photographic film is exposed to light, silver bromide grains are sensitized by interaction with the light. A reasonable assumption is that the number of unsensitized grains decreases to zero by natural decay. The decay rate depends on properties of the light and the silver bromide grains. Photographs taken through telescopes at night require a long exposure.

a. Suppose a small piece of film initially has 1000 silver bromide grains when exposure begins at 1:15 A.M. If 600 grains remain unsensitized at 1:17 A.M., how many will still be unsensitized at 1:20 A.M.?

b. Suppose the optimum exposure time is whatever is needed for 80% of the grains to be sensitized. Assume that an astronomer will measure the percentage p_5 of unsensitized grains after a 5-min exposure. Determine the correct exposure time as a function of p_5.

T 11. Hydrocodone bitartrate is a prescription drug used as a cough suppressant and pain reliever. The drug is eliminated from the body by a natural decay process with a half-life of 3.8 h. The usual oral dose is 10 mg every 6 h.

a. Write down and solve an initial-value problem to model the amount of hydrocodone bitartrate in a patient after a dose. Assume that the amount of drug prior to the dose is Q_0 and that the drug is absorbed immediately.

b. Suppose a patient takes hydrocodone bitartrate for 1 day. Assuming that there is no drug initially in the patient's system, plot the drug amount over a 2-day period. Note that the patient takes 4 doses on the first day and none on the second.

12. Consider the differential equation (5) for capital, where I is taken to be a constant. The equation can be solved by converting it to the decay equation.

a. Recall that an equilibrium solution is a constant function. Find the equilibrium solution K_∞ by substituting $K = K_\infty$ into the differential equation.

b. Let $y(t) = K(t) - K_\infty$. Use this substitution and the result from part *a* to find a differential equation for y.

c. Write down the solution of the y equation (there is no need to solve an equation that we have already solved). Make sure that your solution includes an arbitrary constant.

d. Substitute $y(t) = K(t) - K_\infty$ into the solution from part *c*. The result should be a function of t, with the parameters D and I and an arbitrary constant.

e. Suppose the capital is initially K_0. Use this information to determine the correct value of the arbitrary constant. Write down the solution as a function of t with parameters D, I, and K_0.

f. The problem we have just solved is a mathematical model of capital growth. What assumptions does the model make about the nature of the processes that cause changes in capital? In particular, identify one assumption that is not realistic.

13. Suppose you borrow $12,000 to buy a car. The loan is to be paid in 60 equal monthly installments at an interest rate of 5% per year.

a. Assume that the payments are actually made continuously at whatever rate is needed to pay off the loan in 60 months. Determine the continuous rate per month that would be required. (*Hint:* The problem is easier if you think about it from the lender's point of view. The amount owed begins at $12,000. It increases continuously by a natural growth process and decreases continuously at a fixed rate. In 5 years, the amount owed is zero.)

b. Compare the result of your differential equation model in part *a* with the actual amount. You can find out how much the actual payment would be by consulting with a banker or other investment expert, or you can find appropriate charts in a book or on the Internet. How good is the approximation?

T 14. State lotteries have been described as "a tax on those who are bad at math." Suppose your state has a lottery advertised as having a prize of $1 million. The state pays the winner a total of $1 million over a 20-year period. Assume that the state places an initial sum in a special account on January 1, 2004, and pays the winner at a continuous rate of $50,000 per year out of this account until January 1, 2024. Meanwhile, the account earns investment income for the state.

a. Assume that the initial sum in the account is $1 million. How much money will be left in the account on January 1, 2024, if the account earns interest at a rate of 3% per year?

b. Assume that the state invests y_0 thousands of dollars in an investment that earns interest at a rate of r percent per year. Let Y be the amount, in thousands of dollars, remaining in the account after 20 years. Determine the relationship among y_0, Y, and r. This result can then be used for the remaining parts of this question.

c. How much money will be left in the account on January 1, 2024, if the interest rate is 10% and the initial investment $1 million?

d. Plot a graph of the amount of money left in the account on January 1, 2024, as a function of the interest rate, assuming an interest rate between 2% and 10%.

e. Suppose the state prefers to use the extra money now rather than at the end of 20 years. How much money needs to be invested, assuming 3% interest if the account balance after 20 years is to be $0?

f. Suppose you are teaching a class on money management to high school students. Choose an example and prepare a brief statement explaining why you don't spend *your* money on lottery tickets.

15. Suppose the population of the United States is governed by a natural growth model. Determine the relative growth rate, given that the population was 203 million in the 1970 census and 281 million in the 2000 census. Use these data to predict the population in 2050. Do you think that the actual population in 2050 will be more or less than the predicted amount? Explain your reasoning.

✦ 1.1 INSTANT EXERCISE SOLUTIONS

1. We have $dy/dt = -2 \cos 2t$, so $(dy/dt)(0) = -2$ and $y(0) = 10$. The relative rate of change is
$$\frac{1}{y}\frac{dy}{dt}(0) = -0.2.$$
2. The initial-value problem is
$$\frac{dy}{dt} = -y, \qquad y(0) = 1.$$
The differential equation solutions are again $y = Ae^{-t}$. Only $y = e^{-t}$ satisfies the algebraic condition as well. We want to find t when $y = 1/2$, so we need to solve $e^{-t} = 1/2$, or $e^t = 2$.
3. If we substitute S for T in the differential equation, the left side becomes zero because S is a constant. Since the right side is also zero, the function $T \equiv S$ solves the differential equation.
4. Smaller values of k imply slower decay. The top curve corresponds to the smallest k of the three.
5. From the discussion leading up to Equation (2), $kt_h = \ln 2$. For uranium, $k = 1.537 \times 10^{-10}\ \text{yr}^{-1}$. Hence,
$$t_h = \frac{\ln 2}{1.537 \times 10^{-10}} \approx 4.5 \times 10^9\ \text{yr}.$$
Uranium has a half-life of 4.5 billion yr.

1.2 Differential Equations and Solutions

In this book, we examine a number of differential equations that are in some way representative of real problems. Our aims will be to understand (1) the mathematical principles and solution techniques for these equations and (2) the connection between the differential equations and the real problems. Section 1.1 on natural decay serves as an example. Other important examples include the following:

1. The equation for the motion of a projectile subject to significant air resistance is
$$\frac{d^2\mathbf{r}}{dt^2} + b\left\|\frac{d\mathbf{r}}{dt}\right\|\frac{d\mathbf{r}}{dt} + g\mathbf{k} = \mathbf{0},$$
where $\mathbf{r}(t)$ is the position of an object, g the known surface gravitational constant, and b a known positive constant. This equation is introduced in Section 2.1.
2. The linear oscillator equation for unforced mechanical vibration is
$$m\frac{d^2y}{dt^2} + \beta\frac{dy}{dt} + ky = 0,$$
where $y(t)$ is the deviation of a mass on a spring from its equilibrium position and m, β, and k are the known mass, damping coefficient, and spring constant, respectively. This equation is the focus of Chapter 3.
3. The forced linear oscillator equation for an *RLC* electric circuit is
$$LC\frac{d^2v}{dt^2} + RC\frac{dv}{dt} + v = E(t),$$

where $v(t)$ is the voltage drop across one of the circuit elements; L, R, and C are electrical properties of the circuit; and $E(t)$ is an applied voltage. This equation is introduced in Section 4.1 and studied in detail in Section 4.4.

4. The Lotka–Volterra equations for a simple predator-prey interaction are

$$\frac{dx}{dt} = rx - sxy, \qquad \frac{dy}{dt} = csxy - my,$$

where $x(t)$ and $y(t)$ are the populations of the prey species and the predator species, r is the natural growth rate of the prey, m is the natural death rate of the predator, s is the relative death rate of prey per predator, and c is the conversion factor from prey animals to predator animals. This model is introduced in Section 5.1.

5. A two-compartment model for drug interaction with human tissues consists of the differential equations

$$\frac{dx}{dt} = R - r_1x - k_1x + k_2y, \qquad \frac{dy}{dt} = -r_2y + k_1x - k_2y,$$

where $x(t)$ and $y(t)$ are the amounts of the drug in the blood and the tissues, r_1 and r_2 are the relative rates of decay for whatever processes the body uses to eliminate the drug, and k_1 and k_2 are the relative rates at which the drug moves between blood and tissue. These equations appear in Section 6.1.

6. The one-dimensional wave equation is given by

$$\frac{\partial^2 u}{\partial t^2} = c^2 \frac{\partial^2 u}{\partial x^2},$$

where $u(x, t)$ is the displacement of each point on a vibrating string and c is the known wave speed. This equation is the focus of Chapter 8.

These examples all include one or more derivatives of an unknown function.

A **differential equation** is an equation that contains a derivative (or derivatives) of an unknown function. A differential equation is **ordinary** if all derivatives are with respect to a single independent variable and is **partial** if there are derivatives with respect to two or more independent variables. The **order** of a differential equation is the order of the highest derivative in the equation. A set of n first-order ordinary differential equations with n unknown functions is a **system** of first-order ordinary differential equations; the number n is the **dimension** of the system. An ordinary differential equation or system of ordinary differential equations is **autonomous** if the independent variable does not appear explicitly in the equation.

Number 1 is a system of second-order ordinary differential equations, numbers 2 and 3 are second-order ordinary differential equations, numbers 4 and 5 are systems of two first-order ordinary differential equations, and number 6 is a second-order partial differential equation. The decay equation is a first-order ordinary differential equation. Most differential equations that arise from mathematical models are first-order or second-order, but we will also see some examples of differential equations of higher order. Autonomous equations arise when the physical processes depend only on the state (values) of the dependent variables. The physical process in the equation for number 3 depends on an applied voltage that can change with time. This is the only nonautonomous equation of the six.

Often our task will be to solve a differential equation by determining the functions that satisfy the equation. Sometimes our aim will be merely to obtain qualitative information about the solutions.

Solutions of Ordinary Differential Equations

Chapters 1 through 4 and 7 are almost exclusively about single ordinary differential equations, with systems considered briefly in Chapter 2 and more thoroughly in Chapters 5 and 6 and partial differential equations considered only in Chapter 8. In the remainder of this section, we introduce the key definitions pertaining to solutions of ordinary differential equations.

A function ϕ is a **solution** of an ordinary differential equation on an open interval I if it satisfies the differential equation (makes the equation true) on I. An **equilibrium solution** is a solution that is constant. A problem has a **unique** solution ϕ if every solution of the problem is equal to ϕ.

EXAMPLE 1

The function $\phi(t) = te^t$ is a solution of the differential equation $y'' - 2y' + y = 0$. This is so because $\phi'(t) = te^t + e^t = (t+1)e^t$ and, similarly, $\phi''(t) = (t+2)e^t$; hence, $\phi''(t) - 2\phi'(t) + \phi(t) = 0$. This function ϕ is a solution of the differential equation on any interval in t.

✦ INSTANT EXERCISE 1

Verify that the function $\phi(t) = 1 + 2e^{-t}$ is a solution of the differential equation $y' + y = 1$ by computing $\phi'(t) + \phi(t)$.

EXAMPLE 2

The function $y = 0$ is an equilibrium solution of the natural decay equation of Section 1.1 and the natural growth equation

$$\frac{dy}{dt} = ky, \qquad k > 0. \tag{1}$$

An equilibrium solution is **stable** if solutions that begin sufficiently close to the equilibrium solution tend toward that equilibrium solution as $t \to \infty$. We saw in Section 1.1 that solutions of the decay equation all tend to zero as $t \to \infty$; hence, $y = 0$ is a stable equilibrium solution of the decay equation. In contrast, solutions of the natural growth equation that begin arbitrarily close to the equilibrium solution $y = 0$ move away from that solution; $y = 0$ is an **unstable** equilibrium solution of the natural growth equation.

The idea of stability of solutions is very important, especially for the many differential equations that do not have an elementary solution formula. Study of stability of higher-order equations and systems is deferred to Chapter 5.

Families of Solutions

Consider the family of functions defined by

$$y(t) = Ae^{-3t},$$

where the quantity A can take on any value. This quantity is called an *arbitrary parameter* and is similar to an integration constant in an antiderivative formula. The given family has one such parameter, so it is called a *one-parameter family* of functions.[6] In general, an *n-parameter family* of functions is a family of functions with n arbitrary parameters.[7]

Given a family of functions, we can check to see whether any or all of the functions in the family are solutions of a given differential equation. For example, the formula $y = Ae^{-3t}$ represents a one-parameter family of solutions for the differential equation $dy/dt = -3y$.

How Many Solutions Should We Expect?

So far, we have found one-parameter families of solutions for the natural decay and natural growth equations, both of which are first-order. What about other first-order equations?

EXAMPLE 3

Consider the differential equation

$$\frac{dy}{dt} = \frac{2t}{t^2+1}.$$

This differential equation can be solved by ordinary calculus. The solutions for y are all the functions that are antiderivatives of $2t/(t^2+1)$. Thus,

$$y = \int \frac{2t}{t^2+1}\, dt.$$

This integral can be computed by using substitution. Let $u = t^2 + 1$. Then $du = 2t\, dt$. We therefore have

$$y = \int \frac{1}{u}\, du = \ln|u| + C,$$

where C is an arbitrary integration constant. Substituting $u = t^2 + 1$, which is nonnegative, we get the solution formula

$$y = \ln(t^2+1) + C.$$

---

[6]This concept will be revisited in greater detail in the discussion of linear independence in Chapter 3.

[7]There are some subtleties in the definition of an n-parameter family of functions. We postpone a careful definition until Section 3.2.

✦ INSTANT EXERCISE 2

Solve the differential equation

$$\frac{dy}{dt} = te^{-t^2}.$$

In principle, the simple method of Example 3 should give us solutions for any differential equation of the form

$$\frac{dy}{dt} = f(t). \tag{2}$$

As in the example, any such equation will have a one-parameter family of solutions as long as the function f can be integrated. However, some functions do not have elementary antiderivatives. Can we solve the following differential equation?

$$\frac{dy}{dt} = te^t \arctan t$$

Theorem 1.2.1

Let a be a fixed constant, and let f be a function that is continuous on a closed interval $[t_0, t_1]$, with $t_0 < a < t_1$. Then the formula

$$y = \int_a^t f(\tau)\, d\tau + C \tag{3}$$

defines a one-parameter family of solutions (with C as the arbitrary parameter) of the differential equation

$$\frac{dy}{dt} = f(t)$$

on the interval $[t_0, t_1]$.

- Theorem 1.2.1 is nothing more than a restatement of the fundamental theorem of differential calculus, which says that the derivative of the definite integral $\int_a^t f(\tau)\, d\tau$ with respect to t is the function $f(t)$, where a is any fixed constant and f is continuous on the interval $a \leq \tau \leq t$.
- The constant a can be chosen for convenience, subject to the requirement that f be continuous on an interval containing a. The solution formula (3) looks as if it has two parameters, but that is not the case. The quantity a is taken to be fixed, while C is a parameter.

EXAMPLE 4

Consider again the differential equation

$$\frac{dy}{dt} = \frac{2t}{t^2 + 1}.$$

The function f is continuous for all t, so the choice of a is completely arbitrary. If we take $a = 0$, we obtain the solution

$$y_0 = \int_0^t \frac{2\tau}{\tau^2+1}\,d\tau + C_0;$$

with $a = 1$ we have instead

$$y_1 = \int_1^t \frac{2\tau}{\tau^2+1}\,d\tau + C_1.$$

As we saw in Example 3, we can compute these integrals explicitly because we have a formula for an antiderivative. Thus,

$$y_0 = \ln(\tau^2+1)\Big|_{\tau=0}^{t} + C_0 = \ln(t^2+1) + C_0,$$

which is the same solution formula we found in Example 3. Similarly,

$$y_1 = \ln(\tau^2+1)\Big|_{\tau=1}^{t} + C_1 = \ln(t^2+1) - \ln 2 + C_1.$$

The formulas for y_0 and y_1 are different, but they represent the same one-parameter family of functions, with $C_0 = C_1 - \ln 2$.

EXAMPLE 5

To solve the differential equation

$$\frac{dy}{dt} = \cos t^2,$$

we may use Theorem 1.2.1, with any value of a. Thus, for example, we may write the solution as

$$y = \int_0^t \cos\tau^2\,d\tau + C.$$

✦ INSTANT EXERCISE 3

Construct a solution formula for the differential equation

$$\frac{dy}{dt} = te^t \arctan t.$$

Up to now, we have found solutions for several first-order equations, and each such solution was a one-parameter family. What about higher-order differential equations?

EXAMPLE 6

The differential equation

$$\frac{d^2y}{dt^2} = 1$$

can be integrated once to yield

$$\frac{dy}{dt} = t + C_1.$$

A second integration yields the two-parameter family

$$y = \frac{1}{2}t^2 + C_1 t + C_2.$$

The pattern is clear: each integration introduces an additional integration constant. This suggests that the second-order equation of Example 1 should have a two-parameter family of solutions; however, we only found a single solution $y = te^t$ for the differential equation $y'' - 2y' + y = 0$. Does this differential equation have other solutions?

✦ INSTANT EXERCISE 4

Verify that $y = Ae^t + Bte^t$ is a two-parameter family of solutions for the differential equation $y'' - 2y' + y = 0$.

The examples we have seen suggest a conjecture about the solutions of differential equations.

- A differential equation of order n has an n-parameter family of solutions.

Our conjecture is not strictly true, but it is true for almost all differential equations that you are likely to encounter in practical problems. It serves to give an idea of how large a family of solutions we should expect a differential equation to have. If we are able to construct a one-parameter family for a second-order equation, we should anticipate that there are other solutions we have not yet found.

Initial-Value Problems

Solutions of differential equations usually occur in families, as we have seen. However, real-world situations that we want to model with differential equations often have a unique real-world result. For example, if you drop a bowling ball out of a window, the graph of the height of the bowling ball versus time since its release will always be the same, allowing for experimental error. Mathematically, it is often possible to reduce the number of solutions of a differential equation to exactly one solution by attaching an extra requirement to a differential equation problem.

EXAMPLE 7

Consider the problem

$$\frac{dy}{dt} = -3y, \qquad y(0) = 2.$$

All functions of the form $y = Ae^{-3t}$ satisfy this decay equation. The extra equation $y(0) = 2$ has both an algebraic significance and a geometric significance. The algebraic significance is to specify a value of the function y at a particular value of t. We substitute the solution family into this algebraic condition

$$2 = y(0) = Ae^0 = A,$$

and this becomes an algebraic relationship that determines a unique value of A. To summarize: $y = Ae^{-3t}$ satisfies the differential equation regardless of the value of A, but it only satisfies the extra condition if

$A = 2$. The problem

$$\frac{dy}{dt} = -3y, \qquad y(0) = 2$$

has a unique solution:[8]

$$y = 2e^{-3t}.$$

The geometric significance of the requirement $y(0) = 2$ is that it selects from the family of solutions just the one whose graph passes through the point (0, 2). The graphs of several of the solutions of $dy/dt = -3y$ are illustrated in Figure 1.2.1.

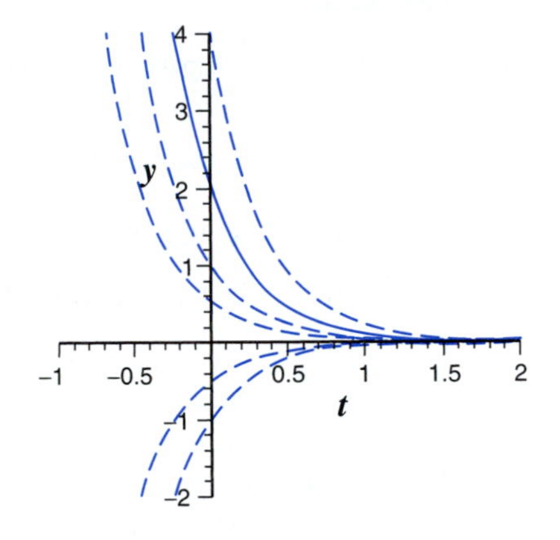

Figure 1.2.1
Several functions in the family $y = Ae^{-3t}$, with the function $y = 2e^{-3t}$ solid and the others dashed.

An **initial-value problem** for a first-order differential equation is a problem of the form

$$\frac{dy}{dt} = f(t, y), \qquad y(t_0) = y_0, \tag{4}$$

where t_0 is any value of the independent variable. The extra condition $y(t_0) = y_0$ is called an **initial condition,** regardless of the value of t_0.

EXAMPLE 8

To solve the problem

$$\frac{dy}{dt} = t, \qquad y(1) = 2,$$

we first find a family of solutions for the differential equation. Since the equation is of the form $dy/dt = f(t)$, we can simply integrate it to get

$$y = \tfrac{1}{2}t^2 + C,$$

[8] We have not actually *shown* this to be true. What we currently know is that we have found only one solution.

with C an arbitrary integration constant. The initial condition yields

$$2 = y(1) = \tfrac{1}{2} + C,$$

so $C = 3/2$ and the solution of the initial-value problem is

$$y = \frac{t^2 + 3}{2}.$$

EXAMPLE 9

Consider the problem

$$\frac{d^2y}{dt^2} + 4y = 0, \qquad y(0) = 1, \qquad \frac{dy}{dt}(0) = -2.$$

Observe that the differential equation has a two-parameter family of solutions,

$$y = A\cos 2t + B\sin 2t.$$

To satisfy the two additional conditions, we must have

$$1 = y(0) = A$$

and

$$-2 = \frac{dy}{dt}(0) = (-2A\sin 2t + 2B\cos 2t)|_{t=0} = 2B.$$

Thus, $A = 1$ and $B = -1$; we have found a solution for the problem:

$$y = \cos 2t - \sin 2t.$$

As Example 9 illustrates, it is generally necessary to have n auxiliary conditions to get a unique solution for an nth-order initial-value problem.

An **initial-value problem** for an nth-order differential equation consists of the differential equation and a set of n algebraic conditions at a common point:

$$y(t_0) = y_0, \qquad \frac{dy}{dt}(t_0) = y_1, \qquad \ldots, \qquad \frac{d^{n-1}y}{dt^{n-1}}(t_0) = y_{n-1}.$$

An Evaporating Raindrop

In arid climates, there is a phenomenon called *phantom rain,* in which the rain that falls from clouds evaporates before it reaches the ground. We can construct two different initial-value problems to model phantom rain, and their study will motivate a final concept relevant to the solutions of differential equations.

MODEL PROBLEM 1.2

A raindrop initially has a radius of 1 mm. If the raindrop takes 4 s to evaporate, determine the volume of the raindrop as a function of time. Assume that the raindrop is spherical.

At first glance, Model Problem 1.2 might seem to be another natural decay problem. However, there is nothing in the narrative to justify the assumption that the relative rate of growth of the volume is a negative *constant*. The key idea in finding a differential equation to describe the evaporation process is that evaporation is a surface phenomenon. Only molecules at the surface of the drop can evaporate. Assume that the rate of evaporation is proportional to the surface area of the raindrop. With V the volume and A the surface area, we have the differential equation

$$\frac{dV}{dt} = -kA(t),$$

where we have been careful to note that the area is a function of time. We do not have a useful model for evaporation yet, because the area and volume are both unknown. However, they are related by geometry. We have

$$V = \tfrac{4}{3}\pi r^3, \qquad A = 4\pi r^2.$$

Solving the volume formula for r and then substituting the result into the area formula yields

$$A = (36\pi)^{1/3} V^{2/3};$$

hence, we have the initial-value problem

$$\frac{dV}{dt} = -(36\pi)^{1/3} k V^{2/3}, \qquad V(0) = \tfrac{4}{3}\pi.$$

There is no advantage to having a complicated factor like $(36\pi)^{1/3}$ multiplied by a constant whose value is not even known, so the model can be conveniently written as

$$\frac{dV}{dt} = -k_1 V^{2/3}, \qquad V(0) = \tfrac{4}{3}\pi. \tag{5}$$

We could solve this problem as we originally solved the decay equation, by thinking of t as a function of V rather than V as a function of t. However, there is a much more clever way to solve the problem. Instead of writing the area in terms of the volume, we can write both area and volume in terms of the radius. From

$$V(t) = \tfrac{4}{3}\pi r^3(t),$$

we have

$$\frac{dV}{dt} = \frac{4}{3}\pi \frac{d}{dt} r^3(t) = \left(\frac{4}{3}\pi\right)\left(3r^2 \frac{dr}{dt}\right) = 4\pi r^2 \frac{dr}{dt}.$$

Substituting this and $A = 4\pi r^2$ into the first differential equation yields a simple initial-value problem for the radius:

$$\frac{dr}{dt} = -k, \qquad r(0) = 1.$$

The radius decreases at a constant absolute rate.

The initial-value problem for r has the solution

$$r = 1 - kt.$$

The constant k is determined by the extra information that the raindrop vanishes after 4 s. To have $r(4) = 0$, we must have $k = 0.25$. Thus,

$$r = 1 - 0.25t.$$

The volume is now given by the volume formula as

$$V = \phi(t) = \tfrac{4}{3}\pi(1 - 0.25t)^3.$$

The graph of $\phi(t)$ appears in Figure 1.2.2. The function decreases to zero at time $t = 4$ as required, but then it becomes negative and continues to decrease. After the time $t = 4$, the function $\frac{4}{3}\pi(1 - 0.25t)^3$ no longer makes sense in the context of volume.

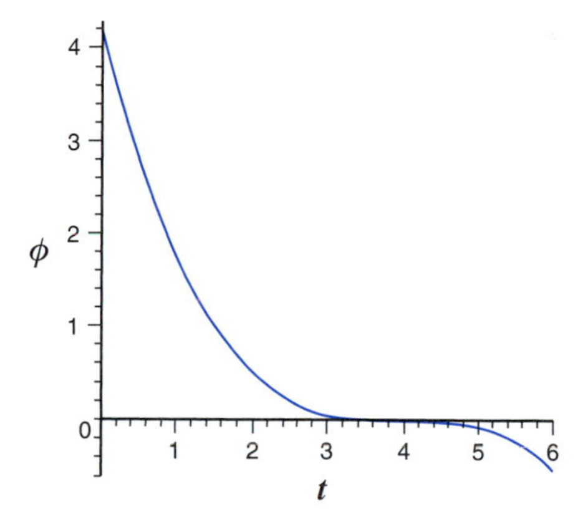

Figure 1.2.2
The function $\frac{4}{3}\pi(1 - 0.25t)^3$.

Although it was not explicitly stated, the derivation of the differential equation for Model Problem 1.2 assumes that there is a raindrop to evaporate. The differential equation only represents the physical problem as long as $V \geq 0$. After $V = 0$, the correct differential equation is $dV/dt = 0$, and the solution is $V = 0$. Considering the full range of physical time, the solution is

$$V(t) = \begin{cases} \frac{4}{3}\pi(1 - 0.25t)^3 & t \leq 4 \\ 0 & t > 4 \end{cases}. \tag{6}$$

Uniqueness

We have already seen that the function ϕ of Figure 1.2.2 solves the initial-value problem

$$\frac{dV}{dt} = -k_1 V^{2/3}, \qquad V(0) = \frac{4\pi}{3}, \qquad k_1 = \left(\frac{9\pi}{16}\right)^{1/3}.$$

Notice that the function V defined by Equation (6) also solves the initial-value problem. This statement requires a careful justification. Clearly V satisfies the differential equation on the intervals $t < 4$ and $t > 4$, and it satisfies the initial condition. Does it satisfy the differential equation at $t = 4$? The function V is continuous at $t = 4$. Differentiating V, we have

$$\frac{dV}{dt} = \begin{cases} -\pi(1 - 0.25t)^2 & t \leq 4 \\ 0 & t > 4 \end{cases},$$

which is continuous at $t = 4$. Thus, $(dV/dt)(4) = 0 = -k_1[V(4)]^{2/3}$, and so the differential equation is satisfied at $t = 4$.

In summary, we can list several conclusions.

- The function ϕ defined by $\phi(t) = \frac{4}{3}\pi(1 - 0.25t)^3$ is defined for all t.
- The function ϕ solves the initial-value problem of Equation (5) for all t.
- The function V defined in Equation (6) also solves the initial-value problem of Equation (5) for all t.
- The initial-value problem of Equation (5) represents Model Problem 1.2 only until such time as the volume becomes zero.
- The function V represents the solution of Model Problem 1.2, provided that the mathematical model accurately describes evaporation.

The Interval of Existence

Model Problem 1.2 is an example of a differential equation that represents the model only for a limited range of the independent variable. Similarly, there are sometimes functions that only solve an initial-value problem for a limited range of the independent variable.

EXAMPLE 10

Consider the initial-value problem

$$\frac{dy}{dt} = 1 + y^2, \qquad y(0) = 0.$$

This differential equation can be solved by the same method we used to solve the decay equation in Section 1.1. Thinking of t as a function of y gives us the initial-value problem

$$\frac{dt}{dy} = \frac{1}{1 + y^2}, \qquad t(0) = 0.$$

Integrating $1/(1 + y^2)$ yields

$$t = \arctan y + C,$$

after which the initial condition requires $C = 0$. Thus, $t = \arctan y$; hence,

$$y = \tan t.$$

The solution is illustrated in Figure 1.2.3.

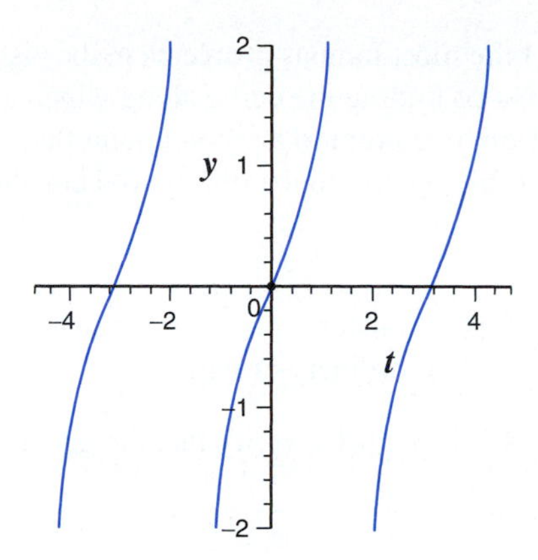

Figure 1.2.3
The function $\tan t$ from Example 10.

The **interval of existence** of a solution of an initial-value problem is the largest interval on which the solution function solves the initial-value problem. In Example 10, the tangent function solves the initial-value problem only on the interval $[-\pi/2, \pi/2]$. It solves the differential equation wherever it is well defined, but it solves the initial-value problem only on the portion of its domain that includes the initial condition.

In summary, we can list several conclusions.

- The tangent function is defined for all t in the intervals $\ldots, [-3\pi/2, -\pi/2], [-\pi/2, \pi/2], [\pi/2, 3\pi/2], \ldots$.
- The tangent function solves the differential equation of Example 10 on all intervals on which the function is defined.
- The tangent function solves the initial-value problem of Example 10 only on the interval $[-\pi/2, \pi/2]$.

A Final Remark

Mathematicians like to emphasize unusual behavior. However, it is important to also emphasize what behavior is more usual. Most nth-order differential equations have an n-parameter family of solutions from which every solution of the differential equation can be obtained by a unique choice of the parameter values. Most initial-value problems have a unique solution, although not necessarily one that is valid for all values of the independent variable.

1.2 Exercises

1. Classify each of the following as ordinary or partial, and indicate the order of the equation.

   a. $y'' + y' = y$

   b. $\dfrac{\partial u}{\partial t} = \dfrac{\partial^2 u}{\partial x^2}$ (This is the equation governing heat flow.)

   c. $\dfrac{d^2 y}{dt^2} + \dfrac{d^2 x}{dt^2} = 0$

2. One of the most famous problems in the history of mathematics is the brachistochrone problem, that of finding the curve along which a particle will require the least amount of time to slide from one point to a second point that is lower than the first, but not directly beneath it. This problem leads to the differential equation

$$(1 + y'^2)y = k,$$

where k is a constant that must be determined later. What is the order of the differential equation? Is it ordinary or partial?

In Exercises 3 through 12, verify that the given function is a solution of the differential equation.

3. $\frac{dy}{dt} - 3y = -2e^{-2t}, \quad \phi = \frac{2}{5}e^{-2t} + 4e^{3t}$

4. $\frac{dy}{dt} + 2y = \cos t, \quad \phi = 2e^{-2t} + \frac{2}{5}\cos t + \frac{1}{5}\sin t$

5. $\frac{d^2y}{dt^2} + y = e^{3t}, \quad \phi = 3\cos t + \frac{1}{10}e^{3t}$

6. $\frac{d^2y}{dt^2} - y = e^{3t}, \quad \phi = 4e^t + 2e^{-t} + \frac{1}{8}e^{3t}$

7. $x^2\frac{d^2y}{dx^2} - 6y = 12, \quad \phi = x^3 - 2$

8. $(1 - x^2)\frac{d^2y}{dx^2} - 2x\frac{dy}{dx} + 6y = 0, \quad \phi = 3x^2 - 1$

9. $\frac{dy}{dx} = \frac{x}{y}, \quad \phi = \sqrt{1 + x^2}$

10. $\frac{dy}{dt} = y - y^2, \quad \phi = \frac{1}{1 + e^{-t}}$

11. $\frac{dy}{dt} + 2y = t, \quad \phi = \frac{1}{2}t - \frac{1}{4} + Ce^{-2t},$ where C is arbitrary

12. $\frac{dy}{dt} + 4y = e^{-4t}, \quad \phi = (C + t)e^{-4t},$ where C is arbitrary

13. Determine the value(s) of r for which each of the differential equations has a solution of the form $y = e^{rt}$.
    a. $y' - 3y = 0$
    b. $y'' + 3y' + 2y = 0$
    c. $y'' + 4y' + 4y = 0$

14. Determine the value(s) of r for which each of the differential equations has a solution of the form $y = x^r$.
    a. $xy' + y = 0$
    b. $x^2y'' - 4xy' - 6y = 0$
    c. $x^2y'' - 2xy' - 6y = 0$

15. *a.* Determine the value(s) of A for which the differential equation $y' + 3y = e^{-2t}$ has a solution of the form $y = Ae^{-2t}$.
    b. Repeat part *a* for the differential equation $y' - 2y = e^{-2t}$.
    c. Repeat part *a* for the differential equation $y' + 2y = e^{-2t}$.
    d. Can you explain why the equation in part *c* is different from the others?

16. *a.* Determine the value(s) of A for which the differential equation $y'' + y = \cos 2t$ has a solution of the form $A \cos 2t$.
    b. Repeat part *a* for the differential equation $y'' - 3y = \cos 2t$.
    c. Repeat part *a* for the differential equation $y' - 3y = \cos 2t$.
    d. Is there a pair of values A and B for which $y' - 3y = \cos 2t$ has a solution of the form $A \cos 2t + B \sin 2t$?

17. The differential equation $y' = 2y^2 - 2y$ has a one-parameter family of solutions $y = 1/(1 + Ce^{2t})$. It also has an equilibrium solution that is not part of the family. Determine this equilibrium solution. This differential equation demonstrates that a first-order differential equation may have one or more solutions in addition to a one-parameter family.

In Exercises 18 through 22, solve the given initial-value problem.

18. $\dfrac{dy}{dt} = 2y, \quad y(0) = 4$

19. $\dfrac{dy}{dt} = 4y, \quad y(0) = 3$

20. $\dfrac{dy}{dt} = \dfrac{t}{\sqrt{1+t^2}}, \quad y(0) = 2$

21. $\dfrac{d^2y}{dt^2} = \dfrac{3}{(1+2t)^2}, \quad y(0) = 1, \quad \dfrac{dy}{dt}(0) = 0$

22. $\dfrac{d^2y}{dt^2} = \dfrac{1}{1+t^2}, \quad y(0) = 1, \quad \dfrac{dy}{dt}(0) = -1$

23. Solve the initial-value problem $\dfrac{dy}{dt} = \tan t$, $y(0) = y_0$ and determine the interval of existence for the solution.

In Exercises 24 through 26, solve the given problem by using a definite integral.

24. $\dfrac{dy}{dt} = e^t \ln(1+t), \quad y(0) = 1$

25. $\dfrac{dy}{dx} = \dfrac{e^{-x}}{x}, \quad y(1) = 3$

26. $\dfrac{d^2y}{dt^2} = -2te^{-t^2}$

27. The error function $\operatorname{erf} x$ is defined by the integral

$$\operatorname{erf} x = \frac{2}{\sqrt{\pi}} \int_0^x e^{-s^2}\, ds.$$

a. Determine an initial-value problem satisfied by $y = \operatorname{erf} x$.

b. Show that the function $y = (\sqrt{\pi}/2)e^{t^2} \operatorname{erf} t$ is a solution of $dy/dt = 2ty + 1$.

28. Consider the initial-value problem

$$\frac{dy}{dt} = \frac{e^{-t}}{\sqrt{\pi t}}, \qquad y(0) = 0.$$

a. Solve the initial-value problem, being careful to check that the improper integral converges.

b. Use a change of variables to write the solution in terms of the error function from Exercise 27.

In Exercises 29 through 34, verify that the given family of functions solves the differential equation. Then find the solution of the initial-value problem and determine its interval of existence.

29. $\dfrac{dy}{dt} = 2y^2, \quad y(0) = 2, \quad y = \dfrac{1}{C - 2t}$

30. $\dfrac{dy}{dt} - 3y = -2e^{-2t}, \quad y(0) = 5, \quad y = \frac{2}{5}e^{-2t} + Ce^{3t}$

31. $\dfrac{dy}{dt} = 2y^2 - 2y, \quad y(0) = 2, \quad y = \dfrac{1}{1 + Ce^{2t}}$

32. $\dfrac{dy}{dt} = (1 - 2t)y^2, \quad y(0) = -\frac{1}{6}, \quad y = \dfrac{1}{C - t + t^2}$

33. $\dfrac{dy}{dt} = y^3, \quad y(0) = -2, \quad y = -\dfrac{1}{\sqrt{C - 2t}}$

34. $\dfrac{dy}{dt} = y^2 \sin t, \quad y(0) = \frac{1}{2}, \quad y = \dfrac{1}{C + \cos t}$

✦ 1.2 INSTANT EXERCISE SOLUTIONS

1. Given $\phi(t) = 1 + 2e^{-t}$, we have

$$\frac{d\phi}{dt} + \phi = -2e^{-t} + (1 + 2e^{-t}) = 1.$$

This calculation confirms that the function ϕ satisfies the differential equation for all values of t.

2. We solve $dy/dt = te^{-t^2}$ simply by computing the antiderivatives of the function te^{-t^2}. Thus,

$$y = \int te^{-t^2}\, dt.$$

Using the substitution $u = -t^2$, we have

$$y = \int -\tfrac{1}{2}\, e^u\, du = -\tfrac{1}{2}\, e^u + C = -\tfrac{1}{2}\, e^{-t^2} + C.$$

3. Since $te^t \arctan t$ has no elementary antiderivative, we construct an antiderivative by using the fundamental theorem of calculus. The function is continuous at $t = 0$, so we choose $a = 0$ for convenience.

The result is

$$y = \int_0^t \tau e^\tau \arctan \tau \, d\tau.$$

4. Suppose $y = Ae^t + Bte^t$. Then

$$y' = Ae^t + B(e^t + te^t) = (A + B)e^t + Bte^t.$$

Similarly,

$$y'' = (A + 2B)e^t + Bte^t.$$

Hence,

$$y'' - 2y' + y = [(A + 2B)e^t + Bte^t] - 2[(A + B)e^t + Bte^t] + (Ae^t + Bte^t) = 0.$$

1.3 Mathematical Models and Mathematical Modeling

You have undoubtedly studied applications of mathematics and solved word problems based on these applications. This is only part of mathematical modeling. The larger view of mathematical modeling is that it is about using mathematics to explore topics outside of mathematics. The intimate connection between science and mathematical modeling is eloquently stated by the great contemporary physicist Stephen Hawking.

> I . . . take the view . . . that a theory of physics is just a mathematical model that we use to describe the results of observations. A theory is a good theory if it is an elegant model, if it describes a wide class of observations, and if it predicts the results of new observations.[9]

Models and Modeling

Mathematical models arise in every field of science. Although the connections between models and physical phenomena in other sciences are not always as strong as in physics, it remains useful to think of theories of science in terms of mathematical models. In the course of this book, we will study mathematical models from various areas of physics and engineering as well as from chemistry, geology, biology, and occasionally nonscience areas such as history.

To begin, we propose the following interpretation of the term *mathematical model* in the context of problems in science and engineering.

> A **mathematical model** is a set of formulas and/or equations based on a quantitative description of real phenomena and created in the hope that the behavior it predicts will resemble the real behavior on which it is based.

With this interpretation, a mathematical model could be as simple as a single formula relating two variables or as complicated as a set of equations describing the relationships between a set of

[9]From Stephen Hawking, "My Position," in *Black Holes and Baby Universes and Other Essays,* Bantam Books, New York, 1993.

unknowns. Note that the goal, according to Hawking, is not to produce a model that is "correct," but rather to elegantly describe and predict classes of observations. Models do not represent reality—at best they represent an oversimplified characterization of reality. Typically, we do not need to use all details of a physical setting to create the model, which leaves additional features to compare with predictions made by the model. A successful mathematical model confirms the scientific understanding used to construct it and makes predictions that can be tested to gain greater understanding.

It is helpful to think of the processes of mathematical description and model creation as being somewhat distinct. We will use the term **conceptual model** to denote an idealized characterization of a real-world situation. The mathematical model is a careful mathematical description of the conceptual model rather than the real situation.

Mathematical modeling is the art and science of constructing mathematical models and using them to gain insight into physical processes or to make predictions concerning physical processes. The science lies in constructing the mathematical model from the conceptual model, and the art lies in determining an appropriate conceptual model.

Outlining the Purpose of a Model

Any particular mathematical model might be very successful for some uses and unsuccessful for others. The key first step in any mathematical modeling project is to decide upon the purpose of the model.

MODEL PROBLEM 1.3

Construct a mathematical model to predict how high a person can throw a ball, given a variety of people and a variety of balls.

Once the purpose is clearly stated, the modeler needs to become familiar enough with the physical setting to understand what elements could be included in the model and how these are described mathematically. We begin with a broad look at mathematical models for moving objects.

Where Do Models Come From?

Suppose a heavy, rigid object is dropped from a height of y_0 ft and falls to the ground. Up to the time that the object hits the ground, its height y will be a function of time that decreases from y_0 to 0. In an elementary physics book, we find the standard formula

$$y = y_0 - 16t^2, \tag{1}$$

where t is the time since the beginning of the fall, measured in seconds. Where does this formula come from?

There are two answers to this question. The original answer is that it comes from experiment. In a modern physics laboratory, we could make detailed measurements of height against time for falling objects. Our results could be plotted as graphs of y versus t, y versus t^2, and so on. With careful measurements, we would discover that the graph of y in feet versus t^2 in seconds squared is approximately a straight line with slope -16 and y intercept y_0.

In practice, things are never so simple as described here. If you actually do the experiment, you will find that the data points do not lie *exactly* on a straight line. Real data cannot be measured with perfect accuracy. We might end up questioning the accuracy of Equation (1), since experiments can only confirm it approximately.

A second answer to the question of where the formula (1) comes from is that it can be derived from a mathematical model. We shall do so, but first we must ask the question, "Where do mathematical models come from?"

Physical Laws

Scientists do not use measurements only to obtain formulas for specific experiments. A bigger goal is to try to determine basic principles that underlie the measured behavior. These basic principles are stated in the form of *physical laws*, such as Newton's second law of motion, which is generally given as

$$F = ma,$$

where F is the net force on the object, m is the object's mass, and a is the acceleration that results from the force.

A **physical law** is a statement that expresses the relationship between quantities closely enough to be taken as exactly true. To qualify as a physical law, a relationship must be observed in a large variety of settings.

The phrase "closely enough to be taken as exactly true" is admittedly vague. We use the expression to indicate a high level of generality and confidence in a formula.

Newton's second law can be written in forms that are more useful. Recall the relationships between acceleration, velocity, and distance:

$$a = \frac{dv}{dt} = \frac{d^2y}{dt^2}.$$

Substituting these expressions into Newton's second law gives the equations

$$m\,\frac{dv}{dt} = F \tag{2}$$

and

$$m\,\frac{d^2y}{dt^2} = F. \tag{3}$$

As a practical matter, we must view Equations (2) and (3) as differential-equations-in-the-making, rather than differential equations, because they are not completely specified. Usually, the mass m is a known quantity, and the function v or y is unknown. To complete the differential equations, we have to make some assumption about the force F. Depending on the situation, F may be a function of y, v, t, or some combination of these.

Motion Due to Gravity

The gravitational force on an object is proportional to the object's mass. A reasonable assumption is that the gravitational force F_g has the simple formula

$$F_g = -mg,$$

where g is a constant, generally taken as $32\,\text{ft/s}^2$ or $9.8\,\text{m/s}^2$. This assumption is not actually correct, but it is a reasonable approximation in most cases. A more accurate formula for the gravitational force is given by Newton's law of gravitation, which is discussed in Exercise 3.[10]

Now assume that the constant gravitational force is the only force on the object. We can use either form of Newton's second law. The second version gives the differential equation

$$m\,\frac{d^2y}{dt^2} = -mg.$$

The constant factor m can be removed to give the simplest model for a falling object:

$$\frac{d^2y}{dt^2} = -g. \tag{4}$$

We could substitute a number for g; however, this requires us to decide which system of measurement to use and how precisely to report the results. Do we use feet per second squared or meters per second squared? If feet, do we use the simpler value 32 or the slightly more accurate value 32.2? Retaining constants such as g as symbols rather than as specific numbers allows us to postpone answering these questions. We can always insert the value for g later.

The differential equation (4) represents the motion in the conceptual model. Integrating the differential equation twice yields a two-parameter family of solutions

$$y = C_1 + C_2 t - \frac{g}{2}t^2.$$

The initial conditions $y(0) = y_0$ and $(dy/dt)(0) = 0$ and the value $g = 32$ yield the result.

Mathematical Models in Science

In general, we obtain formulas relating physical quantities by solving mathematical models, which are, in turn, based on experimental results. Mathematical models begin with physical laws, such as Newton's second law of motion, that are taken to be "exactly" true because they offer a very good description in most circumstances. The physical laws are augmented by simplifying assumptions that describe a conceptual model of the physical setting. Because the mathematical model represents a conceptual model rather than real phenomena, its results must be carefully compared with observations and experiments.

In the conceptual model used here, there are no forces other than gravity, and the force of gravity is independent of height. The real physical problem, however, includes a drag force caused by air resistance, a height-dependent gravitational force, and relativistic effects. Given these differences between the conceptual model we used and the real physical setting, it is pointless to use more than two or three significant digits in the gravitational constant or to report all the digits a calculator gives in a numerical answer.[11] The mathematical model should not be used to solve problems concerning a skydiver drifting under an open parachute or the fall of a feather

[10] Note that it is important to use the correct algebraic sign for any force. A force is positive if it acts to move an object in the positive direction and negative if it acts to move an object in the negative direction. The force of gravity is negative if we follow the standard convention of orienting the vertical axis so that positive is up. It does not matter whether the object is actually moving downward, as its motion could be influenced by inertia or forces other than gravity.

[11] Undue attention to details that are relatively unimportant is a common mistake that we might call the "measure it with your hand, mark it with chalk, cut it with a laser" fallacy.

because the conceptual model on which it is based does not apply to these situations. It is not altogether uncommon for faulty conclusions to be published because they were obtained from an inappropriate model. Every mathematical modeling investigation should carefully consider the validity of the conceptual model.

Mathematical models can even be useful in cases where the mathematical results do not match the real results very well. The actual experimental results suggest changes in the conceptual model, which in turn can yield an improved mathematical model. Over time, conceptual models become better and the physical phenomena become better understood. The connections among observation, theory, model, and mathematical results are illustrated in Figure 1.3.1.

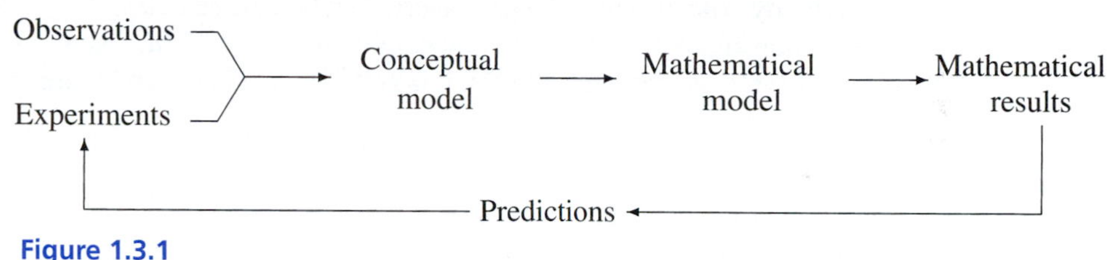

Figure 1.3.1
A schematic diagram of the development of mathematical models for science.

The schematic illustrates a process that has a fixed starting point but no fixed ending point. Observations and experiments lead to the statement of physical laws and suggest reasonable simplifying assumptions. These are combined to create a mathematical model, which is analyzed to determine mathematical results. The model is validated by comparing its results with known observations and experiments. If the comparison is good, the model can be used for making predictions about the physical situation. However, these predictions need to be tested by new experiments to verify that the model is valid in the context of the predictions. The value of mathematical modeling is that it is much less expensive, and often much quicker, to build a mathematical model and perform mathematical experiments than to conduct physical experiments. It is often a wise investment to do mathematical experiments first and to follow up with only those physical experiments that are suggested by the mathematical model.

A First Model for a Thrown Ball

We imagine an athletic contest. There are a variety of participants. Each will throw a golf ball, a baseball, a basketball, and a bowling ball straight up. An observer will measure the height each contestant achieves with each ball, measured from the release point of the throw. The purpose of the model is to predict the maximum height, given a particular thrower and a particular ball.

The Conceptual and Mathematical Models A ball is launched upward from an initial height of zero with initial velocity v_0, a value that is different for each participant. Gravity is the only force on the ball after its release, and the gravitational force is $-mg$, where m is the mass of the ball and g is the gravitational constant in whatever system of measurement we choose. The mathematical model features the differential equation $d^2y/dt^2 = -g$ because the acceleration is equal to the gravitational force on the ball. The ball is characterized by its mass m and the thrower by the initial velocity v_0. Assuming m and v_0 are known, the height $y(t)$ is determined by the initial-value problem

$$\frac{d^2y}{dt^2} = -g, \qquad y(0) = 0, \qquad \frac{dy}{dt}(0) = v_0. \tag{5}$$

We still need to figure out how to determine the maximum height, which we will call Y. There are several different ways to overcome this difficulty. A particularly clever way[12] is to notice that we know the velocity at which the maximum height is achieved, but we do not know the time. Since the height of the ball increases steadily from the beginning of the throw until the maximum height is reached, we can think of the velocity as a function of height rather than time. Let v be the velocity. The differential equation can then be written as $dv/dt = -g$, and we also know that the relationship between y and v, by definition, is $v = dy/dt$. Now assume a function $u(y)$ that gives the velocity for the ball when it is at height y. Then v and u are related by

$$v(t) = u(y(t)).$$

Differentiating this equation gives us

$$\frac{dv}{dt} = \frac{du}{dy}\frac{dy}{dt},$$

which we can rewrite as

$$-g = v\,\frac{du}{dy}.$$

Since the value of u is the same as that of v, we have the differential equation $u\,du/dy = -g$. Along with the differential equation, we have an initial condition $u(0) = v_0$, which says that the velocity is v_0 when the height is zero. We also know that the velocity and height at maximum are 0 and Y, respectively, so $u(Y) = 0$. If we put everything together, the model consists of a first-order differential equation with two extra conditions. The first extra condition is needed to evaluate the integration constant, and the second is needed because there is an additional unknown (Y) in the problem. In summary, we have

$$u\,\frac{du}{dy} = -g, \qquad u(0) = v_0, \qquad u(Y) = 0. \tag{6}$$

Our goal is to determine Y.

Analysis of the Mathematical Model The differential equation can be solved by a clever manipulation. Let $w(y) = u^2(y)$. Then $dw/dy = 2u\,du/dy$. Stated otherwise,

$$\frac{du^2}{dy} = 2u\,\frac{du}{dy}.$$

[12]An even more clever way to do the problem is to apply the principle of conservation of energy. We use a mathematical technique because we are using this simple model problem to illustrate mathematical modeling.

Multiplying the differential equation (6) by 2 allows us to write it as

$$\frac{du^2}{dy} = -2g.$$

This equation is easily integrated, with the result

$$u^2 = -2gy + C.$$

Thus,

$$u = \pm\sqrt{C - 2gy}.$$

Next we apply the initial condition $u(0) = v_0$ to obtain the result $C = v_0{}^2$. The initial-value problem consisting of the first two equations in (6) has the unique solution

$$u = \sqrt{v_0{}^2 - 2gy},$$

which is valid from the beginning of the throw, when $y = 0$, until the peak of the throw, when $u = 0$. (Note that the requirement $u \geq 0$ determines the correct sign.) The peak height Y is determined by the equation $u(Y) = 0$, which gives us $v_0{}^2 = 2gY$. The final result is a formula for Y in terms of known parameters:

$$Y = \frac{v_0{}^2}{2g}. \tag{7}$$

The result (7) gives us a simple prediction for the height of a thrown ball in terms of the parameters of the problem. In particular, the result says that the maximum height of a ball should be the same for every ball. The result is wrong in the context of the purpose of the model. Obviously a given person can throw a baseball higher than a bowling ball, but our model says that the mass of the ball has no effect. The problem lies in our choice of parameters. The mass m is clearly a property of the ball. A little thought should bring you to realize that the initial velocity v_0 is a property of *both* the thrower and the ball. This means that changing the ball changes both m and v_0, so the result (7) is difficult to interpret. The model could be improved by replacing the parameter v_0 with a parameter that is a property of the thrower and independent of the object being thrown. The model is corrected in Exercise 7.

Components of Mathematical Models

Mathematical quantities in models can be classified as variables, constants, parameters, and input functions. An **independent variable** is a quantity that takes on a range of values. Usually, independent variables are measures of time or position. The set of all possible values of the independent variables is the **domain** of the problem. A **dependent variable** is a quantity that changes during a given problem, depending on the value(s) of the independent variable(s). A **constant** is a quantity that has a single fixed value. In our example, there is an independent variable (t), a dependent variable (y), and a constant (g). We introduced a new dependent variable (u) to simplify the model.

A **parameter** is a quantity whose value is fixed throughout the domain of the model but can be varied to give a family of related problems. Parameters are in a sense the most important component of a mathematical model. Look again at the original statement of the problem. The goal is to determine the effect of the input parameters m and v_0 that characterize the ball and the thrower on the maximum height Y. In a sense, a mathematical model is like a function that prescribes values to the output parameters, Y in this case, for given values of the input parameters, here m and v_0.

Figure 1.3.2 is a schematic diagram illustrating the relationships between the components of the example model. The first part of the modeling process deals with the internal problem of the determination of the dependent variables in terms of the independent variables. The parameters are no different from constants at this stage. Once the model has been "solved," in the sense that the output parameters are clearly connected to the input parameters, the modeler deals with the functional relationship between the parameters. At this stage of the process, it is the parameters that serve as the variables. In our example model, the solution of the model is Equation (7).

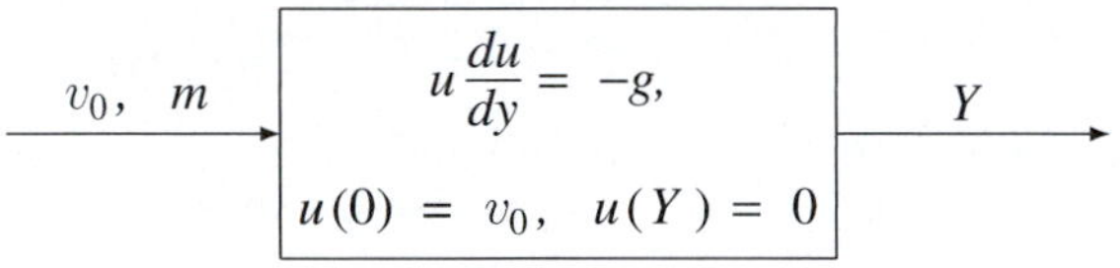

v_0, m → $u\dfrac{du}{dy} = -g$, $u(0) = v_0$, $u(Y) = 0$ → Y

Figure 1.3.2
A schematic diagram of the model for the height of a thrown ball.

The Modeling Process

Although it is not reasonable to try to prescribe a definite list of steps needed in modeling, the following outline can serve as a guide. Often consecutive steps are done together rather than in sequence.

1. Determine the purpose of the mathematical model. Often the purpose is to study a family of related experiments or observations and determine how the differences in the details lead to differences in the results.
2. Outline a conceptual model to describe an idealized representation of the situation being modeled.
3. Define symbols to represent the various quantities in the conceptual model. Be careful to keep track of which are independent variables, dependent variables, constants, and parameters.
4. Derive mathematical equations describing the relationships between the quantities, as determined by the conceptual model. As far as possible, make sure that the full set of equations forms a well-posed problem. Often this means nothing more at this stage than counting the number of algebraic conditions supplementing the differential equations.
5. Make simplifications, if possible, by defining new quantities in terms of the old ones or by combining equations. Such changes are worthwhile if they make the model easier to analyze. Often, there is no compelling reason to retain the original dependent variable.
6. Analyze the model by using analytical, graphical, and/or numerical techniques.
7. Critique the model. Do its predictions agree with everyday experience, common sense, or experimental data? If not (we assume here that you have corrected any mistakes in the calculations), is it because your intuition is wrong or because the model is flawed? Either of these answers is possible.

1.3 Exercises

1. A body is subjected to two different forces. One of these is proportional to the height y and opposite in direction to y. The other is a force of air resistance that is proportional to the velocity v and opposite in direction to v. Write down a differential equation for the height y, being careful to make sure that all parameters in the equation are positive. Note that all variables must be given in terms of t, y, and derivatives of y. What is the order of the differential equation?

2. Experiments with falling bodies suggest that in many cases the force of air resistance is proportional to the square of the velocity, rather than the velocity itself, as in Exercise 1.
   a. Explain why the simplifying assumption $F = -bv^2$ is not correct for this case. (*Hint:* does the formula work for all values of v or only for some values?)
   b. Determine the correct formula for F.

3. The gravitational force of the earth on a falling body actually depends on the height of the body according to the *inverse square law.* Let y be the height above sea level, and let R be the radius of the earth. Then
$$F_g = \frac{-A}{(R+y)^2},$$
where A is some positive constant.
   a. Explain in words why the denominator of the inverse square law involves the factor $R + y$ rather than y.
   b. Determine the constant A in terms of known parameters, given that you know the force of gravity at the earth's surface is $-mg$.
   c. Write down a differential equation to represent the height of an object dropped from a weather balloon. Be sure to state the assumptions you are using for F. Can you solve this equation simply by integrating both sides?

4. A basketball hoop is 10 ft above the ground. A small child is going to throw a basketball underhanded from an initial point just above the ground. Determine approximately the minimum velocity that the ball needs to reach a height sufficient for it to go into the basket. Assume that the only force on the ball is the force of gravity. Begin with the conceptual model, and work the problem from first principles rather than by using formulas we obtained in the analysis of the model.

5. Solve the initial-value problem
$$y\frac{dy}{dt} = -e^t, \qquad y(0) = y_0$$
and determine the interval of existence for the solution in terms of the initial value y_0.

6. The interval of existence for an initial-value problem depends in general on both the differential equation and the initial condition.
   a. Find the solution formula(s) and interval of existence for $2y\,dy/dt = \cos t,\;\; y(0) = 2$.
   b. Find the solution formula(s) and interval of existence for $2y\,dy/dt = \cos t,\;\; y(0) = 1/2$.
   c. Find the solution formula(s) and interval of existence for $2y\,dy/dt = \cos t,\;\; y(-\pi/6) = 0$.
   d. Find the values of y_0 for which the interval of existence of the solution of $2y\,dy/dt = \cos t$, $y(0) = y_0$, is $(-\infty, \infty)$.

Exercises 7 and 8 examine improved mathematical models for the problem of determining the height of a thrown ball.

7. Assume that a specific person is capable of imparting a given momentum p to an object, regardless of the size of the object. Rework Model Problem 1.3, using p rather than v_0 as a characterization of the contestant. Note that momentum is the product of mass and velocity. You should obtain the result

$$Y = \frac{p^2}{2gm^2}.$$

Write a paragraph explaining in simple terms what this result implies about the effect of changing either the ball or the thrower.

8. Consider a conceptual model for the upward motion of a ball of mass m thrown vertically with speed v_0. The ball is subject to the force of gravity and a drag force $F_d = -bv^2$ caused by air resistance.
   - *a.* Determine the correct differential equation for the velocity $v(t)$.
   - *b.* Convert the differential equation to a differential equation for $u(y)$, the velocity as a function of height.
   - *c.* Convert the differential equation for $u(y)$ to an equation for $y(u)$. What is the initial condition for this problem?
   - *d.* Solve the initial-value problem for $y(u)$.
   - *e.* Use the solution for $y(u)$ to obtain a formula for the maximum height Y.
   - *f.* A tennis ball has mass 0.06 kg and drag coefficient $b \approx 0.00037$ kg/m. Plot the maximum height for a tennis ball as a function of initial velocity, given initial velocities up to approximately 100 mi/h. (The top players on the men's professional tennis circuit can hit a serve at about 120 mi/h.) Plot the result (7) from the model without drag for comparison. How important is it to include drag?

CASE STUDY 1 Scientific Detection of Art Forgery

Measurements of radioactive decay can sometimes be used to determine the age of an object. The best-known example of dating by radioactivity is carbon-14 dating, which is used to date objects that were manufactured from living materials. The method was first described by Willard Libby and colleagues at the University of Chicago.[13] Carbon-14 dating will be considered in Exercises 4 through 6. Here we focus on the radioactive dating of lead-based paint. The problem considered here is based on the work of Bernard Keisch of Carnegie Mellon University[14] and has also been discussed by M. Braun.[15]

---

[13] W. F. Libby, E. C. Anderson, and J. R. Arnold, Age determination by radiocarbon content: World-wide assay of natural radiocarbon, *Science,* 109(1949), pp. 227–228.

[14] Bernard Keisch, Dating works of art through their natural radioactivity: Improvements and applications, *Science,* 160(1968), pp. 413–415.

[15] M. Braun, *Differential Equations and Their Applications,* 3rd ed, Springer, New York, 1983.

The Historical Context

Long ago, some unscrupulous artist discovered that there was money to be made by forging works of art. Of course it is much easier to get away with a forgery when the painting appears to have been made by a person long since deceased. Until the mid-20th century, the only way to distinguish between modern forgeries and authentic old master paintings was to have the work judged by art experts. Unfortunately for the true lover of art history, expert opinions are not always correct. A celebrated case is that of the Van Meegeren forgeries. H. A. Van Meegeren was a little-known Dutch painter of the 1930s and 1940s. At the close of World War II, Van Meegeren was arrested under the charge of collaborating with the Nazis by selling to them some paintings by the 17th-century Dutch master Jan Vermeer. Collaborating with the Nazis was a very serious charge; Van Meegeren's defense was that the paintings he had sold to the Nazis were not really 17th-century art treasures: he had actually painted them himself! The unfortunate Van Meegeren was then charged with forgery and died in jail shortly thereafter.

The Van Meegeren paintings were subjected to study by art historians, who concluded that all were forgeries with the possible exception of a painting called "Disciples at Emmaus," about which opinions were mixed. A chemical analysis showed that the painting contained small amounts of a chemical unknown in the 17th century. However, some experts maintained that the painting was too good to be a forgery. Although he could not conclusively date the painting to the 1930s, Keisch demonstrated that the paint used in the painting could not have been manufactured in the time of Vermeer, thereby offering conclusive evidence of a forgery and strong evidence of the painting having been the work of Van Meegeren.

The Uranium Decay Series

The scientific basis behind the method is the series of radioactive isotopes resulting from the decay of uranium-238, the most common isotope of uranium. The key stages in the uranium series are listed in Figure C1.1 along with the half-lives of the decays. The elements are uranium, thorium, radium, lead, and polonium.

$$\text{U-238} \xrightarrow{4.5\text{ Gyr}} \text{U-234} \xrightarrow{250\text{ kyr}} \text{Th-230} \xrightarrow{80\text{ kyr}} \text{Ra-226} \xrightarrow{1600\text{ yr}} \text{Pb-210} \xrightarrow{22\text{ yr}} \text{Po-210} \xrightarrow{138\text{ d}} \text{Pb-206}$$

Figure C1.1
The uranium series of radioactive isotopes (Gyr means 1 billion yr, kyr means 1000 yr, and d means days).

The series also includes several short-lived species not indicated in the figure; these can be omitted without noticeable error. Note that each of the isotopes listed in the series has a shorter half-life than the previous one. A sample that was originally pure uranium-238 will, given enough time, contain all these isotopes in measurable amounts.

Radioactive Isotopes in White Lead

Naturally occurring lead contains small amounts of radium as an impurity. The amounts can vary considerably, depending on the source of the lead. In the natural state, lead-210 is in equilibrium with radium-226; that is, the rate at which lead-210 decays is balanced by the rate at which it is produced by decay of radium. The radium decays slowly enough that the quantity of radium is virtually unaffected by its radioactivity. When the lead ore is refined into white lead, most of the radium is removed, but the lead-210 remains. This disturbs the equilibrium by drastically decreasing the rate at which lead-210 is produced without changing the rate at which it decays. Over a period of approximately 200 to 300 yr, the lead-210 decays to a point where it is in equilibrium with the reduced amount of radium-226.

Determining the Ratio of Lead-210 to Radium-226

Consider a sample of lead that contains r_0 atoms of radium-226 and x_0 atoms of lead-210 at some time $t = 0$. Let $r(t)$ and $x(t)$ be amounts of radium-226 and lead-210, respectively. Let $y(t) = x(t)/r(t)$ be the ratio of lead-210 to radium-226, and let $z(t)$ be the ratio of the radium decay rate to the polonium decay rate. Determine y and z. Use measurements of z to determine the authenticity of "Disciples at Emmaus."

We begin with the initial-value problems for the two isotopes. Changes in the amount of radium are due only to the decay of radium, while changes in the amount of lead are due to both the decay of lead and the decay of radium. Thus, we have the initial-value problems

$$\frac{dr}{dt} = -ar, \qquad r(0) = r_0, \tag{1}$$

$$\frac{dx}{dt} = ar - bx, \qquad x(0) = x_0, \tag{2}$$

where a and b are the relative decay rates for radium and lead decay. These rate constants can be determined from the half-lives given in Figure C1.1. They are approximately

$$a = 0.00043\,\text{yr}^{-1}, \qquad b = 0.0315\,\text{yr}^{-1}.$$

The initial conditions r_0 and x_0 are unknown. They are highly dependent on the source of the ore used for the lead; however, the ratio of lead to radium in the ore is generally known, as we shall see. The initial value of y depends primarily on the quality of the refinement operation, and this is much more predictable than the initial concentrations in the ore. This is one reason why we choose to work with y rather than x.

Now we apply the chain rule to the equation $y = x/r$ to obtain a differential equation for y:

$$\frac{dy}{dt} = \frac{r\,dx/dt - x\,dr/dt}{r^2}.$$

We can use Equations (1) and (2) to replace the derivatives; thus,

$$\frac{dy}{dt} = \frac{r(ar - bx) - x(-ar)}{r^2} = \frac{ar + (a - b)x}{r}.$$

Replacing x by ry yields the initial-value problem

$$\frac{dy}{dt} = a - (b - a)y, \qquad y(0) = y_0. \tag{3}$$

This equation can be stated in a more familiar and useful form by defining two new parameters to replace a and b. Let the parameters B and y_∞ be defined by

$$B = b - a \approx 0.031, \qquad y_\infty = \frac{a}{b - a} \approx 0.014. \tag{4}$$

Replacing $b - a$ by B and a by By_∞ puts the problem into the form

$$\frac{dy}{dt} = B(y_\infty - y), \qquad y(0) = y_0. \tag{5}$$

This is the same differential equation as that for Newton's law of cooling. The solution can easily be found, as in Section 1.1:

$$y = y_\infty + (y_0 - y_\infty)e^{-Bt}.$$

Now define the refinement factor R by

$$R = \frac{y_0 - y_\infty}{y_\infty}.$$

With Ry_∞ replacing $y_0 - y_\infty$, the final result for the lead/radium ratio is

$$y = y_\infty\left(1 + Re^{-Bt}\right). \tag{6}$$

This final result is a family of functions, with y_∞ a known constant and R a parameter that depends on the process used to refine the white lead. Refinement of lead increases the lead/radium ratio by removing radium. Suppose, for example, that the radium is reduced to 5% of its initial value. This means that r is decreased by a factor of 20, so $y_0 = 20y_\infty$, and therefore $R = 19$. Higher values of R result from better manufacturing processes. A typical contemporary value for R is about 100; with less sophisticated manufacturing operations in the 17th century, R would have been somewhat less, but probably not less than 40.

Dating Old Paint

In practice, the lead/radium ratio is not what is actually measured. The amount of radioactive impurities in white lead is very small, and it is easier to measure the radioactive decay rates than the isotope amounts. It turns out to be most useful to measure the radioactivity of radium-226 and polonium-210, rather than lead-210. Let z be defined as the ratio of the radium decay rate to the polonium decay rate. It turns out that z is very closely related to y:[16]

$$z \approx \frac{y_\infty}{y} = \frac{1}{1 + Re^{-Bt}}. \tag{7}$$

The radium/polonium decay ratios with the extreme values $R = 100$ and $R = 40$ are shown in Figure C1.2. Given experimental uncertainty, the measured z value for the painting "Disciples at Emmaus" is somewhere in the range of 0.05 to 0.15. These two values are indicated in the plot.

[16] This approximation is not at all obvious. Exercise 2 takes you through the correct calculation of z.

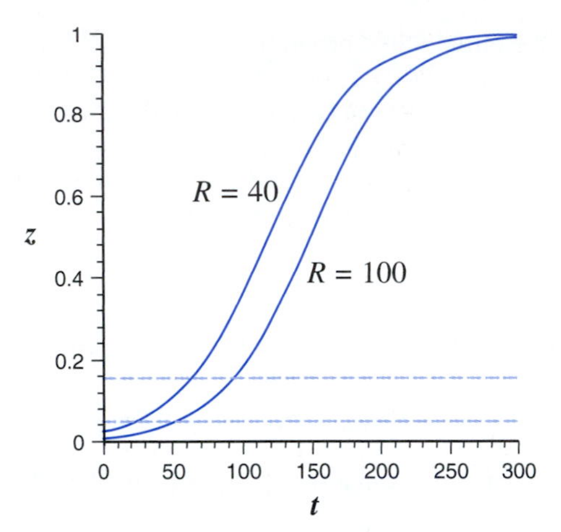

Figure C1.2
The radium/polonium decay ratio z along with the experimental data from "Disciples at Emmaus."

Discussion

The data depicted in Figure C1.2 clearly show that "Disciples at Emmaus" could not have been painted during the life of Jan Vermeer, the artist to whom the painting was originally attributed. At best, the paint could have been only 90 years old (in 1967), so it could not have been manufactured before about 1880. The paint was most likely 30 to 60 years old, which is consistent with the premise that it was actually painted by Van Meegeren. The assumption of equilibrium between polonium-210 and lead-210 makes it difficult to accurately date lead that is less than about 100 years old by this method, but it is certainly clear that the painting "Disciples at Emmaus" is a recent work and not from the 17th century.

In general, the method developed by Keisch and his coworkers is very well suited for distinguishing "old" lead (200 years old or more) from "new" lead (100 years old or less). It cannot be used for fine distinctions. Of course the mathematics is very accurate. Modern measurements are sufficiently accurate for fine distinctions, and the independence of the method from uncertainties in the composition of the original lead sample also suggests the possibility of high accuracy. However, the usefulness of the method is limited by the large uncertainty in the refinement factor R. Radium was not even discovered before 1898, and there were no measurements of R for lead manufacture until recently. The dating of old lead paint is an example in which a very accurate mathematical model is limited in value by very large uncertainty in data, yet this uncertainty is not so large as to prevent the coarse distinction necessary to distinguish 17th-century paintings from 20th-century forgeries.

Case Study 1 Exercises

Exercises 1 through 3 consider an improved model for the dating of old paint without making the approximation $z \approx y_\infty/y$. The analysis begins in Exercise 1 with the derivation of a differential equation to determine $w = 1/z$. This equation is solved in Exercise 2 to determine z for a particular refinement ratio R. It is also solved approximately in Exercise 3 by using a quantifiable approximation. The results are then compared with the approximation $z \approx y_\infty/y$.

1. *a.* Let p be the amount of polonium-210 in a sample of lead. Write down the differential equation for p, assuming that it decays with relative decay rate c. (Remember that polonium-210 is produced by the decay of lead-210.)
   b. Let $w = cp/ar$. (Note that $w = 1/z$.) Differentiate this equation, keeping in mind that r is not a constant, to obtain the differential equation
$$\frac{dw}{dt} = \frac{cb}{a}y(t) - (c - a)w.$$
   c. The initial value of w is not known, but it can be reasonably estimated. Prior to refinement of the lead ore, $w = w_\infty = 1$. What should the initial condition for w be if the refinement factor for polonium is the same as that for radium?

T 2. *a.* The value of c is easily determined from the half-life data in Figure C1.1, and we have already obtained values for a and b and a solution for $y(t)$. Use a computer algebra system to solve the initial-value problem for w (from Exercise 1) with $R = 40$ and again with $R = 100$. Compute z from w.
   b. Plot the graphs of z as determined in part a along with the approximate z given in Equation (7). Is the approximation $z \approx y_\infty/y$ justified? Does the answer to this question depend on the time at which values of z are needed?

T 3. *a.* Substitute the result (6) for y into the differential equation for w (from Exercise 1). Use the approximation that a is much less than either b or c to obtain the approximate differential equation
$$\frac{dw}{dt} = c + cRe^{-Bt} - cw.$$
   b. Assume $w(t) = e^{-ct}W(t)$. Use the product rule to obtain an initial-value problem for $W(t)$. Note that the differential equation is of the form $dW/dt = f(t)$.
   c. Solve the initial-value problem for $W(t)$ and then determine $w(t)$.
   d. Plot the improved z of this exercise for $R = 40$ and $R = 100$ along with the approximate z from the text. At what point in time can we use the simpler formula without significant error?

Radiocarbon Dating

Naturally occurring carbon consists primarily of the stable isotopes carbon-12 and carbon-13, but the interaction of sunlight with nitrogen in the atmosphere produces the radioactive isotope carbon-14. The rate of carbon-14 production and the rate of carbon-14 decay back to nitrogen are approximately in balance, so the fraction of carbon-14 in the carbon of the atmosphere is essentially constant. Samples of living tissue have very nearly the same proportion of carbon-14 as the atmosphere, owing to the metabolism of the living organism. After the organism dies, the carbon-14 in its tissues continues to decay, but no new carbon-14 is introduced. Hence, the amount of carbon-14 (and also the decay rate of carbon-14) satisfies the decay equation with initial conditions given at some time 0 having unknown calendar date. Either the amount of carbon-14 in a sample or the rate of carbon-14 decay is measured to yield the necessary data for the dating calculations. The half-life of carbon-14 is 5730 yr, making radiocarbon dating suitable for objects from about 500 to 50,000 yr old.

4. Mass spectrometry measurements in 1988 indicated that the cloth of the Shroud of Turin contained 91% as much carbon-14 as would be contained in a new sample of the same cloth.
   a. Assuming that radioactive decay is the only process that has acted to change the amount of carbon-14 since the manufacture of the cloth, estimate the date of the Shroud. Note that the calculation gives the date of the cloth, not the object made from the cloth.
   b. For many years, the half-life of carbon-14 was thought to be 5568 yr. What is the approximate date of the Shroud of Turin if we use this value for the half-life?
   c. What other assumptions are implied by the mathematical model that have not been explicitly stated?
   d. Let F be the fraction of carbon-14 atoms in normal living tissue today. Let F_0 be the fraction of carbon-14 atoms in the Shroud of Turin at the time its cloth was manufactured, and assume that F_0 is not necessarily the same as F. Suppose the Shroud of Turin was actually 1950 yr old in 1988, when the fraction of carbon-14 was measured to be $0.91F$. Determine F_0 in terms of F.

5. In 1977, a charcoal fragment from Stonehenge was found to have about 8.2 carbon-14 disintegrations per minute per gram.
   a. The decay rate of carbon in living tissues is 15.3 carbon-14 disintegrations per minute per gram. Assuming that the charcoal was made by the builders of Stonehenge, estimate the date at which Stonehenge was built.
   b. Nuclear weapons tests in the 1950s introduced carbon-14 into the atmosphere. Prior to these tests, the decay rate in living tissues was 13.5 carbon-14 disintegrations per minute per gram. Estimate the date of Stonehenge, using this old value for the natural decay rate.
   c. Which of the estimates of parts a and b do you think is better supported by the data? Why?

6. A check on the accuracy of the radiocarbon dating method is possible for artifacts that can be dated accurately by nonscientific methods. The first scientific study to employ radiocarbon dating studied a wooden artifact from the tomb of Sneferu, who historians can date from written Egyptian records to have died between 2700 B.C. and 2550 B.C.
   a. In 1949, the decay rate of the wood in the artifact was measured to be 6.6 to 7.2 disintegrations per minute per gram. Use these values to obtain a range for the age of the artifact.
   b. Although we now accept 13.5 disintegrations per minute per gram as the correct figure for the pre-nuclear weapon disintegration rate, numerous measurements on fresh wood samples in the same 1949 study yielded an estimate of 12.5 disintegrations per minute per gram for the decay rate of living tissue. Use this value to revise the calculations for the age range of the artifact.
   c. Can you find any reasons why the results in part b are better than those in part a?

chapter 2

Basic Concepts and Techniques

In this chapter, we consider a variety of concepts and techniques that give the reader some basic tools to use in analyzing first-order differential equations. Other techniques for first-order equations appear in Sections 4.2, 4.3, 4.5, A.1, and A.3.

Section 2.1 introduces a collection of mathematical models from a variety of disciplines in science and engineering, including chemistry, biology, fluid flow, and Newtonian mechanics. Some additional modeling techniques are discussed in the context of these models.

The method of separation of variables is discussed in Section 2.2. This method can be used to obtain solution formulas for many first-order equations. The formulas cannot always be solved explicitly, so the issue of implicit solutions needs to be addressed as well.

Section 2.3 introduces the slope field, a method for visualizing the solutions of first-order differential equations that is based on interpreting the differential equation not as a problem to be solved, but as a statement about the rate of change of a function.

Some conceptual and theoretical issues are raised in Section 2.4, focusing on the related questions of whether an initial-value problem has a unique solution and whether new functions can be defined by initial-value problems.

Sections 2.5 and 2.6 deal with the question of how to obtain approximate solutions for differential equations that cannot be solved exactly. Since most differential equations are in this category, these sections are of significant importance to practitioners of mathematical modeling. The most basic method is discussed in Section 2.5, primarily for its conceptual value. Section 2.6 develops some methods that are quite useful in practice.

2.1 A Collection of Mathematical Models

In this section, we introduce a collection of mathematical models from a variety of disciplines. Some additional concepts of mathematical modeling arise from the discussion.

Population Dynamics of Diseases

Mathematical models can be used to study the population dynamics of diseases. These models begin with a classification of the population into groups according to their status. A complicated model might subdivide the population into those who are susceptible, those who have been exposed but are not yet infective, those who are infective, those who are recovered and immune, and perhaps those who are asymptomatic carriers. Then it is necessary to determine the pathways between the classes (e.g., only exposed individuals can become infective) and model the mechanisms for each pathway.

MODEL PROBLEM 2.1a

Construct a simple mathematical model to study the population dynamics of infectious diseases.

The simplest model might be for a disease in which everyone can be classified as either infective or susceptible. A conceptual model for our idealized endemic disease might assume the following:

1. The host population is constant in size, with no births or deaths.
2. Individuals in the population encounter other individuals randomly at a fixed rate.[1] This means that the number of encounters per day is fixed, but the identity of the encountered individual is chosen randomly.
3. A fixed proportion of contacts between healthy and infected individuals results in the transmission of the infection.
4. Infected individuals recover at a constant relative rate.[2]

Let N be the total population, and let $I(t)$ be the population of infected individuals. The population I increases because of infection transmission and decreases because of recovery. To determine the rate of transmission, consider a single healthy individual. The individual encounters others at a rate proportional to the population, say kN. Of these encounters, only a fraction I/N will be with infected individuals. Our healthy individual therefore encounters those who can transmit the infection at a rate of kI. There are a total of $N - I$ healthy individuals, each of whom has the same rate of contact with infected individuals, so the overall rate of contacts that can transmit the disease is $kI(N - I)$. Of these, some proportion p results in new infections, so the transmission rate is $pkI(N - I)$. The recovery rate is proportional to the infected population, or rI for some positive constant r. Thus, the population of infected individuals changes according to the differential equation

$$\frac{dI}{dt} = pkI(N - I) - rI. \tag{1}$$

As it currently stands, there are four parameters in the model: p, k, r, and N. If, as in Section 1.3, our goal is to understand the role that parameters play in the behavior of the disease, it would be nice to have fewer parameters. Of course we can reduce the number by 1 simply by

[1]This is, of course, a great oversimplification. What would be an excellent assumption regarding gas molecules in a container sounds silly when applied to humans in a population. However, excellent results can sometimes be obtained in spite of oversimplification.

[2]This assumption is also problematic. It fails to distinguish between those who have been sick for a day and those who have been sick for a week.

noticing that p and k only occur in the combination pk. We can reduce the number of parameters further by the process of **nondimensionalization,** in which dimensional quantities, such as the time t, are replaced by new dimensionless quantities. Nondimensionalization is accomplished in two steps, with dimensional variables replaced by dimensionless counterparts in the first step and parameters grouped into a smaller number of dimensionless parameters in the second step. It is helpful to further subdivide the variable replacement step into substeps:

1. Replace the dimensionless variables with nondimensional counterparts.
   a. Choose a reference quantity x_r for each dimensional variable X. The reference quantities must be parameters in the model or products/quotients of parameters and must be dimensionally consistent with the dimensional variable. For example, if X is a length, then x_r must be a length also. Ideally one should try to choose **scales,** which are reference quantities that are of a magnitude comparable to the expected values of the variables.[3]
   b. Define a dimensionless quantity x for each dependent variable X by X/x_r, where x_r is the reference quantity chosen for X.
   c. Systematically replace the dimensional dependent variables with their dimensionless counterparts.
   d. Define a dimensionless quantity τ for each independent variable t, as was done with the dependent variables.
   e. Systematically replace the dimensional independent variables with their dimensionless counterparts. Use the chain rule to replace derivatives.
2. Rearrange the equations in the model so that the remaining parameters appear in dimensionless combinations. Define dimensionless parameters to replace the combinations of dimensional parameters. To the extent possible, one should try to choose dimensionless parameters that have a clear meaning in the conceptual model.

The disease model has I as the only dependent variable. The dimension of I is *people.* The parameter N also has the dimension *people.* We can define a new variable i by

$$i = \frac{I}{N}.$$

Note that i is the fraction of the population that are infected. Since the fraction infected can in principle range from 0 to 1, N is a scale for the population I.

All instances of I are now replaced by Ni, keeping in mind that N is a constant. The resulting equation is

$$N\frac{di}{dt} = pkNi(N - Ni) - rNi,$$

which simplifies to

$$\frac{di}{dt} = pkNi(1 - i) - ri.$$

What choices do we now have for reference times? For the sake of discussion, assume that some public health measure, such as a vaccination program, can be enacted to reduce the transmission

[3]For example, in many population dynamics models, a convenient time scale would be the average life span of the population.

probability p to zero. The differential equation is then simplified to $di/dt = -ri$. Over a short interval of time, this can be approximated as $\Delta i/\Delta t \approx -ri$, or

$$\Delta t \approx -\frac{1}{r}\frac{\Delta i}{i}.$$

Roughly speaking, $1/r$ is then the amount of time needed for significant changes in i. So we can think of $1/r$ as a rough measure of the duration of the illness. Similarly, retaining the infection process and ignoring the recovery process mean that, at worst, $di/dt = pkNi$. By the same reasoning as before, we identify $1/pkN$ as a rough measure of the time required for a new epidemic to get established.

Either $1/r$ or $1/pkN$ could be used for a reference time. Perhaps $1/r$ is slightly better because it is probably easier to measure the typical duration of the disease than the infection time. We therefore define the dimensionless time by

$$\tau = rt.$$

Now we can think of i as a function of the new time variable $\tau(t)$. In other words,

$$i = i(\tau(t)).$$

Differentiating with respect to t yields

$$\frac{di}{dt} = \frac{di}{d\tau}\frac{d\tau}{dt} = r\frac{di}{d\tau},$$

where we have used the equation $\tau = rt$ to determine $d\tau/dt$. With this change, the differential equation becomes

$$r\frac{di}{d\tau} = pkNi(1-i) - ri.$$

We have now completed the first step in the nondimensionalization process. The *variables* are dimensionless, but the *equation* is not. However, we can divide the equation by r to obtain the dimensionless equation

$$\frac{di}{d\tau} = \frac{pkN}{r}i(1-i) - i.$$

The nondimensionalization process has created the dimensionless parameter pkN/r, which we can identify as the ratio of the infection rate to the recovery rate when i is small. The final form of the model, after simplification, is

$$\frac{di}{d\tau} = i[(R_0 - 1) - R_0 i], \qquad R_0 = \frac{pkN}{r}. \tag{2}$$

Note that the differential equation has two equilibrium solutions, $i \equiv 0$ and $i \equiv (R_0 - 1)/R_0$. The first corresponds to the disease-free state. The second corresponds to an endemic[4] disease equilibrium, provided $R_0 > 1$. If $R_0 < 1$, the second equilibrium solution has no meaning in the context of the model. Here we have a very important result obtained with very little mathematics.

[4]An **endemic** disease is one that is always present to some extent.

Apparently,[5] the disease cannot get established if $R_0 < 1$. Another interpretation of R_0 is that it is a rough measure of the average number of new infections caused by one infected individual when the infection is just getting started. In this context, R_0 is called the **basic reproductive number.**

The result we obtained for Model Problem 2.1a is indicative of the value and limitation of modeling with elementary models. The model is too simple to yield testable quantitative predictions. It surely misses many important qualitative features of real diseases. However, it provides the motivation for the definition of the basic reproductive number, a concept that can be extended to realistic models. The formula for R_0 in terms of the original parameters in a more realistic model will certainly be different from the formula we got for the simplest model; however, the qualitative features of our formula are still important. We can see, for example, that R_0 can be reduced by shortening the typical recovery time $1/r$, which means that it might be possible to change a disease from endemic to sporadic simply by devising a way to shorten the recovery time. This is a fact that probably was not so obvious before our mathematical investigation.

Water Drainage

Fluid flow usually requires models that are too complicated to consider in a first course in differential equations. However, the problem of determining the height of water in a leaking bucket is simple enough to include in a first course.

MODEL PROBLEM 2.1b

Consider a cylindrical bucket of constant cross-sectional area A, with a hole of cross-sectional area a, as shown in Figure 2.1.1. The small hole is plugged, and the bucket is filled to depth h_0. A clock is started as the plug is removed and the water begins to leak out of the hole. Construct a model to determine the height $h(t)$.

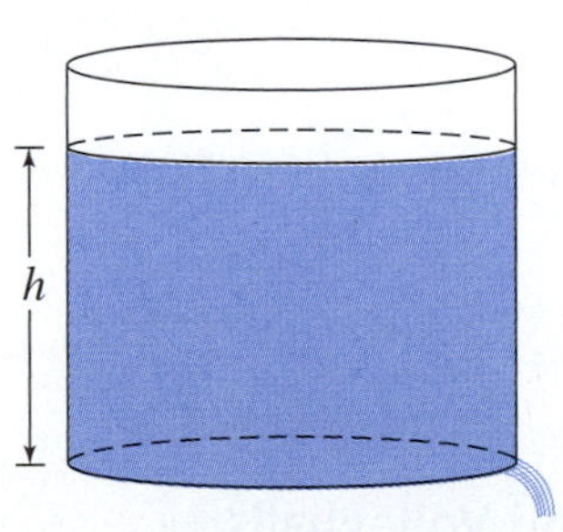

Figure 2.1.1
A leaking bucket.

The differential equation for a leaking bucket does not appear in most introductory physics textbooks. In the 21st century, this is not a significant difficulty for a classical problem such as

[5] This is confirmed in Section 2.3, Exercise 21.

this. A simple Internet search for *leaking bucket* yields several websites that contain the necessary information. The key idea is that the water flow is a response to differences in pressure between the top and bottom of the water column. The pressure difference (ΔP) is related to the velocity of the water column by Bernoulli's law,

$$\Delta P = \frac{1}{2}\rho u^2,$$

where ρ is the density of the water and u is the velocity. The pressure difference is also given by the weight per unit area of the water column

$$\Delta P = \rho g h,$$

where g is the gravitational constant. By combining these formulas, we obtain the velocity of the water exiting from the bucket as

$$u = \sqrt{2gh}.$$

The rate of change of the water volume V is easily determined from the velocity of the water; specifically, the flow rate (volume/time) is the product of the velocity (length/time) and the cross-sectional area, so

$$\frac{dV}{dt} = -au = -\sqrt{2gh}\,a.$$

The model is completed by noting the simple relationship

$$V = Ah$$

among volume, height, and cross-sectional area. Differentiating this last equation and combining the result with the previous equation yields a differential equation for the height:

$$A\frac{dh}{dt} = -\sqrt{2gh}\,a.$$

We can simplify the model further by combining the parameters into a single parameter:

$$\frac{dh}{dt} = -k\sqrt{h}, \qquad k = \sqrt{2g}\,\frac{a}{A}. \tag{3}$$

Rockets and Volleyballs

In Section 1.3, we created a mathematical model for motion caused by the force of gravity. The motion was limited to one dimension, and we made a number of simplifying assumptions.

MODEL PROBLEM 2.1c

Create a mathematical model for the position of a projectile moving through a resistive medium (such as air) and acted on by gravity.

In general, models of motion are derived from a vector version of Newton's second law of motion, familiar in one dimension as $F = ma$:

$$m\frac{d^2\mathbf{r}}{dt^2} = \mathbf{F},$$

where $\mathbf{r}$ is the position vector of the object and $\mathbf{F}$ is the vector of all forces acting on the object. Aside from geometric considerations, models in Newtonian mechanics are distinguished by the variety of forces that can be included in different settings. We consider some of the forces that might be present.

Gravitational Forces In Section 1.3, we assumed that the gravitational force was proportional to the mass of the object and was directed downward. In vector notation, this is

$$\mathbf{F}_g = -mg\mathbf{k},$$

where g is the gravitational constant and the vector $-\mathbf{k}$ is the unit vector that identifies the force as being directed downward. This approximation can be used when the changes in position of the object are sufficiently small. In general, however, the gravitational force on a falling body actually depends on the height of the body from the center of the earth. Let z be the height above sea level, and let R be the radius of the earth. Then

$$\mathbf{F}_g = -mg\frac{R^2}{(R+z)^2}\mathbf{k}.$$

Note that the magnitude of the force is just mg if $z = 0$, and it differs only slightly from mg if z is much smaller than R. For an object located far from the earth's surface, the gravitational force is significantly reduced.

Damping Forces Imagine a scene from an old Western movie. A cowboy is sitting at one end of a bar and calls out to the bartender for a whiskey. The bartender pours the whiskey and then slides the glass across the bar so that it stops right in front of the cowboy. The glass stops because of the damping force of friction.

> A **damping force** is a force that acts to oppose an object's motion.

The whiskey glass is moving toward the cowboy, but the force of damping acts in the direction away from the cowboy. This makes the glass slow down. Eventually, the velocity of the glass reaches zero, at which point the damping force acting on it also vanishes.

To model damping forces, we need a formula that gives the force as a function of the velocity. For damping caused by friction, as in the Western movie, it is often reasonable to assume that the magnitude of the force does not actually depend on the velocity of the object. This is best explained by noting that the velocity of the object acts in a horizontal direction, while the force of gravity that causes the friction acts in a vertical direction. If we assume that the frictional force is proportional to the weight mg of the object, we have the formula

$$F_d = -\phi mg \operatorname{sgn}(v),$$

where ϕ is a constant that depends on the roughness of the surfaces and sgn (v) is the sign of the velocity ($+1$ if positive, -1 if negative, and 0 if the object is not moving).[6] The given formula assumes one-dimensional motion in a horizontal direction.

Objects traveling through a fluid experience a damping force called **drag.** The drag force can be considered to be the net force resulting from friction between the object and the fluid particles. It is common to assume

$$\mathbf{F}_d = -\beta \mathbf{v}$$

for drag forces, where $\mathbf{v}$ is the velocity vector and β is a constant. This is a crude approximation made as much for mathematical convenience as for quantitative accuracy. It is often most accurate to assume that the magnitude of the drag force is proportional to the square of the speed. This yields a force vector with magnitude $\beta \|\mathbf{v}\|^2$, where $\|\mathbf{r}\|$ is the magnitude of any vector $\mathbf{r}$, and with direction given by the unit vector $-\mathbf{v}/\|\mathbf{v}\|$; hence,

$$\mathbf{F}_d = -\beta \|\mathbf{v}\| \mathbf{v}.$$

Restoring Forces Now replay the bar scene with two big differences. First, imagine that the friction is somehow eliminated. Second, imagine that the bartender has looped an elastic band around the glass. The bartender slides the glass toward the cowboy. This time, the glass slows as it reaches the cowboy, but it stops only momentarily and then slides back along the bar to the bartender. The elastic band exerts a restoring force on the glass.

A **restoring force** is a force that acts to oppose an object's displacement.

The glass slows because the force acts in the direction opposite to its motion, as in the damping case. However, the magnitude of the force is an increasing function of displacement rather than speed. The force increases as the glass slows down, so the glass eventually changes direction and returns toward its starting position.

For a restoring force, we need a formula that gives the force as a function of the displacement. A reasonable approximation is that the force is proportional to the displacement, so we have

$$\mathbf{F}_r = -k(\mathbf{r} - \mathbf{r}_0),$$

where k is a positive constant that measures the strength of the restoring force and $\mathbf{r} - \mathbf{r}_0$ is the point at which the restoring force is zero. As with the damping force, we are making a simplifying assumption rather than using an exact result. The formula given here is known as *Hooke's law,* but it must be considered as a rough approximation. For most cases, Hooke's law is a good approximation as long as the displacements are not too great.

Motion of a Projectile Suppose a projectile is subject to the force of gravity and to air resistance. Assume that the horizontal component of the motion is always in the direction of the unit vector **i.** We may then assume that the motion is two-dimensional, with no y component. If we use the

[6] The function sgn is called the *signum* function. This function can sometimes be thought of as sgn $(v) = v/|v|$, but this formula is not correct when $v = 0$.

more complicated model for the force of gravity and the assumption that the magnitude of the air resistance is proportional to the square of the speed, we have

$$m\,\frac{d^2\mathbf{r}}{dt^2} = -\beta\,\|\mathbf{v}\|\,\mathbf{v} - mg\,\frac{R^2}{(R+z)^2}\mathbf{k},$$

where the displacement vector $\mathbf{r}$ has horizontal component x and vertical component z. Substituting $\mathbf{v} = d\mathbf{r}/dt$ and dividing by m, we have

$$\frac{d^2\mathbf{r}}{dt^2} + b\left\|\frac{d\mathbf{r}}{dt}\right\|\frac{d\mathbf{r}}{dt} + g\frac{R^2}{(R+z)^2}\mathbf{k} = \mathbf{0}, \qquad b = \frac{\beta}{m}. \tag{4}$$

This equation is more complicated than necessary for most problems. If the object is far enough from earth that the inverse-square gravitation formula is needed, then air resistance is certainly negligible for most of the motion. Hence, there are two useful simplifications of the model for projectile motion. For the flight of objects that travel large vertical distances, we can often neglect the air resistance altogether and use

$$\frac{d^2\mathbf{r}}{dt^2} + g\frac{R^2}{(R+z)^2}\mathbf{k} = \mathbf{0}. \tag{5}$$

For projectile motion with air resistance but small vertical distances relative to the radius of the earth, the appropriate model is

$$\frac{d^2\mathbf{r}}{dt^2} + b\left\|\frac{d\mathbf{r}}{dt}\right\|\frac{d\mathbf{r}}{dt} + g\mathbf{k} = \mathbf{0}. \tag{6}$$

The discussion of projectile motion indicates the need to think about what model is appropriate for a given set of circumstances. In Section 5.3, we examine the flight of a spacecraft as it leaves the moon's gravity field. The correct model is Equation (5). The variation of the gravitational force with height must be included. The moon has no atmosphere; even if it did, one could probably justify omitting air resistance. Case Study 2 models the serve of a volleyball. We'll see that it is necessary to include air resistance for the model to be useful. The best model to choose for this problem is Equation (6). Equation (4) is more accurate, but the error made in neglecting the height dependence of the gravitational force is surely smaller than the error made in neglecting the spin of the ball. It is foolish to retain features in a model that one expects to be less important than some omitted feature.

Chemical Kinetics

Chemical kinetics is the area of chemistry that is concerned with the *rates* of chemical reactions; hence, it is natural to anticipate that the area will offer many opportunities for mathematical modeling. We focus here on the key ideas and one interesting problem that is worthy of study but often neglected.

MODEL PROBLEM 2.1d

A chemical reaction occurs between compounds A and B. The reaction yields a product C and liberates a quantity of heat, which raises the temperature and influences the reaction rate.

Model Problem 2.1d suggests a general model rather than a model with a specific physical setting in mind. As we develop the model, we will make whatever assumptions are needed in the conceptual model to try to keep the model simple but still incorporate the effects of temperature changes. Only when the model is complete will we know the full conceptual model to which it corresponds. A careful description of the conceptual model would be helpful to a chemist interested in testing the predictions of the model. At best, we would expect the model to yield useful predictions only when the assumptions in the conceptual model are approximately valid.

Suppose the initial amount of reactant A is x_0 and the initial amount of reactant B is y_0. The simplest case occurs when the reaction is caused by collision of one molecule of A with one molecule of B. With these assumptions, we get the equations

$$\frac{dx}{dt} = -kxy, \qquad x(0) = x_0, \qquad \frac{dy}{dt} = -kxy, \qquad y(0) = y_0,$$

where $x(t)$ is the amount of A remaining, $y(t)$ is the amount of B remaining, and k is the relative rate of reaction at room temperature T_0. Note that $d(y-x)/dt = 0$, so $y - x = y_0 - x_0$, or $y = y_0 - (x_0 - x)$. Here we make another assumption to simplify the conceptual model. If y_0 is much larger than x_0, then the amount of B is essentially unchanged during the course of the reaction. This assumption allows us to take y as a constant in the differential equation for x.

So far, we have neglected the effect of temperature on the reaction rate. A convenient way to incorporate the effect of temperature is to replace the relative rate constant ky_0 with the rate "constant" $ky, a(T)$, where $a(T_0) = 1$ and $a' > 0$. Thus, a is the ratio of the reaction rate at temperature T to the reaction rate at temperature T_0. We defer for the moment the task of specifying the function a. With temperature dependence included, the differential equation is

$$\frac{dx}{dt} = -Ka(T)x, \qquad x(0) = x_0, \qquad K = ky_0.$$

Rather than use the amount of reactant as the dependent variable, it is more convenient to use the progress variable $z(t)$, which we define as the ratio of the amount of reactant used to the amount of reactant initially available. Thus,

$$z(t) = \frac{x_0 - x(t)}{x_0}.$$

We then have

$$\frac{dz}{dt} = -\frac{1}{x_0}\frac{dx}{dt}, \qquad x = x_0(1-z),$$

leading to the initial-value problem

$$\frac{dz}{dt} = Ka(T)(1-z), \qquad z(0) = 0.$$

Suppose the reaction of 1 unit of A and 1 unit of B releases enough heat to raise the temperature by s degrees. The total amount of temperature change if all of A is consumed will then be sx_0. When a fraction z of the reaction is complete, the temperature change will have been sx_0z; thus,

$$T = T_0 + sx_0z,$$

and we obtain the model

$$\frac{dz}{dt} = Ka(T_0 + sx_0 z)(1 - z), \qquad z(0) = 0. \tag{7}$$

There is another important feature of the conceptual model to be discerned at this point. In assuming that the temperature increase is proportional to the reaction progress, we are assuming that all the heat produced by the reaction stays in the mixture.[7] In the conceptual model, the reaction vessel insulates the reaction so that no heat can escape. This is not technically possible, of course; but perhaps it is possible to insulate the reaction sufficiently well that the heat loss is unimportant. The mathematical model also neglects spatial variations of temperature, which is tantamount to assuming that the reacting mixture remains well mixed. It is sufficient to include in the conceptual model the requirement that the chemical species involved in the reaction be gases that are initially well mixed.

It remains to choose a function a. This is where the model gets messy. We might be tempted to assume that the reaction rate a is simply proportional to the temperature, as we have already assumed that the reaction rate is proportional to the amount of A and the temperature is proportional to the reaction progress. However, those assumptions are based on plausible arguments. There is no reason to think that doubling the temperature should double the reaction rate. In fact, experiments have led to the formula

$$Ka(T) = re^{-E/RT},$$

where R is a universal constant, E a property of the chemical reaction called the activation energy, and r a rate constant for the particular reaction. Use of this equation requires that T be the absolute temperature in kelvins, rather than the relative temperature in degrees Celsius. (Temperatures given in degrees Celsius are converted to kelvins by adding 273.2°.) By definition, K is the rate constant at temperature T_0, and $a(T_0) = 1$, so

$$K = re^{-E/RT_0}.$$

Dividing these last two equations yields a formula for a:

$$a(T) = e^{E/RT_0}\, e^{-E/RT}. \tag{8}$$

Equations (7) and (8) constitute a complete model, but the model should be made dimensionless before analysis is begun. Ultimately, the number of parameters in the model is reduced from five to two, with the model taking the final form

$$\frac{dz}{d\tau} = (1 - z)\, e^{\rho z/(1+\beta z)}, \qquad z(0) = 0, \tag{9}$$

where τ is a dimensionless time and β and ρ are dimensionless parameters defined by

$$\beta = \frac{sx_0}{T_0}, \qquad \rho = \frac{E}{RT_0}\frac{sx_0}{T_0}.$$

[7] If not, then more heat is lost at higher temperatures.

 **INSTANT EXERCISE 1**

Nondimensionalize the model given by Equations (7) and (8).

2.1 Exercises

1. Discuss the conceptual model for the endemic disease.
   a. Explain why each assumption simplifies the mathematical model as compared with a more realistic assumption.
   b. For each assumption, suggest realistic circumstances for which that assumption might be a reasonable approximation.
   c. Explain why this model is not suitable for chicken pox. (*Hint:* think about the classes of individuals.)
   d. Do you think this model should give good results for the common cold? Explain your reasoning.

2. Suppose $y(t)$ is the proportion of people in a city who have heard a particular rumor. Suppose the rate at which the rumor is spread, relative to the population of people who have not yet heard the rumor, is proportional to the proportion of people who have heard the rumor. Derive the differential equation for the spread of the rumor. This model is also used for the spread of technological innovations within a given society.

3. Consider the leaking bucket problem of Equation (3). Assume that the initial height of the water is h_0 and that the bucket drains until it is empty.
   a. Determine a reference time by applying the argument in the text used to identify $1/r$ as a reference time for the disease model. [Arrange the difference equation for Δt so that it has the form $\Delta t = X(\Delta h/h)$, where X is whatever is left after you factor out the relative change. Then replace any dependent variables in X with their reference quantities, and argue that $|X_r|$ is a reasonable choice for a reference time.]
   b. Think of time in the dimensional model (3) as a function of height, and write down the corresponding differential equation.
   c. Determine the time required for the bucket to drain by integrating the differential equation from part *b* with respect to h from the initial height h_0 at time 0 to the final height 0 at time t_e.
   d. Based on the result of part *c*, what does the reference time from part *a* actually represent?
   e. The result of part *d* confirms that the reference time of part *a* is a scale for time. Nondimensionalize the model, using that reference time and the "obvious" reference length.

4. Suppose the cross-sectional area A of a leaking bucket is actually a function of h rather than a constant. Derive the differential equation for this more general version of the leaking bucket problem. (What part of the derivation is changed by having A depend on h?)

5. Develop a mathematical model for a pendulum with linear damping (Fig. 2.1.2).

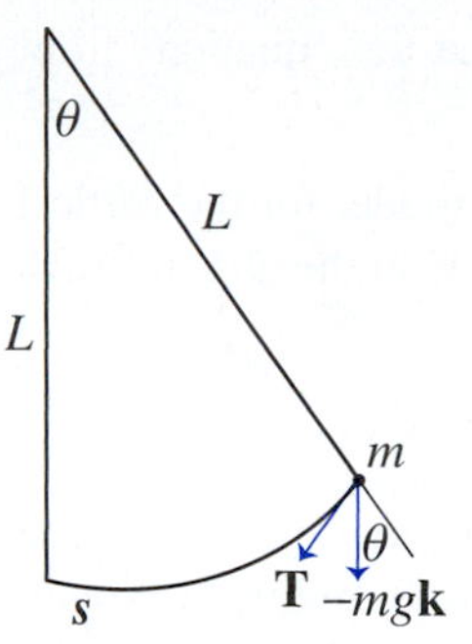

Figure 2.1.2
Notation for the pendulum of Exercise 5.

a. Begin with the differential equation

$$\frac{d^2\mathbf{r}}{dt^2} + b\,\frac{d\mathbf{r}}{dt} + g\mathbf{k} = \mathbf{0}.$$

Let θ be the angle the pendulum rod makes with respect to a vertical line, and let s be the arc of the circle of radius L measured from the bottom of the circle to the pendulum bob. Let $\mathbf{T}$ be a unit vector that is tangent to the circle. Use the geometry of the pendulum to determine the component in the $\mathbf{T}$ direction of each term in the differential equation. You should get a second-order differential equation of the form $s'' = f(\theta, s')$, where derivatives are with respect to time.

b. Use the geometric relationship between s and θ to obtain an equation of the form $\theta'' = g(\theta, \theta')$.

c. Nondimensionalize the differential equation. The angle θ is already dimensionless. In choosing a reference time, do not use the parameter b because the damping can be expected to be relatively unimportant. Think about the dimensions of the other parameters in the differential equation.

6. Suppose an object moves under the influence of gravity through a resistive medium with the horizontal component of the motion limited to one direction. Equation (6) can be recast as a system of two first-order differential equations for this case. Let $x(t)$ be the distance along the horizontal axis, and let $z(t)$ be the distance along the vertical axis. Let $s(t)$ be the speed of the motion, and let $\theta(t)$ be the angle between the velocity vector and the horizontal axis.

a. Resolve the vector equation (6) into horizontal and vertical components.

b. Write down an equation giving s^2 in terms of dx/dt and dz/dt. Differentiate this equation with respect to t, and use the equations from part *a* to remove the second derivatives. Use algebraic manipulation to reduce the result to a differential equation of the form $ds/dt = f(s, \theta)$.

c. Write down equations for dx/dt and dz/dt in terms of s and θ. Differentiate the equations with respect to t. Use the equations from part *a* to remove the second derivatives. Use algebraic manipulation to get two differential equations with derivatives ds/dt and $d\theta/dt$.

d. Combine the equations from parts b and c to obtain a first-order differential equation for θ.

7. Consider a model for the vertical flight of a projectile launched from earth from the ground with speed V in the absence of air resistance. By Equation (5), this is

$$\frac{d^2z}{dt^2} = -\frac{gR^2}{(R+z)^2}, \qquad z(0) = 0, \qquad \frac{dz}{dt}(0) = V.$$

a. Suppose the projectile is a rocket that will be used to launch a satellite. Explain why R is a good reference length. Nondimensionalize the model, using R as the reference length. For the reference time, use the time required to travel the distance R at the constant velocity V. Your dimensionless model should include one dimensionless parameter.

b. Dimensionless parameters can be thought of as a ratio of two competing reference quantities. Use your dimensionless parameter from part a to find an alternative reference length.

c. Nondimensionalize the model again, this time using the alternative reference length. Choose the reference time in the same manner as in part a. You will have to do some algebraic manipulation to get the model to appear with all parameters grouped into a single dimensionless combination.

d. Models can sometimes be reduced to simpler models by using an **asymptotic approximation.** The idea is that if all reference quantities are scales, then the relative magnitudes of the terms in the equations can be determined just from the sizes of the dimensionless parameters. What do you have to assume about your dimensionless parameter to reduce the scaled problem to $Z'' = -1$?

e. What has to be true about the initial velocity V so that the dependence of gravitational force on height can be neglected?

8. Consider the spread of a disease through a region. The model is similar to the model (1) for the spatially mixed population, except that I and N are now population densities, in individuals per unit area, and I now depends on location as well as time. For simplicity, we assume that the disease is spreading only in the x direction and that the recovery rate is negligible. With these assumptions, we have the model

$$\frac{\partial I}{\partial t} = D\frac{\partial^2 I}{\partial x^2} + pkI(N-I).$$

The equation reflects the assumption that changes in the density of infected individuals are caused by spatial spread and transmission to susceptible individuals. The coefficient D is a measure of the speed of spatial spread. Nondimensionalize the model.

9. Assume that the chemical reaction discussed in the text proceeds from $z = 0$ to $z = Z$, where Z is some fixed value with $0 < Z < 1$. Let τ_Z be the time required for the process. Determine an integral formula for τ_Z by writing the chemical reaction model (9) as a differential equation for τ in terms of z and then integrating the equation from $z = 0$ to $z = Z$.

T 10. Fix $\beta = 1$. Plot $\tau_{0.99}$ as a function of ρ for $0 < \rho < 10$, where τ_Z is defined in Exercise 9.

11. Show that the integral τ_1, where τ_Z is defined in Exercise 9, representing the time required for the chemical reaction in the chemical kinetics model (9) to be completed, diverges. This

shows that the model predicts that the reaction is not completed in finite time. You will not be able to calculate the integral. Instead, find a constant value that represents the minimum of the exponential factor in the integral. Replacing the exponential function with its minimum value makes the integrand smaller. The convergence or divergence of the simplified integral is easy to determine.

12. Interpret the parameters β and ρ in the chemical kinetics model.
    a. Both the numerator and the denominator of β have a clear meaning in the model. Use these meanings to explain the meaning of β.
    b. The meaning of ρ is harder to determine. One idea is to consider the reaction rate for small values of z, which can be done by replacing the right-hand side of Equation (7) by its linear approximation at the point $z = 0$. Then consider the rate of change of the reaction rate z'.

Chemical Mixing

Chemical mixing problems involve a chemical solution of fixed quantity but variable composition. The constant volume is maintained dynamically by having rates of inflow and outflow that are equal. Solute may be created by a chemical reaction in the solution or introduced with the inflow, or it may simply be present at the beginning of the mixing process. The solution is considered to be well mixed, so that the composition of the outflow is the same as that of the solution. If all processes are maintained in time, the system tends toward a steady state in which the rate at which solute is created internally or introduced via inflow is exactly balanced by the rate at which solute exits with the solution. Exercises 13 through 17 consider different scenarios for which chemical mixing is a suitable model. Recall that differential equations of the form $y' = -ky + c$, with k and c constant, can be solved by converting them to decay equations. See, for example, Exercise 12 of Section 1.1.

13. A lake of fixed volume V contains an amount Q of a pollutant. At time zero, the factory (or factories) responsible for the pollutant installs systems that remove nearly all the pollutant from the wastewater discharged into the lake.
    a. Assuming that fresh water flows into the lake at a rate r and lake water flows out at the same rate, determine a differential equation for the quantity of pollutant. (*Hint:* The mass decreases by the mechanism of outflow of lake water of concentration Q/V and flow rate r. Check your differential equation for dimensional consistency. Note that the flow rate has dimensions of volume per time.)
    b. Solve the differential equation and apply the initial condition to obtain a formula for $Q(t)$ with r, V, and Q_0 as parameters.
    c. Determine how much time is required for the pollution in the lake to be reduced to 10% of its original value. Compute the actual times required for Lake Erie (volume of 460 km^3 and flow rate of 180 km^3/yr) and for Lake Superior (volume of 12,000 km^3 and flow rate of 65 km^3/yr).
    d. Discuss the significance of the difference in results between Lake Erie and Lake Superior.

14. Cigarette smoke contains carbon monoxide as well as other noxious compounds. Consider a small room of volume V m^3. Let f be the rate of air exchange in cubic meters per minute. Let $m(t)$ be the mass in milligrams of carbon monoxide in the room.

    a. Write an initial-value problem for the mass of carbon monoxide in the room. Assume that the initial amount is m_0 mg. No more carbon monoxide is introduced in the room after time zero. Carbon monoxide leaves the room via the air exchange between the room and its environment; the concentration of carbon monoxide in the room in milligrams per cubic meter can be determined from the mass of carbon monoxide and the room volume.

    b. Solve the initial-value problem from part a.

    c. Concentrations of trace chemicals are often measured in parts per million (ppm). For example, 50 ppm means that every 1 million molecules includes 50 of carbon monoxide, which translates to a volume equal to 50 millionths of the room volume. The OSHA standard for maximum allowable carbon monoxide concentration in a room where people will spend 8 h/day is 50 ppm. Let $P(t)$ be the concentration of carbon monoxide in parts per million. Use the answer from part b to determine $P(t)$ for the case where m_0 corresponds to a concentration of 50 ppm. Use the fact that 1 kg of carbon monoxide gas occupies a volume of 0.86 m^3 at standard room temperature and atmospheric pressure.

    d. Consider a small office of volume 25 m^3. Prior to 1999, the EPA regulations for a large office building required a flow rate of 0.14 m^3/min per person. (In 1999, the requirement was increased by a factor of 4.) Our hypothetical small office is an appropriate size for one person. Assuming this flow rate, how long does it take for the concentration to drop from 50 ppm down to a relatively safe level of 5 ppm?

T 15. There is a flaw in the model of Exercise 13: It does not account for differences in the rates at which the lakes accumulate pollution.

    a. Assume that the lake is initially pollution-free. At time zero, the inflow begins to carry the pollutant at a concentration (quantity of pollutant per unit volume) C_0. Determine the differential equation for the quantity of pollutant.

    b. Solve the equation. (*Hint:* note the similarity with Newton's law of cooling.)

    c. Suppose the polluting process lasts for 20 yr and is followed by a 20-yr cleanup process as described in Exercise 13. Let $C = Q/V$. Then C/C_0 is a measure of the pollution in the lake relative to the pollution level in the original inflow. Determine formulas for C/C_0 for $0 \le t \le 20$ and $20 \le t \le 40$.

    d. Plot the pollution severity ratio C/C_0 for Lake Erie and Lake Superior. Discuss the results.

16. Exercise 14 considers the problem of cleaning out a room with a high concentration of carbon monoxide. In this exercise, we consider a situation that could lead to the one considered there.

    According to the Federal Trade Commission 2000 Report on Tar, Nicotine, and Carbon Monoxide, which is available on the World Wide Web, a typical cigarette releases about 14 mg of carbon monoxide. Assume that a cigarette burns in 7 min.

    a. Suppose 5 cigarettes are lit in the room of Exercise 14*d* and left to burn. Each time a cigarette burns out, a new one is immediately lit to replace it. How long does it take for the concentration of carbon monoxide in the room to reach 50 ppm, assuming no carbon monoxide is initially present?

b. Repeat part *a* for the case where 10 cigarettes are burning at a time. Why is the time required with 10 cigarettes less than one-half that required for 5 cigarettes?
c. In light of your calculations, do you think carbon monoxide is the greatest danger of secondhand cigarette smoke? Why or why not?

T 17. Consider a medicine, such as a chemotherapy drug, that has toxic side effects. Assume that the severity of the side effects depends on the maximum amount in the patient's blood. If a single dose is large enough to cause the side effects, the drug can be administered over a time interval rather than all at once.

a. Suppose an amount Q_0 of a drug is administered at a constant rate over a time interval T and that the drug is eliminated from the patient's system by a natural decay process with rate constant k. Write down an initial-value problem for this situation, assuming that there is no drug in the patient before the dose is administered.
b. Solve the initial-value problem of part *a*. What would be the maximum amount of the drug in the patient if the drug were to continue to be administered at the same rate?
c. Suppose that Q_T is the maximum tolerable amount of drug in the patient. Use the result of part *b* to determine an appropriate administration time T to guarantee that the amount of the drug never exceeds Q_T.
d. After the administration of the drug has been completed, the drug will continue to be eliminated from the patient's system by the same process as before. Solve the initial-value problem for the postadministration period. Note that the initial condition is at time T rather than time 0 and that the initial amount is $Q(T)$ as determined from the solution you found in part *b*.
e. The solutions obtained in parts *b* and *d*, taken together, give the drug content of the patient during and after administration of the drug. Let $\tau = t/t_h$, $y = Q/Q_0$, and $S = Q_0/Q_T$. Rewrite the solutions, using τ and y rather than t and Q. Note that the results do not contain any parameters other than S.
f. Plot $y(\tau)$ for the cases $S = 1$ and $S = 2$. (Your plot should extend to about 3 to 4 half-lives after the end of the drug administration.) Discuss the results. What is the physical significance of S? (In other words, use the definition of S to explain what it means.) What effect does S have on the graph?

18. Consider an alternative conceptual model for an endemic disease. (This model has been used to study disease transmission in prison populations.) The model includes assumptions 2, 3, and 4 of the endemic disease model in the text, but it replaces assumption 1 with the following:

1. The host population is constant in size. Randomly selected individuals are removed from the population at a constant rate, but healthy new individuals arrive at the same rate.

Let ρ be the constant rate of arrival and removal, and note that the proportion of removals who have the disease is the same as the proportion of the population who have the disease. Derive a mathematical model corresponding to the new conceptual model.

✦ 2.1 INSTANT EXERCISE SOLUTION

1. In the differential equation (7), the progress variable z is already dimensionless, and the function a is dimensionless because it represents a ratio of the rate at temperature T to that at temperature T_0. The only quantities having dimensions are the time t and the constant k, which has dimensions of time^{-1}. Thus, we define a new dimensionless time variable τ by

$$\tau = kt.$$

The differential equation is put in terms of τ rather than t by using the chain rule. Instead of thinking of z as $z(t)$, think of it as $z(\tau(t))$. Then

$$\frac{dz}{dt} = \frac{dz}{d\tau}\frac{d\tau}{dt} = k\frac{dz}{d\tau}.$$

This substitution into the differential equation (7) eliminates the parameter k:

$$\frac{dz}{d\tau} = a(T_0 + sx_0z)(1 - z), \qquad z(0) = 0.$$

Now

$$a(T_0 + sx_0z) = \exp\left(\frac{E}{RT_0} - \frac{E}{RT_0 + Rsx_0z}\right) = \exp\left[\frac{Esx_0z}{RT_0(T_0 + \beta T_0z)}\right] = \exp\left(\frac{\rho z}{1 + \beta z}\right).$$

2.2 Separable First-Order Equations

Many first-order differential equations are of a type called *separable*. These equations can, in principle, be solved by a two-step method, in which the equation is first put into *separated form* and then both sides of the equation are integrated.

EXAMPLE 1

Consider the differential equation

$$3y^2\frac{dy}{dt} = 1.$$

It is very tempting to "multiply" the equation by dt to get

$$3y^2\,dy = dt$$

and then write

$$\int 3y^2\,dy = \int dt.$$

Integrating, we have

$$y^3 = t + C,$$

where we have added an arbitrary integration constant to just one side. Taking the cube root gives us a solution formula:

$$y = (t + C)^{1/3}.$$

✦ INSTANT EXERCISE 1

Verify that the function obtained in Example 1 actually does solve the differential equation.

Example 1 raises some interesting questions. In "multiplying" the equation by dt, we are treating the derivative dy/dt as though it were a fraction. However, the symbol dy/dt represents a single entity, the derivative, rather than a fraction, and it is not clear that either dy or dt in the derivative notation has a clear meaning independent of the derivative. And yet the procedure produced the correct answer.

Development of the Method

The ad hoc procedure of Example 1 worked for that one example, and it works for other examples. Use it enough and it is easy to become convinced that it always works. This is not a satisfactory conclusion in mathematics. There is always the chance that a method fails for a certain class of problems, and we just never tested one of those. It is particularly troublesome to have a method that is based on a sloppy interpretation of notation. Instead of simply using a method that seems to work, mathematicians want to find a justification of the method and a set of conditions that guarantees that the method works. Our aim in this section is to develop a method that is essentially that of Example 1 and then to show that the method can be used for a broad class of problems.

MODEL PROBLEM 2.2

Obtain a symbolic formula for the solution of the differential equation

$$(t+1)\frac{dy}{dt} = 1 - y. \tag{1}$$

In general, we can get rid of a derivative by integration, so it makes sense to try to integrate both sides of the differential equation (1). We must integrate with respect to the independent variable t, so we have (keeping in mind that y is a function of t)

$$\int (t+1)\frac{dy}{dt}\,dt = \int [1 - y(t)]\,dt. \tag{2}$$

This integral form of the differential equation is useful only if we can actually compute the integrals on both sides. If we already knew the function y, then we could substitute it into the integrals and perform the integration. However, we cannot integrate either side of the integral equation (2) until *after* we have found the solution of the differential equation (1). So this attempt to integrate the differential equation fails.

Returning to the original differential equation, we can divide both sides by $t+1$ and $1-y$. The equation then reads as

$$\frac{1}{1-y}\frac{dy}{dt} = \frac{1}{t+1}. \tag{3}$$

Observe that the coefficient of dy/dt in Equation (3) does not depend *explicitly* on t (it depends on t implicitly, only because y is a function of t), while the function on the right side of the equation does not depend on y at all.

Now suppose we try to integrate both sides of Equation (3) with respect to t. This time we have

$$\int \frac{1}{1-y(t)}\frac{dy}{dt}\,dt = \int \frac{1}{t+1}\,dt. \tag{4}$$

The integral on the right side of Equation (4) is not a problem, but we still have to find a way to integrate the left side when we don't yet know the function $y(t)$.

Integrating by Substitution

To see how to integrate Equation (4), it helps to review briefly the method of integration by substitution. Let's say we want to compute

$$\int \frac{\cos t}{1-\sin t}\,dt.$$

We define a new integration variable, say u, by $u = \sin t$. Then we have $du = \cos t\,dt$, and the integral becomes

$$\int \frac{1}{1-u}\,du.$$

We can then integrate with respect to u.

Notice that the integral

$$\int \frac{2t}{1-t^2}\,dt$$

also reduces to $\int 1/(1-u)\,du$, this time using the substitution $u = t^2$. We can create lots of other integrals that reduce to $\int 1/(1-u)\,du$. All we have to do is pick a function y and then construct the integrand by dividing dy/dt by $1-y(t)$. In general, we have

$$\int \frac{1}{1-y(t)}\frac{dy}{dt}\,dt = \int \frac{1}{1-u}\,du,$$

using the substitution $u = y(t)$. Since u and y in this context are equivalent, there is no harm in using the symbol y rather than u:

$$\int \frac{1}{1-y(t)}\frac{dy}{dt}\,dt = \int \frac{1}{1-y}\,dy. \tag{5}$$

Equation (5) shows that the integral on the left side of Equation (4) can be simplified by the substitution of y for t as the integration variable *even though we don't yet know the function* y. This substitution gives us the equation

$$\int \frac{1}{1-y}\,dy = \int \frac{1}{t+1}\,dt.$$

Now the left side is an integral of a function of y with respect to y, and the right side is an integral in t. We have

$$-\ln|1-y| = \ln|t+1| + C, \qquad y \neq 1, \qquad t \neq -1. \tag{6}$$

(We only need to include an integration constant on one side of the equation.) This result is an **implicit solution formula,** so called because it is a relation between the independent and dependent variables but is not explicitly solved for the dependent variable.

An Explicit Solution Formula

An implicit solution formula such as Equation (6) is not an ideal way to present the solution of a differential equation. In general, an explicit solution formula is preferable. In this example, we can solve the implicit solution formula for y to obtain an explicit solution formula. We begin by multiplying by -1 and then raising both sides over the exponential base e:

$$|1-y| = e^{-\ln|t+1|-C} = e^{-C}e^{-\ln|t+1|} = \frac{e^{-C}}{|t+1|},$$

or

$$1-y = \pm\frac{e^{-C}}{t+1}.$$

Thus we get the explicit solution formula

$$y = 1 - \frac{\pm e^{-C}}{t+1}.$$

This result is messier than it needs to be. The constant C is arbitrary, so the constant $-(\pm e^{-C})$ is also arbitrary. Let $A = -(\pm e^{-C})$; then we have the simpler formula

$$y = 1 + \frac{A}{t+1}. \tag{7}$$

Note that A is not *completely* arbitrary, according to our derivation. There is no real number C for which $e^{-C} = 0$, so A cannot be 0 if it is defined in terms of the real number C. Nevertheless, the constant function $y \equiv 1$, corresponding to $A = 0$, *is* a solution of the original Equation (1). Thus, all functions of the family (7) are solutions, even though our solution procedure did not actually yield the solution $y \equiv 1$.

We lost the solution $y \equiv 1$ because the solution method involves division by zero when $y = 1$. In theory, we should take this much care with possible values for arbitrary constants for all problems that we solve by separation of variables. In practice, it is equally valid, and much easier, simply to check any solution formula we obtain. The real test of a solution formula is not whether we can derive it, but whether it satisfies the differential equation.

✦ INSTANT EXERCISE 2

Find an explicit solution formula for the equation $dx/dt = xt$.

The General Procedure

Now that we have worked out a method for a model problem, it is time to generalize the method to a class of problems. We need to determine what properties of the model problem were sufficient to allow the method to work. The key point is that we needed to be able to make the substitution of y for t on the left side of Equation (4). The specific functions $1/(1-y)$ and $1/(t+1)$ don't matter, but the absence of t from the left side of Equation (3) and the absence of y from the right side of the equation are necessary. In general, a first-order differential equation of the form

$$g(y)\frac{dy}{dt} = f(t) \tag{8}$$

is said to be in **separated form.** A **separable equation** is a first-order equation that can be written in separated form.

EXAMPLE 2

The differential equation

$$\frac{dy}{dt} = 2y^2 - 5t^2$$

cannot be written in separated form. We can move the $2y^2$ to the left side by dividing both sides by $2y^2 - 5t^2$, but that yields

$$\frac{1}{2y^2 - 5t^2}\frac{dy}{dt} = 1.$$

The right side looks good, but now there is a problem with the left side. Alternatively, we could have written the original equation as

$$\frac{dy}{dt} - 2y^2 = -5t^2,$$

but that is not of the form "function of y times derivative of y equals function of t." This equation is not separable and cannot be solved by the method of separation of variables.

We solve a separable first-order differential equation by writing it in separated form and then integrating by substitution. Then we try to obtain an explicit solution formula by using algebra.

EXAMPLE 3

Solve

$$\frac{dy}{dt} = 1 - y^2.$$

Dividing by $1 - y^2$ puts the equation in separated form:

$$\frac{1}{1-y^2}\frac{dy}{dt} = 1.$$

Next we integrate with respect to t,

$$\int \frac{1}{1-y^2}\frac{dy}{dt}\,dt = \int dt,$$

and use substitution on the left side to get

$$\int \frac{1}{1-y^2}\,dy = \int dt.$$

The integral on the left side can be found in a table of integrals or converted (by decomposition into partial fractions) to

$$\frac{1}{2}\int \left(\frac{1}{1+y} + \frac{1}{1-y}\right) dy.$$

Integrating gives the implicit solution formula

$$\frac{\ln|1+y| - \ln|1-y|}{2} = t + C.$$

Combining terms on the left and then raising over e yields

$$\left|\frac{1+y}{1-y}\right| = e^{2t+2C},$$

which we rewrite as before to get

$$\frac{1+y}{1-y} = Ae^{2t},$$

where $A = \pm e^{2C}$, and simplify to get the final answer

$$y = \frac{Ae^{2t}-1}{Ae^{2t}+1}.$$

This answer is valid for any value of A, on any time interval for which the denominator never vanishes. It is interesting to note that this solution formula does not include all solutions of the differential equation. The constant function $y \equiv 1$ satisfies the differential equation, so it is an equilibrium solution even though we cannot obtain $y \equiv 1$ from the one-parameter family of solutions by choosing a value of A.

A family of solutions of a differential equation is called a **general solution** if the family includes all solutions of the differential equation. A solution that is not part of the family is called a **singular solution.** In Example 3, the differential equation has a singular solution; therefore, the one-parameter family of solutions is not a general solution. It is important to keep in mind that not all differential equations have a general solution.

EXAMPLE 4

Solve

$$\frac{dy}{dx} = e^x y^3 - xe^x y^3, \qquad y(0) = -1.$$

We first factor the right side so as to obtain

$$\frac{dy}{dx} = (1-x)e^x y^3.$$

This equation is separable; in separated form we have

$$y^{-3}\frac{dy}{dx} = (1-x)e^x.$$

Integrating with respect to x and then replacing $(dy/dx)\,dx$ by dy yields

$$\int y^{-3}\,dy = \int (1-x)e^x\,dx.$$

Integrating both sides gives the implicit solution formula

$$-\tfrac{1}{2}y^{-2} = (2-x)e^x + C.$$

Substituting the initial condition into this result yields

$$-\tfrac{1}{2} = 2 + C;$$

hence, $C = -5/2$. Substituting this value into the implicit formula and then multiplying by -2 gives us

$$y^{-2} = 5 - 2(2-x)e^x,$$

or

$$y^2 = \frac{1}{5 - 2(2-x)e^x}.$$

We must be careful in taking the square root to obtain y. The quantity on the right has both a positive square root and a negative one. The correct one is the one that is consistent with the initial condition; thus,

$$y = -\frac{1}{\sqrt{5 - 2(2-x)e^x}}.$$

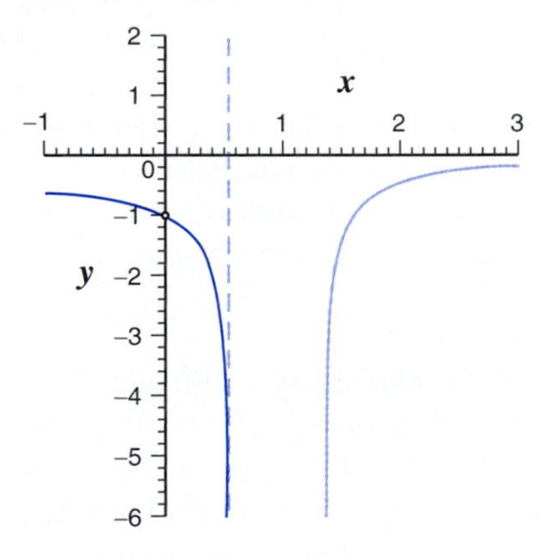

Figure 2.2.1
The solution of Example 4 (darker curve), along with the vertical asymptote marking the right boundary of the interval of existence and the other portion of the function defined by $y(x) = -[5 - (2-x)e^x]^{-1/2}$.

This is an explicit solution formula. The graph of this function appears in Figure 2.2.1, with the solution of the initial-value problem consisting of just the darker portion of the curve. Note that the function is not defined for all x because the quantity inside the square root can be negative for some values of x. As we discussed in Section 1.2, only the portion of the graph that is connected to the initial condition represents the solution of the initial-value problem. Thus, the solution is defined only for $x < x_1$, where x_1 is the smaller root of the equation

$$5 - (2-x)e^x = 0.$$

Using a computer, we obtain the approximate value $x_1 = 0.533$. The vertical asymptote $x = x_1$ appears as a dashed line in Figure 2.2.1.

EXAMPLE 5

Consider the differential equation

$$\frac{dy}{dt} = \frac{1}{1 + e^y}.$$

This equation can be written in separable form as

$$(1 + e^y)\frac{dy}{dt} = 1.$$

Integration with respect to t yields

$$\int (1 + e^y)\frac{dy}{dt}\, dt = \int 1\, dt.$$

Substituting y as the integration variable on the left side and then integrating yields the implicit solution formula

$$y + e^y = t + C.$$

We cannot solve this relation for y explicitly, but we can still check to see if it is a solution formula. Differentiating both sides, we have

$$\frac{dy}{dt} + e^y \frac{dy}{dt} = 1,$$

or

$$\frac{dy}{dt} = \frac{1}{1 + e^y}$$

this confirms that the solution formula is correct on any interval in t for which it defines a function $y(t)$.

A relation $F(t, y) = 0$ is an **implicit solution** of a differential equation for $y(t)$ on an interval I if it makes the equation true for t in the interval I. It may be difficult to think of an equation such as $y + e^y = t + C$ as defining a function y when we can't solve it to get an explicit formula for y. However, the graph of the function defined implicitly by $y + e^y = t$ in Figure 2.2.2 confirms that there really is a function y corresponding to the implicit relation.

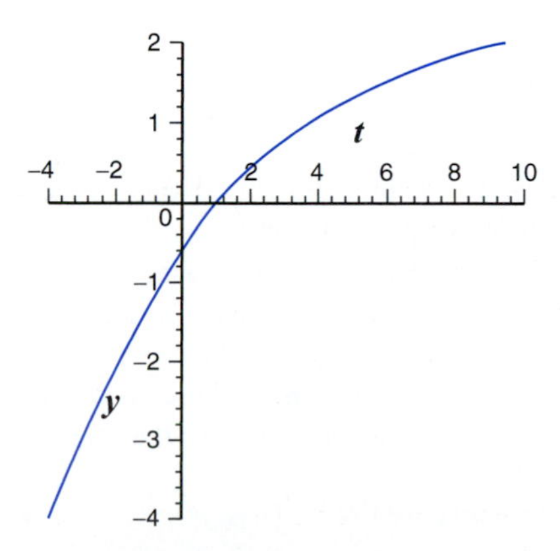

Figure 2.2.2
The function y defined by the relation $y + e^y = t$.

In general, suppose we have a differential equation

$$g(y)\frac{dy}{dt} = f(t),$$

and we know a function G whose derivative is g. Then the formula

$$G(y) = \int f(t)\,dt$$

gives an implicit solution of the differential equation. We can derive this solution in the general case, using the integration technique, but it is simpler to check the result directly using the chain rule. Differentiating both sides of the implicit formula with respect to t, we have

$$\frac{dG}{dy}\frac{dy}{dt} = f(t),$$

or

$$g(y)\frac{dy}{dt} = f(t).$$

Since we can use Theorem 1.2.1 to construct an antiderivative G for any continuous function g and an antiderivative for any continuous function f, we can now say that any separable equation with continuous factors f and g can be solved by the method developed here, given that an implicit solution counts as a solution.

2.2 Exercises

In Exercises 1 through 12, determine whether the differential equation is separable. If it is, solve the differential equation or initial-value problem. Put your solution in explicit form if possible.

1. $\dfrac{dy}{dt} = t - 1, \quad y(0) = 3$
2. $\dfrac{dy}{dt} + y = e^t$
3. $\dfrac{dy}{dt} = y^3, \quad y(0) = 1$
4. $\dfrac{dy}{dt} = \dfrac{1+t^2}{t(1+2y)}, \quad y(1) = 0$
5. $\dfrac{dy}{dt} = te^t y^2$
6. $\dfrac{dy}{dx} = \dfrac{y-1}{x-1}$
7. $x\dfrac{dy}{dx} = e^{-y}\ln x$
8. $\dfrac{dy}{dt} = ty + 1, \quad y(0) = 1$
9. $y\dfrac{dy}{dx} = -2x$
10. $\dfrac{dy}{dt} = \dfrac{y\cos t}{1+2y^2}, \quad y(0) = 1$
11. $y^2\dfrac{dy}{dx} = (x + xy^3)e^{x^2}$
12. $\dfrac{dy}{dt} = ye^{t^2}, \quad y(0) = 2$

In Exercises 13 through 18, find the solution in explicit form, plot the graph of the solution, and determine (at least approximately) the interval of existence.

13. $\dfrac{dy}{dt} = 2ty^2, \quad y(0) = 5$
14. $\dfrac{dy}{dt} = \dfrac{2\sin t}{1-y}, \quad y(0) = -1$
15. $\dfrac{dy}{dt} = y^2\cos t, \quad y(0) = 1$
16. $t + ye^{-t}\dfrac{dy}{dt} = 0, \quad y(0) = 1$

17. $xy\dfrac{dy}{dx} = 1 + y^2, \quad y(1) = -2$

18. $\dfrac{dy}{dt} = y^3 \cos t, \quad y(0) = 2$

In Exercises 19 through 22, (*a*) verify that the given relation is an implicit solution for the given problem and (*b*) plot the solution.

19. $y = x^2 + \ln(1 + y), \quad y\dfrac{dy}{dx} = 2x(1 + y)$

20. $y^2 - xy = e^{-x}, \quad (2y - x)\dfrac{dy}{dx} = y - e^{-x}$

21. $e^y + xy = x^2, \quad \dfrac{dy}{dx} = \dfrac{2x - y}{e^y + x}$

22. $xy^2 - \sin y = 5x^2 - 1, \quad \dfrac{dy}{dx} = \dfrac{10x - y^2}{2xy - \cos y}, \quad y(0) = \dfrac{\pi}{2}$

23. *a.* Solve the differential equation

$$\frac{dy}{dt} = \frac{1}{y - 2}.$$

b. Let a be a real number. Solve the initial-value problem

$$\frac{dy}{dt} = \frac{1}{y - 2}, \qquad y(0) = a.$$

Determine the interval of existence of the solution as a function of a.

24. Solve the initial-value problem

$$\frac{dy}{dt} = \frac{1}{3y^2 - 1}, \qquad y(0) = 0$$

and determine the interval of existence. (*Hint:* look for points where the solution curve has a vertical tangent. Alternatively, plot t as a function of y and ask on what interval the function has an inverse.)

Orthogonal Trajectories

If two straight lines are perpendicular to each other, the slope of one is the negative reciprocal of the slope of the other. Similarly, two families of curves are said to be *mutually orthogonal* if the product of the slopes is -1 at every point of intersection between curves of each family. Orthogonal trajectories can be determined from differential equations. For example, consider the family of curves $y = cx$. Differentiating gives $dy/dx = c$. We can eliminate c by substitution and obtain the differential equation $dy/dx = y/x$. Curves orthogonal to these curves must satisfy the differential equation $dy/dx = -x/y$. By separation of variables, we obtain the implicit solution formula $x^2 + y^2 = k$. Thus, the family of concentric circles centered at the origin is orthogonal to the family of straight lines through the origin. See Figure 2.2.3.

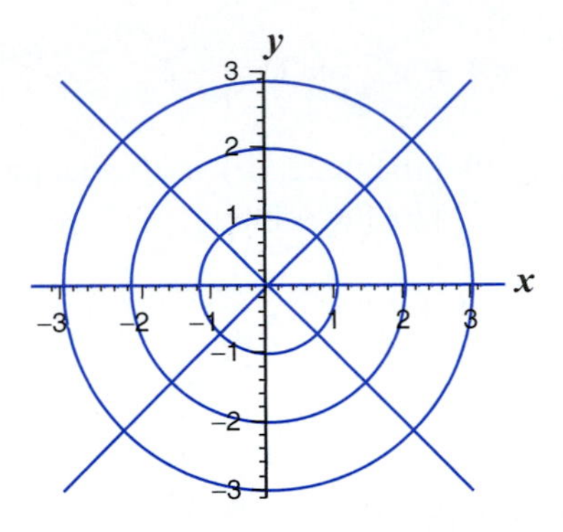

Figure 2.2.3
Orthogonal trajectories.

In Exercises 25 through 28, determine the family of curves that are orthogonal to the given family. Plot members of the orthogonal families on a common set of axes.

25. $y = Ae^x$ 26. $x^2 - y^2 = c^2$ 27. $y = kx^2$ 28. $x^2 + 4y^2 = c^2$

Special Substitution Techniques

Equations of the form

$$\frac{dy}{dt} + cy = f(t)$$

can be solved by a special substitution technique. Let

$$v(t) = e^{ct} y(t).$$

Then we can differentiate this equation to get

$$\frac{dv}{dt} = e^{ct}\frac{dy}{dt} + ce^{ct} y = e^{ct} f(t).$$

The solution for y is easily found after v is determined by integration. (This substitution is a special case of a general method that appears in Section 4.5.)

Equations of the form

$$\frac{dy}{dx} = F\left(\frac{y}{x}\right)$$

can be solved by a different substitution. Let

$$v(x) = \frac{y(x)}{x}.$$

Then we can differentiate the equation $y(x) = xv(x)$ to get $dy/dx = x\,dv/dx + v$. Comparing this result with the differential equation gives a differential equation for v:

$$x\frac{dv}{dx} + v = F(v).$$

The equation for v is separable; once it is solved, the original variable y can be determined from $y = xv$. As an example, consider the equation $y' = (x + 2y)/(2x - y)$. With

$v(x) = y(x)/x$, we have $y' = xv' + v$ or $xv' + v = (x + 2y)/(2x - y) = (1 + 2v)/(2 - v)$. After simplification, we have $xv' = (1 + v^2)/(2 - v)$. After separation of variables, this is

$$\int \frac{2 - v}{1 + v^2}\, dv = \int \frac{1}{x}\, dx.$$

This equation integrates to yields $2 \arctan v - [\ln(1 + v^2)]/2 = \ln|x| + C$. Substitution of $v = y/x$ and algebraic manipulation eventually yield the relation $4 \arctan(y/x) = \ln(x^2 + y^2) + 2C$.

Equations of the form

$$\frac{dy}{dx} = F(ay + bx + c),$$

where a, b, and c are constants, can be solved by the substitution

$$v(x) = ay(x) + bx + c.$$

In Exercises 29 through 38, solve the given differential equation by using one of the special substitution techniques.

29. $\dfrac{dy}{dt} + y = 2t + 2$
30. $\dfrac{dy}{dt} + y = t^2$
31. $\dfrac{dy}{dt} + y = 2t + t^2$
32. $\dfrac{dy}{dt} - y = 2t - t^2$
33. $x\dfrac{dy}{dx} = x + y$
34. $x\dfrac{dy}{dx} = x + 2y$
35. $xy\dfrac{dy}{dx} = y^2 - x^2$
36. $(x + y)\dfrac{dy}{dx} = 3x - y$
37. $\dfrac{dy}{dx} = \dfrac{x - y}{2}$
38. $\dfrac{dy}{dx} = (y + x)^2$
39. The differential equation

$$y\frac{dy}{dx} = x^3 - xy^2$$

can be solved by the substitution

$$v(x) = y^2(x) - x^2.$$

Use the chain rule to differentiate v in terms of x, y, and dy/dx. Solve the resulting separable differential equation for v, and use the result to obtain the solution for the original equation.

40. Solve the differential equation

$$y\frac{dy}{dx} = \frac{x}{y^2 - x^2},$$

using the method of Exercise 39.

41. Determine the class of differential equations for which the substitution $v(x) = y^2(x) - x^2$ yields a separable equation for v.

✦ 2.2 INSTANT EXERCISE SOLUTIONS

1. Given $y = (t + C)^{1/3}$,
$$\frac{dy}{dt} = \frac{1}{3}(t + C)^{-2/3}.$$
Thus,
$$3y^2\frac{dy}{dt} = 3(t + C)^{2/3}\left[\frac{1}{3}(t + C)^{-2/3}\right] = 1.$$

2. The equation, in separated form, is
$$\frac{1}{x}\frac{dx}{dt} = t.$$
Integrating with respect to t on both sides yields
$$\int \frac{1}{x}\frac{dx}{dt}\,dt = \int t\,dt,$$
or
$$\int \frac{1}{x}\,dx = \int t\,dt.$$
Integrating, we have
$$\ln|x| = \frac{t^2}{2} + C.$$
Raising both sides as exponents over e gives us
$$|x| = e^{t^2/2+C} = e^C e^{t^2/2}.$$
The constant e^C can take on any positive value. We can remove the absolute value bars from $|x|$ by allowing the constant to take any nonzero value. We can also check that the zero function solves the differential equation. Hence, the solutions are given by
$$x = Ae^{t^2/2}.$$

2.3 Slope Fields

The **solution curves**[8] of a differential equation $dy/dx = f(x, y)$ are the graphs of the solutions in the xy plane. Differential equations give indirect information about a family of functions by giving direct information about the relationship of the functions and their derivatives. This direct information can be used to sketch solution curves without the need of a solution formula.

MODEL PROBLEM 2.3

Sketch the solution curves of the differential equation
$$\frac{dy}{dx} = \frac{x - y}{2} \tag{1}$$
without using solution formulas.

[8]Some authors use the term **integral curves** instead.

In Section 2.2, we interpreted a differential equation as a problem to be solved for an unknown function. We can also interpret a differential equation as a formula for computing the slope of the solution curves. As an example, consider the point $(x_1, y_1) = (0, -2)$. From the slope formula (1), we calculate the slope to be $(x_1 - y_1)/2 = 1$. This means that the graph of a solution $y(x)$ through the point $(0, -2)$ has slope 1 at that point. We can indicate this fact by sketching a short horizontal line segment of slope 1 centered at the point $(0, -2)$. Similarly, we can compute the slopes of solution curves at any point (x, y) and sketch line segments at (x, y) to indicate the slope. A short line segment centered at a point and having a slope consistent with the differential equation at that point is called a **minitangent.** Figure 2.3.1 shows the minitangents for all points at integer coordinates in the range $-2 \leq x, y \leq 2$.

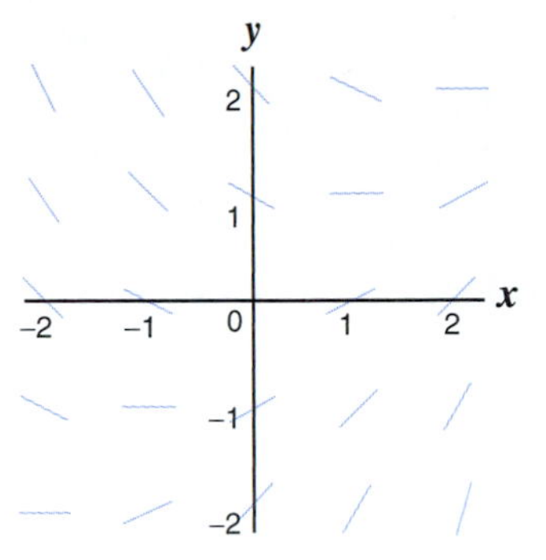

Figure 2.3.1
Some minitangents for $dy/dx = (x - y)/2$.

✦ INSTANT EXERCISE 1

Sketch the minitangents for $dy/dx = (x + y)/2$ at the points (x, y) where x and y are the integers $-2, -1, 0, 1$, and 2.

The **slope field** of a differential equation is the set of all the minitangents.

Notice that we cannot draw the *complete* slope field, because there would have to be a minitangent at every point. Instead, we can *sketch* the slope field by choosing a grid and drawing only those minitangents that pass through grid points. Each minitangent calculation is simple, but it takes a lot of minitangents to make a meaningful sketch. Fortunately, computers can perform repetitive calculations quickly, so we will generally want to use them to sketch slope fields.[9] Figure 2.3.2 shows a computer-generated sketch of the slope field of the differential equation (1) for a portion of the xy plane.

✦ INSTANT EXERCISE 2

Use Maple, Mathematica, or some other computer software to sketch the slope field for the differential equation $dy/dx = (x + y)/2$. Compare the sketch with the one you produced by hand in Instant Exercise 1.

[9]In Section 5.2, we will see that the slope field for an *autonomous* differential equation can be sketched easily by hand.

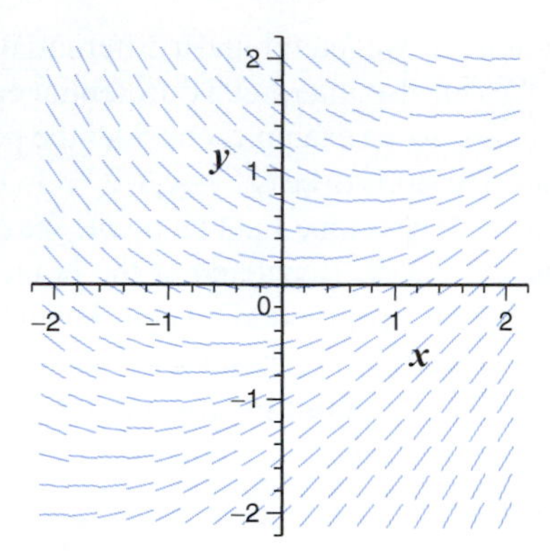

Figure 2.3.2
The slope field for $dy/dx = (x - y)/2$.

The Slope Field and Solution Curves

The differential equation (1) is not separable, but it can be solved by the general methods of Sections 4.5 and A.1; it can also be solved by a substitution (Section 2.2, Exercise 37). The solution, in any case, is

$$y = x - 2 + Ae^{-x/2}.$$

This formula can be used to sketch solution curves. Figure 2.3.3 shows the solution curves with A given as 0, 1, 2, 3, and 4 along with the slope field from Figure 2.3.2. Note that each time a solution curve passes near the center of a minitangent, it passes through with the same slope as the minitangent; in other words, the solution curves are tangent to the minitangents. This property follows from the two different ways of interpreting the differential equation.

- The functions that solve a differential equation have slopes that can be calculated by the differential equation.

This point is obvious once you think about it, and it nicely summarizes the value of a slope field sketch.

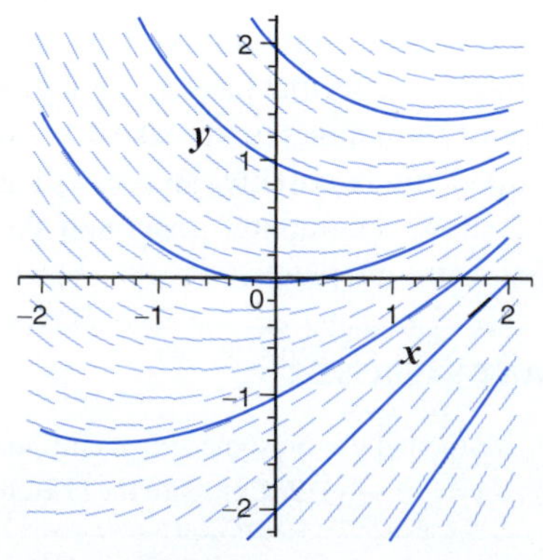

Figure 2.3.3
The slope field and some solution curves for $dy/dx = (x - y)/2$.

✦ INSTANT EXERCISE 3

Use the slope field in Figure 2.3.3 to sketch the solution curve that passes through the point $(-2, 0)$. Then use the solution formula to check the accuracy of your sketch.

EXAMPLE 1

The differential equation

$$(t+1)\frac{dy}{dt} = 1 - y$$

was examined in Section 2.2. Note that the slope field is not well defined at every point in the ty plane for this equation. At any point with $t = -1$, the derivative is undefined. This does not necessarily mean that the slope of the solution curve at such a point is undefined. The limit as $(t, y) \to (-1, 0)$ of dy/dt is $\pm\infty$, which tells us that the solution curves are vertical at $(-1, 0)$. The same is true for all other points $(-1, y)$ except the point $(-1, 1)$. The function $(1 - y)/(t + 1)$ does not have a limit (not even an infinite limit) at the point $(-1, 1)$. A sketch of the slope field should not show any minitangent at $(-1, 1)$, but it can show a vertical minitangent at all other points on the line $t = -1$. The slope field appears in Figure 2.3.4.

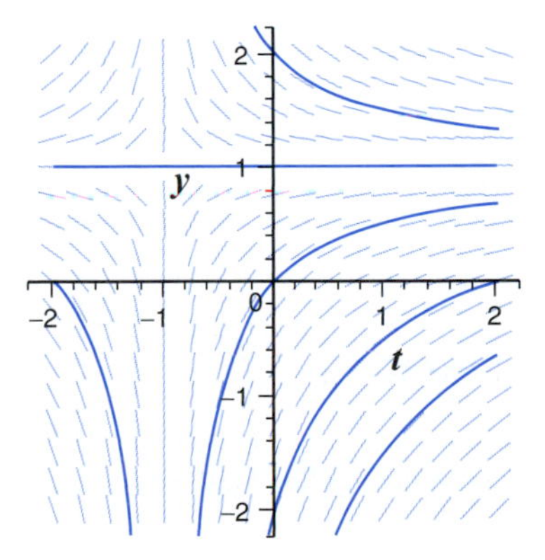

Figure 2.3.4
The slope field and some solution curves for Example 1.

Most differential equations cannot be solved. Randomly create a function of both x and y for dy/dx, using the standard functions of algebra, precalculus, and trigonometry, and it is highly unlikely that an explicit solution formula can be found for that differential equation. Differential equations that cannot be solved can still be studied by a variety of mathematical tools, such as the slope field, that do not require a solution formula. Even differential equations that *can* be solved may have solutions that are inconvenient because they are implicit or in the form of a definite integral. In cases such as this, it is often easier to use slope fields to plot solution curves than it is to use a solution formula.

The Zero Isocline

Consider again the differential equation $dy/dx = (x - y)/2$. Suppose we set the derivative equal to zero. The result is the curve $y = x$. This curve consists of precisely those points at which the solution curve has a horizontal tangent. Furthermore, $dy/dx > 0$ if $y < x$ and $dy/dx < 0$

if $y > x$. The curve $y = x$ is not a solution of the differential equation, but it has graphical significance in that it divides the plane into regions where the solutions are increasing and regions where the solutions are decreasing. The **zero isocline**[10] is a curve consisting of points for which the slopes of solution curves are zero. The curve $y = x$ is the zero isocline for the differential equation of Model Problem 2.3. Figure 2.3.5 shows solution curves for Model Problem 2.3 along with the zero isocline. Note how each curve has a horizontal tangent at the point where it crosses the zero isocline. The zero isocline therefore passes through the local extrema of all the solution curves.

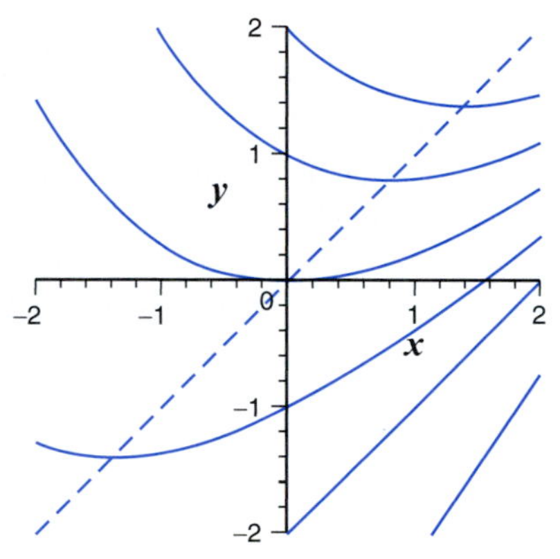

Figure 2.3.5
Solution curves and the zero isocline (dashed) for the differential equation $dy/dx = (x - y)/2$.

✦ INSTANT EXERCISE 4

Show that $y = -x - 2 + Ae^{x/2}$ is a family of solutions for $dy/dx = (x + y)/2$. Sketch several of these solutions along with the isocline for the differential equation.

Slope Fields for Autonomous Equations

The slope field for the decay equation

$$\frac{dy}{dt} = -y$$

appears in Figure 2.3.6. The graph illustrates the qualitative properties of the solutions. Each solution is always positive, always negative, or always zero; in all cases, the function approaches zero in the limit as t approaches infinity. As we observed in Section 1.1, the constant function defined by $y(t) \equiv 0$ is a stable equilibrium solution. These qualitative properties could have been found directly from the slope field even if we had no solution formula. Notice also that the minitangents on a given horizontal line are all parallel. This is so because y' is a function of y alone. Any differential equation in which the independent variable does not appear explicitly is

[10]Some other authors use the term **null isocline** or **nullcline**.

autonomous. Because y' does not depend on t for an autonomous equation, it is always the case that all minitangents on a given horizontal line have the same slope. Similarly, equations of the form $y' = f(t)$ have the property that all minitangents on a given vertical line have the same slope.

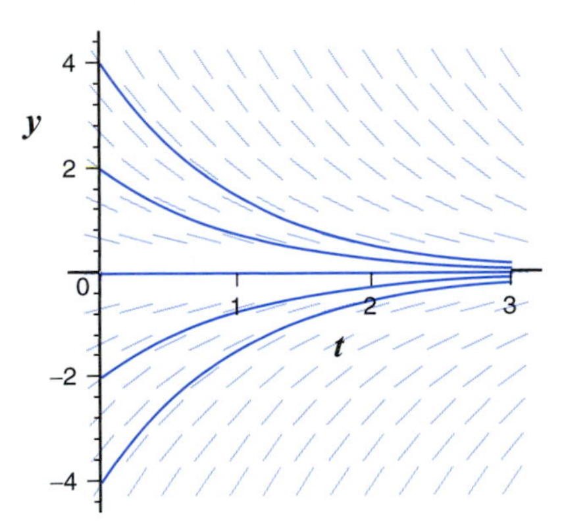

Figure 2.3.6
The slope field for $y' = -y$, along with some solution curves.

EXAMPLE 2

Consider the differential equation

$$\frac{dy}{dt} = \frac{1}{1+e^y},$$

for which we found the implicit solution formula

$$y + e^y = t + C$$

in Example 5 of Section 2.2. The equation is autonomous, so the slopes of all minitangents on any given horizontal line are the same. The slope field appears in Figure 2.3.7, along with several solution curves.

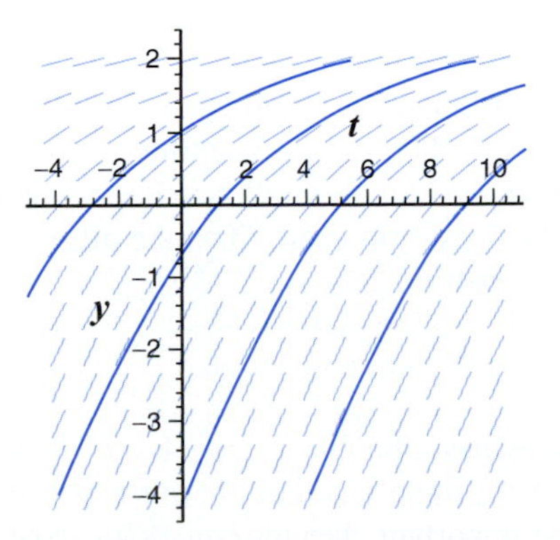

Figure 2.3.7
The slope field for $y' = 1/(1+e^y)$, along with some solution curves.

Long-Term Behavior of Solutions

Sometimes solutions to differential equations have simple approximations for large values of the independent variable. In the case of Model Problem 2.3 the solutions are given by

$$y = x - 2 + Ae^{-x/2}.$$

As $x \to \infty$, the function $e^{-x/2}$ vanishes, so the solutions approach the linear function $y = x - 2$. This behavior is not clear from the figures we have seen so far, but only because these figures look at the interval $-2 < x < 2$. Figure 2.3.8 illustrates the long-term behavior of the same solutions that are shown in Figures 2.3.3 and 2.3.5.

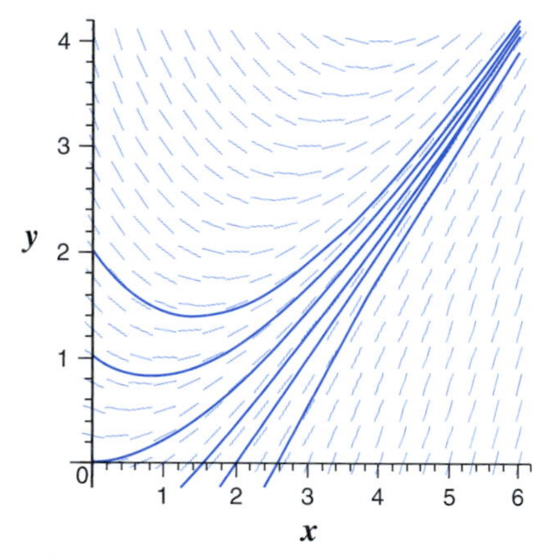

Figure 2.3.8
Long-term behavior of solutions of $dy/dx = (x - y)/2$.

In this case, we discovered the long-term behavior from the solution formulas. It would have been possible, but more difficult, to discover the behavior directly from the slope field. There is another way to examine long-term behavior by approximating the differential equation.

Long-term behavior of solutions tends to be simpler than the full behavior of solutions because some terms in the solution are less important for large values of the independent variable. It may also be true that some terms in the differential equation are also less important. The differential equation of Model Problem 2.3 has three terms in its differential equation: dy/dx, $x/2$, and $y/2$. If we guess that one of these terms is unimportant for large x, there are three choices. One possibility is that $y/2$ is least important. Then the differential equation can be approximated for large x by the simpler equation

$$\frac{dy}{dx} = \frac{x}{2}.$$

This equation has solutions $y = x^2/4 + C$. However, this solution contradicts the assumption that $y/2$ is the least important term, because $(x^2/4 + C)/2$ is bigger than $x/2$. Similarly, if the term $x/2$ is least important, then the equation can be approximated by

$$\frac{dy}{dx} = -\frac{y}{2}.$$

This equation is separable, so it is easily solved, and the solution is $y = Ae^{-x/2}$. However, we assumed that $x/2$ is less important than $y/2$, and this assumption is contradicted by the result $y = Ae^{-x/2}$. Finally, we suppose that the derivative term is the least important. Then we have the approximate equation

$$0 = \frac{x - y}{2},$$

with solution $y = x$. In this case, the term we are assuming to be least important turns out to be zero for the approximate solution. This approximation is consistent with the assumptions.

Although the approximation $y \approx x - 2$ that we got from the solution formula is better than the one we got by omitting a term from the differential equation, the approximation $y = x$ is still quite good for x large enough. In cases where we want to know the long-term behavior but we do not have a solution formula, the combination of the slope field and this approximation technique can sometimes yield a good answer.

2.3 Exercises

1. Sketch the solution curves that pass through the points (0, 0) and (0, 1) in Figure 2.3.9.

2. Sketch the solution curves that pass through the points (0, 0), (0, 0.5), (0, 1), (0, 1.5) and (0, 2) in Figure 2.3.10.

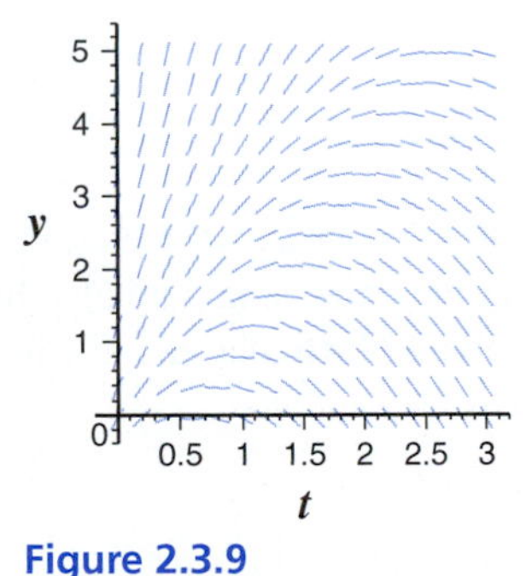

Figure 2.3.9
The slope field for Exercise 1.

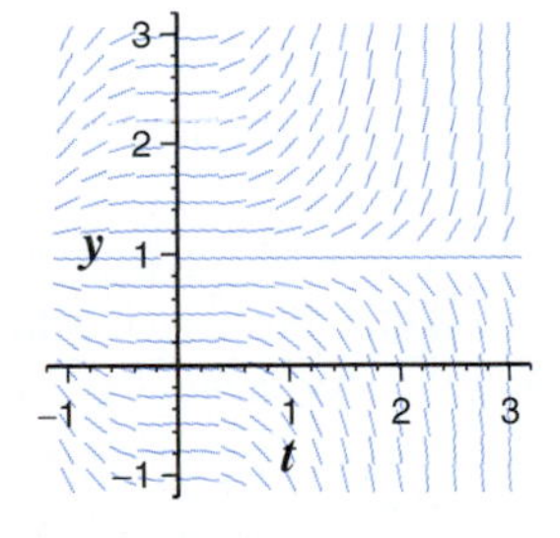

Figure 2.3.10
The slope field for Exercise 2.

In Exercises 3 through 8, sketch the slope field along with the solution curves through the indicated points. Sketch the zero isocline. If possible, determine the behaviors of solutions as the independent variable approaches ∞. Compare with the solution formulas if possible.

T 3. $\dfrac{dy}{dt} + y = 2t + 2$, $(0, -4)$, $(0, 0)$, $(0, 4)$ (Section 2.2, Exercise 29)

T 4. $\dfrac{dy}{dt} + y = t^2$, $(0, -4)$, $(0, 0)$, $(0, 4)$ (Section 2.2, Exercise 30)

T 5. $\dfrac{dy}{dt} + y = 2t + t^2$, $(0, -4)$, $(0, 0)$, $(0, 4)$, $(0, 8)$ (Section 2.2, Exercise 31)

T 6. $\dfrac{dy}{dt} - y = 2t - t^2$, $(0, -1)$, $(0, 0)$, $(0, 2)$ (Section 2.2, Exercise 32)

T 7. $\dfrac{dy}{dt} = t - 0.2y^2$, $(0, -2)$, $(0, 0)$, $(0, 2)$, $(0, 4)$

T 8. $\dfrac{dy}{dt} = \dfrac{y \cos t}{1 + 2y^2}$, $(0, -1)$, $(0, 0)$, $(0, 1)$ (Section 2.2, Exercise 10)

In Exercises 9 through 14, sketch the slope field along with the solution curves through the indicated points. Sketch the zero isocline. If possible, determine the behaviors of solutions as the independent variable approaches ∞. Compare with the solution formulas as indicated.

T 9. $\dfrac{dy}{dt} = 2ty^2$, $y(0) = 5$ (Section 2.2, Exercise 13)

T 10. $\dfrac{dy}{dt} = \dfrac{2 \sin t}{1 - y}$, $y(0) = -1$ (Section 2.2, Exercise 14)

T 11. $\dfrac{dy}{dt} = y^2 \cos t$, $y(0) = 1$ (Section 2.2, Exercise 15)

T 12. $t + ye^{-t} \dfrac{dy}{dt} = 0$, $y(0) = 1$ (Section 2.2, Exercise 16)

T 13. $xy \dfrac{dy}{dx} = 1 + y^2$, $y(1) = -2$ (Section 2.2, Exercise 17)

T 14. $\dfrac{dy}{dt} = y^3 \cos t$, $y(0) = 2$ (Section 2.2, Exercise 18)

In Exercises 15 through 20, sketch the slope field along with three representative solution curves. Also sketch the zero isocline. If possible, determine the behaviors of solutions as the independent variable approaches ∞. Compare with the solution formulas as indicated.

T 15. $x \dfrac{dy}{dx} = x + y$ (Section 2.2, Exercise 33)

T 16. $x \dfrac{dy}{dx} = x + 2y$ (Section 2.2, Exercise 34)

T 17. $xy \dfrac{dy}{dx} = y^2 - x^2$ (Section 2.2, Exercise 35)

T 18. $(x + y) \dfrac{dy}{dx} = 3x - y$ (Section 2.2, Exercise 36)

T 19. $\dfrac{dy}{dx} = x + y$ (Use the method of Section 2.2, Exercise 37)

T 20. $\dfrac{dy}{dx} = (y + x)^2$ (Section 2.2, Exercise 38)

21. Plot the zero isoclines for the endemic disease model of Equation (2) of Section 2.1, assuming $r < pkN$. Sketch the direction field by making use of the property that all minitangents on any given horizontal line are parallel. Use the sketch to determine the long-term behavior of solution curves that begin with $I > 0$.

22. Plot the zero isocline for the chemical kinetics model of Equation (9) of Section 2.1. Determine the minitangent slopes for $z = 0$, 0.2, 0.4, 0.6, and 0.8 for the case $\beta = 1$, $\rho = 1$, and use these to sketch the slope field. What happens to the solution curve that begins at $z = 0$?

✦ 2.3 INSTANT EXERCISE SOLUTIONS

1.

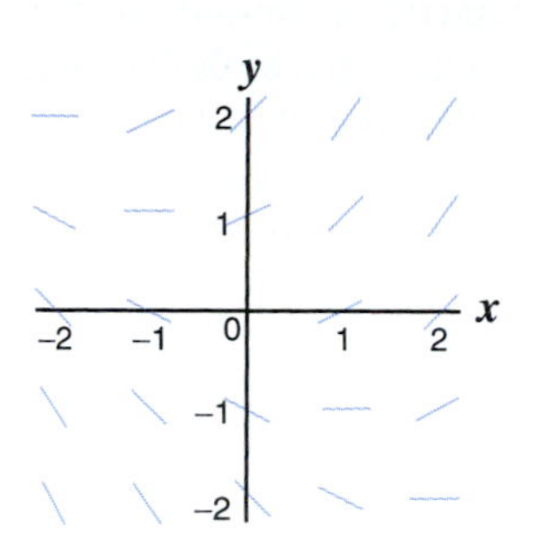

2.

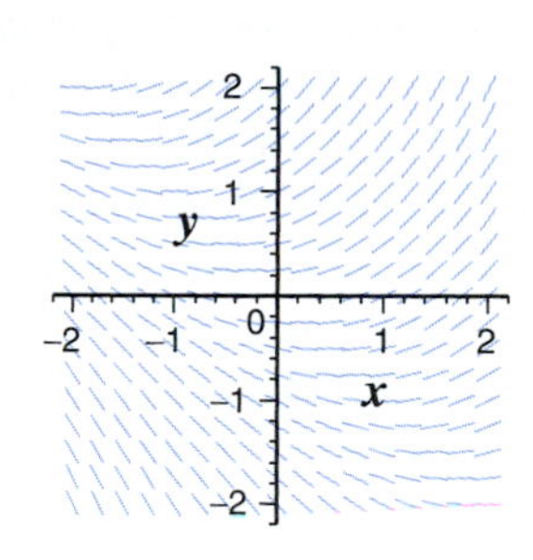

3.

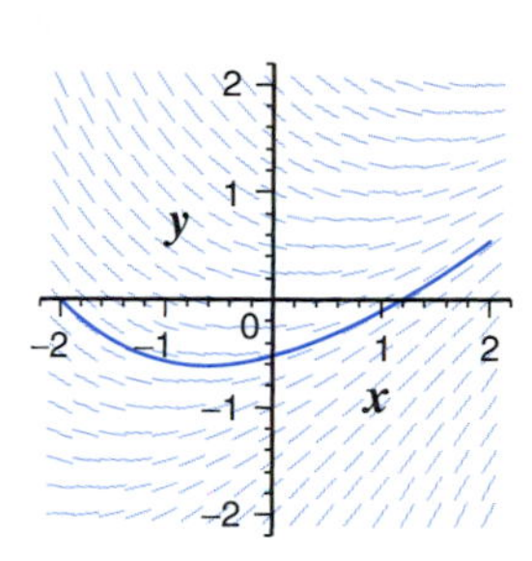

4. Given $y = -x - 2 + Ae^{x/2}$, we have $dy/dx = -1 + (A/2)e^{x/2}$ and $y + x = -2 + Ae^{x/2}$. Thus, $dy/dx = (x + y)/2$.

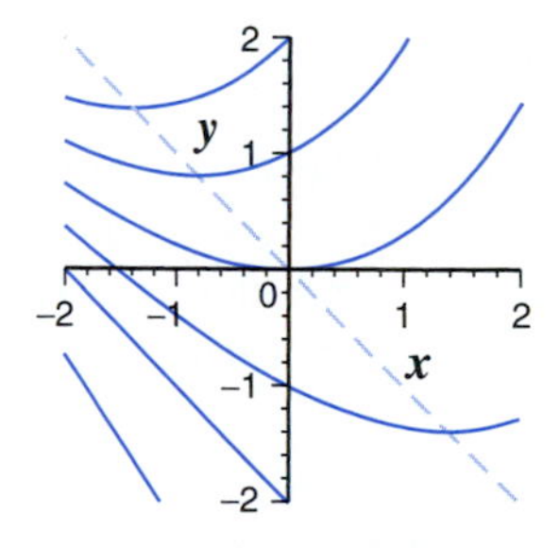

2.4 Existence of Unique Solutions

Up to this point, we have found solutions for a number of differential equations. In almost all cases, we have found a one-parameter family of solutions for a first-order differential equation and a single solution for an initial-value problem of first order. Our inability to find additional solutions for a given initial-value problem suggests that there is only one solution to be found. The question of whether a problem has a unique solution is of mathematical interest, but it is also of practical interest. If a scientific experiment yields the same result every time, then any mathematical model for that experiment should have a unique solution.

One application of uniqueness is that any problem that has a unique solution can be used to define a function. The reader is undoubtedly familiar with all the *elementary functions,* namely, power functions, exponential functions, logarithmic functions, and trigonometric functions. These functions are gradually introduced into the precalculus curriculum as the subjects needed for their definitions are introduced. There are many other named functions that cannot be written in terms of elementary functions. Some of these *special functions* can be defined by definite integrals. Others solve simple initial-value problems. Once a function has been defined as the unique solution of an initial-value problem, an algorithm for calculating the function values can be programmed into a calculator or a computer algebra system. After that, the function is accessible, even though it is not an elementary function.

In some sense, the exponential and natural logarithm functions are actually special functions. Power functions are defined in terms of arithmetic operations, so it is easy to think of these functions as well understood. Trigonometric functions can be defined by geometric constructions, so again it is easy to think of these as well understood. But how does one define the function e^x? One can calculate 10^n for any integer n by ordinary multiplication. Quantities such as $10^{5/3}$ can be defined by multiplication and roots. But 10^π cannot be defined by multiplying π factors of 10. Trying to define e^π by multiplication is even worse because of the added complication of understanding the number e.

From a careful mathematical point of view, the "standard" way to define the exponential function is to first define the natural logarithm function as

$$\ln x = \int_1^x \frac{1}{t}\,dt.$$

Then the exponential function is defined to be the inverse function of the natural logarithm: e^x is defined to be the number y for which $\ln y = x$. Given these definitions, the exponential and logarithm functions really should be considered as special functions rather than elementary functions. They are elementary only in that we are used to them and that they have some simple properties.

The function e^x satisfies the initial-value problem $y' = y$, $y(0) = 1$. If the initial-value problem can be shown to have a unique solution, then we could define the exponential function by using the initial-value problem rather than the inverse of the natural logarithm function. One could certainly argue that such a definition would make the function seem more familiar than the standard definition.

MODEL PROBLEM 2.4

1. Can the initial-value problem

$$\frac{dy}{dx} = y, \qquad y(0) = 1 \tag{1}$$

be used to define the function e^x?

2. Can one use the differential equation

$$\frac{d^2y}{dx^2} = xy \tag{2}$$

to define a function?

If we are to use the initial-value problem of Equation (1) to define a function, we have to assume that we do not know anything about the function we are defining other than that it solves the initial-value problem. The key property that we need to establish is that the initial-value problem has exactly one solution. We start by showing that it cannot have more than one solution.

Let $y(x)$ and $z(x)$ be any two solutions of the initial-value problem (1). Our task is to show that these functions are the same, using only the fact that these functions satisfy the differential equation and initial condition. Let $w(x) = y(x)/z(x)$. If we can show that $w \equiv 1$, that will prove that y and z are the same function. We start by checking the derivative of w:

$$\frac{dw}{dx} = \frac{z(x)y'(x) - z'(x)y(x)}{z^2(x)} = \frac{z(x)y(x) - z(x)y(x)}{z^2(x)} = 0,$$

where we have used the differential equation to substitute for y' and z'. The calculation of w' shows that w is constant. We also know that $y(0) = z(0) = 1$; hence, $w \equiv 1$, which completes the proof that $y \equiv z$.

Actually, this demonstration neglects one annoying detail. The definition of w makes sense only if $z(x)$ is not zero for any finite value of x. This is not obvious. It can, however, be demonstrated by using only the information that is given. To avoid complicating details, consider only the interval $x \geq 0$. Since $z' > 0$ for $z > 0$, and $z(0) = 1$, z must be an increasing function, and thus $z > 1$ for $x \geq 0$. More is needed to show that $z > 0$ for $x < 0$.

We have now demonstrated that the initial-value problem (1) can have only one solution. Proving that a problem actually has a solution is generally more difficult. It is not difficult when a solution can be exhibited. Here, for example, if we first define the exponential function in the standard way, then we can exhibit it as the solution of the problem. If we want to use the problem to *define* the exponential function, then we need a much more subtle argument to establish existence.

The Existence and Uniqueness Theorem for First-Order Equations

To show that an initial-value problem defines a function, there are two distinct properties that must be shown:

1. The initial-value problem has a solution.
2. The solution of the initial-value problem is unique.

Sometimes we can demonstrate the existence of a solution for a differential equation simply by solving the equation; however, we often need to study differential equations for which we do not have a solution formula. In these cases, existence is difficult to demonstrate. Although uniqueness is easier to demonstrate than existence, it is still difficult. Fortunately, there are standard theorems that can be used to demonstrate the existence of a unique solution for most initial-value problems. Theorem 2.4.1 applies to first-order initial-value problems of the form

$$\frac{dy}{dt} = f(t, y), \qquad y(t_0) = y_0; \tag{3}$$

the proof of the theorem is discussed in Section A.2.

Theorem 2.4.1 **Existence of Unique Solutions (First-Order)** Let R be a rectangle in the ty plane that contains the initial point (t_0, y_0) in its interior. If the functions f and $\partial f/\partial y$ are continuous throughout R, then the initial-value problem (3) has a unique solution on some interval $t_l < t < t_r$ that contains t_0.

It is important to be careful about the interpretation of Theorem 2.4.1.

- Geometrically, Theorem 2.4.1 says that each point in the region R has exactly one solution curve passing through it; thus, solution curves cannot cross within R.
- Theorem 2.4.1 does not guarantee a large interval of existence.
- Theorem 2.4.1 does not address cases for which the hypotheses are not met. Suppose f is not continuous at (t_0, y_0). There may be a unique solution through the point, there may be multiple solutions through the point, or there may be no solutions through the point.
- In practice, it is often important to know what initial conditions are guaranteed to work for a given differential equation. The easiest way to address this issue may be to look for points where f and $\partial f/\partial y$ are not both continuous.

EXAMPLE 1

Consider the differential equation

$$(1+t)\frac{dy}{dt} = 1 - y.$$

For this equation, we have

$$f(t, y) = \frac{1-y}{1+t}, \qquad \frac{\partial f}{\partial y} = -\frac{1}{1+t}.$$

Both of these functions are continuous at all points not on the line $t = -1$. We can use Theorem 2.4.1 to assert the following: Any initial-value problem

$$(t+1)\frac{dy}{dt} = 1 - y, \qquad y(t_0) = y_0, \qquad t_0 \neq -1$$

has a unique solution on some interval containing t_0. Any additional properties, such as the behavior of solutions and the existence or uniqueness of solutions through points on the line $t = -1$, can only be determined from other information.

To get definitive information about solutions other than what is available from Theorem 2.4.1, we need a solution formula. The differential equation of Example 1 served as Model Problem 2.2; we found the solution family

$$y = 1 + \frac{A}{1+t}.$$

Some of the solution curves for this differential equation were displayed in Figure 2.3.4 along with the slope field. Notice that there actually is a solution through the point $(-1, 1)$, namely, the constant solution $y = 1$. Theorem 2.4.1 does not guarantee such a solution, but neither does it rule it out. Solution curves through most points exist on the interval $(-\infty, -1)$ or the interval $(-1, \infty)$; however, solutions with initial condition $y(t_0) = 1$ exist for all t.

EXAMPLE 2

Consider the differential equation

$$\frac{dV}{dt} = -3V^{2/3},$$

which we derived in Section 1.2 as the differential equation for the volume of an evaporating raindrop (we have taken $k_1 = 3$ for algebraic convenience). By separating variables, we obtain the solution family

$$V = (C - t)^3.$$

These solutions have the interval of existence $(-\infty, \infty)$. However, the differential equation still has an obvious singular solution $V \equiv 0$. Theorem 2.4.1 is again of some help. We have

$$f(t, V) = -3V^{2/3}, \qquad \frac{\partial f}{\partial V} = -2V^{-1/3}.$$

The function $\partial f/\partial V$ is not continuous at $V = 0$. Theorem 2.4.1 guarantees a unique solution through any point not on the t axis, but it has nothing to say about points on the t axis. In fact, we can show that any function of the forms

$$V_1 = \begin{cases} (C - t)^3 & t \le C \\ 0 & C < t \end{cases}, \qquad V_2 = \begin{cases} 0 & t < C \\ (C - t)^3 & C \le t \end{cases}, \qquad V_3 = \begin{cases} (C_1 - t)^3 & t \le C_1 \\ 0 & C_1 < t < C_2 \\ (C_2 - t)^3 & C_2 \le t \end{cases}$$

is also a solution. Figure 2.4.1 illustrates some of these solutions.

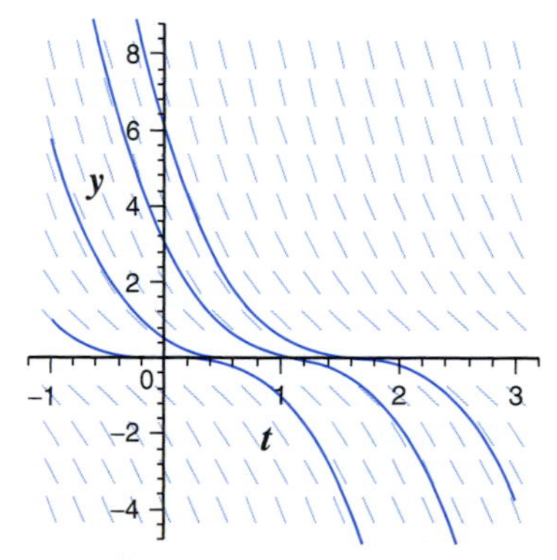

Figure 2.4.1
Some solutions of $dV/dt = -3V^{2/3}$.

✦ INSTANT EXERCISE 1

In Example 2, show that V_1 is a solution of the differential equation for all t.

The Existence and Uniqueness Theorem for Linear Equations

A first-order **linear** differential equation is a differential equation that can be written in the form

$$\frac{dy}{dt} + p(t)y = g(t). \tag{4}$$

Similarly, a second-order **linear** differential equation can be written in the form

$$\frac{d^2y}{dt^2} + p(t)\frac{dy}{dt} + q(t)y = g(t); \tag{5}$$

in general, an nth-order **linear** differential equation can be written in the form

$$\frac{d^n y}{dt^n} + p_1(t)\frac{d^{n-1}y}{dt^{n-1}} + p_2(t)\frac{d^{n-2}y}{dt^{n-2}} + \cdots + p_{n-1}(t)\frac{dy}{dt} + p_n(t)y = g(t).$$

Linear differential equations enjoy some special properties that make them very important. These include special methods for solution, which are discussed in Chapters 3 and 4, as well as a special existence and uniqueness theorem that is more predictive than Theorem 2.4.1.

Theorem 2.4.2

Existence of Unique Solutions (Linear) Let I be an open interval (t_l, t_r), and let t_0 be a point in the interval I. If the functions $p_i (i = 1, \ldots, n)$ and g are continuous throughout I, then the initial-value problem

$$\frac{d^n y}{dt^n} + p_1(t)\frac{d^{n-1}y}{dt^{n-1}} + p_2(t)\frac{d^{n-2}y}{dt^{n-2}} + \cdots + p_{n-1}(t)\frac{dy}{dt} + p_n(t)y = g(t), \tag{6}$$

$$y(t_0) = y_0, \qquad \frac{dy}{dt}(t_0) = y_0', \qquad \ldots, \qquad \frac{d^{n-1}y}{dt^{n-1}}(t_0) = y_0^{n-1} \tag{7}$$

has a unique solution throughout the interval I.

Note that Theorem 2.4.2 applies to Example 1. That differential equation can be rewritten in the form

$$\frac{dy}{dt} + \frac{y}{1+t} = \frac{1}{1+t}.$$

The only problem with the coefficient functions occurs at $t = -1$. If the initial condition is given at a point $t_0 > -1$, then Theorem 2.4.2 guarantees a unique solution on the interval $(-1, \infty)$; if the initial condition is given at a point $t_0 < -1$, then the guaranteed interval of existence is $(-\infty, -1)$. As with Theorem 2.4.1, Theorem 2.4.2 does not give any information if the initial condition is at $t_0 = -1$. Theorem 2.4.2 does not apply to Example 2 because that differential equation is not linear.

Some Special Functions

There are a lot of differential equations that cannot be solved in terms of elementary functions. Many of these can be shown by Theorem 2.4.1 or 2.4.2 to have unique solutions, when augmented by a suitable set of initial conditions. The initial-value problem can then be used to define the function that solves it uniquely. Such functions can be studied, tabulated, and programmed into software packages so that they are readily accessible when needed. Similarly, some definite integrals cannot be evaluated symbolically because there is no elementary antiderivative for the integrand. These can also be defined as special functions and made accessible. A few of the most important special functions are presented here, and others will be introduced in Chapter 3.

Error Functions The error function (erf) and the complementary error function (erfc) arise in probability and heat flow. They are defined by definite integrals:

$$\operatorname{erf} x = \frac{2}{\sqrt{\pi}} \int_0^x e^{-t^2}\, dt, \qquad \operatorname{erfc} x = 1 - \operatorname{erf} x = \frac{2}{\sqrt{\pi}} \int_x^{\infty} e^{-t^2}\, dt. \tag{8}$$

The value of $\operatorname{erf} x$ is the area under the curve of the function $(2/\sqrt{\pi})e^{-t^2}$ under a region that extends from $t = 0$ to $t = x$ (see Figure 2.4.2). The factor $2/\sqrt{\pi}$ is included in the definition because then $\operatorname{erf}(\infty) = 1$.

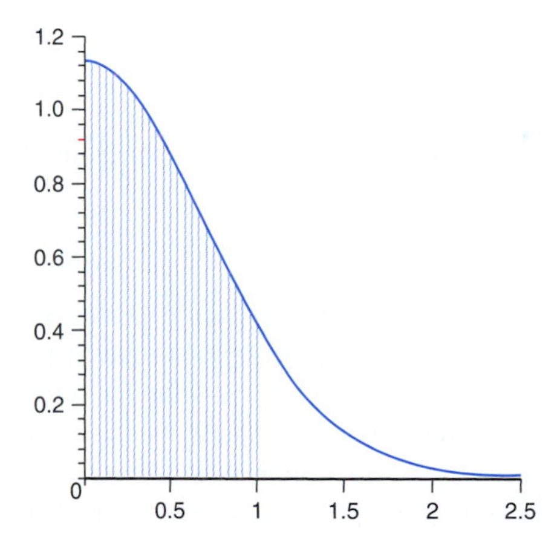

Figure 2.4.2
The value of $\operatorname{erf}(1)$ visualized as an area.

The error function has some very nice properties: It is defined for all values of x, bounded between -1 and 1, and always increasing. See Figure 2.4.3.

Functions defined by definite integrals on a finite interval can also be thought of as having been defined by initial-value problems. For example, $\operatorname{erf} x$ is clearly the unique solution to

$$\frac{dy}{dx} = \frac{2}{\sqrt{\pi}} e^{-x^2}, \qquad y(0) = 0.$$

Error functions arise more commonly as solutions of the second-order equation

$$\frac{d^2y}{dx^2} + 2x\frac{dy}{dx} = 0. \tag{9}$$

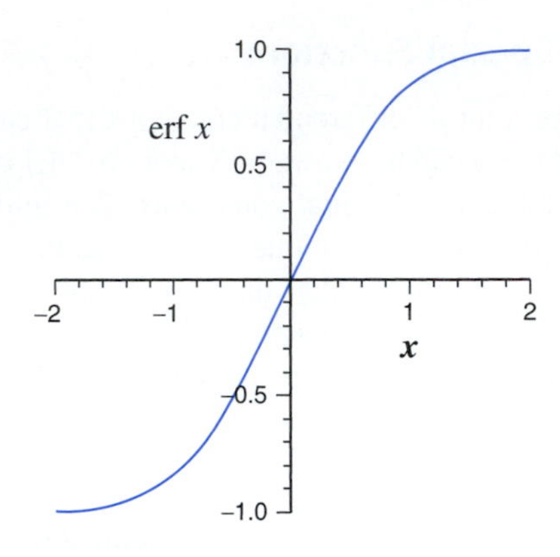

Figure 2.4.3
The error function erf x.

✦ INSTANT EXERCISE 2

Show that $y = C \operatorname{erf} x$ solves the family of initial-value problems

$$\frac{d^2y}{dx^2} + 2x\frac{dy}{dx} = 0, \qquad y(0) = 0.$$

Airy Functions Consider now the differential equation

$$\frac{d^2y}{dx^2} - xy = 0 \tag{10}$$

of Model Problem 2.4. This equation is known as the **Airy equation.** The Airy equation is linear, and the only coefficient function, x, is continuous everywhere. Theorem 2.4.2 guarantees that any initial-value problem for this equation has a unique solution that is valid for all x. However, the Airy equation does not have solutions that can be written in terms of elementary functions.

Since the Airy equation is second-order, it has a two-parameter family of solutions. Therefore, we could define two *special functions* as solutions of the equation with specific initial conditions. The general solution of the equation is then a sum of these functions multiplied by arbitrary constants. It is customary to define the **Airy functions** Ai and Bi as solutions of Equation (10) with initial conditions

$$\mathrm{Ai}\,(0) = a_1, \qquad \frac{d\mathrm{Ai}}{dx}(0) = -a_2, \qquad \mathrm{Bi}\,(0) = \sqrt{3}\,a_1, \qquad \frac{d\,\mathrm{Bi}}{dx}(0) = \sqrt{3}\,a_2, \tag{11}$$

where a_1 and a_2 are constants given approximately as

$$a_1 = 0.3550280539, \qquad a_2 = 0.2588194038.$$

You might wonder why the Airy functions are defined with these strange constants rather than 1 and 0. The answer is that the Airy functions were originally defined as definite integrals:

$$\mathrm{Ai}\,(x) = \frac{1}{\pi}\int_0^\infty \cos\left(\frac{t^3}{3} + xt\right)dt, \quad \mathrm{Bi}\,(x) = \frac{1}{\pi}\int_0^\infty \left[\exp\left(-\frac{t^3}{3} + xt\right) + \sin\left(\frac{t^3}{3} + xt\right)\right]dt.$$

The definite integral definitions are convenient for some purposes, but they are difficult to use to compute values because they involve improper integrals of functions that do not vanish as $t \to \infty$. Standard numerical methods, such as Simpson's rule, cannot be used to evaluate the Airy functions from these integral definitions. The general solution of the Airy equation is written as

$$y = c_1 \operatorname{Ai}(x) + c_2 \operatorname{Bi}(x).$$

Although the Airy functions seem rather strange, they are part of the standard definition package for computer algebra systems. They can be used just as more familiar functions are. The Airy functions are plotted in Figure 2.4.4.

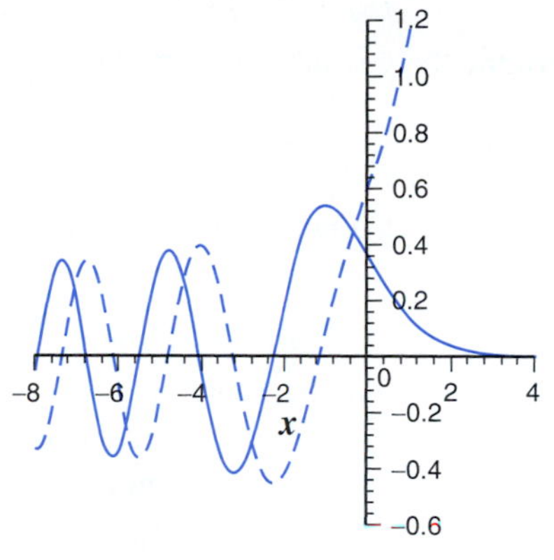

Figure 2.4.4
The Airy functions Ai (solid) and Bi (dashed).

2.4 Exercises

In Exercises 1 through 4, state the region(s) in the ty plane where Theorem 2.4.1 guarantees a unique solution through any given initial point in the region.

1. $(x+1)\dfrac{dy}{dx} = y^{1/3}$

2. $\dfrac{dy}{dx} = \dfrac{x+y}{x-y}$

3. $\dfrac{dy}{dt} = \dfrac{\sqrt{25 - t^2 - y^2}}{t+y+1}$

4. $\dfrac{dy}{dt} = \sqrt{e^t - y^2}$

In Exercises 5 through 8, use Theorem 2.4.2 to determine the guaranteed interval of existence for solutions with initial conditions at $x = 0$.

5. $(x-2)(x+1)\dfrac{d^2y}{dx^2} + x\dfrac{dy}{dx} + y = 0$

6. $(x-2)(x+1)\dfrac{d^2y}{dx^2} + (x+1)\dfrac{dy}{dx} + (x-2)y = 0$

7. $\cos x \dfrac{d^2y}{dx^2} + e^x y = x$

8. $x(x+1)\dfrac{d^2y}{dx^2} + \sin xy = 0$

9. Determine the guaranteed interval of existence for solutions of

$$\frac{d^2y}{dx^2} + 2x\frac{dy}{dx} + \ln(1-x) = 0, \qquad y(0) = 1, \qquad \frac{dy}{dx}(0) = 0.$$

10. Determine the guaranteed interval of existence for solutions of

$$x^2\frac{d^2y}{dx^2} + 2x\frac{dy}{dx} + 4x = 0, \qquad y(1) = 1, \qquad \frac{dy}{dx}(1) = 0.$$

11. *a.* Determine the guaranteed interval of existence for the initial-value problem

$$(1-x^2)\frac{d^2y}{dx^2} - 2x\frac{dy}{dx} + n(n+1)y = 0, \qquad y(0) = 1, \qquad \frac{dy}{dx}(0) = 0$$

without solving the problem.

b. Are there initial conditions for which a unique solution is not guaranteed?

c. For each of $n = 0, 1, 2, 3$, find a polynomial of degree n that solves

$$(1-x^2)\frac{d^2y}{dx^2} - 2x\frac{dy}{dx} + n(n+1)y = 0, \qquad y(1) = 1.$$

d. Is there a contradiction between theory and actual result for the solutions[11] in part *c*?

T 12. *a.* Determine the guaranteed interval of existence for the initial-value problem

$$x\frac{d^2y}{dx^2} + \frac{dy}{dx} + xy = 0, \qquad y(0) = 1, \qquad \frac{dy}{dx}(0) = 0$$

without solving the problem.

b. Use a computer algebra system to solve the problem and plot the solution.[12]

13. Consider the leaking bucket problem [Section 2.1, Eq. (3)]

$$\frac{dh}{dt} = -2\sqrt{h}, \qquad h(0) = 4.$$

a. Solve the problem using separation of variables.

b. Observe that the solution is defined for all t, but that beyond $t = 2$ it stops being a realistic description of the height of water in a bucket. What physical event occurs at $t = 2$?

c. Write down a function that uses the solution of part *a* up to the time $t = 2$ and then uses a realistic solution for $t > 2$.

---

[11]The differential equation is the Legendre equation of order n, and the polynomials that solve it with the condition $y(1) = 1$ are the Legendre polynomials.

[12]The differential equation is the Bessel equation of order 0, and the solution with the given initial conditions is the Bessel function of the first kind of order 0, denoted by J_0.

d. Show that the solution of part c is valid for all t.
e. Do these results contradict Theorem 2.4.1?
f. Plot the solutions of parts a and c together with the slope field. Note that it is easy to plot the slope field of an autonomous equation by hand because each slope value can be used for all minitangents on a horizontal line.

T 14. Consider the problem

$$\frac{dy}{dt} = 3y^{2/3}.$$

a. Find a family of solutions.
b. Find another family of solutions that is valid for all t. (*Hint:* see Example 2.)
c. Compare this problem with Example 2.
d. Do the results contradict Theorem 2.4.1?

T 15. Consider the problem

$$(x-1)\frac{dx}{dt} = t, \qquad x(0) = 1.$$

a. What does Theorem 2.4.1 say about this problem?
b. Try to solve the problem.
c. Use a computer algebra system to sketch the slope field and some solution curves.
d. Is there a contradiction between the results of parts a and b?

T 16. Consider the differential equation

$$\frac{dy}{dx} = 3\frac{y-1}{x-1}.$$

a. Solve the differential equation and graph some of the solutions.
b. What does Theorem 2.4.2 say about this equation with initial condition at $x = 1$?
c. Are there any values of y_0 for which there is a solution passing through $(1, y_0)$? If so, are the solutions unique?
d. What is the guaranteed interval of existence for the solution passing through a point $(0, y_0)$, where y_0 is any real number?

17. In this exercise, we show that the exponential function as defined by Model Problem 2.4 is in fact the inverse of the natural logarithm function. To do this, it is necessary to prove two identities:

$$\ln y(x) = x, \qquad \text{and} \qquad y(\ln x) = x, \qquad x > 0.$$

We are allowed to use all the properties of the natural logarithm, but the only properties of the exponential function that we can use are the differential equation $y' = y$ and initial condition $y(0) = 1$ that it satisfies.

a. To show that $\ln y(x) = x$, let $f(x) = \ln y(x)$. Differentiate this equation, using the chain rule along with the differential equation for y. This yields a differential equation for f. Use the initial condition for y to determine an initial condition for f. Solve the initial-value problem for f to complete the demonstration.

b. To show that $y(\ln x) = x$, let $z(x) = \ln x$ and $f(x) = y(\ln x)$. Then $f(x) = y(z(x))$. Use the chain rule along with the differential equation for y and the properties of the natural logarithm to derive a differential equation for f. Obtain an initial condition for this equation by evaluating f at $x = 1$. Solve the initial-value problem for f to complete the demonstration.

18. Two functions s and c can be defined by the initial-value problems

$$s'' + s = 0\,, \qquad s(0) = 0\,, \qquad s'(0) = 1$$

and

$$c'' + c = 0\,, \qquad c(0) = 1\,, \qquad c'(0) = 0\,.$$

Define the functions S and C by $S = s'$ and $C = c'$.

a. We can determine the function S by constructing an initial-value problem for it. Obtain the differential equation for S by differentiating the differential equation for s and then making the appropriate substitutions. The initial condition for S follows immediately from that for s', and the initial condition for S' follows from evaluating the differential equation for s at 0 and then making the appropriate substitutions. Now determine S in terms of s and c by comparing its initial-value problem with those for s and c.
b. Repeat part *a* to determine the function C in terms of s and c.
c. Define a new variable f by $f = s^2 + c^2$. Differentiate this equation and make the appropriate substitutions to show that f is a constant. Use the initial conditions to determine the correct constant.
d. What familiar functions are defined by the initial-value problems?

19. Show that

$$y = \operatorname{erf} \frac{x}{2\sqrt{kt}}$$

is a solution of the heat equation

$$\frac{\partial y}{\partial t} = k \frac{\partial^2 y}{\partial x^2},$$

where k is a constant. [*Hint:* Let $z(x, t) = x/(2\sqrt{kt})$. Differentiate $y = \operatorname{erf}(z(x, t))$, using the chain rule.]

20. Consider the equation

$$\frac{d^2 y}{dx^2} - 2x \frac{dy}{dx} - 2y = 0.$$

a. What conclusions can you draw about this equation without solving the equation?
b. Show that $y = e^{x^2} \operatorname{erf} x$ is a solution of the equation.

21. Use the fact that Ai and Bi are solutions of the Airy equation to show that any function of the form $y = c_1 \operatorname{Ai}(x) + c_2 \operatorname{Bi}(x)$ also solves the Airy equation.

T 22. *a.* Show that $\mathrm{Ai}\,(k^{1/3}x)$ is a solution of the equation

$$\frac{d^2y}{dx^2} - kxy = 0.$$

Conclude that the general solution is $y = c_1\,\mathrm{Ai}\,(k^{1/3}x) + c_2\,\mathrm{Bi}\,(k^{1/3}x)$.

b. Plot or sketch the graphs of the solutions $\mathrm{Ai}\,(-x)$ and $\mathrm{Bi}\,(-x)$ of $d^2y/dx^2 + xy = 0$.

✦ 2.4 INSTANT EXERCISE SOLUTIONS

1. For $t < C$, we have $V = (C - t)^3$. Thus, $dV/dt = -3(C - t)^2 = -3V^{2/3}$. Hence V_1 solves the differential equation for $t < C$, and obviously it also solves the differential equation for $t > C$. At $t = C$, we have

$$\lim_{t \to C^-} \frac{dV_1}{dt} = \lim_{t \to C^-} [-3(C - t)^2] = 0, \qquad \lim_{t \to C^+} \frac{dV_1}{dt} = \lim_{t \to C^+} 0 = 0.$$

The limits agree, so the derivative exists and satisfies the differential equation. Hence, the interval of existence for the solution family V_1 is $(-\infty, \infty)$.

2. With $y = C\,\mathrm{erf}\,x$, we have

$$\frac{dy}{dx} = \frac{2C}{\sqrt{\pi}}e^{-x^2}, \qquad \frac{d^2y}{dx^2} = -\frac{4Cx}{\sqrt{\pi}}e^{-x^2} = -2x\,\frac{dy}{dx}.$$

2.5 Euler's Method

In Section 2.3, the interpretation of a differential equation as a formula for the rate of change of a function led to a graphical technique for visualizing solution curves without finding solution formulas. If we want to approximate solutions, rather than merely visualize them, we need a computational technique. Euler's method, named after the 18th-century Swiss mathematician Leonhard Euler,[13] is based on the slope field.

MODEL PROBLEM 2.5

Approximate the solution of

$$\frac{dy}{dt} = \frac{8e^{-t}}{3 + y}, \qquad y(0) = 0$$

without using the solution formula.

Figure 2.5.1 shows the slope field for the differential equation of Model Problem 2.5. In the absence of a solution formula, some method is needed to add the solution curves to a slope

[13]The correct pronunciation is "Oy-ler."

field sketch. The problem of adding solution curves to the slope field is a numerical problem; in other words, some computational approximation is needed to use the slope field data to construct approximate solution curves.

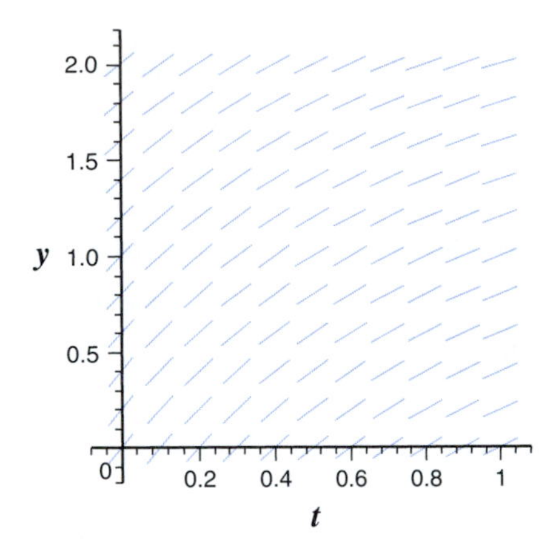

Figure 2.5.1
The slope field for $dy/dt = (8e^{-t})/(3+y)$.

A First Approximation

To begin, suppose we just want to approximate the solution on the interval $0 \le t \le 0.2$. There are only two pieces of information about the solution curve that we know for sure. First, it passes through the initial point (0, 0). Second, its slope at the point (0, 0) can be calculated by using the differential equation: $y'|_{(0,0)} = 8e^{-t}/(3+y)|_{(0,0)} = 8/3 \approx 2.667$. Graphically, this means that the solution curve has slope 8/3 at the point (0, 0), as depicted by the minitangent in the slope field. The simplest way to approximate the solution on $0 \le t \le 0.2$ is to draw the straight-line segment through the point (0, 0) in the direction of the minitangent and ending at a point where $t = 0.2$. Figure 2.5.2 shows this line segment along with the slope field.

The y coordinate of the right end of the line segment can be easily determined from

$$\Delta y = (\text{slope})\,(\Delta t) = \left(\tfrac{8}{3}\right)(0.2) \approx 0.533.$$

Thus, the right endpoint of the line segment approximation has

$$y \approx 0 + 0.533 = 0.533.$$

This calculation is illustrated in Figure 2.5.3. The technique of approximating the solution on intervals $[t_n, t_n + \Delta t]$ by using line segments whose slopes are determined at $t = t_n$ is **Euler's method.**

Now suppose we want to approximate the solution on the interval $0 \le t \le 0.4$. We can use the line segment just calculated, up to $t = 0.2$, and then we can use the same procedure on the interval $0.2 \le t \le 0.4$. We sketch a line segment from the point (0.2, 0.533) to $t = 0.4$, using the slope of the minitangent at (0.2, 0.533), which is approximately 1.854. To find the y coordinate of

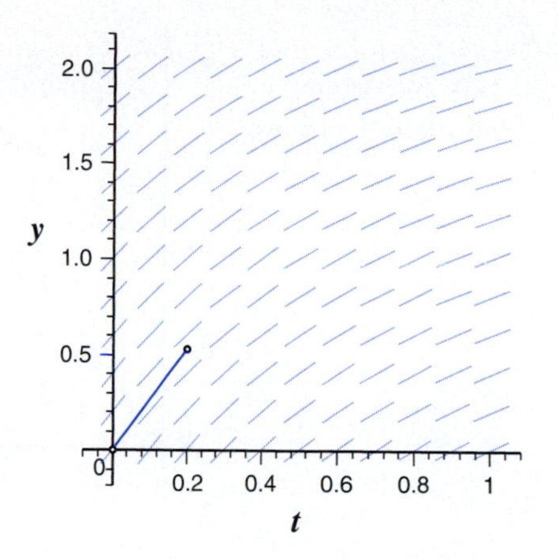

Figure 2.5.2
A line segment approximation for $dy/dt = 8e^{-t}/(3+y)$, $y(0) = 0$ on the interval [0, 0.2], along with the slope field.

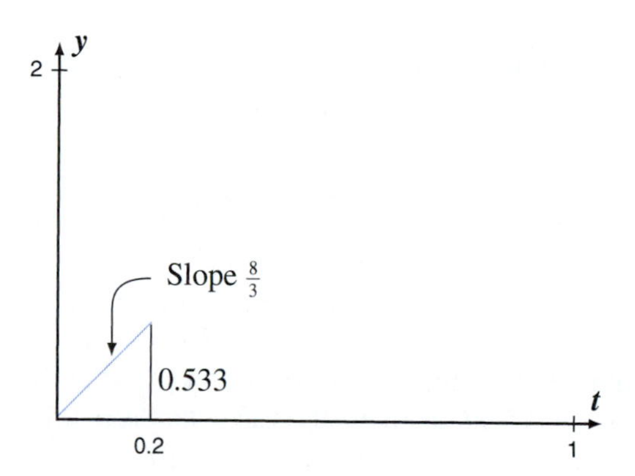

Figure 2.5.3
The line segment from (0, 0) to $t = 0.2$ in the approximation for Model Problem 2.5.

the right end, we have

$$y = 0.533 + \Delta y = 0.533 + (1.854)(0.2) \approx 0.904.$$

Figure 2.5.4 shows this approximation along with the slope field.

We can continue to use this same procedure to get more line segments approximating portions of the solution curve.

✦ INSTANT EXERCISE 1

Determine the next point ($t = 0.6$) on the graph of the straight-line approximation to the problem

$$\frac{dy}{dt} = \frac{8e^{-t}}{3+y}, \qquad y(0) = 0.$$

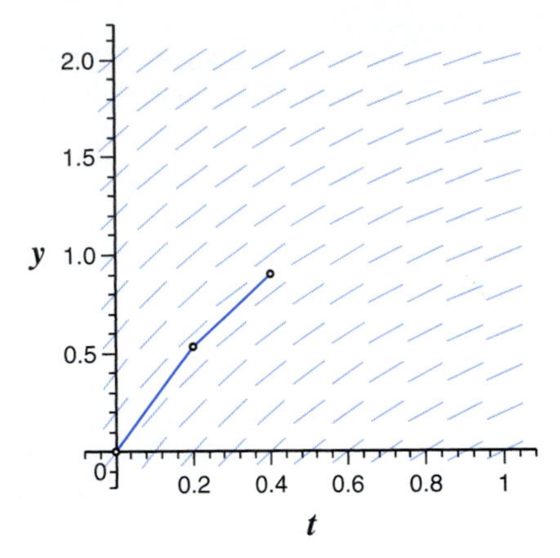

Figure 2.5.4
A line segment approximation for $dy/dt = 8e^{-t}/(3+y)$, $y(0) = 0$, on the interval [0, 0.4], along with the slope field.

The full set of points obtained in this manner up to $t = 1$ is given in Table 2.5.1. All values are rounded off to 3 decimal digits.

t	0	0.2	0.4	0.6	0.8	1.0
y	0	0.533	0.904	1.179	1.389	1.553
dy/dt	2.667	1.854	1.374	1.051	0.819	

Table 2.5.1 Approximate values for y and dy/dt for Euler's method applied to Model Problem 2.5 on the interval [0, 1], with $\Delta t = 0.2$

Figure 2.5.5 shows the approximate solution along with the slope field. Clearly the approximation is not perfect. A close look at the figure shows that each line segment is parallel to the direction field at its left end, but not at its right end. This is so because the solution curve is really *curved*; that is, the slope of the correct solution curve changes continuously, while the slope of the approximate solution curve changes only at specific points separated by a distance $\Delta t = 0.2$.

Normally, it is difficult to know the accuracy of an approximation. In this case, there is a solution formula:

$$y = \sqrt{25 - 16e^{-t}} - 3.$$

This function is included with the approximation and the slope field in Figure 2.5.6.

✦ INSTANT EXERCISE 2

Solve Model Problem 2.5 by using the method of separation of variables.

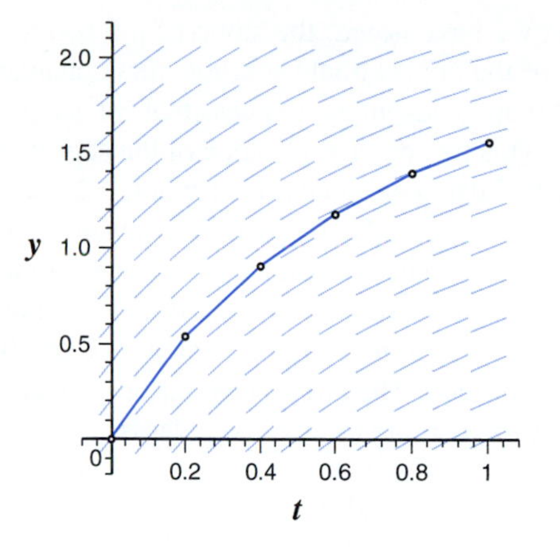

Figure 2.5.5
The Euler approximation for $dy/dt = 8e^{-t}/(3 + y)$, $y(0) = 0$, with $\Delta t = 0.2$, along with the slope field.

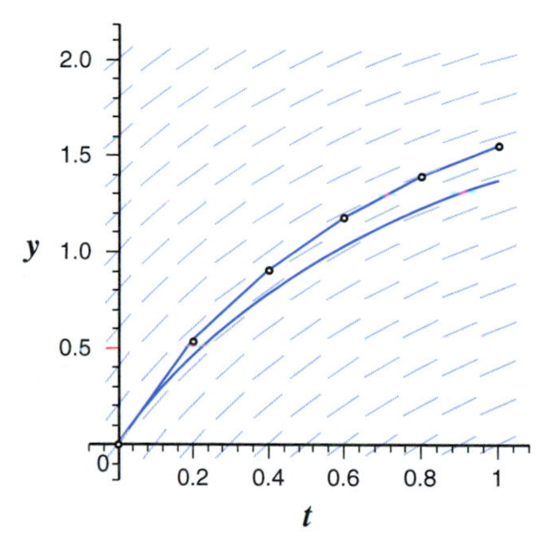

Figure 2.5.6
The Euler approximation for $dy/dt = 8e^{-t}/(3 + y)$, $y(0) = 0$, with $\Delta t = 0.2$, along with the slope field and the exact solution, $y = \sqrt{25 - 16e^{-t}} - 3$.

Improving the Results

The results of the calculations do not seem to be very good, and a careful examination of Figure 2.5.6 shows why. The point (0, 0) is on the correct solution curve, but the next approximation point (0.2, 0.533) is not. This point is, however, on a different solution curve. If we could avoid making any further error beyond $t = 0.2$, the numerical approximation would follow *a* solution curve from then on, but it would not be the *correct* solution curve. The real situation is more complicated, because additional error is introduced at each step of the approximation. The amount of error ultimately caused by the error in the first approximation step can become larger or smaller at later times, depending on the relationship between the different solution curves.[14]

---

[14]See Section A.3.

As we have noted, the slope of the correct solution curve changes continuously, while the slope of the approximate solution curve changes only at specific points separated by a distance Δt. It seems reasonable to guess that we can get a better approximation by using a smaller value of Δt. This means that the slope of the approximate solution curve will be corrected more often. Instead of using subintervals of width 0.2, we could try using subintervals of width 0.1 or less. We cannot just make Δt extremely small without introducing other difficulties. Using smaller subintervals means that the slope is changed more often, but at a cost of more calculations. (Taking $\Delta t = 10^{-9}$, for example, would require calculation of 1 billion data points to obtain the graph.) The approximations using the values 0.2 and 0.1 are illustrated in Figure 2.5.7, along with the correct solution. The circles near the correct solution curve were obtained with $\Delta t = 0.025$; these are almost accurate to visual standards.

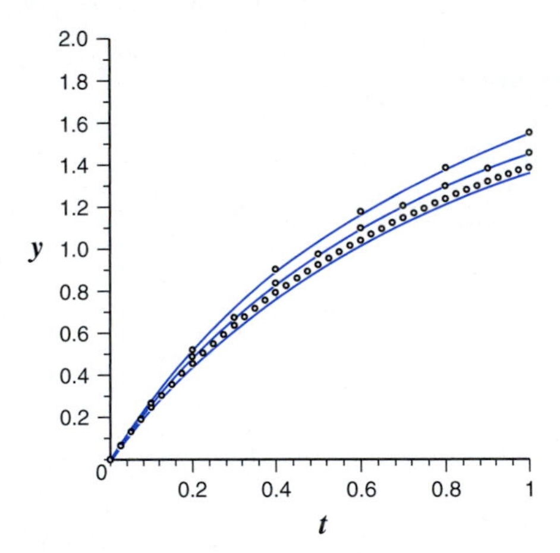

Figure 2.5.7
The Euler approximations using $\Delta t = 0.2$ and $\Delta t = 0.1$, for $dy/dt = 8e^{-t}/(3+y)$, $y(0) = 0$, along with the correct solution $y = \sqrt{25 - 16e^{-t}} - 3$ and data points obtained with $\Delta t = 0.025$.

The error in each new approximation seems to be roughly one-half of the error in the previous approximation. (See Table 2.5.2.) We can keep improving the accuracy, to a point, by choosing smaller subintervals. Some better methods are discussed in Section 2.6.

	Approximation					**Error**			
	Δt				**Solution**	Δt			
t	**0.2**	**0.1**	**0.05**	**0.025**	$y(t)$	**0.2**	**0.1**	**0.05**	**0.025**
0.2	0.533	0.488	0.468	0.459	0.450	0.083	0.038	0.018	0.009
0.4	0.904	0.837	0.807	0.792	0.778	0.126	0.059	0.029	0.014
0.6	1.179	1.099	1.062	1.045	1.027	0.152	0.072	0.035	0.018
0.8	1.389	1.301	1.259	1.240	1.220	0.169	0.081	0.039	0.020
1.0	1.553	1.458	1.414	1.393	1.372	0.181	0.086	0.042	0.021

Table 2.5.2 Comparison of results for Euler's method applied to Model Problem 2.5 on the interval [0, 1] with several different step sizes

Euler's Method in General

We want to be able to apply Euler's method to any first-order initial-value problem. The general form for a first-order initial-value problem is

$$\frac{dy}{dt} = f(t, y), \qquad y(t_0) = y_0. \tag{1}$$

Here we allow the function representing the slope of $y(t)$ to depend on both y and t, and we have allowed for any point (t_0, y_0) in the ty plane to provide the initial condition. We assume that we want to approximate the problem for the range $t_0 \le t \le t_f$, with t_f any value of t larger than t_0. (In practice, the method works equally well for $t_f \le t \le t_0$, with $t_f < t_0$.)

Assume that we want to divide the interval $[t_0, t_f]$ into N equal subdivisions, and that we want to approximate $y(t)$ at each of the subdivision points. The step size is

$$\Delta t = \frac{t_f - t_0}{N}$$

and the subdivision points are

$$t_n = t_0 + n\,\Delta t, \qquad n = 0, 1, \ldots, N. \tag{2}$$

In particular, the last point is $t_N = t_f$. Notice that the t coordinates of the approximation points are all determined in advance by the choices of initial condition, final point, and number of steps. The y coordinates of the approximation points must be determined sequentially, from left to right, by assuming that the slope of each line segment is that of the minitangent at the left endpoint.

Initially, we know only y_0. On the interval $[t_0, t_1]$, we approximate the slope by evaluating it at the left endpoint (t_0, y_0). The approximate slope is $f(y_0, t_0)$. A line segment that extends from (t_0, y_0) to $t = t_1$ with the given slope has a change of y given by

$$\Delta y = f(y_0, t_0)\,\Delta t.$$

The right endpoint of the line segment is the point $(t_1, y_0 + \Delta y)$. We want y_1 to approximate the y coordinate at the point t_1, so we choose

$$y_1 = y_0 + f(y_0, t_0)\,\Delta t.$$

The same calculation works for the next subinterval and, in turn, all the others. We have the result

$$y_{n+1} = y_n + f(y_n, t_n)\,\Delta t, \qquad n = 0, 1, \ldots. \tag{3}$$

The full Euler approximation is obtained by calculating all points (t_n, y_n) with Equations (2) and (3) and then connecting the points with line segments. See Figure 2.5.8.

A Test Problem

Consider the family of problems

$$y\frac{dy}{dt} = \frac{\pi}{2}\cos \pi t, \qquad y(0) = y_0 \approx 1, \tag{4}$$

which has the solution

$$y = \sqrt{y_0^2 + \sin \pi t}. \tag{5}$$

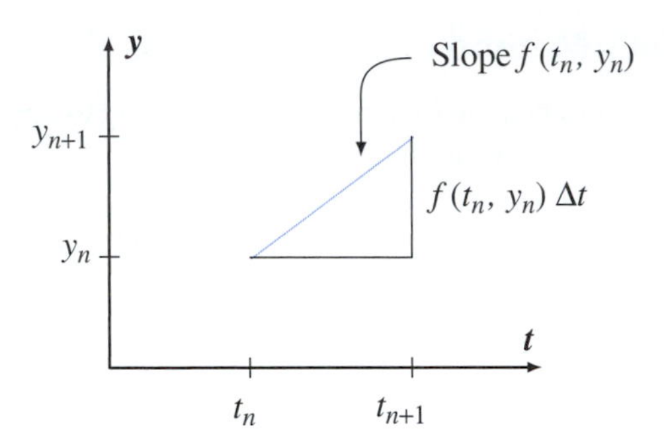

Figure 2.5.8
A line segment in Euler's method for the general case.

✦ INSTANT EXERCISE 3

Derive the solution formula for the test problem (4).

The test problem is interesting for two reasons. First, the solution is periodic, with $y(n) = y_0$ for any positive integer n. This gives us a convenient check on quantitative accuracy over a long interval. Second, the interval of existence of the solution[15] depends critically on y_0. If $y_0 > 1$, then the solution exists for all t. If $y_0 = 1$, then the solution exists only until the first point at which $\sin \pi t = -1$, which is $t = 3/2$. If $y_0 < 1$, the solution ceases to exist a bit sooner than $t = 3/2$. Error in the method may lead to failure in these qualitative results. It could be that solutions will cease to exist for some y_0 values greater than 1 or will continue to exist for some y_0 values less than 1. We cannot expect a numerical method to give an exact answer, but we must be alert to the danger that a numerical answer is qualitatively wrong.

Figure 2.5.9 shows the exact solution to the test problem with $y_0 = 1$ on the interval $[0, 2]$ along with the Euler approximations using 100 steps and 200 steps. The exact solution drops to exactly 0 at $t = 1.5$ and then rebounds. The numerical approximations are visually acceptable up to $t = 1$, but soon afterward they drift away from the exact solution. Curiously, the approximate solutions come back close to the exact solution after that, although the same problem will occur approaching $t = 3.5$, the next time the exact solution reaches 0.

Now suppose $y_0 = 0.9$. In this case, the solution should cease to exist when y reaches 0 at $t \approx 1.30$. Figure 2.5.10 shows the approximations on the interval $[0, 2]$ with 200 and 400 time steps. The solution does not stop at $t = 1.30$, but it clearly exhibits some strange behavior near that point. The approximate curves quickly move away from $y = 0$, but the direction in which they move depends on the step size. What is happening?

We can draw one conclusion simply from the observation that doubling the number of steps makes a drastic change in the numerical result. Doubling the number of steps should make the approximation look more like the correct solution, in which case we should not be seeing a qualitative difference between the two approximations. This qualitative difference should make us suspicious even if we do not have a solution formula or any other detailed knowledge. This

[15] Section 1.2.

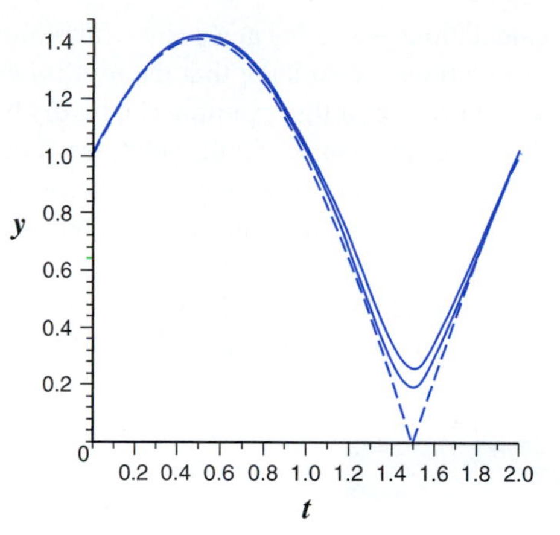

Figure 2.5.9
The Euler approximations for $y\,dy/dt = (\pi/2)\cos \pi t$, $\quad y(0) = 1$, using 100 and 200 time steps on the interval [0, 2], along with the exact solution (dashed).

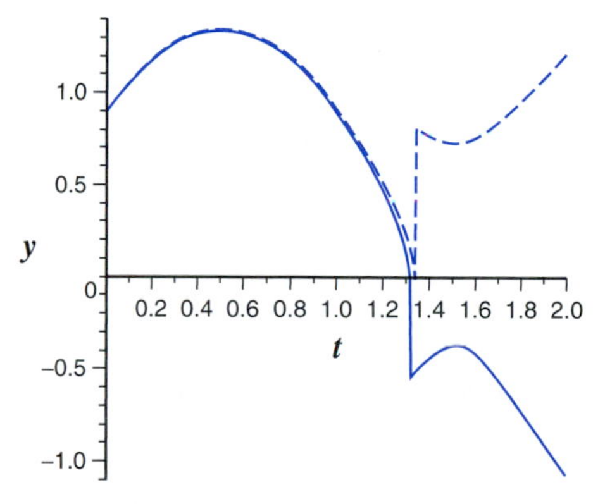

Figure 2.5.10
The Euler approximations for $y\,dy/dt = (\pi/2)\cos \pi t$, $y(0) = 0.9$, using 200 and 400 time steps on the interval [0, 2].

sort of behavior indicates that the method is not working, and it suggests that there is something unusual about the solution at that point.

Careful examination of the differential equation also helps explain the problem. In the form $dy/dt = f(t, y)$, the differential equation is

$$\frac{dy}{dt} = \frac{\pi \cos \pi t}{2y}.$$

The function that gives the slope is unbounded as $y \to 0$. As y becomes small, the calculated slopes become large. If y is negative at such a point, we will get a large negative slope; otherwise we will get a large positive slope. A large slope magnifies the error made in using a line segment to approximate the solution curve. Each of the sudden jumps in Figure 2.5.10 is the result of one slope calculation that yields a result large in magnitude. It is simply a coincidence that one of these calculations was done at a point with positive y and the other with negative y. The errors in these calculations are so large that the results beyond these points are meaningless.

The importance of this example can hardly be overstated. Computers are a wonderful tool for approximating and visualizing the solutions of differential equations. However, computers cannot tell whether something is wrong with their calculations unless they are programmed to do so. Professional numerical differential equation solvers generally avoid difficulties such as these, but no numerical scheme is so good as to be immune to trouble. The educated user needs to be able to look at numerical results and spot indications of trouble. Difficulties such as we have seen here are caused by a combination of method error[16] and *ill-conditioning* of the differential equation.[17]

2.5 Exercises

1. Use Euler's method (do not use a computer program) with step size 0.1 to approximate $y(0.1)$, $y(0.2)$, $y(0.3)$, and $y(0.4)$ for the problem

$$\frac{dy}{dt} + y = 2t + 2, \qquad y(0) = 2.$$

Compare the result with the exact solution $y = 2t + 2e^{-t}$.

2. Use Euler's method (do not use a computer program) with step size 0.25 to approximate $y(1.25)$, $y(1.5)$, $y(1.75)$, and $y(2)$ for the problem

$$x\frac{dy}{dx} = 2y, \qquad y(1) = 2.$$

Compare with the exact solution.

T 3. Consider the initial-value problem

$$\frac{dy}{dt} = 2ty^2, \qquad y(0) = 5$$

(Section 2.2, Exercise 13, and Section 2.3, Exercise 9).

a. Use Euler's method with step size of 0.02 to approximate $y(0.1)$. Compare the approximation with the correct solution and determine the percentage error.

b. Use Euler's method with step size of 0.02 to approximate $y(0.4)$. Compare the approximation with the correct solution and determine the percentage error. Compare this result with that of part a.

c. Repeat parts a and b using step sizes of 0.01 and 0.005. What effect does halving the step size have on the error?

d. Plot the approximations and the exact solution.

T 4. Consider the initial value problem

$$\frac{dy}{dt} = \cos(2\pi t)\, y, \qquad y(0) = 1.$$

[16]In the context of numerical approximations, the word *error* does not mean that the approximation scheme or computer is making a mistake. Rather, it refers to errors that are a natural consequence of the numerical approximation.

[17]These topics are explored further in Section A.3.

a. Use Euler's method with step size of 0.1 to approximate $y(0.5)$. Compare the approximation with the correct solution and determine the percentage error.

b. Use Euler's method with step size of 0.1 to approximate $y(2)$. Compare the approximation with the correct solution and determine the percentage error. Compare this result with that of part *a*.

c. Repeat parts *a* and *b*, using step sizes of 0.05 and 0.025. What effect does halving the step size have on the error?

d. Plot the approximations and the exact solution.

T 5. Consider the initial-value problem

$$\frac{dy}{dt} + y = 2t + 2, \qquad y(0) = -4$$

(Section 2.2, Exercise 29, and Section 2.3, Exercise 3).

a. Use Euler's method with 10 steps to approximate $y(2)$. Compare the approximation with the correct solution and determine the percentage error.

b. Repeat part *a*, using 20 steps and 40 steps. What effect does halving the step size have on the error?

c. Plot the approximations and the exact solution.

d. Determine a formula to estimate the error in the approximation when the step size is Δt.

T 6. Consider the initial-value problem

$$\frac{dy}{dt} + y = t^2, \qquad y(0) = 2$$

(Section 2.2, Exercise 30, and Section 2.3, Exercise 4).

a. Use Euler's method with 10 steps to approximate $y(2)$. Compare the approximation with the correct solution and determine the percentage error.

b. Repeat part *a*, using 20 steps and 40 steps. What effect does halving the step size have on the error?

c. Plot the approximations and the exact solution.

d. Determine a formula to estimate the error in the approximation when the step size is Δt.

T 7. Consider the initial-value problem

$$\frac{dy}{dt} = \frac{5}{1 + 5e^{-y}}, \qquad y(0) = 0.$$

a. Use Euler's method with 10 steps to approximate $y(2)$.

b. Repeat part *a*, using 20 steps and 40 steps.

c. Plot the approximations and the exact solution.

T 8. Consider the initial-value problem

$$\frac{dy}{dt} = \frac{y \cos t}{1 + 2y^2}, \qquad y(0) = 1$$

(Section 2.2, Exercise 10, and Section 2.3, Exercise 8).

a. Use Euler's method with 10 steps to approximate $y(\pi)$.

b. Repeat part *a*, using 20 steps and 40 steps.

c. Plot the approximations and the exact solution.

T 9. Consider the initial-value problem

$$\frac{dy}{dt} = y^2 \cos t, \qquad y(0) = 1$$

(Section 2.2, Exercise 15, and Section 2.3, Exercise 11).

a. Use Euler's method with 40 steps on the interval [0, 2].
b. Repeat part *a*, using 80 steps and 160 steps.
c. Plot the approximations along with the exact solution. Can you explain the graph?

T 10. Consider the initial-value problem

$$\frac{dy}{dt} = \frac{2 \sin t}{1 - y}, \qquad y(0) = -1$$

(Section 2.2, Exercise 14, and Section 2.3, Exercise 10).

a. Use Euler's method with 18 steps on the interval [0, 1.8].
b. Repeat part *a*, using 36 steps and 72 steps.
c. Plot the approximations along with the exact solution. Can you explain the graph?

T 11. Consider the initial-value problem

$$\frac{dy}{dt} = t - 0.2y^2, \qquad y(0) = 0.$$

a. Use Euler's method with 40 steps to approximate $y(1)$.
b. Repeat part *a*, using 80 steps.
c. Assume that the error in the approximation y_N for $y(1)$ is halved each time the number of steps is doubled. What is the best approximation you can make for $y(1)$?

✦ 2.5 INSTANT EXERCISE SOLUTIONS

1. The current point is $t_2 = 0.4$, $y_2 = 0.904$. The slope at this point is

$$\frac{8e^{-t_2}}{3 + y_2} \approx 1.374.$$

The change in y is determined by

$$\Delta y \approx 1.374\, \Delta t = 0.275.$$

(Note that we are rounding at the third digit because previous values were already rounded at that point.) Thus, $y_3 = 1.179$.

2. The equation in separated form is

$$(3 + y)\frac{dy}{dt} = 8e^{-t}.$$

Integration and substitution yield

$$\int (3 + y)\, dy = 8 \int e^{-t}\, dt,$$

or

$$3y + \frac{y^2}{2} = -8e^{-t} + C.$$

This implicit solution formula can be solved explicitly for y by completing the square. First multiply both sides by 2 to get

$$y^2 + 6y = 2C - 16e^{-t}.$$

Adding 9 to both sides completes the square:

$$y^2 + 6y + 9 = (2C + 9) - 16e^{-t},$$

or

$$(y + 3)^2 = A - 16e^{-t},$$

where A is an arbitrary constant. This is a convenient point at which to determine the constant. The solution passes through the origin if $A = 3^2 + 16 = 25$. Thus,

$$(y + 3)^2 = 25 - 16e^{-t}.$$

There are two possible solutions to this last equation:

$$y = -3 \pm \sqrt{25 - 16e^{-t}}.$$

The positive root is the correct one because $y(0) > -3$. Thus,

$$y = \sqrt{25 - 16e^{-t}} - 3.$$

3. The equation is already separated, but it is convenient to multiply by 2 before integrating. We have

$$\int 2y\, dy = \int \pi \cos \pi t\, dt.$$

The integration yields

$$y^2 = \sin \pi t + C,$$

and the initial condition yields $C = y_0^2$. Since $y_0 > 0$, the correct square root is the positive one, so the result is Equation (5).

2.6 Runge–Kutta Methods

Our experience with Euler's method raised some questions about numerical approximations for the solutions of differential equations. Are there more clever methods that perform better than Euler's method? In particular, the error in Euler's method is cut approximately in half by doubling the number of steps. It would be useful to find methods in which doubling the number of steps does better than cutting the error in half, if there are such methods.

MODEL PROBLEM 2.6

Obtain a numerical approximation for the solution of

$$y\frac{dy}{dt} = \frac{\pi}{2} \cos \pi t, \qquad y(0) = 1.01$$

with error less than 0.001 at each point.

First note that the performance of Euler's method on this problem is not impressive. Although the solution of this problem

$$y = \sqrt{1.0201 + \sin \pi t}$$

is defined for all t, Euler's method has difficulty approximating this solution when it is near its minimum point at $(1.5, 0.14)$. Figure 2.6.1 shows the Euler approximations with 100 and 200 time steps on the interval $[0, 2]$.

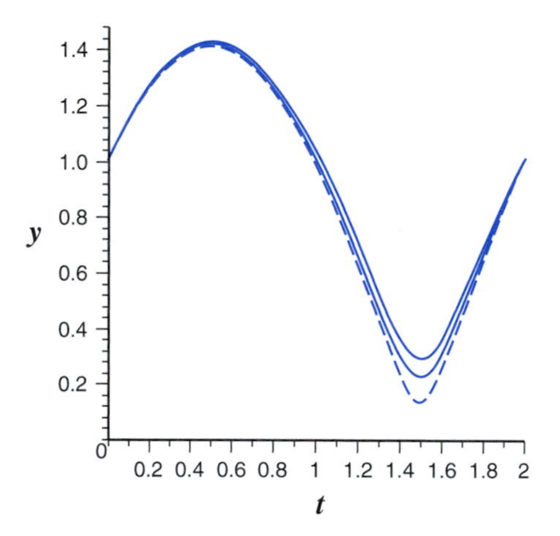

Figure 2.6.1
The Euler approximations for Model Problem 2.6 using 100 and 200 steps (solid) along with the exact solution (dashed).

By trial and error, it turns out that Euler's method requires $N = 2047$ over the interval $[0, 1.5]$ to achieve an error less than 0.01. Since doubling the number of steps roughly halves the error, we can expect to have to increase the number of steps by a factor of 10 to reduce the error by a factor of 10. By this estimate, we would need to use more than 20,000 steps to achieve the desired accuracy with Euler's method.

Numerical Methods from Numerical Integration Rules

We obtained the formula for Euler's method in Section 2.5 by using a tangent line approximation. Other numerical approximations can be obtained from the definite integral form of the differential equation. We can integrate any differential equation of the form

$$\frac{dy}{dt} = f(t, y)$$

over the interval (t_n, t_{n+1}), provided f and $\partial f/\partial y$ are continuous on the interval:

$$\int_{t_n}^{t_{n+1}} \frac{dy}{dt}\, dt = \int_{t_n}^{t_{n+1}} f(t, y(t))\, dt.$$

The integral on the left side can be computed immediately, yielding

$$y(t_{n+1}) - y(t_n) = \int_{t_n}^{t_{n+1}} f(t, y(t))\, dt.$$

In principle, if y_n is known, we can compute y_{n+1} from

$$y_{n+1} = y_n + \int_{t_n}^{t_{n+1}} f(t, y(t))\, dt, \tag{1}$$

but we cannot compute the integral on the right side without knowing the exact solution. However, we can use an approximation formula for the integral and thereby obtain a numerical approximation for the differential equation. One option would be to simply use a left-hand Riemann sum with one subinterval. The integral is then approximated as

$$\int_{t_n}^{t_{n+1}} f(t, y(t))\, dt \approx f(t_n, y_n)\, \Delta t.$$

Combining this with the integral form of the differential equation (1) gives us the approximation formula

$$y_{n+1} = y_n + f(t_n, y_n)\, \Delta t.$$

Thus, Euler's method corresponds to using the left-hand Riemann sum to approximate an integral. One-sided Riemann sums are not very accurate for integrals, so it is not surprising that Euler's method is not sufficiently accurate for careful work. Think of it this way: The rate of change of y with respect to t varies from the point $(t_n, y(t_n))$ to the point $(t_{n+1}, y(t_{n+1}))$, and the difference between the two y values depends on the average derivative; however, Euler's method only uses the value of the derivative at one end of the subinterval.

The Trapezoidal Method

We could make a better method out of the integral formulation (1) by using the trapezoidal rule with one subdivision

$$\int_{t_n}^{t_{n+1}} f(t, y)\, dt \approx \frac{f(t_n, y_n) + f(t_{n+1}, y_{n+1})}{2}\, \Delta t$$

to approximate the integral. This approximation gives us the trapezoidal method for numerical solution of differential equations:

$$y_{n+1} = y_n + 0.5[f(t_n, y_n) + f(t_{n+1}, y_{n+1})]\, \Delta t. \tag{2}$$

We can expect the trapezoidal method to be significantly more accurate than Euler's method. However, it has a difficulty of its own. Since y_{n+1} appears on the right-hand side of the approximation, the equation defines y_{n+1} implicitly. In general, Equation (2) must be solved numerically to determine y_{n+1}.

Similarly, any other method of approximating the integral in Equation (1) is going to have a similar problem, except of course the left-hand Riemann sum.

The Modified Euler Method: The Best of Both Worlds

We would like to have a method that uses a good integral approximation scheme, such as the trapezoidal method, but we would also like a method that allows computation of y_{n+1} using explicit formulas, such as Euler's method. One attempt to achieve these goals is the *modified Euler method.* The idea is that the trapezoidal method could be used more easily if we used an approximation for y_{n+1} on the right side of Equation (2). The modified Euler method uses Euler's method to compute a first approximation for y_{n+1} and then uses the trapezoidal method to calculate y_{n+1} from y_n and the first approximation for y_{n+1}. The method requires two computations in each step. First, an intermediate value is determined by using Euler's method

$$k = y_n + f(t_n, y_n)\,\Delta t; \tag{3}$$

then the intermediate value is used in place of y_{n+1} in a trapezoidlike approximation

$$y_{n+1} = y_n + 0.5[f(t_n, y_n) + f(t_{n+1}, k)]\,\Delta t. \tag{4}$$

We can't expect the modified Euler method to be as good as the trapezoidal method because it evaluates the function f at an approximate point rather than at the exact point; however, it ought to be much better than Euler's method because it uses the essential idea of averaging the derivative between the left and right endpoints of the interval. Figure 2.6.2 shows the modified Euler approximation for Model Problem 2.6 with just 20 time steps. The calculation requires the function f to be evaluated 40 times, so the computational time needed for the approximation is roughly 20% of the time needed for Euler's method with 200 steps. Only the points obtained by the modified Euler method are displayed, along with the exact solution.

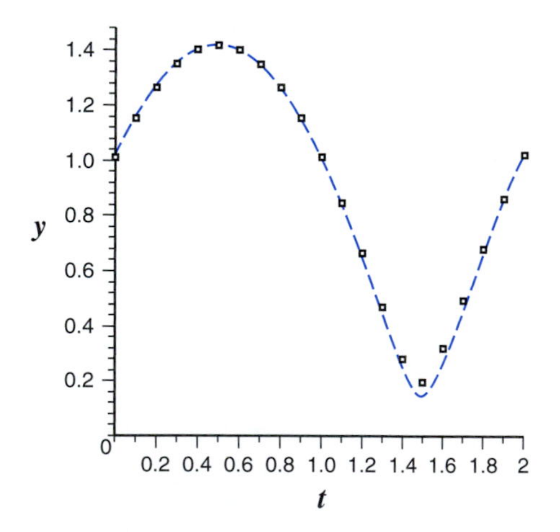

Figure 2.6.2
The modified Euler approximation with 20 steps (points) for Model Problem 2.6 along with the exact solution.

The comparison between the methods is even more striking when one looks at the effort required to achieve a high degree of accuracy. We noted earlier that Euler's method achieves an error less than 0.01 at $t = 1.5$ only with 2047 steps or more. In contrast, the modified Euler method achieves the same accuracy with only 33 steps. Since each step requires two function evaluations, the computer time required for this level of accuracy with the modified Euler method is about 3.2% of that required for Euler's method. Whereas Euler's method was estimated to require more than 20,000 steps to achieve an error less than 0.001, trial and error shows that the modified Euler method accomplishes this goal with just 91 steps, or 182 function evaluations. Thus, the modified Euler method achieves the desired degree of accuracy with slightly fewer function evaluations than were used to compute the better of the two Euler approximations of Figure 2.6.1.

The Order of a Numerical Method

In general, suppose a numerical method is being used to approximate the solution of an initial-value problem on an interval $[t_0, T]$, with the initial condition given at $t = t_0$. Let E_N be the magnitude of the error in the approximation at $t = T$ when the method is applied with N steps. Thus,

$$E_N = |y_N - y(T)|.$$

EXAMPLE 1

Consider the Euler approximations for $y(1.5)$ for Model Problem 2.6. The 100-step and 200-step approximations on the interval $[0, 2]$ yielded the approximations $y_{75} = 0.3044$ and $y_{150} = 0.2404$, respectively (1.5 is three-fourths of the way from 0 to 2), and the correct value is $y(1.5) = 0.1418$. Thus, $E_{75} = |0.3044 - 0.1418| = 0.1626$ and $E_{150} = |0.2404 - 0.1418| = 0.0986$. Similarly, we obtain $E_{300} = 0.0569$ and $E_{600} = 0.0313$. Also note that $E_{150}/E_{75} = 0.61$, $E_{300}/E_{150} = 0.58$, and $E_{600}/E_{300} = 0.55$. As the number of steps is increased, the error is approximately halved when the number of steps is doubled.

In general, the error in Euler's method is given approximately by

$$NE_N = C_E,$$

where the constant C_E depends strongly on the specific problem. For Example 1, we can estimate (using $E_{2047} = 0.01$) $C_E \approx 20.5$. Model Problem 2.6 is particularly troublesome for numerical approximation, so the value of C_E is relatively high.

The error in the modified Euler method changes with step size in a different way. We've seen that the error in an approximation with 91 steps is approximately 0.001. The error in an approximation with 182 steps is approximately 0.00022, which is roughly one-fourth of the error in the 91-step approximation. In general, the modified Euler method has an error given approximately by

$$N^2 E_N = C_{ME},$$

where C_{ME} is a constant. We have deliberately used different symbols for the constants to indicate that there is no reason to think that the constant for one method should be the same as the constant for another. Based on the 182-step approximation, the value of C_{ME} for Model Problem 2.6 at $t = 1.5$ is approximately 7.3.

Although the difference between C_E and C_{ME} has some effect on the difference in the performance of the two methods, the difference between $E_N \propto N^{-1}$ and $E_N \propto N^{-2}$ is much more important. The power of N in the error approximation formula is called the **order** of the numerical method. Euler's method is first-order, and the modified Euler method is second-order.

The modified Euler method requires little effort to program into a computer, so students with a small amount of programming experience can write their own programs to approximate solutions of initial-value problems using the modified Euler method.

The Classical Fourth-Order Runge–Kutta Method

In general, there are a number of numerical approximation methods that use one or more intermediate computations in a scheme to compute y_{n+1} from y_n. These are collectively known as **Runge–Kutta methods.** Two Runge–Kutta methods of fourth order are in common usage. Many standard numerical packages, Maple for example, use a fourth-order method called **rkf4,** or the **Runge–Kutta–Fehlberg method.** This sophisticated method not only computes the approximation, but also estimates the error as it goes, and it adjusts the step size to maintain an acceptable degree of accuracy. The method is excellent for professional software, but rather complicated for the student to program. People who want to write their own numerical approximation algorithm generally use the "classical" **fourth-order Runge–Kutta method,** also called just **rk4.** The word *classical* is used here because there are an infinite variety of fourth-order Runge–Kutta methods, but the one we are using is by far the most common. The rk4 method involves four computations for each step:

$$y_{n+1} = y_n + \frac{k_1 + 2k_2 + 2k_3 + k_4}{6}\Delta t, \tag{5}$$

where

$$\begin{aligned}
k_1 &= f(t_n, y_n),\\
k_2 &= f\left(t_n + \tfrac{1}{2}\Delta t,\, y_n + \tfrac{1}{2}k_1\Delta t\right),\\
k_3 &= f\left(t_n + \tfrac{1}{2}\Delta t,\, y_n + \tfrac{1}{2}k_2\Delta t\right),\\
k_4 &= f(t_{n+1}, y_n + k_3\Delta t).
\end{aligned}$$

This method is difficult to derive, but not too difficult to understand. The quantities k_1 and k_4 approximate the slopes at t_n and t_{n+1}, respectively, and the quantities k_2 and k_3 are two different approximations for the slope at the midpoint of the interval. Given these identifications of the intermediate values, the change in y is computed by using Simpson's rule for the integral in the general approximation scheme (1).

Figure 2.6.3 displays the numerical approximations generated by the rk4 method on Model Problem 2.6, using 8 and 16 steps. The 16-step approximation is very good to visual standards at all values of t. The 8-step approximation is better than the 20-step modified Euler approximation at $t = 1.5$, but its error is greater at $t = 2$. One must take $N \geq 13$ to guarantee error less than 0.01 at $t = 2$, and $N \geq 21$ to guarantee error less than 0.001.

As noted before, computational time is measured as the number of function evaluations rather than the number of steps. To achieve error less than 0.01 on the whole interval [0, 2], the rk4 method uses 52 function evaluations for 13 steps, and the modified Euler method uses 88 function evaluations for 44 steps. For a tolerance of 0.001, the rk4 method uses 84 function evaluations as compared with 242 for the modified Euler method. The difference is slight when low accuracy is sufficient, but it becomes more and more important as the need for accuracy increases.

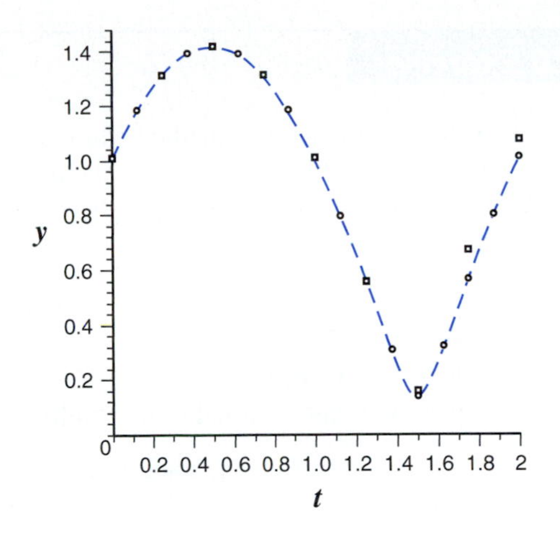

Figure 2.6.3
The rk4 approximations with 8 steps (squares) and 16 steps (circles) for Model Problem 2.6 along with the exact solution.

EXAMPLE 2

Consider the chemical kinetics problem [Equation (9)] from Section 2.1:

$$\frac{dz}{d\tau} = (1 - z)\, e^{\rho z/(1+\beta z)}, \qquad z(0) = 0.$$

In particular, we consider the case $\rho = 1,\ \beta = 1$. An implicit solution formula can be obtained by separation of variables; however, it is easier to get a numerical approximation directly from the initial-value problem than it is to get an approximation from the implicit solution formula. Figure 2.6.4 displays the reaction progress for this case, obtained by using rk4 with 20 steps. Doubling the number of steps produces no visible change in the graph, so the numerical approximation undoubtedly represents the model with greater accuracy than the model represents the real chemical reaction.

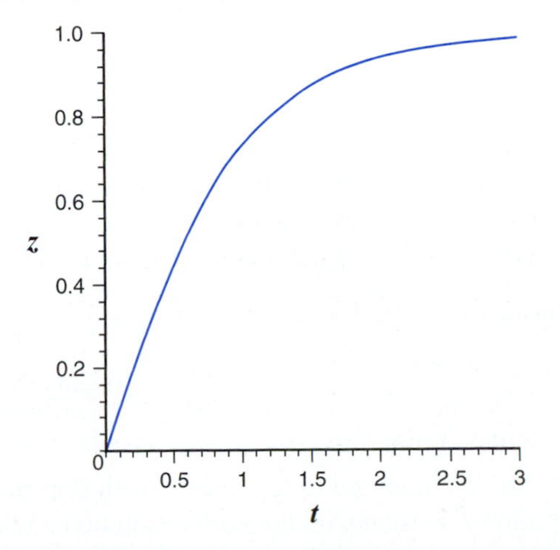

Figure 2.6.4
The progress of the chemical reaction of Example 2, obtained by using the rk4 method.

2.6 Exercises

1. *a.* Use the modified Euler method (do not use a computer program) with step size 0.2 to approximate $y(0.2)$ and $y(0.4)$ for the problem

$$\frac{dy}{dt} + y = 2t + 2, \qquad y(0) = 2.$$

   b. Compare the result with the exact solution $y = 2t + 2e^{-t}$ and the results of Exercise 1 of Section 2.5.

   c. Use the rk4 method, without a computer program, with step size 0.4 to approximate $y(0.4)$. Compare with the modified Euler results.

2. *a.* Use the modified Euler method (do not use a computer program) with step size 0.5 to approximate $y(1.5)$ and $y(2)$ for the problem

$$x\frac{dy}{dx} = 2y, \qquad y(1) = 2.$$

   b. Compare with the exact solution and the results of Exercise 2 of Section 2.5.

   c. Use the rk4 method, without a computer program, with step size 1.0 to approximate $y(2)$. Compare with the modified Euler results.

T 3. Consider the initial-value problem

$$\frac{dy}{dt} = 2ty^2, \qquad y(0) = 5$$

(Section 2.5, Exercise 3).

   a. Use the modified Euler method with step size of 0.05 to approximate $y(0.1)$. Compare the approximation with the correct solution and determine the percentage error.

   b. Use the modified Euler method with step size of 0.05 to approximate $y(0.4)$. Compare the approximation with the correct solution and determine the percentage error. Compare this result with part *a*.

   c. Repeat parts *a* and *b*, using step sizes of 0.025 and 0.0125. What effect does halving the step size have on the error?

   d. Plot the approximations and the exact solution.

   e. Compare with the results of Exercise 3 of Section 2.5.

   f. Use the rk4 method, with step sizes of 0.1 and 0.05, to approximate both $y(0.1)$ and $y(0.4)$. What effect does halving the step size have on the error?

T 4. Consider the initial-value problem

$$\frac{dy}{dt} = \cos(2\pi t)\, y, \qquad y(0) = 1$$

(Section 2.5, Exercise 4).

   a. Use the modified Euler method with step size of 0.2 to approximate $y(0.6)$. Compare the approximation with the correct solution and determine the percentage error.

   b. Use the modified Euler method with step size of 0.2 to approximate $y(2)$. Compare the approximation with the correct solution and determine the percentage error. Compare this result with part *a*.

c. Repeat parts *a* and *b*, using step sizes of 0.1 and 0.05. What effect does halving the step size have on the error?
d. Plot the approximations and the exact solution.
e. Compare with the results of Exercise 4 of Section 2.5.
f. Use the rk4 method, with step sizes of 0.2 and 0.1, to approximate both $y(0.6)$ and $y(2)$. What effect does halving the step size have on the error?

T 5. Consider the initial-value problem

$$\frac{dy}{dt} = y^2 \cos t, \qquad y(0) = 1$$

(Section 2.5, Exercise 9).

a. Use the modified Euler method with 20 steps on the interval [0, 2].
b. Repeat part *a*, using 40 steps and 80 steps.
c. Plot the approximations along with the exact solution.
d. Compare with the results of Exercise 9 of Section 2.5.
e. Repeat part *a*, using the rk4 method with 10, 20, and 40 steps. Plot these approximations and compare with the other results.

T 6. Consider the initial-value problem

$$\frac{dy}{dt} = \frac{2 \sin t}{1 - y}, \qquad y(0) = -1$$

(Section 2.5, Exercise 10).

a. Use the modified Euler method with 9 steps on the interval [0, 1.8].
b. Repeat part *a*, using 18 steps and 36 steps.
c. Plot the approximations along with the exact solution.
d. Compare with the results of Exercise 10 of Section 2.5.
e. Repeat part *a*, using the rk4 method with 9 and 18 steps. Plot these approximations and compare with the other results.

T 7. Use the rk4 method to plot graphs of z versus τ for the chemical reaction model [Equation (9) of Section 2.1], using β values of $0, 1, 2,$ and 4, with $\rho = 2\beta$ in each case. These differences correspond to making systematic changes in the heat liberated by the reaction without changing any of the other basic parameters. Describe and explain the effect of increasing the heat of reaction.

T 8. Consider the family of problems

$$\frac{dy}{dt} = k(y - t^2) + 2t, \qquad y(0) = 0,$$

where k is a real number.

a. Show that $y = t^2$ is the solution of the problem for any value of k. The equations in the family all have the same solution.
b. Determine (by trial and error) how many steps are needed for Euler's method to give an error less than 5% for $y(1)$ with $k = -2$.
c. Repeat part *b* for $k = 2$.
d. Repeat parts *b* and *c* for the modified Euler method.

e. Estimate the number of steps needed to get an error less than 1% for each case.

f. Plot the direction fields for the problems with $k = -2$ and $k = 2$. Compare them with the results of parts *a* and *b*. Use the direction field to explain why numerical methods work better with negative values of k than with positive values. (It may help to sketch several solutions on the direction fields.)

CASE STUDY 2

A Successful Volleyball Serve

The serve in volleyball is delivered by a player who is standing behind the back line of the court on the right side. The ball must travel over the net without bouncing or touching the net. If it lands outside the court, then the serve is not successful. Let's imagine that we have been hired to coach a coed intramural volleyball team. We have observed that many points are lost by the serving team because the ball either hits the net or lands outside the court. Can we use mathematics to help determine a strategy to decrease the number of missed serves?

Mathematics is used to solve problems in science and engineering. There is a key distinction between science and engineering problems. In science, the goal is to understand something; in engineering, the goal is to design something. The thrown ball model of Section 1.3 is an example of a science problem, while the volleyball serve problem we are considering here is an engineering problem. Of course, one has to understand the relevant scientific principles to design something, so engineering problems can often be conveniently divided into a part concerned with making clear connections between the design parameters and the results and a part concerned with determining the effect of design choices on the results.

A careful model of a volleyball serve should take into account the possibility of lateral movement. In this case, we have to keep track of all three spatial coordinates of the ball during its flight. For the sake of simplicity, we consider a conceptual model in which there is no lateral movement. As illustrated in Figure C2.1, we choose a coordinate system with the origin at the service line, the x axis pointing toward the opposite side of the net, and the y axis pointing up. The court is 18 m long, with the net at the 9-m mark. The net height depends on the specific rules being followed, so we take it to be H m. The server hits the ball at a point h m above the service line. The value of h depends on the server's height and serving style. The ball begins its flight with a speed s_0 m/s and a direction given as an angle α above the horizontal. The ball must clear the net at $(9, H)$ and then land at a total distance less than 18 m.

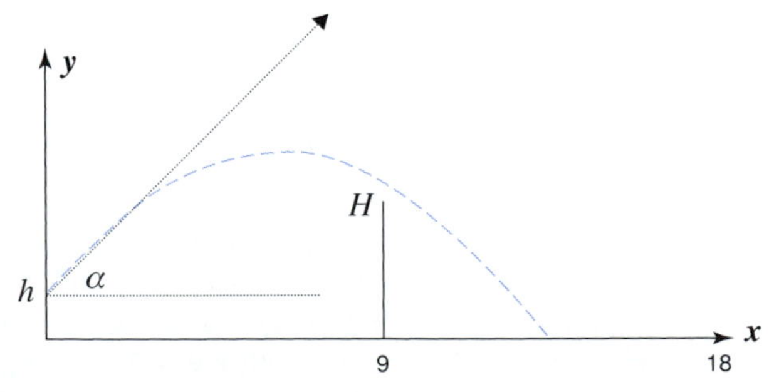

Figure C2.1
The conceptual model for a volleyball serve.

We begin with the problem of finding the connections between the parameters in the problem and the success of the serve. As noted in the discussion of Newtonian mechanics in Section 2.1, the behavior of the model is determined by the forces that we include and the formulas used to model them. For a first approximation, we assume that air resistance can be neglected. Although this assumption is not correct, it makes the problem much easier to solve. The accuracy of this assumption will be challenged at the end of the investigation; if we are lucky, it will prove to make little difference. If air resistance does seem to matter, we can build it into the model and repeat the analysis.

PROBLEM 1 - Conceptual Statement

A ball begins its flight from the point $(0, h)$ with initial speed s_0 at an angle α measured counterclockwise from the x axis. The ball moves under the influence of a constant gravitational force, with no damping forces. Determine the path of the ball.

We can write separate initial-value problems for the horizontal and the vertical motion. There is no acceleration in the x direction because there are no forces in that direction. The only force in the y direction is the constant gravitational force $-mg$. The initial conditions for the motion are determined from the information given in the conceptual statement. Specifically, the initial velocity is given as a vector of magnitude s_0 and direction indicated by the angle α. This vector can be resolved into horizontal and vertical components. A triangle with sides along the coordinate axes and hypotenuse of magnitude s_0 at angle α has adjacent side $s_0 \cos\alpha$ and opposite side $s_0 \sin\alpha$.

PROBLEM 1 - Mathematical Statement

Solve the initial-value problems

$$\frac{d^2x}{dt^2} = 0, \qquad x(0) = 0, \qquad \frac{dx}{dt}(0) = s_0 \cos\alpha \tag{1}$$

and

$$\frac{d^2y}{dt^2} = -g, \qquad y(0) = h, \qquad \frac{dy}{dt}(0) = s_0 \sin\alpha \tag{2}$$

to determine the path of the ball. Find conditions that must be satisfied for the serve to be successful.

The initial-value problems (1) and (2) are easily solved by integrating the equations directly. The results are

$$x = s_0 (\cos\alpha) t, \qquad y = h + s_0 (\sin\alpha) t - \tfrac{1}{2} g t^2. \tag{3}$$

At this point, we have formulas that give us the x and y coordinates at any time. We are not interested in the time for the flight; rather, we are interested in the path of the ball. Since $x(t)$ is one-to-one, we can rewrite the first result to give t as a function of x:

$$t = \frac{x}{s_0 \cos\alpha}.$$

Substituting this time into the solution for y gives us, after some algebraic manipulation, the formula

$$y = h + (\tan\alpha)x - \frac{g\sec^2\alpha}{2s_0^2}x^2. \tag{4}$$

For the ball to clear the net, the height at $x = 9$ must be at least H. Thus, we require

$$h + 9\tan\alpha - \frac{81g\sec^2\alpha}{2s_0^2} > H. \tag{5}$$

The obvious way to determine whether a serve is too long is to solve the equation $y = 0$ and see if the solution is less than 18. However, there is a simpler way to test the length of the serve. Imagine that the floor is removed from the court on the far side of the net. Balls that clear the net continue on to the end line at $x = 18$. The serve is good if the height at $x = 18$ is negative, for that corresponds to a ball that hits the floor. Given a serve that clears the net, the additional condition for the serve to land inside the court is

$$h + 18\tan\alpha - \frac{162g\sec^2\alpha}{s_0^2} < 0. \tag{6}$$

Now we have completed the science part of the problem. We have two simple inequalities that must be satisfied in order for a serve to be successful. It remains to solve the engineering part of the problem—to develop an appropriate strategy for the serving player.

PROBLEM 2 - Conceptual Statement

The success or failure of a serve is dependent on four parameters: net height, service height, service speed, and service angle. Of these parameters, the net height is fixed by the rules, while the other three are not. A player has some choice for the service height, a range of possible service speeds, and a free choice of service angle. A player cannot choose these values exactly, but a player can choose target values and attempt to serve the ball with values close to the target values.

Suppose that we are interested in solving the problem only for one specific net height. We still have three parameters. If we assume that each player has only one comfortable serving height, then there remain two parameters to be chosen for each player. A convenient way to display the information provided by Equations (5) and (6), given values of H and h, is to plot a graph with coordinates given by s_0 and α. Each point in the plane corresponds to a choice of service parameters. We can prepare a plot that shows what points in the plane correspond to a successful serve. We could use either s_0 or α as the horizontal coordinate, but α is a more convenient choice because the inequalities can be solved for s_0.

PROBLEM 2 - Mathematical Statement

Given values for the parameters H and h, determine the region in the αs_0 plane that corresponds to a successful serve. Identify good choices for target values of α and s_0.

We can manipulate inequalities (5) and (6) to obtain a pair of inequalities for s_0:

$$0 < \frac{81g\sec^2\alpha}{2(h - H + 9\tan\alpha)} < s_0^2 < \frac{162g\sec^2\alpha}{h + 18\tan\alpha} \tag{7}$$

Figure C2.2 shows these curves for the case $H = 2.43$ m (the standard net height for mixed volleyball) and a service height of 1 m.

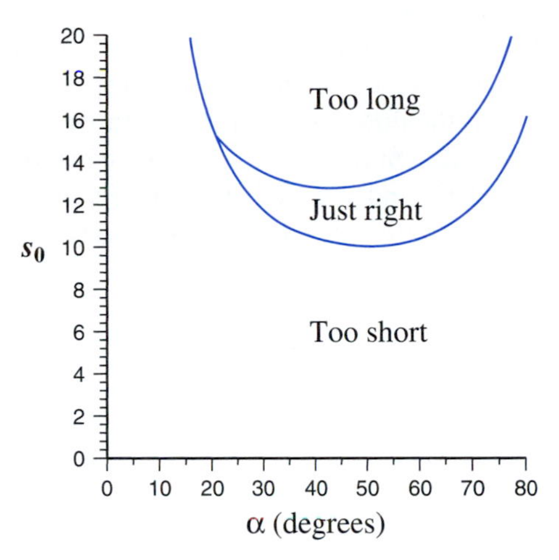

Figure C2.2
The success region for a serve with net height of 2.43 m and service height of 1 m, assuming no air resistance.

There remains the question of the best choice for intended angle and speed of serve. There is no definite answer to this question. Certainly it is desirable to aim for a point in the parameter space that is well within the success region, but there is no clear way to define a best point.

Incorporating Air Resistance

The correct differential equation to use for the volleyball serve with air resistance is Equation (6) from Section 2.1:

$$\frac{d^2\mathbf{r}}{dt^2} + b\left\|\frac{d\mathbf{r}}{dt}\right\|\frac{d\mathbf{r}}{dt} + g\mathbf{k} = \mathbf{0}. \tag{8}$$

It is helpful to separate the components of this vector differential equation. If we use the notation from the problem without air resistance, the horizontal and vertical components of $\mathbf{r}$ are $x(t)$ and $y(t)$, and the quantity $\|d\mathbf{r}/dt\|$ represents the speed $s(t) = \sqrt{x'^2 + y'^2}$. These changes yield the initial-value problems

$$x'' + bsx' = 0, \qquad x(0) = 0, \qquad x'(0) = s_0\cos\alpha$$

and

$$y'' + bsy' + g = 0, \qquad y(0) = h, \qquad y'(0) = s_0\sin\alpha,$$

where

$$s(t) = \sqrt{x'^2(t) + y'^2(t)}.$$

These equations are coupled, meaning that they must be solved simultaneously, and of second order. It is not possible to solve these equations. It is possible to recast the equations with different variables, one of which is the speed s and the other the angle θ between the vector $\mathbf{r}$ and the x axis.[18] Doing this decouples the equations and yields an equation that can be solved to determine the speed as a function of the angle.[19] However, after all that calculation, one is still left with a numerical problem, so there does not seem to be any advantage to that approach rather than a direct numerical approximation of the system.

In the context of the volleyball problem, what we really want to know is the height of the ball when $x = 9$ and $x = 18$. Given that we are not interested in the time, we can simplify the problem by a change of variables to omit the time. First, it is better to write the problem as a system of four first-order equations than a system of two second-order equations. To that end, let u and v be the velocity components in the x and y directions, respectively. The second-order equations for x and y become first-order equations for u and v, respectively. We have

$$\begin{aligned} x' &= u, & x(0) &= 0, \\ y' &= v, & y(0) &= h, \\ u' &= -bu\sqrt{u^2+v^2}, & u(0) &= s_0 \cos\alpha, \\ v' &= -bv\sqrt{u^2+v^2} - g, & v(0) &= s_0 \sin\alpha. \end{aligned}$$

As before, $x(t)$ is always increasing, so we can think of x as the independent variable for y, u, and v. If we have derivatives of these three quantities with respect to x, then we do not need to know the corresponding times for each point. Eliminating time reduces the system from four equations to three, and it has the further advantage of making the success criteria easier to determine. Given y as a function of x, a successful serve is one for which $y(9) > H$ and $y(18) < 0$.

By the chain rule, we have

$$\frac{dy}{dx} = \frac{dy}{dt}\frac{dt}{dx} = \frac{y'}{x'} = \frac{v}{u}.$$

Treating the other variables similarly, we obtain a system of three equations:

$$\begin{aligned} \frac{dy}{dx} &= \frac{v}{u}, & y(0) &= h, \\ \frac{du}{dx} &= -b\sqrt{u^2+v^2}, & u(0) &= s_0 \cos\alpha, \\ \frac{dv}{dx} &= -b\sqrt{u^2+v^2}\,\frac{v}{u} - \frac{g}{u}, & v(0) &= s_0 \sin\alpha. \end{aligned}$$

Note that the initial conditions are unchanged only because $x = 0$ at $t = 0$. One final simplification is suggested by the appearance of the factor v/u, which represents the tangent of the angle that the velocity vector makes with the floor. If we let $w = v/u$, we can replace the third equation

[18] See Exercise 6 of Section 2.1.

[19] See Neville de Mestre, *The Mathematics of Projectiles in Sport*, Cambridge University Press, London, 1990, Section 3.4.

with an equation for w. The final result is

$$\frac{dy}{dx} = w, \qquad y(0) = h, \tag{9}$$

$$\frac{du}{dx} = -bu\sqrt{1+w^2}, \qquad u(0) = s_0 \cos\alpha, \tag{10}$$

$$\frac{dw}{dx} = -\frac{g}{u^2}, \qquad w(0) = \tan\alpha. \tag{11}$$

Is air resistance enough to make a difference in the success region of Figure C2.2? Using a fluid mechanics model to determine the nature of b and data for a volleyball,[20] we can estimate

$$b \approx 0.016\ \text{m}^{-1}.$$

Applying the rk4 numerical approximation method to the model with air resistance, using this value of b, we obtain the data needed to plot Figure C2.3, showing the success region with air resistance taken into account. The difference is significant, so air resistance cannot be neglected in this investigation. Note that the success region is a little larger with air resistance included. Air resistance makes it easier to make a successful serve because the effect of slowing down the ball matters more at 18 m than at 9 m.

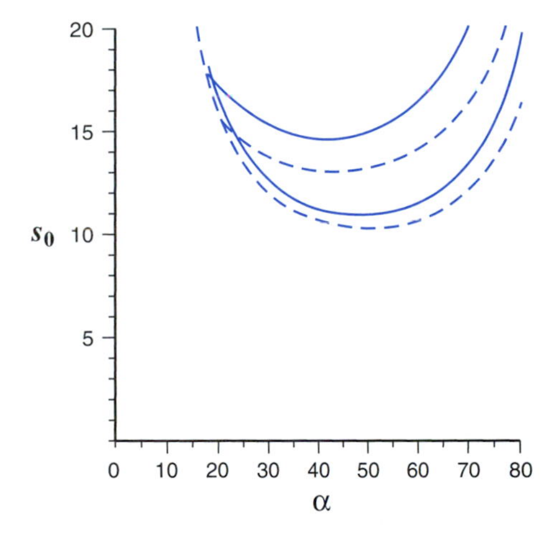

Figure C2.3
The success region for a serve with net height of 2.43 m and service height of 1 m, including the effect of air resistance (solid), as compared with the region without air resistance (dashed).

Case Study 2 Exercises

In Exercises 1 through 6, neglect air resistance.

1. Find the minimum speed of a successful serve, given $H = 2.43$ m and $h = 1$ m.

[20] See Neville de Mestre, *The Mathematics of Projectiles in Sport,* Cambridge University Press, London, 1990, Section 3.4.

2. Find the maximum velocity that guarantees the serve will not be too long given $H = 2.43$ m and $h = 1$ m. What problem is encountered in choosing an angle to go with a velocity slightly greater than this maximum value?
3. The initial height of 1 m corresponds roughly to an underhand serve. Assume that an overhand serve has an initial height of 2 m. Plot the success region for the overhand serve, and compare it with Figure C2.2. Based on the problem of getting the serve to land in the right place, which style of serve do you recommend?
4. Develop a mathematical model to determine successful combinations of speed and angle for a basketball free throw. Does the model explain the observation that the tallest players generally have the greatest difficulty making free throws? Explain why basketball players must practice free throws much more than volleyball players practice serves.
5. Construct a plot similar to that of Figure C2.2 for a tennis serve. You will need to find the appropriate data for a tennis court and net and to make a reasonable estimate for the initial height of the serve.
6. In anticipation of the U.S. Civil War, you have been hired as a consultant by the Secretary of War (now Defense) to determine whether the walls of Fort Sumter are high enough to prevent cannonballs from coming over a wall and landing in the interior. Assume that the cannons are placed no closer than a distance L from the fort. The cannons fire a cannonball at speed s_0 from a height h. The cannoneers can choose the angle α.

   a. Determine a mathematical model for the flight of a cannonball, assuming that the only relevant force on the ball is the force of gravity.
   b. Solve the mathematical model to obtain a formula for the trajectory of the ball's flight.
   c. Assuming that h and s_0 are known values, determine the highest point on the wall of the fort that can be reached by a cannonball.

7. Derive Equations (11).
8. (T) A volleyball serve is launched with initial angle of 50° and initial speed of 11 m/s. Use the model in Equations (9) through (11) to plot the path of the ball. You can use the rk4 method to approximate $y(x)$, but note that there must be values corresponding to k_1, k_2, k_3, and k_4 for each of the three dependent variables. Although you are not going to use the results for u and w, these results are needed at each step to compute y.
9. (T) Given an initial height of 2 m and an angle of 30°, determine the minimum and maximum speeds for a successful serve, incorporating air resistance.

chapter

3

Homogeneous Linear Equations

Most differential equations cannot be solved to yield an explicit solution formula. There is only one large class of differential equations for which an explicit solution can be found, at least in principle, and this is the class of linear equations with constant coefficients. We shall devote a great deal of coverage to these equations, both in this chapter and in Chapter 4. Linear equations with constant coefficients are important not just because we can solve them. They are important because many physical settings can be modeled with them. The theory and solution techniques for linear homogeneous differential equations with constant coefficients serve as essential background for nearly all the more advanced topics in differential equations.

Chapter 3 begins (Section 3.1) with the derivation of the classic *linear oscillator* model, consisting of a homogeneous second-order linear equation with constant nonnegative coefficients, from an idealized conceptual model of mechanical vibration. The linear oscillator model arises in a number of other contexts. Its study motivates almost all the topics in this chapter. Section 3.1 also presents the analysis of the simplest subcase of the linear oscillator model.

Section 3.2 presents a brief exposition of the theory of systems of linear algebraic equations. The reader who is familiar with linear algebra will find this section to be strictly review. The reader who has not previously studied linear algebra will need to study this section in order to have the necessary background for the rest of the chapter.

Section 3.3 presents the theory of homogeneous linear differential equations, with the focus on the development of an algorithm for constructing a general solution formula for a homogeneous linear differential equation from a set of individual solutions. The determination of the individual solutions is deferred to later sections, as the manner for determining solutions depends on the specific type of homogeneous linear equation. The theory developed in this section applies to the broad class of homogeneous linear equations, including many that cannot be solved except by approximation.

The remainder of Chapter 3 deals with solution techniques for certain important homogeneous linear equations. Sections 3.4, 3.5, and 3.6 develop the *characteristic value method* for solving homogeneous linear equations with constant coefficients and use the results to study the properties of the linear oscillator model from Section 3.1. Some other homogeneous linear equations are examined in Section 3.7. The focus in Section 3.7 is on Cauchy–Euler and Bessel equations, both of which arise in mathematical models, and equations that can be recast in terms of these equations. The simplification of certain differential equations by changing variables and the expression of solutions in terms of *special functions* defined by initial-value problems are also emphasized.

Case Study 3 concerns the development and analysis of a simple mathematical model of the digestion process in certain very simple animals. In addition to the specific topic studied in the model, this case study provides an opportunity to discuss the difference between mathematical modeling for science and mathematical modeling for engineering.

3.1 Linear Oscillators

Rhythmic motion occurs in a variety of settings, from the unpleasant motion of a jackhammer to the pleasant vibration of the eardrum when one is listening to music. A first step in the mathematical study of the phenomenon of vibration might be to construct a simple model for a vibrating object. One characteristic common to all vibrating systems is the presence of a force that causes an object to reverse its direction of motion.[1] The simplest conceptual model of vibration would include nothing other than an object subjected to such a force. However, adding more features to models makes the models more interesting, as long as they do not become too complicated for mathematical analysis. The following conceptual model leads to the standard mathematical model of vibration.

One end of a spring is attached to a fixed point overhead. A solid object is hung on the free end of the spring. The weight of the object lengthens the spring from its normal length, and the elongated spring then exerts a restoring force in the upward direction. The motion of the object is determined by the balance of the gravitational and restoring forces along with other forces that might be present, such as a force that resists motion and a force that is caused by a mechanism external to the system.

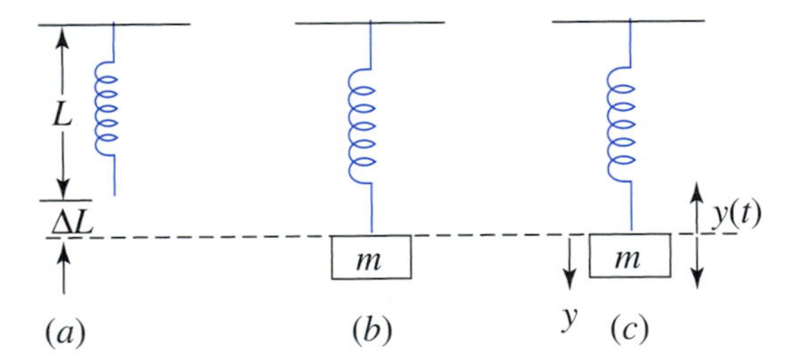

Figure 3.1.1
A conceptual model for vibration: (*a*) an unloaded spring, (*b*) the system at rest, and (*c*) the system in motion.

Figure 3.1.1 illustrates the conceptual model. Figure 3.1.1*a* shows the spring before the solid object is attached. The spring has a natural length L. Figure 3.1.1*b* shows the system after the object is attached, but before it is set into motion. The downward force of gravity is balanced by the upward force of the spring. Figure 3.1.1*c* shows the system in motion. The gravitational force is still constant, but other forces are changing as the object moves, and the system responds accordingly. Note that we have arbitrarily decided to orient the coordinate system with the y axis pointing downward.

[1] See the discussion of restoring forces in Section 2.1.

The Static Case (No Motion) Suppose an object of mass m is attached to the free end of the spring. Assume that the system is allowed to reach a rest position, as depicted in Figure 3.1.1*b*. In this situation, the weight of the object has stretched the spring a distance ΔL beyond its natural length L. The quantity $L + \Delta L$ is the equilibrium length of the spring, and the coordinate y measures the additional lengthening (positive or negative) of the spring beyond this equilibrium length.

There are two forces on the mass in the static situation: a gravitational force exerted by the earth and a restoring force exerted by the spring. The force of gravity acts in the downward direction with magnitude equal to the weight w of the object. With our decision to make the downward direction be positive, we have

$$F_g = w = mg,$$

where g is the appropriate gravitational constant:

- 32 ft/s^2 in the English system (feet-pounds-seconds)
- 9.8 m/s^2 in the mks (meters-kilograms-seconds) system
- 980 cm/s^2 in the cgs (centimeters-grams-seconds) system

In the English system, the primary measurement of the amount of material is weight (pounds), and the mass (slugs) is then determined from $w = mg$. Both sets of units in the metric system have mass (kilograms or grams) as the primary measurement of amount, and the weight (newtons or dynes) is then determined from the mass. If two-digit accuracy is sufficient for a given situation, it is convenient to use $g \approx 10$ in the mks system and $g \approx 1000$ in the cgs system.

The restoring force acts in the upward direction. As in Section 2.1, we assume that the magnitude is $k\,\Delta L$, where k is the spring constant, a parameter that characterizes the stiffness of the spring. Thus,

$$F_r = -k\,\Delta L.$$

We assume that there are no other forces on the system. Note that this means we are neglecting the mass of the spring itself, which is reasonable provided that the mass of the spring is small compared with the mass of the object hung from the free end.

By Newton's second law of motion, the acceleration of an object subjected to a gravitational force and a restoring force is given by $ma = F_g + F_r$. The object is not moving, so its acceleration is zero. Hence, the static situation gives us the result

$$k\,\Delta L = w = mg. \tag{1}$$

This formula provides an excellent way to determine the spring constant k. One simply hangs an object of known mass (if metric) or weight (if English) from the spring and measures the length change.

EXAMPLE 1

If the mass is 0.6 kg and the change in length is 0.4 m, then the weight is approximately $w = mg = 6$ N and the spring constant is approximately $k = w/\Delta L = 15$ N/m.

EXAMPLE 2

If the object weighs 4 lb and the change in length is 6 in, then the mass is approximately $m = w/g = \frac{1}{8}$ slug and the spring constant is approximately $k = w/\Delta L = 8$ lb/ft. Notice that the change in length is taken to be $\frac{1}{2}$ ft rather than 6 in.

Note in these examples that the spring constant has a dimension of force per unit length.

✦ INSTANT EXERCISE 1

A spring of length 25 cm is stretched to a length of 30 cm by the weight of an object of mass 100 grams (g). Determine the spring constant.

The Dynamic Case (Motion) Now suppose we define y as in Figure 3.1.1c. If the system is set into motion, the forces are no longer in equilibrium and the object is not always at its rest position. The object's acceleration is determined by Newton's second law of motion

$$m\frac{d^2y}{dt^2} = F,$$

where F is the sum of all the forces, taking downward forces to be positive and upward forces to be negative. The total force F includes gravitational and restoring forces similar to those in the static case. The conceptual model indicates the presence of a damping force and possibly a force external to the system, such as might be provided by a motor or a magnet.

The gravitational force is the same as in the static case:

$$F_g = mg.$$

The restoring force is a little different from that in the static case because the amount by which the spring is lengthened depends on the displacement y. Specifically, the displacement of the spring from its natural length at any particular time is $\Delta L + y$. Thus, the restoring force, by Hooke's law, is

$$F_r = -k(\Delta L + y).$$

Damping forces F_d are caused by friction or air resistance. As in Section 2.1, the damping force acts to oppose the motion, and we assume it to be proportional to the velocity. Thus,

$$F_d = -b\frac{dy}{dt},$$

where the damping coefficient b is a nonnegative constant. The external force F_e, if present, must be specified for each specific problem.

The sum of all the forces is then

$$F = F_g + F_r + F_d + F_e = mg - k(\Delta L + y) - b\frac{dy}{dt} + F_e = -ky - b\frac{dy}{dt} + F_e,$$

where we have used the static balance (1) to simplify the result. We now substitute this sum of forces into Newton's second law of motion and get

$$m\frac{d^2y}{dt^2} = -ky - b\frac{dy}{dt} + F_e,$$

or

$$my'' + by' + ky = F_e, \tag{2}$$

where the prime symbol denotes derivatives with respect to t.

In general, a mathematical model of the form

$$ay'' + by' + cy = f, \qquad a, c > 0, \qquad b \geq 0, \tag{3}$$

where y is the dependent variable, t is the independent variable, f is some prescribed function, and a, b, and c are constants, is called a **linear oscillator.** A linear oscillator is said to be **forced** if f is not identically zero and **damped** if $b > 0$. In addition to vibration of a mechanical system, linear oscillators occur in models for the current or voltage in an electric circuit and the swinging motion of a pendulum, as well as numerous other settings.

Simple Harmonic Motion

Forced linear oscillators are discussed in Chapter 4, and damped oscillators are discussed in Sections 3.4 through 3.6. Here we focus on unforced, undamped linear oscillators. With $F_e = 0$ and $b = 0$, the mechanical oscillator model (2) reduces to the differential equation

$$my'' + ky = 0. \tag{4}$$

MODEL PROBLEM 3.1

Consider a spring-mass system with mass of 1 unit and spring constant of 9 units. Suppose the mass is raised 0.3 units and released with a downward velocity of 1.2 units. Construct an initial-value problem for this situation, solve it, and describe the properties of the solution.

Model Problem 3.1 is an initial-value problem for the linear oscillator equation

$$y'' + 9y = 0, \qquad y(0) = -0.3, \qquad y'(0) = 1.2$$

We develop a formal method for solving this equation later in this chapter. However, it is not at all difficult to determine the solution by inspection. We begin with a simpler problem.

EXAMPLE 3

Consider the initial-value problem

$$y'' + 9y = 0, \qquad y(0) = 1, \qquad y'(0) = 0.$$

To satisfy the differential equation, we need to find functions having this property: the second derivative is -9 times the function itself. The functions $\cos 3t$ and $\sin 3t$ satisfy this property. Of these two functions, the function $\cos 3t$ also satisfies the initial conditions. The solution is therefore

$$y = \cos 3t.$$

The solution of Model Problem 3.1 is similar to that of Example 3. We have already seen that $\cos 3t$ and $\sin 3t$ solve the differential equation. You can easily check that any multiples of $\cos 3t$ or $\sin 3t$ satisfy the equation as well. Sums of such functions can also be shown to satisfy the differential equation. Thus, there is a two-parameter family of solutions

$$y = c_1 \cos 3t + c_2 \sin 3t.$$

✦ INSTANT EXERCISE 2

Verify that $y = c_1 \cos 3t + c_2 \sin 3t$ satisfies the differential equation $y'' + 9y = 0$ regardless of the values of c_1 and c_2.

Given a two-parameter family of solutions of the differential equation, we now determine whether any member(s) of the solution family satisfies the two initial conditions of Model Problem 3.1 as well. Substituting $t = 0$ into the solution formula yields

$$y(0) = c_1.$$

Thus, the initial condition $y(0) = -0.3$ is satisfied if and only if $c_1 = -0.3$. Similarly, the derivative of the solution formula is

$$y' = -3c_1 \sin 3t + 3c_2 \cos 3t,$$

and so

$$y'(0) = 3c_2.$$

The initial condition $y'(0) = 1.2$ forces $c_2 = 0.4$. Thus, we arrive at the solution

$$y = -0.3 \cos 3t + 0.4 \sin 3t.$$

The graph of the solution appears in Figure 3.1.2. From the graph, we can see that the motion is periodic. Motion described by a sum of sine and cosine functions with the same argument κt, with κ a constant, is called **simple harmonic motion.**

Period, Frequency, and Circular Frequency The coefficient of t in a sinusoidal function is called the **circular frequency.** In Model Problem 3.1, the circular frequency is 3. The circular frequency does not by itself have a physical significance in the linear oscillator model, but it completely determines two related quantities that do have a physical significance. The **period** is defined

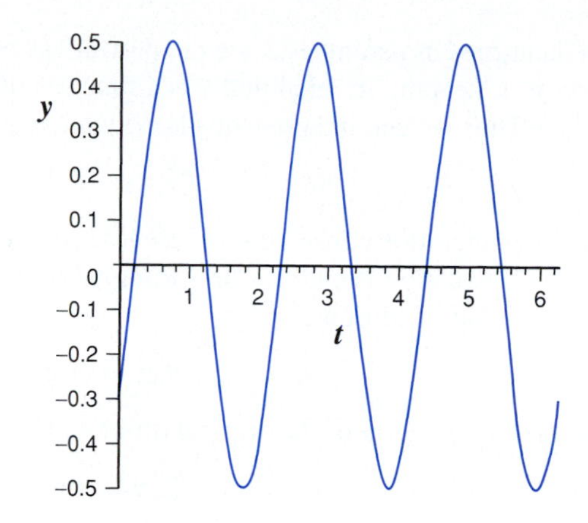

Figure 3.1.2
The solution of $y'' + 9y = 0$, $y(0) = -0.3$, $y'(0) = 1.2$.

to be the smallest time interval over which the graph of a periodic function repeats. The functions $\cos 3t$ and $\sin 3t$ first repeat when $3t = 2\pi$. The period of these functions is therefore $2\pi/3$, and clearly the period of $y = -0.3\cos 3t + 0.4\sin 3t$ is the same. In general, the product of the circular frequency and the period must be 2π. The reciprocal of the period is the **frequency** of the motion. The frequency of periodic motion is usually measured in cycles per second, also known as hertz (Hz), although in certain contexts it may be reported with the unit "rpm" (revolutions per minute).

EXAMPLE 4

Consider the spring-mass system of Example 1. Given $m = 0.6$ and $k = 15$, the system is determined by the differential equation

$$0.6y'' + 15y = 0,$$

or

$$y'' + 25y = 0.$$

Solutions of this equation have a second derivative that is -25 times the function, such as $\cos 5t$ and $\sin 5t$. The circular frequency is therefore 5. The period is the smallest t for which the trigonometric functions repeat; from $5t = 2\pi$, the period is 0.4π. The frequency is $5/(2\pi) \approx 0.796$.

✦ INSTANT EXERCISE 3

Find the circular frequency and period for the spring-mass system of Instant Exercise 1.

Amplitude The **amplitude** A of a sinusoidal function y is the largest value achieved in each period. From Figure 3.1.2, it appears that the amplitude is approximately 0.5. The amplitude can be determined exactly by a clever manipulation of the differential equation. Multiplying the equation $y'' + 9y = 0$ by $2y'$ yields

$$(2y'y'') + 9(2yy') = 0.$$

Both quantities in parentheses are the derivatives of simple functions: The derivative of y^2 with respect to t, keeping in mind that y is a function of t, is $2yy'$, and similarly the derivative of y'^2 is $2y'y''$. Thus we can integrate this last equation and obtain

$$y'^2 + 9y^2 = C,$$

with C an integration constant. A differential equation that is obtained by integration from another differential equation is called a **first integral** of the original equation. The constant is evaluated from the initial conditions:

$$C = v_0^2 + 9y_0^2 = (1.2)^2 + 9(-0.3)^2 = 2.25.$$

Thus, the first integral of the original differential equation is

$$y'^2 + 9y^2 = 2.25.$$

Both terms on the left are nonnegative, so y is largest (in other words, $y = A$) when $y' = 0$. Substituting 0 for y' and A for y on the left side results in the equation

$$9A^2 = 2.25,$$

which yields the result $A^2 = 0.25$, or

$$A = 0.5.$$

Like the period and circular frequency, the amplitude can be determined without even solving the initial-value problem.

✦ INSTANT EXERCISE 4

Suppose the mass in Instant Exercise 1 is pulled down 1 cm and released with an upward velocity of 4 cm/s. Determine the amplitude of the subsequent motion.

Phase Shift Let $x(t) = 0.5\cos 3t$. The function x is a sinusoidal function with the same circular frequency and amplitude as the function y that solves Model Problem 3.1. The graphs of x and y are shown together in Figure 3.1.3. Notice that we could think of the function y as representing a horizontal shift in the function x.

To further clarify this idea, consider the function X defined by

$$X(t) = 0.5\cos\left(3t - \frac{2\pi}{3}\right).$$

This function differs from x by a horizontal shift. Note that $2\pi/3$ is one-third of a period. A function $\cos(\theta - 2\pi/3)$ lags one-third of a period behind the corresponding function $\cos\theta$, so its graph is shifted to the right by $\frac{1}{3}$ period. The graph of X is shown in Figure 3.1.4, along with the original function x. Comparison of Figures 3.1.3 and 3.1.4 suggests that the functions X and y are the same, and in fact they are. Any function of the form

$$c_1 \cos\kappa t + c_2 \sin\kappa t \tag{5}$$

can be written in the form

$$A\cos(\kappa t - \delta). \tag{6}$$

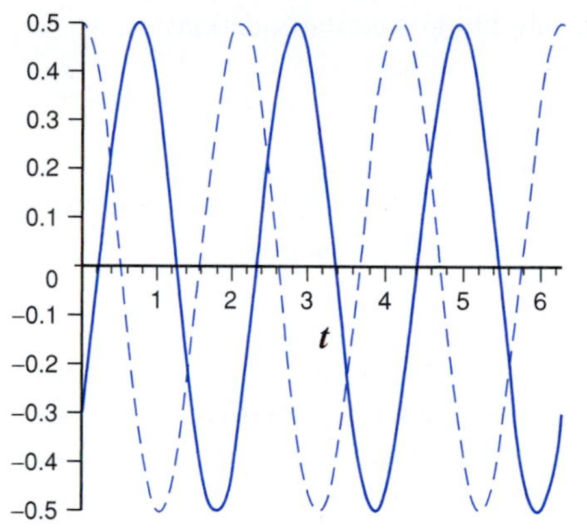

Figure 3.1.3
The solution of $y'' + 9y = 0$, $y(0) = -0.3$, $y'(0) = 1.2$ (solid) along with $x = 0.5\cos 3t$ (dashed).

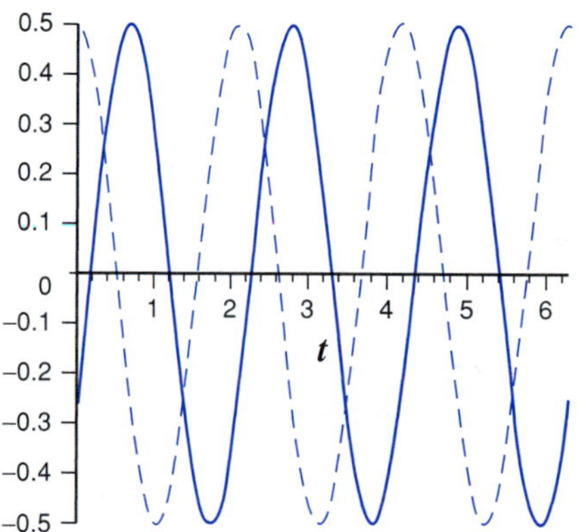

Figure 3.1.4
The functions $X = 0.5\cos(3t - 2\pi/3)$ (solid) and $x = 0.5\cos 3t$ (dashed).

The constant δ is called the **phase shift.** Solutions of initial-value problems for an undamped linear oscillator can be written in either form. Form (5) arises when the problem is solved by applying the initial conditions at time 0 to the general solution of the differential equation. Form (6) is convenient when further analysis is required, because the parameters A and δ have a physical significance that is readily apparent in the graph.

Determining the Phase Shift The relationships among c_1, c_2, A, and δ can be determined by setting the equivalent forms (5) and (6) equal to each other:

$$c_1 \cos \kappa t + c_2 \sin \kappa t = A\cos(\kappa t - \delta).$$

Applying the appropriate trigonometric identity to the right-hand side eventually yields the equation

$$A^2 = c_1^2 + c_2^2 \tag{7}$$

for the amplitude, and the equations

$$\cos\delta = \frac{c_1}{A}, \qquad \sin\delta = \frac{c_2}{A}, \tag{8}$$

to determine δ.

At this stage, a bit of caution is required. In our example, we have $c_1 = -0.3$, $c_2 = 0.4$, and $A = 0.5$. Thus, the problem to be solved for δ is

$$\cos\delta = -0.6, \qquad \sin\delta = 0.8.$$

It is tempting to "solve" the second equation by writing $\delta = \arcsin 0.8 \approx 0.927$. This is not correct, however, because it does not solve the first equation: $\cos 0.927 \approx 0.6$, but we need $\cos\delta = -0.6$.

One must be careful when solving trigonometric equations. The number δ has a positive sine and a negative cosine, and it is therefore in the second quadrant, while the number $\arcsin 0.8$ is in the first quadrant. The correct answer is $\delta = \pi - \arcsin 0.8 \approx 2.21$.[2] Thus, the solution of the initial-value problem

$$y'' + 9y = 0, \qquad y(0) = -0.3, \qquad y'(0) = 1.2$$

is given approximately, in amplitude–phase shift form, as

$$y \approx 0.5\cos(3t - 2.21).$$

EXAMPLE 5

Consider the problem

$$y'' + 3y = 0, \qquad y(0) = -3, \qquad y'(0) = -3.$$

Suppose we want to write the solution in amplitude–phase shift form. We begin by writing the general solution of the differential equation:

$$y = c_1\cos\sqrt{3}t + c_2\sin\sqrt{3}t.$$

Next, we use the initial conditions $-3 = y(0) = c_1$ and $-3 = y'(0) = \sqrt{3}c_2$. So we have

$$y = -3\cos\sqrt{3}t - \sqrt{3}\sin\sqrt{3}t.$$

We can now determine the amplitude from Equation (7) and the phase shift from Equation (8). We have $A^2 = {c_1}^2 + {c_2}^2 = 9 + 3 = 12$, so $A = 2\sqrt{3}$. Also, $\cos\delta = c_1/A = -\sqrt{3}/2$ and $\sin\delta = c_2/A = -\frac{1}{2}$. The correct value of δ is in the third quadrant, but $\arcsin(-\frac{1}{2})$ is in the fourth quadrant. Again, the correct answer is not $\arcsin(-\frac{1}{2})$, but $\delta = \pi - \arcsin(-\frac{1}{2}) = \pi - (-\pi/6) = 7\pi/6$. Hence,

$$y = 2\sqrt{3}\cos\left(\sqrt{3}t - \frac{7\pi}{6}\right).$$

Note that the answer could also be written as

$$y = 2\sqrt{3}\cos\left(\sqrt{3}t + \frac{5\pi}{6}\right).$$

The first answer indicates a shift of $7\pi/6$ radians to the right, while the second answer indicates a shift of $5\pi/6$ radians to the left, and these two shifts give the same result. See Figure 3.1.5.

---

[2]The reader to whom this is not clear needs to review trigonometric equations and inverse trigonometric functions.

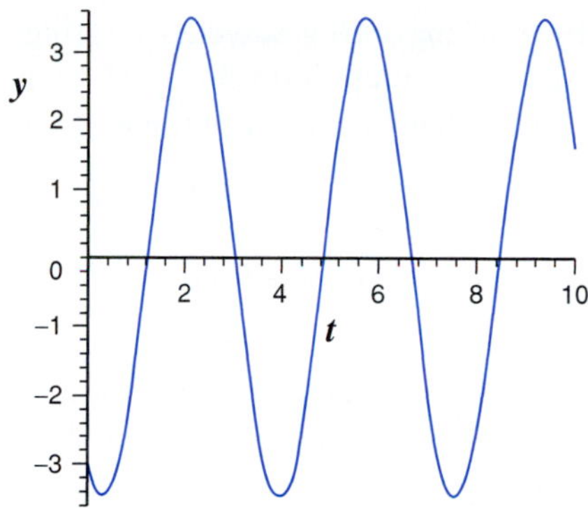

Figure 3.1.5
The solution of Example 5.

✦ **INSTANT EXERCISE 5**

Determine the solution of Instant Exercise 4 in amplitude–phase shift form and graph the motion.

3.1 Exercises

In Exercises 1 through 6, solve the initial-value problem, write the solution in amplitude–phase shift form, and sketch the solution.

1. $y'' + y = 0, \quad y(0) = -1, \quad y'(0) = -\sqrt{3}$
2. $y'' + 9y = 0, \quad y(0) = -1, \quad y'(0) = \sqrt{3}$
3. $y'' + 3y = 0, \quad y(0) = 1, \quad y'(0) = -1$
4. $y'' + 12y = 0, \quad y(0) = -2, \quad y'(0) = 4$
5. T $y'' + 5y = 0, \quad y(0) = -2, \quad y'(0) = -6$
6. T $y'' + 8y = 0, \quad y(0) = 1, \quad y'(0) = 2$
7. An object of mass 10 kg oscillates on the end of a spring with a period of 1 s. What is the spring constant? Determine the period of any resulting oscillation.
8. An object that weighs 10 lb oscillates on the end of a spring with a period of 1 s. What is the spring constant? Determine the period of any resulting oscillation.
9. An object of unknown mass oscillates on the end of a spring with period 5 s. A 2-lb object is attached to the first object, changing the period to 7 s. What is the mass of the first object?
10. An object of unknown mass oscillates on the end of a spring with period 8 s. A 10-kg object is attached to the first object, changing the period to 12 s. What is the mass of the first object?

11. An object of mass 20 g stretches a spring 5 cm. The mass is lifted up 2 cm and then set in motion with an upward velocity of 1 cm/s. Find the initial-value problem of the motion, assuming no damping, and determine the period, amplitude, and phase shift.

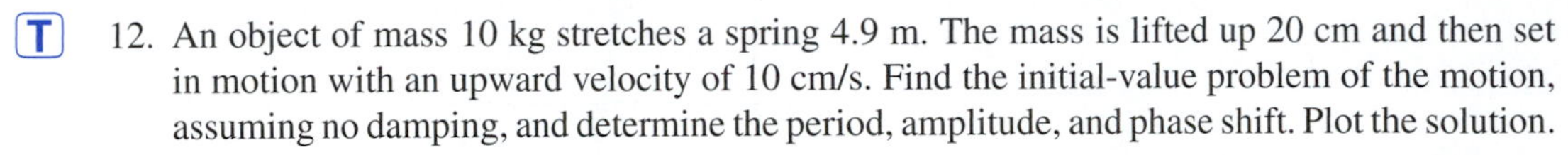

T 12. An object of mass 10 kg stretches a spring 4.9 m. The mass is lifted up 20 cm and then set in motion with an upward velocity of 10 cm/s. Find the initial-value problem of the motion, assuming no damping, and determine the period, amplitude, and phase shift. Plot the solution.

T 13. An object weighing 2 lb stretches a spring 6 in. The mass is pulled down an additional 3 in and then set in motion with a downward velocity of 1 ft/s. Find the initial-value problem of the motion, assuming no damping, and determine the period, amplitude, and phase shift. Plot the solution.

T 14. An object weighing 4 lb stretches a spring 6 in. The mass is pulled down an additional 3 in and then set in motion with a downward velocity of 1 ft/s. Find the initial-value problem of the motion, assuming no damping, and determine the period, amplitude, and phase shift. Plot the solution. Compare with Exercise 13.

The Period of a Pendulum

The differential equation for the motion of an undamped pendulum is

$$\frac{d^2\theta}{dt^2} + \frac{g}{L}\sin\theta = 0,$$

where $\theta(t)$ is the angle of the pendulum rod as measured from the vertical and L is the length of the pendulum as measured to the center of mass. The usual treatment of a pendulum is to assume that the amplitude is small. This results in a linear oscillator model. The goal of Exercise 15 is to determine the error in using the period T_0 of the linear oscillator model as an approximation of the true period T. Let the correction factor F be defined by $T = FT_0$. If, for example, $F = 1.01$, then the true period is 1% larger than the approximation. If one counts the swings of the pendulum to measure time, the use of T_0 instead of T will result in a time measurement that is 1% too small. Such a clock would lose 14.4 min/day.

T 15. *a.* Replace $\sin\theta$ by a linear approximation. Determine the period T_0 of the resulting undamped linear oscillator.

b. Suppose the pendulum begins at angle $\theta = A$ with no initial velocity. The angle changes to zero in one-quarter of a period; hence, the period is determined by the problem

$$\frac{d^2\theta}{dt^2} + \frac{g}{L}\sin\theta = 0, \qquad \theta(0) = A, \qquad \frac{d\theta}{dt}(0) = 0, \qquad \theta\left(\frac{T}{4}\right) = 0.$$

Determine the first integral of this differential equation in a manner similar to that used in the text for the undamped oscillator. Use the initial conditions to evaluate the constant.

c. Manipulate the first integral to obtain a first-order differential equation for $\theta(t)$ [note that $d\theta/dt < 0$ in the time interval $0 < t < T/4$]. Rewrite this equation as a differential equation for t in terms of θ. Complete the problem with two data points of the form $t(\theta_1) = t_1$ taken from the problem of part *b*.

d. Integrate both sides of the equation from part c with respect to θ from A to 0. Combine the result with that of part a to get an integral formula for F.

e. The integral for F cannot be calculated in terms of elementary functions, but it can be converted to a simpler form and evaluated numerically. Use the substitution $x = \cos\theta$ and write F as the integral of a function of x.

f. The integral you obtained in part e is an example of what is called a **complete elliptic integral of the first kind.** If you have access to a computer algebra system, see what happens when you ask it to compute the integral. (If your computer algebra system allows you to specify a range for a parameter, specify the range of A as $0 < A < \pi/2$.)

g. Let a be the degree measure of the amplitude. Plot F versus a and approximate the amplitude corresponding to a relative error of $\frac{1}{60}$. A clock with such an error would be wrong by 1 min out of every hour. What amplitude corresponds to an error of 1 min/day? How about 1 min/week? (Actual pendulum clocks are calibrated with an adjustable pendulum bob; they still need to be reset regularly to maintain the correct time. A longer pendulum, such as in a grandfather clock, is more accurate than a shorter pendulum because it can be designed to operate with a smaller amplitude.)

✦ 3.1 INSTANT EXERCISE SOLUTIONS

1. The length change is $\Delta L = 5$ cm, and the weight is $mg = 98{,}000$ dyn. By the static spring-mass model, the spring constant is $k = mg/\Delta L = 19{,}600$ dyn/cm.

2. Differentiating

$$y = c_1 \cos 3t + c_2 \sin 3t$$

twice yields

$$y' = -3c_1 \sin 3t + 3c_2 \cos 3t$$

and

$$y'' = -9c_1 \cos 3t - 9c_2 \sin 3t = -9y.$$

3. The circular frequency is $\sqrt{k/m} = \sqrt{19{,}600/100} = 14\,\text{s}^{-1}$. The period is then 2π divided by the circular frequency, or $\pi/7 \approx 0.449$ s.

4. We have $y_0 = 1$, $v_0 = -4$, $m = 100$, and $k = 19{,}600$. The differential equation, after simplification, is

$$y'' + 196y = 0.$$

As in Model Problem 3.1, we multiply by $2y'$ and integrate, with the result

$$y'^2 + 196y^2 = C.$$

From the initial conditions,

$$C = (-4)^2 + 196(1)^2 = 212.$$

With $y' = 0$ and $y = A$, the amplitude is

$$A = \sqrt{\frac{C}{196}} = \sqrt{53/49} \approx 1.04\,\text{cm}.$$

5. The solution of the differential equation $y'' + 196y = 0$ is

$$y = c_1 \cos 14t + c_2 \sin 14t.$$

The initial condition $y(0) = 1$ yields $c_1 = 1$, and the condition $y'(0) = -4$ yields $14c_2 = -4$, or $c_2 = -\frac{2}{7}$. The amplitude satisfies

$$A^2 = c_1^2 + c_2^2 = \tfrac{53}{49},$$

or $A = \sqrt{53}/7$, as obtained in Instant Exercise 4. The phase shift δ satisfies the equations

$$\cos\delta = \frac{c_1}{A} = \frac{7}{\sqrt{53}}, \qquad \sin\delta = \frac{c_2}{A} = -\frac{2}{\sqrt{53}}.$$

The combination of positive cosine and negative sine places δ in the fourth quadrant; hence,

$$\delta = \arcsin\frac{-2}{\sqrt{53}} \approx -0.278.$$

The solution appears in Figure 3.1.6.

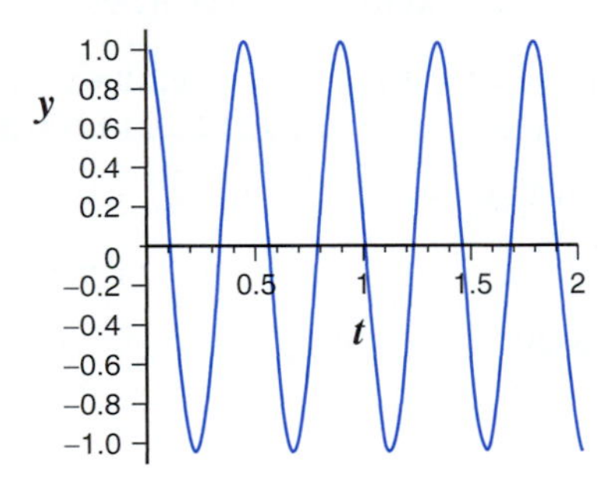

Figure 3.1.6
The approximate solution $y = 1.04\cos(14t + 0.278)$ of Instant Exercise 5.

3.2 Systems of Linear Algebraic Equations

Linear algebra is a subject whose considerable value lies largely in its applications to other areas of mathematics. Many topics in advanced mathematics, differential equations among them, give rise to problems in linear algebra. In particular, the theory and techniques for homogeneous linear differential equations utilize the theory and techniques for systems of linear algebraic equations. This section can serve as a quick review of this material for those who have taken a course in linear algebra or a focused introduction to the topic for those who have not. We begin with the basic definitions.

An n-**vector x** is a column of n numbers $x_1, x_2, \ldots, x_n$. An $n \times n$ **square matrix A** is a rectangular array of numbers with n rows and n columns, with a_{ij} as the entry in row i and column j. The entries $a_{11}, a_{22}, \ldots, a_{nn}$ are called the **main diagonal** of the matrix. Given a square matrix **A** and vector **x,**

$$\mathbf{A} = \begin{pmatrix} a_{11} & a_{12} & \ldots & a_{1n} \\ a_{21} & a_{22} & \ldots & a_{2n} \\ \vdots & \vdots & \vdots & \vdots \\ a_{n1} & a_{n2} & \ldots & a_{nn} \end{pmatrix}, \qquad \mathbf{x} = \begin{pmatrix} x_1 \\ x_2 \\ \vdots \\ x_n \end{pmatrix},$$

the **matrix product Ax** is defined to be the vector

$$\mathbf{Ax} = \begin{pmatrix} a_{11} & a_{12} & \ldots & a_{1n} \\ a_{21} & a_{22} & \ldots & a_{2n} \\ \vdots & \vdots & \vdots & \vdots \\ a_{n1} & a_{n2} & \ldots & a_{nn} \end{pmatrix} \begin{pmatrix} x_1 \\ x_2 \\ \vdots \\ x_n \end{pmatrix} = \begin{pmatrix} a_{11}x_1 + a_{12}x_2 + \cdots + a_{1n}x_n \\ a_{21}x_1 + a_{22}x_2 + \cdots + a_{2n}x_n \\ \vdots \\ a_{n1}x_1 + a_{n2}x_2 + \cdots + a_{nn}x_n \end{pmatrix}. \tag{1}$$

The $n \times n$ **identity matrix**

$$\mathbf{I} = \begin{pmatrix} 1 & 0 & \ldots & 0 \\ 0 & 1 & \ldots & 0 \\ \vdots & \vdots & \vdots & \vdots \\ 0 & 0 & \ldots & 1 \end{pmatrix}$$

is the matrix whose diagonal entries are all 1s and off-diagonal entries are all 0s.

Note that the matrix product of a square matrix and a vector is defined only when the size n of the matrix and vector is the same. The product is a vector of the same size as the original vector **x.** The product **xA** is not defined.

Systems of Linear Equations

A system of n linear equations in n unknowns can be written as

$$\begin{aligned} a_{11}x_1 + a_{12}x_2 + \cdots + a_{1n}x_n &= b_1 \\ a_{21}x_1 + a_{22}x_2 + \cdots + a_{2n}x_n &= b_2 \\ \vdots \qquad\qquad & \quad \vdots \\ a_{n1}x_1 + a_{n2}x_2 + \cdots + a_{nn}x_n &= b_n. \end{aligned}$$

The definition of the matrix product allows the same system to be written conveniently in vector form as

$$\mathbf{Ax} = \mathbf{b},$$

where
$$\mathbf{A} = \begin{pmatrix} a_{11} & a_{12} & \cdots & a_{1n} \\ a_{21} & a_{22} & \cdots & a_{2n} \\ \vdots & \vdots & \vdots & \vdots \\ a_{n1} & a_{n2} & \cdots & a_{nn} \end{pmatrix}, \qquad \mathbf{x} = \begin{pmatrix} x_1 \\ x_2 \\ \vdots \\ x_n \end{pmatrix}, \qquad \mathbf{b} = \begin{pmatrix} b_1 \\ b_2 \\ \vdots \\ b_n \end{pmatrix}.$$

It is also useful to combine all the problem data into a single matrix $(\mathbf{A}|\mathbf{b})$ called the **augmented matrix** for the system and defined as

$$(\mathbf{A}|\mathbf{b}) = \left(\begin{array}{cccc|c} a_{11} & a_{12} & \cdots & a_{1n} & b_1 \\ a_{21} & a_{22} & \cdots & a_{2n} & b_2 \\ \vdots & \vdots & \vdots & \vdots & \vdots \\ a_{n1} & a_{n2} & \cdots & a_{nn} & b_n \end{array}\right).$$

Systems of linear algebraic equations arise when one attempts to find a solution of an initial-value problem from a family of solutions of the corresponding differential equation. As a practical matter, we want to have efficient methods for solving such systems. As a theoretical matter, we want to know what circumstances guarantee that $\mathbf{Ax} = \mathbf{b}$ has a unique solution.

MODEL PROBLEM 3.2

Solve the system

$$3y - 3z = 4, \qquad x + 2y + 4z = 0, \qquad -x + 3y + z = 0.$$

Equivalent Systems

Some systems of linear algebraic equations are easier to solve than others. The system of Model Problem 3.2 consists of three equations with three unknowns. The system is **coupled** because there is no subset of the system that can be solved independently. But suppose we add the second and third equations and use the new equation as a replacement for the second equation. The new system,

$$3y - 3z = 4, \qquad 5y + 5z = 0, \qquad -x + 3y + z = 0,$$

is easier to solve than the original system because it is not fully coupled; one can first solve the subsystem $3y - 3z = 4$, $5y + 5z = 0$ for y and z and then find x from the third equation. We can further simplify the system by dividing the second equation by 5 to get

$$3y - 3z = 4, \qquad y + z = 0, \qquad -x + 3y + z = 0,$$

and then adding 3 times the second equation to the first equation, to get the system

$$6y = 4, \qquad y + z = 0, \qquad -x + 3y + z = 0.$$

This system is **decoupled** because the equations can be solved sequentially: The first equation yields $y = \frac{2}{3}$, then the second yields $z = -\frac{2}{3}$, and the third yields $x = \frac{4}{3}$.

A quick check shows that all the systems in the previous paragraph have the solution $x = \frac{4}{3}$, $y = \frac{2}{3}$, $z = -\frac{2}{3}$. These systems are **equivalent** because they have the same solution.

Row Reduction

We solved Model Problem 3.2 by finding a sequence of equivalent systems that gradually leads to a decoupled system. **Row reduction** is the name given to a more systematic version of the method as applied to the augmented matrix $(\mathbf{A}|\mathbf{b})$. The system can be put in vector form by using $x_1 = x$, $x_2 = y$, and $x_3 = z$ to define the unknown vector **x.** The augmented matrix for the vector form of the system is then

$$(\mathbf{A}|\mathbf{b}) = \left(\begin{array}{ccc|c} 0 & 3 & -3 & 4 \\ 1 & 2 & 4 & 0 \\ -1 & 3 & 1 & 0 \end{array}\right).$$

Row reduction proceeds by a sequence of **elementary row operations;** these are of three types, each corresponding to a basic manipulation of systems of linear algebraic equations. The three operations include

1. Interchanging two rows.
2. Multiplying a row by a nonzero constant.
3. Replacing a row with the sum of that row and a multiple of some other row.

The systematic process of row reduction works by columns from left to right, with the operations performed, if needed, in the order listed. The goal of step 1 is to obtain a system in which column 1 is all 0s except for a 1 in row 1. Similarly, the goal of step k is to obtain a system in which columns 1 through k are all 0s except for 1s in the diagonal positions. If all goes well, step n results in an augmented matrix of the form $(\mathbf{I}|\mathbf{v})$, where $\mathbf{v}$ is some vector and $\mathbf{I}$ is the identity matrix. At this point, the corresponding system of equations is simply $x_1 = v_1$, $x_2 = v_2$, and so on, so that the final augmented matrix is actually $(\mathbf{I}|\mathbf{x})$. Thus, the solution of the system appears on the right side of the final augmented matrix.

Turning to the solution of Model Problem 3.2, we begin work on column 1 of the augmented matrix by switching rows so that the entry in row 1, column 1 is nonzero. There are two ways to do this, as we could interchange row 1 with either of the other two rows. Since our ultimate goal is to have the row 1, column 1 entry be 1, we choose to interchange rows 1 and 2. This yields an equivalent[3] augmented matrix

$$(\mathbf{A}|\mathbf{b}) = \left(\begin{array}{ccc|c} 0 & 3 & -3 & 4 \\ 1 & 2 & 4 & 0 \\ -1 & 3 & 1 & 0 \end{array}\right) \cong \left(\begin{array}{ccc|c} 1 & 2 & 4 & 0 \\ 0 & 3 & -3 & 4 \\ -1 & 3 & 1 & 0 \end{array}\right).$$

Next, we add multiples of row 1 to the other rows, to make the remaining entries in column 1 be 0s. In the present case, the row 2, column 1 entry is already 0. We replace row 3 with the sum of row 3 and row 1 in order to make the row 3, column 1 entry become 0. In summary, we have completed column 1 through the sequence

$$(\mathbf{A}|\mathbf{b}) = \left(\begin{array}{ccc|c} 0 & 3 & -3 & 4 \\ 1 & 2 & 4 & 0 \\ -1 & 3 & 1 & 0 \end{array}\right) \cong \left(\begin{array}{ccc|c} 1 & 2 & 4 & 0 \\ 0 & 3 & -3 & 4 \\ -1 & 3 & 1 & 0 \end{array}\right) \cong \left(\begin{array}{ccc|c} 1 & 2 & 4 & 0 \\ 0 & 3 & -3 & 4 \\ 0 & 5 & 5 & 0 \end{array}\right).$$

[3] It is important to distinguish between augmented matrices that are equal and those that are merely row-equivalent. Here, the new augmented matrix is equivalent to the old one, but not equal to it. We use the symbol $\cong$ to denote row equivalence.

For column 2, the goal is to make the row 2, column 2 entry be 1 and the other two entries 0s without moving row 1.[4] It is a matter of choice whether to interchange the order of rows 2 and 3; here we choose to make this change since it delays the introduction of fractions into the augmented matrix. After the interchange of the two rows, we divide row 2 by 5 to make the row 2, column 2 entry be 1. Then we replace row 1 with the sum of row 1 and −2 times row 2, and we replace row 3 with the sum of row 3 and −3 times row 2 to place 0s in the remaining positions in column 2:

$$\left(\begin{array}{ccc|c} 1 & 2 & 4 & 0 \\ 0 & 3 & -3 & 4 \\ 0 & 5 & 5 & 0 \end{array}\right) \cong \left(\begin{array}{ccc|c} 1 & 2 & 4 & 0 \\ 0 & 5 & 5 & 0 \\ 0 & 3 & -3 & 4 \end{array}\right) \cong \left(\begin{array}{ccc|c} 1 & 2 & 4 & 0 \\ 0 & 1 & 1 & 0 \\ 0 & 3 & -3 & 4 \end{array}\right) \cong \left(\begin{array}{ccc|c} 1 & 0 & 2 & 0 \\ 0 & 1 & 1 & 0 \\ 0 & 0 & -6 & 4 \end{array}\right).$$

Continuing with column 3, we need to make the row 3, column 3 entry be 1 and the others be 0s. To accomplish this, we divide row 3 by −6 and then replace row 1 with the sum of row 1 and −2 times row 3, and we replace row 2 with the sum of row 2 and −1 times row 3:

$$\left(\begin{array}{ccc|c} 1 & 0 & 2 & 0 \\ 0 & 1 & 1 & 0 \\ 0 & 0 & -6 & 4 \end{array}\right) \cong \left(\begin{array}{ccc|c} 1 & 0 & 2 & 0 \\ 0 & 1 & 1 & 0 \\ 0 & 0 & 1 & -\frac{2}{3} \end{array}\right) \cong \left(\begin{array}{ccc|c} 1 & 0 & 0 & \frac{4}{3} \\ 0 & 1 & 0 & \frac{2}{3} \\ 0 & 0 & 1 & -\frac{2}{3} \end{array}\right) = (\mathbf{I}|\mathbf{x}).$$

At this point, the system is completely decoupled. The final augmented matrix immediately yields the solutions x, y, and z.

Note that it is essential not to disturb a column that has already been put into the proper form. For example, when one is working on column 2, it is alright to interchange rows 2 and 3, but not to interchange rows 2 and 1, because that changes the location of the unit entry in column 1. As long as the restriction against disturbing finished columns is not violated, there is some flexibility in the row reduction procedure.

✦ INSTANT EXERCISE 1

Obtain the matrix $(\mathbf{I}|\mathbf{x})$ by row-reducing $\left(\begin{array}{ccc|c} 1 & 2 & 4 & 0 \\ 0 & 3 & -3 & 4 \\ 0 & 5 & 5 & 0 \end{array}\right)$ *without* interchanging rows.

Theory of Systems of Linear Algebraic Equations

The method of row reduction as outlined here does more than outline a solution technique. It also serves to identify the class of problems for which the method gives a unique solution. Model Problem 3.2 has a unique solution because the row reduction algorithm reduces the matrix **A** to the identity matrix **I**. A matrix that is row-equivalent to the identity matrix is said to be **nonsingular.** Not all matrices have this property, just as not all systems of three equations with three unknowns have a unique solution.

[4]Column 1 remains unchanged as long as we do not interchange row 1 with another row.

EXAMPLE 1

Consider the system $x = 1$, $y + z = 0$, $x + y + z = 0$. We can write this system in vector form and row-reduce it, with the final result

$$(\mathbf{A}|\mathbf{b}) = \left(\begin{array}{ccc|c} 1 & 0 & 0 & 1 \\ 0 & 1 & 1 & 0 \\ 1 & 1 & 1 & 0 \end{array}\right) \cong \left(\begin{array}{ccc|c} 1 & 0 & 0 & 1 \\ 0 & 1 & 1 & 0 \\ 0 & 0 & 0 & -1 \end{array}\right).$$

The reduced system includes the equation $0x + 0y + 0z = -1$; therefore, the system has no solutions. Further row reduction is not possible because there is no way to obtain a 1 for the row 3, column 3 entry without disturbing one of the completed columns.

The problem with the system in Example 1 is that the matrix $\mathbf{A}$ does not row-reduce to the identity matrix. If a matrix is not row-equivalent to the identity matrix, then it is instead row-equivalent to a matrix that has a row of 0s. Such a matrix is **singular.**

INSTANT EXERCISE 2

Show that the matrix

$$\mathbf{A} = \begin{pmatrix} 2 & 0 & 4 \\ 0 & 1 & 1 \\ 1 & -1 & 1 \end{pmatrix}$$

is singular.

The theoretical properties of the system $\mathbf{Ax} = \mathbf{b}$, with $\mathbf{A}$ an $n \times n$ square matrix, depend, as indicated in Model Problem 3.2 and Example 1, on whether or not the matrix $\mathbf{A}$ is singular.

Theorem 3.2.1 A linear system of equations $\mathbf{Ax} = \mathbf{b}$ has a unique solution if and only if the matrix $\mathbf{A}$ is nonsingular.

If $\mathbf{A}$ is singular, then either there are no solutions or there are infinitely many solutions, depending on $\mathbf{b}$. The equation $\mathbf{Ax} = \mathbf{0}$, with $\mathbf{A}$ singular, is discussed in Section 6.2, where it is needed for the theory and techniques for systems of homogeneous linear equations.

The Determinant

We have seen how to use row reduction to determine a unique solution for a linear system $\mathbf{Ax} = \mathbf{b}$ and to determine a criterion for the existence of a unique solution. These tasks can also be accomplished with the aid of a function called the *determinant*. The determinant function can be defined in any of several equivalent ways; here we use a definition that prescribes a unique value for the determinant of any square matrix by prescribing the values for the determinants of the possible outcomes of row reduction along with rules for how the determinants of row-equivalent matrices are related.

If **A** has a row of 0s, then the **determinant** of **A,** denoted as $\det(\mathbf{A})$ or $|\mathbf{A}|$, is defined to be 0; also, $\det(\mathbf{I}) = 1$. Row operations change the determinant value in a predictable way: interchanging two rows multiplies the determinant by -1, multiplying a row by a scalar c multiplies the determinant by c, and replacing a row by a sum of that row and a multiple of some other row does not change the value of the determinant at all.

The definition given here does not read as a standard definition does, but it does uniquely define the determinant function.

EXAMPLE 2

Let **A** be the matrix from Model Problem 3.2. Following the same steps in the row reduction,

$$\det(\mathbf{A}) = \begin{vmatrix} 0 & 3 & -3 \\ 1 & 2 & 4 \\ -1 & 3 & 1 \end{vmatrix} = -\begin{vmatrix} 1 & 2 & 4 \\ 0 & 3 & -3 \\ -1 & 3 & 1 \end{vmatrix} = -\begin{vmatrix} 1 & 2 & 4 \\ 0 & 3 & -3 \\ 0 & 5 & 5 \end{vmatrix} = \begin{vmatrix} 1 & 2 & 4 \\ 0 & 5 & 5 \\ 0 & 3 & -3 \end{vmatrix}$$

$$= 5\begin{vmatrix} 1 & 2 & 4 \\ 0 & 1 & 1 \\ 0 & 3 & -3 \end{vmatrix} = 5\begin{vmatrix} 1 & 0 & 2 \\ 0 & 1 & 1 \\ 0 & 0 & -6 \end{vmatrix} = -30\begin{vmatrix} 1 & 0 & 2 \\ 0 & 1 & 1 \\ 0 & 0 & 1 \end{vmatrix} = -30\begin{vmatrix} 1 & 0 & 0 \\ 0 & 1 & 0 \\ 0 & 0 & 1 \end{vmatrix} = -30.$$

The definition of the determinant allows for a restatement of Theorem 3.2.1. Which of the theorem statements is more convenient depends on whether determinants can be computed by methods that are more efficient than row reduction.

Theorem 3.2.2 A linear system of equations $\mathbf{Ax} = \mathbf{b}$ has a unique solution if and only if $\det(\mathbf{A}) \neq 0$.

Computing Determinants

The row reduction procedure is not the only way to compute determinants. In particular, the 2×2 case has a simple formula

$$\det\begin{pmatrix} a_{11} & a_{12} \\ a_{21} & a_{22} \end{pmatrix} = \begin{vmatrix} a_{11} & a_{12} \\ a_{21} & a_{22} \end{vmatrix} = a_{11}a_{22} - a_{12}a_{21} \tag{2}$$

that can be derived from the row reduction definition.

✦ INSTANT EXERCISE 3

Compute $\begin{vmatrix} 3 & 1 \\ 2 & -1 \end{vmatrix}$, once using row reduction and once using Equation (2).

Higher-order determinants can be computed by the technique of **cofactor expansion,** whereby an $n \times n$ determinant is given in terms of a sum of n determinants of size $(n-1) \times (n-1)$. Consider 3×3 determinants. The formula

$$\begin{vmatrix} a & b & c \\ d & e & f \\ g & h & i \end{vmatrix} = a\begin{vmatrix} e & f \\ h & i \end{vmatrix} - b\begin{vmatrix} d & f \\ g & i \end{vmatrix} + c\begin{vmatrix} d & e \\ g & h \end{vmatrix}$$

defines cofactor expansion across the first row. The reader has probably used this formula to compute cross products of vectors. Cofactor expansion down the first column is defined in a similar manner as

$$\begin{vmatrix} a & b & c \\ d & e & f \\ g & h & i \end{vmatrix} = a\begin{vmatrix} e & f \\ h & i \end{vmatrix} - d\begin{vmatrix} b & c \\ h & i \end{vmatrix} + g\begin{vmatrix} b & c \\ e & f \end{vmatrix}.$$

One can apply the same basic principle to any row or any column, providing only that the signs are reversed for the case where the row number or column number is even. For example, the formula for cofactor expansion down the second column is

$$\begin{vmatrix} a & b & c \\ d & e & f \\ g & h & i \end{vmatrix} = -b\begin{vmatrix} d & f \\ g & i \end{vmatrix} + e\begin{vmatrix} a & c \\ g & i \end{vmatrix} - h\begin{vmatrix} a & c \\ d & f \end{vmatrix}.$$

Observe that regardless of which row or column is being used for the expansion, the signs of the terms are given schematically by

$$\begin{vmatrix} + & - & + \\ - & + & - \\ + & - & + \end{vmatrix}.$$

EXAMPLE 3

Let $\mathbf{A}$ be defined by

$$\mathbf{A} = \begin{pmatrix} 1 & 2 & -1 \\ 2 & 0 & 1 \\ 0 & 4 & -3 \end{pmatrix}.$$

Then

$$\det(\mathbf{A}) = \begin{vmatrix} 1 & 2 & -1 \\ 2 & 0 & 1 \\ 0 & 4 & -3 \end{vmatrix}.$$

There are six different choices for cofactor expansion, and there may be reasons to prefer one over another. One may prefer to have the zeros appear as elements in the smaller determinants, or one may take advantage of zeros to have fewer subdeterminants in the expansion. Here, cofactor expansion on column 1 or 2 or row 2 or 3 requires only two 2×2 subdeterminants rather than the usual three. With the first column, for example, we have

$$\det(\mathbf{A}) = \begin{vmatrix} 0 & 1 \\ 4 & -3 \end{vmatrix} - 2\begin{vmatrix} 2 & -1 \\ 4 & -3 \end{vmatrix} = -4 - 2(-2) = 0.$$

EXAMPLE 4

To compute

$$\begin{vmatrix} 3 & 1 & 0 & -1 \\ 2 & 1 & 1 & 2 \\ 1 & 0 & -1 & 0 \\ 0 & -1 & 2 & 1 \end{vmatrix},$$

we can decrease the amount of writing by choosing rows or columns with as many zeros as possible:

$$\begin{vmatrix} 3 & 1 & 0 & -1 \\ 2 & 1 & 1 & 2 \\ 1 & 0 & -1 & 0 \\ 0 & -1 & 2 & 1 \end{vmatrix} = \begin{vmatrix} 1 & 0 & -1 \\ 1 & 1 & 2 \\ -1 & 2 & 1 \end{vmatrix} - \begin{vmatrix} 3 & 1 & -1 \\ 2 & 1 & 2 \\ 0 & -1 & 1 \end{vmatrix}$$

$$= \begin{vmatrix} 1 & 2 \\ 2 & 1 \end{vmatrix} - \begin{vmatrix} 1 & 1 \\ -1 & 2 \end{vmatrix} - \begin{vmatrix} 3 & -1 \\ 2 & 2 \end{vmatrix} - \begin{vmatrix} 3 & 1 \\ 2 & 1 \end{vmatrix} = -15.$$

✦ INSTANT EXERCISE 4

Compute the determinant

$$\begin{vmatrix} 3 & 1 & 0 \\ -2 & 2 & 0 \\ 3 & 0 & 2 \end{vmatrix}$$

by using cofactor expansion.

Often a combination of methods allows for faster computation of a determinant than any one method. In addition to cofactor expansion and row reduction, column reduction steps can be performed. In particular, it is often convenient to replace a column with a sum of that column and a multiple of some other column; this manipulation does not change the value of the determinant.

EXAMPLE 5

Here is another computation of the determinant of Example 4.

$$\begin{vmatrix} 3 & 1 & 0 & -1 \\ 2 & 1 & 1 & 2 \\ 1 & 0 & -1 & 0 \\ 0 & -1 & 2 & 1 \end{vmatrix} = \begin{vmatrix} 3 & 1 & 3 & -1 \\ 2 & 1 & 3 & 2 \\ 1 & 0 & 0 & 0 \\ 0 & -1 & 2 & 1 \end{vmatrix} = \begin{vmatrix} 1 & 3 & -1 \\ 1 & 3 & 2 \\ -1 & 2 & 1 \end{vmatrix} = \begin{vmatrix} 1 & 3 & -1 \\ 0 & 0 & 3 \\ 0 & 5 & 0 \end{vmatrix} = \begin{vmatrix} 0 & 3 \\ 5 & 0 \end{vmatrix} = -15.$$

The first step involved replacing column 3 by the sum of columns 1 and 3. Cofactor expansion on row 3 then reduced the determinant from 4×4 to 3×3. The third step was achieved by row reduction applied to column 1. Cofactor expansion on column 1 then reduced the determinant to 2×2 in preparation for the last step.

3.2 Exercises

In Exercises 1 through 10, use row reduction to either find a unique solution or show that the system does not have a unique solution.

1. $2x + 3y - 4z = -8, \quad x - 2y - 2z = -4, \quad x + 3y = 2$
2. $x - y - z = -1, \quad 4x + 2y - z = 2, \quad y - 3z = 8$
3. $2x + y - 2z = 10, \quad 3x + 2y + 2z = 1, \quad 5x + 4y + 3z = 4$
4. $x + 2y - 3z = -1, \quad 3x - y + 2z = 7, \quad 5x + 3y - 4z = 2$
5. $2x + 6y + z = 8, \quad x + 2y - z = -2, \quad 5x + 7y - 4z = 5$
6. $x - y - z = 0, \quad 2y - z = 0, \quad x + 2y = 12$
7. $x - 2y + 3z = -7, \quad 4x + 3y + z = 5, \quad 2x + 7y - 5z = 19$
8. $x + 2y + 2z = 2, \quad x + y + z = 0, \quad x - 3y - z = 0$
9. $w + x - y = -1, \quad w + x + y + z = 2, \quad -w + x + z = 1, \quad w + 2y + z = 3$
10. $5w + x + 2y + 4z = 1, \quad -w + 2y + 3z = -5, \quad w + x + 6y + z = -11, \quad w - 4z = 1$

In Exercises 11 through 16, compute the indicated determinant.

11. $\begin{vmatrix} 0 & 3 & -2 \\ 1 & 0 & 4 \\ -1 & 0 & -3 \end{vmatrix}$

12. $\begin{vmatrix} 2 & -1 & 0 \\ 3 & 0 & 1 \\ -2 & 1 & 3 \end{vmatrix}$

13. $\begin{vmatrix} 2 & 0 & 3 \\ 1 & 7 & 2 \\ 4 & 6 & 2 \end{vmatrix}$

14. $\begin{vmatrix} 3 & 5 & 1 \\ 0 & 3 & 2 \\ 1 & 2 & 0 \end{vmatrix}$

15. $\begin{vmatrix} 1 & -2 & 1 & 2 \\ 4 & 0 & 5 & 0 \\ 0 & 1 & 6 & 1 \\ 1 & 1 & -1 & 5 \end{vmatrix}$

16. $\begin{vmatrix} -4 & 1 & 3 & 4 \\ 0 & 6 & 2 & 2 \\ 0 & 1 & 0 & 1 \\ 1 & 1 & -1 & 5 \end{vmatrix}$

In Exercises 17 through 20, (*a*) determine the value(s) of c for which the matrix $\mathbf{A}$ of coefficients is singular, and (*b*) for all *other* values of c, determine the unique solution.

17. $x + y = 3, \quad 2x + cy = 7$
18. $3x - 4y = 2, \quad -x + cy = 4$
19. $x + 2y + z = 4, \quad 2x - 3z = 6, \quad cy + 4z = 0$
20. $cx - 2y + 2z = 0, \quad -2x + cy - 2z = 4, \quad 2x - 2y + cz = 0$

In Exercises 21 through 22, find a one-parameter family of solutions of the given system. (Reduce the system as far as possible, choose one of the unknowns to be arbitrary, and solve for the other unknowns in terms of the arbitrary one.)

21. $x - 3y = 0, \quad 2x - 6y = 0$

22. $2x + y - z = 0, \quad x + 2y + z = 0, \quad -x + y + 2z = 0$

23. Derive formula (2) for the 2×2 determinant. This must be done carefully because one or more of the matrix entries could be zero. The derivation must consider as many cases as are necessary to establish the formula for all possible values of the matrix entries.

24. Let $\mathbf{A} = \begin{pmatrix} a & b & c \\ d & e & f \\ g & h & i \end{pmatrix}$. Show that cofactor expansion across the first row and cofactor expansion down the second column both yield the formula

$$\det(\mathbf{A}) = aei - afh + bfg - bdi + cdh - ceg.$$

All six cofactor expansion formulas yield the same result. This does not prove that cofactor expansion yields the same result as row reduction, but it does at least prove that cofactor expansion yields consistent results.

25. An **upper triangular matrix** is a square matrix for which the entry in row i and column j is 0 whenever $j < i$.

   a. Use the text's definition of the determinant to show that the determinant of an upper triangular matrix is the product of the diagonal entries.

   b. Show that the cofactor expansion formula for the determinant is consistent with the result of part *a*.

26. Let $\mathbf{A}$ be as in Exercise 24 with $a = 1$. Show that the formula obtained in Exercise 24 is correct, by using the definition of the determinant given in the text. Note that there are two separate cases that must be checked: $e = bd$ and $e \neq bd$. This exercise can be used as the primary component of a proof of the equivalence of the cofactor expansion formulas with the determinant as defined by row reduction.

✦ 3.2 INSTANT EXERCISE SOLUTIONS

1.
$$\left(\begin{array}{ccc|c} 1 & 2 & 4 & 0 \\ 0 & 3 & -3 & 4 \\ 0 & 5 & 5 & 0 \end{array}\right) \cong \left(\begin{array}{ccc|c} 1 & 2 & 4 & 0 \\ 0 & 1 & -1 & \frac{4}{3} \\ 0 & 5 & 5 & 0 \end{array}\right)$$

$$\cong \left(\begin{array}{ccc|c} 1 & 0 & 6 & -\frac{8}{3} \\ 0 & 1 & -1 & \frac{4}{3} \\ 0 & 0 & 10 & -\frac{20}{3} \end{array}\right) \cong \left(\begin{array}{ccc|c} 1 & 0 & 6 & -\frac{8}{3} \\ 0 & 1 & -1 & \frac{4}{3} \\ 0 & 0 & 1 & -\frac{2}{3} \end{array}\right)$$

$$\cong \left(\begin{array}{ccc|c} 1 & 0 & 0 & \frac{4}{3} \\ 0 & 1 & 0 & \frac{2}{3} \\ 0 & 0 & 1 & -\frac{2}{3} \end{array}\right).$$

2.
$$\begin{pmatrix} 2 & 0 & 4 \\ 0 & 1 & 1 \\ 1 & -1 & 1 \end{pmatrix} \cong \begin{pmatrix} 1 & -1 & 1 \\ 0 & 1 & 1 \\ 2 & 0 & 4 \end{pmatrix} \cong \begin{pmatrix} 1 & -1 & 1 \\ 0 & 1 & 1 \\ 0 & 2 & 2 \end{pmatrix} \cong \begin{pmatrix} 1 & -1 & 1 \\ 0 & 1 & 1 \\ 0 & 0 & 0 \end{pmatrix}.$$

The matrix **A** is row-equivalent to a singular matrix; hence, it is itself singular.

3. By row reduction,

$$\begin{vmatrix} 3 & 1 \\ 2 & -1 \end{vmatrix} = 3\begin{vmatrix} 1 & \frac{1}{3} \\ 2 & -1 \end{vmatrix} = 3\begin{vmatrix} 1 & \frac{1}{3} \\ 0 & -\frac{5}{3} \end{vmatrix} = -5\begin{vmatrix} 1 & \frac{1}{3} \\ 0 & 1 \end{vmatrix} = -5\begin{vmatrix} 1 & 0 \\ 0 & 1 \end{vmatrix} = -5.$$

By Equation (2),

$$\begin{vmatrix} 3 & 1 \\ 2 & -1 \end{vmatrix} = (3)(-1) - (1)(2) = -5.$$

4. Using cofactor expansion on the third column yields

$$\begin{vmatrix} 3 & 1 & 0 \\ -2 & 2 & 0 \\ 3 & 0 & 2 \end{vmatrix} = 2\begin{vmatrix} 3 & 1 \\ -2 & 2 \end{vmatrix} = (2)(8) = 16.$$

3.3 Theory of Homogeneous Linear Equations

In Section 3.1, we considered initial-value problems for the differential equation

$$y'' + 9y = 0.$$

Our efforts can be summarized as follows:

1. We determined (by inspection) two solutions, $y_1 = \cos 3t$ and $y_2 = \sin 3t$.
2. We found that the two solutions y_1 and y_2 could be used to construct a two-parameter family

$$y = c_1 \cos 3t + c_2 \sin 3t$$

of solutions.

3. We found a unique pair of values for c_1 and c_2 such that the solution satisfies the two initial conditions as well as the differential equation.

The discussion of $y'' + 9y = 0$ suggests several questions.

1. If the functions $y_1(t), y_2(t), \ldots, y_n(t)$ are solutions of a differential equation, is $y(t) = c_1 y_1(t) + c_2 y_2(t) + \cdots + c_n y_n(t)$ a family of solutions of the same differential equation?
2. If we have an n-parameter family of solutions of an nth-order differential equation, can we be sure that there is a set of parameter values to satisfy any given set of n initial conditions?
3. If we have found a solution of an initial-value problem, can we be sure that the initial-value problem has no other solutions?

None of these questions can be answered in the affirmative without some restrictions.

MODEL PROBLEM 3.3

Find a three-parameter family of solutions for the differential equation

$$y''' - y' = 0$$

that satisfies the properties of question 2 and question 3.

A family of solutions that satisfies the properties of questions 2 and 3 is the **general solution** of the differential equation. Not all differential equations have a general solution; however, if a differential equation does have a general solution, then any initial-value problem for it can be solved uniquely by applying the initial conditions to the general solution.[5]

The answer to question 3 for Model Problem 3.3 can be determined immediately by using Theorem 2.4.2. The differential equation is linear and third-order, with coefficient functions that are continuous for all t. Theorem 2.4.2 asserts that there is a unique solution defined for all t for the initial-value problem

$$y''' - y' = 0, \qquad y(t_0) = y_0, \qquad y'(t_0) = v_0, \qquad y''(t_0) = w_0,$$

for any numbers t_0, y_0, v_0, and w_0. Thus, the model differential equation has a general solution. It will be easier to answer the other questions after we develop some useful concepts.

Linear Differential Operators

The language of linear differential operators gives a convenient setting in which to introduce the theory of linear differential equations.

A **differential operator** is a rule L that uses derivatives to assign a function $L[y]$ to any sufficiently differentiable function y. The function $L[y]$ is called the **image** of y under L.

EXAMPLE 1

Let L be the differential operator defined by $L[y] = y'' + 9y$. This operator is a rule that says that the output is computed from the input by adding 9 times the input function to its second derivative. For example, $L[e^{-3t}] = 18e^{-3t}$ and $L[\cos 3t] = 0$.

✦ INSTANT EXERCISE 1

Calculate $L[e^{-3t}]$ and $L[\cos 3t]$ for Example 1 to verify the indicated results.

Observe that both y and $L[y]$ are functions of a common variable. An operator is similar to a function in the sense that both are rules that prescribe for each suitable input a unique output. But where the inputs and outputs of a function are numbers or vectors, the inputs and outputs of a differential operator are functions of a common variable. A schematic diagram of a differential

[5]The general solution, if there is one, is unique, but the representation of the solution in terms of a particular set of n solutions is not. See Exercise 21.

operator is shown in Figure 3.3.1, using the specific input function y defined by $y(t) = e^{-3t}$ and operator L defined by $L[y] = y'' + 9y$.

$y = e^{-3t}$ → $L[y] = y'' + 9y$ → $L[y] = 18e^{-3t}$

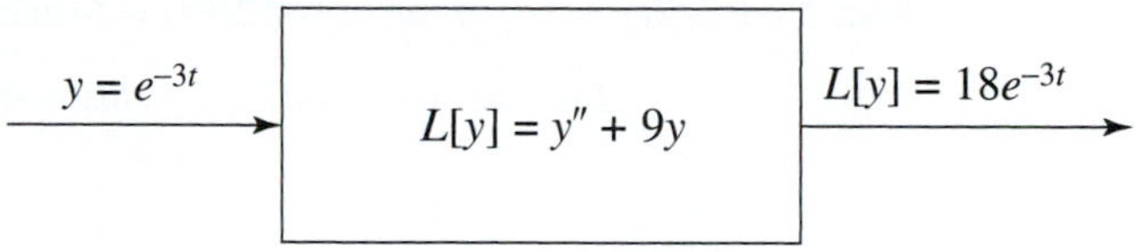

Figure 3.3.1
A schematic diagram of the differential operator $L[y] = y'' + 9y$ applied to $y(t) = e^{-3t}$.

Operators are particularly useful when applied to families of functions.

EXAMPLE 2

Consider the family of functions given by

$$y(t) = c_1 \cos 3t + c_2 \sin 3t$$

and let L be defined as in Figure 3.3.1. Then $L[y]$ is given by

$$y'' + 9y = (-9c_1 \cos 3t - 9c_2 \sin 3t) + 9(c_1 \cos 3t + c_2 \sin 3t) = 0.$$

In operator notation,

$$L[c_1 \cos 3t + c_2 \sin 3t] = 0.$$

Thus, the image of y under L is the zero function.

✦ INSTANT EXERCISE 2

Find the image of $y = c_1 e^t$ under the linear differential operator L defined by $L[y] = y'' + 9y$.

In Section 3.1, we constructed a family of functions $y = c_1 \cos 3t + c_2 \sin 3t$ from the set $\{\cos 3t, \sin 3t\}$ by adding arbitrary multiples of each of the two functions. In general, a function of the form

$$y = c_1 y_1 + c_2 y_2 + \cdots + c_m y_m,$$

where the coefficients c_i are numbers, is a **linear combination** of the functions $y_1, \ldots, y_m$. The full set $\{y = c_1 y_1 + c_2 y_2 + \cdots + c_m y_m | c_i \in \mathbb{R}\}$ is called the **span** of the functions $y_1, \ldots, y_m$.

The application of differential operators to the the span of a set of functions is of particular interest in the study of differential equations.

EXAMPLE 3

Let $y_1(t) = \cos 2t$ and $y_2(t) = \sin 2t$. Let L be the differential operator defined by $L[y] = y'' + 9y$. Let y be the span of y_1 and y_2. We have

$$\begin{aligned} L[y](t) &= (c_1 \cos 2t + c_2 \sin 2t)'' + 9(c_1 \cos 2t + c_2 \sin 2t) \\ &= -4c_1 \cos 2t - 4c_2 \sin 2t + 9c_1 \cos 2t + 9c_2 \sin 2t, \end{aligned}$$

or

$$L[y](t) = 5c_1 \cos 2t + 5c_2 \sin 2t.$$

Also, we have $L[\cos 2t](t) = 5\cos 2t$ and $L[\sin 2t](t) = 5\sin 2t$. Thus, we have the result

$$L[c_1 \cos 2t + c_2 \sin 2t] = c_1 L[\cos 2t] + c_2 L[\sin 2t].$$

The property illustrated in Example 3 says that the arithmetic operations of scalar multiplication and addition can be applied either before or after the operator is applied, with no difference in the result. This property is not a coincidence, nor did it have anything to do with the functions $\cos 2t$ and $\sin 2t$.

✦ INSTANT EXERCISE 3

Let L be defined by $L[y] = y'' + 9y$. Show that $L[c_1 e^t + c_2 e^{2t}] = c_1 L[e^t] + c_2 L[e^{2t}]$.

A differential operator L for which

$$L[c_1 y_1 + c_2 y_2] = c_1 L[y_1] + c_2 L[y_2] \tag{1}$$

for any pair of functions y_1 and y_2 in the domain of the operator and any pair of constants c_1 and c_2 is a **linear differential operator.**

EXAMPLE 4

The differential operator L defined by $L[y] = y'' + 9y$ is linear. To see this, we simply repeat the calculation of Example 3 and Instant Exercise 3 in the general case:

$$\begin{aligned} L[c_1 y_1 + c_2 y_2] &= (c_1 y_1 + c_2 y_2)'' + 9(c_1 y_1 + c_2 y_2) = c_1(y_1'' + 9y_1) + c_2(y_2'' + 9y_2) \\ &= c_1 L[y_1] + c_2 L[y_2]. \end{aligned}$$

EXAMPLE 5

The differential operator L defined by $L[y] = y'' + 9$ is nonlinear.

$$L[c_1 y_1 + c_2 y_2] = (c_1 y_1 + c_2 y_2)'' + 9 = c_1 y_1'' + c_2 y_2'' + 9,$$

but

$$c_1 L[y_1] + c_2 L[y_2] = c_1 y_1'' + c_2 y_2'' + 9c_1 + 9c_2;$$

these quantities are not equal for *all* values of c_1 and c_2.

✦ INSTANT EXERCISE 4

Show that the operator L defined by $L[y] = y''' - y'$ is linear.

Linear differential operators must fit a particular form; specifically, $L[y]$ must be a linear combination of derivatives of y, with coefficients that may depend on t. Thus, the first-order linear differential operators are those operators that can be written in the form

$$L[y] = a\frac{dy}{dt} + qy, \tag{2}$$

where a is a nonzero function of t and q is any function of t. Linear differential operators of order higher than 1 are analogous to the first-order case. Second-order linear differential operators are those that can be written as

$$L[y] = a\frac{d^2y}{dt^2} + p\frac{dy}{dt} + qy, \tag{3}$$

where a, q, and p are functions of t with a not identically zero. Similarly, a linear differential operator of order n has the form[6]

$$L[y] = p_0\frac{d^ny}{dt^n} + p_1\frac{d^{n-1}y}{dt^{n-1}} + \cdots + p_ny, \tag{4}$$

where, as before, p_0 is not identically zero.

Linear Differential Equations

The language of linear differential operators is convenient for stating the theory of linear differential equations. A linear differential equation can be written in the form

$$L[y] = g, \tag{5}$$

where L is a linear differential operator and g is a function of the same variable as the functions in the domain of L.[7] Linear differential operator notation allows us to use the equation $L[y] = g$ to refer to a specific linear differential equation or to an arbitrary linear equation of any order. It also allows us to think of the solutions of linear equations as the set of functions whose image under L is the function g.

The properties of linear differential operators give rise to corresponding properties of linear differential equations. These properties make the solutions of linear differential equations easier

[6]A notation with subscripted coefficients is messier than one with different letters for the coefficients. It is common to use different letters for problems of a fixed order and use subscripted letters for problems of unspecified order.

[7]Linear differential equations were previously defined in Section 2.4 without using the terminology of linear differential operators.

to characterize than nonlinear differential equations. Linear equations arise naturally in many physical models, such as the spring-mass model of Section 3.1. They can also be used to determine properties of nonlinear equations, in much the same way as a tangent line can be used to study nonlinear functions.[8]

Homogeneous Linear Equations

Linear differential equations for which $g \equiv 0$ are called **homogeneous** linear equations. Homogeneous equations occur frequently in mathematical models. They also provide information that is needed to solve the more general equation $L[y] = g$. Given a specific linear differential equation $L[y] = g$, the corresponding homogeneous equation $L[y] = 0$ is sometimes called the **homogeneous equation associated with $L[y] = g$.** For example, $y'' + 9y = 0$ is the homogeneous equation associated with the nonhomogeneous equation $y'' + 9y = \cos t$.

Suppose L is a linear differential operator and $\{y_1, y_2, \ldots, y_m\}$ is a set of solutions of the homogeneous linear equation $L[y] = 0$. By the definition of linear differential operators,

$$\begin{aligned} L[c_1 y_1 + c_2 y_2 + \cdots + c_m y_m] &= L[c_1 y_1] + L[c_2 y_2] + \cdots + L[c_m y_m] \\ &= c_1 L[y_1] + c_2 L[y_2] + \cdots + c_m L[y_m] = 0. \end{aligned}$$

Thus, linear combinations of solutions of $L[y] = 0$ are also solutions. This property, called the **principle of superposition,** is the key to determining a simple structure for the solution sets of linear differential equations.[9]

Theorem 3.3.1 Let L be a linear differential operator. If $y_1, y_2, \ldots, y_m$ are solutions of $L[y] = 0$, then so is $y = c_1 y_1 + c_2 y_2 + \cdots + c_m y_m$. In other words, all linear combinations of solutions of $L[y] = 0$ are also solutions.

✦ INSTANT EXERCISE 5

Observe that $y_1 = 1$, $y_2 = e^t$, and $y_3 = e^{-t}$ are all solutions of $y''' - y' = 0$. Verify the principle of superposition for this example by showing that

$$y = c_1 + c_2 e^t + c_3 e^{-t}$$

is a family of solutions of $y''' - y' = 0$.

Theorem 3.3.1 provides a useful answer to question 1. Families constructed from solutions are themselves solutions, provided the differential equation is linear and homogeneous.

---

[8]See Section 6.7.

[9]The same property holds for other homogeneous linear problems. For example, if $\mathbf{x}_1, \mathbf{x}_2, \ldots, \mathbf{x}_m$ are solutions of the homogeneous linear vector algebraic equation $\mathbf{A}\mathbf{x} = \mathbf{0}$, then so is $\mathbf{x} = c_1\mathbf{x}_1 + c_2\mathbf{x}_2 + \cdots + c_m\mathbf{x}_m$.

EXAMPLE 6

Let $y_1 = t$ and $y_2 = e^t$, and let L be defined by

$$L[y] = (t-1)y'' - ty' + y.$$

Then

$$L[y_1] = (t-1)(0) - t(1) + (t) = 0$$

and

$$L[y_2] = (t-1)(e^t) - t(e^t) + (e^t) = 0.$$

Thus, both y_1 and y_2 are solutions of the homogeneous linear equation

$$(t-1)y'' - ty' + y = 0.$$

By Theorem 3.3.1, the linear combination

$$y = c_1 t + c_2 e^t$$

is a family of solutions.

Satisfying Initial Conditions

We turn now to question 2. In Instant Exercise 5, we found a solution family

$$y = c_1 + c_2 e^t + c_3 e^{-t}$$

for the differential equation $y''' - y' = 0$. Theorem 2.4.2 guarantees a unique solution for any initial-value problem for this differential equation. Suppose we choose three numbers, y_0, v_0, and w_0. Can we always find a member of the solution family that solves the initial-value problem

$$y''' - y' = 0, \qquad y(0) = y_0, \qquad y'(0) = v_0, \qquad y''(0) = w_0?$$

Substituting the solution family into the initial conditions results in three algebraic equations:

$$c_1 + c_2 + c_3 = y_0, \qquad c_2 - c_3 = v_0, \qquad c_2 + c_3 = w_0.$$

These are three linear algebraic equations with three unknowns, and we can write them as a system. Let vectors **b** and **c** and matrix **A** be defined by

$$\mathbf{A} = \begin{pmatrix} 1 & 1 & 1 \\ 0 & 1 & -1 \\ 0 & 1 & 1 \end{pmatrix}, \qquad \mathbf{b} = \begin{pmatrix} y_0 \\ v_0 \\ w_0 \end{pmatrix}, \qquad \mathbf{c} = \begin{pmatrix} c_1 \\ c_2 \\ c_3 \end{pmatrix}.$$

Given these definitions, the equations for the coefficients can be written as

$$\mathbf{Ac} = \mathbf{b}.$$

This is a system of linear algebraic equations; hence, we can examine it by using the methods of Section 3.2. In particular, the system has a unique solution,

$$c_1 = y_0 - w_0, \qquad c_2 = \frac{w_0 + v_0}{2}, \qquad c_3 = \frac{w_0 - v_0}{2}.$$

✦ INSTANT EXERCISE 6

Use the row reduction method to solve the system $\mathbf{Ac} = \mathbf{b}$ from Model Problem 3.3.

Similarly, we can show that there is a unique set of values for c_1, c_2, and c_3 for any initial conditions at $t_0 = 0$ simply by showing that $\det(\mathbf{A}) \neq 0$. Similar results can be obtained for initial conditions at some other t_0.

✦ INSTANT EXERCISE 7

Consider the initial-value problem

$$(t-1)y'' - ty' + y = 0, \qquad y(0) = y_0, \qquad y'(0) = v_0,$$

where y_0 and v_0 are any constants. In Example 6, we found a two-parameter family of solutions

$$y = c_1 t + c_2 e^t$$

for this differential equation. Show that values of c_1 and c_2 can be chosen to satisfy the initial conditions.

Application of the n initial conditions to the n-parameter family of solutions always results in a set of n linear equations with n unknowns. In general, there is not always a unique solution for a set of n linear equations with n unknowns.

EXAMPLE 7

Note that $y_1 = e^t$ and $y_2 = 2e^t$ are both solutions of $y'' - 2y' + y = 0$. By the principle of superposition, all functions $y = c_1 y_1 + c_2 y_2$ are also solutions of $L[y] = 0$. To satisfy the initial conditions $y(0) = 1$ and $y'(0) = 0$, we must find constants c_1 and c_2 such that $c_1 + 2c_2 = 1$ and $c_1 + 2c_2 = 0$. This is clearly impossible.

The Wronskian

Consider the case of a third-order problem,

$$a(t)y''' + p(t)y'' + q(t)y' + r(t)y = 0, \qquad y(0) = y_0, \qquad y'(0) = v_0, \qquad y''(0) = w_0.$$

Assume that we have a solution formula

$$y = c_1 y_1 + c_2 y_2 + c_3 y_3.$$

Substituting this formula into the initial conditions yields a system of three equations in the three unknowns c_1, c_2, and c_3:

$$\begin{aligned} y_1(0)c_1 + y_2(0)c_2 + y_3(0)c_3 &= y_0, \\ y_1'(0)c_1 + y_2'(0)c_2 + y_3'(0)c_3 &= v_0, \\ y_1''(0)c_1 + y_2''(0)c_2 + y_3''(0)c_3 &= w_0. \end{aligned}$$

By Theorem 3.2.2, this system has a unique solution if and only if the determinant of the matrix of coefficients is nonzero. The usefulness of this determinant motivates the following definition.

Suppose $y_1, y_2, \ldots, y_n$ are solutions of a homogeneous nth-order linear differential equation $L[y] = 0$. Let $y_j^{(i)}$ be the ith derivative of y_j. The **Wronskian** is the determinant

$$W[y_1, y_2, \ldots, y_n] = \begin{vmatrix} y_1 & y_2 & \cdots & y_n \\ y_1' & y_2' & \cdots & y_n' \\ \vdots & \vdots & \vdots & \vdots \\ y_1^{(n-1)} & y_2^{(n-1)} & \cdots & y_n^{(n-1)} \end{vmatrix}. \tag{6}$$

EXAMPLE 8

The Wronskian of the solutions $y_1 = 1$ and $y_2 = e^{-t}$ of $y'' + y' = 0$ is given by

$$W[1, e^{-t}](t) = \begin{vmatrix} 1 & e^{-t} \\ 0 & -e^{-t} \end{vmatrix} = -e^{-t}.$$

The Wronskian of the solutions $y_1 = 1$, $y_2 = e^t$, and $y_3 = e^{-t}$ of $y''' - y' = 0$ is given by

$$W[1, e^t, e^{-t}](t) = \begin{vmatrix} 1 & e^t & e^{-t} \\ 0 & e^t & -e^{-t} \\ 0 & e^t & e^{-t} \end{vmatrix} = \begin{vmatrix} e^t & -e^{-t} \\ e^t & e^{-t} \end{vmatrix} = e^t e^{-t} - (-e^{-t})e^t = 2.$$

Note that the Wronskian is a differential operator. Its inputs are n functions of t that solve a homogeneous linear differential equation of order n, and it uses derivatives and algebraic operations to obtain an output that is also a function of t. The function $W[y_1, y_2, \ldots, y_n]$ can be shown to satisfy a simple differential equation.[10]

Theorem 3.3.2

Abel's Theorem Suppose $y_1, y_2, \ldots, y_n$ are solutions of a homogeneous nth-order linear differential equation

$$\frac{d^n y}{dt^n} + p_1(t)\frac{d^{n-1} y}{dt^{n-1}} + \cdots + p_n(t)y = 0$$

on an open interval I where the coefficients are all continuous. Then $W[y_1, y_2, \ldots, y_n]$ satisfies the first-order equation

$$W' = -p_1 W;$$

hence, W is either never zero on I or identically zero on I.

EXAMPLE 9

In Example 8, the Wronskian of solutions for the differential equation $y'' + y' = 0$ was determined to be the function $-e^{-t}$. By Abel's theorem, the Wronskian should satisfy the differential equation $W' = -W$, given $p_1 = 1$, and so it does. Similarly, the Wronskian of the solutions $y_1 = 1$, $y_2 = e^t$, and $y_3 = e^{-t}$ of

[10] The proof of this theorem is considered in Exercises 22 through 24.

$y''' - y' = 0$ was found to be a constant; Abel's theorem indicates that the Wronskian should satisfy the differential equation $W' = 0$.

✦ INSTANT EXERCISE 8

Calculate the Wronskian for the solutions of Example 6, and show that it satisfies the differential equation $W' = -p_1 W$. Note that the given differential equation must be divided by $t - 1$ before identification of p_1.

The theory of systems of homogeneous linear algebraic equations, the definition of the Wronskian, and Abel's theorem combine to produce a test for whether a set of n solutions of an nth-order differential equation is a general solution.

Theorem 3.3.3 Let L be a linear differential operator of order n with coefficients as denoted in (4). Let I be an interval on which all p_i are continuous and $p_0 \neq 0$, and let $t_0 \in I$. Let $\{y_1, y_2, \ldots, y_n\}$ be a set of solutions of the differential equation $L[y] = 0$ on the interval I. Then the unique solution of the initial-value problem

$$L[y] = 0, \qquad y(t_0) = y_0, \qquad y'(t_0) = y_0', \qquad \ldots, \qquad y^{(n-1)}(t_0) = y_0^{(n-1)}$$

can always be found from

$$y = c_1 y_1 + c_2 y_2 + \cdots + c_n y_n,$$

provided $W[y_1, y_2, \ldots, y_n](t_0) \neq 0$. By Abel's theorem, it is sufficient to show that $W \neq 0$ at any point in I.

An equivalent statement of Theorem 3.3.3 is that, given the nonzero Wronskian, the n-parameter family of solutions in the theorem is the general solution of the differential equation.

EXAMPLE 10

The functions $y_1 = 1 + t$, $y_2 = 1 + 2t + t^2$, and $y_3 = 1 - t^2$ are all solutions of the third-order linear differential equation $y''' = 0$. Is $y = c_1 y_1 + c_2 y_2 + c_3 y_3$ the general solution of $y''' = 0$? We have

$$\begin{aligned} W[y_1, y_2, y_3] &= \begin{vmatrix} 1+t & 1+2t+t^2 & 1-t^2 \\ 1 & 2+2t & -2t \\ 0 & 2 & -2 \end{vmatrix} \\ &= (1+t)\begin{vmatrix} 2+2t & -2t \\ 2 & -2 \end{vmatrix} - 1\begin{vmatrix} 1+2t+t^2 & 1-t^2 \\ 2 & -2 \end{vmatrix} \\ &= (1+t)[(2+2t)(-2) - (2)(-2t)] - [(1+2t+t^2)(-2) - (1-t^2)(2)] \\ &= (1+t)(-4) + (2+4t+2t^2) + (2-2t^2) = 0. \end{aligned}$$

By Theorem 3.3.3, the given family is not the general solution of the differential equation. Not all sets of initial conditions can be satisfied using this family.

Linear Independence

The concept of *linear independence* is closely related to the Wronskian.

A set of m functions $y_1, y_2, \ldots, y_m$ is **linearly independent** on an interval I if

$$c_1 y_1 + c_2 y_2 + \cdots + c_m y_m = 0$$

is true for all $x \in I$ only if $c_1 = c_2 = \cdots = c_m = 0$. A set that is not linearly independent is **linearly dependent.**

Linear dependence of the set of solutions of a homogeneous linear differential equation on the interval I is equivalent to $W \equiv 0$ on I; hence, Theorem 3.3.3 can be paraphrased in terms of linear independence:

- $y = c_1 y_1 + c_2 y_2 + \cdots + c_n y_n$ is the general solution of an nth-order linear differential equation $L[y] = 0$ on an interval I if and only if $\{y_1, y_2, \ldots, y_n\}$ is a linearly independent set of solutions of $L[y] = 0$ on I.

EXAMPLE 11

The three solutions of $y''' = 0$ given in Example 10 were such that $W[y_1, y_2, y_3] \equiv 0$; hence, the three solutions form a linearly dependent set. Alternatively, consider the equation

$$c_1 y_1 + c_2 y_2 + c_3 y_3 = 0.$$

We have

$$\begin{aligned} c_1 y_1 + c_2 y_2 + c_3 y_3 &= c_1(1 + t) + c_2(1 + 2t + t^2) + c_3(1 - t^2) \\ &= (c_1 + c_2 + c_3) + (c_1 + 2c_2)t + (c_2 - c_3)t^2. \end{aligned}$$

For this quantity to be identically zero, the coefficients of the three terms must all vanish; hence, the vector **c** must satisfy the algebraic equations

$$c_1 + c_2 + c_3 = 0, \qquad c_1 + 2c_2 = 0, \qquad c_2 - c_3 = 0.$$

These equations can be solved by linear algebra methods, but it is easy enough simply to use the latter two equations to replace c_1 and c_3 in the first, resulting in the equation

$$-2c_2 + c_2 + c_2 = 0.$$

This equation is true for all c_2; in particular, we could choose $c_2 = 1$. Then $c_1 = -2$ and $c_3 = 1$. There exists a set of c_i values, not all zero, for which the linear combination is zero. Hence, the set is linearly dependent according to the definition.

An Algorithm for Solving Homogeneous Linear Equations

Theorem 3.3.3 provides the outline of a basic plan for solving homogeneous linear equations. The theorem reduces the problem of solving $L[y] = 0$ to the problem of finding a linearly independent set of n solutions of $L[y] = 0$. We only need to find n different solutions in order to obtain all the solutions, provided the n solutions are sufficiently distinct that they form a linearly independent set.

Algorithm 3.3.1

Solving Homogeneous Linear Equations Given an nth-order linear differential equation $L[y] = 0$,

1. Find n different solutions on an interval I.
2. Show that the span of the solutions is the general solution on the interval I by demonstrating that the Wronskian of the solutions is nonzero at some point in I.

Step 1 is often difficult. For most homogeneous linear equations, we are unable to determine any solutions in terms of elementary functions. If the coefficients of L are all constant, then the solutions can be found by a method that will be the subject of Sections 3.4 to 3.6. Otherwise, there are a few equations, such as Cauchy–Euler equations (Section 3.7), that can be solved by converting them to constant-coefficient equations; and there are a number of differential equations, such as Bessel equations (Section 3.7), whose solutions can be expressed in terms of *special functions* that are accessible using computers. Any other homogeneous linear equations must generally be solved by numerical methods.

In practice, the same theorems that prescribe methods for step 1 also guarantee the linear independence of the set of solutions determined by the method, so step 2 does not actually need to be performed. The Wronskian is of theoretical importance because it is used to establish such guarantees. It is also used in the determination of solutions for nonhomogeneous equations.

3.3 Exercises

In Exercises 1 through 8, find the image of the given function or family of functions under the given linear differential operator.

1. $y = t^3 - t, \quad L[y] = y'' + 2y$
2. $y = e^{3t}, \quad L[y] = y'' + 2y' - 3y$
3. $y = e^{-t}\cos t, \quad L[y] = y'' + 2y$
4. $y = e^{2t}\sin t, \quad L[y] = y'' + y$
5. $y = c_1 e^{-2t}, \quad L[y] = y'' + y' + 4y$
6. $y = c_1 e^{-2t}, \quad L[y] = y'' + 5y' + 6y$
7. $y = c_1 \cos 2t, \quad L[y] = y'' + 3y' + 4y$
8. $y = c_1 t^3 + c_2 t^2 + c_3 t + c_4, \quad L[y] = y'' + 2y' + 3y$
9. Use the result of Exercise 1 to determine a differential equation of the form $L[y] = g$, where L is the given operator, for which $y = t^3 - t$ is a solution.
10. Use the result of Exercise 6 to determine a differential equation for which $y = c_1 e^{-2t}$ is a one-parameter family of solutions.

In Exercises 11 through 14, (a) show that the given functions are solutions of the given differential equation and (b) use Theorem 3.3.3 to construct a formula for the general solution of the equation.

11. $y''' - 2y'' = 0, \quad y_1 = 1, \quad y_2 = t, \quad y_3 = e^{2t}$

12. $y'' - (m + r)y' + mry = 0, \quad y_1 = e^{rt}, \quad y_2 = e^{mt} \quad (m \neq r)$

13. $y'' - 2y' + y = 0, \quad y_1 = e^t, \quad y_2 = te^t$

14. $y''' - 3y'' + 3y' - y = 0, \quad y_1 = e^t, \quad y_2 = te^t, \quad y_3 = t^2e^t$

In Exercises 15 through 20, (*a*) find the largest interval on which a unique solution is guaranteed by Theorem 2.4.2, (*b*) verify that the given functions solve the differential equation, (*c*) verify that the given solutions form a linearly independent set, (*d*) verify the differential equation for W in Theorem 3.3.2, (*e*) construct the general solution, (*f*) complete the solution of the initial-value problem, and (*g*) determine the actual interval of existence for the solution.

15. $y'' - y' - 6y = 0, \quad y(0) = 0, \quad y'(0) = 1, \quad y_1 = e^{-2t}, \quad y_2 = e^{3t}$

16. $y'' + 7y' + 10y = 0, \quad y(0) = -3, \quad y'(0) = 2, \quad y_1 = e^{-5t}, \quad y_2 = e^{-2t}$

17. $x^2y'' - 3xy' + 4y = 0, \quad y(1) = 1, \quad y'(1) = -1, \quad y_1 = x^2, \quad y_2 = x^2 \ln x$

18. $(t^2 + 2t)y'' + (t^2 - 2)y' - (2t + 2)y = 0, \quad y(1) = 1, \quad y'(1) = -4, \quad y_1 = e^{-t}, \quad y_2 = t^2$

19. $y''' + 2y'' = 0, \quad y(0) = 1, \quad y'(0) = 0, \quad y''(0) = 0, \quad y_1 = 1, \quad y_2 = t, \quad y_3 = e^{-2t}$

20. $y''' - y'' + y' - y = 0, \quad y(0) = 0, \quad y'(0) = 1, \quad y''(0) = 0, \quad y_1 = e^t,$
$y_2 = \cos t, \quad y_3 = \sin t$

21. Consider the differential equation $y'' - y = 0$.
    a. Show that $Y = a_1e^t + a_2e^{-t}$ is a two-parameter family of solutions.
    b. Show that $Z = b_1 \cosh t + b_2 \sinh t$ is a two-parameter family of solutions.
    c. Use the definitions of cosh and sinh to obtain formulas for a_1 and a_2 in terms of b_1 and b_2.
    d. Determine formulas for b_1 and b_2 in terms of a_1 and a_2.
    e. Explain why the results of parts *c* and *d* show that any function in one of the solution families is also in the other. Conclude that Y and Z are different representations of the same set of functions.

22. Derive the formula $W' = -pW$ of Theorem 3.3.2 for the general second-order equation

$$y'' + p(t)y' + q(t)y = 0.$$

23. Derive the formula $W' = -pW$ of Theorem 3.3.2 for the general third-order equation

$$y''' + p(t)y'' + q(t)y' + r(t)y = 0.$$

    a. The derivative of an $n \times n$ determinant is the sum of n $n \times n$ determinants, with the ith term obtained by replacing the entries in row i of the original determinant with their derivatives. Use this fact to show that

$$W'[y_1, y_2, y_3] = \begin{vmatrix} y_1 & y_2 & y_3 \\ y_1' & y_2' & y_3' \\ y_1''' & y_2''' & y_3''' \end{vmatrix}.$$

    b. Use the differential equation to write the entries in the bottom row of the determinant W' in terms of lower-order derivatives.
    c. Complete the derivation by using row reduction to compute the determinant.

24. Generalize the argument of Exercise 22 to homogeneous linear differential equations of order n.

25. Suppose $\{y_1, y_2, \ldots, y_n\}$ is a set of solutions of a homogeneous linear differential equation of order n, and that $W[y_1, y_2, \ldots, y_n]$ is not identically zero. Show that the set $\{y_1, y_2, \ldots, y_n\}$ is linearly independent according to the definition of linear independence.

✦ 3.3 INSTANT EXERCISE SOLUTIONS

1.

$$L[e^{-3t}] = (9e^{-3t}) + 9(e^{-3t}) = 18e^{-3t}.$$

$$L[\cos 3t] = (-9\cos 3t) + 9(\cos 3t) = 0.$$

2.

$$L[c_1e^t] = (c_1e^t) + 9(c_1e^t) = 10c_1e^t.$$

The image of the family c_1e^t under L is the family $10c_1e^t$.

3. We have

$$L[e^t] = (e^t) + 9(e^t) = 10e^t$$

and

$$L[e^{2t}] = (4e^{2t}) + 9(e^{2t}) = 13e^{2t}.$$

We have

$$L[c_1e^t + c_2e^{2t}] = (c_1e^t + 4c_2e^{2t}) + 9(c_1e^t + c_2e^{2t}) = 10c_1e^t + 13c_2e^{2t} = c_1(10e^t) + c_2(13e^{2t});$$

thus,

$$L[c_1e^t + c_2e^{2t}] = c_1L[e^t] + c_2L[e^{2t}].$$

4.

$$\begin{aligned} L[c_1y_1 + c_2y_2] &= (c_1y_1''' + c_2y_2''') - (c_1y_1' + c_2y_2') = c_1(y_1''' - y_1') + c_2(y_2''' - y_2') \\ &= c_1L[y_1] + c_2L[y_2]. \end{aligned}$$

5. Let L be the linear differential operator defined by $L[y] = y''' - y'$. Then

$$L[c_1 + c_2e^t + c_3e^{-t}] = (c_2e^t - c_3e^{-t}) - (c_2e^t - c_3e^{-t}) = 0.$$

6.

$$(\mathbf{A}|\mathbf{b}) = \left(\begin{array}{ccc|c} 1 & 1 & 1 & y_0 \\ 0 & 1 & -1 & v_0 \\ 0 & 1 & 1 & w_0 \end{array}\right) \cong \left(\begin{array}{ccc|c} 1 & 0 & 2 & y_0 - v_0 \\ 0 & 1 & -1 & v_0 \\ 0 & 0 & 2 & w_0 - v_0 \end{array}\right)$$

$$\cong \left(\begin{array}{ccc|c} 1 & 0 & 2 & y_0 - v_0 \\ 0 & 1 & -1 & v_0 \\ 0 & 0 & 1 & \frac{w_0 - v_0}{2} \end{array}\right) \cong \left(\begin{array}{ccc|c} 1 & 0 & 0 & y_0 - w_0 \\ 0 & 1 & 0 & \frac{w_0 + v_0}{2} \\ 0 & 0 & 1 & \frac{w_0 - v_0}{2} \end{array}\right) = (\mathbf{I}|\mathbf{c}).$$

7. Substituting the formulas

$$y = c_1t + c_2e^t, \qquad y' = c_1 + c_2e^t$$

into the initial conditions yields the algebraic equations

$$c_2 = y_0, \qquad c_1 + c_2 = v_0;$$

hence

$$c_2 = y_0, \qquad c_1 = v_0 - y_0.$$

These calculations hold for any constants y_0 and v_0.

8.

$$W = \begin{vmatrix} t & e^t \\ 1 & e^t \end{vmatrix} = (t-1)e^t.$$

Then

$$W' = e^t + (t-1)e^t = te^t = \frac{t}{t-1}W = -p_1 W.$$

3.4 Homogeneous Equations with Constant Coefficients

The general plan for solving homogeneous linear equations is outlined in Algorithm 3.3.1. The key step is to find n solutions with nonzero Wronskian, where n is the order of the differential equation. In Section 3.1, we found the required solutions for the equation $y'' + 9y = 0$ by inspection, but most differential equations do not have obvious solutions. However, there is a formal procedure similar to the method of inspection that allows us to solve any homogeneous linear equation with constant coefficients.

MODEL PROBLEM 3.4

Solve the family of initial-value problems

$$y'' + 4y' + 3y = 0, \qquad y(0) = 2, \qquad y'(0) = v_0$$

and interpret the results in the context of a spring-mass problem.

The differential equation says that we are looking for functions with the property that a particular linear combination of the function and its first two derivatives is zero. This is a remarkable property for a function to have, as it means that the function and its derivatives are not a linearly independent set. The solution could not be the polynomial t^3, for example, because $\{t^3, 3t^2, 6t\}$ is linearly independent. Clearly, the solution cannot be a polynomial. What kind of function could it be?

Exponential functions have the right property because the derivative of an exponential function is a constant multiple of the function. This suggests that homogeneous linear differential equations with constant coefficients might have solutions of the form

$$y(t) = e^{rt}$$

for some specific value(s) of r.

Differentiating $y = e^{rt}$ yields the formulas $y' = re^{rt}$ and $y'' = r^2e^{rt}$. With these substitutions, the differential equation becomes

$$(r^2 + 4r + 3)e^{rt} = 0.$$

Since the factor e^{rt} is never 0, the differential equation reduces to the polynomial equation

$$r^2 + 4r + 3 = 0.$$

This equation is called the **characteristic equation,** and the polynomial on its left side is called the **characteristic polynomial.** We can factor the characteristic polynomial for this problem and obtain two roots,

$$r_1 = -1, \qquad r_2 = -3.$$

Roots of the characteristic polynomial are called **characteristic values** of the operator L. Each characteristic value corresponds to an exponential solution, so we have found two solutions:

$$y_1 = e^{-3t}, \qquad y_2 = e^{-t}.$$

By the principle of superposition, these two solutions generate a two-parameter family of solutions,

$$y = c_1 e^{-3t} + c_2 e^{-t}.$$

Following Algorithm 3.3.1, it remains to check the Wronskian to be sure that this family is the general solution. It is sufficient to check the Wronskian at $t = 0$.

$$W[e^{-3t}, e^{-t}](0) = \begin{vmatrix} e^{-3t} & e^{-t} \\ -3e^{-3t} & -e^{-t} \end{vmatrix}_{t=0} = \begin{vmatrix} 1 & 1 \\ -3 & -1 \end{vmatrix} = 2 \neq 0.$$

Hence,

$$y = c_1 e^{-3t} + c_2 e^{-t}$$

is the general solution for the differential equation.

EXAMPLE 1

Consider the differential equation

$$y'' + y' - 6y = 0.$$

The assumption $y = e^{rt}$ leads to the characteristic equation

$$r^2 + r - 6 = 0,$$

which we can factor to get

$$(r - 2)(r + 3) = 0.$$

Thus, there are two roots, $r = -3$ and $r = 2$, and two corresponding solutions, $y_1 = e^{-3t}$ and $y_2 = e^{2t}$. We have

$$W[e^{-3t}, e^{2t}](0) = \begin{vmatrix} e^{-3t} & e^{2t} \\ -3e^{-3t} & 2e^{2t} \end{vmatrix}_{t=0} = 5 \neq 0.$$

By Algorithm 3.3.1,

$$y = c_1 e^{-3t} + c_2 e^{2t}$$

is the general solution of the differential equation.

✦ **INSTANT EXERCISE 1**

Find the general solution for $y'' - y = 0$ by looking for solutions of the form $y = e^{rt}$ and then applying Algorithm 3.3.1.

Overdamped Linear Oscillators

We now turn to the family of initial-value problems

$$y'' + 4y' + 3y = 0, \qquad y(0) = 2, \qquad y'(0) = v_0.$$

In the context of the spring-mass model of Section 3.1, we can interpret this problem as prescribing the motion of a particular damped spring-mass system with the mass released from a point 2 units below the equilibrium position and with varying initial velocity. Our goal is to determine the motion and how it depends on the initial velocity.

We've already seen that the general solution of the differential equation is

$$y = c_1 e^{-3t} + c_2 e^{-t}.$$

Hence,

$$y' = -3c_1 e^{-3t} - c_2 e^{-t}.$$

Substituting these formulas into the initial conditions yields a pair of algebraic equations for c_1 and c_2:

$$c_1 + c_2 = 2, \qquad -3c_1 - c_2 = v_0.$$

Thus, we have the solution

$$y = \left(3 + \frac{v_0}{2}\right) e^{-t} - \left(1 + \frac{v_0}{2}\right) e^{-3t}. \tag{1}$$

✦ **INSTANT EXERCISE 2**

Solve the algebraic system of equations for c_1 and c_2 to obtain Equation (1).

The graphs of several of these solutions appear in Figure 3.4.1.

As t increases, the second term in the solution formula decays to zero much faster than the first term, unless the first term is absent ($v_0 = -6$). Thus, the solution is eventually given approximately by

$$y \approx \left(3 + \frac{v_0}{2}\right) e^{-t}.$$

If $v_0 < -6$, this approximate solution is negative; in this case, the solution begins at a positive value, reaches zero at some finite time, and then remains negative for all time, approaching zero in the limit $t \to \infty$. If $v_0 \geq -6$, the solution is positive for all time as it decays to zero. When $v_0 = -6$, the solution is $y = 2e^{-3t}$, and the corresponding rate of decay is much faster.

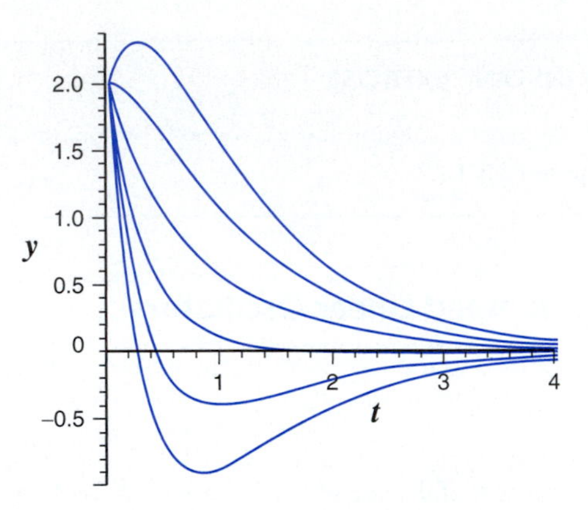

Figure 3.4.1
The solutions of $y'' + 4y' + 3y = 0$, $y(0) = 2$, $y'(0) = v_0$, with $v_0 = 3, 0, -3, -6, -9, -12$.

The General Case

Consider now the general homogeneous linear differential equation of second order

$$ay'' + by' + cy = 0,$$

where $a > 0$, b, and c are given constants. Our method for a constant-coefficient linear equation is to look for solutions of the form

$$y = e^{rt}.$$

Here, this results in the characteristic equation

$$ar^2 + br + c = 0.$$

The quadratic formula yields the roots

$$r = \frac{-b \pm \sqrt{b^2 - 4ac}}{2a}.$$

At present, we are only able to obtain the general solution if we can find two distinct real zeros, and this is true if and only if $b^2 > 4ac$.

If the differential equation represents a linear oscillator, then the coefficients are further restricted by $b \geq 0$ and $c > 0$. The system has two distinct real zeros whenever the damping coefficient b is larger than the critical value $2\sqrt{ac}$. Moreover, the two zeros must be negative because $\sqrt{b^2 - 4ac} < b$. A linear oscillator with distinct real characteristic values is said to be **overdamped.**

In spite of the name *oscillator,* the solutions of an overdamped linear oscillator do not oscillate. The solutions of an overdamped linear oscillator are decaying exponentials, as shown in Figure 3.4.1. This is not the behavior that comes to mind when we think of a spring-mass system because under normal circumstances there is not a lot of damping. However, if we imagine placing our spring-mass system in a vat of maple syrup, we can get a mental picture of the overdamped oscillator. When we pull the mass down from its rest position and let it go, it will slowly drift back toward its rest position.

Another example of an overdamped oscillator from our everyday experience is the shock absorber of a car. When we step on the brake, the chassis stops very quickly. If the body of the car were bolted onto the chassis, then the body would stop just as quickly, and this would be quite annoying to the occupants of the car. Instead, the body and chassis are connected by a shock absorber, which we can think of as a spring. The car body serves as the mass, and its inertia carries it forward beyond the equilibrium position. The motion of the body is then governed (approximately) by an unforced damped linear oscillator equation. A spring-type shock absorber is designed to be overdamped. The result is that the occupants of the car feel a gradual approach to equilibrium, without oscillation.

Homogeneous Linear Equations of Higher Order

Consider a general nth-order homogeneous linear differential equation with constant coefficients. This type of equation has the form

$$L[y] = p_0\frac{d^n y}{dt^n} + p_1\frac{d^{n-1}y}{dt^{n-1}} + p_2\frac{d^{n-2}y}{dt^{n-2}} + \cdots + p_{n-1}\frac{dy}{dt} + p_n y = 0, \tag{2}$$

where $p_0 > 0$.[11] As in the second-order case, we can look for solutions of the form e^{rt}. When we substitute $y = e^{rt}$ into Equation (2), we obtain the characteristic equation

$$P(r) = p_0 r^n + p_1 r^{n-1} + p_2 r^{n-2} + \cdots + p_{n-1} r + p_n = 0. \tag{3}$$

The following definition and theorem summarize what we have learned.

The polynomial P obtained from a linear differential operator $L[y]$ by replacing y by 1, y' by r, and $d^n y/dt^n$ by r^n, in general, is called the **characteristic polynomial** for the differential operator. The equation $P(r) = 0$ is called the **characteristic equation** for the differential equation $L[y] = 0$, and its solutions are called the **characteristic values** of L.

Theorem 3.4.1 Let $L[y] = 0$ be a linear differential equation with constant coefficients. The assumption

$$y = e^{rt}$$

leads to the characteristic equation $P(r) = 0$. Each real characteristic value corresponds to a solution $y = e^{rt}$.

By Algorithm 3.3.1, we need only find a set of n solutions of the differential equation such that the span of the solutions can satisfy all initial conditions. Since an nth-order linear differential equation gives rise to an nth-degree characteristic polynomial, there is a possibility of finding n

[11]There is no loss of generality in assuming that the first coefficient is positive. The quantity p_0 is defined to be the coefficient of the highest-order term in the equation, so it must be nonzero. If it is negative, then the equation can be multiplied by -1 without changing the set of solutions.

distinct characteristic values, each of which corresponds to a solution of the form e^{rt}. If we do find n such solutions, then we don't need to check the Wronskian.

Theorem 3.4.2  Sets of functions of the form e^{rt}, with distinct values of r, are linearly independent.

✦ INSTANT EXERCISE 3

Solve the differential equation $y''' - 4y' = 0$.

Recapitulation

Theorem 3.4.1 reduces the problem of solving a homogeneous linear equation with constant coefficients to the problem of finding roots of a polynomial equation. Theorems 3.4.1 and 3.4.2 together yield a general solution whenever the nth-degree characteristic polynomial has n distinct real zeros. The catch is that nth-degree polynomials do not always have n distinct real zeros. The fundamental theorem of algebra asserts that all polynomials can be reduced to a product of linear factors. Each of these corresponds to a zero of the polynomial; however, the zeros could be complex. It is also possible that not all the linear factors are distinct. Each of these possibilities poses difficulties that must be overcome before we can claim to have a general method for homogeneous linear equations with constant coefficients. In Section 3.5, we see how to manage with complex characteristic values by finding a pair of real-valued solutions corresponding to each pair of complex characteristic values. In Section 3.6, we find a way to form a linearly independent set of m solutions for any characteristic value whose factor appears m times in the characteristic polynomial, and we establish the linear independence of solutions obtained in this manner. Ultimately, we will have a linearly independent set of n solutions for any homogeneous linear differential equation of order n with constant coefficients.

3.4 Exercises

In Exercises 1 through 6, solve the given differential equation.

1. $y'' + 4y' + 3y = 0$
2. $y'' - 5y' + 6y = 0$
3. $y'' - 4y = 0$
4. $4y'' + 5y' + y = 0$
5. $y''' + 5y'' + 4y' = 0$
6. $y''' - 2y'' + y' - 2y = 0$

In Exercises 7 through 10, solve the given initial-value problem, sketch the solution, and determine the long-time approximation.

7. $y'' + 5y' + 6y = 0, \quad y(0) = 2, \quad y'(0) = 0$
8. $2y'' + 3y' + y = 0, \quad y(0) = 0, \quad y'(0) = -2$
9. $y'' + 2y' - 8y = 0, \quad y(0) = 4, \quad y'(0) = 0$
10. $y'' - 4y' + 3y = 0, \quad y(0) = 2, \quad y'(0) = 3$

11. Consider a spring-mass system with spring constant 4, mass 1, and damping coefficient b.
   a. Determine the critical amount of damping.
   b. Suppose the damping coefficient is twice that needed for critical damping. Determine the motion $y(t)$ if the mass is pulled down 1 unit and then released.
   c. Suppose the mass in part b is pulled down 1 unit and released with an upward velocity of $-s$ (note $s > 0$). Determine how large s must be so that the mass reaches its equilibrium position in finite time.

12. Consider a spring-mass system with spring constant 2, mass m, and damping coefficient 4.
   a. Determine the mass that makes the system critically damped.
   b. Suppose the mass is twice that needed for critical damping. Determine the motion $y(t)$ if the mass is pulled down 1 unit and then released.
   c. Suppose the mass in part b is pulled down 1 unit and released with an upward velocity of $-s$ (note $s > 0$). Determine how large s must be so that the mass reaches its equilibrium position in finite time.

13. Show by calculation that Theorem 3.4.2 holds for sets of two exponential functions.

14. Show that Theorem 3.4.2 holds for sets of three exponential functions.
   a. Use the fact that a common factor can be removed from any row or column of a determinant to show that $W[e^{r_1 t}, e^{r_2 t}, e^{r_3 t}]$ is the product of an exponential function and a determinant of constants for which all entries in the top row are 1.
   b. Show that the determinant of constants in part a evaluates to $(r_2 - r_1)(r_3 - r_2)(r_3 - r_1) \neq 0$.

✦ 3.4 INSTANT EXERCISE SOLUTIONS

1. The assumption $y = e^{rt}$ leads to the characteristic equation

$$r^2 - 1 = 0,$$

which we can factor to get

$$(r + 1)(r - 1) = 0.$$

Thus, there are two roots, $r = -1$ and $r = 1$, and two corresponding solutions $y_1 = e^{-t}$ and $y_2 = e^t$. Then

$$W[e^{-t}, e^t](0) = \left| \begin{array}{cc} e^{-t} & e^t \\ -e^{-t} & e^t \end{array} \right|_{t=0} = \left| \begin{array}{cc} 1 & 1 \\ -1 & 1 \end{array} \right| = 2 \neq 0.$$

By Algorithm 3.3.1,

$$y = c_1 e^{-t} + c_2 e^t$$

is the general solution.

2. Adding the two equations, we obtain $-2c_1 = 2 + v_0$. Adding 3 times the first equation to the second yields $2c_2 = 6 + v_0$. Thus, $c_1 = -1 - v_0/2$ and $c_2 = 3 + v_0/2$.

3. By Theorem 3.4.1, the characteristic equation is $r^3 - 4r = 0$, or $r(r-2)(r+2) = 0$. The characteristic values 0, 2, and -2 correspond to solutions e^{rt}. By Theorem 3.4.2, the general solution is

$$y = c_1 + c_2 e^{-2t} + c_3 e^{2t}.$$

(Note that the three terms can be in any order—it does not matter which constant goes with which solution.)

3.5 Real Solutions from Complex Characteristic Values

MODEL PROBLEM 3.5

Solve the initial-value problem

$$y'' + 2y' + 5y = 0, \qquad y(0) = 1, \qquad y'(0) = 0$$

and interpret the results in the context of a spring-mass problem.

The differential equation of Model Problem 3.5 is homogeneous and linear, with constant coefficients. By Theorem 3.4.1, the substitution $y = e^{rt}$ yields the characteristic equation

$$r^2 + 2r + 5 = 0.$$

However, unlike the examples we solved in Section 3.4, this characteristic equation has no real roots. The quadratic formula provides two complex roots:

$$r = \frac{-2 \pm \sqrt{4 - 20}}{2} = -1 \pm 2i.$$

It is tempting to write the solution as

$$y(t) = k_1 e^{(-1+2i)t} + k_2 e^{(-1-2i)t} = e^{-t}\left(k_1 e^{2it} + k_2 e^{-2it}\right),$$

but we don't know whether exponential functions of complex numbers have any meaningful interpretation. This is a question that needs an answer before we can claim to have solved the differential equation.

The Complex Exponential

Our task is to give a meaningful definition of the function e^{it}, after which we can compute e^{2it} by substitution of $2t$ for t.

Consider the initial-value problem

$$y'' + y = 0, \qquad y(0) = 1, \qquad y'(0) = 0.$$

The substitution $y = e^{rt}$ yields the characteristic equation $r^2 + 1 = 0$, which has roots $r = \pm i$. The general solution would thus seem to be

$$y(t) = c_1 e^{it} + c_2 e^{-it}.$$

Using this general solution along with the given initial conditions, we see that $c_1 + c_2 = 1$ and $ic_1 - ic_2 = 0$. Thus, both constants are $\frac{1}{2}$, and we have

$$y = \frac{e^{it} + e^{-it}}{2}.$$

Note that the function $y = \cos t$ also solves the initial-value problem. We know from Theorem 2.4.2 that the initial-value problem we are examining has a unique solution. So if the complex exponential function is to make any sense at all, then the two solution formulas must be different formulas for the same function. Thus, the complex exponential e^{it} must satisfy

$$\cos t = \frac{e^{it} + e^{-it}}{2}. \tag{1}$$

Similar consideration of the initial-value problem

$$y'' + y = 0, \qquad y(0) = 0, \qquad y'(0) = 1$$

leads to the requirement

$$\sin t = \frac{e^{it} - e^{-it}}{2i}. \tag{2}$$

✦ INSTANT EXERCISE 1

Derive the solution $(e^{it} + e^{-it})/2i$ for the problem $y'' + y = 0$, $y(0) = 0$, $y'(0) = 1$ and thereby obtain Equation (2).

We can think of these two identities as algebraic equations to be solved for the functions e^{it} and e^{-it}. Adding i times the second identity to the first identity gives us a formula for e^{it}. The result, called **Euler's formula,** is

$$e^{it} = \cos t + i \sin t.$$

Euler's formula serves as a definition of the complex exponential function. Note that we can also subtract i times the second identity from the first to get

$$e^{-it} = \cos t - i \sin t.$$

We can get a very general formula by replacing t by bt and writing the two formulas together:

$$e^{\pm ibt} = \cos bt \pm i \sin bt. \tag{3}$$

Finally, we can extend the result to any complex number, since

$$e^{a+ib} = e^a e^{ib}.$$

The general result is summarized in Theorem 3.5.1.

Theorem 3.5.1 The exponential function of complex numbers $a \pm ib$ is given by

$$e^{a \pm ib} = e^a(\cos b \pm i \sin b).$$

Real-Valued Solutions

We now have the necessary background to complete the solution of

$$y'' + 2y' + 5y = 0.$$

The characteristic values are $-1 \pm 2i$, so we get

$$y = k_1 e^{(-1+2i)t} + k_2 e^{(-1-2i)t}.$$

Application of Theorem 3.5.1 gives us

$$y = e^{-t}[k_1(\cos 2t + i \sin 2t) + k_2(\cos 2t - i \sin 2t)],$$

or

$$y(t) = e^{-t}[(k_1 + k_2)\cos 2t + i(k_1 - k_2)\sin 2t].$$

Since the constants k_1 and k_2 do not have known values, we have no compelling reason to retain them as they were originally defined. Instead, we can define new constants $c_1 = k_1 + k_2$ and $c_2 = i(k_1 - k_2)$. The solution then appears as

$$y(t) = e^{-t}(c_1 \cos 2t + c_2 \sin 2t).$$

By alternately choosing one constant to be 1 and the other 0, we have two real-valued solutions:

$$y_1(t) = e^{-t}\cos 2t, \qquad y_2(t) = e^{-t}\sin 2t.$$

✦ INSTANT EXERCISE 2

Verify that $y = e^{-t}\cos 2t$ is a solution of $y'' + 2y' + 5y = 0$.

By Algorithm 3.3.1, it remains only to check the Wronskian of the solutions at a point.

$$W[e^{-t}\cos 2t, e^{-t}\sin 2t](0) = \left| \begin{matrix} e^{-t}\cos 2t & e^{-t}\sin 2t \\ e^{-t}(-\cos 2t - 2\sin 2t) & e^{-t}(2\cos 2t - \sin 2t) \end{matrix} \right|_{t=0}$$

$$= \begin{vmatrix} 1 & 0 \\ -1 & 2 \end{vmatrix} = 2 \neq 0.$$

We have found a real-valued general solution formula. With real-valued initial conditions, we will always be able to find real constants c_1 and c_2. For Model Problem 3.5, we have

$$y = e^{-t}(c_1 \cos 2t + c_2 \sin 2t)$$

and

$$y' = e^{-t}[(-2c_1 \sin 2t + 2c_2 \cos 2t) - (c_1 \cos 2t + c_2 \sin 2t)].$$

The initial conditions $y(0) = 1$ and $y'(0) = 0$ reduce to the equations

$$c_1 = 1, \qquad 2c_2 - c_1 = 0,$$

with the results $c_1 = 1$ and $c_2 = \frac{1}{2}$. The unique solution of Model Problem 3.5 is

$$y = e^{-t}(\cos 2t + \tfrac{1}{2}\sin 2t).$$

This solution could have been obtained from the initial conditions and the general solution formula

$$y = k_1 e^{(-1+2i)t} + k_2 e^{(-1-2i)t};$$

however, the constants k_1 and k_2 are complex-valued. If we apply the initial conditions to this general solution formula, we get a result that appears to be complex; algebraic simplification, including an application of Euler's formula, eventually converts the solution to the real-valued form $e^{-t}(\cos 2t + \frac{1}{2}\sin 2t)$. The families $y = k_1 e^{(-1+2i)t} + k_2 e^{(-1-2i)t}$ and $y = e^{-t}(c_1 \cos 2t + c_2 \sin 2t)$ are two different representations of the same general solution; both yield the same real-valued solution when initial conditions are given.

✦ INSTANT EXERCISE 3

Use the characteristic value method, Theorem 3.5.1, and Algorithm 3.3.1 to find a real-valued general solution for

$$y'' + 9y = 0,$$

which we solved by inspection in Section 3.1.

The following theorem saves us the trouble of having to apply the complex exponential formula and check the Wronskian every time we get a problem with complex characteristic values.

Theorem 3.5.2

Nonreal roots of a real polynomial come in pairs of the form $\alpha \pm \beta i$, where α is any real number and without loss of generality $\beta > 0$. Corresponding to the nonreal characteristic value pair $\alpha \pm \beta i$ are two real-valued solutions: $e^{\alpha t}\cos \beta t$ and $e^{\alpha t}\sin \beta t$. Furthermore, a set of distinct functions of the forms $e^{\alpha t}\cos \beta t$, $e^{\alpha t}\sin \beta t$, and e^{rt} is always linearly independent.

EXAMPLE 1

Consider the differential equation

$$y''' - 5y'' + 4y' + 10y = 0.$$

The substitution $y = e^{rt}$ leads to the characteristic equation

$$r^3 - 5r^2 + 4r + 10 = 0.$$

Like the quadratic formula, there is a general formula for the roots of a cubic polynomial; however, the formula is unwieldy. Instead, it is just as well to look for any integer or rational roots and to use a numerical approximation if none are found. A graph of the polynomial (see Figure 3.5.1) suggests that there is a real root $r = -1$, and a quick check by substituting $r = -1$ into the characteristic equation confirms the result. A root of $r = -1$ means that the polynomial has a factor $r + 1$, and we can determine the remaining quadratic factor by setting up a simple algebra problem.[12] Any quadratic factor must be of the form $ar^2 + br + c$ for some correct choices of a, b, and c, so we assume a factorization

$$r^3 - 5r^2 + 4r + 10 = (r + 1)(ar^2 + br + c).$$

[12]The traditional way to approach the problem of finding the quadratic factor is to apply the long division algorithm to divide $r^3 - 5r^2 + 4r + 10$ by $r - 1$. The method presented here is an alternative that requires only a simple multiplication followed by a simple algebra problem.

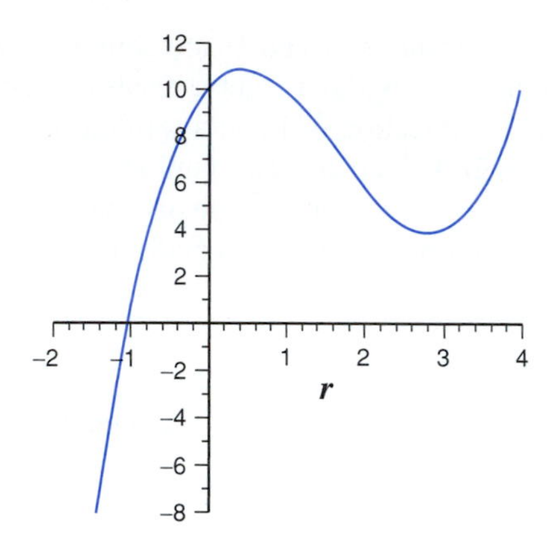

Figure 3.5.1
The characteristic polynomial $r^3 - 5r^2 + 4r + 10$ from Example 1.

Before we multiply out the right-hand side in full, notice that the r^3 term will be ar^3 and the constant term will be c. Thus, we have immediately that $a = 1$ and $c = 10$. We may therefore assume the factorization to be

$$r^3 - 5r^2 + 4r + 10 = (r + 1)(r^2 + br + 10).$$

Multiplying out the right-hand side gives

$$r^3 - 5r^2 + 4r + 10 = r^3 + (b + 1)r^2 + (b + 10)r + 10.$$

Comparing coefficients, we have the equations $b + 1 = -5$ and $b + 10 = 4$, either of which yields the result $b = -6$. The quadratic factor is then $r^2 - 6r + 10$, which is irreducible. By the quadratic formula, we obtain the complex pair $r = 3 \pm i$. We therefore have one real root and one complex pair:

$$r_1 = -1, \qquad r_2, r_3 = 3 \pm i.$$

The real root gives us a solution $y_1 = e^{-t}$, and the complex pair gives us a pair of solutions $y_2 = e^{3t} \cos t$ and $y_3 = e^{3t} \sin t$. By Theorem 3.5.2, the general solution of the differential equation is

$$y(t) = c_1 e^{-t} + e^{3t}(c_2 \cos t + c_3 \sin t).$$

Underdamped Linear Oscillators

Model Problem 3.5 can be interpreted as prescribing the motion of a particular spring-mass system with the mass released from a point 1 unit below the equilibrium position and with no initial velocity. The solution appears in Figure 3.5.2.

The solution oscillates and has zeros at regular intervals determined by the period of the trigonometric factors. Given nonzero damping, the solution is not actually periodic, but rather decays to zero, as illustrated by Figure 3.5.2. A linear oscillator with complex characteristic values is **underdamped.**

The Envelope of the Solution In Section 3.1, we saw that a sum of cosine and sine functions with the same circular frequency can be combined into a single function with a phase shift. The same computation is useful for the solutions of underdamped oscillators. As in the undamped

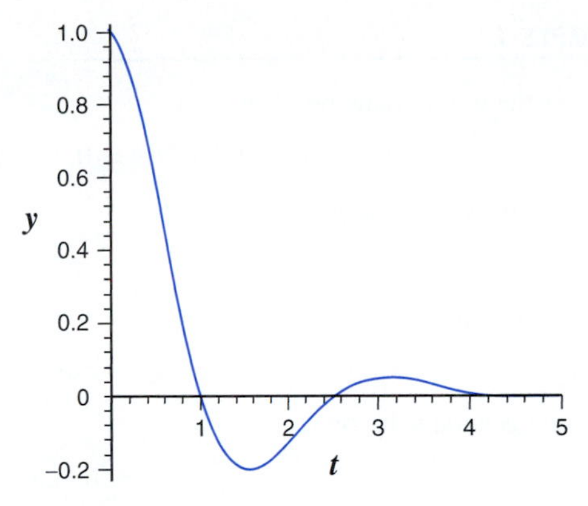

Figure 3.5.2
The solution of $y'' + 2y' + 5y = 0$, $y(0) = 1$, $y'(0) = 0$.

case, the solution of Model Problem 3.5 can be written in the form

$$y = Ae^{-t}\cos(2t - \delta),$$

where the amplitude and phase shift are determined by [Equations (7) and (8) from Section 3.1]

$$A^2 = 1^2 + \tfrac{1}{2}^2 = \tfrac{5}{4}, \qquad \cos\delta = \frac{1}{A} = \frac{2}{\sqrt{5}}, \qquad \sin\delta = \frac{\frac{1}{2}}{A} = \frac{1}{\sqrt{5}}.$$

Thus, $A = \sqrt{5}/2$ and $\delta = \arcsin(1/\sqrt{5}) \approx 0.464$. The functions

$$y_e = \pm Ae^{-t}$$

are called the **envelope** of the solution because $|y| \leq |y_e|$ with y_e tangent to y at a point near the top and bottom of each oscillation.[13] The solution and its envelope are shown together in Figure 3.5.3.

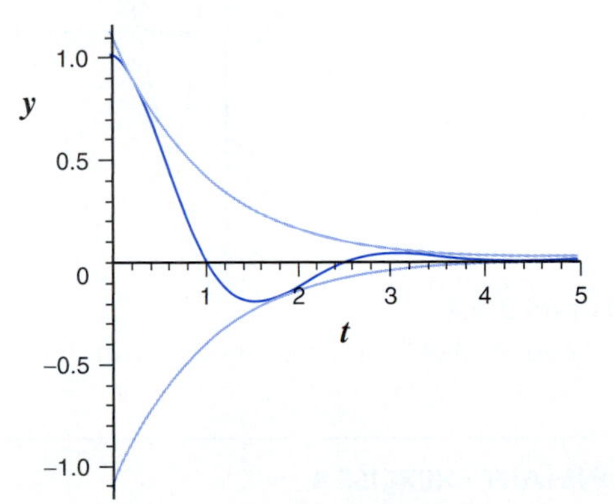

Figure 3.5.3
The solution of $y'' + 2y' + 5y = 0$, $y(0) = 1$, $y'(0) = 0$ and its envelope.

Decay Rate and Quasi-Period Solutions of underdamped oscillators show varying amounts of oscillation.

[13] The graph of the function just "fits inside" the envelope.

EXAMPLE 2

Consider the initial-value problem

$$y'' + 2y' + 26y = 0, \qquad y(0) = 1, \qquad y'(0) = 0.$$

The characteristic equation

$$r^2 + 2r + 26 = 0$$

has characteristic values

$$r = -1 \pm 5i;$$

hence, the general solution is

$$y = e^{-t}(c_1 \cos 5t + c_2 \sin 5t).$$

The initial condition for y gives us $c_1 = 1$, and then the initial condition for y' yields $c_2 = \frac{1}{5}$. The solution is therefore

$$y = e^{-t}(\cos 5t + \tfrac{1}{5} \sin 5t),$$

which can be written in amplitude–phase shift form as

$$y = \frac{\sqrt{26}}{5} e^{-t} \cos(5t - \delta), \qquad \delta = \arcsin \frac{1}{\sqrt{26}} \approx 0.197.$$

The solution and its envelope are shown in Figure 3.5.4.

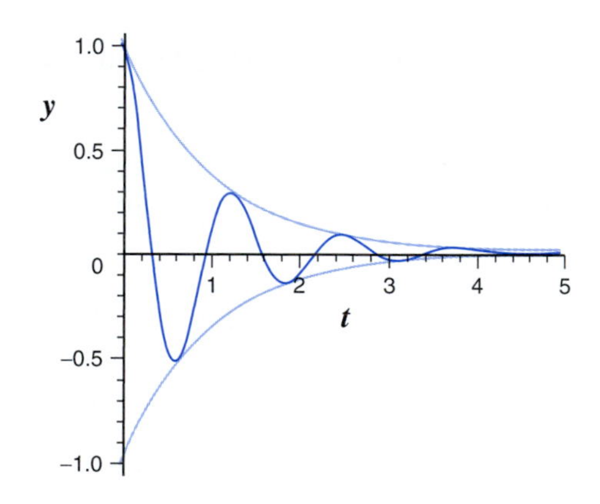

Figure 3.5.4
The solution of $y'' + 2y' + 26y = 0$, $y(0) = 1$, $y'(0) = 0$ and its envelope.

✦ INSTANT EXERCISE 4

Derive the results for the amplitude and phase shift in Example 2.

The differences between the qualitative features of the solutions of Model Problem 3.5 and Example 2 stem from the different characteristic values associated with the differential equations. A characteristic value pair $-\alpha \pm i\beta$ corresponds to a solution of the form $Ae^{-\alpha t} \cos(\beta t - \delta)$. The period of the trigonometric function is $2\pi/\beta$. The solution is not periodic, but it does have zeros that are periodic. The motion is said to be **quasi-periodic,** and the quantity $2\pi/\beta$ is called the

quasi-period. The quantity α, which has units of inverse time, is conveniently thought of as the rate at which the envelope decays. When $t = 1/\alpha$, the envelope has the value $Ae^{-1} \approx 0.37A$, so $1/\alpha$ is the amount of time needed for the envelope to decrease to about 37% of its original value. One can show (see the exercises) that $4.6/\alpha$ is approximately the time required for the envelope to be decreased to 1% of its original value. Both Model Problem 3.5 and Example 2 have $\alpha = 1$, so the envelopes exhibit the same decay rate and the time interval $0 \leq t \leq 5$ is sufficient to show the envelope decay to just less than 1%. The ratio β/α is a measure of the degree of oscillation visible in the graph. The higher degree of oscillation in Figure 3.5.4 compared with Figure 3.5.3 indicates the difference between β/α ratios of 5 and 2.

The General Case

The search for solutions of the form

$$y = e^{rt}$$

for a linear oscillator model

$$ay'' + by' + cy = 0$$

leads to the result

$$r = \frac{-b \pm \sqrt{b^2 - 4ac}}{2a}.$$

The oscillator is underdamped if $b^2 < 4ac$. The quantity inside the square root is negative, so the characteristic values are complex. Specifically, we can write the characteristic values as

$$r = -\frac{b}{2a} \pm i\frac{\sqrt{4ac - b^2}}{2a}.$$

The solution for an underdamped oscillator is the product of a decaying exponential function and a sinusoidal function.

3.5 Exercises

In Exercises 1 through 6, solve the given differential equation.

1. $y'' + 5y = 0$
2. $y'' - 6y' + 13y = 0$
3. $y'' - 4y' + 8y = 0$
4. $2y'' + 6y' + 5y = 0$
5. $y''' - y' = 0$
6. $y''' + y' = 0$

In Exercises 7 through 10, solve the given initial-value problem, sketch the solution, determine the envelope, and sketch the solution with the envelope.

7. $y'' + 4y' + 5y = 0, \quad y(0) = 2, \quad y'(0) = 0$
8. $y'' - 2y' + 10y = 0, \quad y(0) = 0, \quad y'(0) = -2$
9. $y'' + 2y' + 6y = 0, \quad y(0) = 0, \quad y'(0) = 2$
10. $y'' + 4y' + 10y = 0, \quad y(0) = 2, \quad y'(0) = 0$
11. Determine $W[e^{\alpha t} \cos \beta t, e^{\alpha t} \sin \beta t]$.

12. Determine $W[e^{rt}, e^{\alpha t}\cos\beta t, e^{\alpha t}\sin\beta t]$.

T 13. Consider a spring-mass system with spring constant 4, mass 1, and damping coefficient b. Suppose the damping coefficient is one-half of that needed for critical damping. Determine the motion $y(t)$ if the mass is pulled down 1 unit and then released. Sketch the solution and its envelope.

T 14. Consider a spring-mass system with spring constant 2, mass m, and damping coefficient 4. Suppose the mass is one-half of that needed for critical damping. Determine the motion $y(t)$ if the mass is pulled down 1 unit and then released. Sketch the solution and its envelope.

15. The most common derivation of Euler's formula uses Taylor series. Write down the Taylor series for e^x centered at $x = 0$, obtain the Taylor series for e^{ix} by substitution, simplify using $i^2 = -1$, collect real and imaginary parts, and identify the real and imaginary parts as the Taylor series for cosine and sine functions.

✦ 3.5 INSTANT EXERCISE SOLUTIONS

1. The general solution, $y(t) = c_1 e^{it} + c_2 e^{-it}$, has already been found. Substitution into the initial conditions yields the two equations $c_1 + c_2 = 0$ and $ic_1 - ic_2 = 1$. Thus, $c_1 = 1/(2i)$ and $c_2 = -1/(2i)$, and the solution is $y = (e^{it} - e^{-it})/2i$.

2. Let $y = e^{-t}\cos 2t$. Then $y' = e^{-t}(-2\sin 2t) + (-e^{-t}\cos 2t) = e^{-t}(-\cos 2t - 2\sin 2t)$. And $y'' = e^{-t}(2\sin 2t - 4\cos 2t) + (-e^{-t})(-\cos 2t - 2\sin 2t) = e^{-t}(-3\cos 2t + 4\sin 2t)$. Hence,

$$y'' + 2y' + 5y = e^{-t}[(-3\cos 2t + 4\sin 2t) + 2(-\cos 2t - 2\sin 2t) + 5\cos 2t] = 0.$$

3. The substitution $y = e^{rt}$ yields the characteristic equation

$$r^2 + 9 = 0.$$

Thus, $r^2 = -9$, so the characteristic values are $r = \pm 3i$. By Theorem 3.5.1, we have solutions $\cos 3t$ and $\sin 3t$. Then

$$W[\cos 3t, \sin 3t] = \begin{vmatrix} \cos 3t & \sin 3t \\ -3\sin 3t & 3\cos 3t \end{vmatrix} = 3\cos^2 3t + 3\sin^2 3t = 3.$$

The general solution is

$$y = c_1\cos 3t + c_2\sin 3t.$$

3.6 Multiple Solutions for Repeated Characteristic Values

MODEL PROBLEM 3.6

Solve the differential equation

$$y'' + 2y' + y = 0$$

and interpret the results in the context of a spring-mass problem.

The substitution $y = e^{rt}$ into the differential equations yields the characteristic equation

$$0 = r^2 + 2r + 1 = (r + 1)^2.$$

Thus, we have a single characteristic value $r = -1$ that corresponds to two factors of $r + 1$ in the characteristic polynomial. We have one solution $y_1(t) = e^{-t}$, but what are we to choose for a second solution?

A first thought would be to simply choose a multiple of the first solution. But a set of solutions where one is a constant multiple of another is not a linearly independent set.

A Second Solution by Reduction of Order

There is a method called **reduction of order** that can be used to find the second solution of an equation $L[y] = 0$ when one solution y_1 is known. The idea is to look for a solution of the form $y_2 = v(t)y_1(t)$, where v is an unknown function. Theorem 2.4.2 guarantees that $y'' + 2y' + y = 0$ has a general solution, so there must certainly be such a function v. Furthermore, the set $\{y_1, y_2\}$ is linearly independent as long as v is not a constant.

✦ INSTANT EXERCISE 1

Use the Wronskian to show that two nonzero solutions y_1 and vy_1 of a second-order differential equation are linearly independent as long as v is not a constant.

The real question is whether it is any easier to find v than it is to find y_2.

In the current problem, we have

$$y_1 = e^{-t},$$

so we assume

$$y_2 = e^{-t}v(t).$$

Then

$$y_2' = e^{-t}(v' - v), \qquad y_2'' = e^{-t}(v'' - 2v' + v).$$

Substituting these formulas into the differential equation yields

$$0 = y_2'' + 2y_2' + y_2 = e^{-t}(v'' - 2v' + v) + 2[e^{-t}(v' - v)] + e^{-t}v = e^{-t}v''.$$

Thus, v must satisfy $v'' = 0$. The general solution of this equation is $v = c_1 + c_2t$. Any such function v can be used, as long as it is not a constant. The simplest choice is $v = t$, and this yields a second solution

$$y_2 = te^{-t}.$$

The general solution of $y'' + 2y' + y = 0$ is then

$$y(t) = e^{-t}(c_1 + c_2t).$$

✦ INSTANT EXERCISE 2

Verify by direct computation that te^{-t} is a solution of $y'' + 2y' + y = 0$.

Figure 3.6.1 illustrates some of the solutions of Model Problem 3.6.

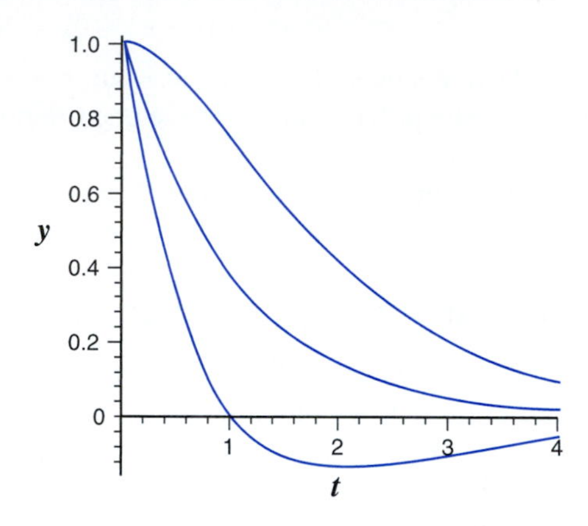

Figure 3.6.1
Some solutions [$y = e^{-t}(1 + c_2 t)$ with $c_2 = 1, 0, -1$] of Model Problem 3.6.

The method of reduction of order is so called because it leads, in general, to a first-order equation for v'.

EXAMPLE 1

Consider the Legendre equation of order 1,

$$(1 - x^2)y'' - 2xy' + 2y = 0.$$

Observe that $y_1(x) = x$ is a solution of this equation. A second solution with an interval of existence of at least $-1 < x < 1$ can now be found by reduction of order. We look for a second solution of the form $y = xv(x)$, where v is a nonconstant function to be determined. Differentiating gives $y' = xv' + v$ and $y'' = xv'' + 2v'$. Substituting into the original equation yields

$$(1 - x^2)(xv'' + 2v') - 2x(xv' + v) + 2xv = 0,$$

which we can simplify to

$$(x - x^3)v'' + (2 - 4x^2)v' = 0.$$

This latter equation is a first-order equation for v'. To solve it, let $u = v'$. Then we have

$$(x - x^3)u' + (2 - 4x^2)u = 0.$$

Separation of variables yields the integral equation

$$\int u^{-1}\, du = -\int \frac{2 - 4x^2}{x - x^3}\, dx.$$

The integral on the right is tricky. Note that it can be rewritten so that the numerator is the derivative of the denominator:

$$\frac{2 - 4x^2}{x - x^3} = \frac{2x - 4x^3}{x^2 - x^4};$$

thus,

$$\int u^{-1}\, du = -\int \frac{2x - 4x^3}{x^2 - x^4}\, dx.$$

Integrating this equation, we have

$$\ln |u| = -\ln |x^2 - x^4| = \ln \frac{1}{|x^2 - x^4|}.$$

Note that we do not need to include an integration constant because we need find only one function u. We may also resolve the absolute value (given that we are looking for a solution with domain $|x| < 1$) to obtain

$$u = \frac{1}{x^2 - x^4}.$$

Now, returning to $u = v'$, we have

$$v' = \frac{1}{x^2 - x^4}.$$

Again we have a nasty integral that can be simplified by algebra, with the final result

$$v = -\frac{1}{x} + \frac{1}{2} \ln \frac{1+x}{1-x}.$$

Note, as before, that we do not need an integration constant because we only need one function v. Finally, $y_2 = xv$, so

$$y_2 = -1 + \frac{x}{2} \ln \frac{1+x}{1-x}.$$

Some solutions of Legendre's equation of order 1 appear in Figure 3.6.2. The solution $y_1 = x$ has an interval of existence that spans the real line, but all other solutions are confined to the interval $-1 < x < 1$ guaranteed by theory. Note that these curves do cross, but that does not contradict Theorem 2.4.2. Unique solutions are guaranteed for each *set* of initial conditions, and sets of initial conditions for a second-order equation include values of y and y' at a given point. Uniqueness guarantees only one solution through a given point with a given slope.

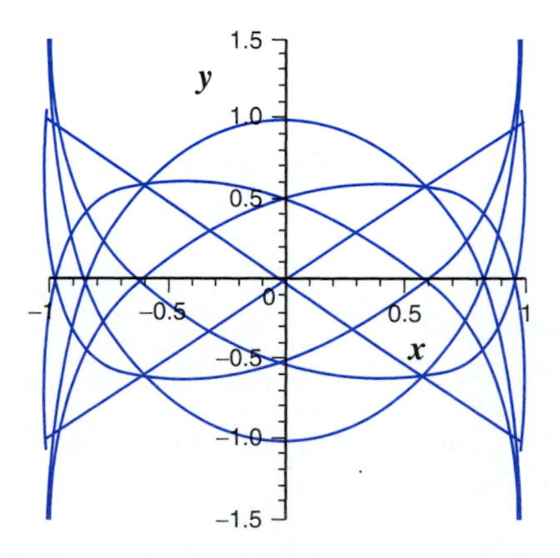

Figure 3.6.2
Some solutions of $(1 - x^2)y'' - 2xy' + 2y = 0$.

✦ INSTANT EXERCISE 3

Obtain the function v in Example 1 by integrating v'.

Critically Damped Linear Oscillators

We have seen in the previous two sections that the character of the motion of a linear oscillator depends on the relative magnitudes of b^2 and $4ac$, with decaying oscillatory behavior if $b^2 < 4ac$ and decaying nonoscillatory behavior otherwise. In the special case $b^2 = 4ac$, we have a single characteristic value $r = -b/(2a)$ of multiplicity 2. The solution is then

$$y = (c_1 + c_2 t)e^{-bt/(2a)}.$$

This behavior is very much like the overdamped case. It should be understood that this case cannot really occur in practice. It is not physically possible to build a system in which b is *exactly* equal to $2\sqrt{ac}$. Even if we can control design parameters to an extremely high degree of accuracy, we cannot completely eliminate all discrepancies between designed values and actual values.

At the same time, the distinction between underdamped and critically damped oscillators is actually less than might be expected. It is true that the former exhibits oscillation while the latter does not, but in practice this difference can be small. An example serves to illustrate the point.

EXAMPLE 2

Consider the family of linear oscillators defined by

$$y'' + by' + y = 0, \qquad y(0) = 0, \qquad y'(0) = 1.$$

The characteristic equation is

$$r^2 + br + 1 = 0;$$

by the quadratic formula, the characteristic values are

$$r = \frac{-b \pm \sqrt{b^2 - 4}}{2}.$$

The critical case is $b = 2$. Then we have a single root, $r = -1$, of multiplicity 2. Hence, the general solution is

$$y = (c_1 + c_2 t)e^{-t}.$$

Differentiating gives

$$y' = (c_2)e^{-t} + (c_1 + c_2 t)(-e^{-t}) = (c_2 - c_1 - c_2 t)e^{-t}.$$

Thus, $y(0) = c_1$ and $y'(0) = c_2 - c_1$. Comparing these results with the initial conditions yields $c_1 = 0$ and $c_2 = 1$. Thus,

$$y = te^{-t} \qquad \text{with} \qquad b = 2.$$

Now suppose instead we take $b = 1.8$, which is 90% of the critical value. The characteristic values are then

$$r = -0.9 \pm \sqrt{0.19}i.$$

Thus, the general solution is

$$y = e^{-0.9t}[c_1 \cos(\sqrt{0.19}\,t) + c_2 \sin(\sqrt{0.19}\,t)].$$

The first initial condition forces $c_1 = 0$, and thus

$$y = c_2 e^{-0.9t} \sin(\sqrt{0.19}\,t).$$

Then

$$y' = c_2 e^{-0.9t}[-0.9 \sin(\sqrt{0.19}\,t) + \sqrt{0.19}\cos(\sqrt{0.19}\,t)].$$

The second initial condition then gives $c_2 = 1/\sqrt{0.19}$, so the solution is

$$y = \frac{1}{\sqrt{0.19}} e^{-0.9t} \sin(\sqrt{0.19}\,t) \qquad \text{with} \qquad b = 1.8.$$

The graphs of these two solution formulas are compared in Figure 3.6.3. Although the solution formulas are very different, the graphs of the functions are quite similar.

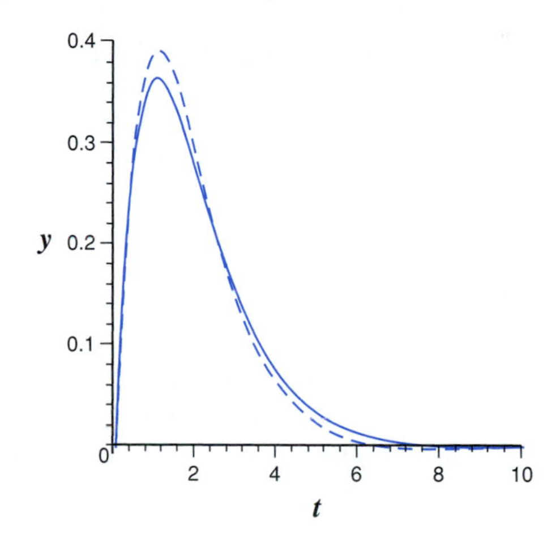

Figure 3.6.3
The solutions of $y'' + by' + y = 0$, $y(0) = 0$, $y'(0) = 1$, with $b = 2$ (solid) and $b = 1.8$ (dashed).

Repeated Characteristic Values in General

EXAMPLE 3

The characteristic polynomial for the differential equation

$$y^{(4)} + y'' = 0$$

is

$$P(r) = r^4 + r^2 = r^2(r^2 + 1).$$

The characteristic value $r = 0$ appears twice in the characteristic polynomial, so we need two solutions to correspond to it. One is $y_1 = e^{0t} = 1$. Based on Model Problem 3.6, perhaps the second solution is

$y_2 = t$. A quick check confirms this guess. There is also a pair of complex characteristic values $r = \pm i$, and from these we have solutions $y_3 = \cos t$ and $y_4 = \sin t$. Thus, the general solution is

$$y = c_1 + c_2 t + c_3 \cos t + c_4 \sin t,$$

provided that the Wronskian of this set of solutions is nonzero. We have

$$W[1, t, \cos t, \sin t](0) = \begin{vmatrix} 1 & t & \cos t & \sin t \\ 0 & 1 & -\sin t & \cos t \\ 0 & 0 & -\cos t & -\sin t \\ 0 & 0 & \sin t & -\cos t \end{vmatrix}_{t=0} = \begin{vmatrix} 1 & 0 & 1 & 0 \\ 0 & 1 & 0 & 1 \\ 0 & 0 & -1 & 0 \\ 0 & 0 & 0 & -1 \end{vmatrix}$$

$$= \begin{vmatrix} 1 & 0 & 1 \\ 0 & -1 & 0 \\ 0 & 0 & -1 \end{vmatrix} = \begin{vmatrix} -1 & 0 \\ 0 & -1 \end{vmatrix} = 1.$$

In Model Problem 3.6, the factor $r + 1$ appeared twice in the characteristic polynomial. We say that $r = -1$ is a characteristic value of multiplicity 2. The **multiplicity** of a root of a polynomial is defined to be the number of factors in the polynomial that correspond to that root. Thus, in Example 3, the root 0 is a characteristic value of multiplicity 2, and the complex pair $\pm i$ are characteristic values of multiplicity 1.

In general, we can find a linearly independent set of m solutions for any real characteristic value of multiplicity m. We state the result here without proof.

Theorem 3.6.1 Let $L[y] = 0$ be a homogeneous linear differential equation of order n with constant coefficients. If r is a real characteristic value of multiplicity m, then the functions $e^{rt}, te^{rt}, \ldots, t^{m-1}e^{rt}$ form a linearly independent set of m solutions.

EXAMPLE 4

Consider the equation

$$y''' - 3y'' + 3y' - y = 0.$$

The characteristic polynomial for this equation is $P(r) = r^3 - 3r^2 + 3r - 1 = (r - 1)^3$. By Theorem 3.6.1, the functions e^t, te^t, and t^2e^t form a linearly independent set of three solutions of the differential equation; thus, the general solution is

$$y = (c_1 + c_2 t + c_3 t^2)e^t.$$

✦ INSTANT EXERCISE 4

Verify by computation that $y_2 = t^2e^t$ is a solution of $y''' - 3y'' + 3y' - y = 0$.

A similar theorem applies to repeated nonreal characteristic value pairs. Note that a pair $\alpha \pm \beta i$ of multiplicity m should give rise to $2m$ solutions.

Theorem 3.6.2 Let $L[y] = 0$ be a homogeneous linear differential equation of order n. If $r = \alpha \pm \beta i$ is a nonreal pair of characteristic values of multiplicity m, then the functions $e^{\alpha t} \cos \beta t$, $e^{\alpha t} \sin \beta t$, $te^{\alpha t} \cos \beta t$, $te^{\alpha t} \sin \beta t, \ldots, t^{m-1}e^{\alpha t} \cos \beta t$, $t^{m-1}e^{\alpha t} \sin \beta t$ form a linearly independent set of $2m$ solutions.

EXAMPLE 5

The differential equation

$$y^{(4)} + 2y'' + y = 0$$

has characteristic polynomial

$$P(r) = r^4 + 2r^2 + 1 = (r^2 + 1)^2.$$

Each factor $r^2 + 1$ yields a complex pair of characteristic values $\pm i$, so these characteristic values are of multiplicity 2. By Theorem 3.6.2, we have a linearly independent set of four solutions: $y_1 = \cos t$, $y_2 = t \cos t$, $y_3 = \sin t$, and $y_4 = t \sin t$. Thus,

$$y = (c_1 + c_2 t) \cos t + (c_3 + c_4 t) \sin t.$$

✦ INSTANT EXERCISE 5

Verify that $y_2 = t \cos t$ is a solution of $y^{(4)} + 2y'' = y = 0$.

The Complete Method for Homogeneous Linear Equations with Constant Coefficients

We now have expressions for m solutions for any real characteristic value of multiplicity m and $2m$ solutions for any complex characteristic value pairs of multiplicity m. Are all these solutions enough? Theorem 3.6.3 guarantees that they are.

Theorem 3.6.3 Let L be a linear differential operator with constant coefficients, and let P be the corresponding characteristic polynomial. The sum of the multiplicities of all the characteristic values is n; furthermore, the set of n solutions of $L[y] = 0$ constructed from Theorems 3.6.1 and 3.6.2 is linearly independent.

We can now solve any homogeneous linear equation with constant coefficients, provided we can factor the characteristic polynomial into individual linear and irreducible quadratic factors. Here is a complete algorithm for the method.

Algorithm 3.6.1

Solving Homogeneous Linear Equations with Constant Coefficients

Given an nth-order linear differential equation with constant coefficients $L[y] = 0$:

1. Find the characteristic polynomial P by substituting $y(t) = e^{rt}$ into the differential equation.
2. Find the characteristic values and their multiplicities by factoring $P(r)$ into linear and irreducible quadratic factors.
3. Find m solutions for each real characteristic value of multiplicity m, using Theorem 3.6.1, and $2m$ solutions for each complex pair of characteristic values of multiplicity m, using Theorem 3.6.2.
4. The family of linear combinations of the solutions from step 3 is guaranteed by Theorem 3.6.3 to be an n-parameter family of solutions, and is therefore the general solution.

EXAMPLE 6

Consider the differential equation

$$y''' + 4y'' + 4y' = 0.$$

The substitution $y = e^{rt}$ leads to the characteristic equation $r^3 + 4r^2 + 4r = 0$, or $r(r + 2)^2 = 0$. Thus, 0 is a root of multiplicity 1, and -2 is a root of multiplicity 2. The general solution, following Algorithm 3.6.1, is

$$y(t) = c_1 + (c_2 + c_3 t)e^{-2t}.$$

It sounds as if we have made a great accomplishment over the last three sections in developing methods for solving differential equations, but careful reflection puts our achievement in a proper perspective. Given a differential equation, we must require the equation to be linear and homogeneous just to be able to use Algorithm 3.3.1. These requirements are already rather restrictive. However, we cannot generally solve even homogeneous linear equations without additional restrictions. To be able to use Algorithm 3.6.1, we must require also that the coefficients be constants rather than functions of t. We must also require, for the time being, that the equation be homogeneous, although we shall see in Chapter 4 that there are many nonhomogeneous problems $L[y] = g$ for which the hardest part of the problem is finding the solution of $L[y] = 0$. The class of equations that we can solve, at least in principle, by the method presented here is small indeed. In practice, we must also worry about the additional difficulty of factoring the characteristic polynomial, so that we can only be *certain* of success with homogeneous first-order and second-order linear equations with constant coefficients.

3.6 Exercises

1. Solve $y'' - 4y' + 4y = 0, \quad y(0) = 2, \quad y'(0) = 1.$
2. Solve $y'' + 6y' + 9y = 0, \quad y(0) = 3, \quad y'(0) = -1.$

3. Solve $y''' - 6y'' + 13y' = 0, \quad y(0) = 3, \quad y'(0) = 6, \quad y''(0) = 10.$

4. Solve $y''' - y'' - y' + y = 0.$

In Exercises 5 through 12, verify that the given function is a solution of the given differential equation, and use the method of reduction of order to find the general solution. Note that you may need to leave your answer as a definite integral. You may also find it useful to use the error function erf that was defined in Section 2.4.

5. $xy'' - 2(x+1)y' + (x+2)y = 0, \quad y_1 = e^x$

6. $(x+1)y'' + (2x+1)y' + xy = 0, \quad y_1 = e^{-x}$

7. $y'' - (e^x + 2)y' + (e^x + 1)y = 0, \quad y_1 = e^x$

8. $y'' - \cos x y' + (\cos x - 1)y = 0, \quad y_1 = e^x$

9. $x^2y'' + xy' + (x^2 - \frac{1}{4})y = 0, \quad y_1 = x^{-1/2} \sin x$

10. $xy'' - y' + 4x^3y = 0, \quad y_1 = \cos x^2$

11. $y'' - 2xy' - 2y = 0, \quad y_1 = e^{x^2}$

12. $y'' + 2(x+1)y' + (2x+1)y = 0, \quad y_1 = e^{-x}$

13. Consider the differential equation $y''' + 3y'' + 3y' + y = 0$. Observe that $y_1 = e^{-t}$ is a solution. Derive the general solution for the differential equation by finding a three-parameter family of functions v such that $y = vy_1$.

14. The Legendre equation of order n,

$$(1 - x^2)y'' - 2xy' + n(n+1)y = 0,$$

has a solution P_n that is called the Legendre polynomial of degree n.

a. Assume a second solution $y_n = P_n v_n$ and derive a differential equation for v_n in terms of P_n and P_n'.

b. After you divide through by v_n', the remaining equation is integrable (without specifying P_n). Derive the formula

$$v_n' = \frac{1}{(1 - x^2)P_n^2}.$$

c. The Legendre polynomial of degree 2 is $(3x^2 - 1)/2$. Solve the equation for v_2 and determine y_2. You should get

$$y_2 = \frac{P_2}{2} \ln \frac{1+x}{1-x} + Ax$$

for some constant A.

d. Given the pattern of y_2 and $P_3 = (5x_3 - 3x)/2$, try to find a solution of the Legendre equation of order 3 that is of the form

$$y_3 = \frac{P_3}{2} \ln \frac{1+x}{1-x} + Ax^2 + B$$

for some constants A and B.

✦ 3.6 INSTANT EXERCISE SOLUTIONS

1.
$$W[y_1, vy_1] = \begin{vmatrix} y_1 & vy_1 \\ y_1' & vy_1' + v'y_1 \end{vmatrix} = y_1(vy_1' + v'y_1) - vy_1y_1' = v'y_1 \neq 0.$$

2. Let $y = te^{-t}$. Then $y' = e^{-t} + t(-e^{-t}) = (1-t)e^{-t}$ and $y'' = (-1)e^{-t} + (1-t)(-e^{-t}) = (t-2)e^{-t}$. Hence,
$$y'' + 2y' + y = (t-2)e^{-t} + 2(1-t)e^{-t} + te^{-t} = 0.$$

3. The integrand can be decomposed into partial fractions:
$$\frac{1}{x^2 - x^4} = \frac{A}{x} + \frac{B}{x^2} + \frac{C}{1+x} + \frac{D}{1-x},$$
where the coefficients A, B, C, and D are to be determined. Once these constants are known, the integration will yield the formula
$$v = \int \frac{1}{x^2 - x^4}\,dx = A\ln|x| - \frac{B}{x} + C\ln(1+x) - D\ln(1-x),$$
where we have assumed $|x| < 1$. To find the coefficients, we multiply the equation for them by the common denominator $x^2 - x^4 = x^2(1+x)(1-x)$:
$$1 = Ax(1+x)(1-x) + B(1+x)(1-x) + Cx^2(1-x) + Dx^2(1+x).$$
When evaluated at $x = 0$, $x = -1$, and $x = 1$, respectively, this equation reduces to
$$1 = B, \qquad 1 = 2C, \qquad 1 = 2D;$$
hence, $B = 1$ and $C = D = \frac{1}{2}$. To find A, we can compare the third-degree coefficients on both sides of the full polynomial equation
$$0 = -A - C + D = -A,$$
and thus $A = 0$. We arrive at the formula presented in Example 1.

4. With $y = t^2e^t$, we have $y' = 2te^t + t^2e^t = (2t + t^2)e^t$, $y'' = (2+2t)e^t + (2t+t^2)e^t = (2 + 4t + t^2)e^t$, and $y''' = (4+2t)e^t + (2+4t+t^2)e^t = (6+6t+t^2)e^t$. Then
$$y''' - 3y'' + 3y' - y = (6+6t+t^2)e^t - 3(2+4t+t^2)e^t + 3(2t+t^2)e^t - t^2e^t = 0.$$

5. With $y = t\cos t$, we have $y' = \cos t - t\sin t$, $y'' = -t\cos t - 2\sin t$, $y''' = -3\cos t + t\sin t$, and $y^{(4)} = t\cos t + 4\sin t$. So
$$y^{(4)} + 2y'' + y = t\cos t + 4\sin t + 2(-t\cos t - 2\sin t) + t\cos t = 0.$$

3.7 Some Other Homogeneous Linear Equations

Some physical problems can be modeled with linear equations that have variable coefficients. A variety of methods are available for this type of problem. Some can be solved by looking for polynomial solutions or by conversion to constant-coefficient equations. Others have solutions that can be written in terms of **special functions,** which are functions defined by integrals or initial-value problems. These functions cannot be given as simple formulas in terms of standard functions of precalculus; but they are well known to mathematicians and scientists, and they are built-in functions for computer algebra systems. Of course these problems can also be solved numerically if necessary, and their solutions can be approximated near the initial conditions by using power series.

MODEL PROBLEM 3.7*a*

A long steam pipe of radius 3 cm is wrapped in insulation out to a radius of 10 cm. The temperature of the pipe is 110°C, and the temperature of the outside of the insulation is 25°C. Determine the temperature of the insulation as a function of the radius r.

Steady-State Heat Flow in a Tube

For Model Problem 3.7*a*, we need a mathematical model of steady-state heat flow in a tube. Consider a long cylindrical tube, with inside radius r_1 and outside radius r_2. See Figure 3.7.1. Suppose the temperature in the tube depends only on the radial coordinate r. Fourier's law says that the net heat flow out of the tube at r_2 is

$$-KA_2\frac{du}{dr}(r_2),$$

where $u(r)$ is the temperature, A_2 the surface area on the outside of the cylinder, and K is a material property called the *thermal conductivity*. Thermal conductivity has dimensions of energy per length per time per temperature, and the temperature derivative has dimensions of temperature per length; thus, the product has dimensions of energy per time. The sign is negative because heat flows out if the temperature gradient at r_2 is negative. Similarly, the heat flow out of the region at r_1 is $KA_1(du/dr)(r_1)$. The surface areas, given length L, of the outside and inside boundaries

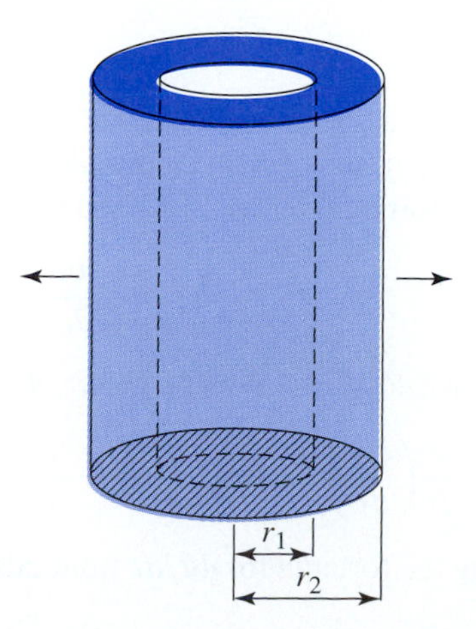

Figure 3.7.1
Heat flow in a tube.

are $2\pi Lr_2$ and $2\pi Lr_1$, so the net heat flow out of the tube is

$$-2\pi Lr_2K\frac{du}{dr}(r_2)+2\pi Lr_1K\frac{du}{dr}(r_1)=0.$$

Removing common factors, we have

$$r_2\frac{du}{dr}(r_2)-r_1\frac{du}{dr}(r_1)=0.$$

To convert this result to a differential equation, we make use of the derivative formula

$$\frac{d}{dr}\left(r\frac{du}{dr}\right)=r\frac{d^2u}{dr^2}+\frac{du}{dr}.$$

Integrating both sides yields a corresponding integral formula

$$\int_{r_1}^{r_2}\left(r\frac{d^2u}{dr^2}+\frac{du}{dr}\right)dr=r\frac{du}{dr}\bigg|_{r_1}^{r_2}=r_2\frac{du}{dr}(r_2)-r_1\frac{du}{dr}(r_1).$$

Thus, steady-state heat flow in the cylinder satisfies the equation

$$\int_{r_1}^{r_2}\left(r\frac{d^2u}{dr^2}+\frac{du}{dr}\right)dr=0.$$

It is particularly noteworthy that the integral is zero regardless of the values of r_1 and r_2. This implies that the integrand itself must be zero; hence, the temperature distribution must satisfy the differential equation

$$r\frac{d^2u}{dr^2}+\frac{du}{dr}=0. \tag{1}$$

This equation can be solved by converting it to a constant-coefficient equation. Define a new independent variable t by

$$r=e^t. \tag{2}$$

Think of u not as specifically a function of t, but as a dependent variable that could be written as a function of any appropriate independent variable, either r or t as needed. By the chain rule, we can derive a formula relating the t and r derivatives of u

$$\frac{du}{dt}=\frac{du}{dr}\frac{dr}{dt}=r\frac{du}{dr}, \tag{3}$$

where the last equality follows from $dr/dt=e^t=r$. Similarly,

$$\frac{d^2u}{dt^2}=\frac{d}{dt}\left(\frac{du}{dt}\right)=\frac{d}{dt}\left(r\frac{du}{dr}\right)=\frac{d}{dr}\left(r\frac{du}{dr}\right)\frac{dr}{dt}=\left(r\frac{d^2u}{dr^2}+\frac{du}{dr}\right)r=r^2\frac{d^2u}{dr^2}+r\frac{du}{dr}.$$

Subtracting the formula for du/dt from this last result gives

$$r^2\frac{d^2u}{dr^2}=\frac{d^2u}{dt^2}-\frac{du}{dt}. \tag{4}$$

Combining the substitution formulas (3) and (4), we have

$$r^2\frac{d^2u}{dr^2} + r\frac{du}{dr} = \left(\frac{d^2u}{dt^2} - \frac{du}{dt}\right) + \frac{du}{dt} = \frac{d^2u}{dt^2}.$$

The left side of this last equation is zero; hence, the temperature profile equation (1) reduces to

$$\frac{d^2u}{dt^2} = 0.$$

This equation has the general solution

$$u = c_1 + c_2 t,$$

which in the original variables is

$$u = c_1 + c_2 \ln r.$$

To complete the problem, note that the temperature is given at both $r = 3$ and $r = 10$. Thus,

$$110 = c_1 + c_2 \ln 3, \qquad 25 = c_1 + c_2 \ln 10.$$

The solution of these equations is

$$c_2 = -\frac{85}{\ln(10/3)}, \qquad c_1 = 110 + \frac{85\ln(3)}{\ln(10/3)}.$$

Thus,

$$u = 110 + \frac{85\ln 3}{\ln(10/3)} - \frac{85\ln r}{\ln(10/3)} = 110 - 85\frac{\ln(r/3)}{\ln(10/3)}.$$

The temperature profile is illustrated in Figure 3.7.2.

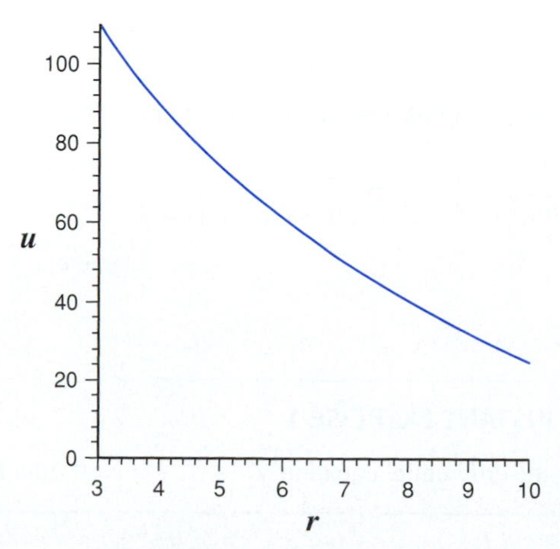

Figure 3.7.2
The temperature profile for Model Problem 3.7*a*.

Cauchy–Euler Equations

Equations of the form

$$x^2 \frac{d^2y}{dx^2} + px \frac{dy}{dx} + qy = G(x), \qquad x \neq 0, \tag{5}$$

where p and q are constants, are called *Cauchy–Euler equations* of order 2. We consider here only the homogeneous case $G \equiv 0$. Nonhomogeneous Cauchy–Euler equations will be considered in Chapter 4.

Cauchy–Euler equations can always be converted to constant-coefficient equations by a change of variables[14]

$$x = e^t. \tag{6}$$

As in the heat flow problem, this substitution results in formulas for the derivative terms in the Cauchy–Euler equation:

$$x \frac{dy}{dx} = \frac{dy}{dt}, \quad x^2 \frac{d^2y}{dx^2} = \frac{d^2y}{dt^2} - \frac{dy}{dt}. \tag{7}$$

Formulas (6) and (7) allow us to rewrite any Cauchy–Euler equation of order 2 or less as a constant-coefficient equation with the independent variable t. For a Cauchy–Euler equation of higher order, we could develop analogous substitution formulas for higher-derivative terms.

EXAMPLE 1

Consider the differential equation

$$x^2 \frac{d^2y}{dx^2} + 2x \frac{dy}{dx} - 2y = 0.$$

The substitution formulas give

$$\left(\frac{d^2y}{dt^2} - \frac{dy}{dt}\right) + 2\frac{dy}{dt} - 2y = 0,$$

or

$$\frac{d^2y}{dt^2} + \frac{dy}{dt} - 2y = 0.$$

The solution of this equation is easily found to be

$$y = c_1 e^t + c_2 e^{-2t}.$$

Substitution of x for e^t gives the final result

$$y = c_1 x + c_2 x^{-2}.$$

✦ INSTANT EXERCISE 1

Solve the differential equation $y'' + y' - 2y = 0$ from Example 1.

---

[14] We are assuming the domain $x > 0$; if instead $x < 0$, then the correct substitution is $x = -e^t$.

Example 1 suggests that Cauchy–Euler equations can be solved by looking for solutions of the form $y = x^r$. This is true, but complex roots for r have to be handled with care. (The change-of-variables method is often more convenient for nonhomogeneous problems as well.)

EXAMPLE 2

Consider the differential equation

$$x^2 \frac{d^2y}{dx^2} + 7x\frac{dy}{dx} + 10y = 0, \qquad x > 0.$$

The substitution formulas give

$$\left(\frac{d^2y}{dt^2} - \frac{dy}{dt}\right) + 7\frac{dy}{dt} + 10y = 0,$$

or

$$\frac{d^2y}{dt^2} + 6\frac{dy}{dt} + 10y = 0.$$

The solution of this equation is

$$y = e^{-3t}(c_1 \cos t + c_2 \sin t).$$

Substitution of x for e^t gives the final result

$$y = x^{-3}[c_1 \cos(\ln x) + c_2 \sin(\ln x)].$$

✦ INSTANT EXERCISE 2

Verify the solution of Example 2 by solving

$$\frac{d^2y}{dt^2} + 6\frac{dy}{dt} + 10y = 0.$$

Also check directly that $y = x^{-3}\cos(\ln x)$ is a solution.

Bessel Equations

Bessel equations are differential equations of the form

$$x^2 \frac{d^2y}{dx^2} + x\frac{dy}{dx} + (x^2 - \nu^2)y = 0, \tag{8}$$

where ν is a nonnegative constant. The spatial coordinate x is generally assumed to be nonnegative, but this depends on the context.

Bessel equations cannot be solved in terms of elementary functions. However, they can be solved by numerical methods or by approximation with power series. The solutions have proved to be useful, so mathematicians have defined **Bessel functions of the first kind** J_ν and **Bessel functions of the second kind** Y_ν as solutions of these equations with certain prescribed initial conditions. As with the Airy functions of Section 2.4, Bessel functions are among the special functions that are incorporated into most computer algebra systems as part of the standard set of functions. We may write the general solution of the Bessel equation of order ν as

$$y = c_1 J_\nu(x) + c_2 Y_\nu(x). \tag{9}$$

Both kinds of Bessel functions are slowly decaying oscillatory functions as x increases. The Bessel functions of the first kind are well behaved at $x = 0$, but the Bessel functions of the second kind approach $-\infty$ as $x \to 0$. Figures 3.7.3 and 3.7.4 show the functions J_0, J_1, Y_0, and Y_1. Bessel functions of other orders have graphs similar to those of these examples.

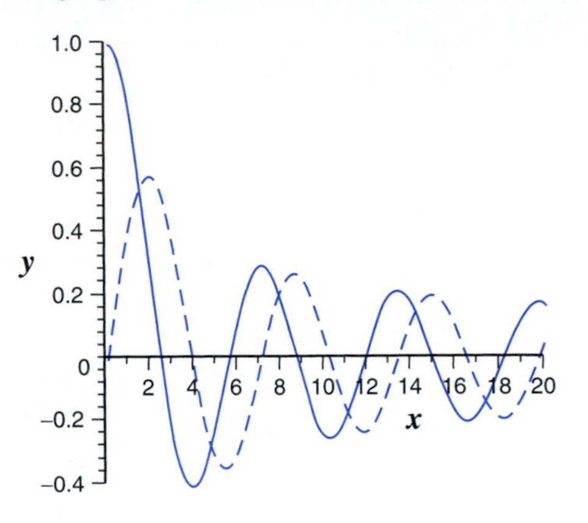

Figure 3.7.3
The Bessel functions J_0 (solid) and J_1 (dashed).

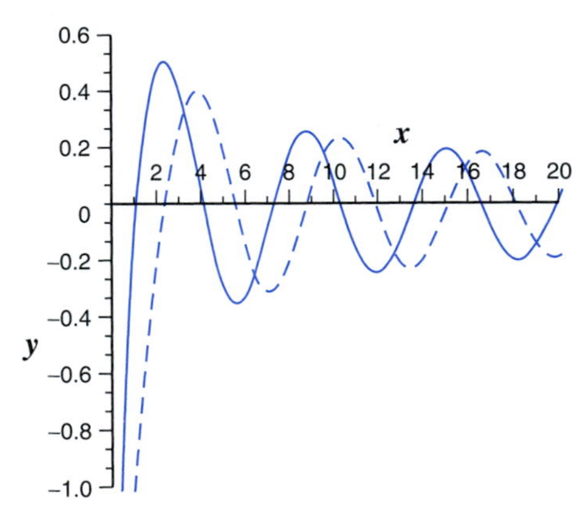

Figure 3.7.4
The Bessel functions Y_0 (solid) and Y_1 (dashed).

As a practical matter, engineers and scientists consider a problem solved if its solution can be stated in terms of Bessel functions. Among the more useful properties of Bessel functions are the following:

$$J_0'(x) = -J_1(x), \qquad Y_0'(x) = -Y_1(x), \qquad W[J_\nu, Y_\nu](x) = \frac{2}{\pi x}, \tag{10}$$

where $W(J_\nu, Y_\nu)$ is the Wronskian.[15]

Bessel functions occur frequently in partial differential equations with axial symmetry, such as in the heating of a solid by an embedded electric wire. There are also some applications of Bessel equations in ordinary differential equations.

---

[15] See Section 3.3.

MODEL PROBLEM 3.7*b*

A spring is made out of a material whose stiffness decreases with time according to the formula $k = e^{-0.02t}$. A unit mass is attached to the spring; then the spring is stretched an additional 1 unit and released. Determine the subsequent motion of the system.

The Aging Spring Problem

Model Problem 3.7*b* is the initial-value problem

$$\frac{d^2y}{dt^2} + e^{-0.02t}y = 0, \qquad y(0) = 1, \qquad \frac{dy}{dt}(0) = 0.$$

As in the Cauchy–Euler equations, we make a change in the independent variable. Here, the time variable t is replaced by a new variable which we define by[16]

$$x = 100e^{-0.01t}.$$

As with any change of independent variable, we need to relate the derivatives with respect to t and those with respect to x, using the chain rule. In particular, note that $dx/dt = -e^{-0.01t} = -0.01x$. Then

$$\frac{dy}{dt} = \frac{dy}{dx}\frac{dx}{dt} = -0.01x\frac{dy}{dx}$$

and

$$\frac{d^2y}{dt^2} = \frac{d}{dx}\left(\frac{dy}{dt}\right)\frac{dx}{dt} = \frac{d}{dx}\left(-0.01x\frac{dy}{dx}\right)(-0.01x) = 0.0001x\frac{d}{dx}\left(x\frac{dy}{dx}\right).$$

Meanwhile, $x^2 = 10{,}000e^{-0.02t}$, so $e^{-0.02t} = 0.0001x^2$. In terms of x rather than t, the aging spring equation is

$$x\frac{d}{dx}\left(x\frac{dy}{dx}\right) + x^2y = 0,$$

or

$$x^2\frac{d^2y}{dx^2} + x\frac{dy}{dx} + x^2y = 0.$$

This is the Bessel equation of order 0, so the general solution is

$$y = c_1J_0(x) + c_2Y_0(x).$$

It is best to evaluate the constants before returning to the original independent variable. The initial point $t = 0$ corresponds to $x = 100$, and the derivative $(dy/dt)(0)$ is the same as $-(dy/dx)(100)$. Hence, the initial conditions are

$$y(100) = 1, \qquad \frac{dy}{dx}(100) = 0;$$

c_1 and c_2 must satisfy

$$J_0(100)c_1 + Y_0(100)c_2 = 1, \qquad J_0'(100)c_1 + Y_0'(100)c_2 = 0.$$

[16]This choice for x will be justified later.

These equations can be solved for c_1 and c_2; then

$$y = \frac{Y_0'(100)J_0(x) - J_0'(100)Y_0(x)}{W[J_0, Y_0](100)}.$$

This result can be simplified by using Equations (10):

$$y = 50\pi\,[J_1(100)Y_0(x) - Y_1(100)J_0(x)].$$

Substituting $x = 100e^{-0.01t}$ yields the solution of Model Problem 3.7*b*:

$$y = 50\pi\left[J_1(100)Y_0\left(100e^{-0.01t}\right) - Y_1(100)J_0\left(100e^{-0.01t}\right)\right].$$

The solution appears in Figure 3.7.5. As the spring weakens, the oscillations increase in both period and amplitude.

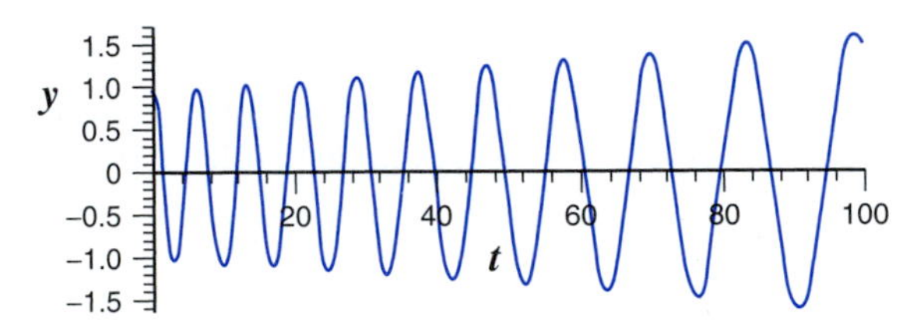

Figure 3.7.5
The solution of Model Problem 3.7*b*.

Solving Certain Equations with Bessel Functions

Up to this point, no justification has been given for the substitution $x = 100e^{-0.01t}$ that solved the aging spring problem. Consider a differential equation of the form

$$\frac{d^2y}{dt^2} + k(t)y = 0.$$

If the equation can be converted by some change of variables to a Bessel equation, then it can be solved by using Bessel functions, much as we can solve Cauchy–Euler equations by conversion into constant-coefficient equations. We would like to know what functions k allow conversion to a Bessel equation and what variable $x(t)$ needs to be used for a given k. In general,

$$\frac{dy}{dt} = \frac{dy}{dx}\frac{dx}{dt} = x'\frac{dy}{dx},$$

where we use the prime symbol to represent derivatives with respect to t. Similarly,

$$\begin{aligned}\frac{d^2y}{dt^2} &= \frac{d}{dt}\left(x'\frac{dy}{dx}\right) = x'\frac{d}{dt}\left(\frac{dy}{dx}\right) + x''\frac{dy}{dx}\\ &= x'\frac{d}{dx}\left(\frac{dy}{dx}\right)\frac{dx}{dt} + x''\frac{dy}{dx} = (x')^2\frac{d^2y}{dx^2} + x''\frac{dy}{dx}.\end{aligned}$$

With these substitutions, the differential equation becomes

$$(x')^2 \frac{d^2y}{dx^2} + x'' \frac{dy}{dx} + k(t)y = 0.$$

Bessel equations have x^2 as the first coefficient and x as the second, so it is necessary for this last differential equation to have the same coefficients, after multiplying the equation by a suitable constant. The easiest way to proceed, since we don't know what constant would be needed, is to require that the first coefficient be x times the second; thus,

$$(x')^2 = xx''.$$

This equation can be integrated by rewriting it as

$$\frac{x''}{x'} = \frac{x'}{x}.$$

Thus, $\ln|x'| = \ln|x| + C$, or $x' = ax$. Hence, the only substitution that can yield a Bessel equation is

$$x = ce^{at},$$

where $a \neq 0$ and $c > 0$, the latter being necessary so that $x'' > 0$.

Now that x has been determined up to two unknown constants, it can be substituted back into the differential equation. It is most convenient simply to substitute $x' = ax$ and $x'' = a^2x$, with the result

$$a^2x^2 \frac{d^2y}{dx^2} + a^2x \frac{dy}{dx} + k(t)y = 0,$$

or

$$x^2 \frac{d^2y}{dx^2} + x \frac{dy}{dx} + \frac{k(t)}{a^2} y = 0.$$

To get a Bessel equation, it is still necessary that the y coefficient be of the form $x^2 - \nu^2$; thus,

$$k(t) = a^2(c^2e^{2at} - \nu^2).$$

In practice, it is necessary to determine x from k rather than the other way a round, so let $\alpha = ac$ and $\beta = a\nu$. With these changes, we have

$$k(t) = \alpha^2e^{2at} - \beta^2, \qquad x = \frac{\alpha}{a} e^{at}.$$

To summarize:

Any differential equation of the form

$$\frac{d^2y}{dt^2} + (\alpha^2e^{2at} - \beta^2)y = 0, \qquad a \neq 0, \qquad \alpha > 0 \tag{11}$$

can be converted to a Bessel equation by the substitution

$$x = \frac{\alpha}{a} e^{at}.$$

Some other equations that can be converted to Bessel equations appear in Exercises 9, 11, and 12.

3.7 Exercises

In Exercises 1 through 4, find the general solution of the given initial-value problem. Sketch the graph of the solution, and describe its behavior as $x \to 0$.

1. $2x^2y'' + xy' - 3y = 0, \quad y(1) = 1, \quad y'(1) = 4$
2. $4x^2y'' + 8xy' + 17y = 0, \quad y(1) = 2, \quad y'(1) = -3$
3. $x^2y'' - 3xy' + 4y = 0, \quad y(-1) = 2, \quad y'(-1) = 3$
4. $x^2y'' + 3xy' + 5y = 0, \quad y(1) = 1, \quad y'(1) = -1$
5. Find all values of β for which all solutions of $x^2y'' + \beta y = 0$ vanish as $x \to 0$.
6. Find all values of γ for which the solution of

$$x^2y'' - 2y = 0, \qquad y(1) = 1, \qquad y'(1) = \gamma$$

is bounded as $x \to 0$.

7. Consider the spring-mass equation from Section 3.1. Define a new time variable τ and a dimensionless parameter β by

$$\tau = \sqrt{\frac{k}{m}}\, t, \qquad \beta = \frac{b}{2\sqrt{mk}}.$$

Rewrite the differential equation with τ as the independent variable and β as the only parameter. What is the significance of β in the model?

T 8. The problem

$$\frac{d^2y}{dt^2} + e^{0.02t}y = 0, \qquad y(0) = 1, \qquad \frac{dy}{dt}(0) = 0$$

might represent a spring that gets stiffer with age. Solve the problem and plot the solution. Does the solution look like what you would expect of a spring that gets stiffer with time?

9. **Parametric Bessel equations** are

$$r^2\frac{d^2y}{dr^2} + r\frac{dy}{dr} + (\lambda^2r^2 - \nu^2)y = 0,$$

where λ is a nonnegative constant.

a. Solve the equation for the case $\lambda = 0$.
b. Solve the equation for the case $\lambda > 0$ by using the substitution $x = \lambda r$.

10. The Bessel equation of order $\frac{1}{2}$,

$$x^2y'' + xy' + \left(x^2 - \tfrac{1}{4}\right)y = 0,$$

is the only Bessel equation whose solutions can be written in terms of elementary functions.

a. Assume that the solution has the form $y = x^r f(x)$, where the function f and the number r are to be determined. Derive the differential equation

$$x^2 f'' + (2r+1)xf' + \left(x^2 + r^2 - \tfrac{1}{4}\right) f = 0.$$

b. Choose r so that all terms that do not contain the factor x^2 vanish. Solve the resulting constant-coefficient equation, and determine the general solution of the original Bessel equation.

11. Solve the equation

$$xy'' - y' + xy = 0$$

by using a substitution $y = x^r f(x)$ [both r and f need to be determined] to convert the equation to a Bessel equation.

12. Solve the equation

$$xy'' - by' + xy = 0$$

by using a substitution $y = x^r f(x)$ [both r and f need to be determined] to convert the equation to a Bessel equation.

✦ 3.7 INSTANT EXERCISE SOLUTIONS

1. The substitution $y = e^{rt}$ yields the characteristic equation $r^2 + r - 2 = 0$ or $(r+2)(r-1) = 0$. The solution follows from the roots $r = -2$ and $r = 1$.

2. The substitution $y = e^{rt}$ yields the characteristic equation $r^2 + 6r + 10$, and the roots of this equation are given by the quadratic formula as $r = -3 \pm i$. Hence the solutions are $e^{-3t} \cos t$ and $e^{-3t} \sin t$ by Theorem 3.5.2.

Given $y = x^{-3} \cos(\ln x)$, we have

$$\frac{dy}{dx} = -x^{-3} \sin(\ln x) x^{-1} - 3x^{-4} \cos(\ln x) = x^{-4}[-3\cos(\ln x) - \sin(\ln x)]$$

and

$$\begin{aligned}\frac{d^2t}{dx^2} &= x^{-4}[3\sin(\ln x)x^{-1} - \cos(\ln x)x^{-1}] - 4x^{-5}[-3\cos(\ln x) - \sin(\ln x)] \\ &= x^{-5}[11\cos(\ln x) + 7\sin(\ln x)].\end{aligned}$$

Thus,

$$\begin{aligned}x^2 \frac{d^2y}{dx^2} + 7x\frac{dy}{dx} + 10y = x^{-3}\{&[11\cos(\ln x) + 7\sin(\ln x)] \\ &+ 7[-3\cos(\ln x) - \sin(\ln x)] + 10\cos(\ln x)\} = 0.\end{aligned}$$

CASE STUDY 3 How Long Should Jellyfish Hold Their Food?

Some creatures eat by taking food into their bodies, digesting this food for a period of time, and then ejecting whatever undigested material remains. How long should such a creature hold its food? Too long means that time is spent trying to squeeze the last bit of nutrition out of food that has become nutrition-depleted. Too short means that the creature has to search for more food sooner than ought to be the case. It seems reasonable that there should be some optimal length of time, given specific information about the digestive process. This hypothesis can be explored with a simple mathematical model. Of course we cannot expect to get a quantitatively useful answer from the model, but perhaps we can get a qualitative idea of whether the optimal time hypothesis makes sense and what factors contribute to the determination of the optimal time.

How do we model digestion? We are going to have to settle for a crude model because it is not reasonable to try to write a model that accounts correctly for any of the subtle details. As discussed in Chapter 1, we begin with a conceptual model, a simplified description of the phenomena of interest that can then be translated to a mathematical model. What should the conceptual model look like?

In choosing a conceptual model, the most important task is to make sure that key processes are included. In our digestion model, the key idea is that nutrients are gained by digestion and spent in locomotion, feeding, digestion, and growth. Natural selection favors individuals who grow fastest, so the measure of success is the net nutrient gain per unit time. The model needs to keep a budget for whatever resource is in limited supply. Digestion consists of two distinct processes—the process of extracting the nutrients from the food and the process of absorbing the extracted nutrients—so nutrients in food and nutrients absorbed have to be counted separately.

These features seem essential in the model. Anything else can probably be simplified or neglected on a first attempt. For example, random variations in the size of a meal and the time required to find it can be neglected, and the details of the various processes can be simplified. These considerations are met in the following conceptual model.

1. Individuals eat one food item at a time according to a predictable cycle. They eat an item at the beginning of the cycle. They digest the item for a certain amount of time, after which the remains of the item are quickly excreted. Then the individuals begin to forage for the next item of food.
2. Digestion consists of two processes, a process whereby the nutrients are extracted from the food and a process whereby the free nutrients are absorbed into the tissues. Each of these processes occurs at a constant rate relative to the quantity available for the process.
3. The duration of the digestion phase in the cycle is chosen by natural selection to maximize the net nutrient intake per unit time. Some of the nutrients gained by digestion are spent in the physiological processes of digestion and foraging.
4. Each food item contains the same amount of nutrients.
5. It always takes the same amount of time to acquire a food item.

Each of these assumptions plays a role in making the model manageable. Assumption 1 separates the digestion and food search operations so that the durations of each can be summed to determine the total amount of time in the cycle. Assumption 2 provides the information needed to write equations to track the amounts of nutrients in the food, free nutrients, and absorbed nutrients. Assumption 3 provides a way to assign a score to each possible duration choice and dictates that the highest score indicates the value to be chosen. Assumptions 4 and 5 serve two purposes. First, they eliminate random variations, which would make the model more complicated but not affect the qualitative features that we want to discover. Second, we can choose to measure nutrient amounts in terms of a *standard food equivalent* rather than using actual measurements, and similarly we can measure time in terms of the *standard foraging time* rather than an arbitrary standard such as minutes or hours. Arbitrary standards are convenient when it is time to make measurements, but standards that are inherent in a process are more convenient for mathematical modeling. We may now state the problem as follows:

Construct a mathematical model from the conceptual model of digestion given above, and analyze it to determine the optimal digestion duration time, relative to the standard foraging time.

Derivation of the Mathematical Model

Let $S(t)$ and $P(t)$ be the amount of nutrients in the food item and the amount of free nutrients. Let k and r be the relative rates of digestion and absorption of nutrients, respectively. Assume that the cost of digestion is C units of nutrient per unit time and that the total cost of finding food is F units of nutrient. Note that the foraging cost can be fixed because the time requirement is outside the control of natural selection; however, the digestion cost is given per unit time because the organism "chooses" the total digestion time. Let T be the amount of time spent digesting each item, let $A(T)$ be the amount of nutrients absorbed during one cycle, and let $I(T)$ be the net nutrient intake per unit time. These quantities are taken to be functions of T because the point of the model is to determine the effect of different values of T.

We need differential equations for S and P for each of the two phases of the cycle and initial conditions to go with these. We also need to find formulas for A and I. The simple diagram in Figure C3.1 shows the processes, indicated by arrows, that change S and P and the rates at which these changes occur.

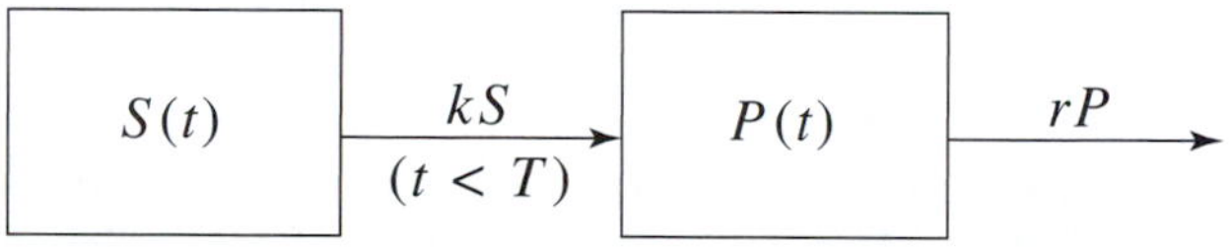

Figure C3.1
A schematic diagram of nutrient movement in the digestive cycle of the model.

Note that the initial ingestion takes place between cycles, so there is no input pathway for the S box. The pathway from S to P is used only during the digestion phase of the cycle because the conceptual model indicates that the food is not present for the foraging phase. The pathway out of P is used for both phases because the absorption process continues during foraging. Using the information displayed in the diagram for the differential equations, we have problems for S and P:

$$S' = -kS \qquad (0 < t < T), \qquad S(0) = 1, \tag{1}$$

$$S(t) = 0 \qquad (T < t < T + 1),$$

$$P' = kS - rP, \qquad P(0) = P(T + 1). \tag{2}$$

The differential equation for P is augmented by an algebraic condition of a different type from what we have seen so far. The initial value of P is not known. What *is* known is that the process is cyclic, with all the cycles identical. The value of P at the end of one cycle is $P(T+1)$, and this must be the same as the value of P at the beginning of the next cycle. This kind of condition is called a **periodicity condition.** The theory of problems with periodicity conditions is much more complicated than that for problems with initial conditions. In practice, we can attempt to use the periodicity condition to evaluate an integration constant, just as we do with initial conditions.

Turning now to the amount of nutrients absorbed in the cycle, we need only note that the rate of nutrient absorption is rP. The total amount absorbed is the integral of the rate; hence,

$$A(T) = r\int_0^{T+1} P(t)\,dt. \tag{3}$$

The net intake of nutrients differs from the total amount absorbed in that the costs of digestion and foraging must be deducted. Thus, the net intake of nutrients is $A(T) - CT - F$. The net intake per unit time is then

$$I(T) = \frac{A(T) - CT - F}{T+1}. \tag{4}$$

In mathematical terms, the problem now reads as follows:

Maximize $I(T)$, where I is defined by Equation (4), A is defined by Equation (3), and S and P are the solutions of the differential equation problems (1) and (2).

Solving for P

The equation for S can be solved easily enough. However, once we have substituted the result for S into the equation for P, we will be faced with a nonhomogeneous equation. Such equations can be solved (Sections 4.5 and A.1), but it is perhaps easier to instead combine the two differential equations into a single homogeneous linear equation of second order for P. The idea is to differentiate Equation (2) to obtain a second-order differential equation for P. The quantities S and S' can then be eliminated from the set of equations consisting of this second-order equation and the original differential equations (1) and (2). The result of these manipulations is the differential equation

$$P'' + (r+k)P' + rkP = 0 \qquad (0 < t < T). \tag{5}$$

The auxiliary conditions become

$$P(0) = P(T+1), \qquad P'(0) + rP(0) = k. \tag{6}$$

The full derivation of this problem is left as Exercise 1. On the time interval $T < t < T+1$, the problem for P follows simply from substituting $S = 0$:

$$P' + rP = 0. \tag{7}$$

The initial condition for this equation comes from the solution of the differential equation (5).

Both differential equations (5) and (7) are linear and homogeneous, with constant coefficients. The substitution e^{at} yields the general solution

$$P = \begin{cases} c_1 e^{-kt} + c_2 e^{-rt} & 0 < t < T \\ c_3 e^{-rt} & T < t < T+1 \end{cases}, \tag{8}$$

where we are assuming $r \neq k$. The constants c_1, c_2, and c_3 are determined by the remaining conditions:

$$c_1 + c_2 = c_3\, e^{-r(T+1)}, \qquad rc_1 - kc_1 = k, \qquad c_1\, e^{-kT} + c_2\, e^{-rT} = c_3\, e^{-rT}. \tag{9}$$

The first two of these correspond to the initial conditions (6), and the last one comes from the requirement that the two solutions for P be equal at time T. The second of the equations gives a solution for c_1, and it is convenient to think of c_2 and c_3 as multiples of c_1. Let $b_2 = c_2/c_1$ and $b_3 = c_3/c_1$. Thus,

$$P = c_1 \begin{cases} e^{-kt} + b_2\, e^{-rt} & 0 < t < T \\ b_3\, e^{-rt} & T < t < T+1 \end{cases}, \tag{10}$$

where

$$c_1 = \frac{k}{r-k}, \qquad 1 + b_2 = b_3\, e^{-rT-r}, \qquad e^{rT-kT} + b_2 = b_3. \tag{11}$$

We postpone solving for b_2 and b_3.

The Net Gain of Nutrients

Now that we have a formula for P, we are finished with the differential equations part of the problem. There is still some more mathematical work and all the interpretation. Here is the remaining problem:

> Maximize $I(T)$ where I is defined by Equation (4), A is given by Equation (3), and P is given by Equations (10) and (11).

The next step in solving the problem is to compute A from Equation (3). Substitution of the solution (10) for P into the definition (3) for A yields

$$A(T) = c_1 \left[\frac{r}{k} - \frac{r}{k} e^{-kT} + (b_3 - b_2)\, e^{-rT} + b_2 - b_3\, e^{-rT-r} \right]. \tag{12}$$

The details of this calculation are left as Exercise 2.

Now it turns out that we do not need to compute b_2 and b_3 because both $b_3 - b_2$ and $b_2 - b_3 e^{-rT-r}$ are easily found from Equations (11). After the calculation is completed, we have the surprisingly simple result

$$A(T) = 1 - e^{-kT}. \tag{13}$$

The net gain of nutrients over one cycle is

$$A(T) - CT - F = 1 - F - CT - e^{-kT}, \tag{14}$$

and the net gain per unit time is

$$I(T) = \frac{1 - F - CT - e^{-kT}}{T + 1}. \tag{15}$$

The Optimal Digestion Time for One Example

Figure C3.2 shows the net gain per unit time for the parameter values $k = 0.1$, $F = 0.2$, and $C = 0.02$. (The values for F and C are justified in Exercise 6.) It is clear from the graph that the optimal digestion time for this set of parameters is a little bit less than 9 times the amount of time required to catch food.[17] The graph of I is fairly flat near the maximum, indicating that the net nutrient gain per unit time is not very sensitive to the duration of the digestion phase. Any value between 7 and 12 is almost as efficient as the optimal value. It should, of course, be remembered that no real creature could follow the results of this model because real food differs in nutrient content and real hunting can take varied amounts of time. It is interesting to note that the value of r makes no difference. The rate of absorption affects the graphs of P and S, but it does not affect the outcome of the cycle.

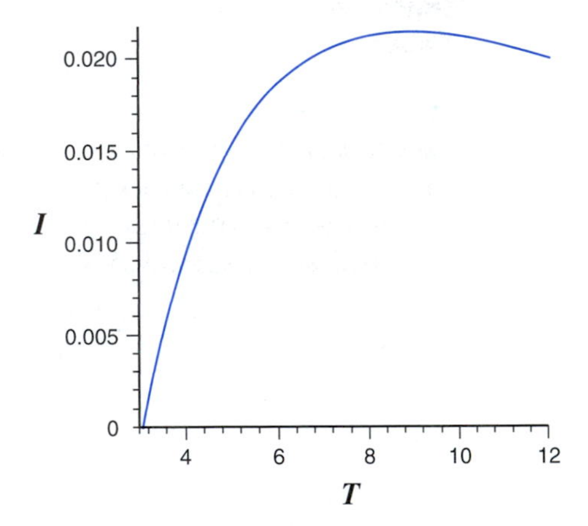

Figure C3.2
The net gain of nutrients per unit time for the case $k = 0.1$, $F = 0.2$, and $C = 0.02$.

Case Study 3 Exercises

1. Derive Equations (5) and (6).
2. Derive Equations (12) and (13).
3. *a.* Show that $-e^{-kT} < kT - 1$ for $T > 0$. (*Hint:* Compare the function values for $T = 0$ and the slopes for $T > 0$.)
   b. Use the tangent line approximation for e^{-kT} near $T = 0$ to obtain an approximation for the net gain of nutrients during one cycle.
   c. Argue that this approximation gives an estimate that is larger than the correct result.
   d. Use the approximation to show that survival requires $C < k$.

[17] There is no point in reporting a precise answer because the model was not designed to be quantitatively accurate.

T 4. *a.* Derive an equation to determine critical points of I. Write the equation in the form $e^{kT} = mT + b$, where m and b do not depend on T.

b. Plot the exponential and linear functions from the equation in part *a* on the same graph, using the data for Figure C3.2. Find the approximate value of T that solves the equation, and compare it with the solution found in the text.

c. Show that the equation from part *a* always has exactly one positive solution T_0 whenever the parameters correspond to a real organism.

d. Use the values $I(0)$ and $\lim_{T\to\infty} I$ to show that T_0 locates the maximum value of I on the set $T > 0$.

5. Find the time for which the net gain of nutrients reaches its maximum. Use logical reasoning (rather than calculations) to determine whether the food should be held for a longer amount of time or a shorter amount of time than this value. Explain your reasoning.

6. Use assumptions 3, 4, and 5, and the definitions of C and F to explain why you should expect C and F to be significantly less than 1. In particular, what do the values $F = 0.2$ and $C = 0.02$ used in the example say about the nutrient requirements for food gathering and digestion compared with the nutrient content of the food?

T 7. Let T_0 be the optimal value of T. (The existence of T_0 is guaranteed by Exercise 4.) For convenience, let $k = 1$ and $C = 0$. The goal of this exercise is to understand how the optimal digestion time depends on the nutrient cost of food gathering.

a. Derive an equation that relates F and T_0 from the requirement that T_0 is a critical point of I.

b. Plot a graph of T_0 as a function of F. Note that you cannot solve the equation for T_0, but you can solve it for F. Then you can plot the parametric curve $(F(T_0), T_0)$.

c. Discuss the effect of F on T_0. In particular, explain why the T_0 intercept is positive, why the slope is positive, and why there is a vertical asymptote. (Note that there is a difference between a *description* and an *explanation*. A description is just a verbal statement about the features of the curve. An explanation connects the features of the curve with the conceptual model from which the curve was obtained.)

chapter

4

Nonhomogeneous Linear Equations

In Chapter 3, we studied the theory of linear differential equations, and we learned to solve selected homogeneous linear equations. In this chapter, we develop additional tools needed to solve some nonhomogeneous equations. The first tool is a theoretical tool. The structure of solutions for nonhomogeneous equations is simple, and because of the simple structure, the problem of solving a nonhomogeneous linear differential equation $L[y] = g$ is reduced to the problem of solving the associated homogeneous equation $L[y] = 0$ along with the problem of finding a single solution of $L[y] = g$. In principle, any differential equation $L[y] = g$ is a candidate for solution as long as the solution of $L[y] = 0$ is given. In practice, it actually *is* possible to construct a solution formula for a nonhomogeneous equation when the corresponding homogeneous equation has been solved, provided of course that a solution exists. Both calculational tools developed in this chapter are aimed at determining particular solutions of linear equations. These are the methods of undetermined coefficients and variation of parameters. Each of these methods exemplifies a broad class of methods for a variety of differential equation problems. Both techniques utilize a substitution, but for different purposes. The method of undetermined coefficients is an example of a technique that solves differential equations by using differentiation to reduce the differential equation problem to an algebra problem. The method of variation of parameters is an example of a technique that solves differential equations by reducing them to integration problems. As a general rule, the former are limited in scope while the latter are more tedious; hence, it is often necessary to learn techniques of both types for the same class of problems.

Section 4.1 begins with the development of the linear oscillator model from a conceptual model for a water flow problem. The conceptual model is also useful, by analogy, for a common type of electric circuit. Section 4.2 develops the key idea that nonhomogeneous problems of the form $L[y] = g$, where L is a linear differential operator, are solved by finding a *particular solution* y_p and the solution of the related homogeneous equation $L[y] = 0$. Section 4.2 also provides an informal introduction to the method of undetermined coefficients for determining y_p. A complete study of the method of undetermined coefficients appears in Section 4.3, followed by a section that uses the method of undetermined coefficients to study the properties of forced linear oscillators, with emphasis placed on finding that part of the system behavior that persists for all time. Included in Section 4.4 is a discussion of the distinction between *mathematical resonance* and *physical resonance*. Physical resonance is typically explained by mathematical resonance, but the phenomena are actually different. Section 4.5 presents the method of variation of parameters for solving nonhomogeneous first-order equations. The method of variation of parameters for second-order nonhomogeneous equations appears in Section 4.6. The chapter concludes with a case study on the design of a tuning circuit for a radio.

4.1 More on Linear Oscillator Models

Electric circuits provide an excellent source of practical problems involving nonhomogeneous linear differential equations; examples are a model of an automobile ignition circuit that appears in Exercise 9 of this section and a model of a radio tuner that appears as Case Study 4. The difficulty in using electric circuit models in a differential equations text is that students who are not knowledgeable in electric circuit theory have little intuition about electric circuits. Our intuition is based on our everyday experience, and nobody can see electricity! On the other hand, modeling of electric circuits illustrates the great power of conceptual models as starting points for mathematical models. We can model any process that is based on rules for determining rates of change, as long as someone who has the correct intuition about the process gets us started by providing a good conceptual model.

Electric current flowing through a wire behaves, in some ways, as water flowing through a hose does. The analogy is only good to a point, as some aspects of electric circuits have no clear analog in water flow. Nevertheless, we can begin with a conceptual model for water flow, and then we can add features corresponding to electric circuits even when they do not quite make sense for water flow.

A Model for Flow of Water in a Pump, Tank, and Pipe System

Figure 4.1.1 illustrates a device consisting of a pump (E), a storage tank (C), a pair of narrow rigid pipes (R_i), and a switch (S) that can open either of the pipes. The pump pushes water up pipe R_1 with a constant amount of pressure. If that pipe is open, water rises through it into the storage tank. Similarly, if the switch is set so that pipe R_2 is open, then water drains from the tank under the force of its own weight. Both pipes can be closed, but the switch does not permit both pipes to be open at the same time.

Given this simple conceptual model, we can use basic principles of water flow to obtain a mathematical model. Let $V(t)$ be the volume of water in the tank. Let $i(t)$ be the flow rate of the water through either open pipe. The flow rate is the rate of change of the volume, so $V' = i$. Let $E(t)$ be the pressure produced by the pump, let $v_C(t)$ be the pressure of the water in the tank, and let $v_R(t)$ be the pressure associated with the movement of water in whichever pipe is open. We have to determine the connections among the three pressures, and we have to connect the pressures v_C and v_R with the volume of the tank or the flow rate through the pipe.

We begin with the relationship of the pressures. The pressure in an open pipe results from the combination of the pressure from below and the pressure from above. Let v_E be the pressure from below ($v_E = E$ when pipe R_1 is open and $v_E = 0$ when pipe R_2 is open). Then the net pressure from below is the pressure from below minus the pressure from above. Using the notation of the

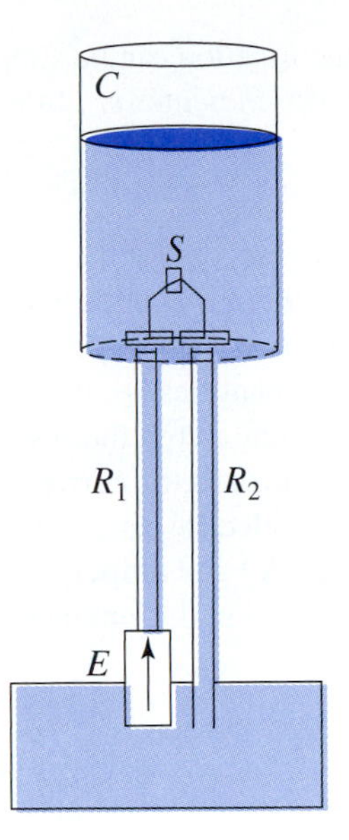

Figure 4.1.1
A conceptual model for a water flow problem.

model, we have $v_R = v_E - v_C$, or

$$v_R + v_C = v_E \tag{1}$$

The pressure of the water in the tank is proportional to the height of the water column, which is simply the volume divided by the constant cross-sectional area. We have

$$v_C = \frac{V}{C}, \tag{2}$$

where C is the cross-sectional area multiplied by a proportionality constant. Given two tanks of different cross-sectional areas, each containing the same volume, the one with the smaller cross-sectional area will produce a greater pressure because the water column will rise higher.

The flow of water in either pipe is caused by pressure in the pipe. In particular, the flow rate should be proportional to the pressure. Let R be a constant that measures the resistance of the pipe to flow. Then $i = v_R/R$, or

$$v_R = Ri = RV'. \tag{3}$$

We can combine Equations (2) and (3) to obtain a formula for v_R in terms of v_C:

$$v_R = RV' = RCv_C'.$$

Substituting this result into the pressure balance equation (1) yields the differential equation

$$RCv_C' + v_C = v_E.$$

At this point, the notation can be simplified by dropping the subscript from v_C. *From here on, v without a subscript indicates v_C.* In its final form, we have the model

$$RCv' + v = v_E. \tag{4}$$

The conceptual model for the flow of water can also be thought of as a conceptual model for an ***RC* series circuit,** as depicted in Figure 4.1.2. The water volume V corresponds to the electric charge q. The pipe and tank correspond to circuit elements called a **resistor** and a **capacitor,** respectively. The quantities v_C, v_R, E, and i represent the **voltage** measured across the capacitor, the voltage measured across the resistor, the voltage (or **electromotive force**) produced by the power supply, and the electric current. The properties R and C are the **resistance** and **capacitance,** respectively, of the electric circuit, measured in **ohms** (Ω) and **farads** (F). Voltage and current are measured in **volts** (V) and **amperes** (A). Whatever results we obtain for the differential equation (4) can be interpreted in the language of electric circuits or the language of water flow.

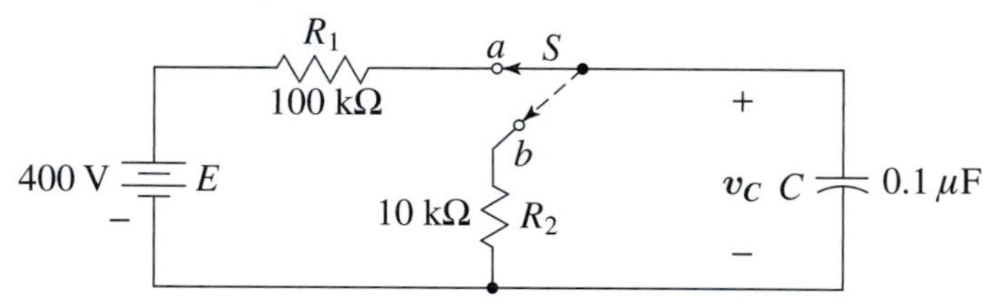

Figure 4.1.2
The electric circuit corresponding to Figure 4.1.1, with parameter values for Example 1.

EXAMPLE 1

Consider the water flow scenario of Figure 4.1.1 and the equivalent electric circuit scenario of Figure 4.1.2. Suppose the switch is initially set so that the tank is not open to either pipe,[1] and there is initially no water in the tank.[2] At time 0, the switch is set so that pipe R_1 is open; thus, $V_E = E$. At $t = 0.03$, the switch is reset so that pipe R_2 is open, corresponding to $V_E = 0$. We want to determine the behavior of the system, as given by $v(t)$ and $i(t)$.

The initial-value problem for the tank pressure before $t = 0.03$ is

$$R_1Cv' + v = E, \qquad v(0) = 0.$$

This differential equation is a decay equation,[3] so it can be solved by using the translation $y = v - E$ to yield the homogeneous problem

$$R_1Cy' + y = 0, \qquad y(0) = -E.$$

[1]Since the water flow model is more accessible to the average student's intuition, we use the water flow language. However, to build intuition about electric circuit models, we use parameter values that are relevant for the electric circuit interpretation. Note that 1 μF $= 10^{-6}$ F.

[2]The pipes are full of water, or else there would be air in them.

[3]Section 1.1.

This problem has the solution

$$y = -Ee^{-t/(R_1 C)},$$

or

$$v = E(1 - e^{-t/(R_1 C)}).$$

Using the parameter values $R_1 = 10^5$, $C = 10^{-7}$, and $E = 400$, as given in Figure 4.1.2, we get

$$v = 400(1 - e^{-100t}).$$

This solution holds as long as the switch is open to the pump. In particular, let v_1 be the pressure at $t = 0.03$; then

$$v_1 = v(0.03) = 400(1 - e^{-3}) \approx 380.$$

At $t = 0.03$, the movement of the switch changes the relevant problem. The pressure now satisfies the initial-value problem

$$R_2 C v' + v = 0, \qquad v(0.03) = v_1.$$

This problem is homogeneous, and easily solved, with the result

$$v = v_1 e^{-(t-0.03)/(R_2 C)}.$$

Given the parameter values, this is

$$v = 400(1 - e^{-3})e^{-1000t+30}.$$

Altogether, we have a complete formula for the pressure:

$$v = 400 \begin{cases} 1 - e^{-100t} & t < 0.03 \\ (1 - e^{-3})e^{-1000t+30} & t > 0.03 \end{cases}.$$

Differentiating this formula gives the result

$$i = Cv' = 0.04 \begin{cases} 0.1e^{-100t} & t < 0.03 \\ -(1 - e^{-3})e^{-1000t+30} & t > 0.03 \end{cases}$$

for the flow rate.

The solution is illustrated in Figures 4.1.3 and 4.1.4. The tank stores water, and since the pressure is proportional to the amount of water, we can think of the tank as storing pressure to be converted to flow at some later time. In the two stages of the problem, the tank is first charged and then drained, with each stage accomplished through the flow of water. The resistance to flow is less in the second pipe than in the first; hence, the draining process is faster, and the flow rate during draining is larger in magnitude. Similarly, the capacitor in the electric circuit stores charge and voltage; we can think of the stored voltage as a potential for future current. The capacitor is charged and then discharged, with the rates of these processes, and the associated currents, dependent on the relative resistances of the associated portions of the circuit.

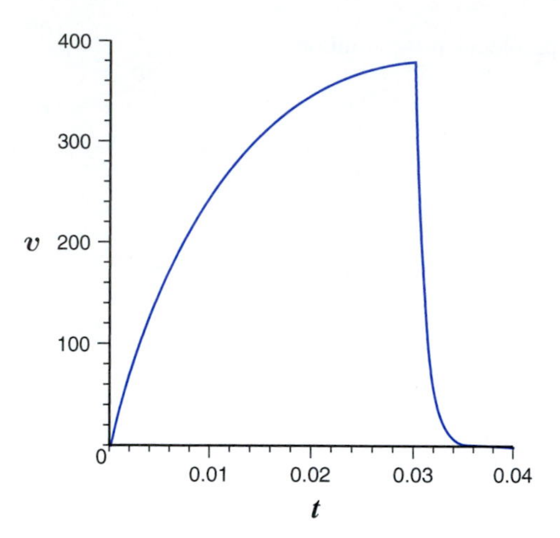

Figure 4.1.3
The tank pressure for the flow problem of Example 1.

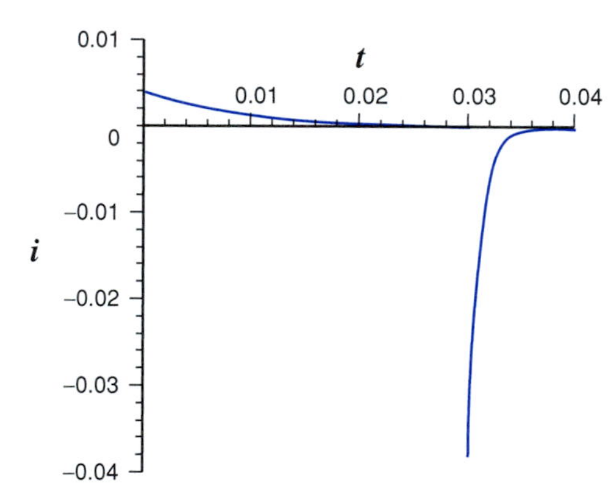

Figure 4.1.4
The flow rate for the flow problem of Example 1.

An Electrical Oscillator

An ***RLC* series circuit** consists of a resistor, an inductor, and a capacitor, along with an optional power source (see Figure 4.1.5), arranged in a loop with the elements in any order. We are already acquainted with resistors and capacitors, which are devices that impede current flow and store charge (potential current). Imagine a device that consists of a paddle wheel in a tube, similar to a revolving door. When the flow changes speed, the wheel changes the pressure that drives the flow. A circuit element characterized by this property is called an **inductor.** Specifically, we assume that the pressure induced by the change in flow rate is given by

$$v_L = Li' = LCv'', \tag{5}$$

where the **inductance** L is measured in **henrys** (H).

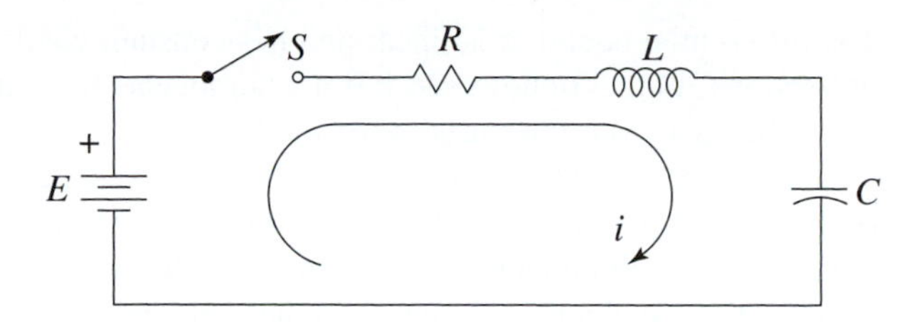

Figure 4.1.5
A schematic diagram of an *RLC* series circuit with applied direct-current power—the circuit elements can be arranged in any order.

Recall that the current in an *RC* circuit is driven by the surplus voltage $v_R = E - v_C$. Similarly, in a circuit with an inductor, the voltage of the inductor decreases the voltage available for producing current; hence, $v_R = E - v_C - v_L$, or

$$v_L + v_R + v_C = E. \tag{6}$$

Substituting from Equations (2), (3), and (5) yields the differential equation for the capacitor voltage in an *RLC* series circuit

$$LCv'' + RCv' + v = E. \tag{7}$$

Generally, initial-value problems for the *RLC* circuit include the initial conditions

$$v(0) = v_0, \qquad Cv'(0) = i_0, \tag{8}$$

where v_0 is the initial voltage stored in the capacitor and i_0 is the initial current in the circuit.

RLC Electric Circuits with Direct-Current Forcing

The two common varieties of electric power are **direct current,** in which the applied voltage is constant, and **alternating current,** in which the applied voltage is a sinusoidal function. Some electric devices require direct current, and others require alternating current. Batteries can be used to supply direct current, and electric generators create alternating current. Transformers are used as needed to convert between the two types. We consider here the *RLC* circuit model for direct current given by Equations (7) and (8), with E a given constant. Analysis of circuits powered by alternating current is discussed in Section 4.4.

MODEL PROBLEM 4.1

Use the *RLC* series circuit model to determine the capacitor voltage and current in a circuit with $L = 10\,\text{H}$, $R = 1000\ \Omega$, $C = 10\ \mu\text{F}$, and $E = 3\,\text{V}$, assuming that the circuit is initially at rest.[4]

The parameters in the model correspond to the initial-value problem

$$0.0001v'' + 0.01v' + v = 3, \qquad v(0) = 0, \qquad v'(0) = 0.$$

[4]A circuit at rest is one with no current flowing and no charge on the capacitor.

The differential equation is linear and has constant coefficients, as in the differential equations of Sections 3.4 to 3.6; however, it is not homogeneous. This difficulty can be eliminated with the same change of variables used in Example 1.

Let $y = v - 3$. Then $y' = v'$ and $y'' = v''$. The differential equation for y is almost the same as the differential equation for v, except that replacing v in the last term on the left with $y + 3$ removes the nonhomogeneous term from the differential equation. The initial condition for $v(0)$ also needs to be changed, resulting in the problem

$$0.0001y'' + 0.01y' + y = 0, \qquad y(0) = -3, \qquad y'(0) = 0.$$

This differential equation is homogeneous as well as linear, and so it can be solved by the characteristic value method of Chapter 3. The solution is

$$y = -e^{-50t}\left[3\cos(50\sqrt{3}\,t) + \sqrt{3}\sin(50\sqrt{3}\,t)\right].$$

✦ INSTANT EXERCISE 1

Derive the solution $y(t)$.

Given $y = v - 3$, the capacitor voltage of Model Problem 4.1 is

$$v = 3 - e^{-50t}\left[3\cos(50\sqrt{3}\,t) + \sqrt{3}\sin(50\sqrt{3}\,t)\right].$$

The current is then

$$i = Cv' = \frac{\sqrt{3}}{500}e^{-50t}\sin(50\sqrt{3}\,t).$$

These solutions appear in Figures 4.1.6 and 4.1.7.

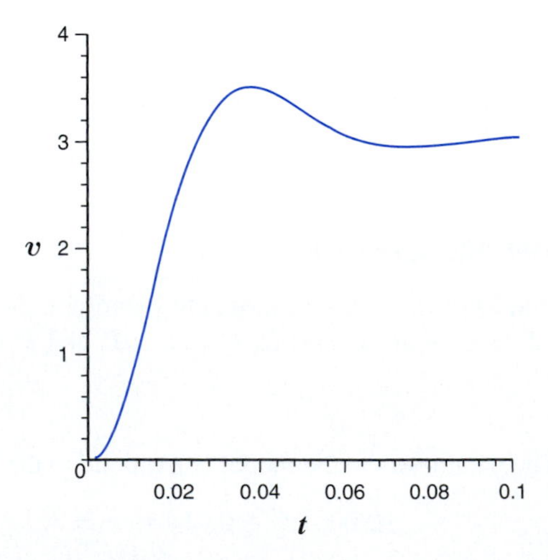

Figure 4.1.6
The capacitor voltage for Model Problem 4.1.

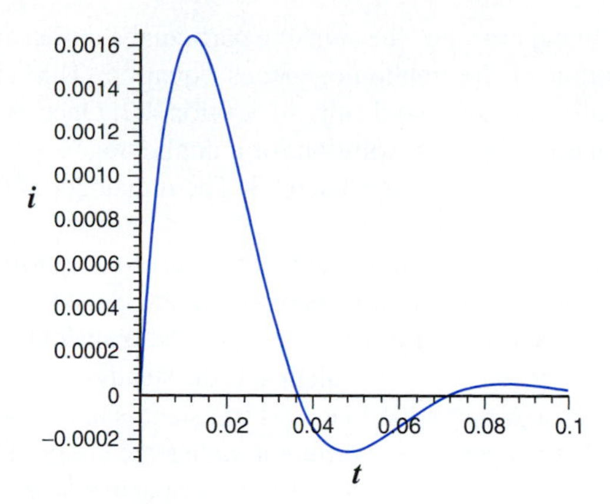

Figure 4.1.7
The current for Model Problem 4.1.

Solution Components

The general solution of the differential equation for y, as determined in Instant Exercise 1, is

$$y = e^{-50t}\left[c_1 \cos(50\sqrt{3}\,t) + c_2 \sin(50\sqrt{3}\,t)\right].$$

This corresponds to a general solution for the original differential equation:

$$v = 3 + e^{-50t}\left[c_1 \cos(50\sqrt{3}\,t) + c_2 \sin(50\sqrt{3}\,t)\right].$$

The solution for v is conveniently split into two portions—a portion that includes the arbitrary constants and a portion that does not. Here, these portions are

$$v_p = 3, \qquad v_c = e^{-50t}\left[c_1 \cos(50\sqrt{3}\,t) + c_2 \sin(50\sqrt{3}\,t)\right].$$

The subscripts p and c are standard; the portions of the solution are called the *particular* and *complementary* solutions, respectively.[5]

Let L be a linear differential operator of order n. A **particular solution** of

$$L[y] = g \tag{9}$$

is any function y_p that satisfies the differential equation. The **complementary solution** of the differential equation is the n-parameter general solution for the associated homogeneous equation $L[y] = 0$.

[5] Do not confuse v_c with v_C. The latter represents the voltage drop across the capacitor, which we generally write without the subscript, while the former indicates the complementary solution of a nonhomogeneous differential equation for a function v.

In our example, the sum of a particular solution and the complementary solution is the general solution of the nonhomogeneous equation. This general property of nonhomogeneous linear equations is discussed fully in Section 4.2. Once established, this property reduces the problem of finding a general solution for a nonhomogeneous equation to two subproblems, one of which was discussed fully in Chapter 3. The remainder of this chapter deals with techniques for finding particular solutions.

For applications, it is sometimes useful to partition solutions into *steady-state* and *transient* components, where the **steady-state** component is the part that does not decay to zero as time increases, if there is such a part, and the **transient** component is the part that does decay to zero as $t \to \infty$. In Model Problem 4.1, the steady-state component is v_p, and the transient component is v_c. The observation that v_c is transient is useful because it suggests the answer to the question of what happens when different values are chosen for L, R, C, and E. Given a constant applied voltage E, the voltage stored in the capacitor eventually approaches E. The voltage drop across the resistor, the current through the resistor, and the voltage produced by the inductor all approach zero. In Model Problem 4.1, the capacitor is charged to a value of 3 V. The capacitor voltage actually exceeds 3 V for part of the charging process because of the oscillatory factors in the solution formula.

4.1 Exercises

In Exercises 1 through 6, determine the current for the circuit with the given parameters and initial conditions. (Note that $1\text{ mF} = 10^{-3}\text{ F}$.)

1. $L = 0$, $R = 1\,\Omega$, $C = 1\text{ mF}$, $E = 1.5\text{ V}$, $v(0) = 0$
2. $L = 10\text{ H}$, $R = 1000\,\Omega$, $C = 10\,\mu\text{F}$, $E = 300\text{ V}$, $v(0) = 0$, $i(0) = 0.2$
3. $L = 4\text{ H}$, $R = 40\,\Omega$, $C = 10\text{ mF}$, $E = 10\text{ V}$, $v(0) = 0$, $i(0) = 0$
4. $L = 10\text{ H}$, $R = 20\,\Omega$, $C = 10\text{ mF}$, $E = 100\text{ V}$, $v(0) = 0$, $i(0) = 0$
5. $L = 100\text{ H}$, $R = 20\,\Omega$, $C = 10\text{ mF}$, $E = 20\text{ V}$, $v(0) = 20$, $i(0) = 1$
6. $L = 3\text{ H}$, $R = 40\,\Omega$, $C = 10\text{ mF}$, $E = 100\text{ V}$, $v(0) = 0$, $i(0) = 0$

Exercise 7 develops a model for an *RL* series circuit driven by direct current. This model is used in Exercises 8 and 9.

7. *a.* Derive an initial-value problem for an *RL* series circuit, using the current as the dependent variable.
   b. Convert the problem to a homogeneous equation by defining a new dependent variable $y = i - k$, where k is appropriately chosen.
   c. Solve the initial-value problem for y.
   d. Determine the current.
   e. What is the steady-state current in the circuit?
   f. Compare the result for the *RL* circuit with the result of Model Problem 4.1. What feature of the *RLC* circuit cannot be obtained without the capacitor? What feature of the *RL* circuit cannot be retained if a capacitor is added to the circuit?

8. In the circuit shown in Figure 4.1.8, the switch S_1 has been closed long enough for the circuit to reach a steady-state condition. Determine the circuit behavior (i, v_R, and v_L) after the switch S_2 is closed.

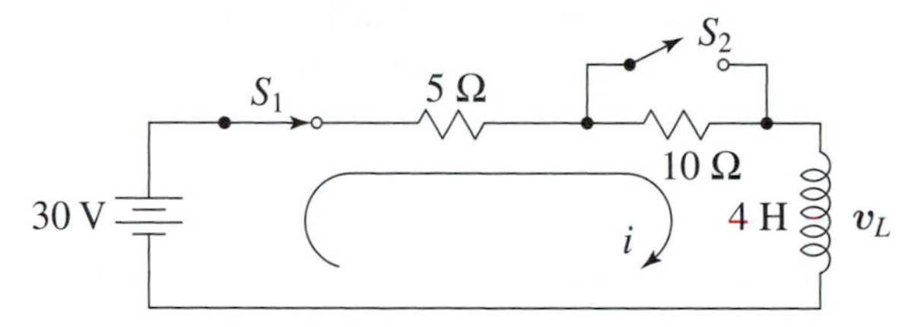

Figure 4.1.8
The circuit for Exercise 8.

9. Figure 4.1.9 illustrates an automobile ignition circuit. The switch is normally open, but in each cycle of the engine, the switch is briefly closed. The spark plug is ignited by a voltage spike produced in the ignition coil. This voltage spike has a magnitude proportional to the voltage in the primary winding.

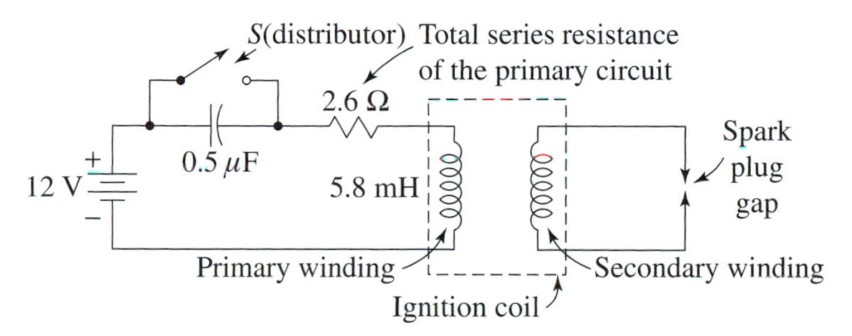

Figure 4.1.9
The circuit for Exercise 9.

a. When the switch is open, the circuit is an *RLC* series circuit. After some time, the capacitor is fully charged and the current drops to zero. When the switch is closed, the effect is to discharge the capacitor immediately and to effectively remove it from the circuit. Thus, the circuit is an *RL* series circuit when the switch is closed. Determine the voltage in the primary winding after the switch is closed, assuming that the current is initially zero.

b. Assume that the steady-state values for the voltage in the primary winding and the current are obtained before the switch opens. Determine the voltage in the primary winding after the opening of the switch.

c. Compare the results of parts a and b. Determine at what point in the cycle the spark plug ignites. (Keep in mind that the switch is closed only briefly and that the ignition is caused by a spike in v_L.)

10. Repeat the calculations of Model Problem 4.1, using resistance values of 0 Ω, 1000 Ω, and 4000 Ω. Discuss how the voltage and current depend on the resistance in the circuit. In particular, why is the current sometimes negative in some circuits and not in others? What is the general effect of increasing the resistance in a circuit?

11. To study the effect of the inductor on an *RLC* series circuit, consider such a circuit initially at rest, with $R = 2$, $C = 1$, and $E = 1$. (These values are chosen to simplify computations.)
    a. Determine the value $L = L_0$ for which the system is critically damped.
    b. Determine the current for the cases $L = 0$, $L = 2L_0$, $L = 5L_0$, and $L = 10L_0$.
    c. Plot the current for each of the cases in part *b*.
    d. Discuss the effect of the inductor on the behavior of the circuit.

✦ 4.1 INSTANT EXERCISE SOLUTION

1. Assume $y = e^{rt}$. Substituting this into the differential equation yields the characteristic equation $0.0001r^2 + 0.01r + 1 = 0$, or $r^2 + 100r + 10{,}000 = 0$. Using the quadratic formula, we see the roots are $r = -50 \pm 50\sqrt{3}\,i$. Thus, the general solution is $y = c_1 e^{-50t}\cos(50\sqrt{3}\,t) + c_2 e^{-50t}\sin(50\sqrt{3}\,t)$. The derivative of the solution is $y' = (-50c_1 + 50\sqrt{3}\,c_2)e^{-50t}\cos(50\sqrt{3}\,t) + (-50\sqrt{3}\,c_1 - 50c_2)e^{-50t}\sin(50\sqrt{3}\,t)$. The initial conditions yield the equations $-3 = c_1$ and $0 = -50c_1 + 50\sqrt{3}\,c_2$. Hence, $c_1 = -3$ and $c_2 = -3/\sqrt{3} = -\sqrt{3}$.

4.2 General Solutions for Nonhomogeneous Equations

Recall that linear differential equations were written in Chapter 3 as $L[y] = g$, where the left side is the image of the unknown function y under a linear differential operator L and the right side is a given function of the independent variable. We turn now to the problem of finding solutions for such differential equations. As stated in Chapter 3, linear differential operators satisfy the linearity property

$$L[c_1 y_1 + c_2 y_2] = c_1 L[y_1] + c_2 L[y_2] \tag{1}$$

and can be written as

$$L[y](t) = p_0(t)\frac{d^n y}{dt^n} + p_1(t)\frac{d^{n-1} y}{dt^{n-1}} + \cdots + p_n(t)y, \tag{2}$$

where n is the order of the highest derivative in L.

There are several different methods for finding solutions of nonhomogeneous linear equations $L[y] = g$, where g is not identically zero. The case where g is constant was treated in Section 4.1, but the method of that section does not generalize to the case where g is not constant. In this section, we begin to develop the general method of undetermined coefficients, which can be used when g belongs to a particular class of functions. The main focus of this section is the structure of solutions of nonhomogeneous linear equations rather than solution methods. Knowledge of the structure of solutions makes it easier to find solutions.

MODEL PROBLEM 4.2

Find the general solution of the differential equation

$$y' + 3y = 4e^{-t}.$$

Finding a Particular Solution

We can solve Model Problem 4.2 in the absence of a formal method by reading the equation carefully. It is of the form $L[y] = g$, where $L[y] = y' + 3y$ and $g(t) = 4e^{-t}$. Given any function y, $L[y]$ is the sum of the derivative and 3 times the function. Solutions of $L[y] = g$ are functions for which the linear combination $L[y]$ is the correct multiple of e^{-t}. Notice that $L[e^{-t}] = (-e^{-t}) + 3(e^{-t}) = 2e^{-t}$. The derivative of e^{-t} is a constant multiple of e^{-t}, and this property carries over to $L[e^{-t}]$ as well.

We haven't found a solution of $L[y] = g$, but we have found a function that has the right general behavior, and that is sufficient because of the linearity of L. Suppose we try Ae^{-t} instead of e^{-t}, where A is a constant to be determined later. Then

$$L[Ae^{-t}] = AL[e^{-t}] = A(2e^{-t}) = 2Ae^{-t}.$$

While $L[e^{-t}]$ is specifically 2 times e^{-t}, $L[Ae^{-t}]$ can be made to equal *any* multiple of e^{-t}. In particular, we obtain $L[Ae^{-t}] = 4e^{-t}$ by choosing $A = 2$. We have therefore found a particular solution

$$y_p = 2e^{-t},$$

as defined in Section 4.1.

Note the role each number appearing in the differential equation plays in determining the appropriate family (Ae^{-t}) in which a particular solution can be found. The correct family, at least for this example, depends only on the factor (-1) in the forcing function $4e^{-t}$. The coefficient 3 in the operator L and the coefficient 4 of g contributed to the determination of the coefficient A but not to the choice of the family Ae^{-t}.

 **INSTANT EXERCISE 1**

Find a particular solution for the differential equation

$$y' + 3y = 18e^{3t}.$$

Finding the General Solution

Model Problem 4.2 is first-order, so it should have a one-parameter family of solutions. A clue for finding the other solutions comes from Model Problem 4.1.

EXAMPLE 1

Consider the differential equation

$$0.0001v'' + 0.01v' + v = 3$$

from Model Problem 4.1. This differential equation clearly has a particular solution $v_p = 3$. The complementary solution, as defined in Section 4.1, is the general solution of the associated homogeneous differential equation

$$0.0001v'' + 0.01v' + v = 0.$$

As we saw earlier, the complementary solution is

$$v_c = e^{-50t}\left[c_1 \cos(50\sqrt{3}\,t) + c_2 \sin(50\sqrt{3}\,t)\right].$$

Let

$$v = v_p + v_c = 3 + e^{-50t}\left[c_1 \cos(50\sqrt{3}\,t) + c_2 \sin(50\sqrt{3}\,t)\right].$$

All these functions are solutions of the nonhomogeneous differential equation.

Perhaps the same idea works for Model Problem 4.2. Note that c_1e^{-3t} is the general solution of $y' + 3y = 0$. Let $y = 2e^{-t} + c_1e^{-3t}$. For this function, $y' = -2e^{-t} - 3c_1e^{-3t}$, and

$$L[2e^{-t} + c_1e^{-3t}] = (-2e^{-t} - 3c_1e^{-3t}) + 3(2e^{-t} + c_1e^{-3t}) = 4e^{-t} + 0e^{-3t} = g(t).$$

Thus, $y = 2e^{-t} + c_1e^{-3t}$ is a one-parameter family of solutions of $L[y] = g$.

At first it may seem surprising that $y = 2e^{-t} + e^{-3t}$, for example, can be a solution when $y_p = 2e^{-t}$ is a solution. Adding the extra term e^{-3t} doesn't seem to matter. This is not a coincidence, however, but a consequence of two properties:

1. L is linear.
2. $y = e^{-3t}$ is a solution of $L[y] = 0$.

Using the notation $y_1 = e^{-3t}$, we can rewrite the relevant calculation as

$$L[y_p + c_1y_1] = L[y_p] + c_1L[y_1] = g + (c_1)(0) = g.$$

✦ INSTANT EXERCISE 2

Find a one-parameter family of solutions for the differential equation

$$y' + 3y = 18e^{3t}.$$

The Structure of Solutions of $L[y] = g$

We now have a one-parameter family of solutions of $y' + 3y = 4e^{-t}$. Are there any solutions that are not in this family? It is not difficult to answer this question in the negative. Suppose y is any solution of $y' + 3y = 4e^{-t}$. Let $u = y - 2e^{-t}$. Then

$$L[u] = L[y - 2e^{-t}] = L[y] - L[2e^{-t}] = 4e^{-t} - 4e^{-t} = 0.$$

Thus, u must be a solution of $L[y] = 0$. More generally, a similar calculation shows that the difference between any two solutions of $L[y] = g$ is a solution of $L[y] = 0$. This is a general property of nonhomogeneous linear equations.

✦ INSTANT EXERCISE 3

Use the linearity property of L to prove that $L[z - y] = 0$ for any two solutions y and z of a linear differential equation $L[y] = g$.

Theorem 4.2.1 summarizes the result of Instant Exercise 3.

Theorem 4.2.1

Let $L[y] = g$ be a nonhomogeneous linear differential equation, and let I be an interval on which any initial-value problem for this equation has a unique solution. If y_p is a particular solution of $L[y] = g$ and y_c is the general solution of the associated homogeneous equation $L[y] = 0$, then the general solution of the equation $L[y] = g$ is $y = y_p + y_c$.

Theorem 4.2.1 reduces the problem of finding the[6] general solution of $L[y] = g$ to two subproblems. We must find the general solution of $L[y] = 0$ and a particular solution of $L[y] = g$. The problem of finding the general solution of $L[y] = 0$ was the primary subject of Chapter 3. In the remainder of this chapter, we develop methods for finding particular solutions for equations where the complementary solution is known.

Two Simple Cases of the Method of Undetermined Coefficients

The solution technique illustrated by Model Problem 4.2 is called, appropriately enough, the *method of undetermined coefficients.* Finding a particular solution by this method involves three steps.

1. Find a **trial solution,** a family of functions Y that includes a correct particular solution. A trial solution is not a particular solution, but rather a family that includes one; generally only one set of coefficients in a correct trial solution yields a particular solution. *In general, trial solutions are based primarily on the general form of the forcing function* g. The coefficients of L play a role that will be determined later. The coefficients in g never play any role in determining a correct trial solution.
2. Find $L[Y]$, the image of the trial solution Y under the operator L.
3. Determine the correct coefficient values in the trial solution by setting $L[Y]$ equal to g.

We now consider two simple cases of the method of undetermined coefficients.

Exponential Forcing Our choice of trial solution for the model problem was based on two properties of the differential equation:

1. The left side of the equation is a linear operator L with constant coefficients.
2. The right side of the equation is an exponential function.

Here is an additional example.

EXAMPLE 2

Consider the equation

$$y'' + 4y = e^{-3t} = g(t).$$

[6]Keep in mind that there may be more than one formula for the general solution; however, the different formulas represent the same general solution.

As before, we guess that there is a particular solution in the family

$$Y = Ae^{-3t}.$$

Substituting this result into the left side gives

$$Y'' + 4Y = 9Ae^{-3t} + 4Ae^{-3t} = 13Ae^{-3t}.$$

The image of Y is g, provided $13A = 1$. Thus,

$$y_p = \tfrac{1}{13}e^{-3t}.$$

To complete the general solution, we need to solve the homogeneous equation

$$y'' + 4y = 0.$$

We can solve this equation by assuming solutions of the form $y = e^{rt}$, but by now the reader should recognize that the solution is

$$y_c = c_1 \cos 2t + c_2 \sin 2t.$$

By Theorem 4.2.1, the general solution is

$$y = \tfrac{1}{13}e^{-3t} + c_1 \cos 2t + c_2 \sin 2t.$$

So far, exponential forcing has been true to a simple pattern: Ae^{rt} is a trial solution for any equation $L[y] = we^{rt}$. It would be nice if the trial solution could always be so easily determined, but that is not the case, as Examples 3 and 4 illustrate.

EXAMPLE 3

Consider the differential equation

$$x^2y'' - 2xy' + 2y = e^x.$$

We might think that there should be a particular solution that is a multiple of e^x, but this does not work:

$$x^2(Ae^x) - 2x(Ae^x) + 2(Ae^x) = A(x^2 - 2x + 2)e^x,$$

so there is no choice of the constant A for which $L[Ae^x] = e^x$.

The method we are developing depends on being able to predict a function whose image under L differs from g only by a constant factor. As demonstrated by Example 3, this is likely only when the coefficients in the linear differential operator L are constants.

EXAMPLE 4

Consider the equation

$$y' + 3y = 4e^{-3t}.$$

As before, we assume a trial solution $Y = Ae^{-3t}$. Substituting this form into the left side of the differential equation gives us

$$Y' + 3Y = -3Ae^{-3t} + 3Ae^{-3t} = 0.$$

There is no choice of A that makes $0 = 4e^{-3t}$. The family Ae^{-3t} does not contain a particular solution.

In Example 4, the expected Ae^{rt} is not a trial solution for $L[y] = we^{rt}$, even though L is linear and has constant coefficients. This time, the problem is that $L[Ae^{rt}] = 0$. Ultimately, we need a general rule that prescribes a trial solution for $g(t) = we^{rt}$ even if $L[e^{rt}] = 0$. This issue is addressed in the exercises and settled in Section 4.3. For now, we have the following rule that addresses the standard case.

Let L be a linear differential operator with constant coefficients, and let r and w be constants. The differential equation $L[y] = we^{rt}$ has a trial solution $Y = Ae^{rt}$, provided $L[e^{rt}] \neq 0$.

EXAMPLE 5

Consider the differential equation

$$L[y] = y'' + 6y = e^{2t}.$$

As before, it seems likely that we can use the trial solution

$$y_p = Ae^{2t}.$$

We then have

$$L[y_p] = y_p'' + 6y_p = 4Ae^{2t} + 6Ae^{2t} = 10Ae^{2t}.$$

We want to have $L[y_p] = e^{2t}$, so we must have $10A = 1$, or $A = \frac{1}{10}$. We have found the particular solution

$$y_p = \tfrac{1}{10}e^{2t}.$$

The method also works when the right side is a constant, since a constant is a multiple of the exponential function e^{0t}.

EXAMPLE 6

For the differential equation

$$y'' + 2y = 6,$$

we try the trial solution

$$y_p = A.$$

Then

$$y_p'' + 2y_p = 0 + 2A = 2A.$$

Setting this result equal to 6 yields the particular solution

$$y_p = 3.$$

Example 6 is similar to the problems we solved in Section 4.1. In the method of that section, we find the particular solution $y_p = 3$ and then look for solutions of the form $y = 3 + y_c$ by writing down the differential equation for y_c and solving it. The procedure we are developing now is a little different in that we construct the particular and complementary solutions separately and add them. Of course the results are the same.

Sinusoidal Forcing Many practical problems involve forcing by a sinusoidal function; for example, an *RLC* circuit with an alternating-current (ac) power source is represented by a linear differential equation with constant coefficients and sinusoidal forcing. A first approximation to the mechanical vibration in an automobile caused by a rough road might also be a linear differential equation with constant coefficients and sinusoidal forcing.

EXAMPLE 7

Consider the differential equation

$$L[y] = y' + 4y = 5\cos 2t.$$

If we take $y_p = A\cos 2t$, we get

$$L[y_p] = y_p' + 4y_p = (-2A\sin 2t) + 4(A\cos 2t) = 4A\cos 2t - 2A\sin 2t.$$

There is no constant A that makes $L[y_p] = 5\cos 2t$. Taking $A = 5/4$ will give us the right coefficient for the cosine term, but then we have the extra sine term. This means that $y_p = A\cos 2t$ is *not* a correct trial solution.

The clue to solving the problem of Example 7 is to notice that the image of a sine or cosine function under a linear differential operator L with constant coefficients is a sinusoidal function that generally includes both sine and cosine terms. This suggests that a trial solution for a problem like that of Example 7 should have both kinds of sinusoidal functions.

EXAMPLE 8

Consider again the differential equation

$$y' + 4y = 5\cos 2t.$$

We try the trial solution

$$y_p = A\cos 2t + B\sin 2t.$$

Substituting into the left side of the differential equation, we have

$$y_p' + 4y_p = (-2A\sin 2t + 2B\cos 2t) + 4(A\cos 2t + B\sin 2t) = (2B + 4A)\cos 2t + (4B - 2A)\sin 2t.$$

Setting this result equal to $5\cos 2t$ results in the algebraic equation

$$(2B + 4A)\cos 2t + (4B - 2A)\sin 2t = 5\cos 2t.$$

The cosine terms match if $2B + 4A = 5$, and the sine terms match if $4B - 2A = 0$. The second equation gives the result $A = 2B$, whence the first becomes $2B + 8B = 5$. Thus, $B = \frac{1}{2}$ and $A = 1$. We have found the particular solution

$$y_p = \cos 2t + \tfrac{1}{2}\sin 2t.$$

The complementary solution is $y_c = c_1 e^{-4t}$; hence, the general solution is

$$y = \cos 2t + \tfrac{1}{2}\sin 2t + c_1 e^{-4t}.$$

INSTANT EXERCISE 4

Find a particular solution and the general solution for $y' + 2y = 5\cos t$.

As with the exponential case, the method does not work for all problems.

EXAMPLE 9

Consider the differential equation

$$L[y] \equiv y'' + 9y = \cos 3t.$$

Following the method of the previous examples, we should try

$$y_p = A\cos 3t + B\sin 3t.$$

However,

$$L[A\cos 3t + B\sin 3t] = 0,$$

indicating that a particular solution cannot be found in the indicated family of functions.

The results are summarized in the following rule:

> Let L be a linear differential operator with constant coefficients, and let ω, c, and s be constants, with $\omega \neq 0$ and s and c not both zero. The differential equation $L[y] = c\cos\omega t + s\sin\omega t$ has a particular solution of the form $y_p = A\cos\omega t + B\sin\omega t$, provided $L[\cos\omega t] \neq 0$.

Note that all our examples have had $s = 0$; however, it should make no difference whether the function g is a cosine, a sine, or a sum of cosine and sine, as long as the circular frequency ω is the same for both terms. Note also that either both of the functions $\sin\omega t$ and $\cos\omega t$ are solutions of $L[y] = 0$ or neither of them is,[7] so the rule need only mention one of the functions as possible solutions.

Looking Ahead

So far we have developed only a portion of the method of undetermined coefficients. The full method works whenever it is possible to prescribe a trial solution that is guaranteed to include a

[7]Both solutions come from the same complex conjugate pair of characteristic values.

particular solution, given the correct choice of coefficients. This turns out to be possible whenever the linear operator L has constant coefficients and the function g is a polynomial, an exponential, a sinusoidal, a product of such functions, or a sum of products of such functions. Section 4.3 completes the development of the method by constructing a set of general rules for determining trial solutions.

4.2 Exercises

In Exercises 1 through 8, find the general solution of the differential equation.

1. $y'' - 4y = e^t$
2. $y' + 4y = e^{-3t}$
3. $y'' + 6y' + 9y = e^{2t}$
4. $y'' + 6y' + 10y = e^{-3t}$
5. $y' + 4y = \cos 3t$
6. $y'' + y' = \cos t$
7. $y'' + 4y = \sin 3t$
8. $y'' + 6y' + 10y = \sin t$

In Exercises 9 through 12, solve the initial-value problem.

9. $y' + 2y = 3e^t, \quad y(0) = 3$
10. $y'' + y = 1, \quad y(0) = 0, \quad y'(0) = 0$
11. $y' + 5y = 3\sin 2t, \quad y(0) = 3$
12. $y'' + 2y' + y = \sin t, \quad y(0) = 0, \quad y'(0) = 0$
13. *a.* Try to find a particular solution for $y' + 3y = \cos 2t - \sin 2t$ by using the trial solution $y_p = A\cos 2t + B\sin 2t$.
    b. Try to find a particular solution for $y' + 3y = \cos 2t - \sin 4t$ by using the trial solution $y_p = A\cos 2t + B\sin 4t$.
    c. Find a particular solution for $y' + 3y = \cos 2t - \sin 4t$ by using an appropriate trial solution. (*Hint:* the trial solution should have four undetermined coefficients.)
14. Consider the differential equation $L[y] = y' + 2y = e^{-2t}$.
    a. Determine $L[e^{-2t}]$.
    b. Determine $L[te^{-2t}]$.
    c. Use the results of parts a and b to find a particular solution.
    d. Conjecture the form of the correct rule for choosing a trial solution for $L[y] = we^{rt}$, with r and w constants and L a linear differential operator with constant coefficients. The rule has to have two cases, depending on whether or not e^{rt} is a solution of $L[y] = 0$.
15. Consider the differential equation $L[y] = y' + 2y = 1 + t^2$.
    a. Determine $L[1]$.
    b. Determine $L[t]$.
    c. Determine $L[t^2]$.
    d. Use the results of parts a, b, and c to find a particular solution.
    e. Conjecture the form of the correct rule for choosing a trial solution for $y' + py = C(t)$, with $p \neq 0$ and C a polynomial of degree k.

✦ 4.2 INSTANT EXERCISE SOLUTIONS

1. Because a linear combination of an exponential function and its derivatives is an exponential function with the same rate factor, it seems reasonable to try $y = e^{3t}$. This yields $L[e^{3t}] = 6e^{3t}$. Thus, $L[Ae^{3t}] = AL[e^{3t}] = 6Ae^{3t}$. With $A = 3$ we get the desired result, so we have found a particular solution $y_p = 3e^{3t}$.
2. We already have a particular solution $y_p = 3e^{3t}$, and we have the general solution $y_c = c_1e^{-3t}$ of $L[y] = 0$. All functions in the family $y = 3e^{3t} + c_1e^{-3t}$ are solutions of $L[y] = g$ because $L[y_p + y_c] = L[y_p] + L[y_c] = g + 0 = g$.
3. Let y and z be solutions of $L[y] = g$ on an interval I. Then $L[z - y] = L[z] - L[y] = g - g = 0$.
4. To get $L[y] = 5\cos t$, we need a general enough trial function to yield $L[y] = 5\cos t + 0\sin t$ with correct choice of coefficients. As in the previous example, we try $y_p = A\cos t + B\sin t$. Now, $L[y_p] = (-A\sin t + B\cos t) + 2(A\cos t + B\sin t) = (2A + B)\cos t + (-A + 2B)\sin t$. The image is correct if we can choose A and B so that $2A + B = 5$ and $-A + 2B = 0$. The second equation yields $A = 2B$, whence the first becomes $5B = 5$. Thus $B = 1$ and $A = 2$. We have $y_p = 2\cos t + \sin t$, and by Theorem 4.2.1, $y = 2\cos t + \sin t + c_1e^{-2t}$ is the general solution.

4.3 The Method of Undetermined Coefficients

The method of undetermined coefficients is a method for finding particular solutions for some differential equations of the form $L[y] = g$. The method consists of three steps.

1. Find a trial solution Y, a family of functions that is guaranteed to include a correct particular solution.
2. Find $L[Y]$, the image of the trial solution Y under the operator L.
3. Determine the correct parameter values in the trial solution by setting $L[Y]$ equal to g.

Notice that none of the steps in the method actually requires integration, and this is a big advantage of the method. Unfortunately, the class of problems for which the method is applicable is somewhat limited. In practice, the method can be used for many differential equations of practical interest. Mastery of the method of undetermined coefficients involves three different skills:

1. Knowing whether or not a given problem can be solved by the method.
2. Determining the correct trial solution for problems for which the method works.
3. Being able to find the correct particular solution, given the correct trial solution.

We consider each of these issues, beginning with the simplest, after a brief overview.

Overview of the Method

Let Y and G represent two families of functions, with the forcing function g a member of the family G. The method of undetermined coefficients works when it is possible to choose a family Y that is sufficiently general so that $L[Y]$ contains the family G. In mathematical notation, we write $G \subseteq L[Y]$ or $L[Y] \supseteq G$. This is illustrated schematically in Figure 4.3.1. Intuitively, this is more likely to occur when the image $L[Y]$ is especially simple. This suggests that both the linear differential operator L and the forcing function g have to be "nice" in some way. In particular, the method always works when the linear differential operator L has constant coefficients and g is a function that solves some homogeneous linear equation with constant coefficients.

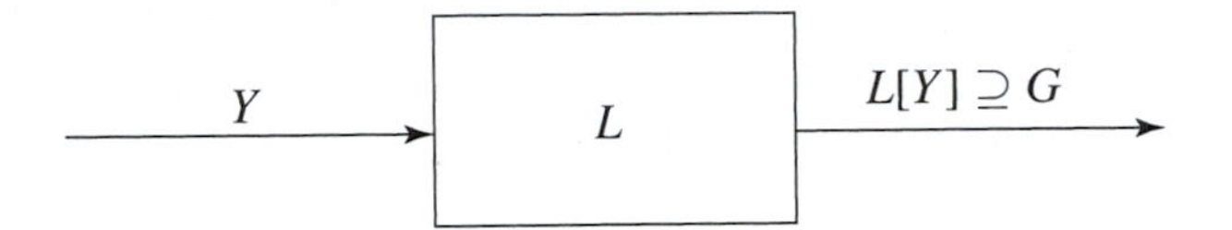

Figure 4.3.1
A schematic diagram showing the operator L acting on a trial solution and producing a family of images that contains the family of forcing functions.

Once it is determined that the linear differential operator L and the forcing function g are of the kinds for which the method works, the crucial step is to determine a correct trial solution Y for the problem. There is a standard case, in which the obvious guess for the trial solution works, and a special case, for which the obvious guess needs to be modified. Finally, it is helpful to show that a function g can be divided into components such that the correct trial solution for g is the sum of the trial solutions for the components.

Finding a Correct Particular Solution, Given a Correct Trial Solution

As in the examples of Section 4.2, finding the correct particular solution is a mechanical procedure if the correct trial solution is known.

EXAMPLE 1

Consider the equation

$$y'' - 9y = 12e^{3t}.$$

Suppose that steps 1 and 2 of the procedure have already been completed, with the result that a correct trial solution is

$$Y = Ate^{3t}.$$

Thus,

$$Y' = Ae^{3t} + At(3e^{3t}) = (3At + A)e^{3t},$$

$$Y'' = (3A)e^{3t} + (3At + A)(3e^{3t}) = (9At + 6A)e^{3t},$$

and

$$Y'' - 9Y = (9At + 6A)e^{3t} - 9(Ate^{3t}) = 6Ae^{3t}.$$

We can now set $L[Y]$ equal to g to obtain

$$6Ae^{3t} = 12e^{3t}.$$

Clearly, $A = 2$. Thus, a particular solution is

$$y_p = 2te^{3t}.$$

The complementary solution is $y_c = c_1e^{-3t} + c_2e^{3t}$, so the general solution is

$$y = 2te^{3t} + c_1e^{-3t} + c_2e^{3t} = c_1e^{-3t} + (2t + c_2)e^{3t}.$$

The action of the linear differential operator is illustrated in Figure 4.3.2. A particular solution can be found for any $G = we^{3t}$ by using the trial solution $Y = Ate^{3t}$ because $6Ae^{3t} = we^{3t}$ can be solved for A given any value of w.

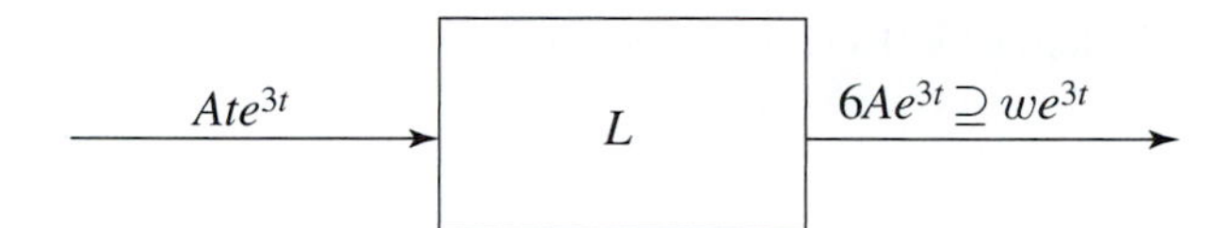

Figure 4.3.2
A schematic diagram showing the operator L of Example 1 acting on the trial solution Y and producing an image that contains the family G.

We deliberately began with a problem for which a trial solution was not obvious in order to make a point. Once a correct trial solution has been determined, the method is completely routine. The real work lies in finding a correct trial solution.

Generalized Exponential Functions

Recall the possibilities for solutions of $L[y] = 0$, where L is a linear differential operator with constant coefficients. As we saw in Chapter 3, such problems are solved by finding constant values r (real or complex) for which e^{rt} solves $L[y] = 0$. A real characteristic value r of multiplicity m gives rise to solutions $e^{rt}, te^{rt}, \ldots, t^{m-1}e^{rt}$ (Theorem 3.6.1). A non-real characteristic value pair $\alpha \pm \beta i$ of multiplicity m gives rise to solutions $e^{\alpha t}\cos\beta t, e^{\alpha t}\sin\beta t, te^{\alpha t}\cos\beta t, te^{\alpha t}\sin\beta t, \ldots, t^{m-1}e^{\alpha t}\cos\beta t, t^{m-1}e^{\alpha t}\sin\beta t$ (Theorem 3.6.2). The functions listed here are the only functions, along with linear combinations of them, that can be solutions of a homogeneous linear differential equation with constant coefficients. These functions are the ones that, as g, enable the use of the standard rules for the method of undetermined coefficients. It is unfortunate that no standard terminology has been devised to describe this collection of functions. Because they arise from an exponential trial solution but include functions that are not exponential, a reasonable name for them is *generalized exponential functions.*

A **generalized exponential function** is a function that can be written as

$$g(t) = e^{\gamma t}[C(t)\cos\omega t + S(t)\sin\omega t], \tag{1}$$

where γ and ω are real numbers and C and S are polynomials. If $\omega = 0$, then the form simplifies to

$$g(t) = C(t)e^{\gamma t}. \tag{2}$$

The two basic properties of generalized exponential functions are the **characteristic value(s)** $r = \gamma \pm i\omega$ and the **degree** k, which is the larger of the degrees of the polynomials C and S for Equation (1) or simply the degree of C for Equation (2).

EXAMPLE 2

The generalized exponential function $\sin 2t$ has characteristic values $\pm 2i$ and degree 0; $(t^2 - 3)e^{-t}$ has characteristic value -1 and degree 2; $e^{2t}(\sin t + t \cos t)$ has characteristic values $2 \pm i$ and degree 1; $e^{2t} + e^t$ is not a generalized exponential function, although it is a sum of two generalized exponential functions with characteristic values 2 and 1 and degree 0.

Solutions of homogeneous linear equations with constant coefficients are either generalized exponential functions or sums of generalized exponential functions.

EXAMPLE 3

The solution of $y'' - 9y = 0$ (Example 1) is

$$y = c_1 e^{-3t} + c_2 e^{3t}.$$

Special cases of this solution family include

$$y = e^{-3t} + e^{3t},$$

which is a sum of two generalized exponential functions, and

$$y = -2e^{3t},$$

which is a generalized exponential function with characteristic value 3.

EXAMPLE 4

The characteristic equation for $y'' + 2y' + 2y = 0$ is $r^2 + 2r + 2 = 0$, which has a complex pair of roots, $r = -1 \pm i$. The solutions $y_1 = e^{-t} \cos t$ and $y_2 = e^{-t} \sin t$, and any linear combination of these, such as $y = e^{-t}(\cos t - \sin t)$, are generalized exponential functions with characteristic value pair $-1 \pm i$.

Examples 3 and 4 illustrate the point that sums of functions having the same characteristic value(s) count as one generalized exponential function, while sums that have different characteristic values count as two different generalized exponential functions.

EXAMPLE 5

The characteristic equation for $y''' + 3y'' + 3y' + y = 0$ is $r^3 + 3r^2 + 3r + 1 = 0$, or $(r + 1)^3 = 0$. Recall that this means that the characteristic value $r = -1$ is of multiplicity 3. The general solution of the equation, by Theorem 3.6.1, is $y = (c_1 + c_2 t + c_3 t^2)e^{-t}$. Assuming $c_3 \neq 0$, these solutions are the family of generalized exponential functions with characteristic value -1 and degree 2.

Example 5 illustrates the correlation between the multiplicity of characteristic values and the maximum degree of the corresponding generalized exponential solutions.

Theorem 4.3.1 A generalized exponential function with characteristic value(s) $\gamma \pm i\omega$ and degree k is a solution of any equation $L[y] = 0$ for which $\gamma \pm i\omega$ is a real characteristic value or a pair of complex characteristic values of L, with multiplicity $m > k$.

In Example 5, the characteristic value -1 of L has multiplicity 3. Thus, all generalized exponential functions with characteristic value -1 and degree less than 3 are solutions of $L[y] = 0$.

Images of Generalized Exponential Functions

The key to determining a correct trial solution is to understand the way the operator L acts on generalized exponential functions to produce an image. If we can predict the form of $L[Y]$, then we ought to be able to use that information to choose Y so that $G \subseteq L[Y]$. We begin with some examples.

EXAMPLE 6

Let L_1 and L_2 be defined by $L_1[y] = y' + 3y$ and $L_2[y] = y' + 2y$. Let $y_1 = e^{-3t}$ and let $y_2 = te^{-3t}$. Then $y_1' = -3e^{-3t}$ and $y_2' = e^{-3t} - 3te^{-3t} = (1 - 3t)e^{-3t}$. Hence,

$$L_1[e^{-3t}] = 0, \qquad L_1[te^{-3t}] = e^{-3t}, \qquad L_2[e^{-3t}] = -e^{-3t}, \qquad L_2[te^{-3t}] = (1 - t)e^{-3t}.$$

These are illustrated schematically in Figure 4.3.3.

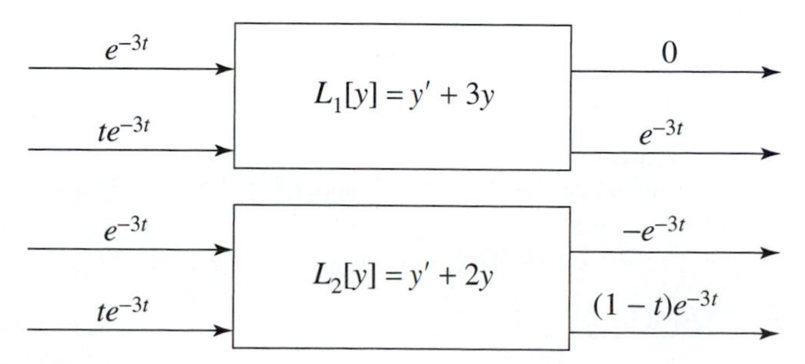

Figure 4.3.3
The images of e^{-3t} and te^{-3t} under $L_1[y] = y' + 3y$ and $L_2[y] = y' + 2y$.

Now consider a more general case, obtained by taking linear combinations of the results of Example 6.

EXAMPLE 7

Let L_2 be defined as in Example 6. Let $y = (A + Bt)e^{-3t}$. Then, using the results of Example 6,

$$L_2[(A + Bt)e^{-3t}] = AL_2[e^{-3t}] + BL_2[te^{-3t}] = A(-e^{-3t}) + B(1 - t)e^{-3t} = (B - A - Bt)e^{-3t}.$$

Notice the pattern. The family of input functions includes all products of polynomials of degree 1 (or less) with e^{-3t}. The images of these functions are also products of polynomials of degree 1 (or less) with e^{-3t}. More importantly, every polynomial of degree 1 or less can be made from the formula $B - A - Bt$, by choosing the appropriate values of A and B.

✦ INSTANT EXERCISE 1

Define L_1 as in Example 6. Determine the family of images $L_1[(A + Bt)e^{-3t}]$.

The results of Example 7 and Instant Exercise 1 are displayed in Figure 4.3.4. What is interesting about the results is the relationship between a family of input functions and the family of its images. The image of a family Y of generalized exponential functions is a family $L[Y]$ of generalized exponential functions with the same characteristic value and the same or lesser degree. The degree is decreased when the characteristic value of Y is a characteristic value of the linear differential operator and is unchanged otherwise.

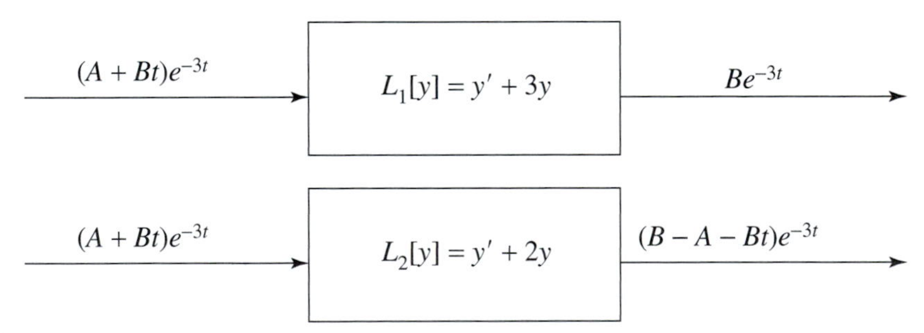

Figure 4.3.4
The images of $(A + Bt)e^{-3t}$ under $L_1[y] = y' + 3y$ and $L_2[y] = y' + 2y$.

EXAMPLE 8

Let $y = At^2e^{-3t}$ and define L by $L[y] = y'' + 6y' + 9y$. Observe that y is a generalized exponential function with characteristic value -3 and degree 2. Calculations show that -3 is a characteristic value of multiplicity 2 of L and $L[At^2e^{-3t}] = 2Ae^{-3t}$. The degree of the image plus the multiplicity of the characteristic value is the degree of the input function.

✦ **INSTANT EXERCISE 2**

Verify the claims in Example 8 by factoring the characteristic polynomial of L and computing $L[At^2e^{-3t}]$.

The observations of Examples 7 and 8 hold true for any generalized exponential function. The results are summarized in Theorem 4.3.2.

Theorem 4.3.2 Let Y be the family of generalized exponential functions with characteristic value $\gamma \pm i\omega$ and degree k. Let L be a linear differential operator with constant coefficients. If $\gamma \pm i\omega$ is a characteristic value of L, then let m be its multiplicity; otherwise, let $m = 0$. If $k \geq m$, then $L[Y]$ is the family of generalized exponential functions with characteristic value $\gamma \pm i\omega$ and degree $k - m$.

 **INSTANT EXERCISE 3**

What happens in the setting of Theorem 4.3.2 if $k < m$?

Determining a Correct Trial Solution

MODEL PROBLEM 4.3

Find particular solutions for each of the following differential equations:

$$y' + 3y = 2e^{-3t}\cos t$$

$$y' + 3y = 18t\cos 3t + 6\sin 3t$$

$$y' + 3y = 4te^{-3t}$$

$$y' + 3y = 2e^{-3t}\cos t + 4te^{-3t}$$

Case I: The Characteristic Value of g Is Not a Characteristic Value of L Consider the first equation in Model Problem 4.3. The operator L defined by $L[y] = y' + 3y$ has a single characteristic value, -3. The forcing function g has characteristic value $-3 \pm i$ and degree 0. Since $-3 \pm i$ is not a characteristic value of L, Theorem 4.3.2 indicates that a family of generalized exponential functions with characteristic value $-3 \pm i$ and degree 0 will have an image with the same characteristic value and degree. Thus, we choose the trial solution

$$Y = e^{-3t}(C\cos t + S\sin t).$$

Note that it does not matter whether g includes only a cosine term, only a sine term, or both terms. Each of these is a generalized exponential function with characteristic value $-3 \pm i$ and degree 0.

Thus,

$$Y' = -3e^{-3t}(C\cos t + S\sin t) + e^{-3t}(-C\sin t + S\cos t)$$
$$= e^{-3t}[(S - 3C)\cos t + (-3S - C)\sin t].$$

Hence,

$$L[Y] = e^{-3t}(S\cos t - C\sin t).$$

The results are depicted in Figure 4.3.5.

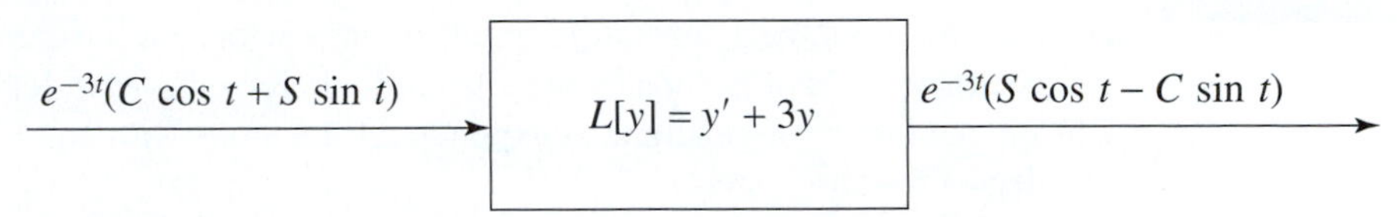

Figure 4.3.5
The image of $e^{-3t}(C\cos t + S\sin t)$ under $L[y] = y' + 3y$.

As predicted by Theorem 4.3.2, the image has the right characteristic value and degree. Setting $L[y]$ equal to $g(t)$, we obtain the equations $S = 2$ and $-C = 0$. We have found the particular solution

$$y_p = 2e^{-3t}\sin t.$$

Had we omitted the sine term from the trial solution, we would have gotten the equations $0 = 2$ and $-C = 0$, and there would have been no solution. The only way to be sure that the desired function g is in the image $L[Y]$ is to include in Y all functions having the appropriate characteristic value and degree.

The second equation in Model Problem 4.3 is similar. The characteristic value $\pm 3i$ of g is not a characteristic value of L. This time, however, the degree of g is 1 rather than 0. The obvious choice for a trial solution is the family of all functions having characteristic value $\pm 3i$ and degree 1. Thus, we try

$$Y = (C_0 + C_1 t)\cos 3t + (S_0 + S_1 t)\sin 3t.$$

From this trial solution, we have

$$L[Y] = [(3C_0 + C_1 + 3S_0) + (3C_1 + 3S_1)t]\cos 3t + [(3S_0 + S_1 - 3C_0) + (3S_1 - 3C_1)t]\sin 3t.$$

✦ INSTANT EXERCISE 4

Verify the last result by computing $L[Y]$ with $L[y] = y' + 3y$ and $Y = (C_0 + C_1 t)\cos 3t + (S_0 + S_1 t)\sin 3t$.

Although the formula we got for $L[Y]$ looks messy, it has a simple structure. It has two terms, each of which is just a first-degree polynomial multiplied by a sinusoidal function. To finish the problem, we have to set the result for $L[Y]$ equal to g:

$$[(3C_0 + C_1 + 3S_0) + (3C_1 + 3S_1)t]\cos 3t + [(3S_0 + S_1 - 3C_0) + (3S_1 - 3C_1)t]\sin 3t$$
$$= 18t\cos 3t + 6\sin 3t.$$

The two quantities in this equation have to be equal for all t, and this requires that corresponding coefficients be equal. We therefore have a set of four equations:

$$3C_0 + C_1 + 3S_0 = 0, \qquad 3C_1 + 3S_1 = 18, \qquad 3S_0 + S_1 - 3C_0 = 6, \qquad 3S_1 - 3C_1 = 0.$$

Four equations in four unknowns can be quite difficult, but these are linear equations and they are not fully coupled. The second and fourth equations can be solved first, with the result

$$C_1 = S_1 = 3.$$

The remaining equations are then

$$3C_0 + 3S_0 = -3, \qquad 3S_0 - 3C_0 = 3.$$

These equations are best solved by taking linear combinations. Adding the equations gives the equation $6S_0 = 0$; hence, $S_0 = 0$. With this substitution we obtain $C_0 = -1$. We have finally arrived at a particular solution:

$$y_p = (-1 + 3t)\cos 3t + 3t \sin 3t.$$

These two model problems illustrate the following rule:

When the characteristic value of g is not a characteristic value of L, the trial solution should be the most general family of functions having the same characteristic value and degree as g.

EXAMPLE 9

Suppose we want a particular solution for

$$y'' + y = t \sin 2t.$$

We have $P(r) = r^2 + 1$, so the characteristic value of L is $r = \pm i$. The trial solution is then

$$Y = (A + Bt)\cos 2t + (C + Dt)\sin 2t.$$

Thus,

$$L[Y] = [-3Bt + (-3A + 4D)]\cos 2t + [-3Dt + (-4B - 3C)]\sin 2t.$$

Setting $L[Y]$ equal to $t \sin 2t$ yields the particular solution

$$y_p = -\frac{4}{9}\cos 2t - \frac{1}{3}t \sin 2t.$$

✦ INSTANT EXERCISE 5

Verify the computations of $L[Y]$ and the coefficients of y_p in Example 9.

Case II: The Characteristic Value of g Is a Characteristic Value of L Consider the third equation of Model Problem 4.3. The forcing function has characteristic value -3, and the operator

L also has characteristic value -3. The trial solution $(A + Bt)e^{-3t}$ does not work for the operator L defined by $L[y] = y' + 3y$ because of the correspondence of characteristic values. By Theorem 4.3.2, the degree of the image family is 1 less than the degree of the trial solution. It is therefore logical to think that we require a trial solution of degree 2 to get an image of degree 1. As we raise the degree of the trial solution, we do not need to add extra terms because these terms will have an image of 0. Hence, we have the trial solution

$$Y = (At + Bt^2)e^{-3t}.$$

From this trial solution, we get[8]

$$Y' = (A + 2Bt)e^{-3t} - 3(At + Bt^2)e^{-3t} = [A + (2B - 3A)t - 3Bt^2]e^{-3t}.$$

Thus,

$$L[Y] = Y' + 3Y = (A + 2Bt)e^{-3t}.$$

As expected, the image of the trial solution is the family of generalized exponential functions with the same characteristic value, but with degree 1 less. Setting this result equal to $4te^{-3t}$, we find $B = 2$ and $A = 0$. Thus, the particular solution is

$$y_p = 2t^2e^{-3t}.$$

As in the model problems and examples, Theorem 4.3.2 gives us all the information we need to find the right trial solution when g is a generalized exponential function. We begin with the family $Y = G$, where G is the family of generalized exponential functions having the same characteristic value and degree as g. If the characteristic value of g is also a characteristic value of multiplicity m of L, then we use $Y = t^m G$.

Theorem 4.3.3

Let g be a generalized exponential function with characteristic value $\gamma \pm i\omega$ and degree k, and let L be a linear differential operator with constant coefficients. If $\gamma \pm i\omega$ is a characteristic value of L, then let m be its multiplicity; otherwise, let $m = 0$. The appropriate trial solution for the equation $L[y] = g$ is

$$Y = t^m e^{\gamma t}[A(t)\cos\omega t + B(t)\sin\omega t],$$

where A and B are arbitrary polynomials of degree k.

✦ INSTANT EXERCISE 6

Use Theorem 4.3.3 to determine the correct trial solution for the differential equation of Example 1.

---

[8]Here, as elsewhere, it is best to simplify formulas to the standard form of generalized exponential functions. This makes the subsequent calculations easier.

EXAMPLE 10

Consider the equation

$$y'' + 6y' + 9y = 6e^{-3t}.$$

The characteristic value of g is again -3. The characteristic polynomial of L is $r^2 + 6r + 9 = (r + 3)^2$, so -3 is a characteristic value of multiplicity 2. Theorem 4.3.3 gives us the trial solution

$$Y = t^2 e^{-3t}[A_0],$$

or

$$Y = At^2 e^{-3t}.$$

✦ INSTANT EXERCISE 7

Verify the conclusion of Example 10 by computing $L[At^2 e^{-3t}]$. Determine the corresponding particular solution.

Case III: g Is a Sum of Generalized Exponential Functions Consider the final equation in Model Problem 4.3:

$$y' + 3y = 2e^{-3t}\cos t + 4te^{-3t}.$$

The forcing function is a sum of two generalized exponential functions. Since these functions have different characteristic values, two distinct trial solutions are required. We have already found the correct trial solutions and the corresponding coefficients; we define these as

$$y_{p_1} = 2e^{-3t}\sin t, \qquad y_{p_2} = 2t^2 e^{-3t}.$$

Clearly, the desired particular solution is

$$y_p = 2e^{-3t}\sin t + 2t^2 e^{-3t};$$

then

$$L[y_p] = L[2e^{-3t}\sin t] + L[2t^2 e^{-3t}] = 2e^{-3t}\cos t + 4te^{-3t}.$$

This problem illustrates a general theorem, which is simply a restatement of the linearity property.

Theorem 4.3.4

Let L be a linear differential operator with constant coefficients. Let $g_1, g_2, \ldots, g_j$ be forcing functions, and let $y_{p_1}, y_{p_2}, \ldots, y_{p_j}$ be particular solutions of the equations $L[y] = g_1$, $L[y] = g_2, \ldots, L[y] = g_j$. Then a particular solution of

$$L[y] = g_1 + g_2 + \cdots + g_j$$

is

$$y_p = y_{p_1} + y_{p_2} + \cdots + y_{p_j}.$$

EXAMPLE 11

Consider the differential equation

$$y'' + y = 2e^t + t^2 + 2.$$

The forcing function can be thought of as having two parts:

$$g_1(t) = 2e^t, \qquad g_2(t) = t^2 + 2,$$

each corresponding to a different characteristic value. Since neither 1 nor 0 is a characteristic value of L, we have particular solutions of the form

$$Y_1 = Ae^t, \qquad Y_2 = B + Ct + Dt^2.$$

Then $L[Y_1] = 2Ae^t$ and $L[Y_2] = (B + 2D) + Ct + Dt^2$, leading to particular solutions $y_{p_1} = e^t$ and $y_{p_2} = t^2$. Hence, the full particular solution is

$$y_p = e^t + t^2$$

and the general solution is

$$y = e^t + t^2 + c_1 \cos t + c_2 \sin t.$$

Nonhomogeneous Cauchy–Euler Equations

The method of undetermined coefficients can also be used for some nonhomogeneous Cauchy–Euler equations.

EXAMPLE 12

Consider the differential equation

$$x\frac{dy}{dx} + 2y = \frac{1}{x^2}, \qquad x > 0.$$

This is a nonhomogeneous Cauchy–Euler equation. Recall (Section 3.7) that homogeneous Cauchy–Euler equations can be converted to constant-coefficient equations by the substitution $x = e^t$. As in Section 3.7, we have

$$x\frac{dy}{dx} = \frac{dy}{dt},$$

and we also have

$$\frac{1}{x^2} = e^{-2t}.$$

With these substitutions, the differential equation becomes

$$\frac{dy}{dt} + 2y = e^{-2t},$$

which has constant coefficients on the left side and a generalized exponential function on the right side. The solution of this equation is

$$y = (c_1 + t)e^{-2t}.$$

The solution of the original problem then follows from substituting $\ln x$ in place of t:

$$y = (c_1 + \ln x)x^{-2}.$$

4.3 Exercises

In Exercises 1 and 2, determine if the given functions are generalized exponential functions. If so, state the characteristic value and degree.

1. *a.* $te^{3t} + e^{3t}$
   b. $t^3 - 3$
   c. $t\cos 3t + e^t \cos 3t$
   d. $te^t \cos 3t - e^t \sin 3t$

2. *a.* $te^{3t} + e^{2t}$
   b. $e^{-t}\cos 3t + e^t \cos 3t$
   c. $\cos 3t - (1 + t^2)\sin 3t$
   d. e^{-t^2}

In Exercises 3 through 6, compute the image of the given generalized exponential function under the given operator defined by $L[y] = y'' + y' - 2y$, and verify the prediction of Theorem 4.3.2.

3. $Y = te^{3t}$
4. $Y = \cos 2t$
5. $Y = te^t$
6. $Y = e^{-2t}\sin t$

In Exercises 7 through 18, find a particular solution by using the method of undetermined coefficients.

7. $y'' - y = (-6 + 4t)e^{-t}$
8. $y'' + 2y' + y = 4e^{-t}$
9. $y'' + y' - 2y = te^{3t}$
10. $y'' + y' - 2y = \cos 2t$
11. $y'' + y' - 2y = 4e^t$
12. $y'' + y' - 2y = e^{-2t}\sin t$
13. $y'' + 4y = e^t \cos 2t$
14. $y'' + 6y = t\sin 2t - \cos 2t$
15. $y'' + 3y' + 2y = 3t^2 e^t$
16. $y'' + 3y' + 2y = 3te^{-t}$
17. $y'' - 4y = e^{2t} + te^t$
18. $y'' + 3y' - y = \sin t + e^t \sin t$

In Exercises 19 through 20, solve the given initial-value problems.

19. $y'' + 9y = 6e^{3t}, \quad y(0) = 0, \quad y'(0) = 1$
20. $y'' + 2y' + y = 4\cos t, \quad y(0) = 2, \quad y'(0) = 0$

In Exercises 21 through 24, solve the given differential equations by using the substitution $x = e^t$, as in Section 3.7, followed by the method of undetermined coefficients.

21. $x^2 y'' - 2xy' + 2y = x^4, \quad x > 0$
22. $x^2 y'' - xy' + y = \dfrac{1}{x}, \quad x > 0$
23. $x^2 y'' - 2xy' + 2y = x^3 \ln x, \quad x > 0$
24. $x^2 y'' - xy' + y = \ln x, \quad x > 0$
25. This exercise illustrates the point that we should expect the method of undetermined coefficients to work only when L has constant coefficients. Consider the operator $L[y] = y' + t^2 y$.
    a. Let $Y = A + Bt + Ct^2$, with $C \neq 0$. Compute $L[Y]$. Note that $L[Y]$ is always a fourth-degree polynomial.

b. Use the result of part *a* to try to find particular solutions for $y' + t^2 y = 2t + 5t^2 + t^4$ and $y' + t^2 y = t + 5t^2 + t^4$.

c. Explain why it is not reasonable to expect to be able to find values of A, B, and C to make $L[A + Bt + Ct^2]$ equal to any given fourth-degree polynomial. (*Hint:* look at the relationship between the set of all fourth-degree polynomials and the image of $A + Bt + Ct^2$.)

✦ 4.3 INSTANT EXERCISE SOLUTIONS

1. From $y = (A + Bt)e^{-3t}$, we have $y' = (B - 3A - 3Bt)e^{-3t}$ and $L_1[(A + Bt)e^{-3t}] = (B - 3A - 3Bt)e^{-3t} + 3(A + Bt)e^{-3t} = Be^{-3t}$.
2. The characteristic polynomial is $r^2 + 6r + 9 = (r + 3)^2$; therefore $r = -3$ is an eigenvalue of multiplicity 2. With $y = At^2 e^{-3t}$, we have $y' = (-3At^2 + 2At)e^{-3t}$ and $y'' = (9At^2 - 12At + 2A)e^{-3t}$. Thus, $L[y] = (9At^2 - 12At + 2A)e^{-3t} + 6(-3At^2 + 2At)e^{-3t} + 9At^2 e^{-3t} = 2Ae^{-3t}$.
3. If $k < m$, then $L[y] = 0$; in other words, y is a solution of the homogeneous equation.
4. We have $Y = (C_0 + C_1 t)\cos 3t + (S_0 + S_1 t)\sin 3t$, and $Y' = C_1 \cos 3t + (-3C_0 - 3C_1 t)\sin 3t + S_1 \sin 3t + (3S_0 + 3S_1 t)\cos 3t = (C_1 + 3S_0 + 3S_1 t)\cos 3t + (S_1 - 3C_0 - 3C_1)\sin 3t$. The final result follows from $L[y] = y' + 3y$.
5. From the given trial solution, $Y' = (B + 2C + 2Dt)\cos 2t + (D - 2A - 2Bt)\sin 2t$ and $Y'' = (4D - 4A - 4Bt)\cos 2t + (-4B - 4C - 4Dt)\sin 2t$. Then $L[Y] = (-3Bt - 3A + 4D)\cos 2t + (-3Dt - 3C - 4B)\sin 2t$. Setting $L[Y]$ equal to $t \sin 2t$ yields four equations for the coefficients:
$$-3B = 0, \qquad -3A + 4D = 0, \qquad -3D = 1, \qquad -4B - 3C = 0.$$
These immediately yield $B = 0$ and $D = -\frac{1}{3}$, after which we have $A = -\frac{4}{9}$ and $C = 0$.
6. The forcing function is a generalized exponential function with characteristic value 3 and degree 0. This characteristic value is also a characteristic value of L of multiplicity 1. Hence, the trial solution given by Theorem 4.3.3 is $Y = (te^{3t})(A_0)$, which we may rewrite as $Y = At3^{3t}$.
7. We have $Y = At^2 e^{-3t}$, $Y' = (-3At^2 + 2At)e^{-3t}$, and $Y'' = (9At^2 - 12At + 2A)e^{-3t}$, so $L[Y] = Y'' + 6Y' + 9Y = 2Ae^{-3t}$. Also $L[Y] = 6e^{-3t}$ requires $A = 3$. Thus, $y_p = 3t^2 e^{-3t}$.

4.4 Forced Linear Oscillators

Recall that a forced linear oscillator[9] is a mathematical model consisting of a nonhomogeneous linear second-order differential equation with constant nonnegative coefficients. Regardless of the physical setting for such a model, one can write the mathematical model as

$$y'' + \beta y' + \kappa^2 y = f,$$

where $\kappa > 0$, $\beta \geq 0$, and f is a function of time, possibly constant or zero. The specific interpretation of the parameters β and κ depends on the physical model. The undamped case ($\beta = 0$) may not be physically realistic, as the viscous damping in a spring-mass system cannot be completely

[9] Section 3.1.

eliminated. Sometimes there is no fundamental distinction between a small amount of damping and none at all; in these cases, it is nice to have the simplification of taking $\beta = 0$.

In this section, we focus on linear oscillators with sinusoidal forcing. There are several reasons to be particularly interested in such problems, for both mechanical and electrical systems. Some devices that are used as shock absorbers for vehicles are spring-mass systems. Movement of the vehicle over a rough surface creates forces that act as external forces on the shock absorber. The magnitude of such forces will depend on the random fluctuations in the surface. Although it is not generally reasonable to assign a formula to the function describing the road surface, it is reasonable to characterize a road surface as oscillating about some mean height with some average frequency. For greater realism, it is possible to better approximate continuous periodic functions by using sums of sine and cosine functions.[10] Alternating-current power sources provide sinusoidal forcing in the *RLC* circuit model introduced in Section 4.1. Radio signals are approximately sinusoidal, with a specific frequency assigned to each radio station. These signals serve as forcing functions for a tuning circuit inside the radio receiver. The design of a tuning circuit is the subject of Case Study 4.

Often, the problem of interest is to look for the steady-state solution component, which is the portion of the solution that persists as $t \to \infty$. The exponential transients are generally short-lived compared with the amount of time over which the forcing persists. Sometimes it is desirable to write the steady-state solution component in amplitude–phase shift form, as in Section 3.1, using the identity

$$C \cos \omega t + S \sin \omega t = A \cos(\omega t + \delta), \tag{1}$$

where

$$A^2 = C^2 + S^2, \qquad A \cos \delta = C, \qquad A \sin \delta = S.$$

MODEL PROBLEM 4.4

Determine the long-time behavior of spring-mass systems with mass 1, spring constant 10, damping coefficient 2, and sinusoidal forcing function of frequency 4 and amplitude 5.

Damped, Sinusoidally Forced Oscillation

Model Problem 4.4 is not quite completely specified. The operator L is defined by $L[y] = y'' + 2y' + 10y$. The forcing function g could be $5 \cos 4t$, $5 \sin 4t$, $3 \cos 4t - 4 \sin 4t$, or any other sinusoidal function with the appropriate frequency and amplitude. The initial conditions are also unspecified. First we solve one such problem.

EXAMPLE 1

Consider the initial-value problem

$$y'' + 2y' + 10y = 5 \cos 4t, \qquad y(0) = 0, \qquad y'(0) = 0.$$

The characteristic polynomial $\lambda^2 + 2\lambda + 10$ is an irreducible quadratic; from the quadratic formula, the characteristic values are

$$\lambda = -1 \pm 3i;$$

[10]This is the idea of Fourier series, a subject treated in Chapter 8.

hence, the complementary solution is

$$y_c = e^{-t}(c_1 \cos 3t + c_2 \sin 3t).$$

To find a particular solution, we first note that the forcing term $5\cos 4t$ is a generalized exponential function with characteristic values $\pm 4i$. This characteristic value is not a root of the characteristic polynomial for the differential equation; hence, a particular solution, by Theorem 4.3.3, has the form

$$Y = C\cos 4t + S\sin 4t,$$

where C and S are constants to be determined. From this expression, we compute Y' and Y'' and substitute both into the left-hand side of the differential equation to get

$$L[Y] = (-6C + 8S)\cos 4t + (-8C - 6S)\sin 4t.$$

Setting this expression equal to $5\cos 4t$ yields a pair of algebraic equations for C and S:

$$-6C + 8S = 5, \qquad -8C - 6S = 0.$$

We can eliminate S from this pair of equations by adding 6 times the first equation to 8 times the second. This gives $-100C = 30$, so $C = -\frac{3}{10}$. From the second equation, we have $S = (-8/6)C = \frac{4}{10}$. Thus, the particular solution is

$$y_p = -\tfrac{3}{10}\cos 4t + \tfrac{4}{10}\sin 4t.$$

Now we add the complementary and particular solutions and apply the initial conditions to get the full solution of the initial-value problem. The reader should perform these steps and verify this result:

$$y(t) = -\tfrac{3}{10}\cos 4t + \tfrac{4}{10}\sin 4t + e^{-t}\left(\tfrac{3}{10}\cos 3t - \tfrac{13}{30}\sin 3t\right).$$

The amplitude of the steady-state component y_p is

$$A = \sqrt{\left(\tfrac{4}{10}\right)^2 + \left(\tfrac{3}{10}\right)^2} = \tfrac{1}{2}.$$

The phase shift of the steady-state solution is determined by

$$\cos\delta = -0.6, \qquad \sin\delta = 0.8.$$

The algebraic signs indicate that the phase shift is in the second quadrant. Thus,

$$\delta = \pi - \arcsin 0.8 \approx 0.7\pi.$$

Figure 4.4.1 shows the graphs of the particular solution $y_p(t)$ and the full solution $y(t)$ from Example 1. The particular solution, like the forcing function, is a sinusoidal function with period $\pi/2$. By the end of two periods, the full solution is indistinguishable from the particular solution owing to the rapid decay of the exponential factor e^{-t}. In this example, the steady-state solution is the particular solution. This will always be the case for a damped linear oscillator with sinusoidal forcing. In summary, the specific choice of forcing function and initial condition yields a steady-state solution that is sinusoidal, with period $\pi/2$, amplitude $\frac{1}{2}$, and phase shift 0.7π. By our

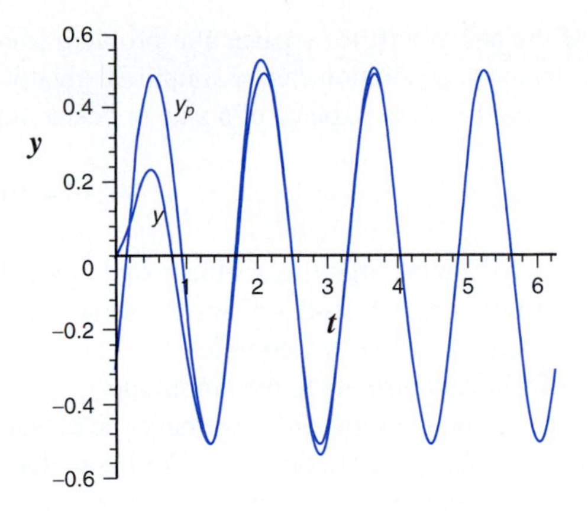

Figure 4.4.1
The full solution and the steady-state solution from Example 1.

definition, the phase shift is the amount by which the sinusoidal function precedes the cosine function. Thus, the steady-state solution runs 35% ahead of the forcing function.[11]

Although we obtained the steady-state solution for Example 1 by solving an initial-value problem completely, we could have simply determined the particular solution without the complementary solution. Sinusoidal forcing functions have purely imaginary characteristic values, while damped linear operators have complex characteristic values with nonzero real part. We do not need the complementary solution to know that the correct form for the particular solution does not include any extra factors of t.

✦ INSTANT EXERCISE 1

Find the steady-state solution for

$$y'' + 2y' + 10y = 5 \sin 4t$$

by calculating y_p. Write the solution in amplitude–phase shift form, verifying that the amplitude of the steady-state solution is the same as in Example 1, but the phase shift is different.

Using the Complex Exponential for Sinusoidal Forcing

A lot of calculations went into the solution of the problem presented in Example 1. The amplitude and phase shift of the particular solution were the two important quantities that required calculation. As we saw with Instant Exercise 1, the result for the amplitude is the same regardless of whether the forcing function is a cosine or sine function.

[11] An advantage of writing the phase shift as a multiple of π is that it allows a simple geometric interpretation.

If the phase shift for a particular problem is unimportant, the calculations for the amplitude of a steady-state solution can be simplified greatly by using a complex exponential function[12] for the forcing function in place of a sine or cosine function. In our example, we could solve

$$y'' + 2y' + 10y = 5e^{4it}.$$

It seems reasonable that this problem should yield a particular solution with the same amplitude as that for Example 1. Both e^{4it} and $\cos 4t$ are sinusoidal functions with circular frequency 4. The only difference is that the complex exponential function has a real part and an imaginary part, both of which contribute to the amplitude.

For the forcing term $5e^{4it}$, we have the characteristic value $4i$, but we can treat it as though it were a real characteristic value. We know that $4i$ is not a characteristic value of $y'' + 2y' + 10y$ because purely imaginary characteristic values correspond to differential equations with no damping. Thus, we have the form

$$Y = A_0 e^{4it},$$

where A_0 is an undetermined complex coefficient. Further, we know that the complementary solution for a damped oscillator decays, so the steady-state solution is the same as the particular solution. To find the amplitude of the steady-state solution, we need only determine A_0 and then determine its magnitude $|A_0|$. Substituting Y into the differential equation gives

$$A_0(-16 + 8i + 10)e^{4it} = 5e^{4it},$$

from which we get the result

$$A_0 = \frac{5}{-6 + 8i}.$$

For a real number, the magnitude is just the absolute value, but for a complex number the magnitude is given by the formula

$$|\alpha + \beta i| = \sqrt{\alpha^2 + \beta^2} \tag{2}$$

It is convenient to also have a formula for the magnitude of a complex number that has the same form as A_0:

$$\left| \frac{c}{\alpha + \beta i} \right| = \frac{|c|}{|\alpha + \beta i|} = \frac{|c|}{\sqrt{\alpha^2 + \beta^2}}. \tag{3}$$

Thus, the amplitude of the particular solution is $5/\sqrt{36 + 64} = \frac{1}{2}$. This is the same amplitude that we got from Example 1 and from Instant Exercise 1, but we got it with less calculation by using complex exponential forcing. Theorem 4.4.1 summarizes what we have learned.

---

[12]Section 3.5.

Theorem 4.4.1

Let $L[y] = y'' + \beta y' + \kappa^2 y$, with $\beta, \kappa > 0$, and let g be a sinusoidal function with frequency ω and amplitude W. The steady-state solution of $L[y] = g$ has frequency ω and amplitude $|A_0|$ determined from the particular solution $y_p = A_0 e^{i\omega t}$.

Undamped, Sinusoidally Forced Oscillation

Theorem 4.4.1 does not summarize all oscillator models with sinusoidal forcing, but is instead limited to problems with nonzero damping. Some very interesting behavior can occur when there is no damping. Here is a typical example.

EXAMPLE 2

Consider the problem

$$y'' + 4y = \cos 5t, \qquad y(0) = 0, \qquad y'(0) = 0.$$

The complementary solution is

$$y_c = c_1 \cos 2t + c_2 \sin 2t$$

and the particular solution is

$$y_p = -\tfrac{1}{21} \cos 5t.$$

Thus, the general solution of the differential equation is

$$y = -\tfrac{1}{21} \cos 5t + c_1 \cos 2t + c_2 \sin 2t.$$

The initial conditions then yield the solution

$$y = \tfrac{1}{21} \cos 2t - \tfrac{1}{21} \cos 5t.$$

This solution can be rewritten by using the identity

$$\cos\alpha - \cos\beta = 2 \sin \frac{\beta + \alpha}{2} \sin \frac{\beta - \alpha}{2}$$

as

$$y = \tfrac{2}{21} \sin \tfrac{7}{2}t \sin \tfrac{3}{2}t.$$

✦ INSTANT EXERCISE 2

Fill in the missing details of Example 2.

The solution of Example 2 appears in Figure 4.4.2. In problems with no damping, the complementary and particular solutions are both part of the steady-state solution, leading to complicated patterns which may be periodic, as in the example, or may not be. This behavior is explored in Exercise 11.

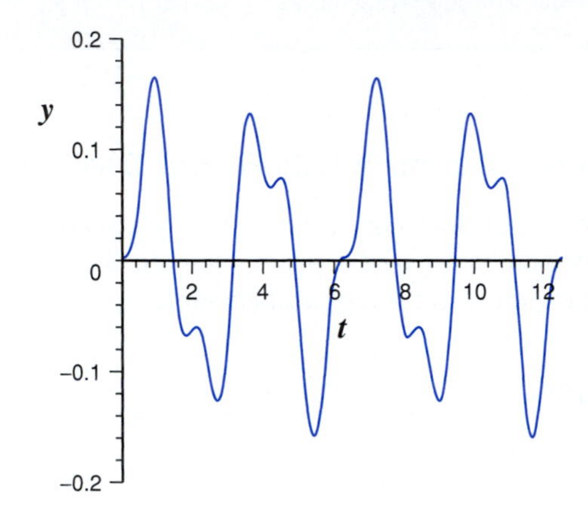

Figure 4.4.2
The solution of $y'' + 4y = \cos 5t$, $y(0) = 0$, $y'(0) = 0$.

Resonance

The situation is qualitatively different when the characteristic values of the operator L and the forcing function are the same.

EXAMPLE 3

Consider the problem

$$y'' + y = \cos t, \qquad y(0) = 0, \qquad y'(0) = 0.$$

The complementary solution is

$$y_c = c_1 \cos t + c_2 \sin t.$$

The forcing function has the same characteristic value as the operator L, so a trial solution is

$$Y = At \cos t + Bt \sin t.$$

Then

$$Y' = (A + Bt) \cos t + (B - At) \sin t$$

and

$$Y'' = (2B - At) \cos t + (-2A - Bt) \sin t.$$

Hence,

$$L[Y] = 2B \cos t - 2A \sin t.$$

Setting this equal to g gives the results $A = 0$ and $B = \frac{1}{2}$. The general solution is then

$$y = \tfrac{1}{2} t \sin t + c_1 \cos t + c_2 \sin t.$$

The first initial condition immediately gives $c_1 = 0$; thus, the second initial condition determines c_2. With $c_1 = 0$, we have

$$y' = \tfrac{1}{2}t\cos t + \tfrac{1}{2}\sin t + c_2\cos t$$

and so

$$y'(0) = c_2.$$

Hence $c_2 = 0$, and the solution is

$$y = \tfrac{1}{2}t\sin t.$$

The solution of Example 3 appears in Figure 4.4.3.

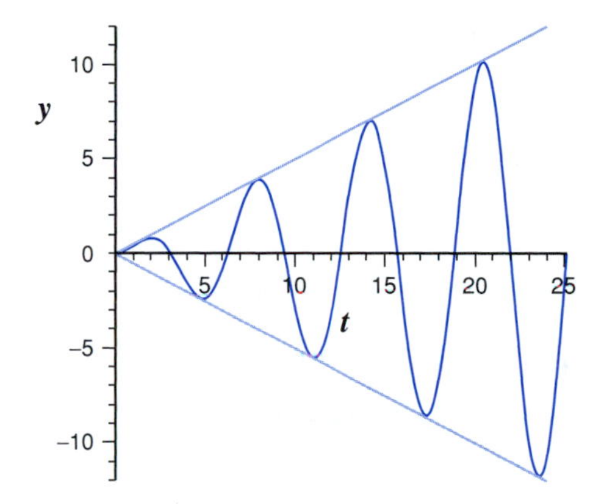

Figure 4.4.3
The solution of $y'' + y = \cos t$, $y(0) = 0$, $y'(0) = 0$, along with its envelope $\pm 0.5t\sin t$.

The key feature of Example 3 is the existence of an unbounded steady-state solution. This feature is often called *resonance* and is used as an explanation for catastrophic vibration. Obviously vibration gets out of control when the amplitude grows linearly with time!

This explanation does not hold up to scrutiny. The linear increase in amplitude with time results from the correspondence of the characteristic value of g with that of L. This is not physically realistic. Real spring-mass systems inevitably have a small amount of damping, and no amount of damping is too small to give the characteristic value of L a real part. Example 3 illustrates **mathematical resonance.** In the context of other problems, mathematical resonance can be important, but another explanation is needed for the phenomenon of **physical resonance,** defined as catastrophic vibration of a physical oscillator. Suppose we alter the problem of Example 3 by adding a small amount of damping; as an example, suppose we add an amount equal to 1% of the amount of damping that would make the oscillator critically damped.

EXAMPLE 4

Consider the problem

$$y'' + 0.02y' + y = \cos t, \qquad y(0) = 0, \qquad y'(0) = 0.$$

The characteristic equation for the complementary solution is $r^2 + 0.02r + 1 = 0$, and the roots of this equation are

$$r = \frac{-0.02 \pm \sqrt{0.02^2 - 4}}{2} = -0.01 \pm i\sqrt{0.9999}.$$

This value is very close to the characteristic value $\pm i$ of g, but it is not *exactly* the same; therefore, the trial solution is

$$Y = A\cos t + B\sin t.$$

Substitution of this trial solution into the left side of the differential equation yields

$$L[Y] = Y'' + 0.02Y' + Y = 0.02B\cos t - 0.02A\sin t.$$

Comparison with the forcing function yields the results $A = 0$ and $B = 50$. We therefore have the particular solution

$$y_p = 50\sin t$$

and the general solution

$$y = 50\sin t + e^{-0.01t}\left[c_1\cos(\sqrt{0.9999}t) + c_2\sin(\sqrt{0.9999}t)\right].$$

The first initial condition gives us $c_1 = 0$, and then

$$y' = 50\cos t + e^{-0.01t}c_2\left[\sqrt{0.9999}\cos(\sqrt{0.9999}t) - 0.01\sin(\sqrt{0.9999}t)\right].$$

The second initial condition then yields the result

$$c_2 = -\frac{50}{\sqrt{0.9999}}.$$

The solution is

$$y = 50\sin t - \frac{50}{\sqrt{0.9999}}e^{-0.01t}\sin(\sqrt{0.9999}t).$$

The solution of Example 4 appears in Figure 4.4.4 along with the envelope of the solution of Example 3.

The addition of a slight amount of damping does not appear to have changed the solution very much, as the graphs in Figures 4.4.3 and 4.4.4 appear virtually identical. Only by comparing the damped solution of Example 4 with the envelope of the undamped solution of Example 3 can we see any difference at all. The peaks of the undamped solution are slightly higher than those of the damped solution, but the difference does not appear to be significant. There doesn't appear to be a significant difference between the mathematical resonance of Figure 4.4.3 and the physical resonance of Figure 4.4.4.

Let us briefly summarize what we have learned. The undamped model of Example 3 is simpler than the damped model of Example 4, but the latter is more realistic. Given the similarity of the curves in Figures 4.4.3 and 4.4.4, it seems reasonable to think that the undamped problem is a good approximation of the damped problem.

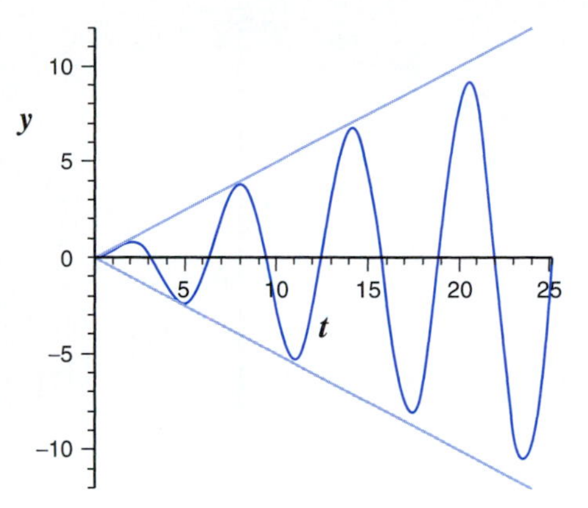

Figure 4.4.4
The solution of $y'' + 0.02y' + y = \cos t$, $y(0) = 0$, $y'(0) = 0$ along with the envelope from Example 3.

The comparison of Figures 4.4.3 and 4.4.4 is somewhat misleading. These graphs include only 4 periods of the oscillating function g. It takes many periods for resonance to reach the full effect, so it is necessary to examine the graphs of the solutions over a much longer time. Figures 4.4.5 and 4.4.6 track the solutions for a time interval equal to 32 periods of the forcing function. Over this time interval, the graphs are significantly different. The amplitude of the solution in the undamped case grows linearly with time and is therefore unbounded. The amplitude of the solution in the damped case is large, but bounded. As with the examples earlier in this section, the solution of the damped problem decays (albeit slowly) to the steady-state solution, which is shown for comparison in Figure 4.4.7. The steady-state solution for the slightly damped case is clearly a better approximation to the full solution than is the solution of the undamped problem. The physical phenomenon of resonance does not correspond to an unbounded amplitude; rather,

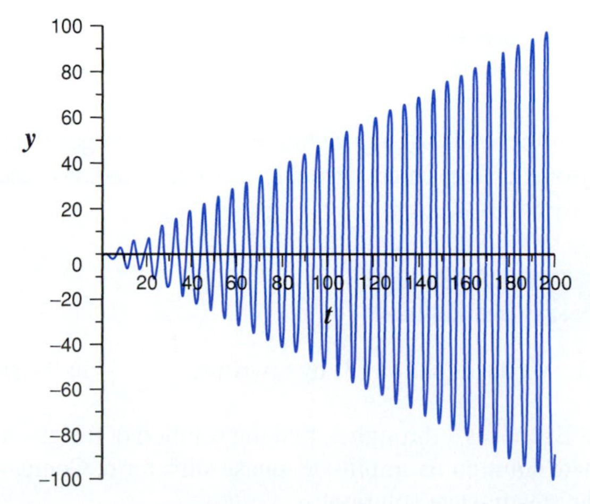

Figure 4.4.5
The solution of $y'' + y = \cos t$, $y(0) = 0$, $y'(0) = 0$.

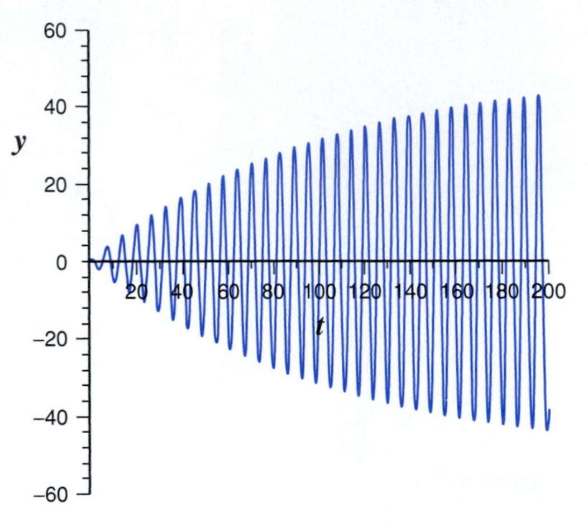

Figure 4.4.6
The solution of $y'' + 0.02y' + y = \cos t$, $y(0) = 0$, $y'(0) = 0$.

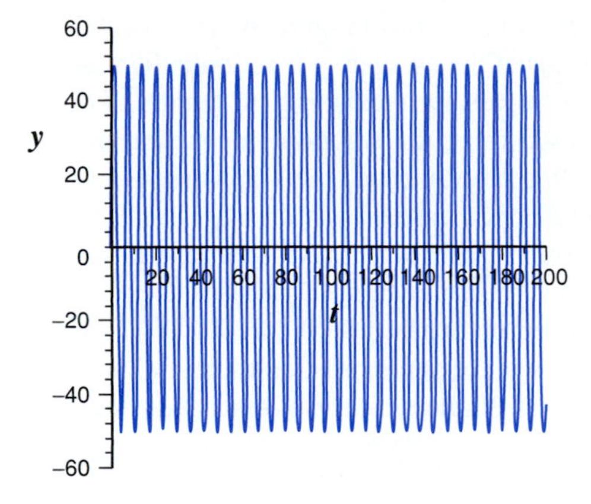

Figure 4.4.7
The steady-state solution ($y = 50 \sin t$) of $y'' + 0.02y' + y = \cos t$.

it corresponds to an amplitude that is far larger than that of the forcing function. This kind of resonance can be catastrophic in some situations, and it can be very useful in others, as seen in Case Study 4.

4.4 Exercises

1. Verify Equation (3) by rewriting $\dfrac{c}{\alpha + \beta i}$ in the standard form $u + vi$.

In Exercises 2 through 6, find the solution of the given initial-value problem and write the steady-state solution in amplitude–phase shift form. Compare the graphs of y and y_p over 4 periods of the steady-state solution.

2. $5y'' + 2y' + 2y = \sin t, \quad y(0) = 1, \quad y'(0) = 0$

3. $y'' + y' + 9y = \cos 3t, \quad y(0) = 0, \quad y'(0) = 0$
4. $4y'' + 5y' + y = 17\cos t, \quad y(0) = 1, \quad y'(0) = 0$
5. $y'' + 3y' + y = \sin 5t, \quad y(0) = 0, \quad y'(0) = 0$
6. $y'' + 2y' + y = 10\sin 3t, \quad y(0) = 0, \quad y'(0) = 1$
7. Find the steady-state solution of $y'' + y' + 9y = e^{3it}$. Compare the amplitude with that of Exercise 3.
8. Find the steady-state solution of $4y'' + 5y' + y = 17e^{it}$. Compare the amplitude with that of Exercise 4.

T 9. Consider the problem

$$y'' + y' + k^2 y = g,$$

where g is any sinusoidal function with period $\pi/2$ and unit amplitude.

a. Find an expression for the amplitude of the steady-state solution.
b. Find the value of k that maximizes the amplitude.
c. Plot the function $A(k)$ and discuss the graph.

T 10. Consider the problem

$$y'' + \beta y' + \omega^2 y = g,$$

where g is any sinusoidal function with circular frequency ω and unit amplitude.

a. Find an expression for the amplitude of the steady-state solution.
b. What happens to the amplitude as the damping approaches 0?
c. Choose ω so that the period of the steady-state solution is 1. Plot the amplitude as a function of the damping coefficient and discuss the graph.

T 11. *a.* Solve the problem $y'' + 4y = \cos 1.4t$, $y(0) = 0$, $y'(0) = 0$.
b. Plot the solution on the interval $[0, 20\pi]$.
c. Show that $\cos 2t - \cos 1.4t$ is periodic with period 10π.
d. Solve the problem $y'' + 4y = \cos(\sqrt{2}\,t)$, $y(0) = 0$, $y'(0) = 0$.
e. Plot the solution on the interval $[0, 20\pi]$.
f. Given the fact that $\cos 2t - \cos(\sqrt{2}\,t)$ is not periodic, suggest a rule for determining which values of β make $\cos 2t - \cos \beta t$ periodic. For the case where it is periodic, suggest a rule for determining the period.

T 12. Pianos have two or three strings for each note. These strings must be exactly in tune for a piano to sound pleasant. Piano tuners learn to tune these unison notes accurately by listening for a phenomenon called **beats.** This phenomenon occurs in an undamped forced oscillator model when the forcing frequency is almost equal to the natural frequency of the vibration.

a. Suppose the natural frequency (1/period) of the oscillator is exactly $1/\pi$ and the frequency of the forcing function is $(1 + \epsilon)/\pi$, where ϵ is a small number, either positive or negative. This corresponds to the model

$$y'' + 4y = \cos(2[1 + \epsilon]t), \qquad y(0) = 0, \qquad y'(0) = 0.$$

Solve the initial-value problem, obtaining the solution

$$y = \frac{1}{4\epsilon + 2\epsilon^2} \sin \epsilon t \sin([2 + \epsilon]t).$$

b. Plot the solution for the case $\epsilon = 0.2$ over the time interval $[0, 10\pi]$. Also plot the functions $y_e = \pm(\sin 0.2t)/0.88$. Observe that the functions y_e are the envelope of the solution.

c. Similarly, plot the solution and its envelope for $\epsilon = 0.1$.

d. When two piano strings of nominally the same pitch are played together, the pitch (frequency) is determined by the fast oscillation, and is therefore the average of the pitches of the two individual strings. The loudness of the sound is the average of the envelope magnitude, and is approximately twice that for a single string. If the strings are almost but not quite in tune, the average person does not notice the amplitude variation, but piano tuners train themselves to hear the periodicity of the amplitude. Suppose a tuner has one string in tune and is tuning a second string to match the first. Explain how the tuner knows that he or she is improving the tuning.

T 13. Suppose the outdoor temperature varies on a daily cycle with a low of 59°F (15°C) and a daily high of 77°F (25°C), with time measured so that the low occurs at time 0 and the high at time 12 each day.

a. Use Newton's law of cooling (see Section 1.1) to write a differential equation for the indoor temperature. Note that the forcing function will have a constant part and a sinusoidal part.

b. Determine the steady-state solution for the differential equation. Write the solution so that the trigonometric part is in amplitude–phase shift form.

c. Choose a value of k corresponding to a half-life of 3 h and plot the solution along with the outdoor temperature.

d. Discuss the relationship between the indoor and outdoor temperatures, paying particular attention to explaining the amplitude and the phase shift.

✦ 4.4 INSTANT EXERCISE SOLUTIONS

1. The characteristic value $4i$ of the forcing function cannot be a characteristic value of the damped oscillator, so the correct trial solution is $Y = C \cos 4t + S \sin 4t$. Then $Y' = 4S \cos 4t - 4C \sin 4t$ and $Y'' = -16C \cos 4t - 16S \sin 4t$. Hence, $Y'' + 2Y' + 10Y = (-6C + 8S) \cos 4t + (-8C - 6S) \sin 4t$. To obtain $5 \sin 4t$, we need $-6C + 8S = 0$ and $-8C - 6S = 5$. Thus, $C = -0.4$ and $S = -0.3$. The particular solution is

$$y_p(t) = -0.4 \cos 4t - 0.3 \sin 4t.$$

The amplitude is $\sqrt{0.4^2 + 0.3^2} = 0.5$, and the phase shift satisfies $\cos \delta = -0.8$ and $\sin \delta = -0.6$. Thus, δ is in the third quadrant and so $\delta = \arcsin(0.6) - \pi \approx -0.8\pi$. Thus,

$$y_p \approx 0.5 \cos(4t + 0.8\pi).$$

2. The characteristic equation is $r^2 + 4 = 0$, so $r = \pm 2i$. Hence, the complementary solution is

$$y_c = c_1 \cos 2t + c_2 \sin 2t.$$

The particular solution, by Theorem 4.3.3, has the form

$$Y = C\cos 5t + S\sin 5t.$$

Then $Y' = 5S\cos 5t - 5C\sin 5t$ and $Y'' = -25C\cos 5t - 25S\sin 5t$. Thus, $Y'' + 4Y = -21C\cos 5t - 21S\sin 5t$. We require $-21C = 1$ and $-21S = 0$, yielding the particular solution $y_p = (-\frac{1}{21})\cos 5t$. Thus, the general solution of the differential equation is

$$y = -\tfrac{1}{21}\cos 5t + c_1\cos 2t + c_2\sin 2t.$$

The initial condition $y(0) = 0$ requires $c_1 = \frac{1}{21}$, and the condition $y'(0) = 0$ requires $c_2 = 0$.

4.5 Solving First-Order Linear Equations

The method of undetermined coefficients allows for the solution of first-order linear equations

$$y' + py = g, \tag{1}$$

provided that g is a generalized exponential function or a sum of such functions and p is constant. The method works for a few other first-order linear equations that can be reduced to equations of the form (1), such as some Cauchy–Euler equations.

There are also two integration methods that can be used to solve first-order linear equations. The better known of these is called the *integrating factor method.* The other method, the *method of variation of parameters,* has two advantages over the integrating factor method. One, it is based on a simpler idea and therefore is easier to remember. Two, the method generalizes nicely to the method of variation of parameters for second-order equations. Both methods have the advantage, compared with the method of undetermined coefficicnts, that they can be used even when the linear equation does not have constant coefficients. As noted above, they have the disadvantage of being generally more difficult than the method of undetermined coefficients. We consider only variation of parameters here. The integrating factor method is described in Section A.1.

MODEL PROBLEM 4.5

Solve

$$y' + (\cos t)y = \sin t\cos t, \qquad y(0) = 0.$$

To apply the method of variation of parameters, we first solve the associated homogeneous equation

$$y' + (\cos t)y = 0.$$

This equation can be solved by separation of variables. We have

$$\frac{dy}{dt} = -(\cos t)y,$$

or

$$\frac{dy}{y} = -\cos t\,dt.$$

Integration yields $\ln y = -\sin t + C$, hence, the complementary solution is

$$y = Ae^{-\sin t}.$$

Now we guess that the solution of the nonhomogeneous problem ought to look something like the solution of the homogeneous problem. Perhaps it would be useful to look for a solution of the form

$$y = e^{-\sin t}\, u(t).$$

Notice that we obtain this form by replacing the parameter A in the homogeneous solution with a function $u(t)$. The replacement of a constant parameter with a nonconstant function is what gives the method its name. From here, we have two ways to complete the solution. We can find a single particular solution y_p by finding one function u, or we can find a one-parameter family of functions u, in which case the result $y = e^{-\sin t}\, u(t)$ is itself the general solution.

With the given form for y and the product rule for derivatives, we get

$$y' = e^{-\sin t}\, u' - (\cos t)e^{-\sin t}\, u.$$

Substituting this into the left side of the differential equation yields

$$y' + (\cos t)y = e^{-\sin t}\, u' - (\cos t)e^{-\sin t}\, u + (\cos t)(e^{-\sin t}\, u) = e^{-\sin t}u'.$$

The differential equation for y becomes the differential equation

$$e^{-\sin t}\, u' = \sin t \cos t,$$

or

$$u' = (\sin t \cos t)(e^{\sin t}).$$

Thus, the substitution $y = e^{-\sin t}\, u(t)$ has reduced the differential equation to a calculus problem.

To complete the solution, we have to find a function $u(t)$ that has the correct derivative. To that end, let $w = \sin t$. Then $dw = \cos t\, dt$, so

$$u = \int (\sin t\, e^{\sin t}) \cos t\, dt = \int w e^{w}\, dw.$$

This expression can be integrated by parts, with the result

$$u = (w - 1)e^{w} + C = (\sin t - 1)e^{\sin t} + C.$$

Without loss of generality, we may choose $C = 0$ and obtain the particular solution

$$y_p = e^{-\sin t}\, u(t) = e^{-\sin t}[(\sin t - 1)e^{\sin t}] = \sin t - 1.$$

Adding this to the complementary solution yields the general solution

$$y = c_1 e^{-\sin t} + \sin t - 1.$$

The initial condition is now used to complete the solution by determining c_1, with the result

$$y = e^{-\sin t} + \sin t - 1.$$

It is curious to note that our motivation for the substitution $y = e^{-\sin t}\, u(t)$ was a belief that the particular solution should somehow resemble the complementary solution; however, the function u has a factor $e^{\sin t}$, and y_p in the end does not look much like y_c. Nevertheless, the method works.

What makes the method work is not the relationship between y_p and y_c, but rather the relationship between y_c and L. It is also interesting that the particular solution is the simple function $\sin t - 1$, which is a sum of generalized exponential functions. However, there is no reasonable way that we could have predicted this by looking at the operator L. Either variation of parameters or the equivalent integrating factor method is necessary for Model Problem 4.5 because there are no families Y and G for which any function in the family G is the image under L of some function in the family Y.

Instead of choosing $C = 0$ in the formula for u, we could have left C unspecified. Then multiplication by $e^{-\sin t}$ yields the general solution

$$y = [C + (\sin t - 1)e^{\sin t}]e^{-\sin t} = Ce^{-\sin t} + \sin t - 1.$$

Only the symbol used for the constant is different.

✦ **INSTANT EXERCISE 1**

Verify that $y = e^{-\sin t} + \sin t - 1$ solves the equation $y' + (\cos t)y = \sin t \cos t$.

The solution of Model Problem 4.5 appears in Figure 4.5.1.

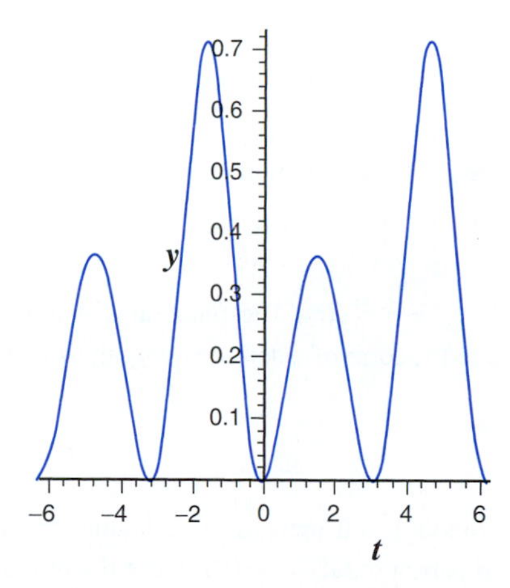

Figure 4.5.1
The solution of $y' + (\cos t)y = \sin t \cos t$, $y(0) = 0$.

The General Case

How general is the method we used to solve Model Problem 4.5? Consider any problem

$$L[y] = y' + py = g.$$

By Theorem 2.4.2, this equation has a unique solution through any point (t_0, y_0) as long as p and g are continuous at t_0. Let y_1 be any nonzero solution of $L[y] = 0$. Now consider $L[y]$ where $y = y_1 u$ for an unknown function u. We have

$$L[y] = L[y_1 u] = (y_1 u' + y_1' u) + p y_1 u = y_1 u' + L[y_1]u = y_1 u'.$$

Substituting this result into the differential equation yields an algebraic equation for u':

$$y_1 u' = g. \tag{2}$$

The method can be summarized as follows:

Theorem 4.5.1 Let p and g be continuous functions over some open interval I, and let y_1 be a solution of $L[y] = y' + py = 0$ on I. If $y = y_1(t)u(t)$, then $L[y] = y_1 u'$; hence, any function $y_1 u$ is a particular solution of $L[y] = g$ provided that $y_1 u' = g$. By the linearity property, an equation $L[y] = g_1 + g_2 + \cdots + g_n$ can be solved by finding a particular solution y_{p_i} for each forcing function g_i.

It is not always possible to *find* an antiderivative formula for u', but one can always *construct* an antiderivative as a definite integral. It appears from Equation (2) that problems could arise if $y_1 = 0$ at some point, but it can be shown that this cannot occur (see Exercise 37).

EXAMPLE 1

Consider the initial-value problem

$$y' + 2y = \tan t, \qquad y(0) = 1.$$

The complementary solution is $y_c = c_1 e^{-2t}$; following the method outlined in Theorem 4.5.1, the problem for y_p reduces to the integration problem

$$u' = e^{2t} \tan t,$$

after which $y_p = e^{-2t}u(t)$. The function u' has no elementary antiderivative, so we construct an antiderivative in the form of a definite integral, as in Theorem 1.2.1. The derivative of any function

$$\int_a^t e^{2w} \tan w \, dw$$

is (by the fundamental theorem of calculus) $e^{2t} \tan t$, regardless of the choice of the constant a. In particular, it is best to take $a = 0$ because the initial condition is given at the point $t = 0$:

$$u = \int_0^t e^{2w} \tan w \, dw.$$

We can now add the particular solution $y_p = y_1 u$ to the complementary solution, yielding the general solution

$$y = ce^{-2t} + e^{-2t} \int_0^t e^{2w} \tan w \, dw.$$

Next we apply the initial condition. Upon substitution of $t = 0$ and $y = 1$, we have

$$1 = c + \int_0^0 e^{2w} \tan w \, dw.$$

The integral must be 0 because it is the integral of a bounded function over an interval of width 0. This is what we gained by choosing $a = 0$. Clearly we must have $c = 1$, so the solution of the initial-value problem is

$$y = e^{-2t} + e^{-2t} \int_0^t e^{2w} \tan w \, dw.$$

✦ INSTANT EXERCISE 2

Verify that the solution of Example 1 satisfies the differential equation.

EXAMPLE 2

Consider the problem

$$y' - 2ty = 1, \qquad y(0) = 1.$$

The associated homogeneous equation

$$y' - 2ty = 0$$

can be written in separated form as

$$\frac{dy}{y} = 2t \, dt.$$

Integration gives us

$$\ln |y| = t^2 + C,$$

from which we obtain the solution

$$y = Ae^{t^2}.$$

This solution is easily checked. Now we look for a solution of the form

$$y_p = e^{t^2} u(t),$$

where u is a nonconstant function. Then

$$y_p' = e^{t^2} u' + 2te^{t^2} u = e^{t^2} u' + 2ty_p.$$

The differential equation becomes

$$u' = e^{-t^2}$$

We can write a definite integral for the antiderivative of e^{-t^2} that appears on the right side. Alternatively, we can use the function erf that was defined in Section 2.4.

$$u = \frac{\sqrt{\pi}}{2} \operatorname{erf} t.$$

This results in the general solution

$$y = Ce^{t^2} + \frac{\sqrt{\pi}}{2} e^{t^2} \operatorname{erf} t.$$

The initial condition, along with the property $\operatorname{erf}(0) = 0$, yields $C = 1$. The final result is

$$y = e^{t^2} + \frac{\sqrt{\pi}}{2} e^{t^2} \operatorname{erf} t.$$

EXAMPLE 3

Consider the differential equation

$$x\frac{dy}{dx} + 2y = \frac{1}{x^2}, \qquad x > 0,$$

which was Example 12 from Section 4.3. The associated homogeneous equation $xy' + 2y = 0$ separates to $dy/y = -2\,dx/x$, from which we obtain the solution $y_1 = x^{-2}$. Given $y_p = x^{-2}u$, we have

$$L[y_p] = x(x^{-2}u' - 2x^{-3}u) + 2(x^{-2}u) = x^{-1}u'.$$

Note that we did not get $L[y_p] = y_1 u'$, which is true only if the lead coefficient of the differential equation is 1. From $L[y_p] = x^{-2}$, we have

$$u' = \frac{1}{x}.$$

We may take $u = \ln x$ and then $y_p = x^{-2}\ln x$. Hence, we obtain the same general solution as before:

$$y = (C + \ln x)x^{-2}.$$

Bernoulli Equations

Bernoulli equations are equations of the form

$$\frac{dy}{dx} + q(x)y = g(x)y^n, \tag{3}$$

where n is any real number. The cases $n = 0$ and $n = 1$ require no special treatment. Other cases can also arise in mathematical models, particularly those that involve fluid mechanics. Bernoulli equations are solved by reduction to linear equations. An example serves to demonstrate the technique.

EXAMPLE 4

The first historical documentation of a Bernoulli equation was for an equation that arises from models for two-dimensional projectile motion under the influence of gravity and air resistance.[13] Using the assumption that air resistance is proportional to the square of velocity, the model for projectile motion reduces to a differential equation relating the speed $s(t)$ of the projectile with the angle of inclination $\theta(t)$

$$\frac{ds}{d\theta} - s\tan\theta = bs^3\sec\theta,$$

[13] See Section 2.1, especially Exercise 6.

where b is a constant that measures the amount of air resistance. To begin solving this equation, we multiply by s^{-3}, which makes the right-hand side independent of s:

$$s^{-3}\frac{ds}{d\theta} - s^{-2}\tan\theta = b\sec\theta. \tag{4}$$

Now define a new dependent variable by

$$w = s^{-2}.$$

Then

$$\frac{dw}{d\theta} = -2s^{-3}\frac{ds}{d\theta},$$

which is the same as the first term of the s equation except for the constant factor. Multiplying Equation (4) by the factor -2, we have the equation

$$\frac{dw}{d\theta} + 2w\tan\theta = -2b\sec\theta,$$

which is a first-order linear equation.

The solution of the w equation is

$$w = (\cos^2\theta)u,$$

where

$$u' = -2b\sec^3\theta.$$

Once w is known, s is recovered from $w = s^{-2}$, so we may use the solution for u directly to obtain a formula for s:

$$s = \frac{\sec\theta}{\sqrt{u}}.$$

Integration of u' yields the result

$$u = b(-\sec\theta\tan\theta - \ln|\sec\theta + \tan\theta| + C).$$

We therefore arrive at the solution

$$s = \frac{\sec\theta}{\sqrt{b[C - \sec\theta\tan\theta - \ln(\sec\theta + \tan\theta)]}},$$

where we have omitted the absolute value bars because the range of possible θ values ($|\theta| < \pi/2$) guarantees $\sec\theta + \tan\theta > 0$.

✦ INSTANT EXERCISE 3

Supply the omitted details of Example 4.

Choosing a Method for a Given First-Order Linear Equation

Most solution techniques for ordinary differential equations ultimately come down to one of two broad classes of methods. Some methods, such as separation of variables and the variation of parameters method of this section, are based on reducing a differential equation to an integration problem. Other methods, such as the characteristic value method and the undetermined coefficient method for homogeneous linear equations with constant coefficients, are based on predicting the functional form of the solution, substituting that form into the differential equation, and thereby reducing the differential equation to an algebra problem.

Methods that reduce a differential equation to an algebra problem can only be applied to a limited class of problems. One might wonder whether such methods are needed at all. The answer is that reduction methods are generally easier to use than integration methods. There are exceptions. Sometimes it is necessary to solve a number of differential equations of the form $y' + py = g(t)$, where the constant p is fixed but the function g varies. The variation of parameters method might be preferable in this case because the problem reduces to

$$y = e^{-pt}u(t), \qquad u' = e^{pt}g(t)$$

regardless of the choice of the forcing function g. In contrast, the trial solution for the method of undetermined coefficients is different for different forcing functions.

Example 5 illustrates that variation of parameters may be preferable for cases where the characteristic value of g is also a characteristic value of L.

EXAMPLE 5

The first step in the solution of

$$y' + 2y = (1 + 4t)e^{-2t}$$

by either undetermined coefficients or variation of parameters is to find the solution $y_1 = e^{-2t}$ of $y' + 2y = 0$.

Continuing by the method of undetermined coefficients, we observe that the forcing function is a generalized exponential function with characteristic value -2 and degree 1, and that the multiplicity of this characteristic value of L is 1. The trial solution, by Theorem 4.3.3, is $Y = (At + Bt^2)e^{-2t}$. Thus, $Y' = (A + 2Bt)e^{-2t} + (At + Bt^2)(-2e^{-2t}) = [A + (2B - 2A)t - 2Bt^2]e^{-2t}$ and $Y' + 2Y = (A + 2Bt)e^{-2t}$. We have $L[Y] = g$ if $A = 1$ and $B = 2$; therefore, a particular solution is $y_p = (t + 2t^2)e^{-2t}$.

Continuing instead by variation of parameters, we look for a solution $y_p = e^{-2t}u$. Thus, $e^{-2t}u' = (1 + 4t)e^{-2t}$. The factor e^{-2t} conveniently appears on both sides of the equation, so we end up with $u' = 1 + 4t$, for which we can choose the antiderivative $u = t + 2t^2$. The corresponding particular solution is again $y_p = (t + 2t^2)e^{-2t}$.

The correspondence of the forcing function e^{-2t} with the solution of the homogeneous equation $L[y] = 0$ made the method of undetermined coefficients more difficult because of the extra factor of t in the trial solution, but it made the method of variation of parameters easier by eliminating the factor $1/y_1$ from u'.

4.5 Exercises

In Exercises 1 through 6, solve the given differential equations, using the method of variation of parameters and the method of undetermined coefficients.

1. $y' - 4y = te^{t}$
2. $y' + 2y = (t - 1)e^{-t}$
3. $y' - 4y = te^{4t}$
4. $y' + y = (t - 1)e^{-t}$
5. $y' - 3y = \cos t$
6. $y' + y = \sin 2t$

In Exercises 7 through 16, solve the given differential equation.

7. $y' + y = 1/(1 + e^t)$

8. $y' + y = 1/(1 + e^{2t})$

9. $y' + ty = 2t$

10. $y' - 4ty = t$

11. $y' + 2ty = e^{-t^2} \cos t$

12. $y' + 2ty = 2t^3$

13. $ty' + (1 + t)y = 1$

14. $ty' + (1 + t)y = e^t$

15. $y' - y \tan t = 1$

16. $y' + y \tan t = t \sin 2t$

In Exercises 17 through 24, solve the given initial-value problem and indicate the interval on which the solution is valid. You may find it convenient to use the error function, as in Example 2.

17. $ty' - 2y = 6t^5, \quad y(1) = 0$

18. $ty' + y = t \cos t, \quad y(\pi) = 0$

19. $(1 + t^2)y' - 2ty = 1 + t^2, \quad y(0) = 0$

20. $(1 + t^2)y' + 4ty = 2t/(1 + t^2)^2, \quad y(0) = 1$

21. $y' + 2y \tan t = 1, \quad y(0) = 1$

22. $y' + y \tan t = \sin t, \quad y(0) = 0$

23. $y' - 8ty = 1, \quad y(0) = 1$

24. $y' - 2y = e^{-t^2}, \quad y(0) = 0$

In Exercises 25 through 28, solve the given differential equation.

25. $t^2 y' + 2ty = y^3, \quad t > 0$

26. $\theta^2 \dfrac{dr}{d\theta} = r^2 + 2r\theta$

27. $p' = rp(1 - p/K), \quad r > 0, \quad K > 0$. This is the logistic equation for population growth, developed in Section 5.1.

28. $y' = \epsilon y - \sigma y^3, \quad \epsilon > 0, \quad \sigma > 0$. This equation arises in fluid flow.

29. Consider the differential equation

$$ty' + my = kt^n, \quad t > 0,$$

where m and n are integers and k is any nonzero real number.

a. Find the general solution for the case where $n = -m$.
b. Find the general solution for the case where $n \neq -m$.
c. Suppose m and n are both positive. Explain why it is possible that the solution could be defined for all t. Does this contradict the existence and uniqueness theorem (Theorem 2.4.2)?
d. Suppose m and n are both negative. Is it possible that the solution could be defined for all t?

30. Let f be continuously differentiable, let g be any continuous function, and let y_0 be any real number. Derive a formula for the solution of the initial-value problem

$$fy' + f'y = g(t), \qquad y(0) = y_0.$$

What additional requirement must the function f satisfy to guarantee that this solution is valid?

31. Consider the family of initial-value problems

$$y' + y \tan t = \sin t, \qquad y(0) = k.$$

a. Find a formula for the solution and the interval on which the solution is valid.
b. Find the limiting value of the solution as it approaches the endpoints of its interval of validity.
c. Determine which values of k result in a solution that vanishes at some point in the interior of the interval of validity.
d. Determine which values of k result in a solution whose maximum occurs at $t = 0$. (*Hint:* It is best not to use the solution of the differential equation—instead, differentiate the differential equation implicitly with respect to t in order to evaluate the second derivative at the initial point $t = 0$.)

32. Consider the family of initial-value problems

$$ty' + (1 + t)y = e^{-t}, \qquad y(0) = y_0$$

a. What does the existence and uniqueness theorem (Theorem 2.4.2) say about this initial-value problem?
b. Find the general solution of the differential equation.
c. Find the unique value of y_0 for which the initial-value problem has a unique solution.
d. Does the answer to part c contradict the answer to part a?

Piecewise-Continuous Forcing

Although the existence and uniqueness theorem (Theorem 2.4.2) assumes that the forcing function g in the linear equation $y' + qy = g$ is continuous, it is more generally true that the forcing function needs only to be piecewise continuous. (A function is said to be **piecewise continuous** if it is continuous and bounded at all but a finite number of points in its domain.) Suppose, for example, the forcing function is discontinuous at some point $T > 0$. To solve an initial-value problem with initial data given at $t = 0$, it is necessary to solve the differential equation separately for $t < T$ and $t > T$. The integration constant for the first portion is determined from the initial condition, and that for the second portion is determined by requiring the solutions to agree at $t = T$. Exercises 33 and 34 deal with differential equations with discontinuous forcing.

33. Solve the initial-value problem

$$y' + y = g, \qquad y(0) = 0,$$

where

$$g(t) = \begin{cases} 1 & t \leq T \\ 0 & t > T \end{cases},$$

with $T > 0$ a constant.

34. Solve the initial-value problem

$$y' + y = g, \qquad y(0) = 0,$$

where

$$g(t) = \begin{cases} f(t) & 0 \leq t \leq 1 \\ 0 & t > 1 \end{cases},$$

with f a continuous function.

T 35. A population has a periodic relative growth rate that varies from a minimum rate at $t = 0$ months to a maximum rate at $t = 6$ months, and is approximately sinusoidal. The minimum growth rate corresponds to a loss of one-half of the population in 6 months, while the maximum growth rate corresponds to a doubling of the population in 3 months. Additionally, the population is harvested at a rate of R per month, where the population is scaled so that the population is 1 unit at the beginning of the year.

a. Derive the initial-value problem

$$\frac{dP}{dt} = \frac{\ln 2}{12}\left(1 - 3\cos\frac{\pi t}{6}\right)P - R, \qquad P(0) = 1.$$

b. Determine a solution P_1 of the associated homogeneous equation.
c. Determine a particular solution P_p. Note that P_p must be written as a definite integral.
d. Evaluate the integration constant to obtain a formula for the population that still contains the parameter R but is otherwise determined.
e. The population is sustained at the same yearly cycle if R is chosen so that $P(12) = 1$. Determine (numerically) the correct value of R. How much of the population is harvested each year relative to the population level at the beginning of the year?
f. Plot the population over a 2-year cycle. Explain why the resulting graph makes sense in terms of the model.

T 36. A patient takes a dose of 1 unit of medicine once per day, at times $t = 0$, $t = 1$, and so on. Suppose the medicine enters the bloodstream at a constant rate over a time interval of length k, where k is on the order of a few minutes. Assume that the medicine is removed from the body at a rate proportional to the amount, with relative rate 1. This is an example of a problem with periodic forcing, in which the action of the forcing function occurs very quickly in each period. In this exercise, we obtain an approximate solution for such problems.

a. Use the method of variation of parameters to find an expression for the solution of the initial-value problem

$$y' + y = g, \qquad y(0) = 0$$

in terms of the as yet unspecified function g.
b. Sketch the forcing function g, assuming that k is about 0.01 (1% of a day).

c. Using the forcing function described above, determine the solution $y(t)$ when $k < t \le 1$. Note that the integration interval must be split into portions, with g a positive constant in some portions and zero in others.

d. Show that $\lim_{k \to 0} k^{-1}(e^k - 1) = 1$. Conclude that the solution y can be approximated as e^{-t} on the interval $0 < t < 1$.

e. Repeat part *c* for times in the range $1 + k \le t \le 2$. Use the limit result of part *d* to obtain the approximate solution $(1 + e)e^{-t}$ on the interval $1 < t < 2$.

f. Similarly, obtain an approximate solution for the interval $2 < t < 3$.

g. Plot the approximate solution on the interval $0 < t < 3$.

37. The solution method for $y' + py = g$ eventually led to Equation (2). We glossed over a potential difficulty with this equation. If $y_1 = 0$ at some point, then u may not be defined at that point, since $u' = g/y_1$. Let $P(t)$ be an antiderivative of $p(t)$. Then we can write $L[y] = 0$ as $y' + P'y = 0$. Solve this differential equation for y in terms of P. Use the result to explain why the possibility $y_1 = 0$ can be ignored.

✦ 4.5 INSTANT EXERCISE SOLUTIONS

1. With $y = e^{-\sin t} + \sin t - 1$, we have $y' = -(\cos t)e^{-\sin t} + \cos t$. Then $y' + (\cos t)y = -(\cos t)e^{-\sin t} + \cos t + (\cos t)(e^{-\sin t} + \sin t - 1) = \cos t \sin t$.

2. We have

$$y = e^{-2t} + e^{-2t} \int_0^t e^{2w} \tan w \, dw.$$

Then

$$\begin{aligned} y' &= -2e^{-2t} + \left(-2e^{-2t} \int_0^t e^{2w} \tan w \, dw\right) + e^{-2t} \frac{d}{dt} \int_0^t e^{2w} \tan w \, dw \\ &= -2y + e^{-2t}(e^{2t} \tan t) = -2y + \tan t. \end{aligned}$$

Thus, $y' + 2y = \tan t$.

3. The homogeneous equation

$$\frac{dw}{d\theta} + 2w \tan\theta = 0$$

is separable, and it can be written as $dw/w = -2\tan\theta \, d\theta$. Integration yields $\ln|w| = -2\ln|\sec\theta| + C = \ln(\cos^2\theta) + C$. Thus, $w = A\cos^2\theta$. Now let $w(\theta) = (\cos^2\theta)u(\theta)$. Then $s = w^{-1/2} = (\sec\theta)u^{-1/2}$ and $w' = \cos^2\theta u' - 2\cos\theta \sin\theta u$ and

$$w' + 2w\tan\theta = \cos^2\theta u' - 2\cos\theta\sin\theta u + 2(\cos^2\theta u)\tan\theta = \cos^2\theta u'.$$

The nonhomogeneous differential equation for w therefore becomes $u' = (\sec^2\theta)(-2b\sec\theta) = -2b\sec^3\theta$. The integral can be found in a table of integrals. It can be integrated by parts to yield

$$u = -2b\sec\theta\tan\theta + 2b\int \sec\theta\tan^2\theta \, d\theta = -2b\sec\theta\tan\theta - u - 2b\int \sec\theta d\theta.$$

This yields $u = -b\sec\theta\tan\theta - b\int \sec\theta d\theta$, and it is still necessary to integrate $\sec\theta$. Finally, $\sec\theta + \tan\theta = (1 + \sin\theta)/\cos\theta$, so the algebraic sign of the sum of secant and tangent is determined by the secant term. For $-\pi/2 < \theta < \pi/2$, the cosine, and hence also the secant, is positive.

4.6 Particular Solutions for Second-Order Equations by Variation of Parameters

Suppose we have a linear differential equation of second order. In general, such an equation has the form

$$y'' + py' + qy = g, \tag{1}$$

where p, q, and g are functions. We know from Section 4.2 that the solution can be thought of as the sum of a particular solution and the complementary solution, the latter being the two-parameter family of solutions of the associated homogeneous equation. We also know (Section 4.3) how to find particular solutions for the very special case where p and q are constants and g is a generalized exponential function or a sum of such functions. If these restrictive conditions are not met, then we have not as yet found a method that works. However, the method of variation of parameters from Section 4.5 is easily extended to higher-order equations.

The Idea of Variation of Parameters

Suppose we have a complementary solution

$$y_c = c_1 y_1 + c_2 y_2 \tag{2}$$

for a linear equation of the form (1). The complementary solution is a two-parameter family of functions. As in the first-order case, we might guess that the particular solution should bear some relationship to the complementary solution, so we propose a particular solution of the form

$$y_p = u_1 y_1 + u_2 y_2, \tag{3}$$

where u_1 and u_2 are nonconstant functions to be determined. This substitution works if the problems to be solved for u_1 and u_2 are calculus problems rather than differential equation problems.

MODEL PROBLEM 4.6

Find a particular solution for

$$y'' + 3y' + 2y = 2te^{-t} \tag{4}$$

of the form (3).

Solving the Model Problem

We begin with the complementary solution

$$y = c_1 e^{-2t} + c_2 e^{-t}.$$

✦ INSTANT EXERCISE 1

Derive the solution $y = c_1 e^{-2t} + c_2 e^{-t}$ for the differential equation $y'' + 3y' + 2y = 0$.

Note that the form (3) of y_p has two unknowns, u_1 and u_2. In general, we expect a problem for two unknowns to be specified by two conditions rather than one, but the only condition imposed by the problem is that y_p must satisfy the differential equation (4). It seems reasonable that we are free to choose an additional requirement to be satisfied by u_1 and u_2. This turns out to be correct, and we can choose the additional requirement to simplify the problems to be solved for u_1 and u_2.

We want to proceed by substituting the form for y_p into the differential equation. To that end, we must compute the first and second derivatives of y_p in terms of u_1 and u_2. From $y_p = u_1 e^{-2t} + u_2 e^{-t}$, we have

$$y_p' = e^{-2t}u_1' + e^{-t}u_2' - 2e^{-2t}u_1 - e^{-t}u_2.$$

If this formula is differentiated again, the resulting formula for y_p'' will include second derivatives of u_1 and u_2. Since our goal is to obtain simple first-order equations for u_1 and u_2, it is imperative that we not permit second derivatives of u_1 and u_2. We can prevent them by a judicious choice of the additional requirement:

$$e^{-2t}u_1' + e^{-t}u_2' = 0. \tag{5}$$

This simplifies the formula for y_p'. Of the original four terms in y_p', two sum to 0, so these can be omitted. Thus,

$$y_p' = -2e^{-2t}u_1 - e^{-t}u_2.$$

This last result holds only because of the assumption of Equation (5). Thus, u_1 and u_2 have to satisfy that equation, and they also have to be chosen so that y_p satisfies the original differential equation. Differentiating the simplified formula for y_p' yields

$$y_p'' = -2e^{-2t}u_1' + 4e^{-2t}u_1 - e^{-t}u_2' + e^{-t}u_2.$$

Substituting these formulas, along with the formula for y_p, into the left side of the differential equation yields, after some simplification,

$$L[y_p] = -2e^{-2t}u_1' - e^{-t}u_2'.$$

Given the assumption (5), the original differential equation (4) reduces to

$$-2e^{-2t}u_1' - e^{-t}u_2' = 2te^{-t}. \tag{6}$$

At this point, we have two differential equations [Equations (5) and (6)] to be satisfied by the unknown functions u_1 and u_2. These equations are coupled, meaning that both unknowns appear in each equation; however, they can easily be uncoupled with some algebraic manipulation. Specifically, adding the two equations yields

$$-e^{-2t}u_1' = 2te^{-t},$$

from which we can then obtain the results

$$u_1' = -2te^{t}, \qquad u_2' = 2t.$$

These latter equations are simple calculus problems for the unknowns u_1 and u_2. Integration yields

$$u_1 = (2 - 2t)e^{t} + c_1, \qquad u_2 = t^2 + c_2.$$

If we want only a particular solution, then we can freely choose the integration constants. Alternatively, we can retain them and use the form (3) to obtain the general solution. Choosing $c_1 = c_2 = 0$ yields the particular solution

$$y_p = [(2-2t)e^t]e^{-2t} + t^2e^{-t} = (2-2t+t^2)e^{-t}; \tag{7}$$

leaving c_1 and c_2 unspecified yields the general solution

$$y = [(2-2t)e^t + c_1]e^{-2t} + (t^2+c_2)e^{-t} = c_1e^{-2t} + (c_2+2-2t+t^2)e^{-t}.$$

Note that the sum $c_2 + 2$ is itself an arbitrary constant, so we can simplify the answer by defining a new arbitrary constant $c_3 = c_2 + 2$, resulting in the solution formula

$$y = c_1e^{-2t} + (c_3 - 2t + t^2)e^{-t}. \tag{8}$$

✦ INSTANT EXERCISE 2

Determine u_1 by two different methods. First, integrate $u_1' = -2te^t$, as you would do in a calculus course. Second, try an undetermined-constant method. Guess that u_1 has the form $u_1 = (A + Bt)e^t$. Differentiate and then choose A and B so that the derivative is $-2te^t$. This latter method has the advantage of requiring only differentiation and algebra, while the first method requires an integration technique.

It is interesting to compare this result with the one we obtain by the method of undetermined coefficients. The trial solution, by Theorem 4.3.3, is

$$Y = (At + Bt^2)e^{-t},$$

and the calculations yield the results $A = -2$ and $B = 1$. Hence, the particular solution obtained by the method of undetermined coefficients is

$$y_p = (-2t + t^2)e^{-t} \tag{9}$$

and the general solution is again Equation (8), with the constant c_2 appearing in the place of c_3. The particular solutions obtained by the methods of variation of parameters and undetermined coefficients are not identical. Both are correct, however. The particular solution from variation of parameters includes an extra term, $2e^{-t}$, that can be incorporated into the complementary solution $c_1e^{-t} + c_2e^{-2t}$.

Variation of Parameters in General

It is generally best to learn solution techniques rather than the formulas resulting from them, except for cases where a simple formula results from the application of a cumbersome technique. In the case of variation of parameters, we can derive simple general formulas for u_1' and u_2' to avoid much of the computation. We consider a second-order linear nonhomogeneous differential equation (1), where the coefficient functions p and q and the forcing function g are continuous. We assume that a complementary solution (2) is known and that we desire a particular solution (3). Our goal is to obtain formulas for u_1' and u_2' that can then be integrated to complete the solution.

As in Model Problem 4.6, our plan is to substitute the form (3) into the differential equation (1) to get an equation involving the unknown functions u_1 and u_2 and their derivatives. We get only one equation from this procedure, but we have two unknown functions. This adds some crucial flexibility to the procedure, because it allows us to arbitrarily impose an additional restriction on u_1 and u_2. We choose a restriction that simplifies the equations.

Using the product rule, we have the first derivative of y_p:

$$y_p{}' = u_1 y_1{}' + y_1 u_1{}' + u_2 y_2{}' + y_2 u_2{}'.$$

As it currently stands, the formula for $y_p{}''$ will include the second derivatives of u_1 and u_2, which will then appear in the differential equation after the substitution. Since we are free to impose an additional restriction, we now choose to require

$$y_1 u_1{}' + y_2 u_2{}' = 0. \tag{10}$$

This eliminates two of the terms from $y_p{}'$ and leaves

$$y_p{}' = u_1 y_1{}' + u_2 y_2{}'.$$

Differentiating again yields

$$y_p{}'' = u_1 y_1{}'' + u_1{}' y_1{}' + u_2 y_2{}'' + u_2{}' y_2{}'.$$

Substituting these derivative formulas and the formula for y_p into the differential equation (1) yields, after some rearrangement,

$$u_1[y_1{}'' + p(t)y_1{}' + q(t)y_1] + u_2[y_2{}'' + p(t)y_2{}' + q(t)y_2] + y_1{}'u_1{}' + y_2{}'u_2{}' = g(t).$$

Recall that y_1 and y_2 are solutions of the homogeneous equation $y'' + p(t)y' + q(t)y = 0$; therefore, the quantities inside the brackets in this last equation are both zero, and the full differential equation (1) reduces to

$$y_1{}'u_1{}' + y_2{}'u_2{}' = g(t). \tag{11}$$

We now have two differential equations [Equations (10) and (11)] for the unknown functions u_1 and u_2. [Note that these reduce to Equations (5) and (6) with the correct y_1, y_2, and g for the model problem.] These equations can be thought of as a pair of algebraic equations for u_1' and u_2'. Since they are linear, we can write them as a system of two algebraic equations,

$$\begin{pmatrix} y_1 & y_2 \\ y_1' & y_2' \end{pmatrix}\begin{pmatrix} u_1' \\ u_2' \end{pmatrix} = \begin{pmatrix} 0 \\ g \end{pmatrix}. \tag{12}$$

The matrix in this system of equations is the matrix whose determinant is the Wronskian $W[y_1, y_2]$.[14] Note that this equation is a generalization to the second-order case of the equation $y_1 u' = g$ from the first-order case.

The system of Equation (12) for u_1' and u_2' can be solved in any manner, with the result

$$u_1' = \frac{-y_2 g}{W[y_1, y_2]}, \qquad u_2' = \frac{y_1 g}{W[y_1, y_2]}. \tag{13}$$

[14] Section 3.3.

Equation (13) assumes $W \neq 0$. Is it possible that the method fails because either u_1' or u_2' is zero at some point? The answer comes from Theorem 3.3.2. As long as the Wronskian is computed only on an interval in which all coefficient functions are continuous, the Wronskian is either everywhere zero or nowhere zero. If it is everywhere zero, then the set $\{y_1, y_2\}$ is not linearly independent. But this cannot be when $y = c_1 y_1 + c_2 y_2$ is the general solution.

We now have the complete method for the solution of Equation (1) worked out.

Theorem 4.6.1

Let p, q, and g be continuous functions, with p and q continuous and g continuous or piecewise continuous. Let $y_c = c_1 y_1 + c_2 y_2$ be the general solution of

$$y'' + py' + qy = 0,$$

and let

$$W[y_1, y_2] = y_1 y_2' - y_2 y_1'.$$

The general solution of

$$y'' + py' + qy = g$$

is

$$y = y_c + u_1 y_1 + u_2 y_2,$$

where u_1 and u_2 are any functions such that

$$u_1' = \frac{-y_2 g}{W[y_1, y_2]}, \qquad u_2' = \frac{y_1 g}{W[y_1, y_2]}.$$

EXAMPLE 1

Consider the equation

$$y'' + y = 2\tan t.$$

The complementary solution is

$$y_c = c_1 \cos t + c_2 \sin t.$$

Note that the Wronskian is $W = y_1 y_2' - y_2 y_1' = 1$. Then

$$u_1' = -2\sin t \tan t, \qquad u_2' = 2\sin t.$$

From an integral table, we have

$$u_1 = 2\sin t - 2\ln(\tan t + \sec t) + c_1$$

and of course $u_2 = -2\cos t + c_2$. Thus, a particular solution is

$$y_p = -2\cos t \, \ln(\tan t + \sec t)$$

and the general solution is

$$y = [c_1 - 2\ln(\tan t + \sec t)]\cos t + c_2 \sin t.$$

EXAMPLE 2

Consider the equation

$$y'' - y = 2\tan t.$$

The complementary solution is

$$y_c = c_1 e^{-t} + c_2 e^{t}.$$

Note that the Wronskian is $W = y_1 y_2' - y_2 y_1' = 2$. Then

$$u_1' = -e^{t}\tan t, \qquad u_2' = e^{-t}\tan t.$$

These functions have no elementary antiderivatives, so we must construct antiderivatives using definite integrals. By the fundamental theorem of calculus,

$$u_1 = c_1 - \int_0^t e^{\tau}\tan\tau\, d\tau, \qquad u_2 = c_2 + \int_0^t e^{-\tau}\tan\tau\, d\tau.$$

We may therefore write the general solution as

$$y = \left(c_1 - \int_0^t e^{\tau}\tan\tau\, d\tau\right)e^{-t} + \left(c_2 + \int_0^t e^{-\tau}\tan\tau\, d\tau\right)e^{t}.$$

Sometimes it is convenient to combine the integrals into a single integral by moving the factors $y_1(t)$ and $y_2(t)$ inside the integrals:

$$y = c_1 e^{-t} + c_2 e^{t} + \int_0^t (e^{t-\tau} - e^{\tau-t})\tan\tau\, d\tau.$$

This last result can also be rewritten, using the hyperbolic sine and cosine functions, as

$$y = C_1\cosh t + C_2\sinh t + 2\int_0^t \sinh(t-\tau)\tan\tau\, d\tau.$$

To check the particular solution, suppose we take $C_1 = C_2 = 0$. Then

$$y_p = 2\int_0^t \sinh(t-\tau)\tan\tau\, d\tau,$$

$$y_p' = 2\int_0^t \cosh(t-\tau)\tan\tau\, d\tau + 2\sinh(0)\tan t = 2\int_0^t \cosh(t-\tau)\tan\tau\, d\tau,$$

and

$$y_p'' = 2\int_0^t \sinh(t-\tau)\tan\tau\, d\tau + 2\cosh(0)\tan t = y_p + 2\tan t.$$

4.6 Exercises

In Exercises 1 through 4, use the method of variation of parameters to find a particular solution. Follow the procedure used in the text to solve Model Problem 4.6; do not use the general formulas of Theorem 4.6.1. Compare the answers with those obtained in the corresponding exercise from Section 4.3.

1. $y'' - y = (-6 + 4t)e^{-t}$ (Exercise 7)
2. $y'' + 2y' + y = 4e^{-t}$ (Exercise 8)
3. $y'' + y' - 2y = te^{3t}$ (Exercise 9)
4. $y'' + y' - 2y = \cos 2t$ (Exercise 10)

In Exercises 5 through 8, find the general solution of the given differential equation.

5. $y'' + 4y = \sec 2t, \quad -\pi/4 < t < \pi/4$
6. $y'' + 4y = \sec t, \quad -\pi/2 < t < \pi/2$
7. $y'' + 2y' + y = t^{-p}e^{-t}, \quad t > 0,$ where p is a positive integer
8. $y'' + 2y' + y = e^{-t}/(1 + t^2)$

In Exercises 9 through 12, verify that the given functions y_1 and y_2 satisfy the associated homogeneous equation and determine the general solution.

9. $(1 - x)y'' + xy' - y = xe^{-x}, \quad |x| < 1, \quad y_1 = x, \quad y_2 = e^x$
10. $x^2(1 - \ln x)y'' + xy' - y = (1 - \ln x)^2, \quad x > 0, \quad y_1 = x, \quad y_2 = \ln x$
11. $4x^2y'' + 4xy' + (4x^2 - 1)y = x^{3/2} \cos x, \quad x > 0,$
    $y_1 = x^{-1/2} \cos x, \quad y_2 = x^{-1/2} \sin x$
12. $(1 + \tan x)y'' - 2y' + (1 - \tan x)y = (1 + \tan x)^2, \quad x > 0, \quad y_1 = \cos x, \quad y_2 = e^x$
13. Consider the family of differential equations given by

$$xy'' - (1 + x)y' + y = g(x), \quad x > 0,$$

where $g(x)$ is a polynomial.

a. Show that $y_1 = 1 + x$ and $y_2 = e^x$ are solutions of the associated homogeneous equation.
b. Find particular solutions for the cases $g = 1$ and $g = x^2$. Use your results to argue that there is always a particular solution that is a second-degree polynomial whenever $g = a + bx^2$.
c. Find a particular solution for any case $g = x^p$ for $p \geq 2$, and note that the solution is a polynomial of degree p.
d. Find a particular solution for the case $g = x$. The solution is conveniently written in terms of the exponential integral function, which is defined by

$$E_1(x) = \int_x^{\infty} t^{-1}e^{-t}\,dt, \qquad x > 0.$$

Note that in general it is not correct that there is a particular solution in the form of a polynomial of degree p when g is a polynomial of degree p.

T 14. The forced Airy equation $y'' - xy = 1$ arises in the solution of a problem to determine the shape of the trunk of a Mexican fan palm bent by a strong wind.[15]

[15] See Donald F. Winter, On the stem curve of a tall palm in a strong wind, *SIAM Review,* vol. 35, pp. 567–579, 1993.

a. Use variation of parameters to derive a particular solution in the form of a definite integral, as in Example 2. See Section 2.4 for information about the Airy functions.

b. Verify that the function $y(x) = \int_0^\infty \exp(xt - t^3/3)\,dt$ is also a particular solution of $y'' - xy = 1$.

c. Use a computer algebra system to plot each of the particular solutions of parts a and b. Are they the same function? If not, is that a contradiction?

d. Discuss the practical merits of each of the two solutions. Is one easier to understand than the other (for example, can you look at them and determine by inspection whether they are strictly increasing, strictly decreasing, or neither)? Is there a difference in the amount of time a computer algebra system takes to plot a particular solution from the two solution formulas?

T 15. This problem examines the behavior of an undamped oscillator forced at its natural frequency for a fixed amount of time and left free to oscillate thereafter. Specifically, the aim is to determine the magnitude of the steady-state amplitude.

a. Use variation of parameters to derive a particular solution for any problem

$$y'' + y = g.$$

(Construct u_1 and u_2 with definite integrals from 0 to t, as in Example 2.)

b. Use the result from part a to compute u_1 and u_2 for the case $g(t) = \sin t$. Show that the particular solution in this case satisfies the initial conditions $y(0) = 0$ and $y'(0) = 0$. Conclude that any steady-state solution represents the motion caused by the forcing.

c. Suppose $g(t) = \sin t$ for times $0 < t < T$, and $g(t) = 0$ for times $t > T$, where T is a positive constant that marks the time at which the forcing ends. Use the result from part a to calculate a formula for y_p that is valid for times $t > T$. [*Hint:* For $t > T$, the expressions for u_1 and u_2 must be calculated with two integrals because the formula for $g(\tau)$ is different for different ranges of τ.]

d. Determine the amplitude of the steady-state solution from part c as a function of T. Plot the graph of amplitude versus forcing interval.

e. Plot the solutions to the full problem with initial conditions $y(0) = 0$ and $y'(0) = 0$, for the cases $T = 0.5\pi, 0.75\pi, \pi, 1.25\pi$, and 1.5π. Compare the results.

16. Consider the family of initial-value problems

$$y'' + y = g, \qquad y(0) = y_0, \qquad y'(0) = v_0$$

where g is any continuous function and y_0 and v_0 are given constants.

a. Use variation of parameters to derive a particular solution. (Construct u_1 and u_2 with definite integrals from 0 to t, as in Example 2.)

b. Use a trigonometric identity to rewrite the particular solution as

$$y_p(t) = \int_0^t \sin(t - \tau)\, g(\tau)\, d\tau.$$

c. Compute $y_p(0)$ and $y_p'(0)$, and use these results to solve the initial-value problem.

d. The integral in y_p is called the **convolution** of $\sin t$ and $g(t)$, and it is sometimes denoted by $(\sin t) * g$. Show that $f * g = g * f$ for any functions f and g for which the convolution is defined.

e. Use the property of part d and the formula from part b to find a particular solution for

$$y'' + y = e^{-t}.$$

f. Use the method of undetermined coefficients to find a particular solution for the equation of part *e*.

g. Is there a contradiction between the results of parts *e* and *f*? Explain.

17. Consider a third-order differential equation

$$L[y] = y''' + py'' + qy' + ry = g,$$

where p, q, r, and g are continuous functions. Suppose the complementary solution $y = c_1y_1 + c_2y_2 + c_3y_3$ is known.

a. Write down a form for y_p in terms of the known solutions of the homogeneous problem and unknown coefficient functions u_i.

b. Derive a system of equations analogous to that of Equation (12).

c. Solve the system of part *b* to obtain formulas for u_1', u_2', and u_3'.

In Exercises 18 through 20, use the method of variation of parameters from Exercise 17 to obtain a particular solution of the given differential equation.

18. $y''' + y' = \tan t, \quad 0 < t < \pi$

19. $y''' - 2y'' - y' + 2y = e^t$

20. $y''' - y'' + y' - y = \sin t$

T 21. The deflection of a uniform beam of unit length (think of a diving board) that is clamped in place at one end and free at the other end is governed by the boundary value problem

$$y''''(x) - y''(x) = -w(x), \qquad y(0) = 0, \qquad y'(0) = 0, \qquad y''(1) = 0, \qquad y'''(1) = 0,$$

where w is the weight per unit length of a load placed on top of the beam.

a. Determine the solution of the associated homogeneous equation $y'''' - y'' = 0$, using hyperbolic functions in place of exponential functions.

b. Confirm that the differential equation has a particular solution given by

$$y_p = -\int_0^x s w(s)\, ds + x \int_0^x w(s)\, ds + \int_0^x \sinh(s - x)\, w(s)\, ds.$$

c. Construct the general solution and use the auxiliary conditions (called **boundary conditions**) at $x = 0$ to reduce the solution to a two-parameter family.

d. Now suppose a unit load is divided over the last tenth of the beam ($w = 0$ for $x < 0.9$ and $w = 10$ for $x > 0.9$). Determine the particular solution.

e. Use the remaining boundary conditions to determine the two remaining integration constants in the solution.

f. Plot the solution.

✦ 4.6 INSTANT EXERCISE SOLUTIONS

1. The substitution $y = e^{rt}$ yields the characteristic equation $r^2 + 3r + 2 = 0$, which has roots $r_1 = -2$ and $r_2 = -1$. Thus, the general solution of $y'' + 3y' + 2y = 0$ is $y = c_1e^{-2t} + c_2e^{-t}$.

2. Let $r = -2t$ and $s' = e^t$. Then $r' = -2$ and $s = e^t$. The integration by parts formula yields

$$u_1 = \int -2te^t\,dt = \int rs'\,dt = rs - \int r's\,dt = -2te^t + \int 2e^t\,dt = -2te^t + 2e^t + c_1.$$

Alternatively, if $u_1 = (A + Bt)e^t + c_1$, then $u_1' = Be^t + (A + Bt)e^t = (A + B + Bt)e^t$. To get $u_1' = -2te^t$, it is necessary to have $B = -2$ and $A = 2$, whence $u_1 = (2 - 2t)e^t + c_1$.

CASE STUDY 4

A Tuning Circuit for a Radio

Recall the *RLC* circuit model from Section 4.1

$$LCv'' + RCv' + v = v_e, \tag{1}$$

where v is the voltage on the capacitor; L, R, and C are, respectively, the inductance, resistance, and capacitance; and v_e is the voltage supplied by an external source.

Our task is to study this model with the aim of designing a tuning circuit for a radio. We consider amplitude modulation (AM) radio signals. These signals are sinusoidal functions with fixed frequency ω and variable amplitude $W(t) > 0$. The variation in the amplitude is very slow relative to the period of the oscillation. It is, therefore, a useful simplification to think of the amplitude as being constant. We then have the equation

$$LCv'' + RCv' + v = We^{2\pi i\omega t}, \tag{2}$$

where ω is the frequency in hertz (cycles per second) of an AM station. Using a complex exponential function to represent a sinusoidal function of unit amplitude simplifies the calculations. Note that the factor 2π is needed so that the period of the signal is $1/\omega$.

The solution $v(t)$ approaches the steady-state solution $v_S(t)$ very quickly, so we need to consider only the steady-state solution of Equation (2). There is damping in the model, so the operator on the left side of Equation (2) cannot have a pure imaginary characteristic value. Thus, the steady-state solution of Equation (2) must have the form

$$v_S(t) = A_0e^{2\pi i\omega t}, \tag{3}$$

where A_0 is a complex constant. The circuit response can be described by a single quantity called the **gain,** which we define to be the ratio of the amplitude $A = |A_0|$ of the steady-state voltage across the capacitor to the amplitude W of the input signal. The gain measures the amplification by the *RLC* circuit of the minuscule voltage generated by the antenna.

Finding the Gain in Terms of the Parameters

The parameters L, R, and C are chosen in the design of the circuit, while ω and W are determined by the radio signal. The quantity $|A_0|$, which depends on these parameters, is determined by finding the amplitude of the steady-state solution of Equation (2). We may therefore think of the model as providing a way to determine G from the input parameters, as illustrated in Figure C4.1.

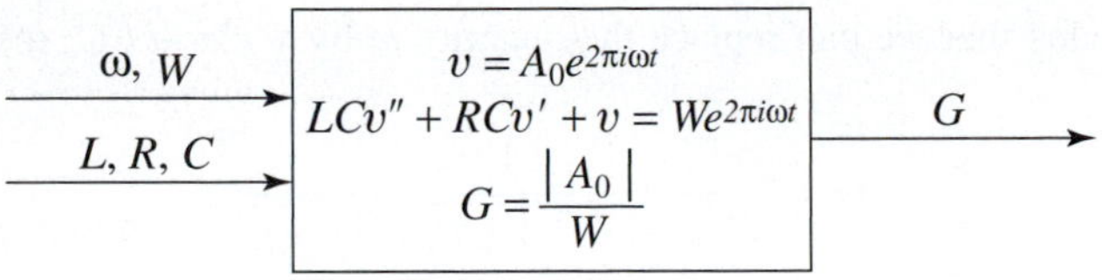

Figure C4.1
A schematic diagram of the tuning circuit model.

Of particular interest is the situation where the circuit parameters are fixed and we want to know how the gain depends on the frequency of the AM station. This is important because the job of the tuning circuit is to discriminate among the different stations so that you can hear only one of them.

Determination of G

Solve the problem indicated in Figure C4.1. Illustrate the dependence of G on ω, with fixed L, R, and C.

Substituting the steady-state solution form (3) into the differential equation (2) ultimately yields the result

$$A_0 = \frac{W}{1 - 4\pi^2 LC\omega^2 + 2\pi i RC\omega}, \tag{4}$$

and thus the gain is

$$G = \frac{1}{\sqrt{(1 - 4\pi^2 LC\omega^2)^2 + (2\pi RC\omega)^2}}. \tag{5}$$

The strength W of the input signal does not appear in the formula for the gain because the factor W is lost in the ratio.

Interpreting the Result for the Gain

Understanding the result (5) is complicated by the dependence of G on four different parameters. We can improve the result by recasting it in dimensionless form. Actually, the result is already a little bit dimensionless because the output amplitude $|A_0|$ appears instead as the dimensionless gain G. The number of parameters is reduced whenever a variable is replaced by a dimensionless counterpart. The next task is to replace the frequency ω with a dimensionless variable. One insight into how to do this comes from the quantity in parentheses inside the square root in Equation (5). To subtract $4\pi^2 LC\omega^2$ from 1, the quantity $4\pi^2 LC\omega^2$ must be dimensionless. We can therefore define a dimensionless frequency γ by

$$\gamma = 2\pi\sqrt{LC}\omega. \tag{6}$$

This definition means that we can replace the quantity ω by $\gamma/2\pi\sqrt{LC}$, thereby rewriting the result as

$$G = \frac{1}{\sqrt{(1-\gamma^2)^2 + (R\sqrt{C/L}\,\gamma)^2}}.$$

Another simplification is now apparent. We define a dimensionless resistance Z by

$$Z = R\sqrt{\frac{C}{L}}.$$

The solution of our problem in dimensionless form is then

$$G = \frac{1}{\sqrt{(1-\gamma^2)^2 + (Z\gamma)^2}}. \tag{7}$$

There are other ways that we could have chosen dimensionless parameters; however, the given way has one big advantage: The parameter Z depends only on the properties of the circuit. The parameter γ is a measure of the frequency of the input signal under consideration. We can get a good idea of the significance of the results by plotting G against γ, for various values of Z. This graph is shown in Figure C4.2.

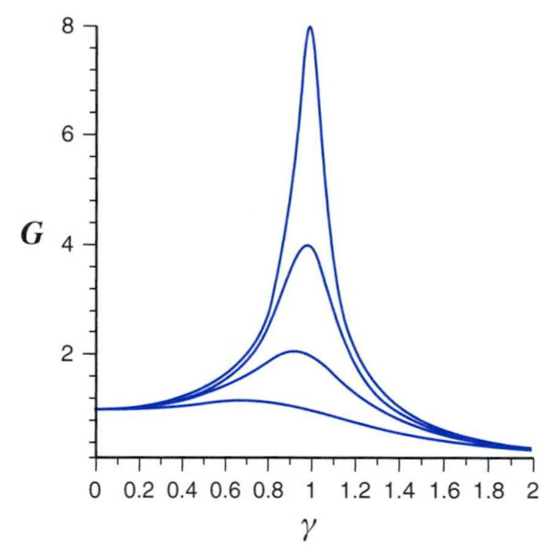

Figure C4.2
The dependence of gain on frequency, with $Z = 1, \frac{1}{2}, \frac{1}{4}, \frac{1}{8}$ from bottom to top.

Tuning Circuit Design

Now we have a mathematical result; the next question concerns how to use the result to design a tuning circuit. We begin with a careful description of the signal that comes to the circuit from a real antenna. AM radio stations have frequencies from 550 kHz to 1600 kHz. Every antenna is continuously bombarded by tiny signals having frequencies in this range. In a populous area, there could well be 100 radio signals, one every 10 kHz, all serving as simultaneous inputs to the antenna. Now suppose the radio is turned on with the dial set at a particular frequency. All the signals are inputs to the antenna, but only one is amplified enough to be heard. Figure C4.2

shows that for each Z, the maximum response is at $\gamma = 1$. This is the value of γ corresponding to the radio station for which the tuner is set. Other values of γ correspond to other stations, and for some the value of γ is not far from 1.

We have to accomplish two goals in the design of the tuning circuit. The circuit must be designed so that the desired signal is amplified many more times than the nearest undesirable signal, and it must be designed so that any frequency from 550 kHz to 1600 kHz can be chosen. To accomplish these goals, we have to choose values of the circuit parameters L, R, and C, with two of these fixed and one controlled by the tuning knob.

All the curves in Figure C4.2 represent values of Z that are much too large to be useful. Consider the top curve, for $Z = 0.125$. There is a clear advantage for the desired station, but it is not enough to block out the nearby stations. A radio with a tuning circuit having $Z = 0.125$ would allow us to hear several different stations interfering with the desired station. A good tuner must have a very sharp curve near the peak at $\gamma = 1$. For a real radio tuner, Z should be small enough that the gain of the closest station is no more than 1% of the gain of the desired station, and probably quite a bit less than that.

The value of ω is predetermined for each station. To set $\gamma = 1$ for any value of ω, we have to be able to change either L or C. In practice, it is easier to make a variable capacitor than a variable inductor. Thus, we design the circuit by choosing constant values of L and R. The value of C is chosen by turning a knob on the radio to tune in the desired station, corresponding to $\gamma = 1$. Given ω and L, we have

$$C = \frac{1}{4\pi^2 L\omega^2}. \tag{8}$$

In practice, a typical value for the inductance L is 250 μH. Given the frequency range

$$0.55 \times 10^6 \leq \omega \leq 1.6 \times 10^6,$$

along with $L = 2.5 \times 10^{-4}$, we have the range

$$3.9 \times 10^{-11} \leq C \leq 3.35 \times 10^{-10}.$$

The variable capacitor must be able to vary from 39 pF[16] to get the station at 1600 kHz to 335 pF to get the station at 550 kHz.

With both the inductance and capacitance now chosen, it remains only to choose the resistance. Since Z is proportional to resistance, and small Z is desirable, it seems appropriate to make the resistance as small as possible, perhaps even zero. In practice, the resistance cannot be made arbitrarily small, both because the wire itself carries a small amount of resistance and because some resistance is necessary to limit the magnitude of the current. The question of how to choose the resistance is considered in Exercise 4.

Case Study 4 Exercises

1. Derive the steady-state solution (4).

2. Calculate the voltage drop across the resistor. Determine the ratio $G_R = A_R/W$, where A_R is the amplitude of the voltage drop across the resistor. Plot G_R versus γ for the same values of Z as in Figure C4.2. Discuss the graph.

[16]The unit is the picofarad (pF), or 10^{-12} F.

3. Show by calculus that the maximum gain as $z \to 0$ occurs at $\gamma = 1$, as suggested by Figure C4.2.

4. *a.* Suppose you want to listen to a station with frequency of 550 kHz and there is a competing station with frequency of 560 kHz. Assuming that the input signals are of equal strength, how small does Z have to be so that the output signal for the competing station is only 1% of that for the desired station? What resistance is necessary to accomplish this much discrimination, given $L = 250\ \mu\text{F}$ and the appropriate value of C?

   b. Repeat part *a* for the case where the desired station has frequency of 1600 Hz and the competing station has frequency of 1590 Hz.

   c. Based on the results of these calculations, what is the largest reasonable value for R?

chapter

5 Autonomous Equations and Systems

An *autonomous* differential equation is a differential equation in which the independent variable does not appear explicitly. The decay equation $y' = -ky$ and the linear oscillator $ay'' + by' + cy = 0$ are examples. A natural process is autonomous if the rates of change depend only on the *state* of the system, and not on the time or position; such processes are modeled by autonomous differential equations. Autonomous problems are worth special study for two reasons: they are common in nature, and the special properties they have are the basis for some useful techniques. This chapter introduces some of these techniques: the phase line and phase plane, the direction field, trajectory analysis, and nullcline analysis.

The modeling in the chapter focuses on population dynamics, including both single populations and systems of two interacting populations. Several population models are introduced in a unified manner in Section 5.1, and these models are discussed as the methods for their study are developed in later sections.

In Section 5.2, we study the *phase line,* a special case of the slope field, for single autonomous first-order equations.

The remaining sections of Chapter 5 focus on autonomous systems of two first-order equations. The *phase plane* interpretation of such systems is introduced in Section 5.3, with the focus on *trajectories* taken by solutions. The *direction field* is introduced in Section 5.4, as are *equilibrium solutions* and the notion of *stability. Nullcline analysis* is presented in Section 5.5; this useful technique is similar to the isocline technique of Section 2.3.

5.1 Population Models

Mathematical models in the physical sciences are based on physical laws that are supported by a wealth of quantitative data. In the life sciences and social sciences, there is seldom such a clear starting point. Population growth, for example, does not seem to be governed by anything as definite as Newton's laws of motion. Instead, models for population growth tend to follow largely from educated guesses. Nevertheless, such models are able to reproduce the qualitative features of observations and can sometimes even produce accurate quantitative results. Population models begin with assumptions about how the relative rate of change[1] of a population depends on the current size of the population and whatever environmental factors, such as food supply or the presence of natural enemies, are considered relevant.

The simplest differential equation model of population growth is based on the assumption that the relative rate of change of a population is constant.[2] Thus,

$$\frac{1}{p}\frac{dp}{dt} = r,$$

or

$$\frac{dp}{dt} = rp,$$

where r is a positive constant with dimension time^{-1}. This **natural growth** equation is the same as the decay equation that we saw in Chapter 1, except that the relative growth rate is positive rather than negative. As before, solutions have the form

$$p = p_0 e^{rt}.$$

The model therefore predicts that population growth will be exponential.[3]

As simple as it is, the natural growth model is still useful for populations that are not significantly affected by resource limitations or crowding. It also serves as a good starting point for discussing population growth. In particular, any better model for population growth has to include a mechanism to account for limited capacity. In the remainder of Section 5.1, we consider several ways to do this.

MODEL PROBLEM 5.1

Construct population models that include reasonable mechanisms for limiting the growth of a rabbit population because of either limited food or predation by carnivores.

Logistic Growth

The best-known mathematical model of population growth to incorporate a limited capacity is the **logistic growth** model.[4] We assume that there is a limiting population value K that represents the environmental capacity. Instead of assuming a constant relative growth rate r, we assume a relative growth rate that is r when the population is small but decreases to zero when the population reaches the value K. The simplest such model assumes that the relative growth rate is a linear

---

[1] Section 1.1.

[2] This model is attributed to Thomas Malthus (1766–1834).

[3] Malthus was not so naive as to predict that any natural population would undergo exponential growth for all time. Rather, he recognized that population increase leads to increases in disease and starvation. His point was that it was not possible to end starvation because abundant food would simply allow the population to grow to a point where food became limited again.

[4] In the military, the term *logistics* is used to describe the process of delivering adequate supplies to troops in the field. In our context, the growth is described as "logistic" because it is limited by availability of supplies. This model is attributed to P. F. Verhulst (1804–1849).

function of the population. The relative growth rate is to have the value r when $p = 0$ and the value 0 when $p = K$ (see Fig. 5.1.1); thus we have

$$\frac{1}{p}\frac{dp}{dt} = r\left(1 - \frac{p}{K}\right),$$

or

$$\frac{dp}{dt} = rp\left(1 - \frac{p}{K}\right). \tag{1}$$

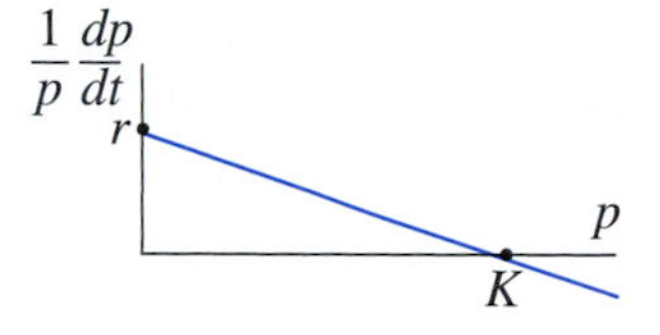

Figure 5.1.1
Relative population growth rate as a function of population in the logistic growth model.

✦ INSTANT EXERCISE 1

Derive the formula $r(1 - p/K)$ from the requirements that the function be linear and yield r for $p = 0$ and 0 for $p = K$.

The **state** of a system is the set of values of the dependent variables in the system. The independent variable t does not appear explicitly in either the natural growth equation or the logistic growth equation; thus, the rates of change depend only on the state of the system. A differential equation or system is **autonomous** if the functions representing the rates of change depend only on the state of the system.

The Lotka–Volterra Model

Another way to limit population growth is to incorporate the effects of competition between populations. Consider populations of rabbits and coyotes living in a grassland. We can think of the coyotes as a mechanism for limiting the population of rabbits. Let $x(t)$ be the rabbit population in a given region, and let $y(t)$ be the coyote population in that region. Both of these functions are unknown, so the model requires two differential equations, each prescribing the growth rate of one of the species. Suppose we assume that the rabbit population exhibits natural growth in the absence of coyotes but decreases in the presence of coyotes. One possibility is

$$\frac{dx}{dt} = rx - sxy, \tag{2}$$

where r is a positive constant with dimension time^{-1} and s is a positive constant with dimension coyotes^{-1}time^{-1}.

Why should we make this choice? In the absence of any fundamental principle or real data, we don't know what to choose, so it makes sense to start with the simplest function that has the right qualitative properties. The function $rx - sxy$ has these properties:

- It is proportional to the rabbit population.
- It assumes that the loss of rabbits due to coyotes is proportional to the product of the rabbit and coyote populations.
- It corresponds to a relative rabbit growth rate that is a linear function of the coyote population.

All these properties seem reasonable. Twice as many rabbits ought to produce a growth rate twice as large. Coyotes eat rabbits some fraction of the time that coyotes and rabbits encounter each other, and it makes sense that the rate at which these encounters occur would be proportional to each population.[5] The encounter rate would be proportional to each population if creatures moved about randomly, as do molecules in a gas. For movement of living creatures, proportionality is only a rough approximation. Nevertheless, linear functions make models simpler, so it makes sense to choose a linear function for a first attempt. We have no way to know in advance whether the choice $rx - sxy$ will produce a successful model. One way to decide this question is to compare the predictions the model makes with observations of real populations.[6] Meanwhile, we should not have much faith in the model, since it is not based on strong evidence.

What about a differential equation to model changes in the coyote population? First we need to make assumptions about the factors most directly responsible for increases or decreases in the number of coyotes. The most logical choice for the cause of coyote population decrease is the starvation that would occur if there were no rabbits. In the absence of rabbits, we might expect that the coyote population would experience a constant relative decay rate, as in a radioactive decay model. The key factor promoting growth in the coyote population is the consumption of rabbits. The differential equation, assuming that these two factors are the only ones needed, is

$$\frac{dy}{dt} = csxy - my, \tag{3}$$

where m is a positive constant with dimension time^{-1} and c is a positive dimensionless constant. The term $csxy$ is the gain in coyote population as a result of the coyote-rabbit interaction; c is the conversion factor of rabbit loss into coyote gain. The constant m represents the per capita mortality rate of the coyotes in the absence of rabbits.

The model given by Equations (2) and (3) is the **Lotka–Volterra**[7] predator-prey model. The specific species that serve as the predator and as the prey are unimportant as long as the designated prey is the principal food source of the designated predator, which in turn is the primary cause of death for the prey. There are also two important assumptions hidden in the model:

- The region is approximately uniform in population densities. Otherwise, it would be necessary to consider spatial variations.
- The region is closed, meaning that predator and prey cannot move in or out of the region.

---

[5]The letter s is chosen for this model because the constant that it represents is a measure of both the *search* rate of the predator and the probability of *success* in hunting.

[6]If the model fails this test, it still requires some thought to determine *which* feature of the model is the most likely problem.

[7]The model is named after the mathematicians Alfred J. Lotka and Vito Volterra, who independently proposed it in 1925 and 1926.

✦ INSTANT EXERCISE 2

In the predator-prey model

$$\frac{dp}{dt} = p(2q - 3), \qquad \frac{dq}{dt} = q(5 - 8p),$$

which is the predator and which is the prey? How do you know?

Mathematically, the Lotka–Volterra model is a **coupled** system of two autonomous nonlinear first-order differential equations because the rate of change of each variable depends explicitly on the other variable. If one of the equations in a system can be solved first without considering the other, the system is **uncoupled.**

EXAMPLE 1

The system

$$\frac{dx}{dt} = -ty^2, \qquad \frac{dy}{dt} = y$$

is an uncoupled system. The second equation can be solved immediately, with solution

$$y = Ae^t.$$

The first equation then becomes

$$\frac{dx}{dt} = -A^2te^{2t}.$$

This can be integrated immediately, with the result

$$x = \frac{A^2}{4}(1 - 2t)e^{2t} + C.$$

INSTANT EXERCISE 3

Integrate the differential equation for x in Example 1 to obtain the final result.

Solution formulas cannot generally be obtained for a coupled system of two nonlinear equations such as the Lotka–Volterra equations. Nevertheless, we will discover methods that can yield a wealth of useful qualitative information about the solutions of such systems.[8] They can also be solved approximately by the rk4 method[9] or other numerical method.

Interacting Populations in General

We could have thought of the rabbit population as being limited by the amount of plant life rather than the number of coyotes. The plants would grow naturally, but be eaten by the rabbits.

[8] This chapter is devoted to such methods. See also Section 6.7.

[9] Section 2.6.

We would then have gotten a predator-prey model in which the rabbits were the predators and the plants were the prey. In this context, the model is often called a consumer-resource model rather than a predator-prey model, and it may be preferable to think of the dependent variables as representing the biomasses of the population, rather than the number of individuals. There are other types of interspecies interactions, and it is interesting to try to construct models for some of these.

Consider the different possibilities for the relative growth rate of a population x to be a linear function of another population y. Assuming that the constants a and b are positive, we have four possibilities:

1. $$\frac{1}{x}\frac{dx}{dt} = a + by,$$
2. $$\frac{1}{x}\frac{dx}{dt} = a - by,$$
3. $$\frac{1}{x}\frac{dx}{dt} = -a + by,$$
4. $$\frac{1}{x}\frac{dx}{dt} = -a - by.$$

Not all these possibilities are reasonable, however. For the population to have a chance to survive, there must be a possibility of a positive growth rate, and this eliminates type 4. There must be some mechanism to limit the population growth, and this requirement eliminates type 1. Linear relative growth rates work only if the terms have opposite signs. In the Lotka–Volterra model, the rabbit equation is of type 2 and the coyote equation is of type 3. Could we get a meaningful model if both equations are of the same type?

Suppose both relative growth rates are of type 2. Each species has a positive growth rate if the other species is absent. Each species has its growth rate decrease in the presence of the other. This is a simple model for competing species. Similarly, suppose both relative growth rates are of type 3. Then each species requires the other for its survival. Clearly this is a simple model of cooperating species. We shall see in the subsequent analysis of these models in Section 5.5 that neither is satisfactory. In both cases, what appears to be a sufficient means of limiting the populations is actually not sufficient. Understanding why the models fail leads to better models for competing or cooperating species.

✦ INSTANT EXERCISE 4

Which of the three types of interacting populations (predator-prey, competing species, cooperating species) could be represented by this system?

$$\frac{da}{dt} = a(3 - 2b), \qquad \frac{db}{dt} = b(1 - 3a).$$

A Complete Family of Examples by Nondimensionalization

Each of the relative growth rate formulas in the interacting species models has two parameters, so our families of simple predator-prey, competing species, and cooperating species models are four-parameter families. Having so many parameters makes it difficult to determine results general enough to apply to the whole family of models. As is typical in mathematical modeling, we can reduce the number of parameters to a reasonable number by nondimensionalization.[10]

Consider the logistic growth equation (1). The dimensions of the parameter K and of the variable p are the same (only a dimensionless number can be subtracted from the dimensionless number 1), so a dimensionless population variable can be defined by

$$P = \frac{p}{K}.$$

The relative growth rate is zero for a population $p = K$ and positive for $p < K$. A small initial population can grow toward K, but cannot exceed K. Thus, K is an upper bound for p. In addition to having the right dimensions, it is a value that is good for comparison. Regardless of what species we are studying with the logistic growth equation or how large the area that the population occupies, $P = 0.5$ will mean that the population is one-half of the upper bound K.

Similarly, note that the relative growth rate is at most r. During a period of maximum growth, we can approximate population changes by

$$\Delta p \approx \frac{dp}{dt}\Delta t \approx rp\,\Delta t.$$

Rearranging this relationship gives us

$$\Delta t \approx \frac{1}{r}\frac{\Delta p}{p}.$$

This shows that the relative growth rate r has units of inverse time, and it also shows that significant population changes require a time period on the order of $1/r$. Thus, $1/r$ is a good reference value for time. Accordingly, we can choose the variable τ defined by

$$\tau = rt$$

as a dimensionless measure of the time.

It remains to convert the original logistic equation to a dimensionless version by replacing p with P and t with τ. We begin by substituting KP for p:

$$\frac{d(KP)}{dt} = rKP(1 - P).$$

The constant K can be removed from the derivative, yielding

$$\frac{dP}{dt} = rP(1 - P).$$

Now to remove t from the equation, note that P is supposed to be a function of τ rather than t. Applying the chain rule, we have

$$\frac{dP}{dt} = \frac{dP}{d\tau}\frac{d\tau}{dt}.$$

The second factor is determined from the equation $\tau = rt$ to be r; hence

$$\frac{dP}{dt} = r\frac{dP}{d\tau}.$$

[10] Section 2.1.

This substitution reduces the model to

$$\frac{dP}{d\tau} = P(1 - P). \tag{4}$$

Note that the new model is similar to the original, but the parameters K and r have disappeared. Instead of a two-parameter family of equations to analyze, we have one single equation. The results produced by the logistic growth model are the same for all species in that the shape of the graph is the same. The only differences are in the amount of actual time that corresponds to $\tau = 1$ and the actual number of individuals that corresponds to $P = 1$.

There are several different ways to nondimensionalize the two-species interaction models. All result in dimensionless counterparts with only one parameter; for example,

Predator-prey

$$X' = X(1 - Y), \qquad Y' = kY(X - 1), \tag{5}$$

Competing species

$$X' = X(1 - Y), \qquad Y' = kY(1 - X), \tag{6}$$

Cooperating species

$$X' = X(Y - 1), \qquad Y' = kY(X - 1), \tag{7}$$

where derivatives are with respect to a suitable time variable τ and the dimensionless parameter k is suitably defined.

5.1 Exercises

1. Consider the logistic growth model

$$p' = 4p\left(1 - \frac{p}{10}\right).$$

Under what circumstances will the population increase? Under what circumstances will the population decrease?

2. Consider the predator-prey model

$$x' = x(1 - 2y), \qquad y' = y(x - 1).$$

a. Are there any values of x and y for which the populations will remain unchanged? If so, what are they?

b. Suppose $x(0) = 1$ and $y(0) = 1$. Will each of these populations be larger than 1 or smaller than 1 shortly after time 0?

3. Which of the following systems are predator-prey models? Indicate which variable represents the predator and which the prey.

a. $x' = -3x + 2xy, \quad y' = -4y + 3xy$

b. $x' = 3x - 2xy, \quad y' = -4y + 3xy$

c. $x' = 3x - 2xy, \quad y' = 4y - 3xy$

4. One of the systems in Exercise 3 represents a model for two cooperative species. Which one is it and how do you know?

5. One of the systems given below represents a model for two species that are competing for the same resources. Which one is it and how do you know?

   a. $x' = x(1 - x - y), \quad y' = y(3 - 2y + x)$
   b. $x' = x(1 - x - y), \quad y' = y(3 + 2y - x)$
   c. $x' = x(1 - x - y), \quad y' = y(3 - 2y - x)$

6. Consider the general Lotka–Volterra model in Equations (2) and (3). Define the new variables

$$X = \frac{csx}{m}, \qquad Y = \frac{sy}{r}, \qquad \tau = rt.$$

   a. Use the chain rule to obtain a set of differential equations for $X(\tau)$ and $Y(\tau)$.
   b. Define the dimensionless parameter k to obtain the dimensionless model (5).
   c. The dimensionless parameter k represents a ratio of two rates. What are these rates? Do you expect $k > 1$ or $k < 1$?

7. Consider the general competition model

$$\frac{dx}{dt} = rx - sxy, \qquad \frac{dy}{dt} = ay - bxy.$$

   Define new dimensionless variables and a dimensionless parameter to obtain the dimensionless model (6).

8. Consider the general cooperation model

$$\frac{dx}{dt} = axy - mx, \qquad \frac{dy}{dt} = bxy - ny.$$

   Define new dimensionless variables and a dimensionless parameter to obtain the dimensionless model (7).

9. Suppose the prey in the predator-prey model is limited by logistic growth in the absence of predators. Incorporate this assumption into the predator-prey model.

10. Suppose $y(t)$ is the proportion of people in a city who have heard a particular rumor. Suppose the rate at which the rumor is spread, relative to the population of people who have not yet heard the rumor, is proportional to the proportion of people who have heard the rumor. Derive the differential equation for the spread of the rumor. This model is also used for the spread of technological innovations within a given society.

11. The standard SIS (susceptible, infective, susceptible) epidemic model divides a population into two classes: the infective class (I), consisting of individuals who are capable of transmitting the disease, and the susceptible class (S), consisting of individuals who are not infective, but could become infective. Let $S(t)$ and $I(t)$ be the populations of the susceptible and infective classes, respectively.

   a. Derive a system of differential equations for the SIS model, using the following assumptions:

      1. Changes of classification for any individual occur by only two mechanisms: A susceptible individual can become infected, and an infected individual can recover to become susceptible again.
      2. The rate at which susceptible people become infected is proportional to the susceptible population and to the infective population, with proportionality coefficient r.

3. The rate at which infective people recover is proportional to the infective population, with proportionality coefficient γ.

b. Explain why assumptions 2 and 3 are reasonable.

12. The standard SIR epidemic model divides a population into three classes: the infective class (I), consisting of individuals who are capable of transmitting the disease; the susceptible class (S), consisting of individuals who are not infective but could become infective; and the removed class (R), consisting of individuals who have had the disease and are no longer able to be infected. Let $S(t)$, $I(t)$, and $R(t)$ be the populations of the susceptible, infective, and removed classes, respectively.

   a. Derive a system of differential equations for the SIR model, using the following assumptions:

   1. Changes of classification for any individual occur by only two mechanisms: A susceptible individual can become infected, and an infected individual can become removed.
   2. The rate at which susceptible people become infected is proportional to the susceptible population and to the infective population, with proportionality coefficient r.
   3. The rate at which infective people become removed is proportional to the infective population, with proportionality coefficient γ.

   b. Explain why assumptions 2 and 3 are reasonable.
   c. Show that the sum of the populations in the three classes is constant. Use this fact to obtain a system of equations for the variables S and I.
   d. Explain why this model is not suitable for the common cold.

13. In this exercise, we derive a model for a waste treatment process that utilizes bacteria.

   a. Let $w(t)$ be the amount of waste in a vessel of volume V, and let $x(t)$ be the population of bacteria in the vessel. Assume that the waste is consumed at the rate kwx, where k is a constant that measures the relative reaction rate of waste per unit amount of bacteria. Let c be a conversion factor for bacteria growth from waste consumption, similar to the parameter c in the Lotka–Volterra predator-prey model. Let m be the relative death rate of the bacteria. Construct the differential equations for the waste and the bacteria.
   b. It is possible to remove all the parameters from the differential equation by nondimensionalization, but it is not obvious what to choose for the reference quantities. Let w_r, x_r, and t_r be the as yet undetermined reference quantities for nondimensionalization. Let $W = w/w_r$, $X = x/x_r$, and $\tau = t/t_r$ be the variables in the dimensionless version of the problem. Determine the differential equations for $W(\tau)$ and $X(\tau)$.
   c. Observe that the differential equations from part *b* include a total of three terms, each of which has a dimensionless coefficient that includes at least one of the reference quantities. Determine the values of the reference quantities needed to make each of the three dimensionless coefficients equal to 1.

14. One of the first mathematical studies of epidemics was done by Daniel Bernoulli in 1760. Bernoulli's goal was to determine the likely effect of a controversial program to immunize young people against smallpox. The model assumes that survivors of the disease have lifetime immunity. Let $P(t)$ be the population of a cohort of individuals who are the same age, with

$t = 0$ corresponding to their birth year. Let $S(t)$ be the population of susceptible individuals from this cohort. Let $m(t)$ be the relative death rate of the cohort in the absence of smallpox. This relative rate of decrease is not constant because the population is to be tracked long enough that mortality must be considered a function of age. We might expect, for example, that $m(t)$ (at least in 1760) was quite large for infants and decreased as the cohort aged toward adulthood. The populations P and S in Bernoulli's model are described by the differential equations

$$S' = -rS - m(t)S, \qquad P' = -brS - m(t)P,$$

where r is a constant and b is the fraction of smallpox cases that result in death.

a. The first term on the right side of the first equation is the rate at which susceptible individuals become infected. For some diseases, this term would have been $-kIS$, where I is the population of infectives. Explain the assumption we are making about smallpox in choosing $-rS$ rather than $-kIS$.

b. The differential equations in the model are impossible to analyze fully without detailed information about $m(t)$. This information is not needed if our goal is simply to determine $y = S/P$, the proportion of individuals who are still susceptible at time t. Let $z = y^{-1} = P/S$. Use the chain rule to differentiate z; substitute the equations of the Bernoulli model to obtain a differential equation that contains only z, along with the parameters b and r. The equation for z should be similar to Newton's law of cooling (Section 1.1).

c. Solve the differential equation for z. Use the initial condition corresponding to the assumption that babies are never born with smallpox. Obtain a formula for y from the formula for z.

d. Bernoulli estimated $b = r = \frac{1}{8}$ from the limited data available. Given these data, what fraction of 20-year-olds is susceptible to smallpox? At what age does susceptibility to smallpox hold for only 5% of the population?

✦ 5.1 INSTANT EXERCISE SOLUTIONS

1. The most general linear function of p is $f(p) = mp + b$, where m is the slope and b the intercept. Given the requirement $f(0) = r$, we have $b = r$. The additional requirement $f(K) = 0$ yields $0 = f(K) = mK + r$, or $Km = -r$. Thus, $m = -r/K$ and $f(0) = -rp/K + r = r(1 - p/K)$.

2. The predator is p because the interaction term pq contributes to the growth of p and the decrease of q.

3. We have

$$x = \int -A^2te^{2t}\,dt = -\frac{A^2}{2}\int t(2e^{2t})\,dt = -\frac{A^2}{2}\int t(e^{2t})'\,dt.$$

Integrating by parts, we have

$$x = -\frac{A^2}{2}\left(te^{2t} - \int e^{2t}\,dt\right) = -\frac{A^2}{2}\left(te^{2t} - \frac{1}{2}e^{2t}\right) + C = \frac{A^2}{4}(1 - 2t)e^{2t} + C.$$

4. The interaction term is negative for both species. Since each decreases because of the other, the system could represent competing species.

5.2 The Phase Line

The autonomous first-order differential equation

$$\frac{dy}{dt} = f(y)$$

represents a process for which changes in the quantity of interest depend only on the level of the quantity and not on the time. In such cases, it might be of interest to know what will eventually happen if the process is allowed to continue indefinitely. We will use the term *longtime behavior* to describe the possible behaviors of a process or system left alone indefinitely.

MODEL PROBLEM 5.2

Identify the possible longtime behaviors for solutions of the equation

$$\frac{dy}{dt} = y(1-y),$$

and determine which longtime behavior results from any given initial condition.

Since we want to know how the solution depends on the initial condition, it makes sense to include the initial data as a parameter in the problem. We therefore consider the one-parameter family of problems

$$\frac{dy}{dt} = y(1-y), \qquad y(0) = y_0. \tag{1}$$

Longtime Behavior and the Limitations of Solution Formulas

Separation of variables yields the solution formula for Model Problem 5.2:

$$y = \frac{y_0}{y_0 + (1-y_0)e^{-t}}. \tag{2}$$

✦ INSTANT EXERCISE 1

Derive the solution of the initial-value problem (1).

To understand the solution, we cannot simply graph the function defined by Equation (2), because it is a one-parameter family of functions. We can graph the solution for any particular y_0, but how can we be sure that the graphs for some specific values represent all the possible behaviors? Instead, we might look at the properties of the family.

Observe that the second term in the denominator of the solution gradually disappears as t increases. Indeed,

$$\lim_{t\to\infty} y = \begin{cases} 1 & y_0 \neq 0 \\ 0 & y_0 = 0. \end{cases}$$

The solution of an initial-value problem may not exist for all t. We have established that the limit as $t \to \infty$ of the *function* defined by the solution formula (2) is always 1 (except for the special case $y_0 = 0$). But this does not by itself mean that solutions of Equation (1) approach $y = 1$. As demonstrated in Section 1.2, functions do not always represent the solution of an initial-value problem over their whole domain.

Consider as an example the case $y_0 = 2$. Here, the solution formula is

$$y = \frac{2}{2 - e^{-t}}.$$

This function is continuous for all $t > 0$, so it defines a solution for all $t > 0$, which therefore approaches 1 as $t \to \infty$. Similarly, the case $y_0 = \frac{1}{2}$ gives the solution formula

$$y = \frac{1}{1 + e^{-t}},$$

which is also continuous for $t > 0$ and so defines a solution that approaches 1 as $t \to \infty$. But now suppose we try $y_0 = -\frac{1}{2}$. This time, the solution formula is

$$y = \frac{1}{1 - 3e^{-t}}.$$

This situation is different. The denominator approaches 0 as $t \to \ln 3$. Thus, the solution of the initial-value problem with $y_0 = -\frac{1}{2}$ exists only on the interval $t < \ln 3$. The three solutions we have already found are plotted together in Figure 5.2.1. Clearly the longtime behavior of a solution of Model Problem 5.2 depends in an important way on the initial condition.

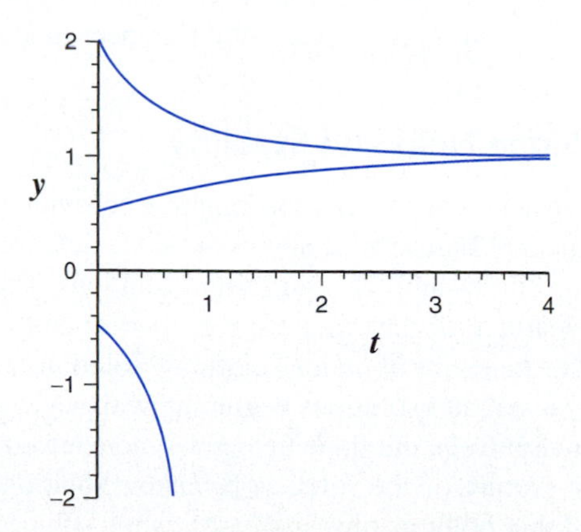

Figure 5.2.1
Some solutions of $y' = y(1 - y)$.

We've seen that solution formulas do not always give a complete picture of the longtime behavior. However, the longtime behavior of an autonomous differential equation can be inferred directly from the differential equation itself, without the need for a solution formula, and we now explore this possibility.

Critical Points and Equilibrium Solutions

The model differential equation is of the form

$$\frac{dy}{dt} = f(y),$$

with

$$f(y) = y(1 - y).$$

The function f represents the rate of change of y as a function of the quantity y. Of particular interest are values of y for which the rate of change is 0. A value of y for which $f(y) = 0$ is called a **critical point** of the differential equation $dy/dt = f(y)$. The function $y(t) = y_c$, with y_c a critical point, is an **equilibrium solution** of the differential equation. The use of the term *critical point* is consistent with its meaning in calculus. In both settings, given a differentiable function, the term refers to points on a graph where the tangent line is horizontal; the difference is that these points are identified by the independent variable in calculus and by the dependent variable for solutions of autonomous differential equations.

The critical points for Model Problem 5.2 are the solutions of

$$y(1 - y) = 0,$$

namely, $y = 0$ and $y = 1$. The functions $y \equiv 0$ and $y \equiv 1$ are equilibrium solutions. The importance of equilibrium solutions stems in part from the consequences of the existence and uniqueness theorem (Theorem 2.4.2). This theorem guarantees a unique solution for $dy/dt = y(1 - y)$ through any point in the (t, y) plane, given that $f(y) = y(1 - y)$ and its derivative are continuous everywhere. This means that solution curves cannot cross in the (t, y) plane. The horizontal lines $y = 0$ and $y = 1$ are solution curves; hence, they divide the (t, y) plane into three distinct regions where there could be significant differences in long-term behavior.

The Slope Field and Stability

A convenient way to study the longtime behavior of solutions of differential equations is to use the slope field.[11] Figure 5.2.2 shows the slope field for Model Problem 5.2, along with the same three solution curves and the equilibrium solutions. The slope field clearly confirms what we have already learned for the cases $y_0 = 2$, $y_0 = \frac{1}{2}$, and $y_0 = -1$. It also gives a complete picture of the longtime behavior of *all* the solutions. Solutions beginning with positive values of y_0 tend toward 1 as $t \to \infty$, and solutions beginning with negative values of y_0 tend toward $-\infty$ in finite time.

Qualitatively, the slope field gives more information than the solution formula because it gives a clear picture of the solution behavior. Solutions that begin near the critical point $y = 1$ tend toward that point as time increases, while solutions beginning near the critical point $y = 0$ tend away from that point as time increases. In this example and in general, some critical points tend to attract nearby solutions, while others repel them. This idea is the basis of the concept of *stability*. The idea is essentially that critical points, and the corresponding equilibrium solutions, are stable if they attract nearby solutions and are unstable if they repel them. Some care is required in the formal definition to give precise meaning to the notions of attraction and repulsion.

[11] Section 2.3.

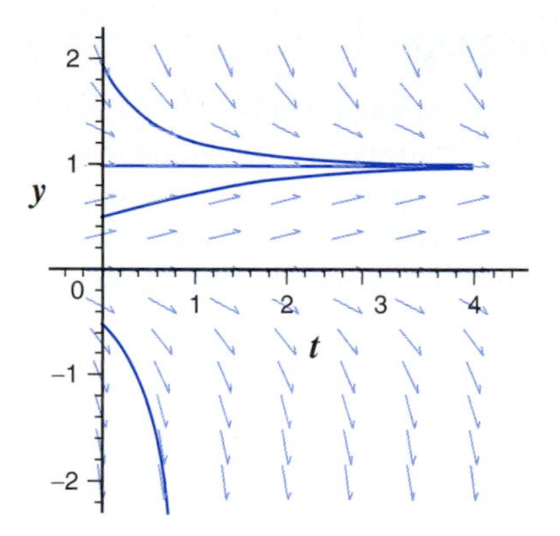

Figure 5.2.2
The slope field for $dy/dt = y(1 - y)$, along with some solution curves.

An equilibrium solution[12] $y \equiv y_c$ for a differential equation $y' = f(y)$ is **asymptotically stable** if there is some open interval $(y_c - \epsilon,\ y_c + \epsilon)$ with the property that $\lim_{t\to\infty} y(t) = y_c$ for any solution that is initially in the interval. An equilibrium solution is **unstable** if there is some open interval $(y_c - \epsilon,\ y_c + \epsilon)$ with the property that $\lim_{t\to\infty} |y(t) - y_c| > \epsilon$ for any solution, other than $y \equiv y_c$, that is initially in the interval.

Consider first the equilibrium solution $y \equiv 1$. If $y_0 > 1$, then the slope of the solution curve through (t, y_0) is

$$f(y_0) = y_0(1 - y_0) < 0.$$

Thus, the solution is decreasing. It must continue to decrease as long as $y > 1$; hence, it must either reach $y = 1$ in finite time or approach $y = 1$ in the limit as $t \to \infty$. The argument is similar for $0 < y_0 < 1$. In summary, we have

$$\lim_{t\to\infty} y = 1 \qquad \text{whenever} \qquad y_0 > 0.$$

Thus, the equilibrium solution $y \equiv 1$ is asymptotically stable.

The equilibrium solution $y \equiv 0$ is unstable by the definition. Solution curves beginning above $y = 0$ move toward $y = 1$, while solutions beginning below $y = 0$ decrease without bound. In either case, solutions cannot stay within a small interval containing the point $y = 0$. The combined stability results completely describe the longtime behavior of the solutions. If $y_0 > 0$, then the solution approaches $y = 1$; if $y_0 < 0$, then the solution decreases without bound.

[12]The terms *critical point* and *equilibrium solution* are almost synonymous, the only distinction being that the critical point is a value of the dependent variable y while an equilibrium solution is a function that takes the same value all the time. Statements about stability can be made using either term; the choice of *equilibrium solution* in this context is a matter of taste.

The Phase Line

Look again at the slope field in Figure 5.2.2. All the minitangents on a given horizontal line are parallel. This is always true for autonomous equations. In the example, the slope is 0 at all points $(t, 1)$ and $\frac{1}{4}$ at all points $(t, \frac{1}{2})$. Now suppose we illustrated only the minitangents at $t = 0$, as in Figure 5.2.3*a*. With fewer arrows and less horizontal space, this plot includes all the essential information that is in the full slope field.

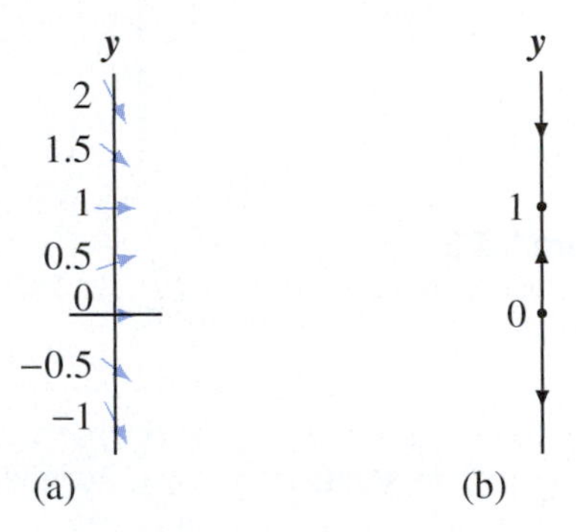

Figure 5.2.3
(*a*) A portion of the slope field for $y' = y(1 - y)$; (*b*) with the phase line.

An even simpler view of the differential equation can be made by placing the arrows directly on the y axis, using them to show only whether the solution curves are increasing or decreasing. This is the **phase line,** as depicted in Figure 5.2.3*b*. All information about the rates of increase or decrease with respect to t is lost in the phase line picture. However, the essential information needed to determine the stability of equilibrium solutions and the longtime behavior of other solutions is efficiently contained in the phase line.

The phase line shows the same properties that we found from the slope field: solution curves beginning below $y = 0$ move off to $-\infty$, solution curves beginning between $y = 0$ and $y = 1$ move up toward $y = 1$, and solution curves beginning above $y = 1$ move down toward $y = 1$. A rough sketch of the solution curves is easily obtained from the phase line.

One advantage of using the phase line is that it is very easy to sketch the phase line from the graph of $f(y)$ for the differential equation $y' = f(y)$. Equilibrium solutions are given by the zeros of f. Regions where f is positive correspond to regions on the phase line where the arrow points up, and regions where f is negative correspond to regions on the phase line where the arrow points down. This interpretation is illustrated in Figure 5.2.4. In this example, the function f vanishes at 0 and 1, and these give the locations of the critical points on the phase line. The function f is positive between 0 and 1; since the differential equation is $y' = f(y)$, this means that y' is positive between $y = 0$ and $y = 1$, and this information gives the direction of the arrowhead between the two critical points on the phase line. The arrowheads in other regions of the phase line are obtained in the same manner. The phase line in Figure 5.2.4 is oriented horizontally to make the connection with $f(y)$ clearer.

✦ INSTANT EXERCISE 2

Sketch the phase line for $y' = y(y - 2)$, and use it to determine the stability of the critical points.

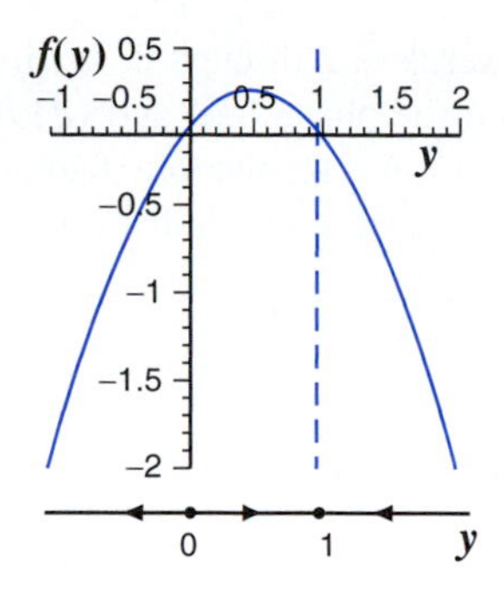

Figure 5.2.4
The function $f(y) = y(1 - y)$ and the phase line for $y' = y(1 - y)$.

The Logistic Growth Equation

The logistic growth model, from Section 5.1, is

$$\frac{dp}{dt} = rp\left(1 - \frac{p}{K}\right). \tag{3}$$

The dimensionless form of the model, with $P = p/K$, is

$$\frac{dP}{d\tau} = P(1 - P),$$

which is exactly the same as Model Problem 5.2. All the results obtained in this section hold for the dimensional logistic growth model (3), with appropriate changes in the variable values. The critical point $p = K$ in Equation (3) corresponds to $y = 1$ in Model Problem 5.2; it is asymptotically stable. The critical point $p = 0$, corresponding to $y = 0$, is unstable. Solution curves beginning with positive values of p tend toward $p = K$, while those beginning with negative values of p, corresponding to initial conditions that are inappropriate for the model, tend toward negative infinity. Of course it is also possible to get these results directly by applying the qualitative methods to the logistic equation (3).

A Final Comment

The solution of an initial-value problem can be thought of as a point that moves along the phase line in time. At any given point in the solution's progress, the phase line shows the value of the dependent variable, but not that of the independent variable. The phase line is only used for autonomous equations $y' = f(y)$. The lack of a time coordinate on the phase line is not very important for an autonomous equation. The current value of y is all that is needed to determine the rate of change of y. We would not want to use the phase line for a general problem $y' = f(t, y)$ because in such a case the current value of y is not sufficient to determine the current rate of change of y.

5.2 Exercises

1. Sketch the function $f(P) = rP(1 - P/K)$, and use it to obtain a sketch of the phase line for the logistic equation. Note that your graphs will be identical to those obtained for Model Problem 5.2 except for the variable values that appear on the axes.

For each of Exercises 2 through 8, (*a*) list the equilibrium solutions, (*b*) sketch y' as a function of y, (*c*) sketch the phase line, and (*d*) determine the stability of all equilibrium solutions. An equilibrium solution is **semistable** if solutions on one side of it tend to approach it while solutions on the other side tend to recede from it.

2. $y' + ky = kS, \quad k > 0, \quad S > 0$
3. $y' = y(y^2 - 4)$
4. $y' = -3y(1 - y)(3 - y)$
5. $y' = 1 - e^{-y}$
6. $y' = \sin y$
7. $y' = -\dfrac{\arctan y}{1 + y^2}$
8. $y' = y^2(1 - y^2)$
9. One of the equations in Exercises 2 through 8 represents a model of a population that can become extinct if it drops below a critical value. Which one is it? What is the critical value? Is the critical value a critical point? If so, what kind is it?
10. Sketch the phase line for the model of rumor spreading from Exercise 10 in Section 5.1. Discuss the behavior of the model.
11. Consider the differential equation $y' = f(y)$. Suppose y_0 is a point satisfying $f(y_0) = 0$ and $f'(y_0) < 0$. Show that y_0 is a stable critical point.
12. The Gompertz model for population growth is

$$p' = kp \ln \frac{M}{p}, \qquad p > 0, \qquad k > 0, \qquad M > 0.$$

a. Sketch the graph of p' as a function of p, sketch the phase line, and determine the stability of any equilibrium solutions.

b. Solve the Gompertz equation by utilizing the substitution $y = \ln(M/p)$.

c. Use the solution from part b to verify the results of part a.

13. The rate at which a drug disseminates into the bloodstream is governed by the differential equation

$$x' = B(A - x),$$

where A and B are positive constants. Find the stable equilibrium value of x. At what time is the concentration one-half of the equilibrium value if there is no drug in the patient's system prior to time 0?

14. Let $x(t)$ be the mass of chemical A in a chemical reactor that initially contains 50 liters (L) of pure water. A solution of A having a concentration 2 kg/L flows into the reactor at a rate of 5 L/min. The solution inside the reactor is kept well mixed and flows out of the reactor at a rate of 5 L/min. The chemical reaction inside the reactor decreases the mass of chemical A at a rate equal to 0.4 times the amount of A present. Determine a differential equation for $x(t)$. Find the equilibrium value and sketch the phase line. Solve the problem to determine $x(t)$.

15. Suppose the population of deer in a forest is governed by a logistic growth equation. Suppose further that the agency charged with managing the deer population grants a limited number of permits to deer hunters. If we assume that the number of deer killed by hunters is proportional to the deer population and the number of hunting permits, we obtain the differential equation

$$\frac{dp}{dt} = rp\left(1 - \frac{p}{K}\right) - Ep,$$

where r and K are as in the logistic equation and E is a parameter with units of time^{-1} that measures the amount of hunting effort and is taken to be proportional to the number of permits issued. This model is called the *Schaefer model;* it is named after the biologist M. B. Schaefer, who applied it to fishery management.

a. The model is easier to analyze if it is scaled by a suitable change of variables. Define new variables y and τ and a new parameter h by

$$y = \frac{p}{K}, \qquad \tau = rt, \qquad h = \frac{E}{r}.$$

Note that y is the population relative to the environmental capacity and h is the hunting effort relative to the natural growth rate. Change variables to arrive at the differential equation

$$y' = y(1 - y) - hy,$$

where the superscript prime represents $d/d\tau$.

b. Assume $E > r$. What will happen to the deer population?

c. Show that there is a positive equilibrium value y_e if $E < r$, and use the phase line to show that y_e is stable.

d. Observe that the value of y_e can be manipulated to achieve any value between 0 and 1 by controlling the number of hunting permits. Sketch a graph of h versus y_e. This graph could be used to determine the appropriate number of permits for any desired equilibrium deer population.

e. Suppose the goal of the management program is to allow a maximum amount of sustainable deer hunting. The hunting is represented by the term hy in the differential equation, so the goal is to maximize the function $Y(h) = hy_e$. Sketch the graph of the function Y.

f. Determine the value of h that maximizes Y.

16. Stefan's law of radiative cooling says that the rate of change of temperature T of a body that interacts with a medium of temperature M is given by

$$T' = K(M^4 - T^4).$$

The temperatures T and M must be given on an absolute temperature scale such as the Kelvin or Rankine, and not in Celsius or Fahrenheit.

a. Sketch the phase line for Stefan's law. Discuss the behavior of an object whose temperature is initially hotter than that of the medium.

b. Suppose the temperatures T and M are roughly comparable. Factor the polynomial $M^4 - T^4$. The resulting equation can be approximated if it is done carefully. Replace T with M in sums, but not in differences; for example, replace $M^2 + T^2$ by $2M^2$. Show that the resulting approximation leads to Newton's law of cooling.

✦ 5.2 INSTANT EXERCISE SOLUTIONS

1. The differential equation is separable, so we can write it as

$$\frac{1}{y(1-y)}\frac{dy}{dt} = 1$$

and integrate with respect to t to get

$$\int \frac{1}{y(1-y)}\,dy = \int dt.$$

The integral in y can be done by a partial fraction decomposition. We need to find the numerators A and B so that

$$\frac{A}{y} + \frac{B}{1-y} = \frac{1}{y(1-y)}.$$

Multiplying both sides of this equation by the common denominator gives

$$A(1-y) + By = 1,$$

or

$$A + (B - A)y = 1 + 0y.$$

For this equation to hold for all y, we must have $A = 1$ and $B - A = 0$. Thus $A = B = 1$. We may therefore rewrite the integral equation as

$$\int \left(\frac{1}{y} + \frac{1}{1-y}\right) dy = \int dt.$$

Now we integrate:

$$\ln|y| - \ln|1-y| = t + C.$$

The two terms on the left combine to give

$$\ln\left|\frac{y}{1-y}\right| = t + C.$$

Raising both sides over e yields

$$\left|\frac{y}{1-y}\right| = e^{t+C} = e^C e^t.$$

Thus,

$$\frac{y}{1-y} = \pm e^C e^t = Ae^t,$$

where A is an arbitrary constant whose value must be determined from the initial condition. Substituting $y = y_0$ and $t = 0$ gives the result $A = y_0/(1 - y_0)$. We now have the solution of the initial-value problem, still in implicit form, as

$$\frac{y}{1-y} = \frac{y_0}{1-y_0} e^t.$$

Multiplying both sides by $1 - y$, $1 - y_0$, and e^{-t} yields

$$(1 - y_0)e^{-t}y = y_0(1 - y).$$

Putting terms with y on the left side and other terms on the right side gives

$$[y_0 + (1 - y_0)e^{-t}]y = y_0,$$

leading to the final answer (2).

2.

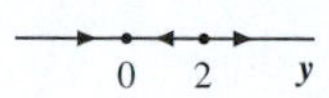

The critical point $y = 2$ is unstable while the critical point $y = 0$ is asymptotically stable. The phase line could just as well have been oriented vertically.

5.3 The Phase Plane

The phase line for an autonomous first-order differential equation is a one-dimensional space with the dependent variable on the axis and arrows to indicate whether solutions increase or decrease. The phase line plot displays many qualitative features of the solutions of the differential equation. Similarly, the **phase plane** for an autonomous second-order differential equation $y'' = f(y, y')$ is a two-dimensional space with one axis for the dependent variable y and the other for its derivative y'. A solution can be represented as a parameterized curve in the phase plane with the independent variable as the parameter; these curves are called **trajectories.** The phase plane for an autonomous system of two equations,

$$\frac{dx}{dt} = f(x, y), \qquad \frac{dy}{dt} = g(x, y), \tag{1}$$

is similarly a two-dimensional space with coordinates for the dependent variables x and y. The use of arrows in the phase plane is discussed in Section 5.4; here, we concentrate on the notion of the phase plane and on trajectories.

EXAMPLE 1

The initial-value problem

$$y'' + 2y' + 26y = 0, \qquad y(0) = 1, \qquad y'(0) = 0$$

has solution (Section 3.5, Example 2)

$$y = e^{-t}(\cos 5t + 0.2 \sin 5t).$$

This initial-value problem could represent the movement of a mass on a spring, in which case the unknown function y represents the distance measured from the rest position. We might also be interested in the velocity, given by

$$v = y' = -5.2e^{-t} \sin 5t.$$

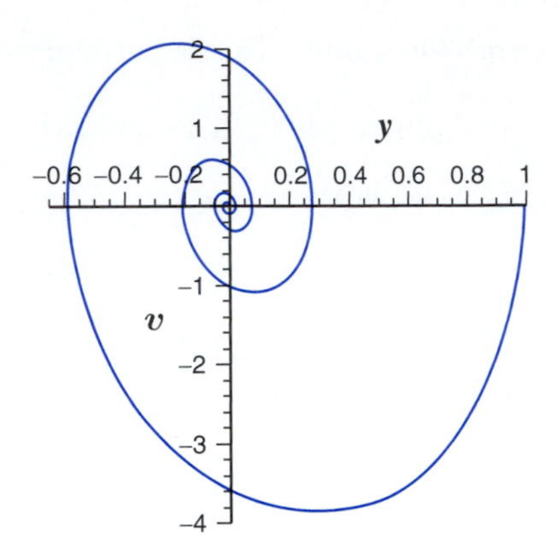

Figure 5.3.1
The solution of Example 1 in the phase plane.

Now suppose we are interested specifically in the relationship between distance and velocity. We could think of the solution as defining a curve with coordinates $(y(t), v(t)) = (e^{-t}(\cos 5t + 0.2 \sin 5t), -5.2e^{-t} \sin 5t)$. By plotting points for various values of t, we obtain the curve shown in Figure 5.3.1. The solution begins at the point $(1, 0)$ and moves around the origin in the clockwise direction. As the velocity becomes more and more negative, the distance decreases from 1 toward 0. Eventually, at the point corresponding to the bottom of the curve, the velocity reaches its most negative value. The distance soon becomes negative, while the velocity becomes less negative. When the point at the left of the curve is reached, the distance is at its most negative value while the velocity is zero, indicating that the mass is momentarily motionless. The remainder of the curve can be similarly interpreted.

✦ INSTANT EXERCISE 1

Let $y(t)$ be the solution of the initial-value problem

$$\frac{d^2y}{dt^2} + y = 0, \qquad y(0) = 1, \qquad \frac{dy}{dt}(0) = 0.$$

Describe how the solution appears in a plot with y on the horizontal axis and $v = y'$ on the vertical axis.

Compare Figure 5.3.1 with Figure 3.5.4. Both graphs show the solution of Example 1 on the time interval $0 < t < 5$, but they show the information in different ways, with different disadvantages. The graph of $y(t)$ in Figure 3.5.4 shows the complete history of the distance y, but it does not show the velocity except by inference. The phase plane graph of Figure 5.3.1 shows the relationship of distance and velocity, but it does not give any indication of the time corresponding to any point on the curve. It does not even give the direction of the motion (clockwise inward spiral or counterclockwise outward spiral) except by inference.

In the case of systems, the phase plane is an alternative to a pair of plots for the two dependent variables against time. The phase plane is useful whenever there is added value to being able to plot the dependent variables in a plane, even though the information about time is lost. Phase plane plots are almost always used when the longtime behavior is the feature of greatest interest.

MODEL PROBLEM 5.3

Suppose a projectile is fired upward with speed v_0 from a point y_0 above the surface of the moon. Determine the motion of the projectile; in particular, determine whether it will return to the moon.

Since Model Problem 5.3 is about motion, we begin with Newton's second law,

$$F = ma,$$

or

$$\frac{dv}{dt} = \frac{F}{m}.$$

For a flight on the moon, there is no air resistance, so the only force that needs to be included in the model is the force of gravity, which is given by[13]

$$F_g = -mg\frac{R^2}{(R+z)^2},$$

where $z(t)$ is the height of the projectile above the moon's surface, R is the radius of the moon, and g is the gravitational constant at the surface of the moon. Thus we have

$$\frac{dv}{dt} = -\frac{gR^2}{(R+z)^2}.$$

The model is not quite complete because there are two unknown functions, v and z. There are two ways to complete the model. One is to rewrite the left-hand side of the equation as d^2z/dt^2 so as to obtain a second-order differential equation for z. This is generally preferable whenever the second-order equation can be solved, as solution methods for higher-order equations are usually more efficient than solution methods for the corresponding system. The other option is to add the equation $dz/dt = v$ and obtain a system of two autonomous first-order equations. This is preferable when a model is to be studied in the phase plane because some phase plane techniques require the problem to be written as a system. Since our plan is to use the phase plane, we have the initial-value problem

$$\frac{dz}{dt} = v, \qquad \frac{dv}{dt} = -\frac{gR^2}{(R+z)^2}, \qquad z(0) = z_0, \qquad \frac{dz}{dt}(0) = v_0, \tag{2}$$

where z_0 is the height from which the projectile is launched and v_0 is the velocity of the projectile at launch.

Writing a Higher-Order Equation as a System

Suppose we had written the differential equation of Model Problem 5.3 as the second-order differential equation

$$\frac{d^2z}{dt^2} = -\frac{gR^2}{(R+z)^2}.$$

Then we could convert the equation to the system (2) by defining $v = dz/dt$ as a second dependent variable and using dv/dt in place of d^2z/dt^2 in the differential equation. Any second-order

[13]Section 2.1.

differential equation

$$y'' = f(t, y, y')$$

can be converted to the system

$$y' = v, \qquad v' = f(t, y, v)$$

in this manner. Differential equations of order $n > 2$ can be similarly converted to higher-dimension systems by defining new variables to represent derivatives up to order $n - 1$. Writing higher-order equations as systems is often convenient. Not only does phase space analysis work only with (autonomous) systems, but also numerical packages are usually designed to work only on first-order equations and systems. When the higher-order equation can be solved by the methods of Chapters 3 and 4, it is almost always better to do so than to try to solve the corresponding system.

 **INSTANT EXERCISE 2**

Write the equation

$$\frac{d^2y}{dt^2} + 2\frac{dy}{dt} + 3y = 0$$

as an autonomous system of two first-order equations.

Nondimensionalization

The differential equations of the current model have two parameters, R and g. We could find numerical values for these parameters and use them. However, it is a better practice to remove the parameters from the problem by nondimensionalization, as was done with the population models in Section 5.1. Results for the dimensionless version of the model will be correct not only on the moon, but on other small astronomical bodies as well.[14] To perform the nondimensionalization, we need to find or construct a reference length, velocity, and time from the parameters R, g, z_0, and v_0. If possible, these quantities should be representative values for z, v, and t so that the problem is properly scaled.

Both the radius R and the initial height z_0 are lengths. Of these, z_0 is not suitable for a reference, because its value could be zero, while R ought to be representative of the height for projectiles that are moving almost fast enough to escape the moon's gravity. Thus, R is an excellent choice for the reference length. There are two reasonable choices for a reference velocity. The initial velocity v_0 is one. The other can be found from a combination of the parameters R and g. Since R has dimension of length and g has dimension of length per time squared, the product gR has dimension of velocity squared. Thus, $\sqrt{gR}$ is a velocity that is somehow representative of the astronomical body. Given that the goal is to determine whether or not the projectile returns to the surface, the moon-based velocity $\sqrt{gR}$ is the better choice for the reference velocity. Given the length R and the velocity $\sqrt{gR}$, a reasonable choice for the reference time is the time required to move a distance R at velocity $\sqrt{gR}$, and this is $\sqrt{R/g}$.

[14] For a large body with a significant atmosphere, the accuracy of the model is affected by the damping of the atmosphere on the motion.

Define dimensionless variables Z, V, and τ by

$$Z = \frac{z}{R}, \qquad V = \frac{v}{\sqrt{gR}}, \qquad \tau = \frac{\sqrt{g}\,t}{\sqrt{R}}.$$

We first replace z by RZ and v by $\sqrt{gR}\,V$. After simplification, this yields

$$\frac{dZ}{dt} = \sqrt{\frac{g}{R}}\,V, \qquad \frac{dV}{dt} = -\sqrt{\frac{g}{R}}\,\frac{1}{(1+Z)^2}.$$

Now we use the chain rule:

$$\frac{dZ}{dt} = \frac{dZ}{d\tau}\frac{d\tau}{dt} = \sqrt{\frac{g}{R}}\,\frac{dZ}{d\tau},$$

and similarly with V, to get the dimensionless model

$$\frac{dZ}{d\tau} = V, \qquad \frac{dV}{d\tau} = -\frac{1}{(1+Z)^2}, \qquad Z(0) = Z_0, \qquad V(0) = V_0, \tag{3}$$

where $Z_0 = z_0/R$ and $V_0 = v_0/\sqrt{gR}$. The dimensionless model is the same as the original model, except that the distance Z is measured in terms of moon radii rather than meters or miles, the velocity V is measured in terms of $\sqrt{gR}$ rather than meters per second or miles per hour, and the time τ is measured in terms of $\sqrt{R/g}$ rather than seconds. Values for these reference quantities are needed to obtain results in familiar units, but not to determine the behavior of the model.

Trajectories and the Phase Portrait

A solution of an autonomous system (1) is a pair of functions $x(t)$ and $y(t)$, which define a trajectory in the phase plane. We can visualize any nonequilibrium solution as a point that moves along some trajectory. Each trajectory represents a family of solutions, because any solution passing through a point on a given trajectory at some time must stay on that trajectory for all time.[15]

The set of all trajectories is the **phase portrait** for the equation or system. Of course one cannot display the entire phase portrait, with curves through every point in a region. Instead, sketches of the phase portrait show a representative set of trajectories. This is analogous to the distinction in Section 2.3 between the slope field, consisting of infinitely many minitangents, and a sketch of the slope field. Standard terminology is to use the term *phase portrait* to denote a *sketch* of the phase portrait; nevertheless, it is important to realize that a phase portrait sketch includes only some of the trajectories. Good visualization of the behavior of a system depends on making a good choice of which trajectories to display in the phase portrait sketch. In Example 1, just one spiral trajectory is sufficient to illustrate the phase portrait. Adding a second trajectory makes the graph more confusing without adding important information. Systems with simpler or more varied trajectories require a larger and thoughtfully chosen sample of trajectories to make a good phase portrait sketch.

For most systems of practical interest, the trajectories have an important property, which is summarized in the following theorem.

[15] This is so because the problem is autonomous. No matter when a solution passes through the point (0, 0), for example, it must follow the same path as other solutions that pass through the same point, since the rates of change depend only on the position in the phase plane and not on the time.

Theorem 5.3.1

Uniqueness of Trajectories If the functions $f(x, y)$ and $g(x, y)$ have continuous first derivatives with respect to both x and y on a region in the xy plane, then the system

$$\frac{dx}{dt} = f(x, y), \qquad \frac{dy}{dt} = g(x, y),$$

has exactly one trajectory through each point (x_0, y_0) in the interior of the region.

Theorem 5.3.1 means that trajectories cannot cross each other except possibly at points where the derivatives dx/dt and dy/dt are not both smooth functions.

The Differential Equation for the Trajectories

It is sometimes convenient to determine a differential equation relating the dependent variables of a system. In Equations (3), suppose we think of V as a function of Z. Since Z is in turn a function of τ, the chain rule gives us

$$\frac{dV}{d\tau} = \frac{dV}{dZ}\frac{dZ}{d\tau}.$$

Substitution from Equations (3) yields a first-order differential equation for V as a function of Z. Along with this differential equation, we have an initial condition that identifies V_0 as the velocity that goes with the height Z_0. Thus, the trajectories are determined by the initial-value problem

$$V\frac{dV}{dZ} = -\frac{1}{(1+Z)^2}, \qquad V(Z_0) = V_0. \tag{4}$$

The differential equation for $V(Z)$ is separable. Integrating both sides with respect to Z and applying the initial condition yields the solution formula

$$V^2 = V_0^2 + \frac{2}{1+Z} - \frac{2}{1+Z_0}. \tag{5}$$

✦ INSTANT EXERCISE 3

Confirm the solution (5) by solving the problem (4) for the trajectories.

We can now sketch the phase portrait by plotting some of the trajectories. The result is shown in Figure 5.3.2. Note that there appear to be two different types of trajectories. Assuming an initial height of 0, corresponding to a launch from the moon's surface, the trajectories that start at $V = 1.2$ and below clearly indicate that projectiles fired at those speeds will reach a maximum height and then fall back to the surface. The trajectories that start at higher initial velocities appear to indicate that projectiles fired at those speeds continue to move away from the surface and eventually escape the moon's gravity. Separating these two types of trajectories is the critical trajectory for which V approaches zero as Z approaches infinity. This happens when $V_0^2 = 2$, as can be seen by taking a limit of the trajectory equation as $V \to 0$ and $Z \to \infty$, given $Z_0 = 0$. To interpret this result, we restate it in terms of the original variables of the problem. The critical

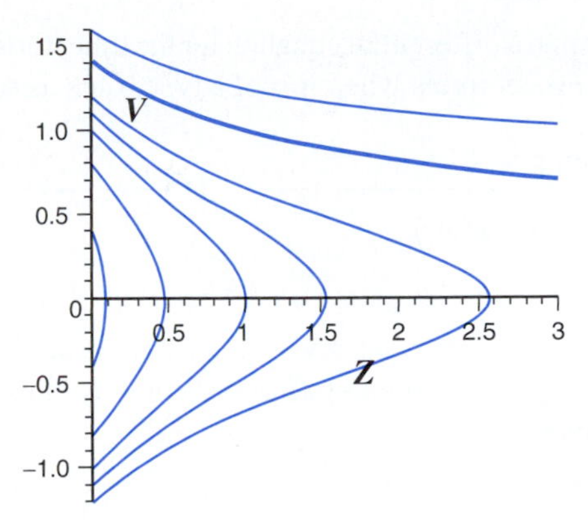

Figure 5.3.2
The phase portrait for Model Problem 5.3.

velocity is $V = \sqrt{2}$, or

$$\frac{v}{\sqrt{gR}} = \sqrt{2}.$$

The parameter

$$v_e = \sqrt{2gR}$$

for any astronomical body is called the **escape velocity.**

Some Conclusions

Note that some information is lost in the phase portrait. One sees the path taken by the solution but cannot see the speed with which the solution progresses along the path. Often the end of the path is approached only in the limit $t \to \infty$. One can also plot separate graphs of the two functions $x(t)$ and $y(t)$. These **time series graphs** are very useful, but they are often difficult to obtain because two-dimensional systems are hard to solve. Much of the important information about solutions of systems can be obtained from the phase plane in the same way that much of the important information about solutions of single first-order equations can be obtained from the phase line. Just as it is easier to sketch the phase line than to solve a scalar differential equation, so is it easier to sketch phase portraits than to solve a system of differential equations. Symbolic solution of autonomous linear systems is the subject of Chapter 6.

Throughout this chapter, we consider methods for obtaining the phase portrait of a two-dimensional system (either a system of two first-order equations or a single second-order equation). We have seen two of these in this section.

- If we have solution formulas for the system, we can sketch in the phase plane the curves represented parametrically by those solution formulas.
- If the differential equation for the solution curves can be solved symbolically, we can sketch the solutions in the phase plane.

Neither of these methods works very often. Solution formulas are generally available only for second-order linear equations with constant coefficients, as in Example 1. We can always determine a differential equation for the trajectories, but we can only solve the differential equation for the trajectories when it is of a type, such as separable, for which solutions can be found.

EXAMPLE 2

Consider the system

$$X' = X(1 - Y), \qquad Y' = Y(X - 1),$$

corresponding to the Lotka–Volterra model from Section 5.1. It is possible to derive a differential equation for the trajectories of the system. Think of Y as a function of X, which in turn is a function of t. By the chain rule,

$$Y' = \frac{dY}{dX}X'.$$

Substituting the differential equations into this equation yields

$$Y(X - 1) = X(1 - Y)\frac{dY}{dX}.$$

This differential equation is separable and leads to the integral form

$$\int \frac{1 - Y}{Y}\, dY = \int \frac{X - 1}{X}\, dX.$$

Integration yields

$$\ln Y - Y = X - \ln X + C,$$

or

$$Y + X + C = \ln XY.$$

This implicit solution formula cannot be solved explicitly for Y, nor does it have an obvious parameterization. Nevertheless, the trajectories given by this equation can be plotted using a computer algebra system. Figure 5.3.3 illustrates some of these trajectories.

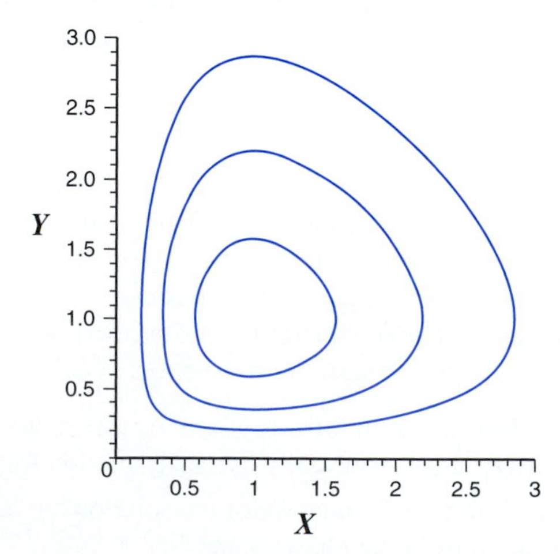

Figure 5.3.3
Trajectories for $X' = X(1 - Y)$, $Y' = Y(X - 1)$.

✦ INSTANT EXERCISE 4

Use the differential equations of Example 2 to determine the direction of solutions traveling on the trajectories of Figure 5.3.3.

Equations of the Form $y'' = f(y)$ In general, a system of two autonomous equations has the form (1). The differential equation for the trajectories is then

$$\frac{dy}{dx} = \frac{g(x, y)}{f(x, y)},$$

and it is unlikely for f and g to be such that the result is separable. The situation is a little better for second-order equations

$$\frac{d^2y}{dt^2} = f\left(y, \frac{dy}{dt}\right).$$

Setting $v = dy/dt$ always yields a system $y' = v$, $v' = f$, so the differential equation for trajectories is

$$v\frac{dv}{dy} = f(y, v).$$

 INSTANT EXERCISE 5

Derive the differential equation for the trajectories of the differential equation

$$\frac{d^2y}{dt^2} = f\left(y, \frac{dy}{dt}\right).$$

The trajectory equation is separable whenever f depends only on y. It is always possible to sketch a phase portrait by solving the differential equation for the trajectories when the original problem, like Model Problem 5.3, has the form

$$\frac{d^2y}{dt^2} = f(y).$$

EXAMPLE 3

For the differential equation $y'' = e^{-y}$, the trajectories are given by the equation

$$v\frac{dv}{dy} = e^{-y}.$$

Integrating this equation yields

$$\int 2v\,dv = \int 2e^{-y}\,dy,$$

or

$$v^2 = C - 2e^{-y}.$$

For any initial conditions $y(0) = y_0$ and $y'(0) = v_0$, the constant C is $C = v_0^2 + 2e^{-y_0}$. Thus,

$$v = \pm\sqrt{v_0^2 + 2(e^{-y_0} - e^{-y})}.$$

Some of the trajectories are illustrated in Figure 5.3.4. The direction of travel along the trajectories must be inferred from the system of equations. For any system derived from a second-order equation, one of the equations in the system is $dy/dt = v$. Hence, the sign of v gives the sign of dy/dt. Trajectories in the upper half-plane show movement to the right, and trajectories in the lower half-plane show movement to the left.

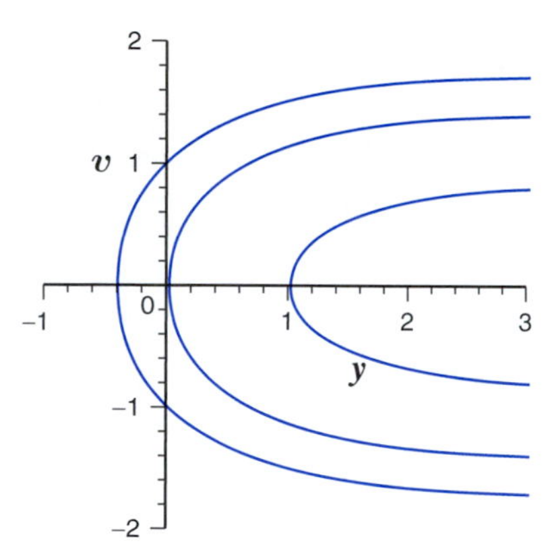

Figure 5.3.4
Trajectories for $y'' = e^{-y}$.

In the final two sections of this chapter, we will develop more general methods for sketching phase portraits.

5.3 Exercises

In Exercises 1 through 6, find and solve the differential equation for the trajectories and sketch the phase portrait.

1. $x' = -xy, \quad y' = -x$
2. $x' = y(1 + x^2), \quad y' = 2xy$
3. $x' = 3y^2, \quad y' = e^x$
4. (T) $x' = e^y, \quad y' = e^{-x}$
5. $y'' + y^2 = 0$
6. (T) $y'' = \dfrac{1}{1 + y^2}$
7. (T) Determine the differential equation for the trajectories of the simple competition model (6) in Section 5.1 with $k = 1$, find an implicit solution formula, and plot some trajectories.

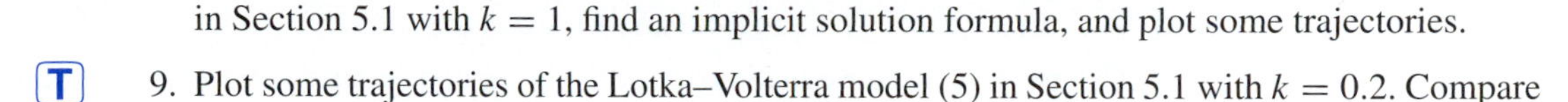

8. Determine the differential equation for the trajectories of the simple cooperation model (7) in Section 5.1 with $k = 1$, find an implicit solution formula, and plot some trajectories.

9. Plot some trajectories of the Lotka–Volterra model (5) in Section 5.1 with $k = 0.2$. Compare with Figure 5.3.3 and explain the significance of the parameter k.

10. Consider the system

$$x' = 2y, \qquad y' = x - y.$$

   a. Determine the differential equation for the trajectories.
   b. Let $v = y/x$. Determine the differential equation for the trajectories in the xv plane.
   c. Solve the differential equation for the trajectories in the xv plane. Then replace v by y/x to obtain the trajectories of the original system. *Note:* With proper simplification, the result should have the form $(x - By)(x + y)^2 = C$, for some constant B.
   d. Plot the phase portrait of the system.

F. W. Lanchester in 1916 proposed a model to predict the results of combat based on the size, efficiency, and character of the opposing forces. The simplest version of the Lanchester model,

$$x' = -y, \qquad y' = -x,$$

assumes that the forces are of equal quality but unequal size. Exercises 11 through 14 concern the Lanchester model and its application to a battle of historical significance.

11. *a.* Explain the assumptions the Lanchester model makes about how losses in combat occur.
   b. Find and solve the differential equation for the trajectories.
   c. If the larger force is initially twice the size of the smaller force and the battle lasts until the smaller force is destroyed, what fraction of the larger force survives?
   d. The British navy won a significant victory over the French navy at the battle of Trafalgar in 1805. The battle began with 27 British ships and 33 French ships, with the ships roughly comparable in combat strength. In a typical naval battle of the time, the ships of each fleet formed two parallel lines so that they could fire cannons from their long sides at the enemy ships. They kept firing until one side conceded the fight. Suppose the battle of Trafalgar had followed standard naval procedure and continued until one fleet was completely destroyed. Who would have won the battle, and how many of their ships would have survived?
   e. Sketch the phase portrait for the simplified Lanchester model, including the trajectory that would have been followed in the battle analyzed in part *d*.

12. Suppose the British commander at Trafalgar, Admiral Lord Nelson, had on his staff a mathematician familiar with the Lanchester model. (Ignore the fact that the battle occurred 111 years before the model was published.) Could this mathematician have devised a winning strategy for the British? One idea would be to divide the battle up into two sub-battles, one favoring each side. Suppose the British navy had divided into a large group and a small group and had managed to arrange the situation so that the larger British group engaged 17 French ships while the smaller British group engaged 16 French ships. Show that both sides have the same number of survivors if the British groups have 23 and 4 ships, respectively.

13. Do equal numbers of survivors from the sub-battles of Exercise 12 guarantee equal chances in the overall battle? This problem analyzes this question.
    a. Use the Lanchester equations to derive a second-order differential equation for $x(t)$, the number of ships on one side of the battle.
    b. Solve the equation from part a, and evaluate the constants, using the initial data $x(0) = x_0$ and $y(0) = y_0$.
    c. Assuming $x_0 < y_0$, show that the time at which the battle ends is given by
    $$t = \operatorname{arctanh} \frac{x_0}{y_0} = \frac{1}{2} \ln \frac{y_0 + x_0}{y_0 - x_0},$$
    where tanh is the hyperbolic tangent function defined by $\tanh u = (e^u - e^{-u})/(e^u + e^{-u})$ and arctanh is the inverse of the hyperbolic tangent function.
    d. Use the result from part c to determine the time that each sub-battle of Exercise 12 ends. If the British had managed to arrange the two sub-battles, what would have happened at Trafalgar?

14. The British strategy in the battle of Trafalgar was indeed to divide the battle into two unequal portions, but they managed to do so in a way that actually did yield an advantage.
    a. Suppose the British managed to arrange a sub-battle of 20 British ships against 12 French ships and a second sub-battle of 7 British ships against 21 French ships. Use the results from Exercises 11 and 13c to determine the number of British survivors for the first battle and the time required for its completion.
    b. Assume that the British admiral ordered the 7 ships in the second sub-battle to try to prolong the fighting with a lot of maneuvering. This would correspond to a change in the model
    $$F' = -aB, \qquad B' = -aF, \qquad F(0) = 21, \qquad B(0) = 7,$$
    where $a < 1$ is a factor that measures the amount of stalling by the British. Solve the model for this sub-battle.
    c. Show that the number of survivors is not changed by a but that the total time for the battle is changed by a. In particular, show that
    $$t = \frac{1}{2a} \ln 2,$$
    for the second sub-battle.
    d. Suppose $a = 0.2$. Which sub-battle ends first? At the time when the first sub-battle ends, approximately how many ships are left for both sides, counting both sub-battles?

T 15. Consider the SIR model of Exercise 12, Section 5.1. This model has three dependent variables, S, I, and R, but R can be determined after S and I by using the fact that the sum of the three variables is constant. Thus, the system can be thought of as a system with two differential equations for the unknowns S and I. Determine the differential equation for the trajectories in the SI plane, and solve it. Explain why epidemics start only if $rS_0 > \gamma$, where S_0 is the initial number of susceptibles. (Note that an epidemic requires a large number of susceptibles but does not require many infectives. This explains why initial exposure of a few individuals to European diseases caused severe epidemics in Native American populations.)

T 16. The equation for motion of a pendulum (suitably scaled) is
$$\theta'' + \sin\theta = 0,$$

where θ is the angle of the pendulum measured from the vertical and derivatives are with respect to time t. (See Exercise 15 of Section 3.1.) Let $\omega = \theta'$ (ω is the **angular velocity**).

a. Determine the equation for the trajectories in the $\theta\omega$ phase plane.
b. Solve the equation for the trajectories.
c. Suppose the pendulum has angular velocity $\omega = \omega_0$ when in the vertical position. Plot the trajectories for a variety of values of ω_0. You should see two distinct types of trajectories.
d. Explain the behavior of the pendulum for each of the two types of trajectories.

17. The simplest model for skydiving must consider two stages: a fast fall with the parachute still closed and a slow fall with the parachute open. Let m be the mass of the skydiver and v_I the maximum safe-impact speed. (The velocity at impact is then $-v_I$.)

a. A skydiving rule of thumb holds that the largest distance a person can fall and safely land on his/her feet is 3 m. Write down a model for a fall from a height of 3 m, assuming constant gravitational force and no damping. Determine the differential equation for the solution curves and solve it. You will need to use the initial conditions appropriate for a free fall from a height of 3 m. Use the solution to determine the velocity at impact, and hence the maximum safe-impact speed.
b. Write down a model for a fall with a closed parachute, assuming constant gravitational force and linear damping, with the damping coefficient $b = 15$ kg/s. Determine the terminal closed-chute velocity v_{cc} for a person of mass 75 kg. (The terminal velocity for a falling object is the stable equilibrium velocity of the differential equation. Note that the terminal velocity is negative in a coordinate system where up is positive.)
c. Skydivers choose a parachute size that yields a damping coefficient of about 1.4 times their mass (in mass units per second). Write down a model for a fall with an open parachute, assuming constant gravitational force and linear damping, with damping coefficient $b = 1.4m$. Determine the terminal open-chute velocity v_{oc}.
d. The damping coefficient $b = 1.4m$ is achieved by selecting the appropriate parachute for each jumper. Explain why this coefficient is a good choice.
e. Determine the differential equation for the solution curves for the open-chute stage. Solve it and plot the phase portrait.
f. Determine the particular trajectory for which the jumper will hit the ground at the maximum safe-impact speed. Now determine the minimum height at which the parachute can safely be opened. You may assume that the jumper is moving with velocity v_{cc} at the moment the chute is opened.

T 18. Consider the dimensionless waste treatment model

$$W' = -WX, \qquad X' = WX - X,$$

where W is the amount of a waste chemical in a given volume of water and X is the population of the bacteria in the water (see Exercise 13 of Section 5.1).

a. Determine the differential equation for the trajectories in the WX plane.
b. Solve the differential equation for the trajectories with initial data $X = X_0$ at $W = W_0$.
c. Plot trajectories in the region $0 < W < 3$, $0 < X < 1$. Describe the behavior of the treatment system, paying particular attention to the direction in which the solutions move along the trajectories. In particular, discuss the difference between trajectories whose W intercept is less than 1 and those whose W intercept is greater than 1.

T 19. Suppose a hoop is placed in a vertical plane so that it just touches the ground. The hoop is then spun at a constant speed with a vertical axis of rotation. (See Fig. 5.3.5.) Suppose further that the hoop is grooved, so that a small bead can slide up or down the hoop as it is spun. Common sense says that the bead will always slide to the bottom of the hoop, as it would if the hoop were not spinning. In this exercise we investigate this prediction.

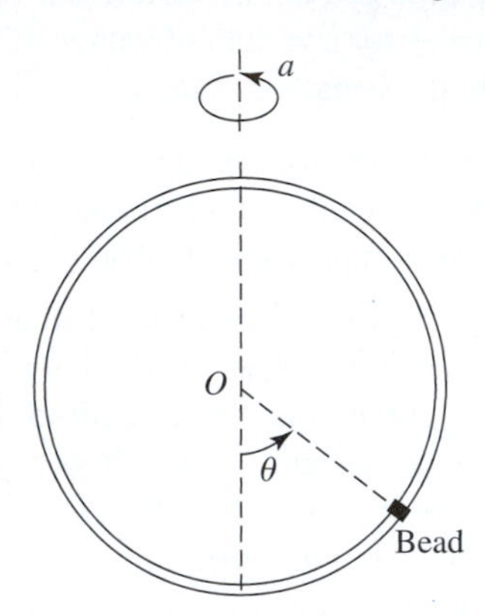

Figure 5.3.5
A bead on a rotating hoop.

Let θ be the angle formed by a vertical ray through the center of the hoop and a ray that connects the center of the hoop to the bead. Thus, $\theta = 0$ corresponds to the position of the bead at the bottom of the hoop. It can be shown that, in the absence of damping forces, the changes in the angle θ are modeled by the differential equation

$$\theta'' = (a^2 \cos\theta - 1)\sin\theta,$$

where the parameter a is proportional to the rotation speed of the hoop. (Note that the case $a = 0$ corresponds to the pendulum model of Exercise 16.)

a. Let $\omega = \theta'$. Determine the differential equation for the trajectories in the $\theta\omega$ phase plane.
b. Solve the differential equation for the trajectories.
c. Plot the phase portrait for the model for the case where $a = 1/\sqrt{2}$.
d. Repeat part *c*, but for the case $a = \sqrt{2}$.
e. Suggest an explanation for the differences between the behavior observed in parts *c* and *d*.

✦ 5.3 INSTANT EXERCISE SOLUTIONS

1. The solution is $y = \cos t$; hence, $v = y' = -\sin t$. By inspection, $y^2 + v^2 = 1$. The solution follows a circle in the yv plane, beginning at $(1, 0)$. The solution moves clockwise; to see this, note that $v'(0) = -\cos 0 = -1$. The solution moves down from the initial point.
2. Let $v = dy/dt$. Then $dv/dt = d^2y/dt^2 = -3y - 2\,dy/dt = -3y - 2v$. The system is

$$\frac{dy}{dt} = v, \qquad \frac{dv}{dt} = -3y - 2v.$$

3. We have

$$2V\,dV = -\frac{2}{(1+Z)^2}\,dZ = -2u^{-2}\,du,$$

where $u = 1 + Z$. Integration yields $V^2 = 2u^{-1} + C = 2/(1 + Z) + C$. The constant is evaluated from the initial condition: $C = V_0^2 - 2/(1 + Z_0)$. The solution follows from the substitution of C into the general solution formula.

4. The solutions move counterclockwise. This can be determined by examining the system of equations in a number of ways. Consider, for example, points with $Y = 1$ and $X < 1$. The equation $Y' = Y(X - 1)$ shows that $Y' < 0$ at these points. Hence, the direction of solutions to the left of the plot in Figure 5.3.3 is primarily downward, corresponding to counterclockwise motion.
5. With $v = y'$, the equation reduces to the first-order equation $v' = y'' = f(y, v)$. Thus, $dv/dy = v'/y' = f(y, v)/v$ or $v(dv/dy) = f(y, v)$.

5.4 The Direction Field and Critical Points

Recall that the slope field[16] of a scalar differential equation

$$\frac{dy}{dt} = f(t, y)$$

is a set of minitangents in the ty plane, each minitangent a line segment centered at some point (t, y) and having the slope dy/dt given by the differential equation. The minitangent at a particular point therefore indicates the direction of a solution curve passing through that point. The slope field can be generated by a computer without having to solve the equation. It can then be used to determine characteristics of the solutions. Autonomous two-dimensional systems can be studied by a similar graphical method.

MODEL PROBLEM 5.4

Describe the trajectories of the systems

$$x' = -2x + 2y, \qquad y' = 2x - 5y,$$

and

$$x' = -2x + 2y, \qquad y' = 2x + y.$$

The Direction Field

The **direction field** for an autonomous system of two equations is the set of arrows in the phase plane that point in the direction taken by a trajectory through the point at its center.

The direction field for an autonomous system is a lot like the slope field of a single nonautonomous differential equation. Both indicate the direction on a graph corresponding to the progress of solutions. The difference is in the coordinates on the axes. The slope field shows the changes in the single dependent variable compared to changes in the independent variable. The direction field shows the path in the two-dimensional phase space that solutions take as time advances. It is not necessary to use arrows to indicate directions in a slope field, because the forward

[16] Section 2.3.

direction for the independent variable is necessarily to the right. The independent variable does not appear on an axis in a direction field, so the direction of forward movement must be indicated with arrows.

The systems of Model Problem 5.4 can be thought of as pairs of formulas that indicate the rates of change of the variables on the axes of a direction field plot. Consider the points that lie on the x axis. For these points, both systems prescribe the rates of change as $x' = -2x$ and $y' = 2x$. All these points have values of x' and y' that are equal in magnitude but opposite in sign. Since both coordinates are changing at the same rate, the direction of motion is along the line $y = -x$. The actual rates of change of position with respect to time are different for each of these points. We could indicate the overall rate of change by using different lengths for direction field arrows. In this example, the arrow at the point $(1, 0)$ could be of length 2 and the arrow at $(2, 0)$ of length 4. A plot in which the lengths of the arrows are relative to the magnitudes of the rate of change is a **vector field.** Vector fields are used sometimes in fluid flow, but they are not often used with other systems of differential equations because the plots can be visually confusing. Direction fields indicate only the direction of motion, not the rate at which the motion occurs.[17]

To sketch the direction field for a system of two autonomous differential equations, one can compute values of x' and y' at points on a grid and then compute the arrow slope from

$$\frac{dy}{dx} = \frac{y'}{x'} \tag{1}$$

It is not practical to do this by hand, but it is easy to use a computer algebra system to sketch direction fields. The direction field for the first system in the model problem appears in Figure 5.4.1. As we computed by hand, the slopes of the arrows at points on the x axis are all -1. Note that the arrowheads point toward the upper left for $x > 0$ and toward the lower right for $x < 0$.

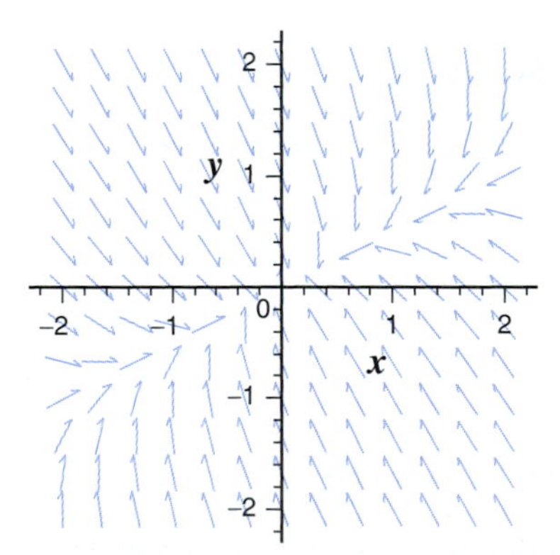

Figure 5.4.1
The direction field for $x' = -2x + 2y$, $y' = 2x - 5y$.

The arrows in the direction field are tangent to the trajectories. Thus, we can use the direction field to generate a sketch of the trajectories. Figures 5.4.2 and 5.4.3 show the phase portraits, including the direction fields, for both systems of Model Problem 5.4.

[17] Some fluid flow maps use the width of the direction field arrows to indicate the speed of the flow, but this practice does not appear to have caught on in mathematics.

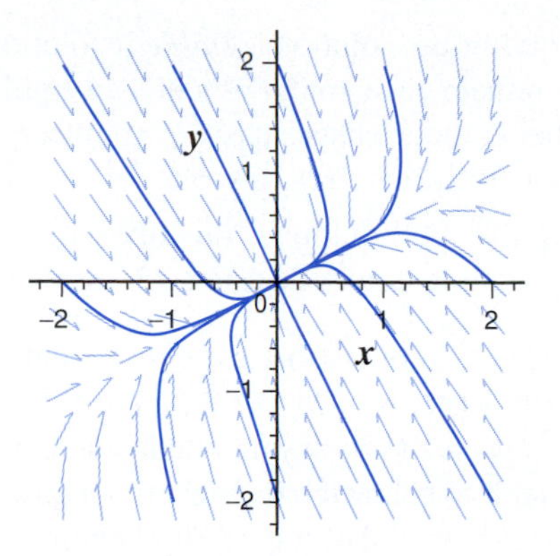

Figure 5.4.2
The phase portrait for $x' = -2x + 2y$, $y' = 2x - 5y$.

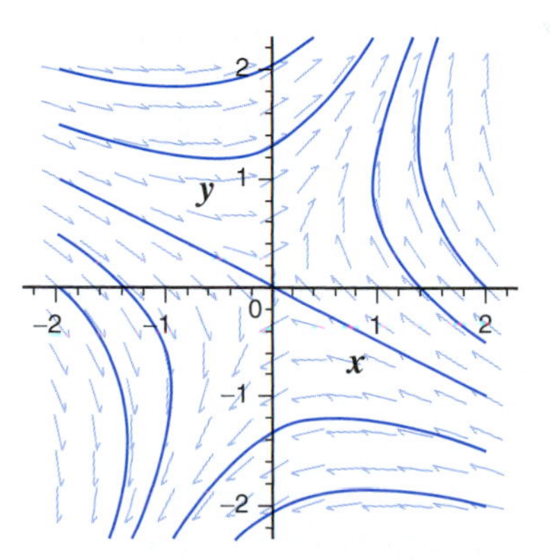

Figure 5.4.3
The phase portrait for $x' = -2x + 2y$, $y' = 2x + y$.

Critical Points and Equilibrium Solutions

Like single autonomous first-order equations, autonomous systems have equilibrium solutions for any values of the dependent variables that make the derivatives zero. A **critical point** for a system $x' = f(x, y)$, $y' = g(x, y)$ is a solution of the algebraic equations $f(x, y) = 0$, $g(x, y) = 0$. The corresponding constant solution is an **equilibrium solution.**

The concept of stability for scalar equations holds equally well for systems, with some modifications. Some care is needed in the definitions. Let (x_c, y_c) be a critical point. For any point (x, y), define the distance $d(x, y)$ by $d(x, y) = \sqrt{(x - x_c)^2 + (y - y_c)^2}$. The distance d allows us to be precise in describing the intuitive ideas of "staying near" a critical point and "approaching" a critical point.

- An equilibrium solution is *asymptotically stable* if all trajectories that come within a distance δ of the critical point enter the critical point in forward time. Formally, an equilibrium solution is **asymptotically stable** if there is a positive number δ small enough that $\lim_{t\to\infty} d(x, y) = 0$ for all initial points (x_0, y_0) for which $d(x_0, y_0) < \delta$.

- An equilibrium solution is *stable* if solution curves that begin close enough to the critical point remain close to it. Formally, an equilibrium solution is **stable** if for any positive number ϵ, there exists a positive number δ small enough that $d(x, y) < \epsilon$ for all initial points (x_0, y_0) for which $d(x_0, y_0) < \delta$.
- An equilibrium solution is **unstable** if it is not stable. The failure of stability occurs when some solution curves that start arbitrarily close to the critical point move away from it.

Note that the requirement for asymptotic stability is stricter than that for mere stability.

For both systems of Model Problem 5.4, the origin is the only critical point. In Figure 5.4.2, the origin appears to be asymptotically stable. The origin in the system of Figure 5.4.3 is a little more difficult to characterize. Most trajectories seem to be moving toward the origin for a while before turning away. The trajectories beginning at the points $(2, -1)$ and $(-2, 1)$, however, appear to head directly into the origin. The origin in Figure 5.4.3 is an example of a *saddle point.* This concept will be more carefully defined in Chapter 6. For now, it is sufficient to think of a saddle point as a critical point for a two-dimensional system that is unstable in spite of having a pair of trajectories that enter it in forward time.

So far, we have considered only systems with just one critical point. The same concepts apply to systems with more than one critical point.

EXAMPLE 1

Consider the predator-prey model

$$X' = X(1 - Y), \qquad Y' = 0.4Y(X - 1).$$

Critical points satisfy $X(1 - Y) = 0$ and $Y(X - 1) = 0$. The first equation requires either $X = 0$ or $Y = 1$. If $X = 0$, then the second equation requires $Y = 0$; if $Y = 1$, then the second equation is satisfied only for $X = 1$. There are two critical points: $(0, 0)$ and $(1, 1)$. The phase portrait, shown in Figure 5.4.4, suggests that the equilibrium solution $x = 0$, $y = 0$ is unstable while the equilibrium solution $x = 1$, $y = 1$ is stable, but not asymptotically stable. The origin appears to be a saddle point because a trajectory enters in along the Y axis. A critical point is a **center** if the nearby trajectories are a set of closed concentric curves; it appears that $(1, 1)$ is a center.

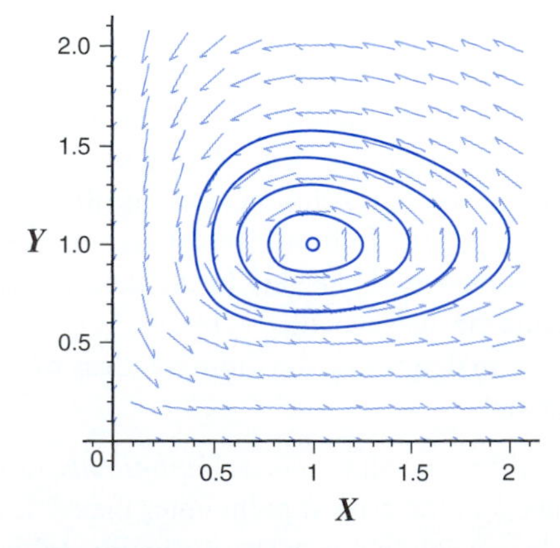

Figure 5.4.4
The phase portrait for $X' = X(1 - Y)$, $Y' = 0.4Y(X - 1)$.

Note the tentative language used in Example 1. Definite conclusions cannot generally be drawn from graphs, particularly graphs generated by numerical methods. A computer-generated phase portrait often provides strong clues for the classification of equilibrium points, but symbolic methods are necessary to confirm the results suggested by the phase portrait.

Separatrices

Systems with multiple critical points often have trajectories that serve to separate regions with different qualitative behavior. These special trajectories are called **separatrices.**

EXAMPLE 2

Consider the competition model

$$X' = X(1 - Y), \qquad Y' = 2Y(1 - X).$$

As in Example 1, there are two critical points: $(0, 0)$ and $(1, 1)$. The phase portrait is shown in Figure 5.4.5. The critical point $(1, 1)$ appears to be a saddle point because most trajectories move away from it, but there is a pair that seems to enter it. The equilibrium solution $x = 0$, $y = 0$ is unstable. All the trajectories are unbounded, but tend to approach one of the axes as time increases. The two trajectories that enter the saddle point divide the first quadrant between the region where trajectories approach the X axis and the region where trajectories approach the Y axis. These two trajectories are separatrices.

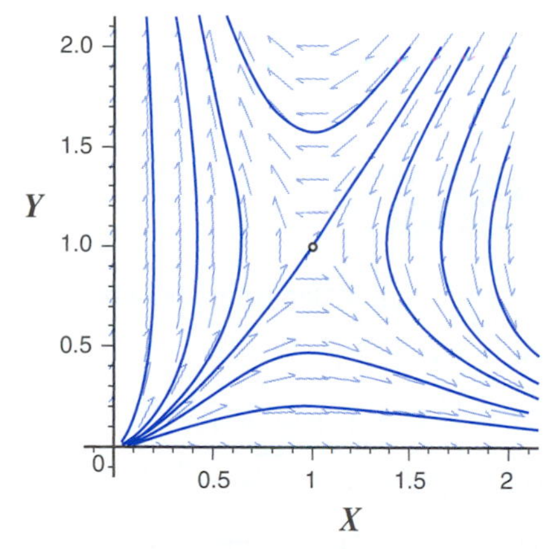

Figure 5.4.5
The phase portrait for $X' = X(1 - Y)$, $Y' = 2Y(1 - X)$.

There are several possible longtime behaviors for a two-dimensional system of autonomous equations.

- Trajectories can terminate in an asymptotically stable equilibrium state.
- Trajectories can approach a closed (stable) trajectory, as in the Lotka–Volterra model (Example 1), where all trajectories are closed.
- Trajectories can be unbounded, as in the simple competition model (Example 2).

Sometimes a system has more than one possible behavior as $t \to \infty$. In these cases, there are portions of the phase plane where the trajectories have one longtime behavior and other portions that have different longtime behavior, and the curves that are the boundaries of these regions are the separatrices.

Separatrices are important because they determine the ultimate fate of a system. In the simple competition model, one species wins and the other loses because the longtime behavior is for one population to increase indefinitely while the other one appears to vanish. Which one wins depends on the initial condition. Initial points to the upper left of the separatrices are on trajectories that ultimately move up and to the left; thus, Y increases and X decreases. Note that the curve at the top of the figure shows an initial decrease in both populations; however, eventually the Y population reaches a minimum value and increases thereafter. Similarly, initial points to the lower right of the separatrices are on trajectories that ultimately move down and to the right, with Y decreasing toward zero and X apparently increasing without bound.

Recapitulation

The examples in this section have illustrated a variety of possibilities for the behavior of autonomous systems. All critical points can be classified by stability. The classification of a critical point can usually be obtained from a computer-generated sketch of the phase portrait, but not with complete reliability. There is also an analytical (symbolic) method of classifying critical points that is the focus of Chapter 6. Saddle points are particularly noteworthy because they have a pair of trajectories that terminate at them even though most trajectories move away from them. Centers are particularly noteworthy because trajectories near them are closed curves. Generally, solutions evolve toward stable equilibrium solutions or closed curves, or are unbounded. Where more than one of these ultimate behaviors is possible, the regions corresponding to each such long-term behavior are separated by trajectories called separatrices.

5.4 Exercises

In Exercises 1 through 4, determine the critical points for the given system.

1. $r' = 3r - 2cr, \quad c' = cr - 4c$
2. $u' = 3u - 2v - 5, \quad v' = u + v - 5$
3. $x' = \sin y, \quad y' = y - x$
4. $x' = y(x^2 + y^2 - 1), \quad y' = -x(x^2 + y^2 - 1)$

In Exercises 5 through 10, use a computer or calculator to sketch the phase portrait; determine whether the equilibrium solution $x = 0$, $y = 0$ is stable or unstable; and determine whether the origin is a saddle point or a center.

T 5. $x' = -x + y, \quad y' = x + y$

T 6. $x' = x - 2y, \quad y' = -2x$

T 7. $x' = -2y, \quad y' = x - 3y$

T 8. $x' = 3x - 2y, \quad y' = 2x - 2y$

T 9. $x' = 2x - 5y, \quad y' = x - 2y$

T 10. $x' = -x + y, \quad y' = -5x + 3y$

In Exercises 11 through 14, (*a*) determine the critical points, (*b*) use a computer or calculator to sketch the phase portrait, (*c*) determine whether each equilibrium solution is stable or unstable, (*d*) identify the saddle point, and (*e*) plot the separatrices. (*Hint:* to plot a separatrix, choose a point that is near the saddle point and appears to lie close to the separatrix. Run the plot backward in time to get a curve that is approximately the separatrix.)

T 11. $x' = x^2 - y, \quad y' = -x + y$

T 12. $x' = y, \quad y' = x - y - x^3$

T 13. $x' = 2y - x, \quad y' = x(2x + y - 5)$

T 14. $x' = x + y, \quad y' = 1 - y^2$

T 15. The growth of a vapor bubble during the boiling process is governed by the differential equation

$$\frac{2}{3}rr'' + (r')^2 = 1 - \frac{1}{r},$$

where r is the radius of the bubble. Define a new variable v by $v = r'$. Write the bubble growth equation as a system, and determine the critical point(s) for this system. Use a computer or calculator to sketch the phase portrait. Determine the stability of the equilibrium solution(s) and identify any saddle points or centers. Note that a bubble radius cannot be negative, so only the half-plane $r \geq 0$ is of interest.

5.5 Qualitative Analysis

Qualitative accuracy means getting the general features right; *quantitative* accuracy means getting agreement to within some satisfactory error tolerance. Both are important. The work we did in Section 5.4 to classify equilibrium solutions was quantitative, as it was based on computer-generated phase portrait sketches. One drawback of quantitative approximation is that it can only yield information about the system for one set of parameter values at a time. Often it is important to determine the longtime behavior of solutions and how that behavior depends on the parameter values. This is the realm of **qualitative analysis.** The phase line of Section 5.2 is an example of a qualitative technique. Nullcline analysis, which is the subject of this section, is a two-dimensional analog of the phase line technique. Sometimes, qualitative analysis is enough to determine the longtime behavior of a system. Other times, it is necessary to use the symbolic quantitative methods that are the subject of Chapter 6.

MODEL PROBLEM 5.5

Use the information given in the differential equations

$$x' = 2y, \qquad y' = x - y$$

to classify the equilibrium solution $x = 0$, $y = 0$ without having to solve the system or rely on numerical approximations.

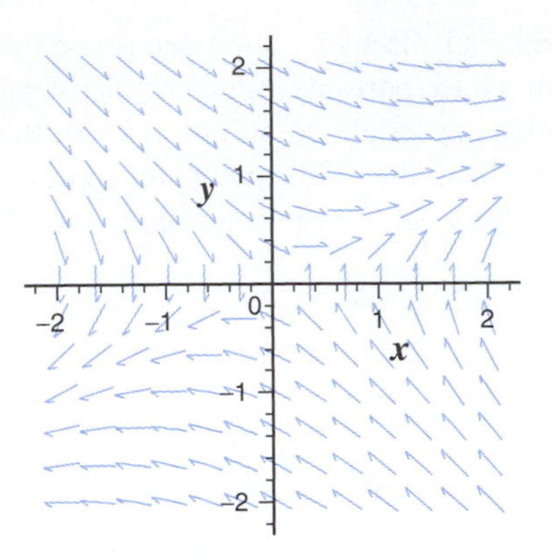

Figure 5.5.1
The direction field for $x' = 2y$, $y' = x - y$.

A plot of the direction field for Model Problem 5.5 appears in Figure 5.5.1. Notice that all the arrows on the x axis point in a vertical direction. This is confirmed by the differential equation for x: $x' = 0$ when $y = 0$. Curves consisting of points at which x' is always zero or points at which y' is always zero are useful tools for qualitative analysis.

A **nullcline** is a curve in the phase plane on which the arrows all point in a direction parallel to a coordinate axis. In the xy plane, for example, an x nullcline is a curve on which $x' = 0$ and a y nullcline is a curve on which $y' = 0$.

The first differential equation in Model Problem 5.5 is $x' = 2y$; thus, $x' = 0$ precisely when $y = 0$. The line $y = 0$ is an x nullcline for the system. Note in the figure that all the arrows in the upper half-plane show x increasing (because the x component of the vector is positive), while all arrows in the lower half-plane show x decreasing. This illustrates an important property of nullclines:

The x nullclines divide the xy phase plane into regions, with each region consisting of points where x is increasing or points where x is decreasing. Similarly, regions set off by y nullclines consist of points where y is increasing or points where y is decreasing.

Note that we can determine which regions have x increasing and which have x decreasing simply by looking at the differential equation for x. We have $x' > 0$ whenever $2y > 0$, and this is in the upper half-plane. A schematic diagram showing the result of the analysis of the equation $x' = 2y$ appears in Figure 5.5.2*a*. Some vertical minitangents appear on the line $y = 0$, which is the x nullcline. Above the nullcline is an arrow indicating that x is increasing, and below it is an arrow indicating that x is decreasing.

Figure 5.5.2*b* shows a similar schematic analysis of the equation $y' = x - y$. The y nullcline is the line $y = x$, on which $y' = 0$. The horizontal minitangents indicate that this line is a y nullcline.

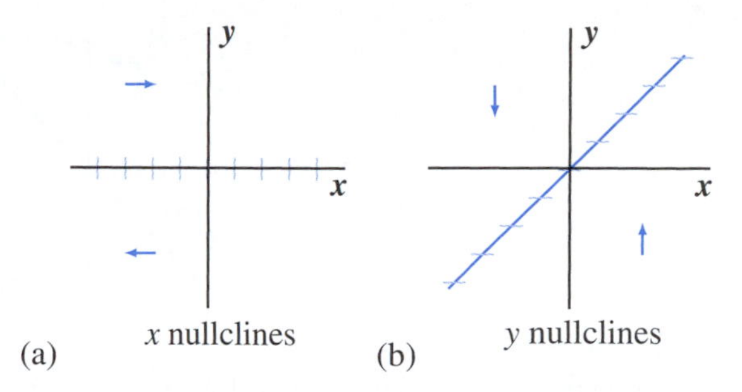

Figure 5.5.2
The individual nullclines for the system $x' = 2y$, $y' = x - y$.

The region $x > y$ lies to the lower right of the nullcline, and this region is marked with an upward arrow to indicate that this is a region where y is increasing. The region on the other side of the nullcline is marked with a downward arrow because $y' = x - y < 0$ for points in that region. Note that the details regarding both nullclines can also be identified in the computer-generated direction field of Figure 5.5.1. All the arrows on the line $y = x$ are indeed horizontal, all arrows above and to the left of this line show y decreasing, and all arrows below and to the right of the line show y increasing.

✦ INSTANT EXERCISE 1

Sketch the individual nullclines (as in Figure 5.5.2) for the system

$$x' = y - x, \qquad y' = -x.$$

Note that the sketches in Figure 5.5.2 do not show arrowheads on the nullcline minitangents. Each sketch was prepared using the information from just one of the differential equations. Information from the x' equation cannot be used to determine whether y is increasing or decreasing on the x nullclines. This information can only be found by studying the differential equations together.

Nullcline Diagrams

Each of the sketches in Figure 5.5.2 is of some value by itself, but a lot more information is obtained by combining the two into a full nullcline diagram. Figure 5.5.3 shows the result. The minitangents on the y nullcline $y = x$ now have arrowheads—these point to the right for the portion above the x axis and to the left for the portion below the x axis. These details follow from the information provided by the x nullclines. Similarly, the minitangents on the x nullcline have arrowheads determined by the analysis of the y nullcline. There is one point where the nullclines intersect. This is the critical point $(0, 0)$. Since critical points are points where neither variable is changing, they are always at the intersection of opposite nullclines. Taken together, the nullclines of Model Problem 5.5 divide the phase plane into four regions, labeled A, B, C, and D in the figure. All points in region A are below the y nullcline and above the x nullcline. Based on our analysis of the nullclines, this means that both x and y are increasing in region A. The symbol consisting of

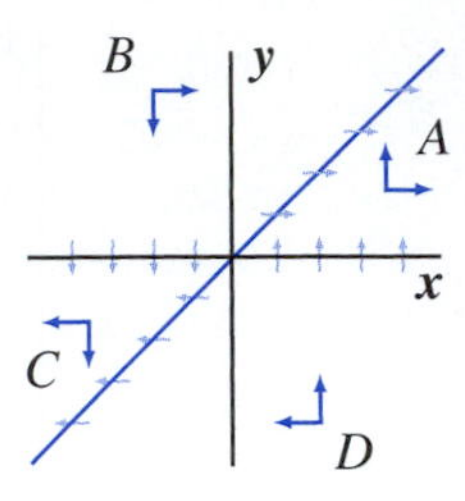

Figure 5.5.3
The nullcline diagram for the system $x' = 2y$, $y' = x - y$.

two connected arrows pointing up and to the right indicates the general trend of the direction field for region A. Similarly, there are symbols in each of regions B, C, and D that indicate the trends of the direction field. The directions of the arrowheads on the nullclines are provided by the trends in the adjoining regions. For example, the nullcline $y = x$ in the first quadrant has horizontal arrows. Both regions A and B, which are adjacent to this nullcline, indicate a general trend to the right; hence the arrows on this portion of the nullcline point to the right. Below the x axis, the arrows on the same nullcline point to the left.

✦ INSTANT EXERCISE 2

Sketch the complete nullcline diagram for the system

$$x' = y - x, \qquad y' = -x.$$

Using Nullclines to Study Critical Points

It is helpful to look at the nullclines in Figure 5.5.3 as providing information about longtime behavior in the same manner as the phase line, but generalized to two dimensions. This analysis is based on the observation that solution curves can only cross a nullcline in the indicated direction.

Consider the behavior of the trajectory through some point in region A. Any solution in region A has both x and y increasing. Stated differently, trajectories in region A can only move up and to the right. What is more important is that they cannot leave region A. The arrows on the boundaries between A and B and between A and D show that solutions can cross these boundaries only to enter region A. Thus, trajectories that begin in A or enter A must thereafter stay in A. The same property holds for region C; trajectories that begin in C must stay in C. Solutions in regions A and C always move away from the origin. This demonstrates that the equilibrium solution $x = 0$, $y = 0$ is unstable.

Now consider regions B and D. Some trajectories in each of these regions move into A, while others move into C. From this information, we can argue that there must be some trajectory that enters the origin. Consider a thought experiment. Mark out all the points in D that are on trajectories that enter A in one color, and mark all the points in D whose trajectories enter C in a different color. There cannot be any white space between these two regions, nor can there be any overlap, because (Theorem 5.3.1) there is a unique trajectory through every point, no matter how close the point is to the origin. Thus, the two regions must have a common boundary that passes through the origin. This boundary is itself a trajectory; indeed, it is a separatrix because it divides points that go into region C from points that go into region A.

Nullcline diagrams can sometimes confirm stability as well as instability, and there are also systems for which no conclusions can be drawn from the nullcline diagram alone.

EXAMPLE 1

Figure 5.5.4 shows the nullclines for

$$x' = -2x + 2y, \qquad y' = 2x - 5y.$$

As in Model Problem 5.5, trajectories in region A or region C must remain in that region. This time, those trajectories move toward the origin, suggesting that the equilibrium solution at the origin is asymptotically stable. To confirm this claim, it is necessary to examine regions B and D. Suppose there is a trajectory in one of these regions that does not approach the origin. Such a trajectory cannot remain in region B or D, but must instead enter either A or C. However, trajectories in A and C move toward the origin. Ultimately, trajectories that begin in any of the four regions must go to the origin, so the origin is indeed asymptotically stable. The phase portrait for this system appears in Figure 5.4.2; trajectories in the phase portrait illustrate the predictions made by nullcline analysis.

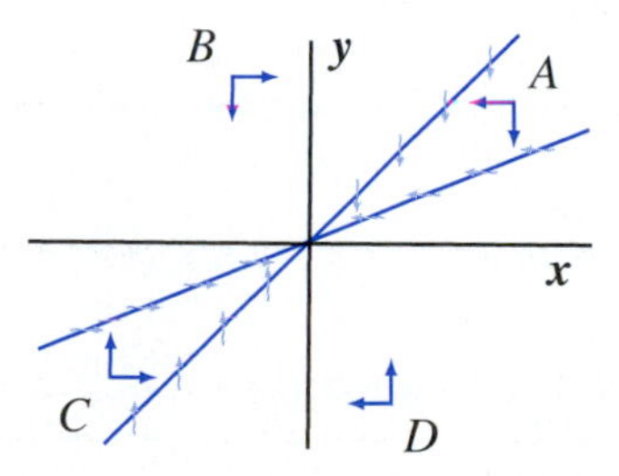

Figure 5.5.4
The nullcline diagram for the system $x' = -2x + 2y$, $y' = 2x - 5y$.

EXAMPLE 2

Figure 5.5.5 shows the nullclines for

$$x' = -2y, \qquad y' = x - 3y.$$

In this case there seems to be a general flow of solution curves from A to B to C to D and so on, which suggests that the solution curves rotate around the origin, but does not offer any evidence concerning stability. The phase portrait in Figure 5.5.6 reveals that the origin is asymptotically stable. Solution curves do not spiral around the origin, but follow paths confined to one or two regions. For example, solution curves beginning in the second quadrant move through region C into region D and then proceed to the origin without leaving region D. The nullclines alone do not give enough information to distinguish this behavior from trajectories that spiral around the origin, such as in Figure 5.3.1, or trajectories that are closed curves, as in Figure 5.3.3.

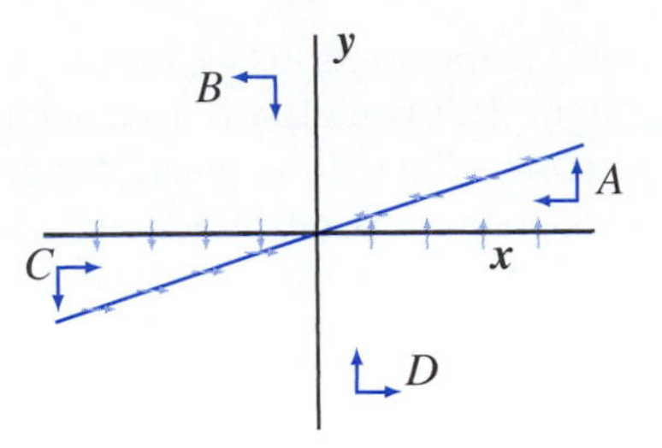

Figure 5.5.5
The nullcline diagram for the system $x' = -2y$, $y' = x - 3y$.

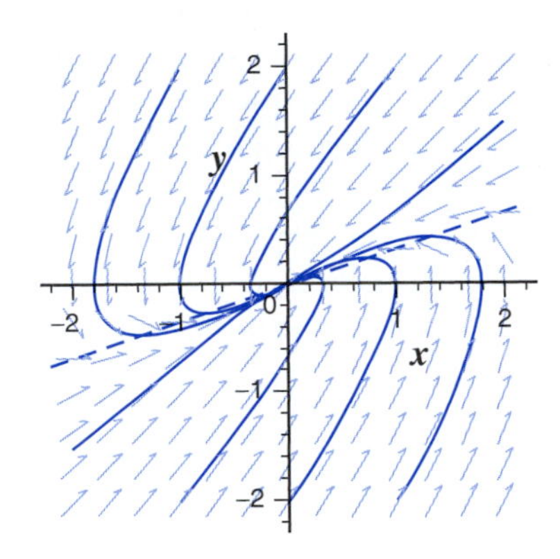

Figure 5.5.6
The phase portrait for $x' = -2y$, $y' = x - 3y$, along with the nullcline $x = 3y$ (dashed).

✦ INSTANT EXERCISE 3

What conclusions, if any, can be drawn from the nullcline diagram for the system

$$x' = y - x, \qquad y' = -x?$$

Nullcline Analysis of Some Population Models

Consider now the three basic models for interacting populations that were developed in Section 5.1. These are the basic predator-prey model

$$X' = X(1 - Y), \qquad Y' = kY(X - 1), \tag{1}$$

the basic competition model

$$X' = X(1 - Y), \qquad Y' = kY(1 - X), \tag{2}$$

and the basic cooperation model

$$X' = X(Y - 1), \qquad Y' = kY(X - 1). \tag{3}$$

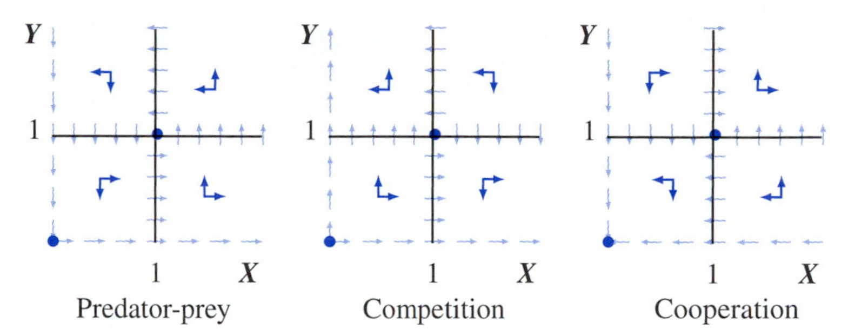

Figure 5.5.7
Nullcline diagrams for the three basic models of interacting populations.

These models show, among other things, the power of abstraction. The nondimensionalization process reduced the number of parameters from four to one, and now we get additional benefits from the change of variables. In all three models, the X nullclines consist of the lines $X = 0$ and $Y = 1$, and the Y nullclines are the lines $Y = 0$ and $X = 1$. The parameter k does not affect the nullclines at all. The nullclines for the three models are shown in Figure 5.5.7. The only difference among the three models is the direction of the arrowheads.

We now use the nullcline diagrams to study the three basic models. In particular, we want to know what happens to the solutions as time increases. The insight of Malthus that populations cannot grow unchecked leads us to require that, in any reasonable model, the solution curves be bounded.

The Basic Predator-Prey Model The nullcline diagram for the basic predator-prey model shows that the origin is a saddle point. This is so because the X axis is a solution that leaves the origin, while the Y axis is a solution that enters the origin. The nullcline diagram does not give enough information to classify the equilibrium point at (1, 1), nor does it guarantee that the solution curves are bounded. Consider, for example, the region $X, Y > 1$. The nullcline diagram shows that X is decreasing in this region while Y is increasing. It appears that the solution curves will cross over the line $X = 1$ and enter into a region where Y is decreasing as well as X. But this is not the only picture consistent with the nullcline diagram. It could be that X remains larger than 1 while Y increases without bound, as would happen if the point (1, 1) were a source.

Of course we have more information than just the nullcline diagram. We also have the computer-generated phase portrait for the case $k = 0.4$, as shown in Figure 5.4.4. This picture appears to confirm that the point (1, 1) is a center and that the solution curves are bounded. This is strong evidence, but it is limited to a single value of the parameter k. The solution of the differential equation for the trajectories, as in Example 2 of Section 5.3, can be used to confirm that solutions are bounded.

One final issue to explore is the significance of the parameter k. Figure 5.5.8 shows the phase portraits for $k = 1$ and $k = 0.5$. The common nullcline diagram guarantees that these systems have similar phase portraits, although there are some quantitative differences.

The Basic Competition Model The nullcline diagram for the competition model clearly indicates that the equilibrium point (1, 1) is a saddle point, with the same argument as in Model Problem 5.5. The origin is unstable also. Both equilibrium solutions are saddles, so all the trajectories

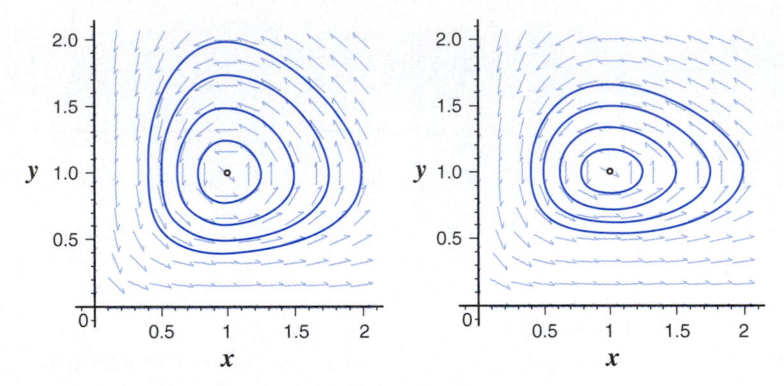

Figure 5.5.8
The phase portrait for the predator-prey model with $k = 0$ (left) and $k = 0.5$ (right).

must be unbounded. Whichever species loses the competition dies out, but then the other grows without bound. This model is too simple to be a useful model for competing species. It can be improved by adding logistic terms to the differential equations. Such models are considered in the exercises.

The Basic Cooperation Model The nullcline diagram for the cooperation model indicates that the point $(1, 1)$ is again a saddle point. The origin, however, appears to be asymptotically stable. Solutions that begin with $X, Y < 1$ move inevitably toward the origin, indicating that the two species die out. Alternatively, solutions that begin with $X, Y > 1$ clearly increase without bound. As in the competition model, the cooperation model requires some additional features in order to work for all starting population levels. The model works only for starting levels too small to sustain the populations and for short-term behavior with larger starting populations.

A Technical Note

Phase portrait plots show good numerical approximations of some example trajectories. Nullclines indicate features that all trajectories in a region must have. Conclusions drawn from nullclines do not rest on numerical approximations, nor are they deduced from examples. They are demonstrated by evidence taken directly from the system of differential equations.

5.5 Exercises

For Exercises 1 through 8, sketch the nullclines for each variable separately, combine them to form a nullcline diagram, and draw what conclusions you can about the nature of the critical points. If there are any separatrices, indicate in which region(s) they lie. Compare the results with a computer-generated phase portrait.

1. $x' = -x + y, \quad y' = x + y$ (Section 5.4, Exercise 5)
2. $x' = x - 2y, \quad y' = -2x$ (Section 5.4, Exercise 6)

3. $u' = 3u - 2v - 5, \quad v' = u + v - 5$
4. $x' = 3x - 2y, \quad y' = 2x - 2y$ (Section 5.4, Exercise 8)
5. $x' = x^2 - y, \quad y' = -x + y$ (Section 5.4, Exercise 11)
6. $x' = y, \quad y' = x - y - x^3$ (Section 5.4, Exercise 12)
7. $x' = 2y - x, \quad y' = x(2x + y - 5)$ (Section 5.4, Exercise 13)
8. $x' = x + y, \quad y' = 1 - y^2$ (Section 5.4, Exercise 14)
9. Prepare a nullcline diagram for the bubble growth equation

$$\frac{2}{3} r r'' + (r')^2 = 1 - \frac{1}{r},$$

from Section 5.4, Exercise 15. Just consider the half-plane $r \geq 0$ corresponding to the physical interpretation of the model. Discuss what the nullcline diagram says about the growth of bubbles of different initial size.

10. Consider the Lotka–Volterra predator-prey model

$$X' = X(1 - Y), \qquad Y' = kY(X - 1).$$

The first study done with this model involved a curious observation made in the 1920s. Italian fishermen at that time caught sharks as well as the more desirable fish. Prior to World War I, about 10 to 12% of the catch consisted of sharks. During World War I, the percentage of sharks rose to as high as 30%. By the mid-1920s the percentage was back down to about 15%. It was obvious that there was much less fishing during the war, but it was not at all clear why that should affect the proportions of the species.

Assume that the model applies to the populations of sharks and food fish during the war, when there was little or no commercial fishing. After the war, the resumption of fishing changes the model by adding an extra term to each equation. Assuming that the rate of fish caught is proportional to the population, the new model is

$$X' = X(1 - Y) - aX, \qquad Y' = kY(X - 1) - bkY,$$

where a and b are positive parameters of fairly small magnitude. Note that the rate of decrease of the shark population (Y) from fishing is written as $-bkY$ rather than $-bY$ so that the parameter k does not affect the nullclines.

a. Sketch the nullcline diagram for the revised model, and compare it with the predator-prey model from Figure 5.5.7. Does the Lotka–Volterra equation adequately explain the observations?

b. The reintroduction of fishing after World War I increased the ratio of food fish to sharks by a factor of 3. Assuming $a = b$, how large must these parameters be in order to change the equilibrium population ratio by this much?

c. The Lotka–Volterra model has also been applied to other problems. Consider a population of rabbits and coyotes, and suppose the local chicken farmers want to hunt coyotes to reduce losses of chickens. Use the Lotka–Volterra model with selective hunting of predators ($a = 0$). What does the model predict will happen to the coyote and rabbit populations if b is increased in an attempt to remove the coyotes? Does this model accurately predict the hunting of predators to extinction that has occurred in many areas?[18]

11. Perhaps the competitive population model would be improved by adding logistic growth. Consider the model

$$X' = X\left(1 - Y - \frac{X}{a}\right), \qquad Y' = Y\left(1 - X - \frac{Y}{b}\right),$$

where $a > 1$ and $b > 1$.

a. Prepare a nullcline diagram. What does the model predict will happen? Discuss whether this model is reasonable as a simple model for competing species.

b. Reconsider the model with $a < 1$ and $b < 1$. What does the model predict will happen? Does the model work for competing species in this case?

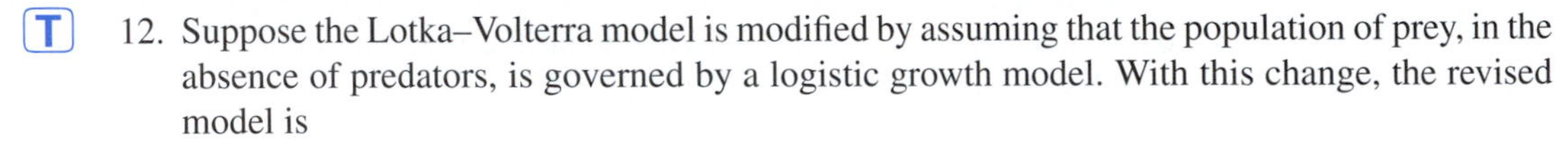

T 12. Suppose the Lotka–Volterra model is modified by assuming that the population of prey, in the absence of predators, is governed by a logistic growth model. With this change, the revised model is

$$X' = X\left(1 - \frac{X}{M} - Y\right), \qquad Y' = kY(X - 1).$$

a. Prepare a nullcline diagram for this model, assuming $M > 1$.

b. Plot the phase portrait, using $k = 1$ and $M = 5$.

c. If we add hunting of predators to the model, we change the Y equation to

$$Y' = kY(X - 1 - b).$$

What does the model predict will happen to the coyote and rabbit populations if b is increased in an attempt to remove the coyotes? Compare with Exercise 10.

5.5 INSTANT EXERCISE SOLUTIONS

1.

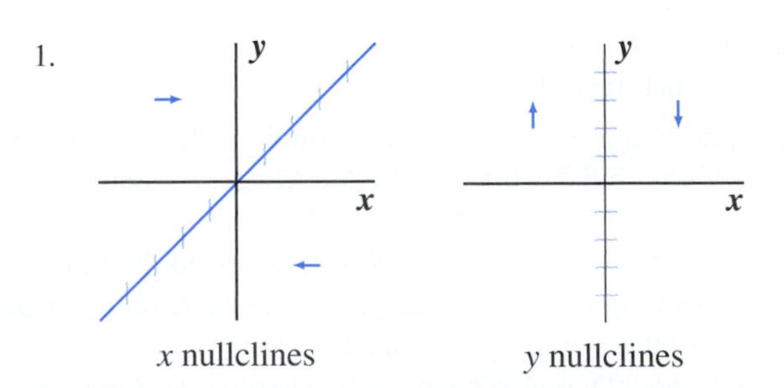

[18] The failure of the Lotka–Volterra model for selective hunting of predators is not a refutation of the model. It is, rather, a warning that models must be fitted to the phenomena to be modeled. The model works well for the shark problem, but it is not sufficient for the coyote problem. It is up to the applied mathematician to determine, through thought experiments and examination of real data, whether a given model is appropriate for a given problem.

2.

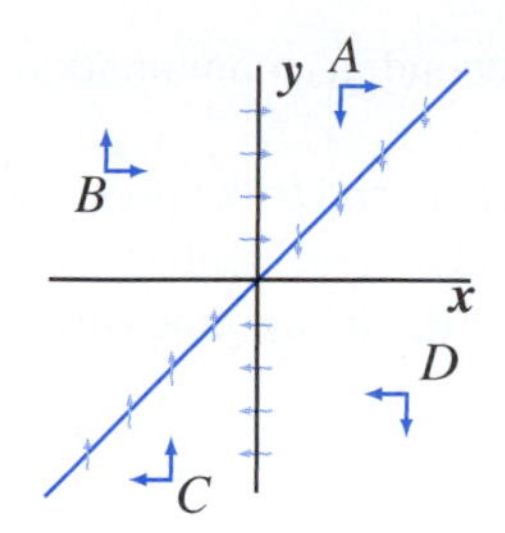

3. No conclusions can be drawn without additional information.

A Self-Limiting Population

CASE STUDY 5

There are many situations in which population growth is limited by environmental factors caused by the population itself. A simpler example would be the population of microorganisms in a closed environment, such as the population of yeast in a commercial fermentation process. Like all other living creatures, yeast produces waste products. In a closed environment, the waste products accumulate, making the environment less hospitable to the yeast.

The Model

We begin with a conceptual model for the problem. Our conceptual model includes a number of assumptions:

1. The microorganisms are distributed uniformly throughout a closed container of fixed volume. Otherwise we would have to consider spatial variations as well as evolution in time.
2. The only relevant quantities in the model are the population $p(t)$ of microorganisms and the amount $w(t)$ of toxic waste material. We are ignoring the possibility that the system could be affected by such things as temperature changes.
3. The rate of waste production is proportional to the population, with rate constant k.
4. In the absence of waste, the population would undergo logistic growth with maximum population M and initial relative growth rate r.
5. The effect of the waste is to kill microorganisms at some rate dependent on both the population and the amount of waste. A reasonable guess is that the death rate relative to the population should be proportional to the amount of waste, with b the constant of proportionality. Intuitively, it makes sense that doubling the amount of waste should double the death rate.
6. Initially, there is a population p_0 that is relatively small compared with the capacity M, and there is no waste.

From these assumptions, we get a pair of differential equations, with initial conditions, for w and p:

$$\frac{dw}{dt} = kp, \qquad w(0) = 0, \tag{1}$$

$$\frac{dp}{dt} = rp\left(1 - \frac{p}{M}\right) - bpw, \qquad p(0) = p_0. \tag{2}$$

The first differential equation follows from assumption 3, while the second is a combination of assumptions 4 and 5.

In its current form, the model has five parameters: the rate constants k, r, and b; the population capacity M; and the initial population p_0. It is very difficult to understand a model with this many parameters. Following the procedure of Section 5.1, we can convert the model to dimensionless form. Using the dimensionless variables

$$P = \frac{p}{M}, \qquad W = \frac{bw}{r}, \qquad \tau = rt,$$

the model takes a simpler form (see Exercise 1).

Self-Limiting Population—Dimensionless Model

Determine the behavior of the system

$$W' = KP, \qquad W(0) = 0, \tag{3}$$

$$P' = P(1 - P - W), \qquad P(0) = P_0, \tag{4}$$

where $K > 0$ is a dimensionless parameter. Of particular interest is the behavior of the system when P_0 is very small.

The dimensionless version of the model has only two parameters, one of which is considerably restricted in its possible values. In this form, the model is much easier to analyze.

Qualitative Analysis

The nullclines for the dimensionless system are obtained in the usual way, by setting each of the derivatives equal to zero. The line

$$P = 0$$

is both a W nullcline and a P nullcline. This means that all points on the W axis are critical points. When the critical points are connected like this, we cannot think of stability in the usual way. The standard concept of stability applies only to isolated critical points. The only other nullcline is the line

$$P + W = 1,$$

which is a P nullcline. The full nullcline diagram appears in Figure C5.1, including only the realistic portion in the first quadrant. Note that neither K nor P_0 appears in the equations for the nullclines; thus, the model has a unique nullcline diagram. The dots along the W axis indicate that all the points on that line are critical points.

The nullcline diagram gives us the general trend of trajectories. All trajectories progress to the right, so we can obtain all of them by considering solutions that begin on the P axis. Those beginning above $P = 1$ move down and to the right, corresponding to decreasing population and

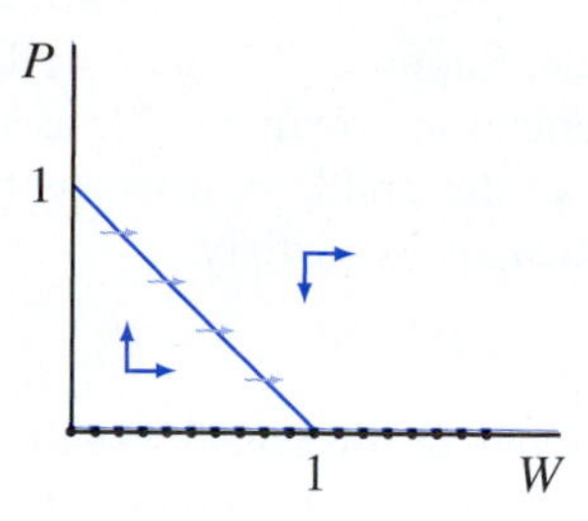

Figure C5.1
The nullcline diagram for $W' = KP$, $P' = P(1 - P - W)$.

increasing waste right from the beginning. Those beginning below $P = 1$ move up and to the right until they reach the line $P + W = 1$, after which they also move down and to the right. The waste increases in all cases. If the starting population is small, it will increase for a time, but eventually it reaches a maximum value and decreases thereafter. The population must approach zero as $t \to \infty$, but it is not clear where on the W axis the trajectories terminate.

The Differential Equation for the Trajectories

Following the usual procedure, we can write the differential equation for the trajectories as

$$\frac{dP}{dW} = \frac{1 - P - W}{K}. \tag{5}$$

This equation is not separable, but it can be solved by a simple change of variables. Let $Z(W)$ be defined by

$$Z = P + W.$$

It can be shown (see Exercise 2) that Z satisfies the differential equation

$$\frac{dZ}{dW} = \frac{1 + K - Z}{K}. \tag{6}$$

In the form (6), the differential equation for the trajectories is autonomous and therefore separable. Before we write the solution of the equation, it is well to note that we can use the phase line to study the autonomous equation for the trajectories. The phase line sketch appears in Figure C5.2.

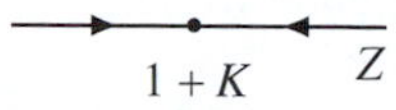

$1 + K$ $\quad Z$

Figure C5.2
The phase line for $dZ/dW = (1 + K - Z)/K$.

If the independent variable were time, the implication of the phase line sketch would be that Z approaches $1 + K$ as time increases without bound. However, the variable W does not increase to infinity, so the best we can say is that Z always changes toward $1 + K$. This is useful information for the trajectories in the WP plane. Suppose Z is initially less than $1 + K$. For the entire life of the population, Z must then remain less than $1 + K$, as indicated by the phase line sketch. Translated to the actual variables of the problem, it means that any trajectories beginning at a point $(0, P_0)$, with $P_0 < 1 + K$, must always satisfy

$$P + W < 1 + K. \tag{7}$$

This result also tells us something about the ultimate fate of the population. The waste W continues to increase, but $P + W$ is bounded by $1 + K$. Eventually, P reaches 0, at which point we have

$$W < 1 + K.$$

Those trajectories that begin at $(0, P_0)$, with $P_0 < 1 + K$ intersect the W axis at a point to the left of $1 + K$.

A Formula for the Trajectories

The solution of the differential equation for the trajectories is

$$Z = 1 + K + Ae^{-W/K}.$$

Since $Z = P + W$, this gives us the result

$$P = 1 + K - W + Ae^{-W/K}. \tag{8}$$

The phase portrait, unlike the nullclines, does depend on the value of the parameter K. We continue now with $K = 0.5$ chosen as an example. With this value for K, the equation for the trajectories is

$$P = 1.5 - W + Ae^{-2W}. \tag{9}$$

The phase portrait appears in Figure C5.3 along with the limiting line $P + W = 1.5$ and the nullcline $P + W = 1$.

Discussion

The problem statement was deliberately open-ended, with the instructions being to "determine the behavior of the system." The analysis used a variety of tools, including nullclines and the differential equation for the trajectories. Analysis of the latter equation required the phase line and symbolic quantitative techniques to obtain the formula for the trajectories. Using a variety of methods has the obvious advantage of giving the largest number of results, but it also has the less obvious advantage of confirming results by consistency. Observe in the phase portrait that the trajectories do indeed cross the nullcline with horizontal slope, and they are indeed bounded above by the inequality $P + W < 1 + K$.

Note that we never solved the original problem of determining the population and waste as functions of time. However, by making use of the autonomy of the model, we have obtained a deep understanding of the history of the self-limiting population represented by the model. The only thing we don't know from our investigation is how rapidly the predicted changes occur.

We might also wonder about the implication of our study for real populations, such as the population of humans on earth. Fortunately, this study says little about the human population. While we create wastes that harm our environment, we live in an environment that is large enough

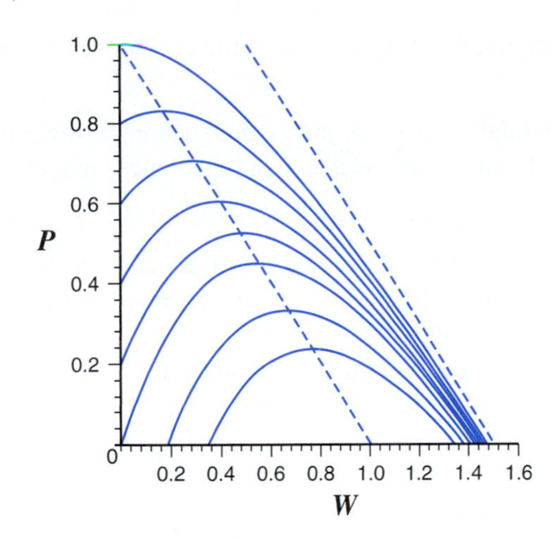

Figure C5.3
The phase portrait for the dimensionless model with $K = 0.5$, along with the nullcline $P + W = 1$ (dashed) and the limiting line $P + W = 1 + K$ (dotted).

and rich enough to have mechanisms for eliminating waste. A more realistic study of human populations would have to include a natural mechanism that decreases the waste.

Case Study 5 Exercises

1. Nondimensionalize the models (1) and (2), using dimensionless variables

$$P = \frac{p}{M}, \qquad W = \frac{bw}{r}, \qquad \tau = rt.$$

Define dimensionless parameters K and P_0 so that you get the dimensionless models (3) and (4). What does K represent?

2. Derive Equation (6) for $Z(W)$.

3. Derive Equation (8) for the trajectories in the WP plane by solving the differential equation (6) for the trajectories in the WZ plane. Note that the equation can be written in the same form as that for Newton's law of cooling from Section 1.1.

T 4. Complete the phase portrait of Figure C5.3 with some trajectories above the line $P + W = 1 + K$.

T 5. Prepare phase portraits with $K = 0.25$ and $K = 1$. Discuss the significance of the parameter K.

6. Results obtained with one tool often suggest investigations to be done with another. For example, we might be interested in knowing the largest population value achieved when the starting population is small. This point can be found in the phase portrait as the intersection of the nullcline with the trajectory that passes through the origin. It can also be found symbolically. Let $K = 0.5$. Set the formula for the trajectory equal to that for the nullcline. Find the waste level at which the maximum population occurs and the corresponding population. How does

the actual maximum population for this case compare with the theoretical limit (that where there is no waste)?

7. Suppose we alter the model to allow for the gradual decrease of waste due to natural environmental renewal. This change results in a new dimensionless model

$$W' = KP - QKW, \qquad P' = P(1 - P - W).$$

   a. Prepare a nullcline diagram for this model. Can any conclusions be drawn from the nullcline sketch regarding the correct classification of the equilibrium points? Does it matter what values are chosen for K and Q?

   b. Take $K = 0.5$ and $Q = 0.5$. Plot the phase portrait for the system. Describe what happens to populations of various initial sizes, assuming there is initially no waste.

chapter

6

Analytical Methods for Systems

We studied symbolic methods for solving homogeneous linear equations in Chapter 3; now we turn our attention to systems of homogeneous linear equations. The methods are similar in that we assume a specific form of solution and use it to reduce the differential equation problem to an algebra problem. However, our algebra problems will be matrix problems rather than scalar problems, and this introduces additional complications.

The chapter begins in Section 6.1 by introducing systems of differential equations and solution techniques without using any tools from linear algebra. Systems arise naturally from problems that can be modeled using the technique of *compartment analysis,* which is introduced in this section.

Section 6.2 begins with a look at *vector spaces* and the study of *homogeneous linear algebraic systems*. These topics are prerequisites to the study of the *eigenvalue problem,* which is the linear algebra problem that arises from the principal method for solving systems of linear algebraic equations. This section serves to provide a minimal knowledge base for students who have not studied linear algebra before and a brief focused review of linear algebra for students who have some background in the subject.[1]

Section 6.3 begins the process of developing the theory and techniques for homogeneous linear systems by appealing to graphical properties of linear systems. Some phase portraits indicate the presence of *linear trajectories* that carry solution curves with simple explicit solution formulas.

Sections 6.4 through 6.6 present the method for solving homogeneous linear systems with constant coefficients. As in the scalar case, the plan is to assume solutions in the form of exponential functions and reduce the differential equation problem to an algebra problem; the difference is that the algebra problem here is the eigenvalue problem. The eigenvalue problem reduces to a problem of finding roots of a *characteristic polynomial,* just as in the scalar case. Sections 6.4, 6.5, and 6.6 consider real roots, complex roots, and repeated roots, respectively.

In Section 6.7, we use the methods for linear systems to study the qualitative behavior of nonlinear systems. The idea is to replace a nonlinear system near a critical point by the best linear approximation to that system. Theorem 6.7.1 indicates the requirements that must be met so that it

[1] See a linear algebra text, such as Keith Nicholson, *Elementary Linear Algebra,* 2nd ed., McGraw-Hill, New York, 2004.

is certain that the behavior of the linear system accurately indicates the behavior of the nonlinear system.

Case Study 6 applies the nonlinear equilibrium point analysis of Section 6.7 to the study of a problem in disease dynamics. After the model is created and used to analyze a stable *endemic* disease, it is modified slightly and used to investigate the problem of first contact between a population in which the disease is established and a population that has not been previously exposed to the disease.

We assume throughout the chapter that matrices have real-valued entries only, unless otherwise noted.

6.1 Compartment Models

Figure 6.1.1 is a schematic diagram depicting the flow of some conserved quantity. As an example, consider the movement of intravenous medication in a human body. The box on the left represents the blood, and the box on the right represents the tissues. The labels inside the boxes indicate that $x(t)$ is the quantity of medication in the blood and $y(t)$ is the quantity of medication in the tissues.[2] Each arrow represents a pathway into the system, out of the system, or between compartments. Medication enters the blood at some rate R. It can move out of each compartment at rates proportional to the amount present. Medication leaves the system through both compartments, either by being flushed out of the blood compartment or by chemical or radioactive breakdown within the blood or tissues. If there is no breakdown of the medication, then $r_2 = 0$.

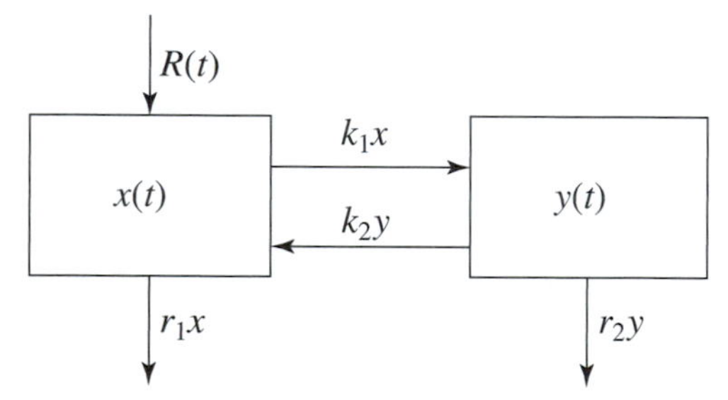

Figure 6.1.1
A schematic diagram of rates of flow in a two-compartment system.

Compartment models such as this are used extensively in models that can be conceptualized as the flow of a quantity between regions. Other examples are pollutants in a lake, chemical reactors, and epidemiology. Sometimes there are more compartments or different sets of connections. Models used for numerical simulations often involve a very large number of compartments.

[2]The study of the interaction of drugs with living systems is called **pharmacokinetics.** In the simplest model, which we are examining here, the liver and kidneys are considered as part of the blood compartment. This is a reasonable assumption, because transfer from blood to these organs is much faster than transfer from blood to other organs.

Derivation of the Model

The quantities r_1, r_2, k_1, and k_2 are rate constants (dimension time^{-1}). The quantity R is an input rate, given in units of the basic quantity per unit time.

The diagram indicates how to write down the differential equations for each of the variables. The equation for y', for example, will have three terms: k_1x is the rate at which y increases by transfer from the other compartment, k_2y is the rate at which y decreases by transfer to the other compartment, and r_2y is the rate at which y is eliminated directly, by whatever mechanism(s) is (are) present. The diagram therefore corresponds to a system of two differential equations:

$$\frac{dx}{dt} = R - r_1x - k_1x + k_2y, \qquad \frac{dy}{dt} = -r_2y + k_1x - k_2y. \tag{1}$$

MODEL PROBLEM 6.1

Suppose the model (1) is used to monitor the levels of a radioactive medicine. The medicine is administered at constant rate R for t_1 hours. Determine the amount of radioactive medicine in the tissues, assuming that the rate at which the medicine is transferred from the blood to the tissues is 4 times the decay rate, and the rate at which the medicine is transferred from the tissues back to the blood is equal to the decay rate. In particular, determine the maximum amount of medicine in the tissues.

We begin by nondimensionalizing the model. There are a number of different choices for the reference quantities. Here, we choose $1/r_1$ for the time scale and R/r_1 for the amounts of medication.[3] The dimensionless quantities are then

$$X = \frac{r_1x}{R}, \qquad Y = \frac{r_1y}{R}, \qquad \tau = r_1t, \qquad T = r_1t_1. \tag{2}$$

With these changes, the model becomes

$$X' = 1 - 5X + Y, \qquad Y' = 4X - 2Y, \qquad 0 < \tau < T, \tag{3}$$

$$X' = -5X + Y, \qquad Y' = 4X - 2Y, \qquad T \le \tau, \tag{4}$$

with the additional requirement that X and Y are both initially zero and must be continuous at $\tau = T$. Note that we have used $k_2 = r_2 = r_1$ and $k_1 = 4r_1$.

✦ INSTANT EXERCISE 1

Use the substitutions of Equation (2) and the model of Equation (1) to derive the dimensionless model of Equations (3) and (4).

---

[3] These represent the time scale of radioactive decay and the amount of medication administered in one unit of dimensionless time.

Graphical Analysis

Figure 6.1.2 is the nullcline diagram for the system of Equation (3) for the region $X, Y \geq 0$ corresponding to the portion of the domain corresponding to the physical setting for the model. There is one equilibrium solution, at $\left(\frac{1}{3}, \frac{2}{3}\right)$. Trajectories in regions A and C can only move to the equilibrium solution, while trajectories in regions B and D must eventually move to region A or C. Thus, the equilibrium solution must be a stable node. The only trajectory that occurs in the model is the one that begins at the initial condition (0, 0). This trajectory moves through region C toward the equilibrium solution. It does not actually reach the equilibrium solution because the system that is sketched in the figure only applies up to time $\tau = T$. After that, the system is given by Equation (4); the nullcline diagram appears in Figure 6.1.3. For this system, the origin is the only equilibrium solution, and it is a stable node.

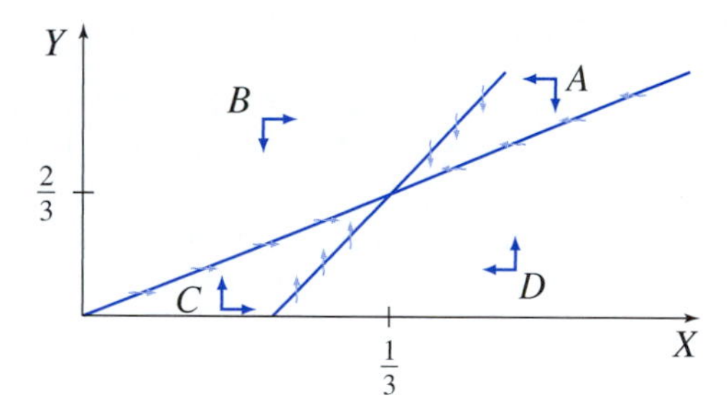

Figure 6.1.2
The nullcline diagram for the system $X' = 1 - 5X + Y$, $Y' = 4X - 2Y$.

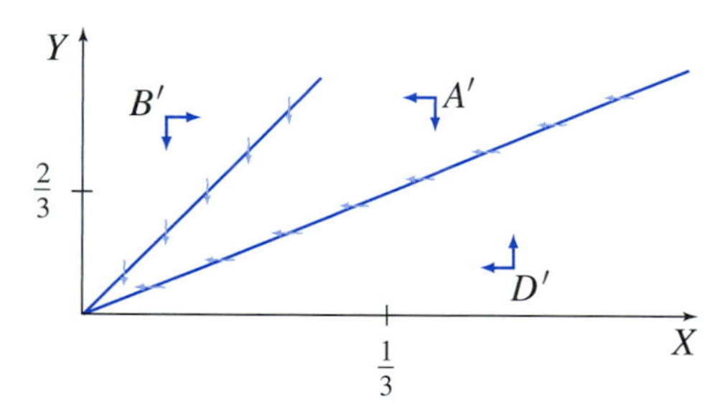

Figure 6.1.3
The nullcline diagram for the system $X' = -5X + Y$, $Y' = 4X - 2Y$.

A number of conclusions can be drawn from the comparative study of Figures 6.1.2 and 6.1.3. As noted before, the trajectory that begins at the origin moves into region C in Figure 6.1.2. It moves toward the equilibrium solution, remaining in the same region. At time T, the nullcline diagram changes to that of Figure 6.1.3. The former region C is wholly contained in region D'. Now the solution of Model Problem 6.1 follows a trajectory that moves from region D' into region A' and on to the origin. Our particular interest is in Y, which represents the amount of medicine in the tissues. This quantity increases throughout the period corresponding to Figure 6.1.2. It continues to increase as long as the trajectory of the solution remains in region D', and it reaches its maximum precisely when the trajectory crosses the Y nullcline into region A'.

Symbolic Analysis

The symbolic analysis of homogeneous linear systems such as that of Equation (4) takes up most of this chapter. A systematic treatment of such problems requires a better background in linear algebra than what has been provided so far. However, autonomous systems of only two linear equations with two unknowns can be recast as second-order linear equations, thereby allowing them to be solved by the methods of Chapters 3 and 4. The idea is to differentiate the differential equation for one of the variables and then eliminate the other variable by substitution. Since our primary interest is in Y, we differentiate the equation for Y, with the result

$$Y'' = 4X' - 2Y'.$$

We now have a second-order equation for Y, but it is still coupled to the equation for X. We have an expression for X' that can be substituted into the new equation, but then we will have (for $\tau < T$)

$$Y'' = 4(1 - 5X + Y) - 2Y',$$

and this equation still contains X. What we really need is an expression for X' that does not include X. To obtain this, we can add 4 times the X' equation of (3) to 5 times the Y' equation of (3). This yields

$$4X' + 5Y' = 4(1 - 5X + Y) + 5(4X - 2Y) = 4 - 6Y,$$

or

$$4X' = 4 - 6Y - 5Y'.$$

Substituting this into $Y'' = 4X' - 2Y'$, after rearrangement, yields

$$Y'' + 7Y' + 6Y = 4, \qquad 0 < \tau < T. \tag{5}$$

The initial conditions are $Y(0) = 0$ and $Y'(0) = 0$, the second following from the initial condition $X(0) = 0$ and the equation for Y'. Similarly, Equations (4) become

$$Y'' + 7Y' + 6Y = 0, \qquad T \le \tau, \tag{6}$$

with initial conditions at $\tau = T$ determined by the solution of Equation (5). The full solution is

$$Y = \frac{2}{15}\begin{cases} 5 - 6e^{-\tau} + e^{-6\tau} & \tau < T \\ 6(e^{T} - 1)e^{-\tau} - (e^{6T} - 1)e^{-6\tau} & \tau > T \end{cases}. \tag{7}$$

✦ INSTANT EXERCISE 2

Obtain the solution (7) by solving Equations (5) and (6).

We can also find the amount of medicine in the blood. The easiest way to compute X is to use the differential equation $Y' = 4X - 2Y$, where Y and Y' are already known. The result is

$$X = \frac{1}{15}\begin{cases} 5 - 3e^{-\tau} - 2e^{-6\tau} & \tau < T \\ 3(e^{T} - 1)e^{-\tau} + 2(e^{6T} - 1)e^{-6\tau} & \tau > T \end{cases}. \tag{8}$$

Finally, the total amount of medicine in the system is given by

$$X + Y = \begin{cases} 1 - e^{-\tau} & \tau < T \\ e^{T-\tau} - e^{-\tau} & \tau > T \end{cases}. \tag{9}$$

The results appear in Figure 6.1.4, with the administration time chosen to be twice the unit time.

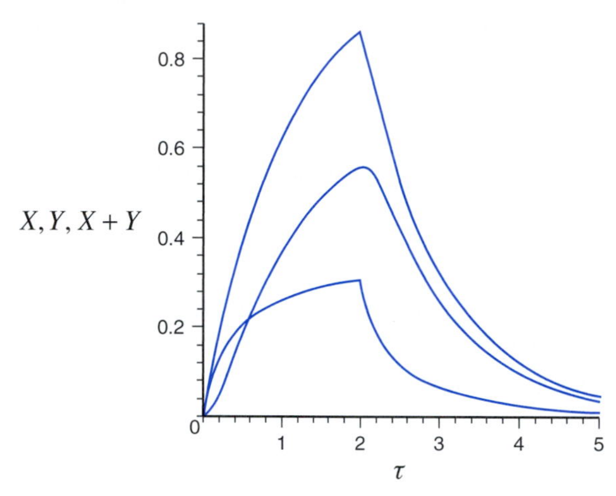

Figure 6.1.4
The amounts of medicine in the blood, tissue, and system for Model Problem 6.1, with $T = 2$.

✦ INSTANT EXERCISE 3

Determine which curve is x and which is y by looking at the qualitative properties of the curves.

Combining the Results

One benefit of having a variety of techniques for studying differential equations is that the results can be combined to get more information and confirmation of results obtained by one of the techniques. In Model Problem 6.1, nullcline analysis produced some predictions for the behavior of the solution trajectory in the phase plane. The solution formulas for X and Y can be used to plot that trajectory. Figure 6.1.5 shows the solution in the XY plane along with the nullclines for both systems. The quantitative results corroborate the conclusions drawn from the nullclines. The solution curve is initially to the right of the Y nullcline and to the left of the X nullcline, and so both X and Y increase in the first phase of the problem. When the administration of medication is stopped, the X nullcline suddenly moves to the other side of the solution trajectory and X immediately begins to decrease. The trajectory is still to the right of the Y nullcline, so Y continues to increase. The increase in Y continues until the solution trajectory reaches the Y nullcline. The solution trajectory crosses the Y nullcline shortly after the administration ends, and both X and Y decrease from that point on toward the stable equilibrium solution at the origin.

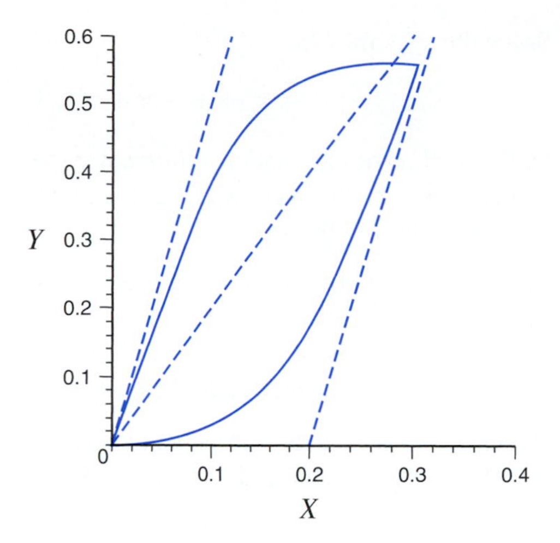

Figure 6.1.5
The solution trajectory and nullclines (dashed) for Model Problem 6.1, with $T = 2$.

6.1 Exercises

In Exercises 1 through 2, solve the indicated problem in the simplest manner (without any second-order equations).

1. Solve the system $x' = 3x$, $y' = x - 2y$.

2. Solve the system $x' = 2x + 3y$, $y' = -2y$.

In Exercises 3 through 8, (a) solve the indicated problem by the elimination method, as in Model Problem 6.1, and (b) plot the solutions in the phase plane along with the nullclines.

3. $x' = -3x - y, \quad y' = x - y$

4. $x' = -2x - y, \quad y' = 4x - 6y$

5. $x' = -2x + y - 1, \quad y' = 2x - y$

6. $x' = x - 2y + 3, \quad y' = -2x + y$

7. $x' = -2x - 3y, \quad y' = 3x - 2y, \quad x(0) = 4, \quad y(0) = 2$

8. $x' = x - 4y, \quad y' = 2x - 3y, \quad x(0) = 0, \quad y(0) = -2$

In Exercises 9 through 12, solve the system by eliminating one of the variables, as in Model Problem 6.1. Note that it may be easier to eliminate one of the variables than the other.

9. $x' = 2y + \cos 3t, \quad y' = -x + 3y$

10. $x' = x - 6y, \quad y' = -x + e^t$

11. $x' = -x + 3y - e^{-2t}, \quad y' = x + y$

12. $x' = -2x + y - e^{-t}, \quad y' = x - 2y$

13. Solve the system

$$x'' + y' - x + y = 1, \qquad x' + y' - x = 0,$$

by the method of elimination. (*Hint:* since the system is second-order in x and first-order in y, it seems reasonable to guess that elimination will result in a third-order differential equation for one of the variables.)

14. Sometimes it is possible to solve a system by defining an appropriate linear combination of the variables. Let $Z(\tau) = X(\tau) + Y(\tau)$ in Model Problem 6.1. Add the differential equations together to get differential equations for Z for $\tau < T$ and $\tau > T$. Solve these first-order equations and verify the result of Equation (9).

T 15. Repeat the solution and subsequent analysis of Model Problem 6.1 for the case $r_2 = 0$, with all other details unchanged. Plot X and Y. How does this change affect the results?

T 16. Repeat the solution and subsequent analysis of Model Problem 6.1 for the case $k_2 = r_1/10$, with all other details unchanged. Plot X and Y. How does this change affect the results?

Exercises 17 and 18 use the compartment model illustrated in Figure 6.1.6 to study the buildup of lead in the human body. Two different approaches are taken, and it is particularly revealing to compare and contrast the results (Exercise 19). The bones are included in the model as a separate compartment because lead interacts differently with bone as compared with other tissues.

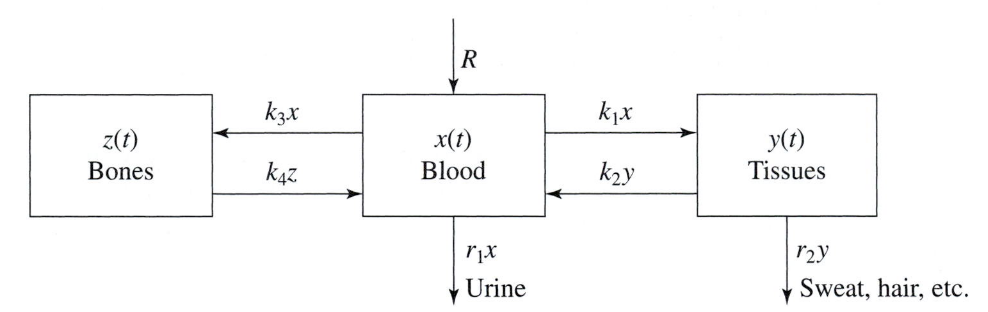

Figure 6.1.6
Lead transport in a vertebrate body.

Rabinowitz, Wetherill, and Kopple[4] collected the following data from a controlled study of an otherwise healthy volunteer, with R in micrograms per day and the various rate constants in days^{-1}:

$$R = 49.3, \qquad r_1 = 0.0211, \qquad r_2 = 0.0162,$$

$$k_1 = 0.0111, \qquad k_2 = 0.0124, \qquad k_3 = 0.0039, \qquad k_4 = 0.000035.$$

T 17. *a.* Write down the system of differential equations corresponding to the compartment diagram.

[4] *Science*, vol. 182, pp. 725–727, 1993.

b. Determine the equilibrium solution for the model.

c. Use a numerical differential equation solver to solve the system on the interval $0 < t < 40{,}000$ with all variables initially zero. Plot all three variables for $0 < t < 400$ on one graph, and plot the lead content of the bones for $0 < t < 40{,}000$.

d. Describe the results of parts *b* and *c* in words. Your description of the results should include general features of the results and should also indicate approximately how long it takes for the lead amounts in each compartment to level off.

e. Discuss the results of parts *b* and *c*. Your discussion should focus on explaining the differences between the results for the blood and tissues and the results for the bones. In particular, what is it about the compartment structure and the parameter values that accounts for the differences?

f. Use the symbolic differential equation solver of a computer algebra system to repeat part *c*. If there are differences in the graphs, which graphs are correct? (*Hint:* compare with the result of part *b*.)

18. Do all parts *a* through *g* if this exercise is not being done with Exercise 17; if you are also doing Exercise 17, do only parts *a* through *e*.

a. Write down the system of differential equations corresponding to the compartment diagram. Approximate the system by removing the term $k_4 z$ from the x equation. Explain why it is a reasonable guess that this term is unimportant.

b. Determine the equilibrium solution for the system of part *a*.

c. The z equation is now decoupled from the xy subsystem. Determine the solution for x, assuming all variables are initially zero, by eliminating y from the xy subsystem.

d. Use the solution for x and the initial condition $z(0) = 0$ to obtain a solution for z.

e. Plot all three variables for $0 < t < 400$ on one graph, and plot the lead content of the bones for $0 < t < 40{,}000$ on a second graph.

f. Describe the results of parts *b* through *e* in words. Your description of the results should include general features of the results and should also indicate approximately how long it takes for the lead amounts in each compartment to level off.

g. Discuss the results of parts *b* through *e*. Your discussion should focus on explaining the differences between the results for the blood and tissues and the results for the bones. In particular, what is it about the compartment structure and the parameter values that accounts for the differences?

19. Discuss the advantages and disadvantages of the approaches used in Exercises 17 and 18. Consider not just the differences in the results, but also the ease of solution and the reliability of the numerical data.

✦ 6.1 INSTANT EXERCISE SOLUTIONS

1. We first make the substitutions $r_2 = r_1$, $k_2 = r_1$, and $k_1 = 4r_1$ to obtain the system

$$\frac{dx}{dt} = R - 5r_1x + r_1y, \qquad \frac{dy}{dt} = 4r_1x - 2r_1y.$$

Next, we replace x and y by RX/r_1 and RY/r_1, respectively. These algebraic substitutions change the system to the form

$$\frac{1}{r_1}\frac{dX}{dt} = 1 - 5X + Y, \qquad \frac{1}{r_1}\frac{dY}{dt} = 4X - 2Y,$$

where we have removed the extra factor of R from all terms in both equations. The substitution $\tau = r_1 t$ corresponds to $d/dt = (d\tau/dt)(d/d\tau) = r_1(d/d\tau)$; hence,

$$\frac{dX}{d\tau} = 1 - 5X + Y, \qquad \frac{dY}{d\tau} = 4X - 2Y.$$

For $\tau > T$, we have only to drop the nonhomogeneous term from the X equation.

2. For both Equations (5) and (6), we have the same complementary solution; however, they need to be written with different names for the parameters since the values of the parameters can be different. Thus, we have

$$Y_c = c_1 e^{-\tau} + c_2 e^{-6\tau}, \qquad \tau < T, \qquad \text{and} \qquad Y = c_3 e^{-\tau} + c_4 e^{-6\tau}, \qquad \tau > T.$$

The particular solution, by the method of undetermined coefficients, is a constant. The substitution $Y = A$ yields the result $6A = 4$, so $A = \frac{2}{3}$. Combining the complementary and particular solutions, we have the general solutions

$$Y = \tfrac{2}{3} + c_1 e^{-\tau} + c_2 e^{-6\tau}, \qquad \tau < T, \qquad \text{and} \qquad Y = c_3 e^{-\tau} + c_4 e^{-6\tau}, \qquad \tau > T.$$

The initial conditions $Y(0) = 0$ and $Y'(0) = 0$ yield a pair of algebraic equations for c_1 and c_2,

$$c_1 + c_2 = -\tfrac{2}{3}, \qquad -c_1 - 6c_2 = 0.$$

These can be solved in the standard manner by constructing an augmented matrix and row-reducing it, as in Section 3.2; but it is simpler just to notice that adding the equations eliminates c_1. Thus, $-5c_2 = -\frac{2}{3}$. From this, we obtain $c_2 = \frac{2}{15}$, and then $c_1 = -6c_2$. This completes the solution for the period $\tau < T$:

$$Y = \tfrac{2}{15}(5 - 6e^{-\tau} + e^{-6\tau}),$$

where we have removed the common factor to avoid having fractions or decimals in the sum. The remaining constants are determined by using the requirement that both Y and Y' be continuous at $\tau = T$. Thus,

$$\tfrac{2}{15}(5 - 6e^{-T} + e^{-6T}) = c_3 e^{-T} + c_4 e^{-6T}$$

and

$$\tfrac{2}{15}(6e^{-T} - 6e^{-6T}) = -c_3 e^{-T} - 6c_4 e^{-6T}.$$

As before, we can determine c_4 by adding the equations and then determine c_3, with the final result given by Equation (7).

3. The function Y has a continuous derivative because the differential equation for Y has a continuous right-hand side. Hence, the graph of Y looks smooth. The graph of X has a sharp corner at $\tau = T$ when the administration is stopped.

6.2 Eigenvalues and Eigenspaces

In Section 3.2, we discussed the theory and solution techniques for nonhomogeneous vector algebraic equations $\mathbf{Ax} = \mathbf{b}$ with $\mathbf{A}$ nonsingular, and we learned techniques for calculating determinants. These topics from linear algebra were necessary to understand the theory of homogeneous linear differential equations and to efficiently determine the values of the integration constants from initial conditions. Most of this chapter is devoted to the theory and techniques for solving systems of linear differential equations. This topic requires more background in linear algebra than what has been presented so far. In particular, understanding of and computational facility with *eigenvalues*[5] and *eigenspaces* are prerequisite to any study of solutions of systems of linear differential equations. This section begins with two topics that are needed as preliminary background for this task. The first of these is the notion of a *vector space,* which applies to solution sets of homogeneous linear differential equations and systems as well as to geometric vectors. The second is the theory and solution techniques for homogeneous vector algebraic equations in which the matrix is singular. These preliminary topics are then applied to the central problem of finding eigenvalues and characterizing eigenspaces.

In this section and throughout the chapter, matrices will be assumed to have only real-valued entries unless specifically stated otherwise.

Vector Spaces

Let V be the set of all vectors in three-dimensional space. One can think of these vectors as arrows with magnitude and direction or as sets of three numbers indicating the components of the vectors in the three mutually perpendicular directions defined by the x, y, and z axes. Along with the vectors themselves, consider the usual operations of addition of vectors and multiplication by real numbers (called **scalars** in the context of vector algebra). Any two vectors can be added by adding the components, and any vector can be multiplied by any scalar componentwise, in each case the result being a vector. The set V and the operations of addition and scalar multiplication have the following fundamental properties:

1. $\mathbf{u} + \mathbf{v} \in V$ and $c\mathbf{u} \in V$ for all $\mathbf{u}, \mathbf{v} \in V$ and scalars c.
2. $\mathbf{u} + \mathbf{v} = \mathbf{v} + \mathbf{u}$ for all $\mathbf{u}, \mathbf{v} \in V$.
3. $\mathbf{u} + (\mathbf{v} + \mathbf{w}) = (\mathbf{u} + \mathbf{v}) + \mathbf{w}$ for all $\mathbf{u}, \mathbf{v}, \mathbf{w} \in V$.
4. There is a vector $\mathbf{0}$ with the property $\mathbf{u} + \mathbf{0} = \mathbf{u}$ for all $\mathbf{u} \in V$.
5. Each vector $\mathbf{u} \in V$ has a corresponding vector $-\mathbf{u} \in V$ such that $\mathbf{u} + (-\mathbf{u}) = \mathbf{0}$.
6. $1\mathbf{u} = \mathbf{u}$ for all $\mathbf{u} \in V$.
7. $a(b\mathbf{u}) = (ab)\mathbf{u}$ for all $\mathbf{u} \in V$ and scalars a and b.
8. $(a + b)\mathbf{u} = a\mathbf{u} + b\mathbf{u}$ for all $\mathbf{u} \in V$ and scalars a and b.
9. $a(\mathbf{u} + \mathbf{v}) = a\mathbf{u} + a\mathbf{v}$ for all $\mathbf{u}, \mathbf{v} \in V$ and scalars a.

Linear algebra is much more than the study of geometric vectors. Many other mathematical settings consist of some set with operations of addition and scalar multiplication that satisfy all nine fundamental properties of vector algebra. The algebra of these other settings is identical to the algebra of geometric vectors.

---

[5]The correct pronunciation is "eye-genvalues," with the stress on the first syllable and a hard g as in *grow*.

A **vector space** is any set on which there are operations of addition and scalar multiplication that satisfy the nine fundamental properties of vector algebra. The elements of the set are, in the context of the vector space, called **vectors.** The scalars are either the real numbers or the complex numbers, depending on the setting.

EXAMPLE 1

Let w, x, y, and z be any functions of a real variable defined for all t on some particular interval and having derivatives of all orders on that interval. Let the vector $\mathbf{x}$ be defined by

$$\mathbf{x} = \begin{pmatrix} w \\ x \\ y \\ z \end{pmatrix}.$$

The set V consisting of all such vectors $\mathbf{x}$ is a vector space using the standard addition and scalar multiplication of functions.

The set of all solutions to a given homogeneous linear differential equation is a vector space, called the **solution space** of the differential equation.

EXAMPLE 2

Let V be the set of all solutions of the differential equation $y'' - y = 0$. Define addition and scalar multiplication of these "vectors," using the standard algebraic operations. The elements of the set V are all the functions of the form $y = c_1e^t + c_2e^{-t}$. Since linear combinations of solutions are also solutions, the operations of addition and scalar multiplication are well defined on the set. All the necessary properties are satisfied (see Exercise 21), so V is a vector space.

The solution space in Example 2 consists of all the linear combinations of the vectors e^t and e^{-t}. Stated otherwise, the solution space of the equation $y'' - y = 0$ is the span[6] of vectors e^t and e^{-t}. The **dimension** of a vector space is the smallest number of vectors that are needed to span the space; the solution space in Example 2 is of dimension 2 because it cannot be spanned by a single vector. In general, a homogeneous nth-order linear differential equation with continuous coefficients has (by Theorem 2.4.2) a solution space of dimension n.

Note that the solution set of a nonhomogeneous equation cannot be a vector space because it does not contain the zero function.

The Algebraic Equation $\mathbf{Ax} = \mathbf{0}$

The algebraic equation $\mathbf{Ax} = \mathbf{b}$ arises in the problem of determining what values the parameters in a general solution should take in order to satisfy initial conditions and was discussed in that

[6]See Section 3.2.

context in Section 3.2. The essential fact in that case is that a unique solution for $\mathbf{x}$ is guaranteed whenever the matrix $\mathbf{A}$ is nonsingular. The equation $\mathbf{Ax} = \mathbf{b}$ also arises in the satisfaction of initial conditions for systems.

Now we consider in detail the special case of the homogeneous linear system[7]

$$\mathbf{Ax} = \mathbf{0}. \tag{1}$$

Consider the analogous scalar equation $ax = 0$. If we think of a as a known scalar and x as an unknown scalar, then the solution depends on whether or not $a = 0$. If $a \neq 0$, then $x = 0$ is the only solution. If $a = 0$, then all numbers x are solutions. Returning to the vector equation (1), we see that the vector $\mathbf{0}$ is always a solution, and all vectors $\mathbf{x}$ are solutions if $\mathbf{A}$ is a matrix consisting of all zeros. In the scalar case, either a or x must be zero. Are there any other possibilities for the vector case?

Suppose we have

$$\mathbf{A} = \begin{pmatrix} 1 & 0 \\ 3 & 0 \end{pmatrix}, \quad \mathbf{B} = \begin{pmatrix} 1 & -1 \\ -2 & 2 \end{pmatrix}, \quad \mathbf{x} = \begin{pmatrix} 0 \\ 1 \end{pmatrix}, \quad \mathbf{y} = \begin{pmatrix} 1 \\ 1 \end{pmatrix}.$$

Notice that the products $\mathbf{Ax}$ and $\mathbf{By}$ are $\mathbf{0}$, but the products $\mathbf{Ay}$ and $\mathbf{Bx}$ are not. Obviously, it is possible for a product of a matrix and a vector to be $\mathbf{0}$ *even if both the matrix and the vector are nonzero*. This happens for some nonzero vectors whenever the matrix $\mathbf{A}$ is singular. As in Theorem 3.2.2, it is convenient to identify singular matrices as matrices whose determinant is 0. We have the following result:

Theorem 6.2.1 The equation $\mathbf{Ax} = \mathbf{0}$ has solutions $\mathbf{x} \neq \mathbf{0}$ if and only if $\det(\mathbf{A}) = 0$.

EXAMPLE 3

Suppose we want to find all solutions of the equation $\mathbf{Ax} = \mathbf{0}$ where $\mathbf{A}$ is given by

$$\mathbf{A} = \begin{pmatrix} -1 & -2 \\ 2 & 2 \end{pmatrix}.$$

The determinant of $\mathbf{A}$ is not zero ($\mathbf{A}$ is nonsingular), so the only solution is $\mathbf{x} = \mathbf{0}$.

Our interest in the equation $\mathbf{Ax} = \mathbf{0}$ is twofold. We have now determined when the equation has nonzero solutions; it remains to find those solutions. An example suffices to illustrate the method.

EXAMPLE 4

Suppose we want to find all solutions of the equation $\mathbf{Ax} = \mathbf{0}$ where $\mathbf{A}$ is given by

$$\mathbf{A} = \begin{pmatrix} 1 & -2 \\ -2 & 4 \end{pmatrix}.$$

[7]Note that the quantity on the right is the *vector* zero, not the scalar zero.

This time, $\det(\mathbf{A}) = 0$, so there are nontrivial solutions. Suppose $\mathbf{x} = \begin{pmatrix} a \\ b \end{pmatrix}$ is such a solution. What restrictions must be placed on a and b? Substituting $\mathbf{x} = \begin{pmatrix} a \\ b \end{pmatrix}$ into $\mathbf{Ax} = \mathbf{0}$, we have

$$\begin{pmatrix} 0 \\ 0 \end{pmatrix} = \mathbf{0} = \mathbf{Ax} = \begin{pmatrix} 1 & -2 \\ -2 & 4 \end{pmatrix} \begin{pmatrix} a \\ b \end{pmatrix} = \begin{pmatrix} a - 2b \\ -2a + 4b \end{pmatrix}.$$

The vector equation corresponds to two scalar equations

$$a - 2b = 0, \qquad -2a + 4b = 0.$$

These equations are redundant, precisely because $\mathbf{A}$ is singular. In effect, there is only one equation rather than two. Any pair of numbers (a, b) satisfies both equations as long as $a = 2b$. The vector $\mathbf{x}_1 = \begin{pmatrix} 2 \\ 1 \end{pmatrix}$ is such a vector, and so are $\mathbf{x} = 2\mathbf{x}_1 = \begin{pmatrix} 4 \\ 2 \end{pmatrix}$ and $\mathbf{x} = -\mathbf{x}_1 = \begin{pmatrix} -2 \\ -1 \end{pmatrix}$. Clearly, any multiple of the vector $\mathbf{x}_1$ is a solution of $\mathbf{Ax} = \mathbf{0}$. The set of solutions of $\mathbf{Ax} = \mathbf{0}$ is the one-parameter family

$$\mathbf{x} = \begin{pmatrix} 2b \\ b \end{pmatrix} = b \begin{pmatrix} 2 \\ 1 \end{pmatrix}.$$

The set of solutions of an equation $\mathbf{Ax} = \mathbf{0}$ with $\mathbf{A}$ singular is a vector space called the **nullspace** of the matrix $\mathbf{A}$. In Example 4, the nullspace of $\mathbf{A}$ has dimension 1. In contrast, the nullspace of the matrix $\mathbf{A} = \begin{pmatrix} 0 & 0 \\ 0 & 0 \end{pmatrix}$ is of dimension 2 because all 2-vectors are in the nullspace.

✦ INSTANT EXERCISE 1

Find all solutions of the equation $\mathbf{Ax} = \mathbf{0}$ where

$$\mathbf{A} = \begin{pmatrix} 1 & 2 \\ 2 & 4 \end{pmatrix}.$$

The Eigenvalue Problem

We turn now to the principal topic of this section.

Given a matrix $\mathbf{A}$, the **eigenvalue problem** is the problem of finding scalars λ and nonzero vectors $\mathbf{x}$ such that

$$\mathbf{Ax} = \lambda\mathbf{x}. \tag{2}$$

The scalars λ are called **eigenvalues;** given an eigenvalue λ, the vectors $\mathbf{x}$ that solve $\mathbf{Ax} = \lambda\mathbf{x}$ are called **eigenvectors corresponding to** λ (simply **eigenvectors** whenever λ is clear from the context). The set of all eigenvectors corresponding to an eigenvalue λ for a matrix $\mathbf{A}$ is a vector space called the **eigenspace corresponding to λ of A.**

A geometric interpretation of $\mathbf{Ax} = \lambda\mathbf{x}$ is useful. Suppose $\mathbf{x}$ is a geometric vector in three-dimensional space. Then both $\mathbf{Ax}$ and $\lambda\mathbf{x}$ are also vectors in three-dimensional space, with $\lambda\mathbf{x}$ the set of all vectors parallel to $\mathbf{x}$. If $\mathbf{Ax} = \lambda\mathbf{x}$, then $\mathbf{Ax}$ is parallel to $\mathbf{x}$. Vectors are usually rotated when they are multiplied by a matrix, but here multiplication by $\mathbf{A}$ does not rotate the vector. In geometric terms, we seek vectors that are not rotated upon multiplication by $\mathbf{A}$.

MODEL PROBLEM 6.2

Determine the eigenvalues and eigenspaces for the matrix

$$\mathbf{A} = \begin{pmatrix} -1 & -2 \\ -2 & 2 \end{pmatrix}.$$

It is harder to make algebraic sense out of Equation (2), which we need to do in order to discover a solution method, but it can be done with a change in notation. Since $\mathbf{Ix} = \mathbf{x}$, Equation (2) can be written as

$$\mathbf{Ax} = \lambda\mathbf{Ix}. \tag{3}$$

This equation requires vectors $\mathbf{x}$ and scalars λ for which multiplication of $\mathbf{x}$ by $\mathbf{A}$ and $\lambda\mathbf{I}$ results in the same vector. This does not mean that $\mathbf{A}$ and $\lambda\mathbf{I}$ are the same matrix, because the equality of $\mathbf{Ax}$ and $\lambda\mathbf{Ix}$ only happens for vectors $\mathbf{x}$ that are eigenvectors corresponding to the eigenvalue λ. More progress is made by rewriting Equation (3) as

$$\mathbf{Ax} - \lambda\mathbf{Ix} = \mathbf{0}.$$

Now we can factor out the vector and get

$$(\mathbf{A} - \lambda\mathbf{I})\mathbf{x} = \mathbf{0}. \tag{4}$$

This is exactly the problem we considered in the previous subsection, but with a one-parameter family of matrices $\mathbf{A} - \lambda\mathbf{I}$ instead of a single matrix $\mathbf{A}$. We want the product $(\mathbf{A} - \lambda\mathbf{I})\mathbf{x}$ to be $\mathbf{0}$ without having $\mathbf{x} = \mathbf{0}$, and this is possible if and only if $\mathbf{A} - \lambda\mathbf{I}$ is singular. Thus, the eigenvalues of $\mathbf{A}$ are the solutions of the equation

$$\det(\mathbf{A} - \lambda\mathbf{I}) = 0. \tag{5}$$

The eigenspace corresponding to an eigenvalue λ is simply the nullspace of the matrix $\mathbf{A} - \lambda\mathbf{I}$; hence, the eigenspace for any eigenvalue can be determined by the same method as in Example 4.

For Model Problem 6.2, we have

$$\det(\mathbf{A} - \lambda\mathbf{I}) = \begin{vmatrix} -1-\lambda & -2 \\ -2 & 2-\lambda \end{vmatrix} = (-1-\lambda)(2-\lambda) - 4 = \lambda^2 - \lambda - 6 = (\lambda - 3)(\lambda + 2).$$

Thus, the eigenvalues are 3 and -2. We find the eigenspace by solving the equation

$$(\mathbf{A} - \lambda\mathbf{I})\mathbf{x} = \mathbf{0}$$

for each of the eigenvalues. For $\lambda = 3$, we get

$$\begin{pmatrix} -4 & -2 \\ -2 & -1 \end{pmatrix}\begin{pmatrix} a \\ b \end{pmatrix} = \begin{pmatrix} 0 \\ 0 \end{pmatrix}.$$

The scalar components of this matrix equation are redundant, so any vector $\begin{pmatrix} a \\ b \end{pmatrix}$ that satisfies $-2a - b = 0$ is an eigenvector corresponding to the eigenvalue $\lambda = 3$. The eigenspace corresponding to $\lambda = 3$ consists of all the vectors that can be written as $c_1\begin{pmatrix} 1 \\ -2 \end{pmatrix}$ for some c_1. The choice of the eigenvector $\begin{pmatrix} 1 \\ -2 \end{pmatrix}$ is arbitrary, subject to the requirement $-2a - b = 0$; we could equally well have written the eigenspace as the set of vectors that can be written as $c_1\begin{pmatrix} -1 \\ 2 \end{pmatrix}$ for some c_1, or we could have used many other formulas.

It is customary to use language that is a little sloppy to summarize the results of the eigenvalue problem. The statement "$\mathbf{x}_1$ is the eigenvector corresponding to λ_1" is typically used in place of the more precise statement "The eigenspace corresponding to λ_1 is the set of vectors $c_1\mathbf{x}_1$."

✦ INSTANT EXERCISE 2

Find the eigenspace corresponding to $\lambda = -2$ for the matrix of Model Problem 6.2.

Eigenvalues and Eigenspaces for Matrices of Higher Dimension

Up to this point, we have solved the eigenvalue problem only for 2×2 matrices. The process is technically more difficult for matrices of higher dimension, although it follows the same logical plan.

EXAMPLE 5

We consider the problem of finding eigenvalues and eigenspaces for the matrix

$$\mathbf{A} = \begin{pmatrix} 0 & 1 & -1 \\ 1 & 1 & 0 \\ -1 & 0 & 1 \end{pmatrix}.$$

We therefore must solve $\det(\mathbf{A} - \lambda\mathbf{I}) = 0$ with

$$\mathbf{A} - \lambda\mathbf{I} = \begin{pmatrix} -\lambda & 1 & -1 \\ 1 & 1-\lambda & 0 \\ -1 & 0 & 1-\lambda \end{pmatrix}.$$

Using the third row for cofactor expansion yields

$$\det(\mathbf{A} - \lambda\mathbf{I}) = \begin{vmatrix} -\lambda & 1 & -1 \\ 1 & 1-\lambda & 0 \\ -1 & 0 & 1-\lambda \end{vmatrix} = -1\begin{vmatrix} 1 & -1 \\ 1-\lambda & 0 \end{vmatrix} + (1-\lambda)\begin{vmatrix} -\lambda & 1 \\ 1 & 1-\lambda \end{vmatrix}$$

$$= -(1-\lambda) + (1-\lambda)(\lambda^2 - \lambda - 1).$$

At this point, note that the factor $1 - \lambda$ is common to both terms. Factoring it out, we have

$$\det(\mathbf{A} - \lambda\mathbf{I}) = (1-\lambda)(\lambda^2 - \lambda - 2) = (1-\lambda)(\lambda - 2)(\lambda + 1).$$

Thus, the eigenvalues are $\lambda_1 = 2$, $\lambda_2 = 1$, and $\lambda_3 = -1$. Now consider the eigenspace corresponding to λ_1. Suppose $\mathbf{x} = \begin{pmatrix} a \\ b \\ c \end{pmatrix}$ is an eigenvector. Then

$$(\mathbf{A} - \lambda_1 \mathbf{I})\mathbf{x} = \begin{pmatrix} -2 & 1 & -1 \\ 1 & -1 & 0 \\ -1 & 0 & -1 \end{pmatrix} \begin{pmatrix} a \\ b \\ c \end{pmatrix} = \begin{pmatrix} 0 \\ 0 \\ 0 \end{pmatrix}.$$

These equations are necessarily redundant because the given value of λ is an eigenvalue of $\mathbf{A}$. A systematic way to proceed is to use the row reduction method.[8] There is no harm in row-reducing the matrix $\mathbf{A} - \lambda\mathbf{I}$ rather than the augmented matrix for the system, because the right-hand portion of the augmented matrix is always a column of zeros. Thus,

$$\mathbf{A} - \lambda_1 \mathbf{I} = \begin{pmatrix} -2 & 1 & -1 \\ 1 & -1 & 0 \\ -1 & 0 & -1 \end{pmatrix} \cong \begin{pmatrix} 1 & -1 & 0 \\ -2 & 1 & -1 \\ -1 & 0 & -1 \end{pmatrix} \cong \begin{pmatrix} 1 & -1 & 0 \\ 0 & -1 & -1 \\ 0 & -1 & -1 \end{pmatrix} \cong \begin{pmatrix} 1 & 0 & 1 \\ 0 & 1 & 1 \\ 0 & 0 & 0 \end{pmatrix}.$$

Row reduction yields a system of two equations,

$$a + c = 0, \qquad b + c = 0,$$

that is equivalent to the original system. Eigenvectors corresponding to λ_1 must have $a = -c$ and $b = -c$; thus, the eigenspace corresponding to λ_1 is the set of vectors

$$\mathbf{x} = \begin{pmatrix} -c \\ -c \\ c \end{pmatrix} = c \begin{pmatrix} -1 \\ -1 \\ 1 \end{pmatrix}.$$

✦ INSTANT EXERCISE 3

The examples treated so far indicate the nature of the equation $\det(\mathbf{A} - \lambda\mathbf{I}) = 0$. Determine the eigenspaces from Example 5 corresponding to the eigenvalues λ_2 and λ_3.

Theorem 6.2.2

If $\mathbf{A}$ is an $n \times n$ matrix of constants, then there exists a polynomial P_A of degree n such that $\det(\mathbf{A} - \lambda\mathbf{I}) = P_A(\lambda)$.

The polynomial P_A of Theorem 6.2.2 is called the **characteristic polynomial** for the matrix $\mathbf{A}$, and the equation $P_A(\lambda) = 0$ is called the **characteristic equation** for $\mathbf{A}$. The eigenvalue problem ultimately reduces to the problem of finding roots of a polynomial.

The calculations required to find eigenvalues of a 2×2 matrix are never difficult, inasmuch as quadratic equations are easily solved with the quadratic formula. There is a possibility that the roots will be complex, a case we defer to Section 6.5. The eigenvalue problem for a 3×3 matrix leads to a cubic equation. In general, the roots of such equations are best approximated by numerical methods, although occasionally, as in Example 2, one can find the roots by identifying a common factor.

[8] See Section 3.2.

Multiple Eigenvalues and Deficient Matrices

EXAMPLE 6

Consider the matrix

$$\mathbf{A} = \begin{pmatrix} -5 & -3 & -3 \\ 3 & 1 & 3 \\ 6 & 6 & 4 \end{pmatrix}.$$

The characteristic polynomial is given by

$$P_A(\lambda) = \begin{vmatrix} -5-\lambda & -3 & -3 \\ 3 & 1-\lambda & 3 \\ 6 & 6 & 4-\lambda \end{vmatrix}.$$

Given the difficulty of solving cubic polynomial equations, it is best to proceed carefully in the hope of finding ways to simplify the result. Using cofactor expansion across the first row, we have

$$P_A(\lambda) = (-5-\lambda)\begin{vmatrix} 1-\lambda & 3 \\ 6 & 4-\lambda \end{vmatrix} + 3\begin{vmatrix} 3 & 3 \\ 6 & 4-\lambda \end{vmatrix} - 3\begin{vmatrix} 3 & 1-\lambda \\ 6 & 6 \end{vmatrix}.$$

Observe that the first term of the expansion will yield a cubic polynomial, while the remaining two terms will be linear functions of λ. In the next step, we combine and simplify the linear second and third terms and the determinant in the first term, but we postpone multiplying out the cubic polynomial:

$$\begin{aligned} P_A(\lambda) &= (-5-\lambda)(\lambda^2 - 5\lambda - 14) + 3(-3\lambda - 6) - 3(6\lambda + 12) \\ &= \cdots = (-5-\lambda)(\lambda^2 - 5\lambda - 14) + (-27\lambda - 54). \end{aligned}$$

If we are going to be able to guess a factor, it will be easiest to do it at this stage. The linear term would be zero if $\lambda = -2$; a quick check shows that $\lambda = -2$ also makes the first term 0. Thus, we can remove a common factor of $\lambda + 2$, leaving a quadratic factor:

$$\begin{aligned} P_A(\lambda) &= -(\lambda+5)(\lambda+2)(\lambda-7) - 27(\lambda+2) = -(\lambda+2)[(\lambda+5)(\lambda-7) + 27] \\ &= -(\lambda+2)(\lambda^2 - 2\lambda - 8) = -(\lambda+2)^2(\lambda-4). \end{aligned}$$

Thus, we have eigenvalues $\lambda = -2$ and $\lambda = 4$.

The eigenspace corresponding to $\lambda = -2$ consists of vectors satisfying the equation

$$\begin{pmatrix} -3 & -3 & -3 \\ 3 & 3 & 3 \\ 6 & 6 & 6 \end{pmatrix}\begin{pmatrix} a \\ b \\ c \end{pmatrix} = \begin{pmatrix} 0 \\ 0 \\ 0 \end{pmatrix}.$$

Since

$$\begin{pmatrix} -3 & -3 & -3 \\ 3 & 3 & 3 \\ 6 & 6 & 6 \end{pmatrix} \cong \begin{pmatrix} 1 & 1 & 1 \\ 0 & 0 & 0 \\ 0 & 0 & 0 \end{pmatrix},$$

eigenvectors need to satisfy the single equation $a + b + c = 0$. Suppose we set $a = -b - c$. Then the eigenvectors are all given by

$$\mathbf{v} = \begin{pmatrix} -b - c \\ b \\ c \end{pmatrix} = \begin{pmatrix} -b \\ b \\ 0 \end{pmatrix} + \begin{pmatrix} -c \\ 0 \\ c \end{pmatrix} = b \begin{pmatrix} -1 \\ 1 \\ 0 \end{pmatrix} + c \begin{pmatrix} -1 \\ 0 \\ 1 \end{pmatrix}.$$

The eigenspace corresponding to $\lambda = -2$ is of dimension 2.

The eigenspace corresponding to $\lambda = 4$ consists of vectors satisfying the equation

$$\begin{pmatrix} -9 & -3 & -3 \\ 3 & -3 & 3 \\ 6 & 6 & 0 \end{pmatrix} \begin{pmatrix} a \\ b \\ c \end{pmatrix} = \begin{pmatrix} 0 \\ 0 \\ 0 \end{pmatrix}.$$

By row reduction,

$$\begin{pmatrix} -9 & -3 & -3 \\ 3 & -3 & 3 \\ 6 & 6 & 0 \end{pmatrix} \cong \begin{pmatrix} -3 & -1 & -1 \\ 1 & -1 & 1 \\ 1 & 1 & 0 \end{pmatrix} \cong \begin{pmatrix} 1 & 1 & 0 \\ 1 & -1 & 1 \\ -3 & -1 & -1 \end{pmatrix} \cong \begin{pmatrix} 1 & 1 & 0 \\ 0 & -2 & 1 \\ 0 & 2 & -1 \end{pmatrix} \cong \begin{pmatrix} 1 & 1 & 0 \\ 0 & -2 & 1 \\ 0 & 0 & 0 \end{pmatrix}$$

Thus, the eigenvectors satisfy $a + b = 0$ and $-2b + c = 0$. Writing a and c in terms of b, we have the eigenspace

$$\mathbf{v} = \begin{pmatrix} -b \\ b \\ 2b \end{pmatrix} = b \begin{pmatrix} -1 \\ 1 \\ 2 \end{pmatrix}.$$

EXAMPLE 7

The characteristic equation for the matrix

$$\mathbf{A} = \begin{pmatrix} -3 & 1 & -1 \\ -2 & 0 & -1 \\ -1 & 1 & -2 \end{pmatrix}$$

is

$$-(\lambda + 1)(\lambda + 2)^2 = 0.$$

Eigenvectors corresponding to $\lambda = -2$ satisfy

$$\begin{pmatrix} -1 & 1 & -1 \\ -2 & 2 & -1 \\ -1 & 1 & 0 \end{pmatrix} \begin{pmatrix} a \\ b \\ c \end{pmatrix} = \begin{pmatrix} 0 \\ 0 \\ 0 \end{pmatrix}.$$

By row reduction,

$$\begin{pmatrix} -1 & 1 & -1 \\ -2 & 2 & -1 \\ -1 & 1 & 0 \end{pmatrix} \cong \begin{pmatrix} -1 & 1 & 0 \\ -2 & 2 & -1 \\ -1 & 1 & -1 \end{pmatrix} \cong \begin{pmatrix} -1 & 1 & 0 \\ 0 & 0 & -1 \\ 0 & 0 & -1 \end{pmatrix} \cong \begin{pmatrix} -1 & 1 & 0 \\ 0 & 0 & 1 \\ 0 & 0 & 0 \end{pmatrix}.$$

The eigenvector equation is equivalent to the two scalar equations $c = 0$ and $-a + b = 0$. Thus, eigenvectors must satisfy $a = b$ and $c = 0$. All eigenvectors can be written as $b \begin{pmatrix} 1 \\ 1 \\ 0 \end{pmatrix}$, so the eigenspace corresponding to $\lambda = -2$ is of dimension 1. The one-dimensional eigenspace corresponding to $\lambda = -1$ can be found in the usual way to be $c\begin{pmatrix} 0 \\ 1 \\ 1 \end{pmatrix}$.

In both Examples 6 and 7, the factor $\lambda + 2$ was squared in the characteristic polynomial; however, the corresponding eigenspace had dimension 2 in one case and dimension 1 in the other.

An eigenvalue λ_1 is said to be of **algebraic multiplicity** $\boldsymbol{m}$ if the factor $\lambda - \lambda_1$ appears m times in the characteristic polynomial $P_A(\lambda)$. An eigenvalue λ_1 is said to be of **geometric multiplicity** $\boldsymbol{p}$ if its eigenspace is of dimension p.

Example 7 illustrates the fact that the geometric multiplicity of an eigenvalue can be less than the algebraic multiplicity of that eigenvalue. The sum of algebraic multiplicities of the eigenvalues of an $n \times n$ matrix is always n, but the sum of geometric multiplicities could be less than n. An $n \times n$ matrix is said to be **deficient** if the sum of the geometric multiplicities of all the eigenvalues is less than n.

6.2 Exercises

Determine the null-space for each of the matrices in Exercises 1 through 4.

1. $\mathbf{A} = \begin{pmatrix} 1 & 0 \\ 2 & 3 \end{pmatrix}$

2. $\mathbf{A} = \begin{pmatrix} 1 & 3 \\ 2 & 6 \end{pmatrix}$

3. $\mathbf{A} = \begin{pmatrix} -1 & 1 & 0 \\ -2 & 2 & 0 \\ 3 & 0 & 1 \end{pmatrix}$

4. $\mathbf{A} = \begin{pmatrix} 1 & 2 & -1 \\ 2 & 0 & 1 \\ 0 & 4 & -3 \end{pmatrix}$

Find the eigenvalues and eigenspaces for each of the matrices in Exercises 5 through 8.

5. $\mathbf{A} = \begin{pmatrix} 1 & 0 \\ 2 & 3 \end{pmatrix}$

6. $\mathbf{A} = \begin{pmatrix} 1 & 3 \\ 2 & 6 \end{pmatrix}$

7. $\mathbf{A} = \begin{pmatrix} -1 & -2 \\ 1 & -4 \end{pmatrix}$

8. $\mathbf{A} = \begin{pmatrix} 3 & -1 \\ 5 & -3 \end{pmatrix}$

Find the eigenvalues and eigenspaces for each of the matrices in Exercises 9 through 18 and note whether or not the matrix is deficient.

9. $\mathbf{A} = \begin{pmatrix} -6 & -1 & 2 \\ 3 & 2 & 0 \\ -14 & -2 & 5 \end{pmatrix}$

10. $\mathbf{A} = \begin{pmatrix} 1 & -2 & -2 \\ -2 & 2 & 3 \\ 2 & -3 & -4 \end{pmatrix}$

11. $\mathbf{A} = \begin{pmatrix} 3 & 2 & 0 \\ 2 & 0 & 0 \\ 1 & 1 & 3 \end{pmatrix}$

12. $\mathbf{A} = \begin{pmatrix} 2 & -2 & -1 \\ 1 & -1 & 1 \\ -1 & 2 & 2 \end{pmatrix}$

13. $\mathbf{A} = \begin{pmatrix} 3 & 2 & 2 \\ 1 & 4 & 1 \\ -2 & -4 & -1 \end{pmatrix}$

14. $\mathbf{A} = \begin{pmatrix} -2 & 0 & -1 \\ 2 & -2 & 0 \\ 1 & 0 & 2 \end{pmatrix}$

15. $\mathbf{A} = \begin{pmatrix} 5 & -6 & -6 \\ -1 & 4 & 2 \\ 3 & -6 & -4 \end{pmatrix}$

16. $\mathbf{A} = \begin{pmatrix} 3 & -1 & 0 \\ 2 & 0 & 0 \\ -1 & 1 & 2 \end{pmatrix}$

17. $\mathbf{A} = \begin{pmatrix} -1 & 1 & 0 \\ -2 & 2 & 0 \\ 3 & 0 & 1 \end{pmatrix}$

18. $\mathbf{A} = \begin{pmatrix} 1 & 2 & -2 \\ 2 & 0 & 6 \\ 0 & 4 & -3 \end{pmatrix}$

19. Show that no vector can be an eigenvector corresponding to two different eigenvalues. Thus, the zero vector is the only vector that can be an element of more than one eigenspace.

20. *a.* Let $\mathbf{A}$ be a 3×3 matrix with eigenvalues λ_1, λ_2, and λ_3. Show that $\det(\mathbf{A}) = \lambda_1\lambda_2\lambda_3$. In general, the product of the eigenvalues of a square matrix, counting eigenvalues of algebraic multiplicity m as m different eigenvalues, is the determinant of the matrix.
    b. The **trace** of a square matrix $\mathbf{A}$ is the sum $a_{11} + a_{22} + \cdots + a_{nn}$. Show that the sum of the eigenvalues of a 3×3 matrix $\mathbf{A}$ is equal to the trace of $\mathbf{A}$. This relationship holds for square matrices of any size.

21. Show that the set V of Example 2 is a vector space.

22. Let $\mathcal{P}_2$ be the set of all polynomials of degree 2 or less. Define addition of polynomials and multiplication of polynomials by a scalar in the usual way. Show that $\mathcal{P}_2$ is a vector space.

23. Let V be a vector space and let $S \subset V$. Prove the following: If S is nonempty and closed under addition and scalar multiplication (property 1), then S is a vector space. In this context, S is a **subspace** of V.

24. Let $\mathbf{A}$ be an $n \times n$ singular matrix, let $\mathbb{R}^n$ be the set of all n-vectors and let $\mathcal{N}(\mathbf{A})$ be the null-space of $\mathbf{A}$. Show that $\mathcal{N}(\mathbf{A})$ is a subspace of $\mathbb{R}^n$. (*Hint:* use Exercise 23.)

25. **Leslie matrices** of size 3×3 are matrices of the form

$$\mathbf{L} = \begin{pmatrix} a & b & c \\ d & 0 & 0 \\ 0 & e & 0 \end{pmatrix}, \qquad a \geq 0,\ \ b \geq 0,\ \ c > 0,\ \ 0 < d \leq 1,\ \ 0 < e \leq 1,$$

that arise in population models for organisms whose life cycle involves three discrete age classes of equal duration. The parameters a, b, and c represent the average number of offspring that are born to first-, second-, and third-generation individuals, respectively; and d and e represent the probability of surviving from the first generation to the second and from the second generation to the third, respectively. The special case $a = 0$ occurs when the first generation does not produce offspring.

a. Show that 3×3 Leslie matrices have exactly one positive eigenvalue.
b. For the special case $a = 0$, show that 3×3 Leslie matrices have a complex pair of

eigenvalues if and only if $4b < 27c^2de^2$. (*Hint:* think about the graph of the characteristic polynomial P_L.)

c. For the special case $a = 0$ show that the positive eigenvalue of a 3×3 Leslie matrix is greater than 1 if and only if $d(b + ce) > 1$.

d. In the context of a population model, having an eigenvalue greater than 1 means that the population grows from generation to generation. Use the meanings of the parameters to explain the condition $d(b + ce)$.

✦ 6.2 INSTANT EXERCISE SOLUTIONS

1. Suppose $\mathbf{x} = \begin{pmatrix} a \\ b \end{pmatrix}$ is a solution. Substituting $\mathbf{x} = \begin{pmatrix} a \\ b \end{pmatrix}$ into $\mathbf{Ax} = \mathbf{0}$, we have

$$\begin{pmatrix} 0 \\ 0 \end{pmatrix} = \mathbf{0} = \mathbf{Ax} = \begin{pmatrix} 1 & 2 \\ 2 & 4 \end{pmatrix} \begin{pmatrix} a \\ b \end{pmatrix} = \begin{pmatrix} a + 2b \\ 2a + 4b \end{pmatrix}.$$

The vector equation corresponds to two scalar equations

$$a + 2b = 0, \qquad 2a + 4b = 0.$$

Any vector $\mathbf{x} = \begin{pmatrix} a \\ b \end{pmatrix}$ satisfies $\mathbf{Ax} = \mathbf{0}$ as long as $a = -2b$. The solutions are the vectors

$$\mathbf{x} = \begin{pmatrix} -2b \\ b \end{pmatrix} = b \begin{pmatrix} -2 \\ 1 \end{pmatrix}.$$

2. With $\lambda = -2$, the eigenvector equation becomes

$$\begin{pmatrix} 1 & -2 \\ -2 & 4 \end{pmatrix} \mathbf{x} = \mathbf{0}.$$

This equation was solved in Example 4. The eigenvectors are the vectors $c\mathbf{x}_1$; where $\mathbf{x}_1 = \begin{pmatrix} 2 \\ 1 \end{pmatrix}$.

3. Suppose $\mathbf{x} = \begin{pmatrix} a \\ b \\ c \end{pmatrix}$ is an eigenvector corresponding to $\lambda_2 = 1$. Then

$$\begin{pmatrix} -1 & 1 & -1 \\ 1 & 0 & 0 \\ -1 & 0 & 0 \end{pmatrix} \begin{pmatrix} a \\ b \\ c \end{pmatrix} = \begin{pmatrix} 0 \\ 0 \\ 0 \end{pmatrix}.$$

By row reduction,

$$\begin{pmatrix} -1 & 1 & -1 \\ 1 & 0 & 0 \\ -1 & 0 & 0 \end{pmatrix} \cong \begin{pmatrix} 1 & 0 & 0 \\ -1 & 1 & -1 \\ -1 & 0 & 0 \end{pmatrix} \cong \begin{pmatrix} 1 & 0 & 0 \\ 0 & 1 & -1 \\ 0 & 0 & 0 \end{pmatrix}.$$

The last matrix corresponds to the system $a = 0$, $b - c = 0$. Thus, $a = 0$ and $b = c$, which means that all eigenvectors have to be of the form $\mathbf{x} = \begin{pmatrix} 0 \\ c \\ c \end{pmatrix} = c \begin{pmatrix} 0 \\ 1 \\ 1 \end{pmatrix}$.

Suppose $\mathbf{x} = \begin{pmatrix} a \\ b \\ c \end{pmatrix}$ is an eigenvector corresponding to $\lambda_3 = -1$. Then

$$\begin{pmatrix} 1 & 1 & -1 \\ 1 & 2 & 0 \\ -1 & 0 & 2 \end{pmatrix} \begin{pmatrix} a \\ b \\ c \end{pmatrix} = \begin{pmatrix} 0 \\ 0 \\ 0 \end{pmatrix}.$$

By row reduction, we have

$$\begin{pmatrix} 1 & 1 & -1 \\ 1 & 2 & 0 \\ -1 & 0 & 2 \end{pmatrix} \cong \begin{pmatrix} 1 & 1 & -1 \\ 0 & 1 & 1 \\ 0 & 1 & 1 \end{pmatrix} \cong \begin{pmatrix} 1 & 0 & -2 \\ 0 & 1 & 1 \\ 0 & 0 & 0 \end{pmatrix}.$$

As always, the equations are redundant; the eigenvectors satisfy the pair of equations $a - 2c = 0$ and $b + c = 0$. The eigenspace is the set of vectors

$$\mathbf{x} = \begin{pmatrix} 2c \\ -c \\ c \end{pmatrix} = c \begin{pmatrix} 2 \\ -1 \\ 1 \end{pmatrix}.$$

6.3 Linear Trajectories

The phase portrait for the linear system

$$x' = -x + 2y, \qquad y' = 4x + y$$

appears in Figure 6.3.1. Although most of the solutions follow curved trajectories, there are two straight lines through the origin that serve as trajectories.

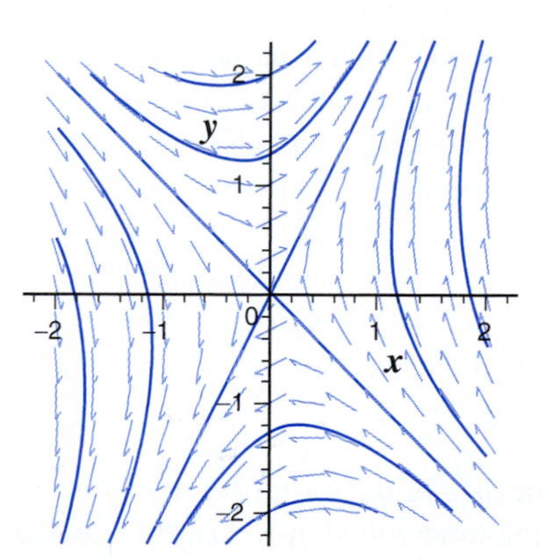

Figure 6.3.1
The phase portrait for $x' = -x + 2y$, $y' = 4x + y$.

A **linear trajectory** is a straight line through the origin that serves as a trajectory for a family of solutions. On a given linear trajectory, either all the solutions move away from the origin, as in the case of the linear trajectory in Figure 6.3.1 with positive slope, or all move toward the origin, as in the case of the other linear trajectory in Figure 6.3.1. Linear trajectories are worth studying for two reasons. First, solutions that follow linear trajectories have simple formulas that are easy to determine. Second, the connection between the linear trajectories and the eigenvectors of the corresponding matrix helps the student better understand eigenvalues and eigenvectors.

MODEL PROBLEM 6.3

Determine solutions for the system

$$x' = -x + 2y, \qquad y' = 4x + y$$

that follow linear trajectories.

Consider a system of differential equations with unknowns $x(t)$ and $y(t)$. The trajectories in the phase portrait of the system are the paths followed by solutions of the equations. These paths are defined parametrically by $(x(t), y(t))$. Given a system of equations such as $x' = -x + 2y$, $y' = 4x + y$, we would expect the movements in the x and y directions to be related in a complicated manner. That these movements can follow a straight line is surprising.

If we define the vector $\mathbf{x}$ and the matrix $\mathbf{A}$ by

$$\mathbf{x}(t) = \begin{pmatrix} x(t) \\ y(t) \end{pmatrix}, \qquad \mathbf{A} = \begin{pmatrix} -1 & 2 \\ 4 & 1 \end{pmatrix},$$

then the system $x' = -x + 2y$, $y' = 4x + y$ corresponds, using matrix multiplication, to the vector differential equation

$$\mathbf{x}' = \mathbf{A}\mathbf{x}. \tag{1}$$

Vector form has several advantages; in particular, it makes a very compact notation that is useful for discussion of systems in general, and it encourages a focus on the geometric properties of systems and their solutions. For example, the system equation (1) is the same for systems of higher dimension. The scalar components of a system of two or three differential equations can be represented as coordinates in the plane or in three-dimensional space, making a clear geometric interpretation of the vector $\mathbf{x}$. Depending on the context, it may be convenient to retain scalar variable names such as $x(t)$ and $y(t)$ as the scalar components of the vector $\mathbf{x}(t)$, or it may be more convenient to simply number the scalar components of the vector $\mathbf{x}(t)$ as $x_1(t)$ and $x_2(t)$.

In vector form, a solution is a function $\mathbf{x}(t)$ specified by a pair of formulas for the scalar components of $\mathbf{x}$. But suppose a particular solution follows a linear trajectory given by a fixed vector $\mathbf{v}$ that passes through the origin. Let $z(t)$ be the length of the vector $\mathbf{x}(t)$, relative to the vector $\mathbf{v}$. Then the solution can be written as

$$\mathbf{x}(t) = z(t)\,\mathbf{v}. \tag{2}$$

The significance of the form (2) is this: instead of two unknown functions $x(t)$ and $y(t)$, we have only one unknown function $z(t)$ and an unknown constant vector $\mathbf{v}$. It is reasonable to guess that it might be easier to calculate a scalar function and a constant vector than to calculate two coupled scalar functions.

Finding Linear Trajectories

It is convenient to work with the general case at first, and to specify the matrix $\mathbf{A}$ only when necessary. To begin, we substitute the linear trajectory formula (2) into the differential equation (1).

We have

$$z'(t)\mathbf{v} = \mathbf{A}z(t)\mathbf{v},$$

or

$$z'(t)\mathbf{v} = z(t)(\mathbf{A}\mathbf{v}). \tag{3}$$

Both sides of Equation (3) are vectors, one parallel to $\mathbf{v}$ and the other parallel to $\mathbf{Av}$. The vectors $\mathbf{v}$ and $\mathbf{Av}$ must therefore be parallel to each other, although their lengths are probably not the same and they need not point in the same direction. Thus, if there is a linear trajectory in the direction of the vector $\mathbf{v}$, then there must be a nonzero scalar λ such that

$$\mathbf{Av} = \lambda\mathbf{v}. \tag{4}$$

A vector $\mathbf{v}$ that denotes a linear trajectory must be an eigenvector of $\mathbf{A}$ corresponding to an eigenvalue λ. Substitution of Equation (4) into Equation (3) leaves us with the equation

$$z'(t)\mathbf{v} = \lambda z(t)\mathbf{v}.$$

Given $\mathbf{v} \neq \mathbf{0}$, this equation reduces to the scalar differential equation

$$z' = \lambda z. \tag{5}$$

The problem of finding a vector $\mathbf{v}$ that indicates the direction of the linear trajectory is now decoupled from the problem of finding the time-dependent factor z that indicates how the solution moves along that trajectory. Furthermore, the equation for z can be solved immediately, with solutions given by

$$z = ce^{\lambda t}. \tag{6}$$

Substituting this result into the linear trajectory formula (2) yields a simpler formula,

$$\mathbf{x} = ce^{\lambda t}\,\mathbf{v}, \tag{7}$$

for solutions that follow linear trajectories. It remains only to solve the eigenvalue problem (4) to determine the eigenvalue-eigenvector pairs for the matrix $\mathbf{A}$.

Returning to the model problem, we first obtain the eigenvalues of $\mathbf{A}$. From

$$0 = \det(\mathbf{A} - \lambda\mathbf{I}) = \begin{vmatrix} -1-\lambda & 2 \\ 4 & 1-\lambda \end{vmatrix} = \lambda^2 - 9 = (\lambda+3)(\lambda-3),$$

the eigenvalues are $\lambda = -3$ and $\lambda = 3$. The eigenvectors corresponding to $\lambda = -3$ must satisfy

$$\mathbf{0} = (\mathbf{A} + 3\mathbf{I})\mathbf{v} = \begin{pmatrix} 2 & 2 \\ 4 & 4 \end{pmatrix}\begin{pmatrix} v_1 \\ v_2 \end{pmatrix}.$$

This vector equation reduces to the scalar equation $2v_1 + 2v_2 = 0$, so we may choose the eigenvector $\begin{pmatrix} 1 \\ -1 \end{pmatrix}$, or any nonzero scalar multiple thereof, for $\mathbf{v}$. Similarly, the eigenvectors corresponding to $\lambda = 3$ must satisfy

$$\mathbf{0} = (\mathbf{A} - 3\mathbf{I})\mathbf{v} = \begin{pmatrix} -4 & 2 \\ 4 & -2 \end{pmatrix}\begin{pmatrix} v_1 \\ v_2 \end{pmatrix}.$$

We may choose the eigenvector $\mathbf{v} = \begin{pmatrix} 1 \\ 2 \end{pmatrix}$, or any of its nonzero scalar multiples, for the eigenvalue $\lambda = 3$.

We have found two one-parameter families of solutions that follow linear trajectories:

$$\mathbf{x} = c\begin{pmatrix} 1 \\ -1 \end{pmatrix} e^{-3t}, \qquad \mathbf{x} = c\begin{pmatrix} 1 \\ 2 \end{pmatrix} e^{3t}.$$

As predicted by the direction field, these solutions stay on the lines $y = -x$ and $y = 2x$, respectively. Notice also that each of these formulas represents a one-dimensional vector space of vector functions; that is, the vectors in the vector space are functions that calculate a time-dependent position vector $\langle x(t), y(t)\rangle$ that moves along a linear trajectory $y = -x$ or $y = 2x$. Thus, a one-dimensional vector space of eigenfunctions corresponds to a one-dimensional vector space of solutions.

Streamlining the Calculations

Let's revisit the calculations that began with the general system problem and ended with the eigenvalue problem and the solution for the scalar multiplier z. At no time in the course of this development did we actually use the entries of the matrix $\mathbf{A}$. This means that the calculation for the scalar multiplier always comes out the same; hence, it does not need to be repeated in every problem. This observation yields a simple method for finding linear trajectories and the solutions that follow them:

1. Assume that the system $\mathbf{x}' = \mathbf{A}\mathbf{x}$ has solutions that follow a linear trajectory. These must be of the form prescribed by Equation (7), where $\mathbf{v}$ is an eigenvector corresponding to the eigenvalue λ.
2. Find the eigenvalues by solving the equation
$$\det(\mathbf{A} - \lambda\mathbf{I}) = 0. \tag{8}$$
3. For each real eigenvalue, find a corresponding eigenvector $\mathbf{v}$ by solving the linear system
$$(\mathbf{A} - \lambda\mathbf{I})\mathbf{v} = \mathbf{0}. \tag{9}$$

EXAMPLE 1

Consider the system

$$\mathbf{x}' = \mathbf{A}\mathbf{x}, \qquad \mathbf{A} = \begin{pmatrix} 7 & -6 \\ 9 & -8 \end{pmatrix}.$$

The direction field, illustrated in Figure 6.3.2, suggests the existence of a linear trajectory $y = x$ with a positive eigenvalue. As in our solution of Model Problem 6.3, the search for linear trajectories results in the eigenvalue problem for $\mathbf{A}$. Thus,

$$0 = \det(\mathbf{A} - \lambda\mathbf{I}) = \begin{vmatrix} 7-\lambda & -6 \\ 9 & -8-\lambda \end{vmatrix} = (\lambda^2 + \lambda - 56) - (-54) = \lambda^2 + \lambda - 2 = (\lambda + 2)(\lambda - 1).$$

The eigenvalues are $\lambda = -2$ and $\lambda = 1$. For $\lambda = -2$, we find an eigenvector from

$$\mathbf{0} = (\mathbf{A} + 2\mathbf{I})\mathbf{v} = \begin{pmatrix} 9 & -6 \\ 9 & -6 \end{pmatrix}\begin{pmatrix} v_1 \\ v_2 \end{pmatrix}.$$

This vector equation gives us the scalar equation $9v_1 - 6v_2 = 0$, or $3v_1 - 2v_2 = 0$, so we may choose the eigenvector $\mathbf{v} = \begin{pmatrix} 2 \\ 3 \end{pmatrix}$. The solutions corresponding to this linear trajectory are given by

$$\mathbf{x} = c \begin{pmatrix} 2 \\ 3 \end{pmatrix} e^{-2t}.$$

This trajectory and that for $\lambda = 1$ are illustrated in Figure 6.3.3. Notice that the presence of this trajectory is *not* apparent in the direction field plot of Figure 6.3.2.

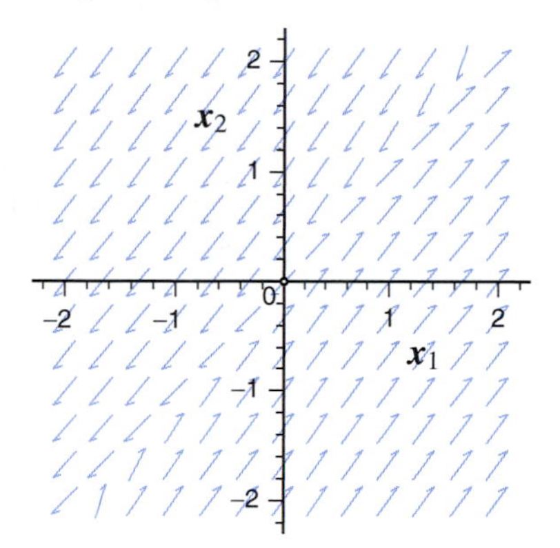

Figure 6.3.2
The direction field for Example 1.

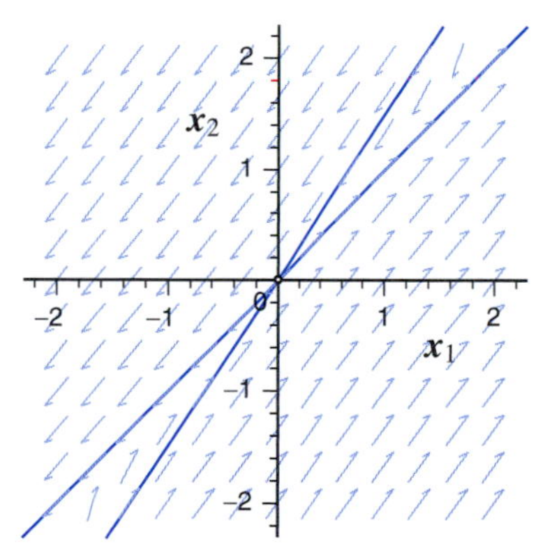

Figure 6.3.3
The direction field for Example 1, along with the linear trajectories.

✦ INSTANT EXERCISE 1

Find the solutions corresponding to the eigenvalue $\lambda = 1$ in Example 1.

A Variety of Possibilities

The characteristic polynomial for an $n \times n$ matrix is always of degree n. This degree indicates the sum of the algebraic multiplicities of the roots, and the geometric multiplicity of any eigenvalue is at least 1 and no more than the algebraic multiplicity. Taken together, this limits the possible number of linear trajectories to a maximum of n. That does not mean that there are always that many.

EXAMPLE 2

Consider the system

$$\mathbf{x}' = \mathbf{A}\mathbf{x}, \qquad \mathbf{A} = \begin{pmatrix} -3 & -1 \\ 1 & -1 \end{pmatrix}.$$

The eigenvalues of $\mathbf{A}$ satisfy $0 = \det(\mathbf{A} - \lambda\mathbf{I}) = \cdots = (\lambda + 2)^2$. There is a single real eigenvalue, $\lambda = -2$. The eigenvectors must satisfy the matrix equation

$$\begin{pmatrix} -1 & -1 \\ 1 & 1 \end{pmatrix}\begin{pmatrix} x_1 \\ x_2 \end{pmatrix} = \begin{pmatrix} 0 \\ 0 \end{pmatrix},$$

so we get the vector $\mathbf{v} = \begin{pmatrix} 1 \\ -1 \end{pmatrix}$ and all its multiples. Solutions corresponding to this linear trajectory are given by

$$\mathbf{x} = c\begin{pmatrix} 1 \\ -1 \end{pmatrix} e^{-2t}.$$

See Figure 6.3.4.

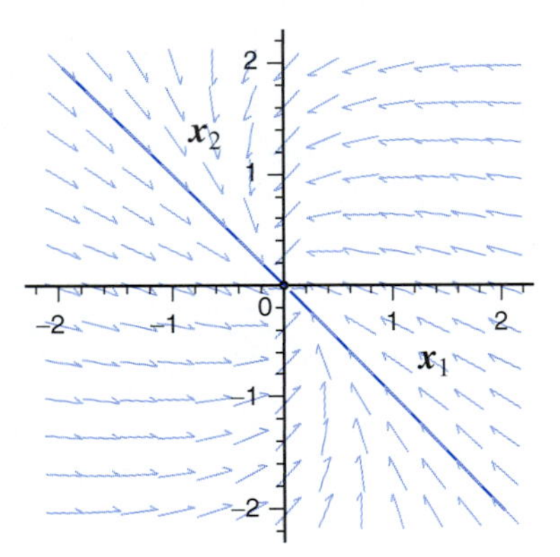

Figure 6.3.4
The linear trajectory and direction field for Example 2.

EXAMPLE 3

Consider the system

$$\mathbf{x}' = \mathbf{A}\mathbf{x}, \qquad \mathbf{A} = \begin{pmatrix} 0 & -1 \\ 1 & 0 \end{pmatrix}.$$

The eigenvalues of $\mathbf{A}$ satisfy $0 = \det(\mathbf{A} - \lambda\mathbf{I}) = \cdots = \lambda^2 + 1$. There is a pair of complex eigenvalues, $\lambda = \pm i$. The phase portrait, shown in Figure 6.3.5, clearly shows that there are no linear trajectories.

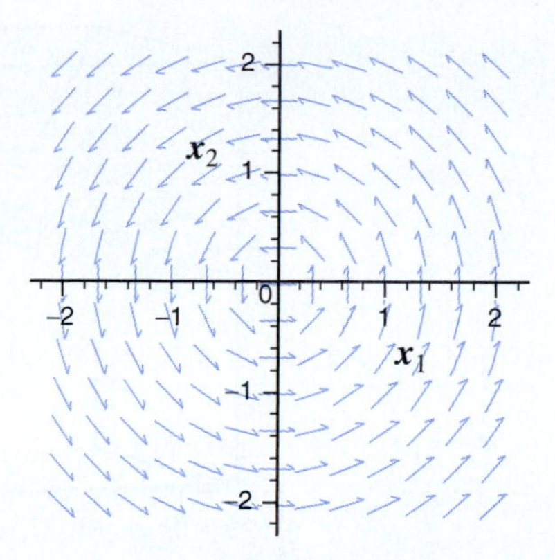

Figure 6.3.5
The direction field for Example 3, showing the absence of linear trajectories.

EXAMPLE 4

Consider the system

$$\mathbf{x}' = \mathbf{A}\mathbf{x}, \qquad \mathbf{A} = \begin{pmatrix} -5 & -3 & -3 \\ 3 & 1 & 3 \\ 6 & 6 & 4 \end{pmatrix}.$$

From Example 6 of Section 6.2, we already know that $\lambda = -2$ is an eigenvalue of algebraic multiplicity 2, with a two-dimensional eigenspace consisting of all the vectors $\mathbf{v}$ such that $v_1 + v_2 + v_3 = 0$, which we can represent by the formula

$$\mathbf{v} = c_1 \begin{pmatrix} -1 \\ 1 \\ 0 \end{pmatrix} + c_2 \begin{pmatrix} -1 \\ 0 \\ 1 \end{pmatrix}.$$

We therefore have a two-dimensional vector space of solutions, given by

$$\mathbf{x}(t) = e^{-2t} \left[c_1 \begin{pmatrix} -1 \\ 1 \\ 0 \end{pmatrix} + c_2 \begin{pmatrix} -1 \\ 0 \\ 1 \end{pmatrix} \right].$$

Instead of a single linear trajectory for this eigenvalue, we have a family of linear trajectories that includes all straight lines that lie in the plane $x_1 + x_2 + x_3 = 0$ and pass through the origin. The solution of the system through any of the points on this plane indicates motion along the line from the point to the origin. Several of these trajectories are illustrated in Figure 6.3.6.

✦ INSTANT EXERCISE 2

Use the results of Example 6 of Section 6.2 to find a solution formula for a linear trajectory that is not in the plane $x_1 + x_2 + x_3 = 0$. Describe how solutions move along this trajectory.

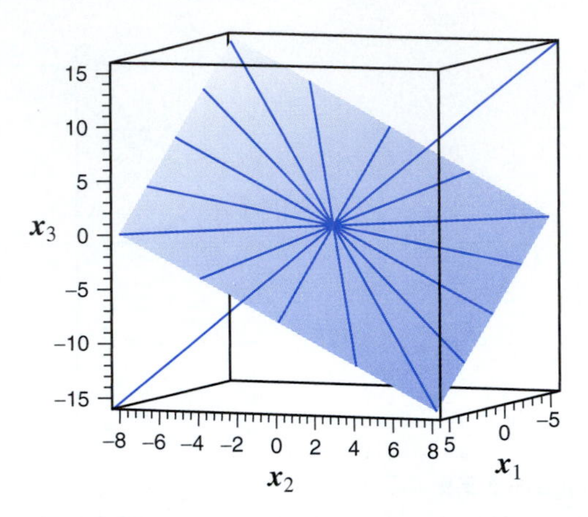

Figure 6.3.6
Some linear trajectories in the plane $x_1 + x_2 + x_3 = 0$ from Example 4 and the linear trajectory from Instant Exercise 2.

EXAMPLE 5

Consider the system

$$\mathbf{x}' = \mathbf{A}\mathbf{x}, \qquad \mathbf{A} = \begin{pmatrix} -2 & 1 \\ 2 & -1 \end{pmatrix}.$$

The search for linear trajectories leads to the characteristic polynomial $P_A(\lambda) = \lambda^2 + 3\lambda$; hence, the eigenvalues are $\lambda = 0$ and $\lambda = -3$. The matrix in this example is singular; the property $\det(\mathbf{A}) = 0$ is equivalent to having an eigenvalue of 0. The eigenspace corresponding to 0 is the set of vectors $c\begin{pmatrix} 1 \\ 2 \end{pmatrix}$, and the eigenspace corresponding to -3 is the set $c\begin{pmatrix} 1 \\ -1 \end{pmatrix}$. There is a linear trajectory given by the equation $y = -x$, corresponding to the second eigenvector; however, the eigenspace corresponding to the eigenvalue 0 does not yield a linear trajectory. From the linear trajectory formula (7), we obtain the solution family

$$\mathbf{x} = c\begin{pmatrix} 1 \\ 2 \end{pmatrix},$$

which represents solutions that are on the straight line $y = x/2$ but that do not move. This example is revisited in Section 6.4.

6.3 Exercises

In Exercises 1 through 4, find and sketch any linear trajectories for the system $\mathbf{x}' = \mathbf{A}\mathbf{x}$.

1. $\mathbf{A} = \begin{pmatrix} 1 & 0 \\ 2 & 3 \end{pmatrix}$ (Section 6.2, Exercise 5)

2. $\mathbf{A} = \begin{pmatrix} 1 & 3 \\ 2 & 6 \end{pmatrix}$ (Section 6.2, Exercise 6)

3. $\mathbf{A} = \begin{pmatrix} -1 & -2 \\ 1 & -4 \end{pmatrix}$ (Section 6.2, Exercise 7)

4. $\mathbf{A} = \begin{pmatrix} 3 & -1 \\ 5 & -3 \end{pmatrix}$ (Section 6.2, Exercise 8)

In Exercises 5 through 12, find and sketch any linear trajectories for the system.

5. $x' = 2x + y, \quad y' = -y$

6. $x' = 2x + y, \quad y' = y$

7. $x' = -2x - 3y, \quad y' = 3x - 2y$

8. $x' = 2x + y, \quad y' = -x + 4y$

9. $x' = -2x + 5y, \quad y' = 2x + y$

10. $x' = 4y, \quad y' = x - 3y$

11. $x' = -3x + 2y, \quad y' = 4x - 5y$

12. $x' = -3x + 2y, \quad y' = 5x - 6y$

In Exercises 13 through 16, find any linear trajectories for the system $\mathbf{x}' = \mathbf{A}\mathbf{x}$.

13. $\mathbf{A} = \begin{pmatrix} -6 & -1 & 2 \\ 3 & 2 & 0 \\ -14 & -2 & 5 \end{pmatrix}$ (Section 6.2, Exercise 9)

14. $\mathbf{A} = \begin{pmatrix} 3 & -1 & 0 \\ 2 & 0 & 0 \\ -1 & 1 & 2 \end{pmatrix}$ (Section 6.2, Exercise 16)

15. $\mathbf{A} = \begin{pmatrix} 5 & -6 & -6 \\ -1 & 4 & 2 \\ 3 & -6 & -4 \end{pmatrix}$ (Section 6.2, Exercise 15)

16. $\mathbf{A} = \begin{pmatrix} 1 & -2 & -2 \\ -2 & 2 & 3 \\ 2 & -3 & -4 \end{pmatrix}$ (Section 6.2, Exercise 10)

17. Prove that every linearly independent system of three homogeneous linear first-order differential equations has at least one linear trajectory.

18. A matrix is **symmetric** if the ij and ji entries are equal for all i and j. Thus, any 2×2 symmetric matrix has the form

$$\mathbf{A} = \begin{pmatrix} a & b \\ b & d \end{pmatrix}.$$

a. Show that every 2×2 system $\mathbf{x}' = \mathbf{A}\mathbf{x}$ with $\mathbf{A}$ symmetric has at least one linear trajectory.
b. Show that every 2×2 system $\mathbf{x}' = \mathbf{A}\mathbf{x}$ with $\mathbf{A}$ symmetric and $b \neq 0$ has exactly two linear trajectories.

✦ 6.3 INSTANT EXERCISE SOLUTIONS

1. For $\lambda = 1$, we find an eigenvector from

$$\mathbf{0} = (\mathbf{A} - \mathbf{I})\mathbf{v} = \begin{pmatrix} 6 & -6 \\ 9 & -9 \end{pmatrix} \begin{pmatrix} a \\ b \end{pmatrix}.$$

This vector equation gives us the scalar equation $a - b = 0$, so $a = b$. We may choose the vector $\mathbf{v} = \begin{pmatrix} 1 \\ 1 \end{pmatrix}$. The corresponding family of solutions is

$$\mathbf{x} = c \begin{pmatrix} 1 \\ 1 \end{pmatrix} e^{t}.$$

2. We previously found the family of eigenvectors

$$\mathbf{v} = b \begin{pmatrix} -1 \\ 1 \\ 2 \end{pmatrix}$$

corresponding to the eigenvalue $\lambda = 4$. The corresponding solutions are

$$\mathbf{x} = c \begin{pmatrix} -1 \\ 1 \\ 2 \end{pmatrix} e^{4t}.$$

These solutions move away from the origin along the linear trajectory given by the eigenvectors.

6.4 Homogeneous Systems with Real Eigenvalues

The theory of systems of linear differential equations is very similar to the theory of scalar linear equations.[9] Here we summarize the key points and then illustrate the solution method for homogeneous linear systems with constant coefficients.

Theory of Linear Systems

In general, consider a system of n linear equations with n unknowns,

$$\begin{aligned} x_1' &= a_{11}(t)x_1 + a_{12}(t)x_2 + \cdots + a_{1n}x_n + g_1(t), \\ x_2' &= a_{21}(t)x_1 + a_{22}(t)x_2 + \cdots + a_{2n}x_n + g_2(t), \\ &\vdots \\ x_n' &= a_{n1}(t)x_1 + a_{n2}(t)x_2 + \cdots + a_{nn}x_n + g_n(t), \end{aligned}$$

[9]See Section 3.3.

and the equivalent vector equation

$$\mathbf{x}' = \mathbf{A}(t)\mathbf{x} + \mathbf{g}(t). \tag{1}$$

An **initial-value problem** for this system is obtained by prescribing a value for $\mathbf{x}$ at some time:

$$\mathbf{x}' = \mathbf{A}(t)\mathbf{x} + \mathbf{g}(t), \qquad \mathbf{x}(t_0) = \mathbf{x}_0. \tag{2}$$

There is an existence and uniqueness theorem for linear systems that is quite similar to that for scalar equations (Theorem 2.4.2).

Theorem 6.4.1

Existence of a Unique Solution (Linear) Let $\mathbf{A}(t)$ be an $n \times n$ matrix, and let $\mathbf{g}(t)$ be a vector function with n components. If all entries of $\mathbf{A}(t)$ and components of $\mathbf{g}(t)$ are continuous on an open interval D containing the initial point t_0, then the initial-value problem (2) has a unique solution that exists throughout the interval D.

The principle of superposition (Theorem 3.3.1) also holds for homogeneous systems.

Theorem 6.4.2

Principle of Superposition If $\mathbf{x}^{(1)}(t), \mathbf{x}^{(2)}(t), \ldots, \mathbf{x}^{(m)}(t)$ are solutions of $\mathbf{x}' = \mathbf{A}(t)\mathbf{x}$, then so is $\mathbf{x} = c_1\mathbf{x}^{(1)} + c_2\mathbf{x}^{(2)} + \cdots + c_m\mathbf{x}^{(m)}$, where $c_1, c_2, \ldots, c_n$ are arbitrary constants.

Theorem 6.4.2 has an equivalent statement that makes use of the language of linear algebra:

- Linear combinations of solutions of a homogeneous linear system are also solutions of that system.

Theorems 6.4.1 and 6.4.2 can be combined to get a rule for the construction of general solutions for homogeneous linear systems that is analogous to Theorem 3.3.3.

Theorem 6.4.3

Let $\mathbf{A}(t)$ be an $n \times n$ matrix of functions that are continuous on an interval I. Let $\{\mathbf{x}^{(1)}(t), \mathbf{x}^{(2)}(t), \ldots, \mathbf{x}^{(n)}(t)\}$ be a set of solutions of the system $\mathbf{x}' = \mathbf{A}\mathbf{x}$ on the interval I, and let $\mathbf{\Psi}(t)$ be the matrix whose jth column is the vector $\mathbf{x}^{(j)}(t)$. Then the unique solution of the initial-value problem

$$\mathbf{x}' = \mathbf{A}\mathbf{x}, \qquad \mathbf{x}(t_0) = \mathbf{x}_0,$$

with $t_0 \in I$, can always be found from

$$\mathbf{x} = c_1\mathbf{x}^{(1)} + c_2\mathbf{x}^{(2)} + \cdots + c_n\mathbf{x}^{(n)},$$

provided $\mathbf{\Psi}(0)$ is nonsingular.

A nonsingular matrix $\mathbf{\Psi}$ is a **fundamental matrix** for the system. The solutions $\{\mathbf{x}^{(1)}(t), \mathbf{x}^{(2)}(t), \ldots, \mathbf{x}^{(n)}(t)\}$ are said to be a **fundamental set** of solutions. The condition that $\mathbf{\Psi}$

be nonsingular is necessary to ensure that the n solutions are sufficient to span $\mathbb{R}^n$ at any time t.[10] As with the Wronskian in the scalar case, the fundamental matrix is either nonsingular throughout the interval I or singular throughout the interval; hence, it suffices to check it at a single point.

Theorem 6.4.3 lays out the steps needed to solve an initial-value problem for a linear system of n equations. We need to find n solutions that form a nonsingular fundamental matrix. The linear combination of those solutions is the general solution, and values can be chosen for the coefficients so as to satisfy any initial condition.

Solving Homogeneous Linear Systems with Constant Coefficients

Homogeneous linear systems with constant coefficients can sometimes be solved quickly with the aid of linear trajectories. Recall from Section 6.3 that each linear trajectory corresponds to a one-parameter family of solutions. Each solution that we find by looking for linear trajectories can be considered as part of the set of solutions needed for Theorem 6.4.3.

MODEL PROBLEM 6.4

Solve the initial-value problem

$$x' = -x + 2y, \qquad y' = 4x + y, \qquad x(0) = 4, \qquad y(0) = -3.$$

In Section 6.3, we found two solution families that follow linear trajectories. In particular, we have two solutions

$$\mathbf{x}^{(1)} = \begin{pmatrix} 1 \\ -1 \end{pmatrix} e^{-3t} \qquad \text{and} \qquad \mathbf{x}^{(2)} = \begin{pmatrix} 1 \\ 2 \end{pmatrix} e^{3t},$$

where $\mathbf{x} = \begin{pmatrix} x \\ - \\ y \end{pmatrix}$. Theorem 6.4.2 guarantees that we can combine these solutions to create a solution formula

$$\mathbf{x} = c_1 \begin{pmatrix} 1 \\ -1 \end{pmatrix} e^{-3t} + c_2 \begin{pmatrix} 1 \\ 2 \end{pmatrix} e^{3t}.$$

We would like to use Theorem 6.4.3 to assert that this is a general solution, so that we are guaranteed unique values of c_1 and c_2 for any initial condition. We can check this by computing $\det(\mathbf{\Psi}(0))$. We have

$$\det(\mathbf{\Psi}(0)) = \begin{vmatrix} 1 & 1 \\ -1 & 2 \end{vmatrix} = 3.$$

Thus, $\mathbf{\Psi}(0)$ is nonsingular and Theorem 6.4.3 guarantees that we have found the general solution of the system.

To complete the solution of Model Problem 6.4, we need to use the initial condition to find the correct values of the coefficients. Substitution of the general solution into the initial condition

[10] See the related discussion about satisfying initial conditions in Section 3.3. Note also that $\mathbb{R}^n$ is the set of vectors with n real components.

gives us

$$c_1\begin{pmatrix}1\\-1\end{pmatrix}+c_2\begin{pmatrix}1\\2\end{pmatrix}=\begin{pmatrix}4\\-3\end{pmatrix}.$$

This system can be solved by writing it in matrix-vector form and row-reducing the augmented matrix, but it is simple enough to solve without going to that much trouble. Combining the terms on the left side yields a pair of scalar equations for c_1 and c_2:

$$c_1+c_2=4, \qquad -c_1+2c_2=-3.$$

Adding these equations gives $3c_2 = 1$, so we have $c_2 = \frac{1}{3}$ and then $c_1 = \frac{11}{3}$. The solution of the initial-value problem is

$$\mathbf{x}=\frac{11}{3}\begin{pmatrix}1\\-1\end{pmatrix}e^{-3t}+\frac{1}{3}\begin{pmatrix}1\\2\end{pmatrix}e^{3t}.$$

In the original scalar notation of Model Problem 6.4, we identify x with the first component of $\mathbf{x}$ and y with the second component. Thus,

$$x=\tfrac{1}{3}e^{3t}+\tfrac{11}{3}e^{-3t}, \qquad y=\tfrac{2}{3}e^{3t}-\tfrac{11}{3}e^{-3t}.$$

This solution is shown in Figure 6.4.1.

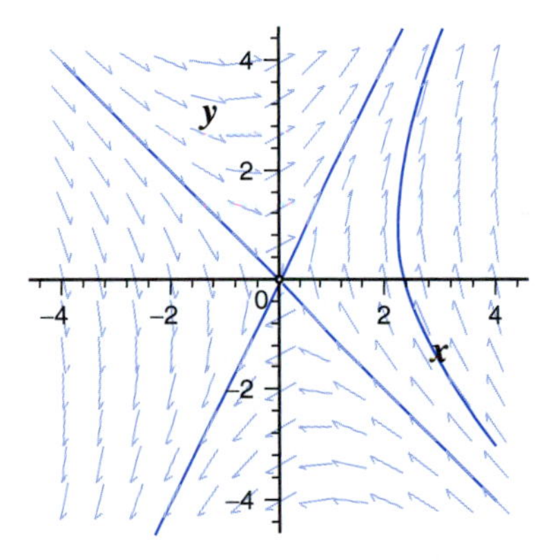

Figure 6.4.1
The solution of Model Problem 6.4, along with the direction field and linear trajectories.

✦ **INSTANT EXERCISE 1**

Check the solution of Model Problem 6.4.

The General Case

The solution of Model Problem 6.4 suggests a general method to solve problems of the form

$$\mathbf{x}'=\mathbf{A}\mathbf{x}, \qquad \mathbf{x}(0)=\mathbf{x}_0, \tag{3}$$

where $\mathbf{A}$ is a matrix of constants.

1. Given the $n \times n$ matrix $\mathbf{A}$, find the eigenvalues λ_j and their algebraic multiplicities m_j.
2. Find the eigenspace for each eigenvalue.
3. Use the eigenvalue-eigenvector pairs to construct solutions $\mathbf{x}^{(1)}, \mathbf{x}^{(2)}, \ldots, \mathbf{x}^{(n)}$ that follow linear trajectories.
4. Check to make sure that $\det(\mathbf{\Psi}(0)) \neq 0$. [Note that $\det(\mathbf{\Psi}(0))$ is simply the determinant whose jth column is the eigenvector used in the solution $\mathbf{x}^{(j)}$.]
5. Construct the solution family

$$\mathbf{x} = c_1\mathbf{x}^{(1)} + c_2\mathbf{x}^{(2)} + \cdots + c_n\mathbf{x}^{(n)}$$

and determine the constants $c_1, \ldots, c_n$ so that $\mathbf{x}(0) = \mathbf{x}_0$.

The method as stated works for many problems, but it misses two important complications.

- Complex pairs of eigenvalues do not correspond to any linear trajectories. This technical complication is addressed in Section 6.5.
- If the matrix $\mathbf{A}$ is deficient,[11] then there are fewer than n eigenvalue-eigenvector pairs from which to construct solutions that follow linear trajectories. This issue is addressed in Section 6.6.

Given a nondeficient matrix with all real eigenvalues, there remains the question of whether the n eigenvectors are sufficient to span $\mathbb{R}^n$. It is unnecessary to check the determinant when the eigenvalues are distinct.[12]

Theorem 6.4.4

If $\lambda_1, \lambda_2, \ldots, \lambda_n$ are distinct real eigenvalues of an $n \times n$ matrix $\mathbf{A}$ with corresponding eigenvectors $\mathbf{v}^{(1)}, \mathbf{v}^{(2)}, \ldots, \mathbf{v}^{(n)}$, then the general solution of the system $\mathbf{x}' = \mathbf{A}\mathbf{x}$ is

$$\mathbf{x} = c_1\mathbf{v}^{(1)}e^{\lambda_1 t} + c_2\mathbf{v}^{(2)}e^{\lambda_2 t} + \cdots + c_n\mathbf{v}^{(n)}e^{\lambda_n t}.$$

✦ INSTANT EXERCISE 2

The linear trajectories for the problem

$$\mathbf{x}' = \begin{pmatrix} 7 & -6 \\ 9 & -8 \end{pmatrix}\mathbf{x}$$

were found in Example 1 and Instant Exercise 1 of Section 6.3. Use these trajectories to find the solution of the problem

$$\mathbf{x}' = \begin{pmatrix} 7 & -6 \\ 9 & -8 \end{pmatrix}\mathbf{x}, \qquad \mathbf{x}(0) = \begin{pmatrix} 0 \\ 1 \end{pmatrix}$$

---

[11] See Section 6.2.

[12] See Exercise 19 of Section 6.2.

EXAMPLE 1

Consider the system

$$\mathbf{x}' = \mathbf{A}\mathbf{x}, \qquad \mathbf{A} = \begin{pmatrix} -5 & -3 & -3 \\ 3 & 1 & 3 \\ 6 & 6 & 4 \end{pmatrix},$$

which we studied in Example 4 of Section 6.3. The matrix has an eigenvalue $\lambda = -2$ that has algebraic and geometric multiplicities of 2. The conditions of Theorem 6.4.4 are not met because of the multiple eigenvalue; however, there are the required three linear trajectory solutions,

$$\mathbf{x}^{(1)}(t) = \begin{pmatrix} -1 \\ 1 \\ 0 \end{pmatrix} e^{-2t}, \qquad \mathbf{x}^{(2)}(t) = \begin{pmatrix} -1 \\ 0 \\ 1 \end{pmatrix} e^{-2t}, \qquad \mathbf{x}^{(3)} = \begin{pmatrix} -1 \\ 1 \\ 2 \end{pmatrix} e^{4t}.$$

We can verify that these solutions are a fundamental set by computing $\det(\mathbf{\Psi}(0))$:

$$\det(\mathbf{\Psi}(0)) = \begin{vmatrix} -1 & -1 & -1 \\ 1 & 0 & 1 \\ 0 & 1 & 2 \end{vmatrix} = \begin{vmatrix} 0 & -1 & 0 \\ 1 & 0 & 1 \\ 0 & 1 & 2 \end{vmatrix} = -\begin{vmatrix} -1 & 0 \\ 1 & 2 \end{vmatrix} = 2.$$

By Theorem 6.4.3, we have the general solution

$$\mathbf{x}(t) = c_1 \begin{pmatrix} -1 \\ 1 \\ 0 \end{pmatrix} e^{-2t} + c_2 \begin{pmatrix} -1 \\ 0 \\ 1 \end{pmatrix} e^{-2t} + c_3 \begin{pmatrix} -1 \\ 1 \\ 2 \end{pmatrix} e^{4t}.$$

As suggested by Example 1, it is not necessary to have n distinct eigenvalues as long as each eigenvalue of algebraic multiplicity m has m distinct eigenvectors. Theorem 6.4.5 summarizes the results.

Theorem 6.4.5

If the sum of the geometric multiplicities of the real eigenvalues of an $n \times n$ matrix $\mathbf{A}$ is n, then the general solution of the system $\mathbf{x}' = \mathbf{A}\mathbf{x}$ can be written as

$$\mathbf{x} = c_1 \mathbf{v}^{(1)} e^{\lambda_1 t} + c_2 \mathbf{v}^{(2)} e^{\lambda_2 t} + \cdots + c_n \mathbf{v}^{(n)} e^{\lambda_n t},$$

where each $\mathbf{v}^{(j)}$ is an eigenvector and each eigenvalue λ_j, not necessarily distinct, is the eigenvalue to which the eigenvector $\mathbf{v}^{(j)}$ belongs.

Sources, Sinks, and Saddles

For many problems involving autonomous systems, the most important information is the set of critical points and their classification by stability and shape. For the autonomous linear system

$$\mathbf{x}' = \mathbf{A}\mathbf{x}, \tag{4}$$

where **A** is a matrix of constants, the critical points are the solutions of the equation

$$\mathbf{Ax} = \mathbf{0},$$

which we investigated in Section 6.2. This equation always has a solution $\mathbf{x} = \mathbf{0}$, which corresponds to a critical point at the origin. If the matrix **A** is nonsingular, then the origin is the only critical point.

To return to Model Problem 6.4, the longtime behavior of solutions follows immediately from the general solution formula

$$\mathbf{x} = c_1 \begin{pmatrix} 1 \\ -1 \end{pmatrix} e^{-3t} + c_2 \begin{pmatrix} 1 \\ 2 \end{pmatrix} e^{3t}.$$

As long as $c_2 \neq 0$, the solution tends toward

$$\mathbf{x} = c_2 \begin{pmatrix} 1 \\ 2 \end{pmatrix} e^{3t}.$$

The origin is unstable[13] because these solutions move away from it. For those solutions with initial conditions on the linear trajectory given by $\begin{pmatrix} 1 \\ -1 \end{pmatrix}$, however, $c_2 = 0$ and the solution has the form

$$\mathbf{x} = c_1 \begin{pmatrix} 1 \\ -1 \end{pmatrix} e^{-3t}.$$

These solutions approach the origin; hence, the origin is a saddle.[14] The phase portrait in Figure 6.4.2 illustrates the characteristic pattern of trajectories near a saddle.

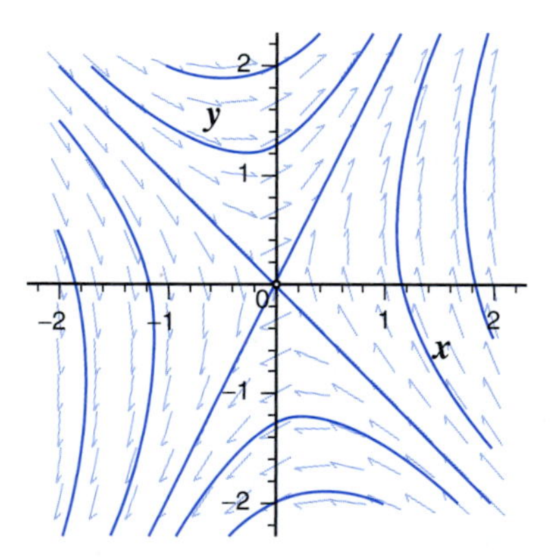

Figure 6.4.2
The phase portrait for Model Problem 6.4.

Note that this analysis of the critical point does not actually make use of the eigenvectors in the solution formula. All that matters is whether the solutions have decaying exponential factors or growing exponential factors. This property follows directly from the eigenvalues.

---

[13] See Section 5.4 for the definitions of the terms *stable*, *asymptotically stable*, and *unstable*.

[14] See Section 5.4.

✦ INSTANT EXERCISE 3

The phase portrait for the system $x' = -2x + 2y, \;\; y' = 2x + y$ appears in Figure 5.4.3. Write the system in vector form, and verify that the origin is a saddle by determining the eigenvalues.

The origin is not always a saddle for 2×2 systems with real eigenvalues.

EXAMPLE 2

Consider the system

$$\mathbf{x}' = \begin{pmatrix} -2 & 2 \\ 2 & -5 \end{pmatrix} \mathbf{x}.$$

The eigenvalues for the system are -1 and -6; thus, all solutions approach the origin. The origin is asymptotically stable and is classified as a **sink.** (See the phase portrait for this system in Figure 5.4.2.)

EXAMPLE 3

Consider the system

$$\mathbf{x}' = \begin{pmatrix} 1 & -2 \\ 1 & 4 \end{pmatrix} \mathbf{x}.$$

The eigenvalues are 2 and 3. The origin is a **source**—all solutions move away from the origin. The phase portrait appears in Figure 6.4.3.

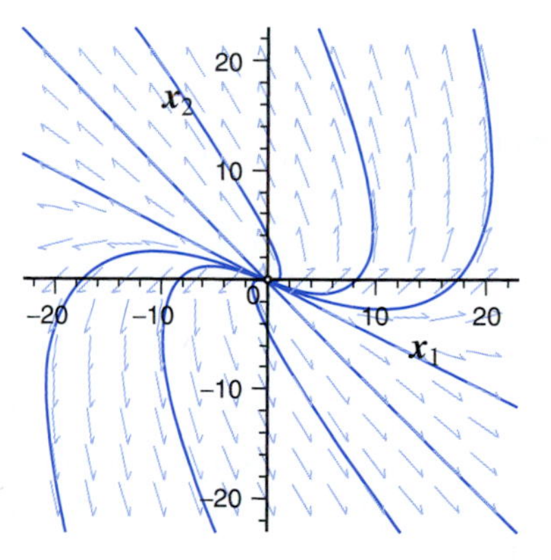

Figure 6.4.3
The phase portrait for Example 3.

✦ INSTANT EXERCISE 4

Solve the system of Example 3.

Let **A** be a 2×2 matrix and consider the origin in the phase portrait for $\mathbf{x}' = \mathbf{A}\mathbf{x}$. Simply put, the origin is a source if the eigenvalues are real and positive, a sink if the eigenvalues are real and negative, and a saddle if the eigenvalues are real, with one positive and the other negative.

Singular Systems

If a 2×2 matrix is singular, then one of its two eigenvalues is 0 and the other is a real number. The eigenspace corresponding to 0 is the null-space of the matrix **A**, and all points in the eigenspace are critical points. Unlike systems with nonsingular matrices, the origin in a system with a singular matrix is not an isolated critical point.

EXAMPLE 4

The system

$$\mathbf{x}' = \mathbf{A}\mathbf{x}, \qquad \mathbf{A} = \begin{pmatrix} -2 & 1 \\ 2 & -1 \end{pmatrix}$$

was examined in Example 5 of Section 6.3. We found a linear trajectory with the corresponding solution family

$$c\begin{pmatrix} 1 \\ -1 \end{pmatrix} e^{-3t}$$

and a family of equilibrium solutions

$$c\begin{pmatrix} 1 \\ 2 \end{pmatrix}$$

corresponding to the eigenvalue 0. Thus, the general solution of the system, by Theorem 6.4.4, is

$$\mathbf{x} = c_1\begin{pmatrix} 1 \\ 2 \end{pmatrix} + c_2\begin{pmatrix} 1 \\ -1 \end{pmatrix} e^{-3t}.$$

The phase portrait for this system appears in Figure 6.4.4.

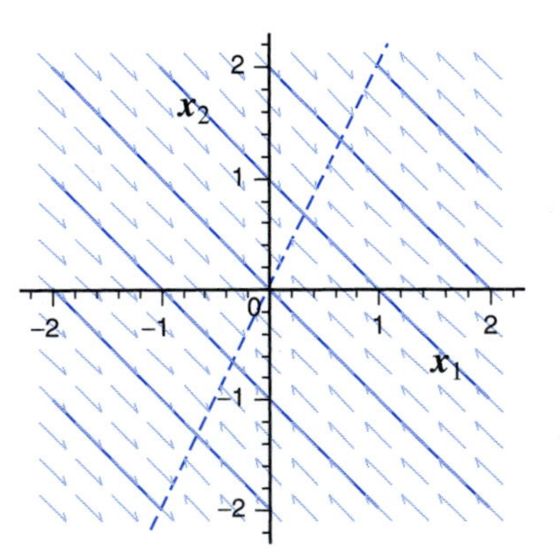

Figure 6.4.4
The phase portrait for Example 4.

All solutions in Example 4 approach the line $y = 2x$, which we can think of as representing something like a stable equilibrium solution, except that it is a line rather than a point. Solutions move in the direction of the eigenvector corresponding to $\lambda = -3$ toward the line $y = 2x$.

6.4 Exercises

In Exercises 1 and 2, solve the initial-value problem.

1. $\mathbf{x}' = \begin{pmatrix} 1 & 0 \\ 2 & 3 \end{pmatrix}\mathbf{x}, \quad \mathbf{x}(0) = \begin{pmatrix} 3 \\ -2 \end{pmatrix}$ (Section 6.3, Exercise 1)

2. $\mathbf{x}' = \begin{pmatrix} -1 & -2 \\ 1 & -4 \end{pmatrix}\mathbf{x}, \quad \mathbf{x}(0) = \begin{pmatrix} 1 \\ 0 \end{pmatrix}$ (Section 6.3, Exercise 3)

In Exercises 3 through 8, use the linear trajectories for the system $\mathbf{x}' = \mathbf{A}\mathbf{x}$ to construct the general solution of the system and determine the appropriate classification for the critical point at the origin; or explain why you cannot. Use all this information to sketch the phase portrait for the system.

3. $\mathbf{A} = \begin{pmatrix} 1 & 0 \\ 2 & 3 \end{pmatrix}$ (Section 6.3, Exercise 1)

4. $\mathbf{A} = \begin{pmatrix} 1 & 3 \\ 2 & 6 \end{pmatrix}$ (Section 6.3, Exercise 2)

5. $\mathbf{A} = \begin{pmatrix} -1 & -2 \\ 1 & -4 \end{pmatrix}$ (Section 6.3, Exercise 3)

6. $\mathbf{A} = \begin{pmatrix} 3 & -1 \\ 5 & -3 \end{pmatrix}$ (Section 6.3, Exercise 4)

7. $\mathbf{A} = \begin{pmatrix} 2 & 1 \\ 0 & -1 \end{pmatrix}$ (Section 6.3, Exercise 5)

8. $\mathbf{A} = \begin{pmatrix} 2 & 1 \\ 0 & 1 \end{pmatrix}$ (Section 6.3, Exercise 6)

In Exercises 9 and 10, use the eigenvalues for the system $\mathbf{x}' = \mathbf{A}\mathbf{x}$ to classify the critical point at the origin. Also sketch the nullclines for each system, and confirm that the information provided by the linear trajectories and the nullclines is consistent. Sketch the phase portrait for the system.

9. $\mathbf{A} = \begin{pmatrix} 2 & 1 \\ 4 & -1 \end{pmatrix}$

10. $\mathbf{A} = \begin{pmatrix} 2 & 1 \\ -1 & 4 \end{pmatrix}$ (Section 6.3, Exercise 8)

In Exercises 11 through 16, use the linear trajectories $\mathbf{x}' = \mathbf{A}\mathbf{x}$ to construct the general solution for the system, or explain why you cannot.

11. $\mathbf{A} = \begin{pmatrix} -6 & -1 & 2 \\ 3 & 2 & 0 \\ -14 & -2 & 5 \end{pmatrix}$ (Section 6.3, Exercise 13)

12. $\mathbf{A} = \begin{pmatrix} 3 & -1 & 0 \\ 2 & 0 & 0 \\ -1 & 1 & 2 \end{pmatrix}$ (Section 6.3, Exercise 14)

13. $\mathbf{A} = \begin{pmatrix} 5 & -6 & -6 \\ -1 & 4 & 2 \\ 3 & -6 & -4 \end{pmatrix}$ (Section 6.3, Exercise 15)

14. $\mathbf{A} = \begin{pmatrix} 1 & -2 & -2 \\ -2 & 2 & 3 \\ 2 & -3 & -4 \end{pmatrix}$ (Section 6.3, Exercise 16)

15. $\mathbf{A} = \begin{pmatrix} 3 & 2 & 0 \\ 2 & 0 & 0 \\ 1 & 1 & 3 \end{pmatrix}$ (Section 6.2, Exercise 11)

16. $\mathbf{A} = \begin{pmatrix} 2 & -2 & -1 \\ 1 & -1 & 1 \\ -1 & 2 & 2 \end{pmatrix}$ (Section 6.2, Exercise 12)

17. Let $\mathbf{A}$ be a 2×2 matrix with $\det(\mathbf{A}) \neq 0$. Show that the origin is a saddle for $\mathbf{x}' = \mathbf{A}\mathbf{x}$ if and only if $\det(\mathbf{A}) < 0$.

18. Consider the system (1) of Section 6.1, with $R = 0$. Show that the origin is a sink whenever $r_1, k_2 > 0$ and $r_2, k_1 \geq 0$.

19. Prove Theorem 6.4.2.

20. Implicit in Theorem 6.4.3 is the statement that a solution matrix $\mathbf{\Psi}$ is always singular or always nonsingular on the interval I. This statement is analogous to Abel's theorem (Theorem 3.3.2). Prove the statement for the 2×2 case. Define

$$\mathbf{A}(t) = \begin{pmatrix} f(t) & b(t) \\ c(t) & g(t) \end{pmatrix}, \qquad \mathbf{x}^{(1)} = \begin{pmatrix} x_1 \\ y_1 \end{pmatrix}, \qquad \mathbf{x}^{(2)} = \begin{pmatrix} x_2 \\ y_2 \end{pmatrix}, \qquad W(t) = \begin{vmatrix} x_1 & x_2 \\ y_1 & y_2 \end{vmatrix},$$

and assume that $\mathbf{x}^{(1)}$ and $\mathbf{x}^{(2)}$ are solutions of $\mathbf{x}' = \mathbf{A}(t)\mathbf{x}$ on an interval I.

a. Show that

$$W' = \begin{vmatrix} x_1' & x_2' \\ y_1 & y_2 \end{vmatrix} + \begin{vmatrix} x_1 & x_2 \\ y_1' & y_2' \end{vmatrix}.$$

b. Determine a differential equation for W.
c. Solve the differential equation to show that $W \equiv 0$ or else $W \neq 0$.

21. Prove Theorem 6.4.4. (*Hint:* use Exercise 19 of Section 6.2.)

✦ 6.4 INSTANT EXERCISE SOLUTIONS

1. Using the results obtained in the text, we have

$$-x+2y=-\left(\tfrac{1}{3}e^{3t}+\tfrac{11}{3}e^{-3t}\right)+2\left(\tfrac{2}{3}e^{3t}-\tfrac{11}{3}e^{-3t}\right)=e^{3t}-11e^{-3t}=x'$$

and

$$4x+y=4\left(\tfrac{1}{3}e^{3t}+\tfrac{11}{3}e^{-3t}\right)+\left(\tfrac{2}{3}e^{3t}-\tfrac{11}{3}e^{-3t}\right)=2e^{3t}+11e^{-3t}=y'.$$

2. Example 1 and Instant Exercise 1 of Section 6.3 found the solutions

$$\mathbf{x}^{(1)}=c_1\begin{pmatrix}2\\3\end{pmatrix}e^{-2t},\qquad \mathbf{x}^{(2)}=c_2\begin{pmatrix}1\\1\end{pmatrix}e^{t}.$$

The general solution, by Theorem 6.4.4, is

$$\mathbf{x}=c_1\begin{pmatrix}2\\3\end{pmatrix}e^{-2t}+c_2\begin{pmatrix}1\\1\end{pmatrix}e^{t}.$$

From the initial conditions, we have the vector equation

$$c_1\begin{pmatrix}2\\3\end{pmatrix}+c_2\begin{pmatrix}1\\1\end{pmatrix}=\begin{pmatrix}0\\1\end{pmatrix},$$

which corresponds to the scalar equations $2c_1+c_2=0$ and $3c_1+c_2=1$. Hence, $c_1=1$ and $c_2=-2$. The solution is

$$\mathbf{x}=\begin{pmatrix}2\\3\end{pmatrix}e^{-2t}-2\begin{pmatrix}1\\1\end{pmatrix}e^{t}=\begin{pmatrix}2e^{-2t}-2e^{t}\\3e^{-2t}-2e^{t}\end{pmatrix}.$$

3. In vector form, the system is $\mathbf{x}'=\mathbf{A}\mathbf{x}$, with $\mathbf{A}=\begin{pmatrix}-2&2\\2&1\end{pmatrix}$. The eigenvalues satisfy the equation

$$0=\det(\mathbf{A}-\lambda\mathbf{I})=\begin{vmatrix}-2-\lambda&2\\2&1-\lambda\end{vmatrix}=\lambda^2+\lambda-6=(\lambda+3)(\lambda-2).$$

Since one eigenvalue is positive and the other negative, the origin is a saddle.

4. The eigenvalues satisfy the equation

$$0=\det(\mathbf{A}-\lambda\mathbf{I})=\begin{vmatrix}1-\lambda&-2\\1&4-\lambda\end{vmatrix}=\lambda^2-5\lambda+6=(\lambda-2)(\lambda-3);$$

hence, the eigenvalues are 2 and 3. If $\mathbf{u}$ is an eigenvector for $\lambda=2$, we have

$$\begin{pmatrix}-1&-2\\1&2\end{pmatrix}\begin{pmatrix}u_1\\u_2\end{pmatrix}=\begin{pmatrix}0\\0\end{pmatrix}.$$

Thus, the eigenvectors must satisfy $u_1 + 2u_2 = 0$; we choose $u_1 = -2$ and $u_2 = 1$ as a representative eigenvector. Similarly, if $\mathbf{u}$ is an eigenvector for $\lambda = 3$, we have

$$\begin{pmatrix} -2 & -2 \\ 1 & 1 \end{pmatrix} \begin{pmatrix} u_1 \\ u_2 \end{pmatrix} = \begin{pmatrix} 0 \\ 0 \end{pmatrix},$$

so we can choose $u_1 = -1$ and $u_2 = 1$. Each eigenvalue-eigenvector pair corresponds to a solution, and their linear combination

$$\mathbf{x} = c_1 \begin{pmatrix} -2 \\ 1 \end{pmatrix} e^{2t} + c_2 \begin{pmatrix} -1 \\ 1 \end{pmatrix} e^{3t}$$

is the general solution.

6.5 Homogeneous Systems with Complex Eigenvalues

In Chapter 3, we considered homogeneous scalar linear equations with constant real coefficients. The assumption of solutions of the form $y = e^{\lambda t}$ leads to a polynomial equation for λ. If there are nonreal roots, we end up with real-valued solutions that include a trigonometric factor as well as an exponential factor.

The same general features occur for homogeneous systems of linear equations with constant real coefficients. We assume a solution of the form $\mathbf{x}(t) = \mathbf{v}e^{\lambda t}$, with $\mathbf{v}$ an unknown constant vector, and get a polynomial equation for λ. If there are nonreal roots, we will again have real-valued solutions that include a trigonometric factor.

MODEL PROBLEM 6.5

Solve the problem

$$x' = -2x - 3y, \qquad y' = 3x - 2y.$$

As in Section 6.3, we first write the problem in matrix-vector form by defining a vector $\mathbf{x}$ whose scalar components x_1 and x_2 correspond to the scalar variables x and y,

$$\mathbf{x}' = \mathbf{A}\mathbf{x}, \qquad \text{where} \quad \mathbf{A} = \begin{pmatrix} -2 & -3 \\ 3 & -2 \end{pmatrix}.$$

The eigenvalues are given by

$$0 = \det(\mathbf{A} - \lambda\mathbf{I}) = \begin{vmatrix} -2-\lambda & -3 \\ 3 & -2-\lambda \end{vmatrix} = \lambda^2 + 4\lambda + 13.$$

From the quadratic formula, we obtain the eigenvalues

$$\lambda = -2 \pm 3i.$$

If $z = \alpha + i\beta$ is one complex number, the related number $\bar{z} = \alpha - i\beta$ is the **complex conjugate** of z. Any quadratic polynomial with real coefficients and no real roots has a pair of complex conjugate roots.

Eigenvectors for Complex Eigenvalues

Suppose $\lambda = \alpha + i\beta$, with $\beta \neq 0$, is an eigenvalue of a matrix $\mathbf{A}$, with $\mathbf{z} = \mathbf{u} + i\mathbf{v}$ an eigenvector corresponding to λ. Here, the scalars α and β and the vectors $\mathbf{u}$ and $\mathbf{v}$ are real-valued. Theorem 6.5.1 reduces the labor of computing the eigenvectors by one-half.

Theorem 6.5.1 If $\mathbf{z}$ is an eigenvector of a matrix $\mathbf{A}$ corresponding to a nonreal eigenvalue λ, then $\bar{\mathbf{z}}$ is an eigenvector of $\mathbf{A}$ corresponding to the eigenvalue $\bar{\lambda}$.

For Model Problem 6.5, define $\lambda = -2 + 3i$. Then

$$\mathbf{A} - \lambda\mathbf{I} = \begin{pmatrix} -3i & -3 \\ 3 & -3i \end{pmatrix},$$

so the eigenvectors are solutions of

$$\begin{pmatrix} -3i & -3 \\ 3 & -3i \end{pmatrix} \begin{pmatrix} z_1 \\ z_2 \end{pmatrix} = \begin{pmatrix} 0 \\ 0 \end{pmatrix}.$$

Only one of the scalar equations contained in this vector equation is needed because λ being an eigenvalue means that $\mathbf{A} - \lambda\mathbf{I}$ has determinant 0, which means that the scalar equations represented by the rows of the matrix are not linearly independent. So (using the top row), we may choose any eigenvector that satisfies the equation

$$-3iz_1 - 3z_2 = 0.$$

Setting $z_1 = 1$ yields $z_2 = -i$; thus,

$$\mathbf{z} = \begin{pmatrix} 1 \\ -i \end{pmatrix}$$

is an eigenvector for the eigenvalue λ. By Theorem 6.5.1,

$$\bar{\mathbf{z}} = \begin{pmatrix} 1 \\ i \end{pmatrix}$$

is an eigenvector for the eigenvalue $\bar{\lambda} = 2 - 3i$.

✦ INSTANT EXERCISE 1

Verify that $\bar{\mathbf{z}} = \begin{pmatrix} 1 \\ i \end{pmatrix}$ is an eigenvector of $\mathbf{A} = \begin{pmatrix} -2 & -3 \\ 3 & -2 \end{pmatrix}$ corresponding to $\bar{\lambda} = -2 - 3i$.

Finding a Two-Parameter Family of Solutions

We now have two eigenvalue-eigenvector pairs for a matrix **A**. As we saw in Section 6.3, each pair corresponds to a solution of the linear system. From λ and $\mathbf{z}$, with the aid of Euler's formula,[15]

$$\mathbf{x} = \begin{pmatrix} 1 \\ -i \end{pmatrix} e^{(-2+3i)t} = \begin{pmatrix} 1 \\ -i \end{pmatrix} e^{-2t}(\cos 3t + i \sin 3t) = e^{-2t} \begin{pmatrix} \cos 3t + i \sin 3t \\ \sin 3t - i \cos 3t \end{pmatrix} \tag{1}$$

is a solution of

$$\mathbf{x}' = \begin{pmatrix} -2 & -3 \\ 3 & -2 \end{pmatrix} \mathbf{x}. \tag{2}$$

We can also construct a second solution using the eigenvalue $\overline{\lambda}$, but this does not turn out to be needed. Observe that the real part of the solution (1) is a vector function

$$\mathbf{x}^{(1)} = e^{-2t} \begin{pmatrix} \cos 3t \\ \sin 3t \end{pmatrix}.$$

Now, we can easily verify that this real-valued function is also a solution. Consider just the upper component of Equation (2). The left side is $x_1' = (e^{-2t} \cos 3t)' = e^{-2t}(-2 \cos 3t - 3 \sin 3t)$, and the right side is $-2x_1 - 3x_2 = e^{-2t}(-2 \cos 3t - 3 \sin 3t)$; thus, the upper equation is satisfied. We can also show that the imaginary part,

$$\mathbf{x}^{(2)} = e^{-2t} \begin{pmatrix} \sin 3t \\ -\cos 3t \end{pmatrix},$$

of Equation (1) is a solution and that

$$\mathbf{x} = e^{-2t} \left[c_1 \begin{pmatrix} \cos 3t \\ \sin 3t \end{pmatrix} + c_2 \begin{pmatrix} \sin 3t \\ -\cos 3t \end{pmatrix} \right] \tag{3}$$

is the general solution.

✦ INSTANT EXERCISE 2

Verify by direct substitution that $\mathbf{x}^{(2)} = e^{-2t} \begin{pmatrix} \sin 3t \\ -\cos 3t \end{pmatrix}$ is a solution of Equation (2), and use Theorem 6.4.3 to show that Equation (3) is the general solution.

The phase portrait for this system appears in Figure 6.5.1, with solution curves corresponding to solutions $\pm\mathbf{x}^{(1)}$ and $\pm\mathbf{x}^{(2)}$. Notice that these solutions do not have a special graphical significance comparable to linear trajectories.

The method used for this example works equally well for any other system with complex eigenvalues. The algebra is often messier; this problem was relatively simple because the components of $\mathbf{z}$ were either strictly real or strictly imaginary. One can always construct the general solution for the complex eigenvalue case by using the same sequence of steps.

[15]See Section 3.5.

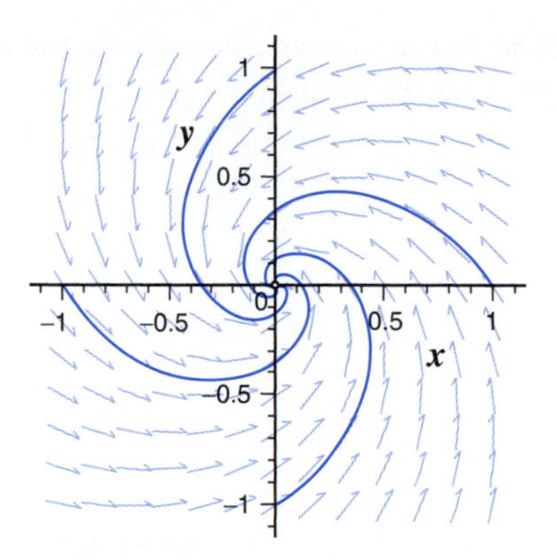

Figure 6.5.1
The phase portrait for Model Problem 6.5.

Algorithm 6.5.1

To find the general solution of a two-dimensional system $\mathbf{x}' = \mathbf{A}\mathbf{x}$, where the eigenvalues of $\mathbf{A}$ are nonreal:

1. Find a complex eigenvector $\mathbf{z}$ corresponding to an eigenvalue λ.
2. Construct a complex-valued solution of the system as $\mathbf{x} = e^{\alpha t}(\cos \beta t + i \sin \beta t)\mathbf{z}$, where $\lambda = \alpha + i\beta$.
3. Separate the solution into a real part $\mathbf{x}^{(1)}$ and an imaginary part $\mathbf{x}^{(2)}$.
4. The general solution of the system is $\mathbf{x} = c_1\mathbf{x}^{(1)} + c_2\mathbf{x}^{(2)}$.

EXAMPLE 1

Consider the problem

$$\mathbf{x}' = \begin{pmatrix} -1 & 4 \\ -2 & 3 \end{pmatrix} \mathbf{x}.$$

We begin by finding the eigenvalues of the matrix. We have

$$0 = \begin{vmatrix} -1-\lambda & 4 \\ -2 & 3-\lambda \end{vmatrix} = (-1-\lambda)(3-\lambda) + 8 = \lambda^2 - 2\lambda + 5.$$

The quadratic formula yields the eigenvalues $\lambda = 1 \pm 2i$.

Next we choose $\lambda = 1 + 2i$ and find an eigenvector. We have

$$\mathbf{A} - \lambda \mathbf{I} = \begin{pmatrix} -2-2i & 4 \\ -2 & 2-2i \end{pmatrix};$$

therefore, the eigenvectors are solutions of

$$\begin{pmatrix} -2-2i & 4 \\ -2 & 2-2i \end{pmatrix} \begin{pmatrix} z_1 \\ z_2 \end{pmatrix} = \begin{pmatrix} 0 \\ 0 \end{pmatrix}.$$

As before, the equations corresponding to the two rows are redundant, so only one is needed. Using the top row, we obtain the equation

$$(-2-2i)z_1 + 4z_2 = 0,$$

or

$$z_2 = \frac{1+i}{2} z_1.$$

We may avoid fractions by choosing $z_1 = 2$; then $z_2 = 1+i$. Hence, we have the eigenvector

$$\mathbf{z} = \begin{pmatrix} 2 \\ 1+i \end{pmatrix}$$

for the eigenvalue λ.

A complex-valued solution is constructed from the formula $\mathbf{x} = e^{\lambda t}\mathbf{z}$ and Euler's formula:

$$\mathbf{x} = \begin{pmatrix} 2 \\ 1+i \end{pmatrix} e^{(1+2i)t} = e^t(\cos 2t + i \sin 2t)\begin{pmatrix} 2 \\ 1+i \end{pmatrix} = e^t \begin{pmatrix} 2\cos 2t + 2i \sin 2t \\ (1+i)\cos 2t + (-1+i)\sin 2t \end{pmatrix}.$$

The real part of this solution is

$$\mathbf{x}^{(1)} = e^t \begin{pmatrix} 2\cos 2t \\ \cos 2t - \sin 2t \end{pmatrix}$$

and the imaginary part is

$$\mathbf{x}^{(2)} = e^t \begin{pmatrix} 2\sin 2t \\ \cos 2t + \sin 2t \end{pmatrix}.$$

Both of these real-valued functions are solutions of the system, and with them we get a general solution

$$\mathbf{x} = e^t \left[c_1 \begin{pmatrix} 2\cos 2t \\ \cos 2t - \sin 2t \end{pmatrix} + c_2 \begin{pmatrix} 2\sin 2t \\ \cos 2t + \sin 2t \end{pmatrix} \right].$$

The phase portrait, including the solutions corresponding to $c_1 = \pm 0.01$, $c_2 = 0$ and $c_1 = 0$, $c_2 = \pm 0.01$ is illustrated in Figure 6.5.2.

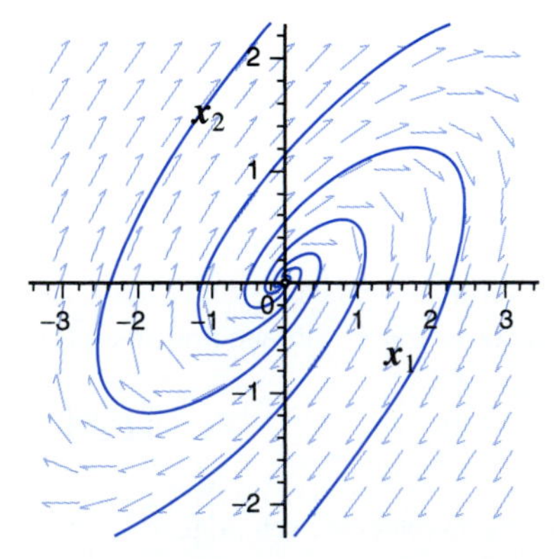

Figure 6.5.2
The phase portrait for Example 1.

Classification of the Critical Point at the Origin

Linear systems with nonreal eigenvalues have nonreal eigenvectors, and these in turn yield real-valued solutions that have an oscillatory trigonometric factor. As in Model Problem 6.5 and Example 1, the trajectories rotate around the critical point at the origin. Critical points with rotating trajectories come in three varieties, depending on the exponential factor $e^{\alpha t}$ that appears in the solution formula. When $\alpha < 0$, as in Model Problem 6.5, the solutions spiral into the origin, and the origin is an **asymptotically stable spiral,** also called a **spiral sink.** When $\alpha > 0$, as in Example 1, the solutions spiral outward and the origin is an **unstable spiral,** or **spiral source.** Finally, when the complex eigenvalue pair is purely imaginary, there is no exponential factor in the solution formula. The solutions oscillate, but neither approach the origin nor recede from it. In this case, the origin is said to be a **center.** Note that the *center* is stable, but not asymptotically stable, while a *spiral sink* is asymptotically stable. The following rule summarizes the results for the stability and classification of critical points in two dimensions.

Theorem 6.5.2

Critical Points for Two-Dimensional Linear Systems Let $\mathbf{A}$ be nonsingular 2×2 matrix of real-valued constants. The linear system $\mathbf{x}' = \mathbf{A}\mathbf{x}$ has a single critical point at the origin, which is classified according to the eigenvalues of $\mathbf{A}$:

- If all eigenvalues have negative real parts, then the critical point is asymptotically stable.
- If at least one eigenvalue has a positive real part, then the critical point is unstable.
- If the eigenvalues are real, then the critical point is a source, a sink, or a saddle, given two positive eigenvalues, two negative eigenvalues, or one of each, respectively.
- If the eigenvalues are a complex conjugate pair with nonzero real part, then the origin is a spiral; in the case of a pure imaginary pair, then the origin is a (stable) center.

Nonreal Eigenvalues in Three-Dimensional Systems

Nonreal zeros of a polynomial with real coefficients can only occur in pairs, so a three-dimensional system with nonreal eigenvalues must also have a real eigenvalue. The three eigenvalues are necessarily distinct when two are a complex pair. Thus, the procedures of this section and Section 6.4 can be used to construct the general solution.

EXAMPLE 2

Consider the system

$$\mathbf{x}' = \mathbf{A}\mathbf{x}, \qquad \mathbf{A} = \begin{pmatrix} -1 & -2 & -3 \\ 5 & -1 & 0 \\ 5 & 0 & -1 \end{pmatrix}.$$

We must first find the characteristic polynomial

$$P_A(\lambda) = \begin{vmatrix} -1-\lambda & -2 & -3 \\ 5 & -1-\lambda & 0 \\ 5 & 0 & -1-\lambda \end{vmatrix}.$$

Cofactor expansion on the second column yields

$$P_A(\lambda) = 2\begin{vmatrix} 5 & 0 \\ 5 & -1-\lambda \end{vmatrix} - (1+\lambda)\begin{vmatrix} -1-\lambda & -3 \\ 5 & -1-\lambda \end{vmatrix} = -10(1+\lambda) - (1+\lambda)(\lambda^2 + 2\lambda + 16).$$

We can make use of the common factor $-(1+\lambda)$ to obtain

$$P_A(\lambda) = -(1+\lambda)(\lambda^2 + 2\lambda + 26).$$

From the first factor, we have the eigenvalue $\lambda_1 = -1$; from the second factor, we have the complex pair

$$\lambda_2 = -1 + 5i, \qquad \overline{\lambda_2} = -1 - 5i.$$

Next, we need to find the eigenvectors. We may choose $\mathbf{v} = \begin{pmatrix} 0 \\ 3 \\ -2 \end{pmatrix}$ as the eigenvector corresponding to $\lambda_1 = -1$. For λ_2, we have

$$\mathbf{A} + (1-5i)\mathbf{I} = \begin{pmatrix} -5i & -2 & -3 \\ 5 & -5i & 0 \\ 5 & 0 & -5i \end{pmatrix} \cong \begin{pmatrix} 5i & 2 & 3 \\ 1 & -i & 0 \\ 1 & 0 & -i \end{pmatrix} \cong \begin{pmatrix} 5i & 2 & 3 \\ i & 1 & 0 \\ i & 0 & 1 \end{pmatrix} \cong \begin{pmatrix} i & 1 & 0 \\ i & 0 & 1 \\ 0 & 0 & 0 \end{pmatrix}.$$

Thus, an eigenvector $\mathbf{z}$ must satisfy the equations $z_2 = z_3 = -iz_1$; alternatively, $z_1 = iz_3$ and $z_2 = z_3$. Hence, we may choose $\mathbf{z} = \begin{pmatrix} i \\ 1 \\ 1 \end{pmatrix}$.

The real eigenvalue-eigenvector pair yields the solution

$$\mathbf{x}^{(1)} = e^{-t}\begin{pmatrix} 0 \\ 3 \\ -2 \end{pmatrix}.$$

The complex eigenvalue-eigenvector pair yields the complex solution

$$\mathbf{x} = e^{-t}(\cos 5t + i \sin 5t)\begin{pmatrix} i \\ 1 \\ 1 \end{pmatrix} = e^{-t}\begin{pmatrix} -\sin 5t + i\cos 5t \\ \cos 5t + i \sin 5t \\ \cos 5t + i \sin 5t \end{pmatrix} = e^{-t}\begin{pmatrix} -\sin 5t \\ \cos 5t \\ \cos 5t \end{pmatrix} + ie^{-t}\begin{pmatrix} \cos 5t \\ \sin 5t \\ \sin 5t \end{pmatrix}.$$

Both the real and imaginary parts of this last solution are real-valued solutions, so the pair of complex eigenvalues results in a pair of real-valued solutions:

$$\mathbf{x}^{(2)} = e^{-t}\begin{pmatrix} -\sin 5t \\ \cos 5t \\ \cos 5t \end{pmatrix}, \qquad \mathbf{x}^{(3)} = e^{-t}\begin{pmatrix} \cos 5t \\ \sin 5t \\ \sin 5t \end{pmatrix}.$$

The general solution is now seen to be

$$\mathbf{x} = e^{-t}\left[c_1\begin{pmatrix} 0 \\ 3 \\ -2 \end{pmatrix} + c_2\begin{pmatrix} -\sin 5t \\ \cos 5t \\ \cos 5t \end{pmatrix} + c_3\begin{pmatrix} \cos 5t \\ \sin 5t \\ \sin 5t \end{pmatrix}\right].$$

Figure 6.5.3 illustrates some of the simpler trajectories in the phase portrait for this system. If $c_2 = c_3 = 0$, the solution lies along the linear trajectory given by the eigenvector corresponding to the real eigenvalue. If $c_1 = 0$, then the solution lies in the plane of the spiral trajectory determined by the complex eigenvalue pair.

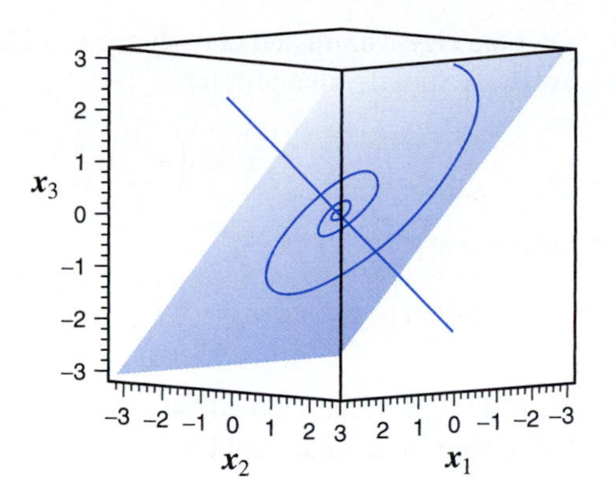

Figure 6.5.3
Some trajectories for Example 2.

✦ INSTANT EXERCISE 3

Find the eigenspace for $\lambda_1 = -1$ in Example 2, showing that $\mathbf{v} = \begin{pmatrix} 0 \\ 3 \\ -2 \end{pmatrix}$ is a suitable eigenvector.

Classification of Equilibrium Points in Higher Dimensions Critical points in three or more dimensions do not necessarily fit the shape classifications for two dimensions. It is, however, a simple matter to at least classify critical points by stability.

Theorem 6.5.3

Critical Points for n-Dimensional Linear Systems Let $\mathbf{A}$ be nonsingular $n \times n$ matrix of real-valued constants. Then the linear system $\mathbf{x}' = \mathbf{A}\mathbf{x}$ has a single critical point at the origin, which is classified according to the eigenvalues of $\mathbf{A}$:

- If all eigenvalues have nonpositive real parts, then the origin is stable.
- If all eigenvalues have negative real parts, then the origin is asymptotically stable.
- If any eigenvalue has a positive real part, then the origin is unstable.

The origin in Example 2 is asymptotically stable because all eigenvalues have real part -1.

6.5 Exercises

1. Consider the system

$$\mathbf{x}' = \begin{pmatrix} 0 & 1 \\ -4 & 0 \end{pmatrix} \mathbf{x}.$$

a. Find the eigenvalues and classify the equilibrium point at the origin.
b. Solve the initial-value problem

$$\mathbf{x}' = \begin{pmatrix} 0 & 1 \\ -4 & 0 \end{pmatrix}\mathbf{x}, \qquad \mathbf{x}(0) = \begin{pmatrix} 1 \\ -2 \end{pmatrix}.$$

2. Consider the system

$$\mathbf{x}' = \begin{pmatrix} 0 & -3 \\ 2 & 0 \end{pmatrix}\mathbf{x}.$$

a. Find the eigenvalues and classify the equilibrium point at the origin.
b. Solve the initial-value problem

$$\mathbf{x}' = \begin{pmatrix} 0 & -3 \\ 2 & 0 \end{pmatrix}\mathbf{x}, \qquad \mathbf{x}(0) = \begin{pmatrix} 1 \\ 2 \end{pmatrix}.$$

3. Consider the system

$$\mathbf{x}' = \begin{pmatrix} 3 & 1 \\ -1 & 3 \end{pmatrix}\mathbf{x}.$$

a. Find the eigenvalues and classify the equilibrium point at the origin.
b. Solve the initial-value problem

$$\mathbf{x}' = \begin{pmatrix} 3 & 1 \\ -1 & 3 \end{pmatrix}\mathbf{x}, \qquad \mathbf{x}(0) = \begin{pmatrix} 2 \\ 1 \end{pmatrix}.$$

4. Consider the system

$$\mathbf{x}' = \begin{pmatrix} -3 & -1 \\ 4 & -3 \end{pmatrix}\mathbf{x}.$$

a. Find the eigenvalues and classify the equilibrium point at the origin.
b. Solve the initial-value problem

$$\mathbf{x}' = \begin{pmatrix} -3 & -1 \\ 4 & -3 \end{pmatrix}\mathbf{x}, \qquad \mathbf{x}(0) = \begin{pmatrix} 1 \\ 3 \end{pmatrix}.$$

5. Consider the system

$$\mathbf{x}' = \begin{pmatrix} -5 & -9 \\ 2 & 1 \end{pmatrix}\mathbf{x}.$$

a. Find the eigenvalues and classify the equilibrium point at the origin.
b. Solve the initial-value problem

$$\mathbf{x}' = \begin{pmatrix} -5 & -9 \\ 2 & 1 \end{pmatrix}\mathbf{x}, \qquad \mathbf{x}(0) = \begin{pmatrix} 1 \\ -2 \end{pmatrix}.$$

6. Consider the system

$$\mathbf{x}' = \begin{pmatrix} 2 & -3 \\ 4 & 2 \end{pmatrix}\mathbf{x}.$$

a. Find the eigenvalues and classify the equilibrium point at the origin.
b. Solve the initial-value problem

$$\mathbf{x}' = \begin{pmatrix} 2 & -3 \\ 4 & 2 \end{pmatrix} \mathbf{x}, \qquad \mathbf{x}(0) = \begin{pmatrix} 2 \\ 1 \end{pmatrix}.$$

7. Consider the system

$$\mathbf{x}' = \begin{pmatrix} 2 & 2 & 0 \\ -4 & 6 & 0 \\ 0 & 0 & -1 \end{pmatrix} \mathbf{x}.$$

a. Find the eigenvalues and determine the stability of the equilibrium point at the origin.
b. Solve the initial-value problem

$$\mathbf{x}' = \begin{pmatrix} 2 & 2 & 0 \\ -4 & 6 & 0 \\ 0 & 0 & -1 \end{pmatrix} \mathbf{x}, \qquad \mathbf{x}(0) = \begin{pmatrix} 1 \\ 0 \\ 1 \end{pmatrix}.$$

8. Consider the system

$$\mathbf{x}' = \begin{pmatrix} -2 & 0 & 2 \\ 1 & -1 & 0 \\ 0 & -1 & -2 \end{pmatrix} \mathbf{x}.$$

a. Find the eigenvalues and determine the stability of the equilibrium point at the origin.
b. Solve the initial-value problem

$$\mathbf{x}' = \begin{pmatrix} -2 & 0 & 2 \\ 1 & -1 & 0 \\ 0 & -1 & -2 \end{pmatrix} \mathbf{x}, \qquad \mathbf{x}(0) = \begin{pmatrix} 0 \\ -1 \\ 3 \end{pmatrix}.$$

9. Prove Theorem 6.5.1.

10. Prove statement 4 in Algorithm 6.5.1.

11. Figure 6.5.4 depicts a mechanical system consisting of two unit masses and three springs with different spring constants. Let $x(t)$ and $y(t)$ be the displacements of the masses from the equilibrium position as shown.

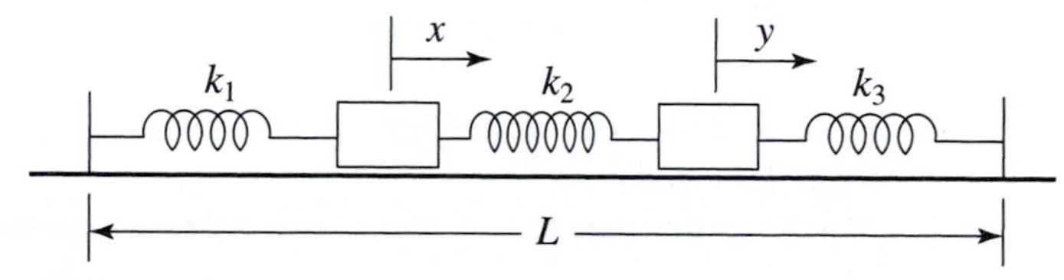

Figure 6.5.4
The mechanical system for Exercise 11.

a. Derive a system of two second-order differential equations for the displacements, assuming no forces other than the restoring forces of the three springs.
b. Let $u = x'$ and $v = y'$, and define a vector $\mathbf{x}$ whose components are, in order, x, y, u, and v. Determine the differential equation for the vector $\mathbf{x}$.

c. Investigate the stability of the equilibrium position for all nonnegative values of k_1, k_2, and k_3.

d. Let $k_1 = k_3 = 1$ and $k_2 = 4$. Find the general solution.

e. Assume that the supports are separated by a distance L. Show that the distance between the masses oscillates, and determine the period of the oscillation. Show that the position of the midpoint between the two masses also oscillates, and determine the period of oscillation.

f. Suppose the mass on the left is displaced 1 unit to the left and the mass on the right is allowed to reach an equilibrium position. At time 0, the mass on the left is released. Determine the subsequent motion. In particular, plot $x(t)$ and $y(t)$. (*Hint:* what is the acceleration of the mass on the right if it is at equilibrium?)

12. Figure 6.5.5 depicts a parallel circuit with three branches. Let $v(t)$ be the voltage across the capacitor, and let $i(t)$ be the current through the inductor. Let $v_1(t)$, $v_2(t)$, $i_2(t)$, and $i_3(t)$ be the voltage across the inductor, the voltage across the resistor that is in series with the inductor, the current through the resistor that is in parallel with the inductor, and the current through the capacitor, respectively. By Kirchoff's current law, the current through the resistor that is in series with the inductor is also $i(t)$, and the sum of the currents out of the node at the left is $i(t) + i_2(t) + i_3(t) = 0$. Similarly, by Kirchoff's voltage law, the voltage across the other resistor is also v, and the sum of voltages through the *RL* series branch is $v_1(t) + v_2(t) = v(t)$. The remaining equations follow from the definition of the circuit components,[16]

$$v = Ri_2, \qquad v_2 = Ri, \qquad Li' = v_1, \qquad Cv' = i_3.$$

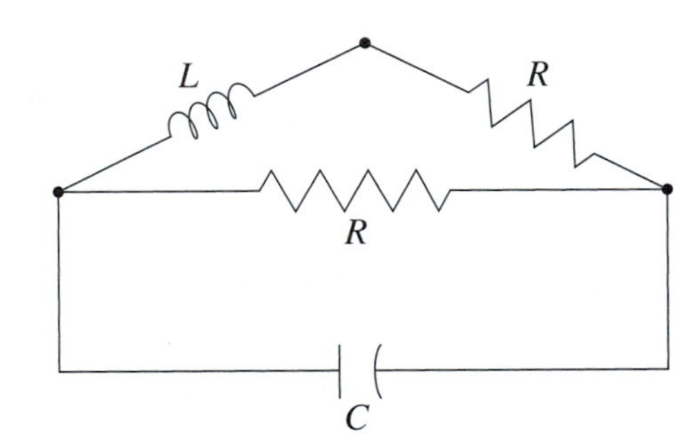

Figure 6.5.5
The electric circuit for Exercise 12.

a. Determine a system of two homogeneous first-order linear differential equations for the variables v and i.

b. Nondimensionalize the equations, using the new quantities

$$V = \frac{v}{v_r}, \qquad I = \frac{\sqrt{L}\,i}{\sqrt{C}\,v_r}, \qquad \tau = \frac{t}{\sqrt{LC}}, \qquad K = R\frac{\sqrt{C}}{\sqrt{L}}.$$

c. Show that the origin is an asymptotically stable critical point for all positive K.

[16] See Section 4.1.

d. Show that the eigenvalues are a complex pair if $K + K^{-1} < 2\sqrt{2}$, and find the corresponding range of K values.

e. Choose the value of K that yields the slowest decay rate for the solution. Solve the problem with that value of K and the initial conditions $V(0) = 1$ and $I(0) = 0$, corresponding to the case where the precharged capacitor was joined to the remainder of the circuit at time 0.

f. Plot the solutions for V and I from part *e*, and discuss the behavior of the circuit.

✦ 6.5 INSTANT EXERCISE SOLUTIONS

1. The simplest way to verify that $\overline{\mathbf{z}}$ is an eigenvector corresponding to $\overline{\lambda}$ is to compute $\mathbf{A}\overline{\mathbf{z}}$ and $\overline{\lambda}\overline{\mathbf{z}}$.

$$\mathbf{A}\overline{\mathbf{z}} = \begin{pmatrix} -2 & -3 \\ 3 & -2 \end{pmatrix}\begin{pmatrix} 1 \\ i \end{pmatrix} = \begin{pmatrix} -2-3i \\ 3-2i \end{pmatrix}, \qquad \overline{\lambda}\overline{\mathbf{z}} = (-2-3i)\begin{pmatrix} 1 \\ i \end{pmatrix} = \begin{pmatrix} -2-3i \\ 3-2i \end{pmatrix}.$$

2. Given $\mathbf{A} = \begin{pmatrix} -2 & -3 \\ 3 & -2 \end{pmatrix}$ and $\mathbf{x}^{(2)} = e^{-2t}\begin{pmatrix} \sin 3t \\ -\cos 3t \end{pmatrix}$,

$$\mathbf{x}^{(2)\prime} = -2e^{-2t}\begin{pmatrix} \sin 3t \\ -\cos 3t \end{pmatrix} + e^{-2t}\begin{pmatrix} 3\cos 3t \\ 3\sin 3t \end{pmatrix} = e^{-2t}\begin{pmatrix} -2\sin 3t + 3\cos 3t \\ 2\cos 3t + 3\sin 3t \end{pmatrix}$$

and

$$\mathbf{A}\mathbf{x}^{(2)} = \begin{pmatrix} -2 & -3 \\ 3 & -2 \end{pmatrix} e^{-2t}\begin{pmatrix} \sin 3t \\ -\cos 3t \end{pmatrix} = e^{-2t}\begin{pmatrix} -2\sin 3t + 3\cos 3t \\ 3\sin 3t + 2\cos 3t \end{pmatrix}.$$

3. For the eigenvalue $\lambda_1 = -1$, we have

$$\mathbf{A} + \mathbf{I} = \begin{pmatrix} 0 & -2 & -3 \\ 5 & 0 & 0 \\ 5 & 0 & 0 \end{pmatrix} \cong \begin{pmatrix} 0 & 2 & 3 \\ 1 & 0 & 0 \\ 0 & 0 & 0 \end{pmatrix}.$$

If $\mathbf{v} = \begin{pmatrix} v_1 \\ v_2 \\ v_3 \end{pmatrix}$ is an eigenvector corresponding to the eigenvalue -1, then $v_1 = 0$ and $2v_2 + 3v_3 = 0$.

6.6 Additional Solutions for Deficient Matrices

The basic plan for solving $n \times n$ linear systems requires us to find a set of eigenspaces whose dimensions add up to n. This can be accomplished whenever the total of the geometric multiplicities[17] of the eigenvalues is n, that is, when every eigenvalue of algebraic multiplicity m

[17] See Section 6.2 for the definitions of algebraic and geometric multiplicity.

has a corresponding m-dimensional eigenspace. As long as the characteristic polynomial can be factored into n distinct linear (possibly complex) factors, we are guaranteed a sufficient set of eigenspaces. Even if the characteristic polynomial has factors of algebraic multiplicity greater than 1, the geometric multiplicities of the corresponding eigenvalues might be large enough. Example 1 of Section 6.4 illustrates this case. However, if any eigenvalue has geometric multiplicity less than its algebraic multiplicity, then the total of the eigenspace dimensions is less than n and there will not be enough solutions based on eigenvalue-eigenvector pairs with which to construct the general solution. Thus, deficient matrices[18] are matrices for which additional solutions must be found by a different method.

MODEL PROBLEM 6.6

Solve the problem $\mathbf{x}' = \mathbf{A}\mathbf{x}$, where

$$\mathbf{A} = \begin{pmatrix} -3 & 1 \\ -1 & -1 \end{pmatrix}.$$

As always, we begin by finding the characteristic equation and solving it for the eigenvalues.

$$0 = \det(\mathbf{A} - \lambda\mathbf{I}) = \begin{vmatrix} -3-\lambda & 1 \\ -1 & -1-\lambda \end{vmatrix} = (-3-\lambda)(-1-\lambda) + 1 = \lambda^2 + 4\lambda + 4 = (\lambda+2)^2.$$

We have a 2×2 system with only one eigenvalue, $\lambda = -2$. The eigenspace is only one-dimensional, with

$$\mathbf{v} = \begin{pmatrix} 1 \\ 1 \end{pmatrix}$$

a suitable eigenvector. This yields a solution,

$$\mathbf{x}^{(1)} = e^{-2t}\begin{pmatrix} 1 \\ 1 \end{pmatrix}.$$

INSTANT EXERCISE 1

Find the eigenspace for $\lambda = -2$ in Model Problem 6.6.

A Second Linearly Independent Solution

Systems of linear equations share many properties with higher-order scalar equations. The methods that work for scalar equations do not always apply to systems, but they can often suggest similar methods for systems.

[18]See Section 6.2.

EXAMPLE 1

Consider the problem

$$y'' + 4y' + 4y = 0.$$

The substitution $y = e^{rt}$ yields the characteristic equation $r^2 + 4r + 4 = 0$, the same characteristic equation as in Model Problem 6.6. We have one solution, $y_1 = e^{-2t}$. The method of reduction of order[19] yields the additional solution $y_2 = te^{-2t}$.

✦ INSTANT EXERCISE 2

Verify that $y_2 = te^{-2t}$ is a solution of $y'' + 4y' + 4y = 0$.

Example 1 suggests that the second solution in Model Problem 6.6 is

$$\mathbf{x}^{(2)} = t\mathbf{x}^{(1)} = te^{-2t}\begin{pmatrix} 1 \\ 1 \end{pmatrix}.$$

However, this is an instance in which the generalization from scalar mathematics to vector mathematics must be made with care. Given $\mathbf{x}^{(2)}$ as defined here,

$$(\mathbf{x}^{(2)})' = \left(t\mathbf{x}^{(1)}\right)' = (-2t+1)e^{-2t}\begin{pmatrix} 1 \\ 1 \end{pmatrix}, \qquad \mathbf{A}\mathbf{x}^{(2)} = \mathbf{A}\left(t\mathbf{x}^{(1)}\right) = -2te^{-2t}\begin{pmatrix} 1 \\ 1 \end{pmatrix}.$$

Instead of being equal, $(\mathbf{x}^{(2)})'$ and $\mathbf{A}\mathbf{x}^{(2)}$ differ by a constant multiple of $\mathbf{x}^{(1)}$. Specifically,

$$\left(t\mathbf{x}^{(1)}\right)' = \mathbf{A}\left(t\mathbf{x}^{(1)}\right) + \mathbf{x}^{(1)}.$$

Although $t\mathbf{x}^{(1)}$ is *not* a solution, it makes sense to look for a solution that is similar to $t\mathbf{x}^{(1)}$. Perhaps the substitution

$$\mathbf{x}^{(2)} = t\mathbf{x}^{(1)} + \mathbf{y}(t)$$

can yield a problem for $\mathbf{y}$ that is easier than the original problem. This substitution yields the results

$$\left(\mathbf{x}^{(2)}\right)' = \left(t\mathbf{x}^{(1)}\right)' + \mathbf{y}' = t\left(\mathbf{x}^{(1)}\right)' + \mathbf{x}^{(1)} + \mathbf{y}' = t\left(\mathbf{A}\mathbf{x}^{(1)}\right) + \mathbf{x}^{(1)} + \mathbf{y}'$$

and

$$\mathbf{A}\mathbf{x}^{(2)} = \mathbf{A}\left(t\mathbf{x}^{(1)}\right) + \mathbf{A}\mathbf{y} = t\mathbf{A}\mathbf{x}^{(1)} + \mathbf{A}\mathbf{y}$$

Equating these results yields a differential equation for $\mathbf{y}$,

$$\mathbf{y}' - \mathbf{A}\mathbf{y} = -\mathbf{x}^{(1)} = -e^{-2t}\mathbf{v}.$$

[19] See Section 3.6.

This equation is a nonhomogeneous linear system. We have not yet studied such systems; however, it is not difficult to suggest a solution based on scalar nonhomogeneous linear equations. Suppose the vector function $\mathbf{y}$ includes a factor e^{-2t}. Then each term in the equation for $\mathbf{y}$ will have the same factor, allowing it to be removed from the equation. Specifically, suppose we look for a solution of the form

$$\mathbf{y}(t) = e^{-2t}\mathbf{w},$$

where $\mathbf{w}$ is a constant vector. This substitution yields the equation

$$-2\mathbf{w} - \mathbf{A}\mathbf{w} = -\mathbf{v},$$

which we can rewrite as

$$(\mathbf{A} + 2\mathbf{I})\mathbf{w} = \mathbf{v} = \begin{pmatrix} 1 \\ 1 \end{pmatrix}.$$

This is a nonhomogeneous linear algebraic equation. The matrix $\mathbf{A} + 2\mathbf{I}$ is singular because -2 is an eigenvalue of $\mathbf{A}$; by Theorem 3.2.1, the system does *not* have a unique solution. Either it has infinitely many solutions, or it has no solutions.

If the components of $\mathbf{w}$ are w_1 and w_2, we have

$$\begin{pmatrix} -1 & 1 \\ -1 & 1 \end{pmatrix} \begin{pmatrix} w_1 \\ w_2 \end{pmatrix} = \begin{pmatrix} 1 \\ 1 \end{pmatrix}.$$

The two scalar equations are redundant, so we may choose any vector $\mathbf{w}$ that satisfies $-w_1 + w_2 = 1$. There is a one-parameter family of solutions, and we may choose any one solution. One choice is $w_1 = 0$, $w_2 = 1$.[20] Putting the results together yields the second solution

$$\mathbf{x}^{(2)} = te^{-2t}\begin{pmatrix} 1 \\ 1 \end{pmatrix} + e^{-2t}\begin{pmatrix} 0 \\ 1 \end{pmatrix} = e^{-2t}\begin{pmatrix} t \\ t+1 \end{pmatrix}$$

Finally, the general solution is obtained from Theorem 6.4.3 as an arbitrary linear combination of $\mathbf{x}^{(1)}$ and $\mathbf{x}^{(2)}$:

$$\mathbf{x} = e^{-2t}\left[c_1\begin{pmatrix} 1 \\ 1 \end{pmatrix} + c_2\begin{pmatrix} t \\ t+1 \end{pmatrix}\right].$$

✦ INSTANT EXERCISE 3

Check the solution $\mathbf{x}^{(2)}$ by verifying that it satisfies the corresponding system of scalar differential equations. Check the general solution by verifying that the matrix $\mathbf{\Psi}(0)$ defined in Theorem 6.4.3 is nonsingular.

[20] Any other choice will yield a *different* $\mathbf{x}^{(2)}$ and a different formula for the *same* general solution.

The phase portrait appears in Figure 6.6.1.

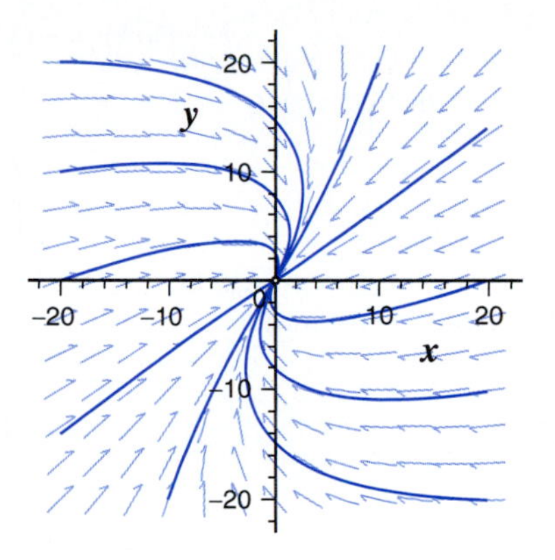

Figure 6.6.1
The phase portrait for Model Problem 6.6.

The General Case

The procedure used in Model Problem 6.6 to find a second solution for eigenvalues of algebraic multiplicity 2 and geometric multiplicity 1 is quite general. To summarize:

Theorem 6.6.1

If $\mathbf{A}$ is a constant real matrix and λ is a real eigenvalue of $\mathbf{A}$ of algebraic multiplicity 2 and geometric multiplicity 1 with $\mathbf{v}$ an eigenvector, then the two-parameter family of solutions of $\mathbf{x}' = \mathbf{A}\mathbf{x}$ corresponding to λ is the linear combination of the solutions

$$\mathbf{x}^{(1)} = e^{\lambda t}\mathbf{v}$$

and

$$\mathbf{x}^{(2)} = e^{\lambda t}(t\mathbf{v} + \mathbf{w}),$$

where $\mathbf{w}$ is any solution of

$$(\mathbf{A} - \lambda\mathbf{I})\mathbf{w} = \mathbf{v}.$$

The vector $\mathbf{w}$ in Theorem 6.6.1 is called a **generalized eigenvector.**

EXAMPLE 2

Consider the problem

$$\mathbf{x}' = \mathbf{A}\mathbf{x}, \qquad \mathbf{x}(0) = \begin{pmatrix} 1 \\ 0 \\ 2 \end{pmatrix}, \qquad \mathbf{A} = \begin{pmatrix} 2 & 1 & -2 \\ -1 & 0 & 0 \\ 0 & 2 & -2 \end{pmatrix}.$$

We begin by finding the characteristic polynomial $P_A(\lambda) = \det(\mathbf{A} - \lambda\mathbf{I})$. After the routine calculation of the determinant, we find

$$P_A(\lambda) = -(\lambda - 2)(\lambda + 1)^2.$$

For $\lambda = 2$, we find an eigenvector

$$\mathbf{u} = \begin{pmatrix} -4 \\ 2 \\ 1 \end{pmatrix};$$

for $\lambda = -1$, we find an eigenvector

$$\mathbf{v} = \begin{pmatrix} 1 \\ 1 \\ 2 \end{pmatrix}.$$

The matrix $\mathbf{A}$ is deficient because the eigenvalue -1 has algebraic multiplicity 2 and geometric multiplicity 1. An additional solution is needed for this eigenvalue, as prescribed by Theorem 6.6.1. To complete the solution, we need to find a generalized eigenvector $\mathbf{w}$ from the equation

$$(\mathbf{A} + \mathbf{I})\mathbf{w} = \mathbf{v},$$

or

$$\begin{pmatrix} 3 & 1 & -2 \\ -1 & 1 & 0 \\ 0 & 2 & -1 \end{pmatrix} \begin{pmatrix} w_1 \\ w_2 \\ w_3 \end{pmatrix} = \begin{pmatrix} 1 \\ 1 \\ 2 \end{pmatrix}.$$

The augmented matrix[21] for this system is

$$(\mathbf{A} + \mathbf{I}|\mathbf{v}) = \left(\begin{array}{ccc|c} 3 & 1 & -2 & 1 \\ -1 & 1 & 0 & 1 \\ 0 & 2 & -1 & 2 \end{array}\right) \cong \left(\begin{array}{ccc|c} -1 & 1 & 0 & 1 \\ 3 & 1 & -2 & 1 \\ 0 & 2 & -1 & 2 \end{array}\right)$$

$$\cong \left(\begin{array}{ccc|c} -1 & 1 & 0 & 1 \\ 0 & 4 & -2 & 4 \\ 0 & 2 & -1 & 2 \end{array}\right) \cong \left(\begin{array}{ccc|c} -1 & 1 & 0 & 1 \\ 0 & 0 & 0 & 0 \\ 0 & 2 & -1 & 2 \end{array}\right).$$

The equations for the scalar components of the generalized eigenvector are $-w_1 + w_2 = 1$ and $2w_2 - w_3 = 2$. Taking $w_1 = 0$, we get $w_2 = 1$ and $w_3 = 0$. Putting everything together, we have solutions

$$\mathbf{x}^{(1)} = e^{2t} \begin{pmatrix} -4 \\ 2 \\ 1 \end{pmatrix}, \qquad \mathbf{x}^{(2)} = e^{-t} \begin{pmatrix} 1 \\ 1 \\ 2 \end{pmatrix}, \qquad \mathbf{x}^{(3)} = e^{-t} \begin{pmatrix} t \\ t+1 \\ 2t \end{pmatrix}.$$

The general solution is therefore

$$\mathbf{x} = c_1 e^{2t} \begin{pmatrix} -4 \\ 2 \\ 1 \end{pmatrix} + e^{-t} \left[c_2 \begin{pmatrix} 1 \\ 1 \\ 2 \end{pmatrix} + c_3 \begin{pmatrix} t \\ t+1 \\ 2t \end{pmatrix} \right].$$

[21] See Section 3.2.

The initial condition becomes

$$c_1 \begin{pmatrix} -4 \\ 2 \\ 1 \end{pmatrix} + c_2 \begin{pmatrix} 1 \\ 1 \\ 2 \end{pmatrix} + c_3 \begin{pmatrix} 0 \\ 1 \\ 0 \end{pmatrix} = \begin{pmatrix} 1 \\ 0 \\ 2 \end{pmatrix},$$

for which the augmented matrix is

$$\left(\begin{array}{ccc|c} -4 & 1 & 0 & 1 \\ 2 & 1 & 1 & 0 \\ 1 & 2 & 0 & 2 \end{array}\right) \cong \left(\begin{array}{ccc|c} 1 & 2 & 0 & 2 \\ -4 & 1 & 0 & 1 \\ 2 & 1 & 1 & 0 \end{array}\right) \cong \left(\begin{array}{ccc|c} 1 & 2 & 0 & 2 \\ 0 & 9 & 0 & 9 \\ 0 & -3 & 1 & -4 \end{array}\right) \cong \left(\begin{array}{ccc|c} 1 & 0 & 0 & 0 \\ 0 & 1 & 0 & 1 \\ 0 & 0 & 1 & -1 \end{array}\right).$$

Thus, $c_1 = 0$, $c_2 = 1$, and $c_3 = -1$. The final solution is

$$\mathbf{x} = e^{-t} \begin{pmatrix} 1 - t \\ -t \\ 2 - 2t \end{pmatrix}.$$

✦ INSTANT EXERCISE 4

Find the characteristic polynomial in Example 2.

✦ INSTANT EXERCISE 5

Find the eigenvector $\mathbf{u}$ for $\lambda = 2$ and the eigenvector $\mathbf{v}$ for $\lambda = -1$ in Example 2.

Higher-order deficiencies can be handled in a manner analogous to the solution of Model Problem 6.6. The methods are illustrated in Exercises 11 and 12.

6.6 Exercises

In Exercises 1 through 9, find the solutions of $\mathbf{x}' = \mathbf{A}\mathbf{x}$ with the given initial condition.

1. $\mathbf{A} = \begin{pmatrix} 3 & 2 \\ -2 & -1 \end{pmatrix}, \quad \mathbf{x}(0) = \begin{pmatrix} 1 \\ 0 \end{pmatrix}$

2. $\mathbf{A} = \begin{pmatrix} -2 & -1 \\ 4 & -6 \end{pmatrix}, \quad \mathbf{x}(0) = \begin{pmatrix} 0 \\ 1 \end{pmatrix}$

3. $\mathbf{A} = \begin{pmatrix} 4 & 3 \\ -3 & -2 \end{pmatrix}, \quad \mathbf{x}(0) = \begin{pmatrix} 2 \\ 5 \end{pmatrix}$

4. $\mathbf{A} = \begin{pmatrix} -2 & 0 \\ 3 & -2 \end{pmatrix}, \quad \mathbf{x}(0) = \begin{pmatrix} 4 \\ 1 \end{pmatrix}$

5. $\mathbf{A} = \begin{pmatrix} 2 & 1 \\ -1 & 4 \end{pmatrix}, \quad \mathbf{x}(0) = \begin{pmatrix} 2 \\ 3 \end{pmatrix}$

6. $\mathbf{A} = \begin{pmatrix} -5 & -2 \\ 2 & -1 \end{pmatrix}, \quad \mathbf{x}(0) = \begin{pmatrix} 1 \\ -2 \end{pmatrix}$

7. $\mathbf{A} = \begin{pmatrix} 2 & 0 & -1 \\ 0 & 2 & 1 \\ 0 & -1 & 0 \end{pmatrix}, \quad \mathbf{x}(0) = \begin{pmatrix} 0 \\ 1 \\ 1 \end{pmatrix}$

8. $\mathbf{A} = \begin{pmatrix} 1 & 0 & 1 \\ -1 & 1 & 0 \\ -4 & 0 & -3 \end{pmatrix}, \quad \mathbf{x}(0) = \begin{pmatrix} 1 \\ 2 \\ 0 \end{pmatrix}$

9. $\mathbf{A} = \begin{pmatrix} -2 & 0 & 1 \\ 2 & -4 & 1 \\ -1 & 0 & -4 \end{pmatrix}, \quad \mathbf{x}(0) = \begin{pmatrix} 1 \\ 0 \\ 1 \end{pmatrix}$

10. The manipulations needed to formulate the characteristic polynomial for a matrix can be done in the general case to avoid the need to compute a determinant for each problem. Let $\mathbf{A}$ be a constant 3×3 matrix. Let $\mathbf{A}_k$ be the matrix obtained from $\mathbf{A}$ by deleting row k and column k. Show that the characteristic polynomial $P_A(\lambda) = \det(\mathbf{A} - \lambda\mathbf{I})$ is given by $-P_A(\lambda) = \lambda^3 + c_1\lambda^2 + c_2\lambda + c_3$, where $c_1 = -\operatorname{tr}(\mathbf{A})$, $c_2 = \sum_{k=1}^{3} \det(\mathbf{A_k})$, $c_3 = -\det(\mathbf{A})$, and the **trace** of an $n \times n$ matrix $\mathbf{A}$ is defined by $\operatorname{tr}(\mathbf{A}) = \sum_{k=1}^{n} a_{kk}$.

11. Consider the system $\mathbf{x}' = \mathbf{A}\mathbf{x}$, where

$$\mathbf{A} = \begin{pmatrix} 0 & 1 & 2 \\ 1 & 0 & 2 \\ -1 & -1 & -3 \end{pmatrix}.$$

a. Show that $\lambda = -1$ is an eigenvalue of algebraic multiplicity 3.

b. Show that the geometric multiplicity of $\lambda = -1$ is 2, and find a set of two vectors $\mathbf{u}$ and $\mathbf{v}$ to represent the eigenspace.

c. Assume that a third solution has the form

$$\mathbf{x}^{(3)} = te^{-t}\mathbf{v} + e^{-t}\mathbf{w}$$

and find a suitable vector $\mathbf{w}$. Note that it does not matter what vector is chosen for $\mathbf{v}$ as long as it is in the eigenspace.

d. Check that the conditions for a general solution, given by Theorem 6.4.3, are met.

e. Solve the initial-value problem

$$\mathbf{x}' = \mathbf{A}\mathbf{x}, \qquad \mathbf{x}(0) = \begin{pmatrix} 1 \\ 0 \\ 0 \end{pmatrix}.$$

12. Consider the system $\mathbf{x}' = \mathbf{A}\mathbf{x}$, where

$$\mathbf{A} = \begin{pmatrix} 0 & 1 & 2 \\ 1 & 0 & 2 \\ 0 & -2 & -3 \end{pmatrix}.$$

a. Show that $\lambda = -1$ is an eigenvalue of algebraic multiplicity 3.

b. Show that the geometric multiplicity of $\lambda = -1$ is 1, and find an eigenvector $\mathbf{u}$ to represent the eigenspace.

c. Assume that a second solution has the form

$$\mathbf{x}^{(2)} = te^{-t}\mathbf{u} + e^{-t}\mathbf{v}$$

and find a suitable generalized eigenvector $\mathbf{v}$.

d. Assume that a third solution has the form

$$\mathbf{x}^{(3)} = \frac{t^2}{2}e^{-t}\mathbf{u} + te^{-t}\mathbf{v} + e^{-t}\mathbf{w}$$

and find a suitable generalized eigenvector $\mathbf{w}$.

e. Check that the conditions for a general solution, given by Theorem 6.4.3, are met.

f. Solve the initial-value problem

$$\mathbf{x}' = \mathbf{A}\mathbf{x}, \qquad \mathbf{x}(0) = \begin{pmatrix} 1 \\ 0 \\ 0 \end{pmatrix}.$$

✦ 6.6 INSTANT EXERCISE SOLUTIONS

1. Eigenvectors $\mathbf{v}$ satisfy

$$\begin{pmatrix} -1 & 1 \\ -1 & 1 \end{pmatrix}\begin{pmatrix} v_1 \\ v_2 \end{pmatrix} = \begin{pmatrix} 0 \\ 0 \end{pmatrix}.$$

The equations are redundant, reducing to the single equation $-v_1 + v_2 = 0$. All eigenvectors are multiples of

$$\mathbf{v} = \begin{pmatrix} 1 \\ 1 \end{pmatrix}.$$

2. If $y = te^{-2t}$, then $y' = (-2t+1)e^{-2t}$ and $y'' = (4t-4)e^{-2t}$. Thus, $y'' + 4y' + 4y = (4t-4)e^{-2t} + 4(-2t+1)e^{-2t} + 4te^{-2t} = 0$.

3. Let $u = x_1$ and $v = x_2$ for convenience. Then the vector differential equation corresponds to the scalar differential equations

$$u' = -3u + v, \qquad v' = -u - v$$

and the solution $\mathbf{x}^{(2)}$ corresponds to

$$u = te^{-2t}, \qquad v = (t+1)e^{-2t}.$$

Then $u' = (-2t+1)e^{-2t} = -3u + v$ and $v' = (-2t-1)e^{-2t} = -u - v$.

From Theorem 6.4.3, $\Psi(0) = \begin{pmatrix} 1 & 0 \\ 1 & 1 \end{pmatrix}$. This matrix is easily seen to be row-equivalent to the identity matrix and is therefore nonsingular. Alternatively, we can compute $\det(\Psi(0)) = 1$.

4. Using cofactor expansion on the first column gives

$$P_A(\lambda) = \begin{vmatrix} 2-\lambda & 1 & -2 \\ -1 & -\lambda & 0 \\ 0 & 2 & -2-\lambda \end{vmatrix} = (2-\lambda)\begin{vmatrix} -\lambda & 0 \\ 2 & -2-\lambda \end{vmatrix} + \begin{vmatrix} 1 & -2 \\ 2 & -2-\lambda \end{vmatrix},$$

or

$$P_A(\lambda) = (2-\lambda)(-2-\lambda)(-\lambda) + (2-\lambda) = (2-\lambda)(\lambda^2+2\lambda+1) = -(\lambda-2)(\lambda+1)^2.$$

5. We have

$$\mathbf{0} = (\mathbf{A} - 2\mathbf{I})\mathbf{u} = \begin{pmatrix} 0 & 1 & -2 \\ -1 & -2 & 0 \\ 0 & 2 & -4 \end{pmatrix}\begin{pmatrix} u_1 \\ u_2 \\ u_3 \end{pmatrix}.$$

The third equation is redundant, so we have $u_2 - 2u_3 = 0$ and $-u_1 - 2u_2 = 0$. Taking $u_3 = 1$ yields $u_2 = 2$ and then $u_1 = -4$. For $\mathbf{v}$, we have

$$\mathbf{0} = (\mathbf{A} + \mathbf{I})\mathbf{v} = \begin{pmatrix} 3 & 1 & -2 \\ -1 & 1 & 0 \\ 0 & 2 & -1 \end{pmatrix}\begin{pmatrix} v_1 \\ v_2 \\ v_3 \end{pmatrix}.$$

By row reduction,

$$\mathbf{A} + \mathbf{I} = \begin{pmatrix} 3 & 1 & -2 \\ -1 & 1 & 0 \\ 0 & 2 & -1 \end{pmatrix} \cong \begin{pmatrix} -1 & 1 & 0 \\ 0 & 2 & -1 \\ 3 & 1 & -2 \end{pmatrix} \cong \begin{pmatrix} 1 & -1 & 0 \\ 0 & 2 & -1 \\ 0 & 4 & -2 \end{pmatrix} \cong \begin{pmatrix} -1 & 1 & 0 \\ 0 & 2 & -1 \\ 0 & 0 & 0 \end{pmatrix}.$$

The eigenvectors must satisfy $-v_1 + v_2 = 0$ and $2v_2 - v_3 = 0$; setting $v_2 = 1$ yields $v_1 = 1$ and $v_3 = 2$.

6.7 Qualitative Behavior of Nonlinear Systems

We have seen in previous sections how to classify the critical point at the origin for linear systems, with the results summarized in Theorems 6.5.2 and 6.5.3. Many useful systems are nonlinear, however. For these systems, we are almost never able to obtain a symbolic formula for the solutions, and we are seldom able to obtain symbolic formulas for the trajectories. However, we can often classify the critical points for autonomous nonlinear systems

$$\mathbf{x}' = \mathbf{f}(\mathbf{x}) \tag{1}$$

by studying the linear system that best approximates the nonlinear system.

A Nonlinear Spring

Consider an unforced spring-mass system as in Section 3.1,

$$my'' + \beta y' + ky = 0, \qquad m, k > 0, \qquad \beta \geq 0.$$

When we derived this model, we assumed that the restoring force of the spring was proportional to the displacement. Suppose instead that the stiffness of the spring increases as the spring is stretched. In particular, we assume that the force per unit displacement has magnitude $k + ay^2$ rather than k; then the restoring force is $F_r = -ky - ay^3$. The result is the model

$$my'' + \beta y' + ky + ay^3 = 0.$$

MODEL PROBLEM 6.7

Determine the properties of the nonlinear oscillator

$$y'' + 0.01y' + y + 0.2y^3 = 0.$$

Ideally, we would like to begin by solving the differential equation. However, like most nonlinear equations, this one cannot be solved. The differential equation is autonomous; therefore, it makes sense to examine it in the phase space. Let $v(t)$ be the velocity of the mass. Then we may write the system as

$$y' = v, \qquad v' = -0.01v - y - 0.2y^3. \tag{2}$$

The origin is the only critical point.[22] We wish to classify this critical point by shape and stability without solving the system. Often the classification is suggested by a computer-generated phase portrait. A computer rendering of the phase portrait for Model Problem 6.7 appears in Figure 6.7.1. It appears from the plot that the origin is a spiral sink (and indeed it is, as we shall see later). However, we must be careful not to jump to that conclusion based on a computer-generated graph. The computer-generated phase portrait is determined by numerical solution of initial-value problems, and numerical computations are approximate. In this case, the spiraling is sufficiently slow that we cannot comfortably conclude that the origin is indeed a spiral sink.

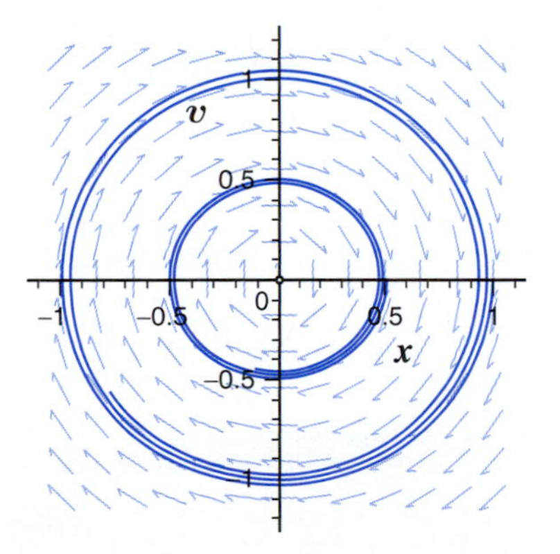

Figure 6.7.1
The phase portrait for Model Problem 6.7.

[22]See Section 5.4 for the definition of *critical point*.

The system of Model Problem 6.7 includes one linear equation and one nonlinear equation. The nonlinearity is contained in the function $-y - 0.2y^3$. In many applications of calculus, it is appropriate to replace a nonlinear function f by its **linear approximation** near a point a; this is defined to be that linear function z that satisfies $z(a) = f(a)$ and $z'(a) = f'(a)$. The linear approximation of $-y - 0.2y^3$ near $y = 0$ is simply $-y$. Replacing $-y - 0.2y^3$ with $-y$ results in the system

$$y' = v, \qquad v' = -0.01v - y. \tag{3}$$

The phase portrait for the linear system (3) appears in Figure 6.7.2.

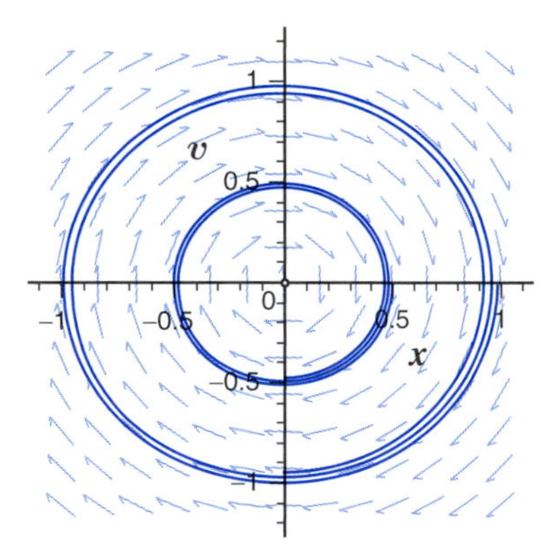

Figure 6.7.2
The phase portrait for the linearized system derived from Model Problem 6.7.

At first glance, the linearized phase portrait appears to be *identical* to the original phase portrait, but a closer look reveals some small differences in the outer solution curves. Note that the curve in Figure 6.7.1 crosses the vertical axis somewhat outside of $v = 1$, while the corresponding curve in Figure 6.7.2 crosses just inside $v = 1$. Note also that the curve in the original phase portrait shows about 2.4 rotations, while the curve in the linearized phase portrait shows 2.25 rotations. The inner solution curves are much more nearly identical; this is what we should expect to see, as it is consistent with the claim that the linearization is exactly correct at the origin.

The linear system, like the original system, appears to have a stable spiral at the origin. In the case of the linear system, this hypothesis can be confirmed. In matrix form, the linear system is

$$\mathbf{x}' = \mathbf{A}\mathbf{x}, \qquad \mathbf{x} = \begin{pmatrix} y \\ v \end{pmatrix}, \qquad \mathbf{A} = \begin{pmatrix} 0 & 1 \\ -1 & -0.01 \end{pmatrix}.$$

The eigenvalues for this system are

$$\lambda = -0.005 \pm i\frac{\sqrt{3.9999}}{2};$$

hence, the origin of the linear system is a spiral sink, by Theorem 6.5.2. This shows that the computer-generated phase portrait for the linear system is qualitatively correct. Given this and the similarity of the phase portraits, it is evidently true that the origin of the nonlinear system is a spiral sink as well.

All uncertainty about the qualitative behavior of the nonlinear system of Model Problem 6.7 can be eliminated by reference to a theorem that appears later in this section. For the moment, we note that there are examples for which the critical point of a nonlinear system has different behavior from the corresponding critical point of the linearized system.

EXAMPLE 1

Consider the differential equation

$$x'' + (x')^3 + x = 0.$$

We can write this equation as a system by defining $v = x'$:

$$x' = v, \qquad v' = -x - v^3.$$

The only critical point for this system is the origin. To linearize the system, we need to replace the right-hand sides of the equations with the linear approximations at the origin. The first equation is already linear. For the second, the function $-x$ yields the same function value and first partial derivatives at the origin as the function $-x - v^3$. Thus, the linear approximation of the system is

$$x' = v, \qquad v' = -x.$$

In matrix form, we have

$$\mathbf{x}' = \mathbf{A}\mathbf{x}, \qquad \mathbf{x} \equiv \begin{pmatrix} x \\ v \end{pmatrix}, \qquad \mathbf{A} = \begin{pmatrix} 0 & 1 \\ -1 & 0 \end{pmatrix}.$$

The eigenvalues of the linear system are

$$\lambda = \pm i,$$

so the origin in the linear system is a center, as shown in Figure 6.7.3. This suggests that the origin is also a center for the nonlinear system. The phase portrait for the nonlinear system (Fig. 6.7.4) seems to suggest that the origin is a spiral sink; however, it is possible that the trajectories nearer to the origin look more like those of the linear system. No matter how close the example trajectories in the phase portrait are to the origin, it is never close enough to overcome the argument that the picture might look better if only it were restricted to a smaller region. Exercise 9 guides the reader through a proof that the origin in the nonlinear system *is* a spiral sink.

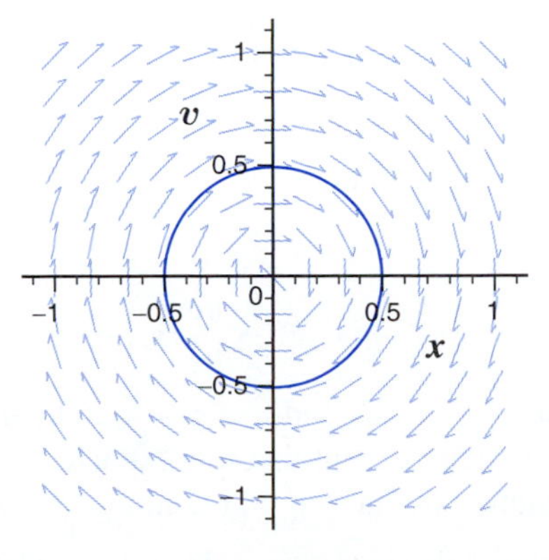

Figure 6.7.3
The phase portrait for the linear system of Example 1.

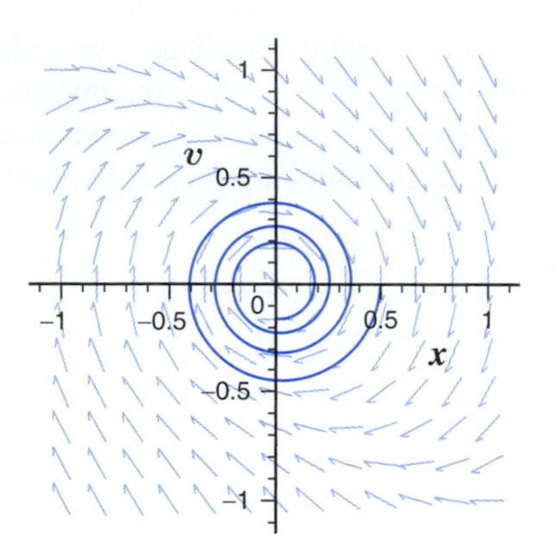

Figure 6.7.4
The phase portrait for the nonlinear system of Example 1.

Model Problem 6.7 serves as an elementary example of the use of linear approximation to study local behavior of nonlinear systems near critical points. Two further developments are needed to establish a systematic method that can be applied to more general problems: an efficient technique for linearizing systems and a theorem that distinguishes between the case where the linear system represents the qualitative behavior of the nonlinear system, as in Model Problem 6.7, and the case where the linear system does not represent the nonlinear system, as in Example 1.

Linearization in General

Consider an autonomous 2×2 nonlinear system

$$x' = f(x, y), \qquad y' = g(x, y), \tag{4}$$

where f and g have continuous first derivatives. The linear approximation of f near any point (x_0, y_0) is

$$f(x, y) \approx f(x_0, y_0) + \frac{\partial f}{\partial x}(x_0, y_0)\,(x - x_0) + \frac{\partial f}{\partial y}(x_0, y_0)\,(y - y_0).$$

If (x_e, y_e) is a critical point of the system, then $f(x_e, y_e) = 0$, so the linear approximation near (x_e, y_e) is

$$f(x, y) \approx \frac{\partial f}{\partial x}(x_e, y_e)\,(x - x_e) + \frac{\partial f}{\partial y}(x_e, y_e)\,(y - y_e).$$

Similarly,

$$g(x, y) \approx \frac{\partial g}{\partial x}(x_e, y_e)\,(x - x_e) + \frac{\partial g}{\partial y}(x_e, y_e)\,(y - y_e).$$

Now define $u = x - x_e$ and $v = y - y_e$. The uv coordinate system is similar to the original xy coordinate system, except that it is "attached" to the critical point of interest. Given that x_e and y_e are constants, we have $u' = x'$ and $v' = y'$. The linear system is conveniently written as

$$u' = \frac{\partial f}{\partial x}(x_e, y_e)\,u + \frac{\partial f}{\partial y}(x_e, y_e)\,v, \qquad v' = \frac{\partial g}{\partial x}(x_e, y_e)\,u + \frac{\partial g}{\partial y}(x_e, y_e)\,v.$$

In vector form, this is

$$\begin{pmatrix} u \\ v \end{pmatrix}' = \begin{pmatrix} \frac{\partial f}{\partial x} & \frac{\partial f}{\partial y} \\ \frac{\partial g}{\partial x} & \frac{\partial g}{\partial y} \end{pmatrix} (x_e, y_e) \begin{pmatrix} u \\ v \end{pmatrix}. \tag{5}$$

The development of Equation (5) generalizes to systems of dimension n.

Let $\mathbf{f} : \mathbb{R}^n \to \mathbb{R}^n$ be a function with continuous first derivatives.[23] The **Jacobian** of the function f is the matrix $\mathbf{J}$ whose entries are given by $\partial f_i / \partial x_j$, where f_i is the ith entry in $\mathbf{f}$ and x_j is the jth independent variable. The **linearized system** at an isolated critical point $\mathbf{x} = \mathbf{x}_e$ of the system $\mathbf{x}' = \mathbf{f}(\mathbf{x})$ is the system $\mathbf{u}' = \mathbf{J}(\mathbf{x}_e)\mathbf{u}$, where $\mathbf{u} = \mathbf{x} - \mathbf{x}_e$.

EXAMPLE 2

Consider the system

$$x' = x(1-x) - xz, \qquad y' = y(1-y) - 5yz, \qquad z' = 2xz + 10yz - z,$$

which could represent an ecological system with two prey species, x and y, and one predator species, z. The prey species are identical in all respects but one, which is that prey y is 5 times as vulnerable to predation as x. The Jacobian matrix at any critical point (x_e, y_e, z_e) is

$$J(x_e, y_e, z_e) = \begin{pmatrix} 1 - 2x_e - z_e & 0 & -x_e \\ 0 & 1 - 2y_e - 5z_e & -5y_e \\ 2z_e & 10z_e & 2x_e + 10y_e - 1 \end{pmatrix}.$$

In particular, note that the point $(1, 1, 0)$ is a critical point, representing the absence of predators. At this critical point, the Jacobian is

$$J(1,1,0) = \begin{pmatrix} -1 & 0 & -1 \\ 0 & -1 & -5 \\ 0 & 0 & 11 \end{pmatrix}.$$

The linearized system is easy to study by computing the eigenvalues for the appropriate Jacobian matrix. Here we have

$$\det(\mathbf{J} - \lambda\mathbf{I}) = \begin{vmatrix} -1-\lambda & 0 & -1 \\ 0 & -1-\lambda & -5 \\ 0 & 0 & 11-\lambda \end{vmatrix} = (-1-\lambda) \begin{vmatrix} -1-\lambda & -5 \\ 0 & 11-\lambda \end{vmatrix} = -(\lambda+1)^2(\lambda - 11).$$

With a positive eigenvalue $\lambda = 11$, the origin in the linear system is unstable. It remains to be seen whether the behavior of the linear system near the origin represents the behavior of the nonlinear system near the critical point $(1, 1, 0)$.

[23] $\mathbf{f} : \mathbb{R}^n \to \mathbb{R}^n$ means that f is a function that produces an n-vector output from an n-vector input, for some particular n.

When Does the Linearized System Represent the Nonlinear System?

Fortunately, there is a large class of problems for which we can be sure that the result obtained for the linearized system applies to the nonlinear system.

Theorem 6.7.1

Suppose $\mathbf{f} : \mathbb{R}^n \to \mathbb{R}^n$ is nonlinear, with continuous first derivatives, and $\mathbf{x}_e$ is a critical point of the nonlinear system $\mathbf{x}' = \mathbf{f}(\mathbf{x})$.

1. If all eigenvalues of the Jacobian matrix $J(\mathbf{x}_e)$ have negative real parts, then the critical point $\mathbf{x}_e$ is asymptotically stable.
2. If any eigenvalue of the Jacobian matrix $J(\mathbf{x}_e)$ has a positive real part, then the critical point $\mathbf{x}_e$ is unstable.

The system of Example 1 is one in which the critical point of the linear system is a center, but that of the corresponding nonlinear system is a spiral sink. This does not violate Theorem 6.7.1 because the theorem does not apply when the Jacobian has purely imaginary eigenvalues. The system of Example 2 satisfies the requirements of Theorem 6.7.1. The origin for that nonlinear system is unstable.

Unstable Equilibria and Limit Cycles

Electronics, such as computers and televisions, utilize what are called *active* circuits that function as negative resistors. Consider an *RLC* circuit, as in Section 4.1, but with the voltage across the "resistor" as $v_R = F(i)$, where i is the current, rather than $v_R = RI$ as in the standard model. The voltages are related by $v_L + v_R + v_C = 0$, with $v_L = Li'$ and $i = Cv_C'$. (The derivatives are with respect to time.) A differential equation for the current is obtained by differentiating the voltage equation and substituting for the voltages. Thus, $0 = v_L' + v_R' + v_C' = Li'' + F'(i)i' + C^{-1}i$. We consider here the specific case $F(i) = ai^3 - bi$; note that for small i, $F(i) \approx -bi$, corresponding to a "negative" resistor. The differential equation for this case is

$$Li'' + (3ai^2 - b)i' + C^{-1}i = 0.$$

The dimensionless form[24] of this equation,

$$u'' + \mu(u^2 - 1)u' + u = 0, \tag{6}$$

is known as the **van der Pol equation.**[25]

The van der Pol equation cannot be solved analytically. However, it is possible to understand the behavior of solutions by using qualitative methods. With $w = u'$, the van der Pol equation is equivalent to the system

$$u' = w, \qquad w' = -u + \mu(1 - u^2)w.$$

---

[24]Here u is the dimensionless current, derivatives are with respect to a dimensionless time, and $\mu = b\sqrt{C/L} > 0$ is a dimensionless parameter that characterizes the nonlinearity of the resistance.

[25]The Dutch physicist Balthasar van der Pol experimented with vacuum tubes in the 1920s and 1930s and developed Equation (6) as a mathematical model to describe electric circuits containing these devices. Vacuum tubes were an essential element in the invention of the radio and other electronic devices that are now made with integrated circuits.

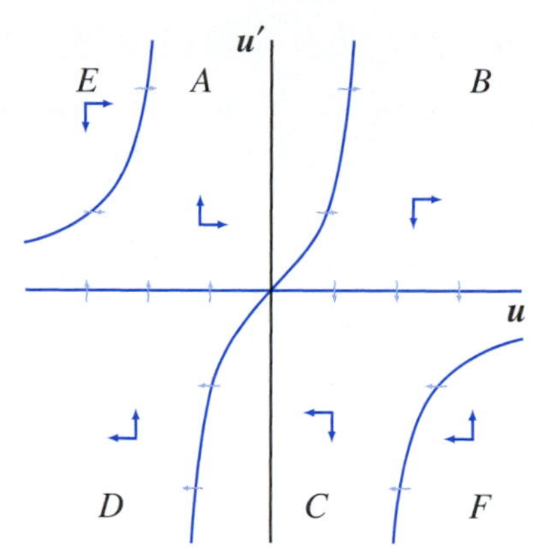

Figure 6.7.5
The nullcline diagram for the van der Pol equation, with $\mu = 1$.

The nullcline diagram for the case $\mu = 1$ is shown in Figure 6.7.5. The trajectories are clearly bounded, with flow possible from region A to B, B to C, C to D, D to A, E to A, and F to C. The critical point is not necessarily a spiral, nor is it necessarily stable.

The eigenvalues of the linearized system are determined from the characteristic polynomial

$$P(\lambda) = \lambda^2 - \mu\lambda + 1$$

to be

$$\lambda = \frac{\mu \pm \sqrt{\mu^2 - 4}}{2}.$$

The origin is a source if $\mu > 2$ and a spiral source if $\mu < 2$. By Theorem 6.7.1, these conclusions also hold for the nonlinear system; in particular, the origin is unstable.

✦ INSTANT EXERCISE 1

Compute the Jacobian for the van der Pol equation and the characteristic polynomial for the corresponding linearized system.

Phase portraits for the van der Pol equation are shown in Figures 6.7.6 and 6.7.7, with $\mu = 1$ and $\mu = 4$. In both cases, solution curves that begin near the origin move away from the origin, the first in a manner similar to a linear spiral source and the second in a manner similar to a linear source. However, these solution curves do not continue their outward motion forever, as in the corresponding linear system. Instead, these solutions approach a closed curve that encloses the origin. The closed curve is a set of points that attracts solutions, in much the same way as a stable equilibrium point attracts solutions; however, it is a solution curve rather than a point. A closed trajectory that attracts solutions is called a **limit cycle.**

The periodic behavior of the solutions is further illustrated by the graphs of the solutions in the tu plane, as shown in Figures 6.7.8 and 6.7.9. In both cases, the solution approaches a periodic function. The approach is oscillatory in the case $\mu = 1$ and nonoscillatory in the case $\mu = 4$; these behaviors are consistent with the classification of the origin as a source in the first case and

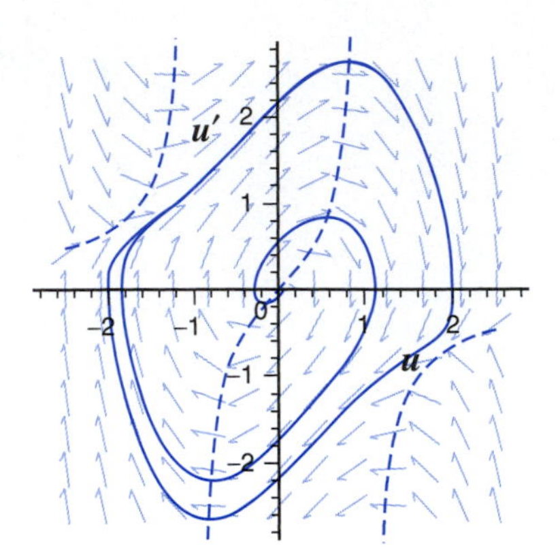

Figure 6.7.6
The phase portrait for the van der Pol equation, with $\mu = 1$, with u' nullclines (dashed).

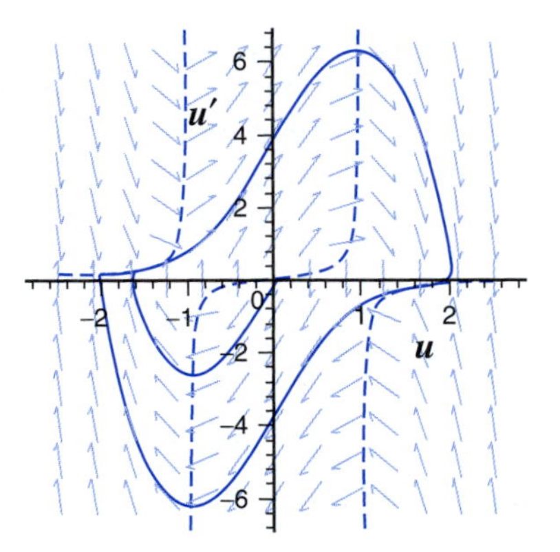

Figure 6.7.7
The phase portrait for the van der Pol equation, with $\mu = 4$, with u' nullclines (dashed).

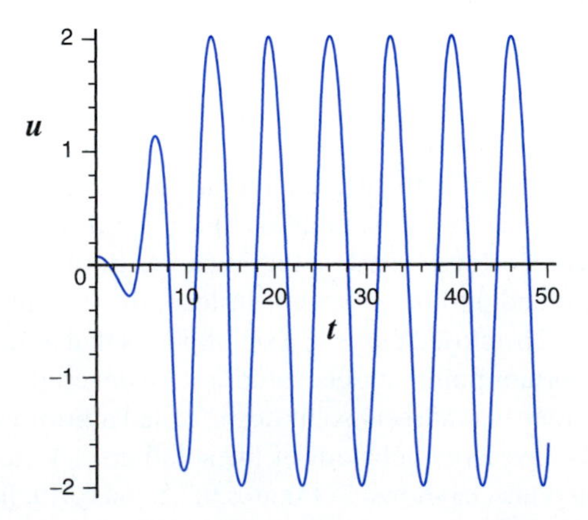

Figure 6.7.8
The solution of the van der Pol equation, with $\mu = 1$ and $u(0) = 0.05$, $u'(0) = 0$.

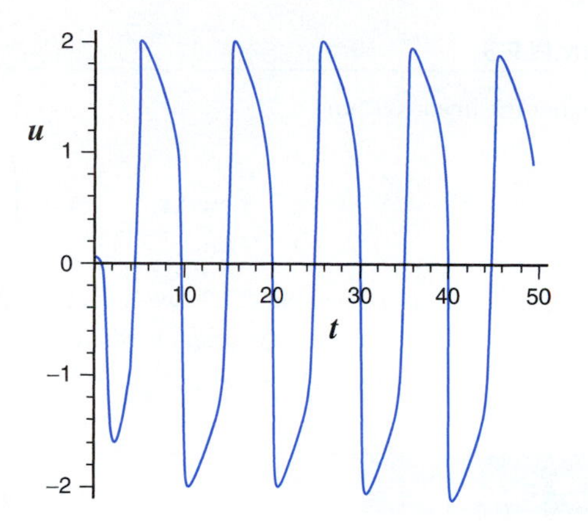

Figure 6.7.9
The solution of the van der Pol equation, with $\mu = 4$ and $u(0) = 0.05$, $u'(0) = 0$.

a spiral source in the second. The period of the motion is different for the two cases, but this could not be predicted by phase plane analysis.

Equilibrium Point Analysis for Three-Dimensional Systems

Equilibrium point analysis is clearly more difficult for a three-dimensional system than for a two-dimensional system. The computations required to find the characteristic polynomial are detailed, and the result is a cubic polynomial that may be difficult to solve. Fortunately, there is a way to tell from the coefficients of the characteristic polynomial whether or not the equilibrium point is asymptotically stable.[26] It is also possible to determine the characteristic polynomial directly from the Jacobian matrix without computing $\det(\mathbf{J} - \lambda\mathbf{I})$.[27]

Theorem 6.7.2 Let (a_1, a_2, a_3) be an isolated critical point for a nonlinear system of three differential equations in three variables. Let $\mathbf{J}$ be the Jacobian of the system at the critical point, let $\mathbf{J}_k$ be the matrix obtained from $\mathbf{J}$ by deleting row k and column k, and define c_1, c_2, and c_3 by

$$c_1 = -\operatorname{tr}(\mathbf{J}), \qquad c_2 = \sum_{k=1}^{3} \det(\mathbf{J}_k), \qquad c_3 = -\det(\mathbf{J}),$$

where the trace $\operatorname{tr}(\mathbf{J})$ is the sum of the diagonal elements of $\mathbf{J}$. Then the critical point of the nonlinear system is asymptotically stable if all three coefficients are positive and $c_1 c_2 > c_3$ and unstable if $c_1 < 0$, $c_3 < 0$, or $c_1 c_2 < c_3$.

---

[26]The necessary conditions for a polynomial to have only roots with negative real parts is called the *Routh–Hurwitz criterion*. The Routh–Hurwitz criterion for a polynomial of degree n can be found in books on the theory of ordinary differential equations, such as W. Kelley and A. Peterson, *The Theory of Differential Equations: Classical and Qualitative*, Pearson Education, Upper Saddle River, NJ, 2004.

[27]See Exercise 10 of Section 6.6.

EXAMPLE 3

Consider the linear system

$$\mathbf{x}' = \mathbf{A}\mathbf{x}, \qquad \mathbf{A} = \begin{pmatrix} 1 & -5 & 0 \\ 0 & -3 & -2 \\ -1 & 4 & 0 \end{pmatrix}.$$

Using the notation of Theorem 6.7.2, we have $c_1 = 2$, $c_2 = 5$, and $c_3 = 2$. All three coefficients are positive and $c_1 c_2 > c_3$; hence, the origin is stable.

6.7 Exercises

In Exercises 1 through 6, (a) determine the critical points, (b) find the Jacobian, (c) determine the stability of the origin for each linearized system, (d) determine the stability of the critical points if permitted by Theorem 6.7.1, (e) sketch the nullclines, (f) use a computer to plot the phase portrait, and (g) summarize your findings.

1. $x' = -3x + 2xy, \quad y' = -4y + 3xy$ (Section 5.1, Exercise 3a)
2. $r' = r(1 - 2c), \quad c' = c(r - 1)$ (Section 5.1, Exercise 2)
3. $x' = x(2 - x - y), \quad y' = y(3 - y - x)$ (This is a competing species model: see Section 5.1.)
4. $x' = x(2 - x - y), \quad y' = y(3 - y - 2x)$ (This is a competing species model: see Section 5.1.)
5. $x' = y, \quad y' = x - y - x^3$ (Section 5.5, Exercise 6)
6. $x' = y - x^3, \quad y' = y - 4x$
7. Determine the stability of the equilibrium point at the origin for the system $\mathbf{x}' = \mathbf{A}\mathbf{x}$, where
$$\mathbf{A} = \begin{pmatrix} -1 & 0 & 1 \\ 1 & -2 & 0 \\ 1 & 2 & -3 \end{pmatrix}.$$
8. Let $\mathbf{A}$ be given by
$$\mathbf{A} = \begin{pmatrix} 1 & -5 & 0 \\ 0 & -3 & -2 \\ -1 & 4 & c \end{pmatrix},$$
where c is a parameter. Use Theorem 6.7.2 to determine which values of c make the origin stable for the system $\mathbf{x}' = \mathbf{A}\mathbf{x}$.
9. Consider the system of Example 1. Let $r(t)$ be the distance from the origin to the point $(x(t), v(t))$. Obtain a differential equation for r by differentiating $r^2 = x^2 + v^2$, and use this equation to argue that the origin is an asymptotically stable equilibrium point.
10. (T) One flaw in the predator-prey models that we have seen so far is that the rate at which predators kill prey is unbounded if the prey population is very large. An improved model was proposed

by Holling in 1959:

$$x' = rx\left(1 - \frac{x}{K}\right) - \frac{pxy}{1 + pqx}, \qquad y' = e\frac{pxy}{1 + pqx} - dy,$$

where r is the maximum prey growth rate, K the prey carrying capacity, d the death rate of the predator, p a measure of the hunting rate of the predator, and q a measure of the time required for the predator to handle each kill.

a. Nondimensionalize the model, using the variables $X = x/K$, $Y = y/(qrK)$, and $\tau = et/q$. The scale for the predator y is the maximum population that could be supported by prey production at the maximum rate rK. Use the parameters $\delta = dq/e$, $\epsilon = e/qr$, and $H = 1/(pqK)$.

b. Determine the critical points for the system. Note that a particular critical point may exist for some parameter values and not for others.

c. Determine the stability of the critical points in the limit $\epsilon \to 0$. Your results will depend on the values of H and δ.

d. Let $H = 0.5$. Choose a value of δ from each of the three regions delineated in part *c*. Plot the phase portrait for each of these cases, with $\epsilon = 0.2$, and superimpose the nullclines on the plots.

e. Discuss the results.

11. Consider the one-parameter family of models with one predator and two prey that is given by the system

$$x' = x(1 - x - z), \qquad y' = ry(1 - y) - yz, \qquad z' = z(2x + 2y - 1).$$

a. Determine the seven meaningful critical points for the system, and note the values of r that permit the existence of each critical point. (*Hint:* only nonnegative values for the variables make sense in the context of the model.) These are conveniently named by the letters of the nonzero variables. For example, a critical point for which y is 0 and x and z are both positive can be named XZ, and the critical point at the origin can be named O.

b. Determine the Jacobian of the system.

c. Show that O, X, Y, and XY are never stable.

d. Show that XZ is asymptotically stable if r is small enough. Similarly, show that YZ is asymptotically stable if r is large enough. (*Hint:* use Theorem 6.7.2.) Explain why this behavior is what we should expect for this system, based on the interpretation of r.

e. Use Theorem 6.7.2 to determine the stability of the remaining equilibrium point, taking care to note that a critical point can be asymptotically stable only if it exists. (*Hint:* The algebra is much easier if you use x_e, y_e, and z_e in the Jacobian matrix rather than the complicated formulas that determine these values as functions of r. With proper simplification, each nonzero entry in the Jacobian can be expressed as a multiple of one of the three critical point coordinates. It is then relatively easy to compute and compare the coefficients of the characteristic polynomial.)

T 12. The **Lorenz system,** defined by the equations

$$x' = -ax + ay, \qquad y' = rx - y - xz, \qquad z' = -bz + xy,$$

where $a, b, r > 0$ and $-\infty < x, y, z < \infty$, is a well-known example of a 3×3 system that looks simple but can have very complicated behavior. The system arises from a mathematical model of the atmosphere, and its behavior offers some insight into the question of why weather is so unpredictable. We explore some of the features of the system in this exercise. In the calculations that follow, assume $a > b + 1$.

a. Determine the ranges of parameter values for which the (0, 0, 0) critical point is asymptotically stable.

b. Determine the other critical points and the range of parameter values for which they exist.

c. Find the characteristic polynomial for the Jacobian at these critical points (they should be the same). Find the requirement that the parameters must satisfy for these critical points to be asymptotically stable. Write the inequality in the form $r < f(a, b)$.

d. The parameters a and b are generally estimated as $a = 10$ and $b = \frac{8}{3}$. The value of r can vary considerably, so we take it to be variable. Solve the problem numerically with $r = \frac{1}{2}$ and initial conditions $x(0) = 1$, $y(0) = 1$, and $z(0) = 1$. Plot $x(t)$. Does the plot agree with the stability results?

e. Solve the problem numerically with $a = 10$, $b = \frac{8}{3}$, $x(0) = 1$, $y(0) = 1$, and $z(0) = 1$, once with $r = 12$ and once with $r = 15$. Plot $x(t)$. Do the plots agree with the stability results?

f. If $r > f(a, b)$, there are no stable equilibrium solutions. Solve the problem numerically with $a = 10$, $b = \frac{8}{3}$, $x(0) = 5$, $y(0) = 5$, and $z(0) = 20$, once with $r = 22$ and once with $r = 25$. Plot $x(t)$ on the interval $0 \leq t \leq 40$. Describe the results.

g. Simple patterns can develop as r is increased further, but the results are unpredictable. To see an example, try $a = 10$, $b = \frac{8}{3}$, $r = 230$, $x(0) = 15$, $y(0) = 15$, and $z(0) = 200$. Plot each of the variables on the interval $0 \leq t \leq 10$. Describe the results.

✦ 6.7 INSTANT EXERCISE SOLUTION

1. The Jacobian is the matrix

$$\mathbf{J} = \begin{pmatrix} 0 & 1 \\ -1 - 2\mu u w & \mu(1 - u^2) \end{pmatrix};$$

at the critical point, we have

$$\mathbf{J}(0, 0) = \begin{pmatrix} 0 & 1 \\ -1 & \mu \end{pmatrix}.$$

The characteristic polynomial is then

$$p(\lambda) = \begin{vmatrix} -\lambda & 1 \\ -1 & \mu - \lambda \end{vmatrix} = \lambda^2 - \mu\lambda + 1.$$

CASE STUDY 6 Invasion by Disease

Until the successes of modern medicine in the 1900s, struggles between civilizations were often influenced by disease, much more so than is generally understood. In particular, the first contact with smallpox has had devastating consequences for many peoples. The fall of the Aztecs can be attributed in part to smallpox acquired from the Spanish Conquistadors. Other tribes throughout the Americas suffered major setbacks from smallpox. Sometimes the smallpox was introduced on purpose, but other times it was introduced simply by casual contact with the newcomers. Although we lack adequate data to fully understand the impact of smallpox on a previously unexposed population, we can get a qualitative sense of the effects with a mathematical model.

Mathematical models for population dynamics with diseases are well known. One of the most common models is the SIR model. The SIR model assumes that a population can be divided into three mutually exclusive classes. Those who are vulnerable to the disease are in the *susceptible* class. Those who can communicate the disease to others are in the *infective* class. Those who are immune are in the *recovered* class. In its standard form, the SIR model considers the short-term event in which a small initial number of infectives are introduced into a population in which the disease is not generally found. Depending on the parameter values, the disease may disappear, or it may bloom into an epidemic. Because the model is intended for short-term events, it does not incorporate births and deaths. The present investigation is intended to look at long-term and short-term events, so births and deaths are included in the model.

The Conceptual Model

We consider two populations. One lives with the specter of a disease that is endemic and a leading cause of death in the population. The other has had no previous exposure to the disease. We can use the same demographic model for both populations, but the investigations we make with the model will be different. We assume that the disease has the following features.

1. It is transmitted by human contact.
2. There is at most a short time interval between the time at which a person is infected and the time at which the person can infect others. The timing of the onset of symptoms makes little difference in the model, as we are interested in events prior to the institution of modern medical treatment, such as quarantine.
3. It has a significant mortality; that is, a noticeable fraction of sufferers die.
4. Recovery from the disease confers immunity to the disease.

Given these assumptions, we can use the SIR model, modified to allow for changes in the population size. Susceptible individuals are at risk of acquiring the disease, infective individuals are capable of passing the disease on to others, and recovered individuals are neither susceptible nor infective.

The conceptual model consists of the classes of individuals and the mechanisms by which membership in a class changes. In particular, there are several processes responsible for membership changes, and these are conveniently grouped into fast and slow processes. The fast processes are those that occur over a period of a month or less. These processes include infection from sources internal to the population, death by the disease, and recovery from the disease. The slow processes are those that occur over a period of years and those that occur over a short time but only infrequently. These include infection from sources external to the population, such as infrequent casual contact with foreigners, death by causes other than the disease, and birth.

Derivation of the Mathematical Model

Let $S(t)$, $I(t)$, and $R(t)$ be the populations of the susceptible, infective, and recovered classes, respectively, and let $N(t)$ be the total population. We assume that every individual is in exactly one of the susceptible, infective, or recovered classes; hence,

$$S + I + R = N. \tag{1}$$

Each of the unknown populations can change with time as individuals move from one class to another. We need only to derive differential equations for three of the quantities, as Equation (1) then determines the fourth.

The key to deriving the differential equations is to model the interactions between classes by using compartment analysis. Individuals enter the susceptible class at birth. Susceptible individuals can become infected, and infected individuals can recover. Individuals in any of the classes can die of causes other than the disease, and infected individuals can also die of the disease. Figure C6.1 illustrates the connections between the classes along with expressions for the rates at which the changes occur. The quantities k_b, k_i, k_r, k_d, and k_m indicate the coefficients of proportionality in the rates of birth, infection, recovery, death from the disease, and mortality due to other causes, respectively. Note that the rates are proportional to the population of the class to which the process applies. The infection rate is also proportional to the population of infectives. The birthrate coefficient is taken to depend on the total population so as to allow for a stable equilibrium solution in the absence of the disease.

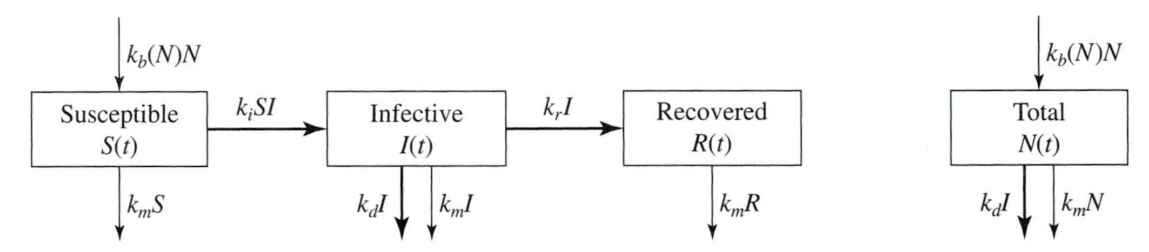

Figure C6.1
A schematic diagram of the long-term SIR model: The heavy arrows indicate processes that take place on the short time scale.

From the schematic diagram of Figure C6.1, we can write down a set of differential equations describing the system. We choose to write equations for the total population N, the susceptible population S, and the infective population I. These are

$$\frac{dN}{dt} = k_b(N)N - k_d I - k_m N. \tag{2}$$

$$\frac{dS}{dt} = k_b(N)N - k_i SI - k_m S, \tag{3}$$

$$\frac{dI}{dt} = k_i SI - (k_d + k_m + k_r)I. \tag{4}$$

To complete the model, we need to know the four rate constants and the birthrate function. Numerical values are problematic, as none of the constants is easy to measure. An example serves to illustrate this point. Let $\mu = k_d/(k_r + k_d)$. This quantity measures the deadliness of the disease; the ratio of the rate of deaths by the disease to the total rate of death and recovery is the same as the fraction of sufferers who die of the disease, ignoring those who die of causes unrelated to the disease during their period of illness. For many diseases, including smallpox, this parameter is quite variable. Smallpox is actually caused by several related viruses, some of which are far more deadly than others. Certain population groups tend to be more severely afflicted by the disease. Based on historical data from Europe and Asia, we can place μ within the range $0.02 \leq \mu \leq 0.4$. This is a wide range, and the behavior of the model may well be different at the two extreme values. A reasonable estimate for the smallpox brought from Europe to the Americas is $\mu = \frac{1}{3}$. The main point to be made is that the lack of precise data with which to determine the rate constants means that it is pointless to aim for a high degree of quantitative precision in the solution. We should feel free to make any approximations that are going to introduce less uncertainty in the results than is automatically introduced by the uncertainty in the parameters.

Rate Constants, Time Scales, and the Birthrate Function

Consider the decay equation $y' = -ky$. In our initial study of this equation in Section 1.1, we identified k as the relative rate of decay. We subsequently argued that $1/k$ is a natural reference time, our main justification being that it has the right dimension. There is another argument, based on a thought experiment, to be made in favor of $1/k$ as a reference time. Imagine the quantity y as consisting of y_0 discrete particles, with the number of undecayed particles at time t determined (up to round-off error) by the decay equation. Suppose we randomly color n of the particles. As the decay process continues, we note the time of decay for each of the colored particles. After the last colored particle has undergone its decay, we compute the average time required for decay. You might guess that the average decay time of a collection of particles should be the same as the half-life of the decay. However, the average decay time is actually $1/k$ (Exercise 1).

This interpretation of the rate constant in the decay equation can be applied to rate constants in other mathematical models, such as our disease model. The constant k_m represents the relative mortality rate in the absence of the disease. Its reciprocal is therefore the life expectancy for a disease-free population. Similarly, the relative rate at which individuals leave the infective class, assuming they do not die of causes other than the disease, is $k_d + k_r$; hence, the mean disease duration is $1/(k_d + k_r)$. The life expectancy and the mean disease duration are the relevant long-term and short-term time scales for the model, and one or the other is the appropriate time scale for the nondimensionalization. In hindsight, the longer scale $1/k_m$ turns out to be more convenient.

The choice of a scale for the populations is more difficult. Consider the differential equation for the total population in the absence of the disease

$$\frac{dN}{dt} = [k_b(N) - k_m]N.$$

A linear function for k_b with negative slope corresponds to the logistic equation[28] (Exercise 2), which has a stable equilibrium value N_0 determined by the equation $k_b(N) = k_m$. This value is the appropriate scale for the populations, and $k_m = k_b(N_0)$ is the appropriate scale for the birthrate.

[28] See Section 5.1.

This last relationship defines one point on the linear function $k_b(N)$. Anticipating nondimensionalization, we define a positive parameter r such that $k_b(0) = (1 + r)k_m$. Figure C6.2 shows the graph of the birthrate function.

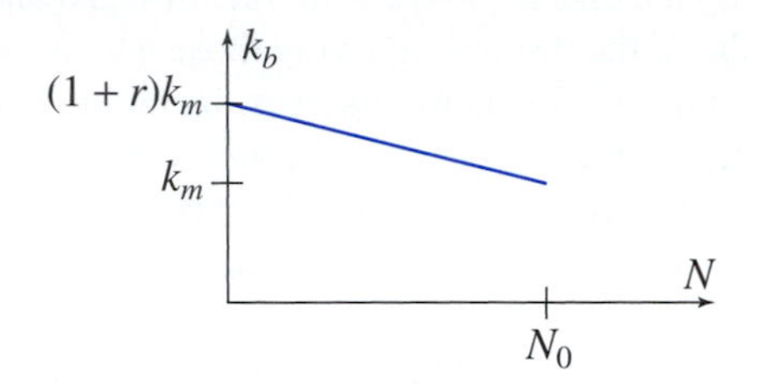

Figure C6.2
The birthrate function $k_b(N)$.

The Dimensionless Model

Nondimensionalization removes the parameter k_m from the model, leaving the constants k_d, k_r, and k_i to be replaced by dimensionless counterparts. To that end, we define

$$\mu = \frac{k_d}{k_d + k_r}, \qquad \alpha = \frac{k_d + k_r}{k_m}, \qquad \beta = \frac{k_i N_0}{k_d + k_r}.$$

The mortality fraction μ has already been discussed. Of the other two parameters, α is the ratio of the long time scale to the short time scale, and is therefore expected to be very large; and β measures the virulence of the disease—it is the ratio of the infection rate of a previously unexposed population to the rate at which the infection ends. With these parameters and the variables $\tau = k_m t$, $n = N/N_0$, $s = S/N_0$, and $i = I/N_0$, we have the dimensionless model

$$n' = rn(1 - n) - \alpha\mu i,$$
$$s' = rn(1 - n) - \alpha\beta si + n - s,$$
$$i' = \alpha\beta si - (\alpha + 1)i.$$

Since α is very large and the parameters are not known to a high degree of precision, it is reasonable to simplify slightly by the approximation $\alpha + 1 \approx \alpha$, resulting in the equation

$$i' = \alpha\beta si - \alpha i.$$

The alert reader might now ask why we do not eliminate other terms that lack the factor α. The reason is that it is the size of the whole term, rather than the coefficients, that matters. Note that all terms with factor α also have factor i. It is quite possible that such terms will *not* be large owing to i being small. By following up on this reasoning, it is logical to rescale i, using a new variable

$$x = \alpha i, \tag{5}$$

and this yields the final version of the model

$$n' = rn(1 - n) - \mu x, \tag{6}$$
$$s' = rn(1 - n) - \beta sx + n - s, \tag{7}$$
$$x' = \alpha(\beta s - 1)x. \tag{8}$$

We use x for calculations and then return to i to interpret the results.

To complete preparations for study of the model, it is helpful to compute the Jacobian of the system.

$$J(n, s, x) = \begin{pmatrix} r - 2rn & 0 & -\mu \\ 1 + r - 2rn & -\beta x - 1 & -\beta s \\ 0 & \alpha\beta x & \alpha(\beta s - 1) \end{pmatrix} \tag{9}$$

The Problem of Endemic Disease

We begin our investigation of the model of Equations (6) through (8) by estimating parameter values and looking for stable equilibrium solutions. Of the four parameters, we have already chosen the value $\mu = \frac{1}{3}$. The parameter α is the ratio of the life expectancy for a disease-free population to the standard illness duration. The rough estimate $\alpha = 500$ corresponds to a life expectancy of about 30 years and a disease duration of about 3 weeks. The remaining parameters, r and β, have to be estimated indirectly.

Estimation of r and β

Assume (*a*) that a population with maximum birthrate and no disease can double in about one-half of the life expectancy time and (*b*) that approximately one-seventh of deaths in the population are due to the disease. Use this information to determine r and β.

With maximum birthrate and no disease, the model simplifies to $n' = rn$, which has the solution $n = n_0 e^{r\tau}$. The population reaches about $2n_0$ at the time $\tau = \frac{1}{2}$. Given these figures, $r = 2\ln 2$. Thus, we estimate $r \approx 1.4$.

Estimation of β is more difficult because a larger portion of the model must be retained for the calculation. The x equation shows that $s = \beta^{-1}$ for a population at steady state with $x \neq 0$. The maximum total population in the absence of the disease is $n = 1$, so it is not possible that the susceptible population in the presence of the disease is as large as 1. Hence the disease can only be endemic if $\beta > 1$. Larger values of β reduce the susceptible population, yielding a lower overall population and a larger percentage of deaths due to the disease. Smaller values of β indicate that the disease must alternate between periods of inactivity, during which the susceptible population builds up due to new births, and brief epidemics, which burn out after the susceptible population has dropped. After some more calculations (see Exercise 3), we obtain the result that the equilibrium population is determined by the equation

$$n^2 + (\gamma - 1)n - \gamma\beta^{-1} = 0, \tag{10}$$

where

$$\gamma = \frac{1}{(\mu^{-1} - 1)r}.$$

It can be shown that Equation (10) has exactly one solution, with $\beta^{-1} \leq n < 1$ (see Exercise 4). Equation (10), along with the information given by the fraction of deaths caused by the disease, gives us the estimate $\beta = 1.7$ in a population in which the disease is endemic and one-seventh of deaths are due to the disease (see Exercise 5).

Given the parameter values

$$\mu = \tfrac{1}{3}, \qquad \alpha = 500, \qquad r = 1.4, \qquad \beta = 1.7,$$

we obtain the equilibrium solution

$$n = 0.881, \qquad s = 0.588, \qquad i = 0.00088. \tag{11}$$

These results indicate that the disease suppresses the total population by about 10%, roughly two-thirds of the population is susceptible, and roughly one out of 1000 people is sick at one time.

With the given data, this equilibrium solution is asymptotically stable (see Exercise 6).

The Problem of First Contact

Let's assume that the parameters $\mu = \frac{1}{3}$, $r = 1.4$, and $\beta = 1.7$ are fixed. Prior to first contact, a previously unexposed population will satisfy the undisturbed equilibrium point $n = 1$, $s = 1$, $i = 0$. This yields an initial-value problem for first contact.

A Problem of First Contact

Determine the stable equilibrium solution and transient properties of the system

$$n' = 1.4n(1-n) - 167i, \qquad n(0) = 1, \tag{12}$$

$$s' = 1.4n(1-n) - 850si + n - (1+\rho)s, \qquad s(0) = 1, \tag{13}$$

$$i' = 850si - 500i + \rho s, \qquad i(0) = 0. \tag{14}$$

The conceptual model of Figure C6.1 has been augmented by an additional pathway leading from the susceptible population to the infective and having the rate ρs. This is a reasonable way to include the effect of contact with members of an outside population that has endemic disease. The parameter ρ represents the average number of transmissions per unit time to susceptible members of the previously unexposed population by infective members of the endemic population. Given a modest amount of contact, the low incidence rate of the disease in the endemic population, and the modest likelihood of transmission through one contact, it is a reasonable guess that ρ is not a large number. Note that it is not possible to have an equilibrium point with $i = 0$ unless $s = 0$ also. Once the disease is introduced, it remains endemic or else the population dies out.[29] It can be shown for this model (see Exercise 7) that the zero population is unstable; that is, the disease may devastate the population, but it cannot cause the population to die out without the influence of other factors such as starvation or war that are not present in the model.

[29]This property occurs because of the assumption that the transmission of the disease between the two populations is an ongoing process. If the disease is introduced by foreigners who then leave, then the model needs to be changed so that the disease might not become endemic.

The Equilibrium Population We first find i and s in terms of n. From the equation $n' = 0$, we have

$$i = 0.00838n(1 - n).$$

To find s, we note that subtracting the second and third differential equations from the first gives us

$$(n - s - i)' = 333i - n + s;$$

at equilibrium, $s = n - 333i = n - 2.8n(1 - n)$, or

$$s = -1.8n + 2.8n^2.$$

Now we can use $i' = 0$ to obtain a formula for ρ in terms of n:

$$\rho = \frac{(500 - 850s)i}{s} = \frac{4.2(1 - n)(1 + 3.06n - 4.76n^2)}{2.8n - 1.8}. \tag{15}$$

This equation defines n implicitly in terms of ρ. A graph of $n(\rho)$ appears in Figure C6.3.

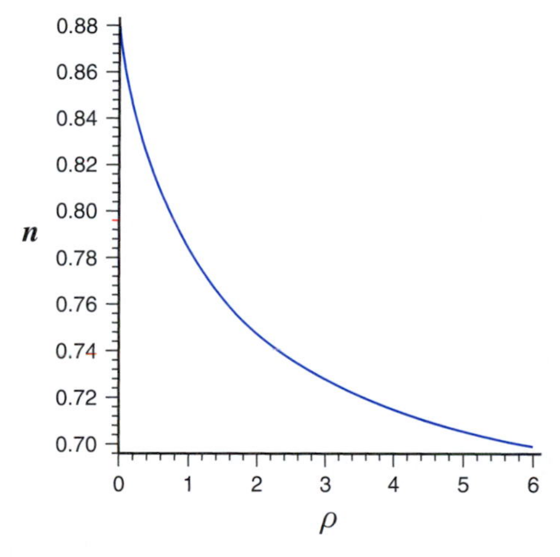

Figure C6.3
The equilibrium population in terms of ρ.

As an example, suppose that the equilibrium population is $n = 0.87$, which is only slightly less than the equilibrium population for an isolated endemic population. This corresponds to $\rho = 0.051$. We deliberately choose a value of ρ that has minimal effect on the equilibrium population to indicate the importance of the transient phenomena. With the given parameter values, the equilibrium solution is asymptotically stable (see Exercise 8). It can be shown to be asymptotically stable no matter how large is the parameter ρ.

Transient Response The equilibrium population of the population of interest is originally $n = 1$. After first contact with an endemic population, the model predicts an equilibrium population of $n = 0.87$. Given time, the population will be reduced by 13% because of the exposure to the disease. This is too small a change to account for the catastrophic effect that first contact with smallpox has had in history. For greater insight into the model, we look at the (transient) solution of the first-contact problem. This problem has to be solved numerically. The solution of this problem is displayed in Figure C6.4. The time axis is $\alpha\tau$, which is the time relative to the short time scale. Thus, $\alpha\tau = 1$ represents 3 weeks after first contact. Of particular interest is $n - i$, which is the number of people who are not currently sick, relative to the initial population.

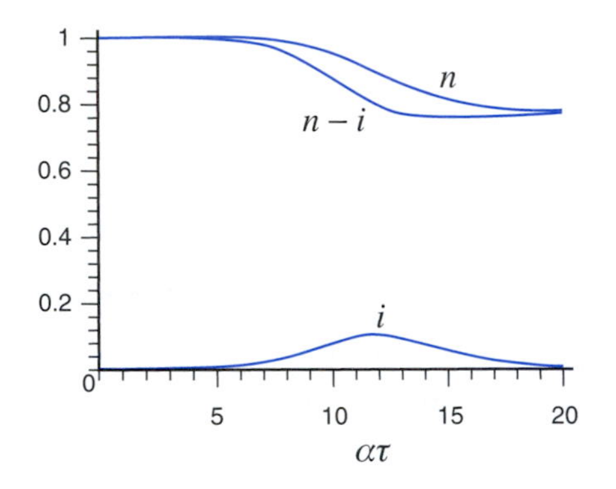

Figure C6.4
The transient response to an introduced disease.

Note that the epidemic does not develop immediately. This is so because the number of people who are infected by the external source is small. The epidemic develops only after the time $t = 6$, which corresponds to about 18 weeks. It takes about twice as long to reach the peak. At its worst, the healthy population is only about three-fourths of its original amount. This is significant, but it is a smaller disaster than is known to have occurred in the history of some Native American tribes. Some other scenarios are explored in Exercises 9 and 10.

Case Study 6 Exercises

1. Show that the average decay time of particles subject to a process governed by the equation $y' = -ky$ is $1/k$.
   a. Begin by writing down the solution of the initial-value problem $y' = -ky$, $y(0) = y_0$ and solving this result explicitly for t.
   b. Now suppose there are n colored particles. If we divide the interval $0 < y < y_0$ into n equal subdivisions, we should expect on average to find one colored particle in each subdivision. Suppose the colored particles have "values" $y_1, y_2, \ldots, y_n$. Write down the formula for the average time.
   c. Replace n in the result from part b, using the definition $\Delta y = y_0/n$. Observe that the result is a Riemann sum,[30] and take a limit as the number of particles approaches infinity.
   d. Compute the average decay time using the formula derived in part c.

2. Let $k_b = b - mN$, where b and m are positive constants. Show that the model reduces to the logistic equation in the case where the disease is absent.

[30]You will probably want to review this topic from elementary calculus.

3. Solve the equation $n' = 0$ for x in terms of n. Then derive Equation (10) by substituting the results for s and x into the equation $s' = 0$, collecting powers of n, and dividing through by the leading coefficient.

4. Show that Equation (10) has a unique solution with $\beta^{-1} < n < 1$. First define the left side of the equation to be a function $F(n)$. Compute $F(\beta^{-1})$ and $F(1)$. These values are opposite in sign. Explain why this shows that there is exactly one solution in the desired interval.

5. In this exercise, we obtain the estimate $\beta \approx 1.7$.
   a. Assume that the ratio of deaths due to the disease versus deaths due to other causes is $\frac{1}{6}$. Use this fact to calculate n without using Equation (10).
   b. Substitute the result for n, as well as the parameters r and μ, into Equation (10) and calculate β.

T 6. Show that the equilibrium solution (11) for the endemic disease problem is asymptotically stable for any value of α.

7. Show that the equilibrium solution $n \equiv 0$, $s \equiv 0$, $x \equiv 0$ for the problem of first contact is unstable, regardless of the value of ρ.

8. Choose $\rho = 0.051$ for the first-contact problem. Show that the equilibrium solution $n \equiv 0.87$, $s \equiv -1.8n + 2.8n^2$, $i \equiv 4.2n(1 - n)$ is asymptotically stable.

T 9. We consider the effect of increasing the rate of transmission from external sources on the equilibrium and transient behavior of the solution.
   a. Suppose ρ is increased to 1, an increase of approximately a factor of 20. How does this change the equilibrium solution? How does this change the time and severity of the infection peak? Is this enough to account for the disasters that befell some of the Native American tribes after first contact?
   b. There are cases in which first contact was initiated deliberately by giving infected blankets to a Native American tribe. Use Equation (15) to show that n cannot be less than $\frac{9}{14}$. Suppose $\rho = 500$, which corresponds to a fast external infection rate. How does this change the transient solution?

T 10. We consider the effect of increasing the rate of transmission from internal sources on the equilibrium and transient behavior of the solution. The parameter β may depend on the behavior of the population with regard to sick individuals. Any kind of isolation, for example, would reduce the value of β. It is reasonable to assume that a society with no prior exposure to the disease will not have developed those good practices that might have been developed in a society where the disease is endemic.
   a. Suppose $\beta = 3.4$. Change the system given by Equations (12) through (14) to reflect this change. Determine the new equilibrium solution.
   b. Determine the effect of this change on the peak in the transient solution.

chapter

7

The Laplace Transform

In Chapters 3 and 4, we learned how to solve homogeneous and nonhomogeneous linear differential equations with constant coefficients by finding a complementary and particular solution. The *Laplace transform* method is entirely different from all the methods we have used so far. This method involves transforming the differential equation for the unknown $y(t)$ into an algebraic equation for an unknown $Y(s)$ that is related to $y(t)$. The equation for Y is easily solved, and then the problem remains of recovering the solution y of the original problem from the solution Y of the transformed problem.

Many phenomena in nature are discontinuous or nearly so. When a switch is thrown on an electric circuit, it is only a matter of milliseconds or less between the moment that the switch stops being "off" to the moment that the switch starts being "on." Instead of trying to carefully model the extremely brief process of switch connection, it is generally more convenient to consider that the switch instantaneously changes from off to on. Similarly, when a mass on a spring is set into motion by a sharp blow, there is an extremely brief interval during which the blow is actually being applied, and it is often more convenient to think of the blow as having been applied instantaneously. Processes can be described using discontinuous or impulsive models, but these pose special problems for standard solution techniques. The Laplace transform method is commonly used by electrical engineers in part because it is well suited to models of processes that are discontinuous or impulsive.

This chapter begins (Sec. 7.1) with the introduction of mathematical notation for discontinuous functions and the solution by methods of Chapters 3 and 4 of a difficult problem that uses discontinuous functions. Section 7.2 introduces the mathematics of the Laplace transform, with an emphasis on the questions of what functions can be transformed and what properties of the transform lead to its utility as a tool for solving differential equations. We use the Laplace transform to solve differential equations in Sections 7.3 and 7.4. Section 7.5 introduces the *convolution integral* and its use in finding particular solutions in the form of a definite integral. Case Study 7 is a classic problem that uses the Laplace transform to study a population model in which the birthrate depends on the distribution of ages in the population.

We do not treat partial fraction decomposition here. Good Laplace transform tables include enough entries to limit the need for this technique. It is also possible to obtain inverse Laplace transform formulas from a computer algebra system, such as Maple or Mathematica. This avoids the computational difficulties of finding inverse Laplace transforms by hand.

The numbers in parentheses are equation numbers while formula numbers without parentheses are Laplace transform table references.

7.1 Piecewise-Continuous Functions

We begin our study of the Laplace transform with a deceptively simple problem. This problem can be solved by a variety of methods, two of which were presented in Chapter 4 and two of which are presented in Sections 7.4 and 7.5. Each method has its drawbacks, and the comparison of the methods for the problem is enlightening.

MODEL PROBLEM 7.1

An *LC* circuit with unit inductance and capacitance is initially at rest. Beginning at time 0, the circuit is forced by an oscillating function that is alternately 1 and -1 over unit intervals of time. Determine the current in the circuit.

The discussion of electric circuits in Section 4.1 allows us to quickly write this problem in mathematical terms. The problem is to find v', where v satisfies the initial-value problem

$$v'' + v = f_{\text{sq}}(t), \qquad v(0) = 0, \qquad v'(0) = 0, \tag{1}$$

with

$$f_{\text{sq}}(t) = \begin{cases} 1 & 0 < t < 1 \\ -1 & 1 < t < 2 \\ f_{\text{sq}}(t-2) & 2 < t < \infty \end{cases}. \tag{2}$$

The *square wave* function f_{sq} is depicted in Figure 7.1.1.

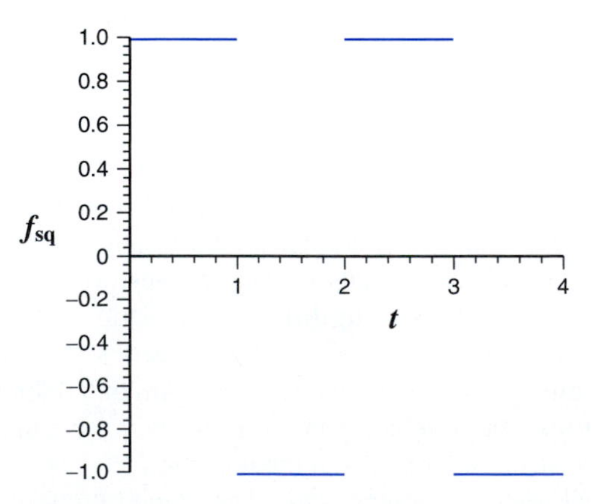

Figure 7.1.1
The square wave function of Model Problem 7.1.

Solution of the Model Problem by the Method of Undetermined Coefficients

Computation of the solution on the interval $0 < t < 2$ points out the plan for obtaining the full solution. The problem for $0 < t < 1$ is

$$v'' + v = 1, \qquad v(0) = 0, \qquad v'(0) = 0,$$

which has solution

$$v = 1 - \cos t, \qquad v' = \sin t.$$

At $t = 1$, we have

$$v(1) = 1 - \cos 1, \qquad v'(1) = \sin 1,$$

and these values serve as the initial conditions for the problem on the interval $1 < t < 2$:

$$v'' + v = -1, \qquad v(1) = 1 - \cos 1, \qquad v'(1) = \sin 1.$$

This problem has solution

$$v = -1 + (-1 + 2\cos 1)\cos t + 2\sin 1 \sin t.$$

The coefficients of $\cos t$ and $\sin t$ are different on each unit interval.

✦ INSTANT EXERCISE 1

Solve the problem

$$v'' + v = -1, \qquad v(1) = 1 - \cos 1, \qquad v'(1) = \sin 1.$$

Rather than continue with one interval at a time, it is more efficient to look for a general result. Let v_n be the solution on the time interval $n - 1 < t < n$. We have already determined the solution on the interval $0 < t < 2$; thus,

$$v_1 = 1 - \cos t, \quad v_2 = -1 + (-1 + 2\cos 1)\cos t + 2\sin 1 \sin t.$$

The differential equation on each subinterval is

$$v_n'' + v_n = (-1)^{n+1},$$

so the solution takes the form

$$v_n = (-1)^{n+1} + a_n \cos t + b_n \sin t, \tag{3}$$

where $a_1 = -1$, $b_1 = 0$, and the remaining a_n and b_n are determined by the requirements

$$v_{n+1}(n) = v_n(n), \quad v_{n+1}'(n) = v_n'(n), \quad n = 1, 2, \ldots. \tag{4}$$

In general, the desired current can be shown to be

$$v'(t) = b_n \cos t - a_n \sin t, \quad \text{for } n-1 < t < n, \tag{5}$$

where

$$a_{n+1} = a_n + 2(-1)^{n+1} \cos n, \quad b_{n+1} = b_n + 2(-1)^{n+1} \sin n, \quad n = 1, 2, \ldots. \tag{6}$$

The solution is illustrated in Figure 7.1.2. It is somewhat repetitive, but it is not truly periodic. The forcing function has period 2, and the oscillator has period 2π (in other words, the complementary solution has period 2π). These periods have no integral least common multiple, so the behavior that combines the oscillations is not periodic.

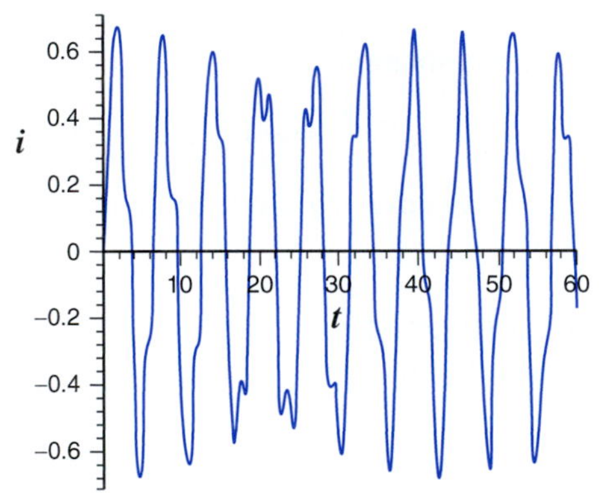

Figure 7.1.2
The solution of Model Problem 7.1.

Piecewise-Continuous Functions

The external voltage in an *RLC* circuit is supplied by batteries, transformers, or other electrical devices. The circuit can, of course, include switches so that components of the voltage turn on or off at a moment in time. For example, the external voltage could be

$$E = \begin{cases} 1 & t < 1 \\ 0 & t > 1 \end{cases},$$

corresponding to a unit voltage that is turned off after 1 time unit. This function E is not continuous, but it fails to be continuous only at one point.

A function f is **piecewise-continuous** on an interval (a, b) if

1. It is continuous at all points except perhaps a finite set of points $t_1, t_2, \ldots, t_n$.
2. Each limit

$$\lim_{t \to t_i^+} f(t), \qquad \lim_{t \to t_i^-} f(t)$$

exists for $i = 1, 2, \ldots, n$.

A function f is piecewise-continuous on the interval $(0, \infty)$ if it is piecewise-continuous on all intervals of the form $(0, A)$, where $A > 0$.

Note the careful definition of piecewise continuity on $(0, \infty)$. It is alright for f to have infinitely many discontinuities as long as there are not infinitely many on a finite interval. Thus, the square wave function f_{sq} is piecewise-continuous.

Physical situations in which the external voltage includes components that are turned on and off instantaneously result in mathematical problems with forcing functions that may be only piecewise-continuous. Such problems are tedious to solve by the standard method for solving linear oscillators. To solve them by the Laplace transform method, it is necessary to write piecewise-defined functions using a single formula. This is accomplished with the **Heaviside function,**[1] a function that represents a switch that turns on at time 0. Specifically, the Heaviside function is defined by

$$H(t) = \begin{cases} 0 & t < 0 \\ 1 & t > 0 \end{cases}. \tag{7}$$

It is easy to see why the Heaviside function is sometimes called the *unit step function.* It takes the value 0 up to the time 0, whereupon it steps up by a unit amount to a value of 1. Many authors assign a value to $H(0)$. Here we choose not to do so because the value assigned at any one point of a piecewise-continuous function makes an inconsequential difference for most practical applications. Another way of saying this is that erasing or moving a single point on the graph of a function changes the function in a way that generally does not matter.

We can use the Heaviside function as a convenient way to produce a formula for other functions that are 0 up to time 0.

EXAMPLE 1

Consider the function

$$f(t) = \begin{cases} 0 & t < 0 \\ \sin t & t > 0 \end{cases}.$$

This function can be written as

$$f(t) = H(t) \sin t.$$

For $t < 0$, we have $H(t) = 0$, so $f(t) = 0$; for $t > 0$, we have $H(t) = 1$ and $f(t) = \sin t$. See Figure 7.1.3.

[1] This function is named after Oliver Heaviside, who pioneered the use of transform methods to solve differential equations.

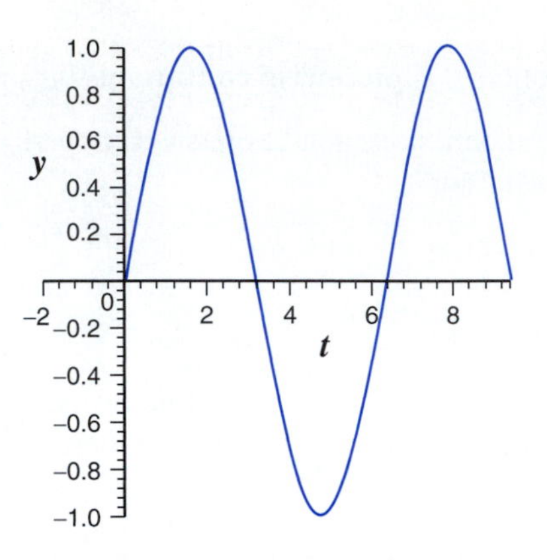

Figure 7.1.3
The function $y = H(t)\sin t$.

We can use a shifted Heaviside function to produce a formula for a function that is 0 up to any particular time.

EXAMPLE 2

Consider the function

$$f(t) = \begin{cases} 0 & t < \dfrac{\pi}{2} \\ \sin t & t > \dfrac{\pi}{2} \end{cases}.$$

This function can be written as

$$f(t) = H\left(t - \frac{\pi}{2}\right)\sin t.$$

The formula for $H(t)$ changes from 0 to 1 at time $t = \pi/2$, so the formula for $f(t)$ changes from 0 to $\sin t$ at the same time. See Figure 7.1.4.

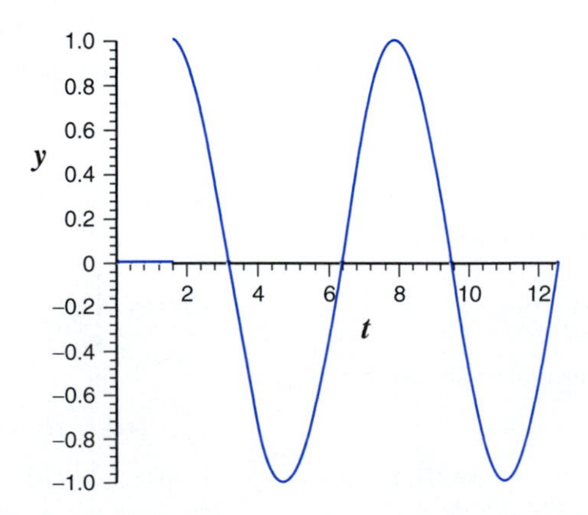

Figure 7.1.4
The function $y = H(t - \pi/2)\sin t$.

In the examples so far, we have looked only at functions that turn on at some time. We can also use the Heaviside function for functions that turn off at a particular time. A switch that turns off at time a is provided by the function

$$1 - H(t-a) = \begin{cases} 1 & t < a \\ 0 & t > a \end{cases}. \tag{8}$$

EXAMPLE 3

The function f defined by

$$f(t) = \begin{cases} e^{-t} & t < 2 \\ 0 & t > 2 \end{cases}$$

can be written as

$$f(t) = [1 - H(t-2)]\, e^{-t}.$$

See Figure 7.1.5.

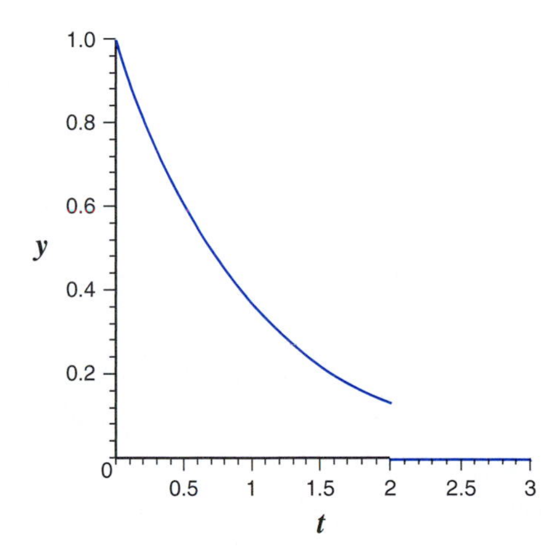

Figure 7.1.5
The function $y = [1 - H(t-2)]\, e^{-t}$.

✦ INSTANT EXERCISE 2

Use the Heaviside function to write the function defined by

$$E = \begin{cases} 1 & t < 1 \\ 0 & t > 1 \end{cases}$$

with a single formula.

Finally, functions that change from one nonzero formula to another can be represented by a combination of terms involving switches. The function $H(t-a) - H(t-b)$, with $a < b$, serves as a filter that selects only the portion of a function that is on the interval $a < t < b$.

EXAMPLE 4

The function f defined by

$$f(t) = \begin{cases} 1 & t < 3 \\ \sin t & t > 3 \end{cases}$$

can be written as

$$f(t) = [1 - H(t-3)] + H(t-3)\sin t.$$

To check this, note that the formula reduces to $f(t) = 1$ for $t < 3$ and $f(t) = \sin t$ for $t > 3$. See Figure 7.1.6.

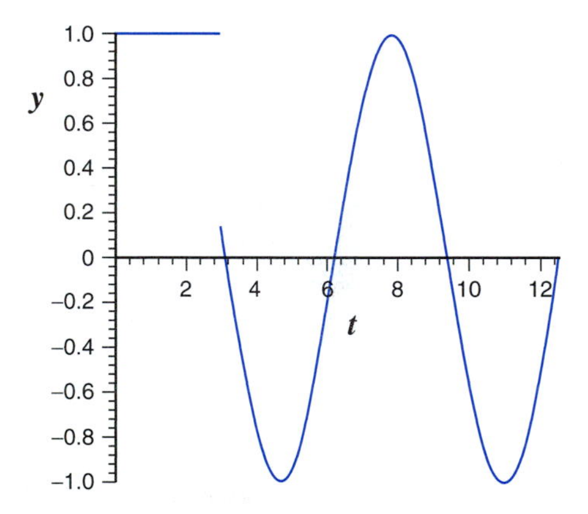

Figure 7.1.6
The function $y = [1 - H(t-3)] + H(t-3)\sin t$.

EXAMPLE 5

The function f defined by

$$f(t) = \begin{cases} 3 & t < 2 \\ t & 2 < t < 5 \\ t^2 & 5 < t \end{cases}$$

can be thought of as a function that turns off at time 2, a function that is active between times 2 and 5, and a function that turns on at time 5. Thus

$$f(t) = 3[1 - H(t-2)] + [H(t-2) - H(t-5)]t + H(t-5)t^2.$$

7.1 Exercises

In Exercises 1 through 10, sketch the function defined by the given formula and rewrite it, using a single formula.

1. $f(t) = \begin{cases} 0 & t < 4 \\ t^2 & t > 4 \end{cases}$

2. $f(t) = \begin{cases} 0 & t < \pi/2 \\ \cos t & t > \pi/2 \end{cases}$

3. $f(t) = \begin{cases} t^2 & t < 1 \\ 0 & t > 1 \end{cases}$

4. $f(t) = \begin{cases} e^{-2t} & t < 1 \\ 0 & t > 1 \end{cases}$

5. $f(t) = \begin{cases} t & t < 2 \\ t^2 & t > 2 \end{cases}$

6. $f(t) = \begin{cases} \sin(\pi t) & t < \pi/2 \\ \ln t & t > \pi/2 \end{cases}$

7. $f(t) = \begin{cases} 0 & t < 2 \\ (t-2)^2 & t \geq 2 \end{cases}$

8. $f(t) = \begin{cases} 0 & t < \pi \\ t - \pi & \pi \leq t \geq 2\pi \\ 0 & t > 2\pi \end{cases}$

9. $f(t) = \begin{cases} 2 - t & t < 2 \\ 0 & 2 < t < 3 \\ t - 3 & 3 < t \end{cases}$

10. $f(t) = \begin{cases} t & t < 1 \\ \dfrac{1}{t} & 1 < t < 2 \\ \cos(\pi t) & 2 < t \end{cases}$

In Exercises 11 through 14, sketch the function defined by the given formula.

11. $f(t) = H(t-1) + 2H(t-3) - 6H(t-4)$
12. $f(t) = (t-\pi)^2 H(t-\pi)$
13. $f(t) = 2tH(t-2)$
14. $f(t) = (t-3)H(t-2) - (t-2)H(t-3)$
15. Write the function f_{sq} from Model Problem 7.1 in terms of the Heaviside function.
16. Derive Equation (6).

✦ 7.1 INSTANT EXERCISE SOLUTIONS

1. The differential equation has particular solution $v_p = -1$ and complementary solution $v_c = c_1 \cos t + c_2 \sin t$; thus, the general solution is $v = -1 + c_1 \cos t + c_2 \sin t$. The initial conditions then yield a pair of equations,

$$-1 + c_1 \cos 1 + c_2 \sin 1 = 1 - \cos 1, \qquad c_2 \cos 1 - c_1 \sin 1 = \sin 1,$$

or

$$(1 + c_1) \cos 1 + c_2 \sin 1 = 2, \qquad (1 + c_1) \sin 1 - c_2 \cos 1 = 0.$$

Multiplying the first equation by $\cos 1$ and the second by $\sin 1$, and then adding the results, yields $1 + c_1 = 2 \cos 1$, whence $c_1 = 2 \cos 1 - 1$. Then $c_2 \cos 1 = (1 + c_1) \sin 1 = 2 \cos 1 \sin 1$, so $c_2 = 2 \sin 1$.

The solution is

$$v = -1 + (2\cos 1 - 1)\cos t + 2\sin 1 \sin t.$$

2. The formula $E = 1$ is correct up to time 1. We can fix this formula by adding a term that turns the value 1 off at time 1. Thus,

$$E = 1 - H(t-1).$$

7.2 Definition and Properties of the Laplace Transform

In the study of linear differential equations, we found it convenient to think of the left side of a linear differential equation as a linear differential operator, that is, as a functionlike rule that produces a unique output function for a given input function. As an example, consider the operator L defined by

$$L[y] = y'' - 2y.$$

Given $y = t^3$, for example, we have $L[t^3] = 6t - 2t^3$.

An **integral transform** is similar to a differential operator. But where the input and output functions of a differential operator are functions of the *same* variable, the input and output functions of an integral transform are functions of *different* variables. A linear differential operator produces a function in the same domain as the input, while an integral transform produces a function in a different domain.

Definition of the Laplace Transform

Let f be a real-valued function defined for $t > 0$. The **Laplace transform** of f is the function F defined by the integral formula

$$F(s) = \mathcal{L}[f](s) = \int_0^\infty e^{-st} f(t)\,dt, \tag{1}$$

provided that the integral converges.

EXAMPLE 1

$$\mathcal{L}[1] = \int_0^\infty e^{-st}\,dt = -\frac{1}{s}e^{-st}\Big|_{t=0}^{t=\infty} = \frac{1}{s}\left(1 - \lim_{t\to\infty} e^{-st}\right) = \frac{1}{s}.$$

Note that the last step of this calculation requires the assumption $s > 0$.

EXAMPLE 2

For any constant a,

$$\mathcal{L}\left[e^{at}\right] = \int_0^\infty e^{(a-s)t}\,dt = \frac{1}{a-s}e^{(a-s)t}\Big|_{t=0}^{t=\infty} = \frac{1}{s-a}\left(1 - \lim_{t\to\infty} e^{(a-s)t}\right) = \frac{1}{s-a},$$

provided $s > a$.

✦ INSTANT EXERCISE 1

Find the Laplace transform of the function defined by $f(t) = t$.

There are several observations that should be made from these examples.

- The evaluation of the antiderivatives removes t from $\mathcal{L}[f]$ but does not remove s; hence, $\mathcal{L}[f]$ is a function of s but not of t. We can say that the function F exists in the s domain rather than the t domain of the function f. The new variable s is not a time or length or any other physical quantity.
- The Laplace transform often produces a function whose character is entirely different from that of the input function. Here, both the constant function and the exponential function have rational functions for their Laplace transform.
- The Laplace transform function is often not defined for all values of s. It is quite common for a particular Laplace transform to be defined only for s larger than some critical value. This detail need not concern us.

The definition of the Laplace transform includes the phrase "provided the integral converges." Not every improper integral converges, so it would be useful to have a way of guaranteeing the existence of the Laplace transform.

MODEL PROBLEM 7.2

What properties determine whether or not a function can be Laplace-transformed? How does the Laplace transform change linear differential equations with constant coefficients?

We can obtain some insight into the first question by examining some functions for which the integral does not converge.

EXAMPLE 3

The (real-valued) function defined by $f(t) = \sqrt{t^2 - 4t + 3}$ does not have a Laplace transform. The problem here is that the function is not defined over the whole interval $t > 0$. It is sometimes possible to integrate a function that is not defined at an isolated point, but it is never possible to integrate a function that is undefined over a portion of the integration interval. See Figure 7.2.1.

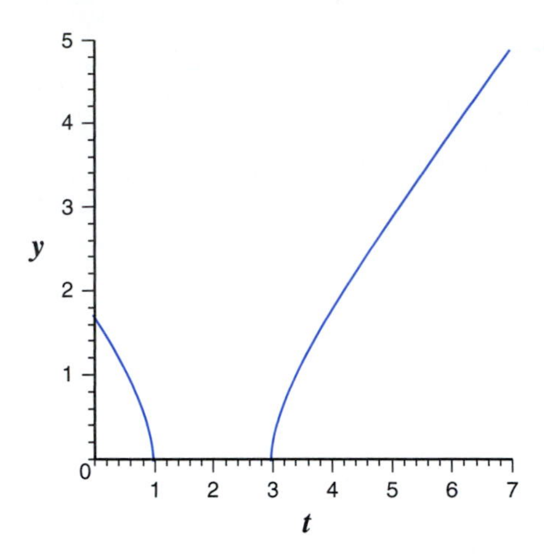

Figure 7.2.1
The function $y = \sqrt{t^2 - 4t + 3}$.

EXAMPLE 4

The function e^{t^2} has no Laplace transform. An integral on the domain $0 < t < \infty$ can exist only if the integrand vanishes as $t \to \infty$. (This condition is not sufficient, but it is certainly necessary.) The integrand of $\mathcal{L}[f]$ is $f(t)e^{-st}$; most of the time, the limit of this integrand is 0 as $t \to \infty$ because of the factor e^{-st}. As an example, suppose $f(t) = e^t$. This function grows quite rapidly, but the product $f(t)e^{-st}$ still vanishes as $t \to \infty$ when $s > 1$. However, e^{t^2} has no Laplace transform because

$$\lim_{t\to\infty} e^{t^2 - st} = \infty$$

for all values of s.

A function f is said to be of **exponential order** if a number s can be chosen large enough[2] that

$$\lim_{t\to\infty} f(t)e^{-st} = 0.$$

The function defined by $f(t) = e^{9t}$ is of exponential order because the limit in the definition is 0 for $s > 9$, but $f(t) = e^{t^2}$ is not of exponential order.

EXAMPLE 5

The integral $\int_0^\infty t^{-1}e^{-st}\,dt$ is improper because the integrand does not exist at the lower limit $t = 0$. Improper integrals sometimes exist, but this one does not (see Exercise 27). Hence, the function f defined by $f(t) = 1/t$ does not have a Laplace transform.

---

[2]If the limit holds for $s = s_1$, then it also holds for $s > s_1$.

Examples 3 through 5 point out that Laplace transformability of a function f depends on the general nature of the function for $(0, \infty)$ and also the rates of growth as $t \to \infty$ or as $t \to 0$. Theorem 7.2.1 gives a set of conditions sufficient to guarantee that a function has a Laplace transform.

Theorem 7.2.1 If a function f is piecewise-continuous on $(0, \infty)$, of exponential order, and $\lim_{t \to 0} |t f(t)| = 0$, then f has a Laplace transform.

EXAMPLE 6

Consider the function defined by $f(t) = t^{-1/2}$. This function is not defined at $t = 0$, but it still satisfies all three conditions of Theorem 7.2.1. It is therefore Laplace-transformable. Specifically, we can calculate $\mathcal{L}\left[t^{-1/2}\right]$, using the substitution $u = \sqrt{st}$ in the defining integral.

$$\mathcal{L}\left[t^{-1/2}\right] = \int_0^\infty t^{-1/2} e^{-st}\, dt = \int_0^\infty \left(e^{-st}\right) t^{-1/2}\, dt = \int_0^\infty \left(e^{-u^2}\right) 2s^{-1/2}\, du = \frac{2}{\sqrt{s}} \int_0^\infty e^{-u^2}\, du.$$

The last integral is well known; its value is $\sqrt{\pi}/2$. Thus, we get the result

$$\mathcal{L}\left[t^{-1/2}\right] = \sqrt{\frac{\pi}{s}}.$$

Two Key Properties of the Laplace Transform

The utility of the Laplace transform depends on two of its most basic properties.

1. The Laplace transform is a linear operation. This means that

$$\mathcal{L}[af + bg] = a\mathcal{L}[f] + b\mathcal{L}[g] \tag{2}$$

   for any Laplace-transformable functions f and g and constants a and b.

2. Suppose the functions f and f' are Laplace-transformable. Then the Laplace transform of f' is

$$\mathcal{L}\left[f'\right] = s\mathcal{L}[f] - f(0). \tag{3}$$

The first of these properties follows directly from the properties of definite integrals. The second is obtained using integration by parts. By definition, the Laplace transform of f' is

$$\mathcal{L}\left[f'\right] = \int_0^\infty f'(t) e^{-st}\, dt.$$

Even though we don't have a specific formula for the function f, we can integrate the formula for $\mathcal{L}\left[f'\right]$ by parts because the integrand has one factor that is easily differentiated and another that is easily integrated. We obtain

$$\mathcal{L}\left[f'\right] = e^{-st} f(t)\Big|_{t=0}^{t=\infty} + s \int_0^\infty f(t) e^{-st}\, dt.$$

As $t \to \infty$, the function $e^{-st} f(t)$ must vanish for any Laplace-transformable function f. Also, the remaining integral is just the Laplace transform of f. Thus we have the result (3). The relationship between the transforms of f and f' is a simple linear algebraic relationship. In a sense, we could say that the relationship between f and f' in the s domain is algebraic even though their relationship in the t domain is "differential."

It is the combination of the two crucial properties (2) and (3) that makes the Laplace transform useful for differential equations. A constant-coefficient linear differential equation, transformed into the s domain, expresses an algebraic relationship between the Laplace transforms of the unknown function and its derivatives. What was a linear differential equation in the t domain becomes a linear algebraic equation in the s domain. The equation in the s domain can then be solved for the Laplace transform of the unknown function. This function of s is used to recover the solution of the original problem. This sounds easy, but there is a hidden difficulty. Finding a function y from its Laplace transform $L[y]$ is problematic. This problem is examined in Section 7.3.

Laplace Transform Tables

Computing Laplace transforms from the definition is time-consuming. Moreover, it is very repetitive, since many different problems can require the Laplace transform of the same function. Accordingly, Laplace transform formulas are collected in tables, much as antiderivative formulas are. A collection of Laplace transform formulas appears on the inside front cover of this book. A more exhaustive list can be found in reference books.[3]

The Laplace Transform of Piecewise-Defined Functions

Suppose f is a function that is 0 up to time $a > 0$ and is given by a formula $g(t)$ after time a. We have

$$f(t) = \begin{cases} 0 & t < a \\ g(t) & t > a \end{cases} = g(t)H(t-a).$$

Now suppose we want to compute the Laplace transform of f. By definition, this is

$$\mathcal{L}[f(t)] = \int_0^\infty e^{-st} f(t)\,dt.$$

The function f is given by 0 on the interval $t < a$ and $g(t)$ on the interval $t > a$, so we can break the integral into two parts:

$$\mathcal{L}[f] = \int_0^a 0\,dt + \int_a^\infty e^{-st} g(t)\,dt = \int_a^\infty e^{-st} g(t)\,dt.$$

Thus, the Laplace transform of f is almost the same integral as the Laplace transform of g. The integral on the right is *not* the transform of g, however, because it has the wrong lower integration limit. This difficulty can be solved by defining a new integration variable. Let $\tau = t - a$ replace t in the integral. The lower limit of the integration is at $t = a$, or $\tau = 0$. The upper limit is at $t = \infty$, which corresponds to $\tau = \infty$. We must also replace the differential dt by $d\tau$, and we

[3]See, for example, D. Zwillinger (ed.), *CRC Standard Mathematical Tables and Formulae,* 30th ed., CRC Press, Boca Raton, FL, 1996.

must replace t by $\tau + a$. Hence, we have

$$\mathcal{L}[f] = \int_0^\infty e^{-s(\tau+a)} g(\tau + a)\, d\tau = e^{-as} \int_0^\infty e^{-s\tau} g(\tau + a)\, d\tau.$$

The choice of symbol for the integration variable is irrelevant, so we can rewrite the result as

$$\mathcal{L}[f(t)] = e^{-as} \int_0^\infty e^{-st} g(t + a)\, dt.$$

The integral in this last formula is the Laplace transform of $g(t + a)$. Thus, we have the final result:

$$\mathcal{L}[g(t)H(t - a)] = e^{-as}\mathcal{L}[g(t + a)], \qquad a > 0. \tag{4}$$

EXAMPLE 7

Let $f(t)$ be given by

$$f(t) = \begin{cases} 0 & t < \dfrac{\pi}{2} \\ \sin t & t > \dfrac{\pi}{2} \end{cases} = H\left(t - \frac{\pi}{2}\right) \sin t.$$

This fits the pattern of Equation (4), with $a = \pi/2$ and $g(t) = \sin t$. Hence,

$$g(t + a) = \sin\left(t + \frac{\pi}{2}\right).$$

To determine the Laplace transform of this function, we must first simplify it. We use the identity

$$\sin(a + b) = \sin a \cos b + \sin b \cos a$$

to obtain

$$\sin\left(t + \frac{\pi}{2}\right) = \cos\frac{\pi}{2} \sin t + \sin\frac{\pi}{2} \cos t = \cos t.$$

Thus,

$$\mathcal{L}[f(t)] = e^{-\pi s/2}\, \mathcal{L}[\cos t] = e^{-\pi s/2} \frac{s}{s^2 + 1}.$$

✦ INSTANT EXERCISE 2

Determine the Laplace transform of the function f defined by

$$f(t) = \begin{cases} 0 & t < 1 \\ e^{-t} & t > 1 \end{cases}.$$

Functions with more complicated switches can be handled in a similar fashion.

EXAMPLE 8

Let $f(t)$ be defined by

$$f(t) = \begin{cases} e^{-t} & t < 2 \\ 0 & t > 2 \end{cases} = [1 - H(t - 2)]\, e^{-t}.$$

Define $g(t)$ by

$$g(t) = e^{-t}.$$

Then

$$f(t) = g(t) - H(t-2)\,g(t).$$

Hence,

$$\mathcal{L}[f(t)] = \mathcal{L}[g(t)] - e^{-2s}\,\mathcal{L}[g(t+2)]\,.$$

Now,

$$g(t+2) = e^{-(t+2)} = e^{-2}\,g(t),$$

so

$$\mathcal{L}[f(t)] = \mathcal{L}[g(t)] - e^{-2s}\,e^{-2}\,\mathcal{L}[g(t)] = \left(1 - e^{-2s-2}\right)\mathcal{L}[g(t)]\,.$$

From the tables, we find

$$\mathcal{L}[g(t)] = \frac{1}{s+1};$$

thus,

$$\mathcal{L}[f(t)] = \frac{1 - e^{-2s-2}}{s+1}.$$

The Laplace Transform of Periodic Functions

It is often necessary to compute the Laplace transform of a periodic piecewise-defined function, such as the function f_{sq}, defined by

$$f_{\text{sq}}(t) = \begin{cases} 1 & 0 < t < 1 \\ -1 & 1 < t < 2 \\ f_{\text{sq}}(t-2) & 2 < t < \infty \end{cases},$$

from Model Problem 7.1. A special formula can be derived to simplify the computation. Suppose f satisfies the property

$$f(t+T) = f(t)$$

for some $T > 0$ and all $t > 0$. Then the Laplace transform of f is

$$\mathcal{L}[f] = \int_0^\infty e^{-st} f(t)\,dt = \sum_{n=0}^\infty \int_{nT}^{(n+1)T} e^{-st} f(t)\,dt = \sum_{n=0}^\infty \int_0^T e^{-s(\tau+nT)} f(\tau+nT)\,d\tau.$$

The factor e^{-nsT} can be removed from the integrals, and the periodicity property can be used to replace $f(\tau+nT)$ by $f(\tau)$; then the integration variable can be renamed t to yield the formula

$$\mathcal{L}[f] = \int_0^\infty e^{-st} f(t)\,dt = \left(\sum_{n=0}^\infty e^{-nsT}\right) \int_0^T e^{-st} f(t)\,dt.$$

Note that we have removed the integral from the sum because it does not depend on n. It remains to compute the infinite series $\sum_{n=0}^\infty e^{-nsT}$. The substitution $r = e^{-sT}$ puts this series into the form of a convergent geometric series,

$$\sum_{n=0}^{\infty} e^{-nsT} = \sum_{n=0}^{\infty} r^n = \frac{1}{1-r} = \frac{1}{1-e^{-sT}}.$$

We thus arrive at the final result,

$$\mathcal{L}[f(t)] = \frac{\int_0^T e^{-st} f(t)\,dt}{1-e^{-sT}} \tag{5}$$

for any function f defined for $t > 0$ and satisfying the property $f(t+T) = f(t)$ for some $T > 0$.

EXAMPLE 9

Let f_{2T} be a square wave of period $2T$ defined by

$$f_{2T}(t) = \begin{cases} 1 & 0 < t < T \\ -1 & T < t < 2T \\ f_{2T}(t-2T) & 2T < t < \infty \end{cases}.$$

We have

$$\int_0^{2T} e^{-st} f_{2T}(t)\,dt = \int_0^T e^{-st} dt - \int_T^{2T} e^{-st}\,dt = -\frac{1}{s} e^{-st}\Big|_0^T + \frac{1}{s} e^{-st}\Big|_T^{2T} = \frac{1}{s}\left(1 - 2e^{-sT} + e^{-2sT}\right).$$

Thus,

$$\mathcal{L}[f_{2T}] = \frac{1-2e^{-sT}+e^{-2sT}}{s(1-e^{-2sT})} = \frac{(1-e^{-sT})^2}{s(1-e^{-sT})(1+e^{-sT})} = \frac{1-e^{-sT}}{s(1+e^{-sT})}.$$

7.2 Exercises

All formula numbers (not in parentheses) refer to the Laplace transform tables on the inside front cover.

1. Use the tables to find the Laplace transform of $t^4 - 3t^3 + t$.
2. Use the tables to find the Laplace transform of $t^2(e^{3t} + 5)$.
3. Use the tables to find the Laplace transform of $\sin 3t + 4\sqrt{t}$.
4. Use the tables to find the Laplace transform of $e^{-t}(\sin 2t - 2\cos 2t)$.
5. Note from formula 24 that

$$\mathcal{L}\left[te^{2t}\right] = \frac{1}{(s-2)^2}.$$

   Derive this result by using the definition of the Laplace transform.
6. Derive formula 22 by using the definition of the Laplace transform. Note the definition

$$\sinh x = \frac{e^x - e^{-x}}{2}.$$

7. Use the formula for the transform of f' and the definition of the Laplace transform to derive the formula for the Laplace transform of f''.

8. Derive Equation (2), using the definition of the Laplace transform.

In Exercises 9 through 24, find the Laplace transform of the given function.

9. $H(t-1) + 2H(t-3) - 6H(t-4)$

10. $f(t) = (t-\pi)^2 H(t-\pi)$

11. $f(t) = 2(t-1)H(t-1)$

12. $f(t) = (t-3)H(t-2) - (t-2)H(t-3)$

13. $f(t) = \begin{cases} 0 & t < 4 \\ t^2 & t > 4 \end{cases}$

14. $f(t) = \begin{cases} 0 & t < \pi/2 \\ \cos t & t > \pi/2 \end{cases}$

15. $f(t) = \begin{cases} t^2 & t < 1 \\ 0 & t > 1 \end{cases}$

16. $f(t) = \begin{cases} e^{-2t} & t < 1 \\ 0 & t > 1 \end{cases}$

17. $f(t) = \begin{cases} t & t < 2 \\ t^2 & t > 2 \end{cases}$

18. $f(t) = \begin{cases} \sin \pi t & t < \pi/2 \\ \ln t & t > \pi/2 \end{cases}$

19. $f(t) = \begin{cases} 0 & t < 2 \\ (t-2)^2 & t \geq 2 \end{cases}$

20. $f(t) = \begin{cases} 0 & t < \pi \\ t - \pi & \pi \leq t \geq 2\pi \\ 0 & t > 2\pi \end{cases}$

21. $f(t) = \begin{cases} 2 - t & t < 2 \\ 0 & 2 < t < 3 \\ t - 3 & 3 < t \end{cases}$

22. $f(t) = \begin{cases} t & t < 1 \\ 1/t & 1 < t < 2 \\ \cos \pi t & 2 < t \end{cases}$

23. $f(t) = \begin{cases} t & 0 < t < 1 \\ f(t-1) & 1 < t < \infty \end{cases}$

24. $f(t) = \begin{cases} \sin t & 0 < t < \pi \\ f(t-\pi) & \pi < t < \infty \end{cases}$

25. Suppose $F = \mathcal{L}[f]$ and both f' and f'' are Laplace-transformable. Use the formula for the transform of derivatives (3) to obtain a formula for $\mathcal{L}\left[f''\right]$ in terms of F.

26. Use the formula for the transform of derivatives (3) and formula 40 to obtain formula 39. Note that the derivation of formula 40 from formula 39 follows immediately from Equation (3).

27. Show that the integral $\int_0^\infty t^{-1}e^{-st}\,dt$ of Example 5 diverges. One way to do this is to establish the inequality $e^{-st} > 1 - st$ for $s > 0$. Applying this inequality to the integral yields an inequality comparing the given integral to a simple divergent integral.

✦ 7.2 INSTANT EXERCISE SOLUTIONS

1. From the definition, $\mathcal{L}[t] = \int_0^\infty te^{-st}\,dt$. Integrating by parts yields the antiderivative formula

$$\int te^{-st}\,dt = -\frac{t}{s}e^{-st} + \frac{1}{s}\int e^{-st}\,dt = -\left(\frac{t}{s} + \frac{1}{s^2}\right)e^{-st}.$$

Thus,

$$\mathcal{L}[t] = \int_0^\infty te^{-st}\,dt = -\left(\frac{t}{s} + \frac{1}{s^2}\right)e^{-st}\Big|_0^\infty = \frac{1}{s^2}.$$

2. We have

$$f(t) = \begin{cases} 0 & t < 1 \\ e^{-t} & t > 1 \end{cases} = H(t-1)\,e^{-t}.$$

This fits the pattern of Equation (4), with $a = 1$ and $g(t) = e^{-t}$. Hence,

$$g(t+1) = e^{-(t+1)} = e^{-1}\,e^{-t}.$$

Thus,

$$\mathcal{L}[f(t)] = e^{-s}e^{-1}\,\mathcal{L}\left[e^{-t}\right] = e^{-(s+1)}\,\frac{1}{s+1}.$$

7.3 Solution of Initial-Value Problems with the Laplace Transform

In this section, we demonstrate how the Laplace transform can be used to solve initial-value problems of the type that can be solved by the method of undetermined coefficients.

MODEL PROBLEM 7.3

Use the Laplace transform to solve the problem

$$y'' + 4y = 8, \qquad y(0) = 2, \qquad y'(0) = 6.$$

In solving this problem, we will make use of the two key Laplace transform properties from Section 7.2. These are summarized here:

$$\mathcal{L}[af+bg] = a\mathcal{L}[f] + b\mathcal{L}[g] \tag{1}$$

for any Laplace transformable functions f and g and constants a and b, and

$$\mathcal{L}\left[f'\right] = s\mathcal{L}[f] - f(0) \tag{2}$$

for any function whose derivative is Laplace-transformable.

Applying the Laplace Transform to an Initial-Value Problem

Formally, we can apply the Laplace transform to the model differential equation simply by writing the transform of both sides:

$$\mathcal{L}\left[y''+4y\right] = \mathcal{L}[8].$$

The transformed equation can now be simplified by using the linearity property (1):

$$\mathcal{L}\left[y''\right] + 4\mathcal{L}[y] = 8\mathcal{L}[1].$$

We now consider each term of the transformed equation.

The term on the right side can be evaluated immediately from the tables. From formula 14, the transform of 1 is $1/s$.

The transform of y cannot be evaluated because we don't know the function y. However, it is not necessary to evaluate this transform, because we will simply carry $\mathcal{L}[y]$ as an unknown. Therefore, we assign the name Y to the transform of y.

To find the transform of y'', we need to use the derivative formula (2). First, we use it to compute $\mathcal{L}\left[y'\right]$:

$$\mathcal{L}\left[y'\right] = s\mathcal{L}[y] - y(0) = sY - 2.$$

Next, we apply the formula again, using y' as the function f:

$$\mathcal{L}\left[y''\right] = s\mathcal{L}\left[y'\right] - y'(0) = s(sY-2) - 6 = s^2Y - 2s - 6.$$

With all the individual transforms determined, we obtain the result

$$\left(s^2Y - 2s - 6\right) + 4Y = \frac{8}{s}. \tag{3}$$

Notice that we cannot transform the initial conditions, nor can we assign initial conditions in the s domain. The initial conditions have not disappeared; they have been incorporated into the equation for Y in the s domain. Furthermore, Equation (3) is an algebraic equation, so no initial conditions are needed.

✦ INSTANT EXERCISE 1

Transform the problem

$$y'' - y = 0, \qquad y(0) = -1, \qquad y'(0) = 3$$

into the s domain.

Finding the Solution in the s Domain

This step is always simple because it requires only the solution of a linear algebraic equation. We just move all terms not containing Y to the right side of the equation and divide by the combined coefficient of Y. Thus, Equation (3) can be rewritten as

$$(s^2 + 4)\, Y(s) = \frac{8}{s} + 2s + 6,$$

and

$$Y(s) = \frac{8}{s(s^2 + 4)} + \frac{2s}{s^2 + 4} + \frac{6}{s^2 + 4}. \tag{4}$$

Note that we have *not* added the three terms in the solution to make one large fraction. The reason for this will be clear when we try to recover the solution $y(t)$ of the original problem from the function $Y(s)$.

 **INSTANT EXERCISE 2**

Find $Y(s)$ if y is the solution of the problem of Instant Exercise 1.

Recovering the Solution of the Original Problem

Suppose you know how to compute derivatives, but you do not know how to compute antiderivatives. Suppose further that you are trying to solve a problem for a function $y(t)$, and that your solution method has yielded instead a formula for the derivative $y'(t)$. Without a method for computing antiderivatives, your only recourse would be to compile a list of known derivatives and then try to use that list to recover the function y from its derivative y'.

The above scenario is not generally a problem once we learn techniques of integration. However, the difficulty posed by the above scenario is exactly what happens when we try to recover a function y from its Laplace transform Y.[4] In Model Problem 7.3, each of the terms in the formula for Y corresponds to an entry in the Laplace transform table, and this allows us to recover y without difficulty. Other problems present a greater challenge.

The calculations are facilitated by defining the inverse Laplace transform.

Suppose a function F is defined by $F = \mathcal{L}[f]$ for some function f. We define the **inverse Laplace transform** $\mathcal{L}^{-1}[F]$ by the formula $f = \mathcal{L}^{-1}[F]$.[5]

[4]There actually *is* a formula that can be used to compute f from $\mathcal{L}[f]$. The formula involves a complicated integral in the complex plane, and its use requires considerable knowledge of complex variable theory.

[5]Strictly speaking, the inverse Laplace transform is not unique. If two functions differ only at a single point, such as the functions $t + 1$ and $(t^2 + t)/t$, then they have the same Laplace transform. The latter function cannot be a solution of a differential equation on any interval including $t = 0$ because the function is not even defined on such an interval. Thus, it is possible to solve a differential equation problem uniquely with the Laplace transform even though the inverse transform is not unique.

EXAMPLE 1

Since $\mathcal{L}[8] = 8/s$,

$$\mathcal{L}^{-1}\left[\frac{8}{s}\right] = 8.$$

✦ INSTANT EXERCISE 3

Find $\mathcal{L}^{-1}[3/(s+2)]$.

We can now view the task of recovering y from Y as a problem of computing the inverse Laplace transform of Y. The inverse Laplace transform operation is linear because of the linearity of the Laplace transform itself, so we can break up sums and factor out constants. Applying the inverse transform to Equation (4) and making use of linearity yields

$$y(t) = 8\mathcal{L}^{-1}\left[\frac{1}{s(s^2+4)}\right] + 2\mathcal{L}^{-1}\left[\frac{s}{s^2+4}\right] + 6\mathcal{L}^{-1}\left[\frac{1}{s^2+4}\right]. \tag{5}$$

We now search the list of formulas in the inside front cover for the appropriate functions in the $F(s)$ column. The first term is an example of formula 33, with $a = 0$ and $k = 2$:

$$\mathcal{L}^{-1}\left[\frac{1}{s(s^2+4)}\right] = \frac{2 - 2\cos 2t}{8} = \frac{1 - \cos 2t}{4}.$$

The second term is given in formula 21, with $k = 2$, and the third term in formula 20, with a suitable change:

$$\mathcal{L}^{-1}\left[\frac{s}{s^2+4}\right] = \cos 2t, \qquad \mathcal{L}^{-1}\left[\frac{2}{s^2+4}\right] = \sin 2t.$$

Thus, we have

$$y(t) = 8\mathcal{L}^{-1}\left[\frac{1}{s(s^2+4)}\right] + 2\mathcal{L}^{-1}\left[\frac{s}{s^2+4}\right] + 3\mathcal{L}^{-1}\left[\frac{2}{s^2+4}\right],$$

or

$$y(t) = 8\,\frac{1 - \cos 2t}{4} + 2\cos 2t + 3\sin 2t.$$

After simplification, we arrive at the final answer:

$$y(t) = 2 + 3\sin 2t. \tag{6}$$

The Laplace Transform Method in General

The Laplace transform method involves the following sequence of steps:

1. Apply the Laplace transform to the initial-value problem. Because of the properties in Equations (1) and (2), this results in a linear algebraic equation in the s domain.
2. Solve the transformed equation to find the unknown transform in the s domain.
3. Determine the function in the t domain whose Laplace transform is the solution from step 2. This function is the solution of the initial-value problem.

Steps 1 and 3 involve the association of functions in the t domain with their transforms. In practice, most of the work needed to determine transforms or inverse transforms is done with the assistance of tabulated formulas.

Model Problem 7.3 can be solved by the method of undetermined coefficients. This requires that several steps be performed in sequence. First one must compute the complementary solution. Then the method of undetermined coefficients is used to compute the particular solution. The general solution is constructed by adding the complementary and particular solutions. Finally, the integration constants in the solution are evaluated from the initial conditions. The full procedure involves a lot of calculations. It might seem that the Laplace transform method is preferable to the standard method.

In the Laplace transform method, the first step is generally not difficult, and the second step is simple algebra. The difficulty lies in the third step. Recovering a function in the t domain from its Laplace transform can be very tedious, as will be seen in the examples that conclude this section. There are some special problems for which the Laplace transform is clearly the method of choice. We will consider some of these problems in Section 7.4.

Inverting Transforms with a Quadratic Denominator

The tables cannot possibly include all useful formulas. Tables are compiled with the idea that users will be able to employ algebraic techniques, such as completing the square, partial fraction decomposition, and the shifted-transform formula, to deal with some cases not explicitly listed in the table. Consider, for example, the situation in which the denominator of a transform F is the quadratic function defined by

$$p(s) = s^2 - 2bs + c,$$

where b and c are real constants. Note the elementary properties of the polynomial. First, its derivative $p'(s) = 2s - 2b$ is zero at the point $s = b$, and the second derivative $p''(s) = 2$ is always positive. This marks $s = b$ as the point locating the minimum of the function. Second, the limits as $s \to \pm\infty$ of p are both $+\infty$. Thus, the number of roots of the polynomial p depends on whether $p(b) = c - b^2$ is positive, negative, or zero. Figure 7.3.1 shows $p(s)$ for the case $b = 2$ with $c = 2, 4, 6$. [The value of c can be read directly from the graph because c is the value $p(0)$.]

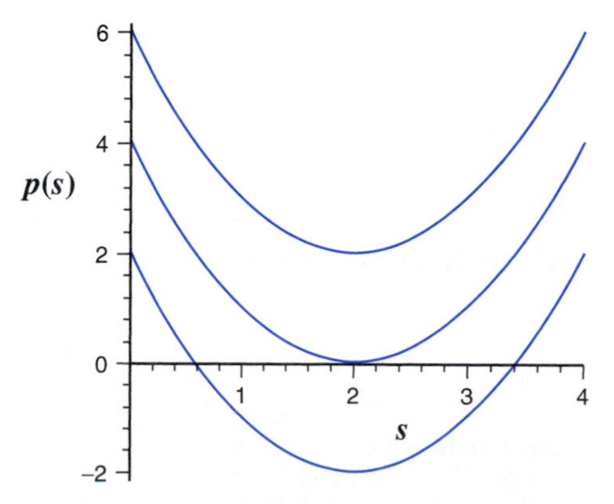

Figure 7.3.1
Functions of the family $s^2 - 4s + c$.

The inversion of a Laplace transform with a quadratic denominator depends on the number of roots, and consequently on the relative values of c and b^2.

1. If $c = b^2$, then $p(b) = 0$. Moreover, b is a double root of p because $p'(b)$ is also zero. Thus, p is a perfect square: $s^2 - 2bs + c = (s - b)^2$. This case fits formulas 24 and/or 29.
2. If $c < b^2$, then $p(b) < 0$ and there are two roots. First the polynomial p must be factored. After that, one can use table entries 30 and/or 31.
3. If $c > b^2$, then there are no real roots. This case corresponds to table entries 25 and 26; however, the denominator must be put into the appropriate form in order to use these entries.

Examples 2 and 3 serve to illustrate the techniques.

EXAMPLE 2

Suppose we want to invert

$$Y(s) = \frac{1}{s^2 + 4s - 5}.$$

The form of Y does not match any of the entries in the table, but we can factor the denominator to obtain

$$Y(s) = \frac{1}{(s - 1)(s + 5)}.$$

This matches formula 30, with $a = 1$ and $b = -5$. Thus,

$$y(t) = \tfrac{1}{6}e^t - \tfrac{1}{6}e^{-5t}.$$

✦ INSTANT EXERCISE 4

Determine the solution of the problem of Instant Exercise 1 by using formulas 30 and 31 to invert the transform $Y(s)$ obtained in Instant Exercise 2.

Completing the Square Returning to the general case for a moment, we suppose that $c > b^2$. The technique of completing the square consists of two steps: Add and subtract b^2, and then group the terms to include the perfect square $s^2 - 2bs + b^2$.

EXAMPLE 3

Consider the problem of inverting

$$Y(s) = \frac{s}{s^2 + 4s + 13}.$$

The denominator cannot be factored, and this means that it must be put in the form of formulas 25 and 26 by completing the square. We have $b = -2$, so we add and subtract 4 and regroup. Thus,

$$s^2 + 4s + 13 = s^2 + 4s + (4 - 4) + 13 = (s^2 + 4s + 4) + (13 - 4) = (s + 2)^2 + 9$$

and so

$$Y(s) = \frac{s}{(s+2)^2+9} = \frac{s+2}{(s+2)^2+9} - \frac{2}{(s+2)^2+9} = \frac{s+2}{(s+2)^2+9} - \frac{2}{3}\frac{3}{(s+2)^2+9}.$$

These terms fit the table entries with $a = -2$ and $k = 3$. Inverting the transform yields

$$y(t) = e^{-2t}\cos 3t - \tfrac{2}{3}e^{-2t}\sin 3t.$$

✦ INSTANT EXERCISE 5

Find $y(t)$ if

$$Y(s) = \frac{3}{s^2-2s+5}.$$

The Shifted-Transform Formula

Suppose we are given a Laplace-transformable function g and its Laplace transform G. Without having a specific formula for g, we can derive a useful formula for the Laplace transform of any function of the form $e^{at}g(t)$:

$$\mathcal{L}\left[e^{at}g(t)\right](s) = \int_0^\infty e^{-st}e^{at}g(t)\,dt = \int_0^\infty e^{-(s-a)t}g(t)\,dt.$$

Now suppose we define a new variable $p = s - a$. Then we have

$$\mathcal{L}\left[e^{at}g(t)\right](p) = \int_0^\infty e^{-pt}g(t)\,dt.$$

This is the integral that defines the Laplace transform, but with p as the transform variable rather than s. Hence, $\mathcal{L}\left[e^{at}g(t)\right] = G(p)$. Substituting back for p gives the shifted-transform formula

$$\mathcal{L}\left[e^{at}g(t)\right] = G(s-a). \tag{7}$$

EXAMPLE 4

Consider again the problem of inverting

$$Y(s) = \frac{s}{s^2+4s+13}.$$

We completed the square in Example 3 to obtain the result

$$s^2 + 4s + 13 = (s+2)^2 + 9.$$

Using this change, we have

$$Y(s) = \frac{s}{(s+2)^2+9}.$$

To use the shifted-transform formula, we need to think of $Y(s)$ as $G(s+2)$ for some function G, which we will determine. So we have

$$G(s+2) = Y(s), \qquad G(s+2) = \frac{s}{(s+2)^2+9}. \tag{8}$$

The first of these equations is inverted with help from the shifted-transform formula:

$$y(t) = e^{-2t} g(t). \tag{9}$$

Once we find the function g, we will have y immediately from Equation (9).

Now we turn to the task of finding g. The plan is to first find the function G and then invert it. The function G is indirectly defined by the second equation in (8). Suppose we substitute $s = p - 2$ for s. The reason for this choice of variable is that we will then have $G(p)$ on the left side of the equation. Systematic substitution of $p - 2$ for s yields

$$G(p) = \frac{p-2}{p^2+9}.$$

This equation holds regardless of the name given to the independent variable. We can therefore change "p" to "s" and obtain

$$G(s) = \frac{s-2}{s^2+9}.$$

The function G can now be inverted by using formulas 20 and 21:

$$g(t) = \mathcal{L}^{-1}\left[\frac{s}{s^2+9}\right] - \tfrac{2}{3}\mathcal{L}^{-1}\left[\frac{3}{s^2+9}\right] = \cos 3t - \tfrac{2}{3}\sin 3t.$$

Thus,

$$y(t) = e^{-2t}\left(\cos 3t - \tfrac{2}{3}\sin 3t\right).$$

It may seem as though we did more work in Example 4 than Example 3; however, the shifted-transform formula often offers a considerable advantage for inversion of rational functions with denominators of third or higher degree.

EXAMPLE 5

Let $F(s)$ be defined by

$$F(s) = \frac{s+6}{s(s^2-2s+5)} = \frac{s+6}{s\left[(s-1)^2+4\right]}.$$

The Laplace transform table in the inside front cover does not have an entry that matches this function. However, if the second factor in the denominator were s^2+4 instead of $(s-1)^2+4$, then we could find a match with formulas 33 and 34. To use the shifted-transform formula, we need to think of $F(s)$ as $G(s-1)$ for some function G, which we will determine. So, we have

$$G(s-1) = F(s), \qquad G(s-1) = \frac{s+6}{s\left[(s-1)^2+4\right]}.$$

Replacing s with $p+1$ in the second equation yields

$$G(p) = \frac{p+7}{(p+1)(p^2+4)}.$$

Thus,

$$g(t) = \mathcal{L}^{-1}\left[\frac{s+7}{(s+1)(s^2+4)}\right] = \mathcal{L}^{-1}\left[\frac{s}{(s+1)(s^2+4)}\right] + 7\mathcal{L}^{-1}\left[\frac{1}{(s+1)(s^2+4)}\right].$$

Substituting in the results from the table, with $a=-1$ and $k=2$, we have

$$g(t) = \frac{-e^{-t}+\cos 2t + 2\sin 2t}{5} + 7\,\frac{2e^{-t} - 2\cos 2t + \sin 2t}{10} = \frac{6}{5}e^{-t} - \frac{6}{5}\cos 2t + \frac{11}{10}\sin 2t.$$

Finally,

$$f(t) = e^t g(t) = \tfrac{6}{5} - \tfrac{6}{5}e^t\cos 2t + \tfrac{11}{10}e^t \sin 2t.$$

7.3 Exercises

In Exercises 1 through 10, find the inverse Laplace transform of the given function $F(s)$.

1. $\dfrac{3}{s^2+4}$
2. $\dfrac{3}{(s-2)^3}$
3. $\dfrac{4}{s^2+3s-4}$
4. $\dfrac{3s}{s^2-s-6}$
5. $\dfrac{2s+2}{s^2+2s+5}$
6. $\dfrac{2s-3}{s^2-4}$
7. $\dfrac{2s+1}{s^2-2s+2}$
8. $\dfrac{8s^2-4s+12}{s(s^2+4)}$
9. $\dfrac{1-2s}{s^2+4s+5}$
10. $\dfrac{2s-3}{s^2+2s+10}$

In Exercises 11 through 14, use the Laplace transform to solve the given initial-value problem.

11. $y''+3y'+2y=0, \quad y(0)=1, \quad y'(0)=0$
12. $y''-2y'+2y=0, \quad y(0)=0, \quad y'(0)=1$
13. $y''''-4y=0, \quad y(0)=1, \quad y'(0)=0, \quad y''(0)=-2, \quad y'''(0)=0$
14. $y''-2y'+2y=e^{-t}, \quad y(0)=0, \quad y'(0)=1$

✦ 7.3 INSTANT EXERCISE SOLUTIONS

1. Using the derivative formula (2), we have $\mathcal{L}[y'] = sY+1$ and then $\mathcal{L}[y''] = s(sY+1)-3 = s^2Y + s - 3$. Thus, the Laplace transform of the problem is

$$s^2Y + s - 3 - Y = 0.$$

2. $Y(s) = \dfrac{-s+3}{s^2-1}$

3. $\mathcal{L}\left[e^{-2t}\right] = 1/(s+2)$, so $\mathcal{L}\left[3e^{-2t}\right] = 3/(s+2)$ and then $\mathcal{L}^{-1}[3/(s+2)] = 3e^{-2t}$.

4. The transform can be rewritten as

$$Y(s) = -\frac{s}{(s-1)(s+1)} + 3\frac{1}{(s-1)(s+1)}.$$

Thus, formulas 30 and 31 apply with $a = 1$ and $b = -1$. The result is

$$y(t) = -\frac{e^t + e^{-t}}{2} + 3\frac{e^t - e^{-t}}{2} = e^t - 2e^{-t}.$$

5. The function $s^2 - 2s + 5$ has the form $s^2 - 2bs + c$ with $b = 1$. Thus, we write $s^2 - 2s + 5 = (s^2 - 2s + 1) + (-1 + 5) = (s-1)^2 + 4$. Thus,

$$Y(s) = \frac{3}{(s-1)^2+4} = \frac{3}{2}\frac{2}{(s-1)^2+4}.$$

Formula 25 applies with $a = 1$ and $k = 2$. Inverting the transform yields

$$y(t) = \tfrac{3}{2}\, e^t \sin 2t.$$

7.4 Piecewise-Continuous and Impulsive Forcing

Recall that nonhomogeneous oscillator problems are often said to be *forced* by the nonhomogeneity. The term *forcing* implies that an oscillator is being driven by some stimulus imposed from outside the oscillator itself. Some situations that are modeled by oscillators involve forcing that cannot be represented by a continuous function. An electric circuit may have a switch whose operation turns forcing circuits on and off; these problems have **piecewise-continuous forcing.** Mechanical vibration problems may include a mechanism that suddenly changes the momentum of the vibrating object; an example of this kind of forcing is a blow by a hammer. It is convenient to represent a hammer blow as an infinite force applied over an infinitesimal amount of time: this is called **impulsive forcing.** In both electrical and mechanical systems there may be **periodic forcing** by functions that are not easily represented with sine and cosine functions, an example being the square wave function in Model Problem 7.1. In all these cases, the Laplace transform method is generally the method of choice. The method of Section 7.3 applies, with the aid of a few additional formulas that we have not yet developed.

MODEL PROBLEM 7.4

An *RLC* circuit with $C = 2$, $L = 0.25$, and $R = 0.1$ is initially at rest. A power source of strength 1 V is applied to the circuit for 10 s. Determine the current in the circuit.

The *RLC* circuit model developed in Section 4.1 applies to this problem. Using v for the voltage across the capacitor, we have the initial-value problem

$$0.5v'' + 0.2v' + v = E(t) = \begin{cases} 1 & t < 10 \\ 0 & t > 10 \end{cases}, \qquad v(0) = 0, \qquad v'(0) = 0.$$

Once we have determined v, the current can be found from $i = Cv'$.

We could solve Model Problem 7.4 by the method of undetermined coefficients, as was done for Model Problem 7.1, but that requires us to solve the problem twice. Instead, we employ the Laplace transform method. The forcing function is turned off by a switch, so it is conveniently written in terms of the Heaviside function as $E = 1 - H(t - 10)$; thus, we have

$$v'' + 0.4v' + 2v = 2[1 - H(t - 10)], \qquad v(0) = 0, \qquad v'(0) = 0.$$

Now define $V(s) = \mathcal{L}[v]$. The rule for the transform of derivatives yields

$$\mathcal{L}\left[v'\right] = s\mathcal{L}[v] - v(0) = sV(s), \qquad \mathcal{L}\left[v''\right] = s\mathcal{L}\left[v'\right] - v'(0) = s^2V(s).$$

The transform of the forcing function is

$$2(\mathcal{L}[1] - \mathcal{L}[H(t - 10)]) = \frac{2(1 - e^{-10s})}{s}.$$

The problem in the s domain is then

$$s^2V + 0.4sV + 2V = \frac{2(1 - e^{-10s})}{s}.$$

As always, the problem in the s domain is algebraic, and we easily obtain the solution

$$V = \frac{2(1 - e^{-10s})}{s(s^2 + 0.4s + 2)}.$$

It remains only to invert the transform to determine the final answer. It is helpful to separate out the effects of the step function, which we do by defining a function F by

$$F(s) = \frac{2}{s(s^2 + 0.4s + 2)}.$$

This allows us to write V as

$$V(s) = F(s) - e^{-10s}F(s). \tag{1}$$

The inversion of F is similar to problems examined in Section 7.3. By completing the square, we obtain

$$f(t) = 1 - e^{-0.2t}\left(\cos 1.4t + \tfrac{1}{7}\sin 1.4t\right). \tag{2}$$

✦ INSTANT EXERCISE 1

Derive Equation (2) by inverting F.

It remains to find $\mathcal{L}^{-1}\left[e^{-10s}F(s)\right]$.

Transform Inversion for Piecewise-Defined Functions

In Section 7.2, we developed a formula for the Laplace transform of a piecewise-defined function:

$$\mathcal{L}[g(t)H(t-a)] = e^{-as}\mathcal{L}[g(t+a)], \qquad a > 0. \tag{3}$$

This formula is not particularly convenient for computing the inverse transform of a function $e^{-as}F(s)$. A more convenient formula is easily derived, however. Define a function $f(t)$ and its Laplace transform by

$$f(t) = g(t+a), \qquad F(s) = \mathcal{L}[f(t)].$$

Then the right side of Equation (3) becomes

$$e^{-as}\mathcal{L}[g(t+a)] = e^{-as}\mathcal{L}[f(t)] = e^{-as}F(s).$$

On the left side, we must find a formula for $g(t)$. Replacing t by $t-a$ in the equation $f(t) = g(t+a)$ gives the result

$$g(t) = f(t-a).$$

Thus, we can rewrite Equation (3) in terms of f:

$$\mathcal{L}[f(t-a)H(t-a)] = e^{-as}F(s), \qquad a > 0. \tag{4}$$

This formula is conveniently used to invert functions in the s domain that include an exponential factor.

EXAMPLE 1

To compute

$$\mathcal{L}^{-1}\left[\frac{4e^{-3s}}{s^3}\right],$$

define $F(s)$ by

$$F(s) = \frac{4}{s^3}.$$

Formula 14 Laplace transform tables gives the transform of t^2 as $2/s^3$, so

$$f(t) = 2t^2.$$

Application of Equation (4) gives the final result:

$$\mathcal{L}^{-1}\left[\frac{4e^{-3s}}{s^3}\right] = \mathcal{L}^{-1}\left[e^{-3s}F(s)\right] = f(t-3)H(t-3) = 2(t-3)^2H(t-3).$$

✦ INSTANT EXERCISE 2

Compute $\mathcal{L}^{-1}\left[\dfrac{e^{-3s}}{s^2+4}\right]$.

Returning to Model Problem 7.4, we have

$$V(s) = F(s) - e^{-10s}F(s).$$

Equation (4) gives us

$$v(t) = f(t) - f(t-10)H(t-10). \tag{5}$$

The function f has already been computed, so we can now write the final result for the voltage:

$$\begin{aligned} v(t) = {} & 1 - e^{-0.2t}\left(\cos 1.4t + \tfrac{1}{7}\sin 1.4t\right) \\ & - \left\{1 - e^{-0.2t+2}\left[\cos(1.4t-14) + \tfrac{1}{7}\sin(1.4t-14)\right]\right\} H(t-10). \end{aligned} \tag{6}$$

The current is given by

$$i(t) = Cv'(t) = 2f'(t) - 2f'(t-10)H(t-10).$$

Differentiating Equation (2), we have

$$f'(t) = \tfrac{10}{7}e^{-0.2t}\sin 1.4t;$$

thus,

$$i(t) = \tfrac{20}{7}e^{-0.2t}\sin 1.4t - \tfrac{20}{7}e^{-0.2t+2}\sin(1.4t-14)H(t-10). \tag{7}$$

Note that we took the derivative of $H(t-10)$ to be zero, which is true except at $t = 10$. The derivative results do not actually hold at $t = 10$, but there is no well-defined function value there either. This is just one point on the graph of a piecewise-continuous function, and this is unimportant in the context of the circuit problem.

✦ INSTANT EXERCISE 3

Derive the formula for $f'(t)$ by differentiating Equation (2).

The solutions for the voltage and current are illustrated in Figures 7.4.1 and 7.4.2. Note that both the voltage and the current are continuous, even at time 10. The differential equation indicates that the discontinuity in the forcing function corresponds to discontinuities in v'' and i'.

Impulsive Forcing and the Dirac Delta Function

Newton's second law of motion can be stated in impulse-momentum form by integrating $F = mv'$ over a time interval:

$$\int_{t_0}^{t_1} F(t)\,dt = \int_{t_0}^{t_1} mv'(t)\,dt = (mv)\Big|_{t_0}^{t_1}.$$

In this form, the law says that the change in momentum caused by a force acting on an interval is equal to the integral of the force. Therefore, a unit change in momentum is accomplished by

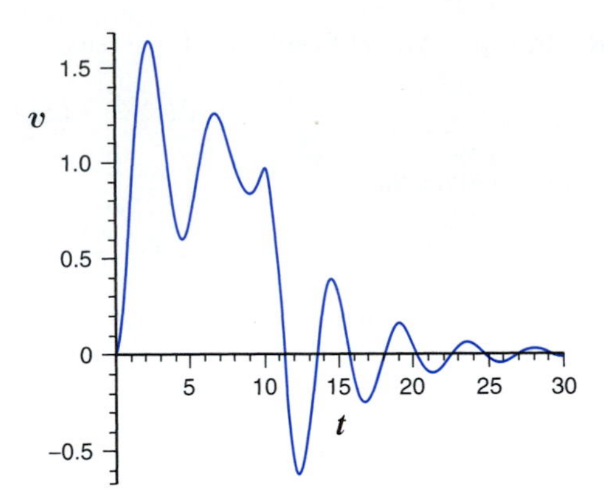

Figure 7.4.1
The output voltage in Model Problem 7.4.

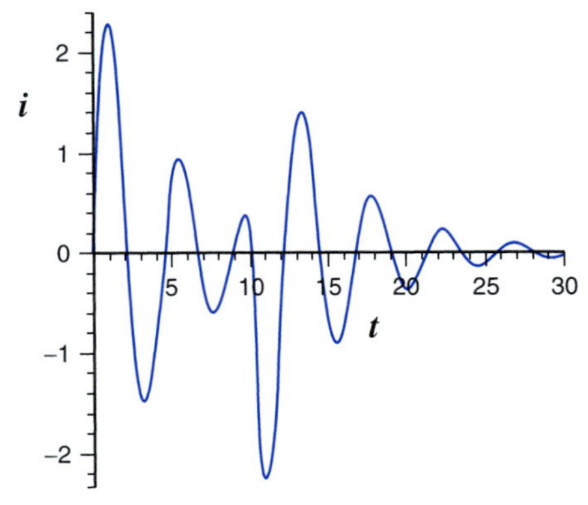

Figure 7.4.2
The current in Model Problem 7.4.

any force whose integral is 1. The integral of force is sometimes called the **impulse;** Newton's second law can thus be stated as "Impulse is equal to change in momentum." The total change in momentum caused by a force is the integral of the force over all time:

$$\Delta(mv) = \int_{-\infty}^{\infty} F(t)\, dt. \tag{8}$$

Suppose a unit impulse is applied beginning at time 0. It seems reasonable to assume that the force is distributed over some short time interval. Assume that the force is distributed uniformly over the time interval $0 \le t \le h$. Then the force is given by a function in the one-parameter family

$$F_h(t) = \begin{cases} \frac{1}{h} & 0 < t < h \\ 0 & \text{otherwise} \end{cases}. \tag{9}$$

The total impulse is

$$\int_{-\infty}^{\infty} F_h(t)\,dt = 1.$$

Figure 7.4.3 illustrates several of these functions.

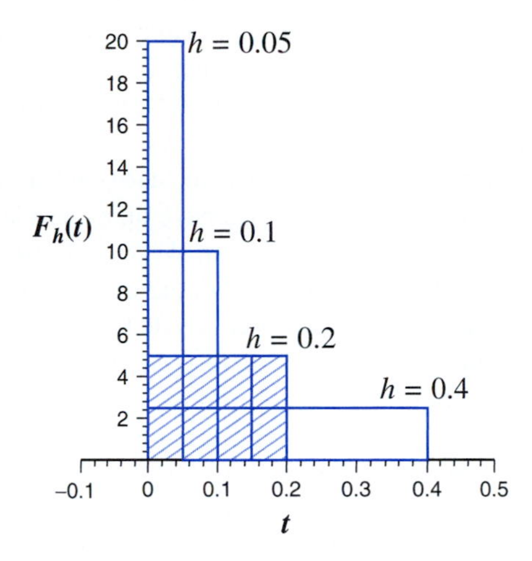

Figure 7.4.3
Functions $F_{0.4}$, $F_{0.2}$, $F_{0.1}$, and $F_{0.05}$, each of which delivers a unit impulse beginning at time 0. The area of the shaded region represents the impulse for $F_{0.2}$.

✦ INSTANT EXERCISE 4

Show that $\int_{-\infty}^{\infty} F_h(t)\,dt = 1$.

Suppose the width of the impulse is made narrower and narrower. It would be nice to define a *unit impulse function* δ as

$$\delta(t) = \lim_{h\to 0} F_h(t), \tag{10}$$

where F_h is defined by Equation (9). For any point $t_1 > 0$, we have $F_h(t_1) = 0$ for any $h < t_1$. For any point $t < 0$, we have $F_h(t_1) = 0$ for all h. Also, we have $\int_{-\infty}^{\infty} F_h(t)\,dt = 0$ for all h. Apparently, the "unit impulse function" must have the properties

$$\delta(t) = 0, \qquad t \neq 0, \tag{11}$$

$$\int_{-\infty}^{\infty} \delta(t)\,dt = 1. \tag{12}$$

No *function* can satisfy both of these properties. This is not as serious a problem as it would appear to be. Mathematicians have often faced situations where the standard set of objects (functions, in this case) is too rigidly defined to allow some desired outcome, and the response is to define a larger standard set of objects. An example is the complex number system. For some problems, such as homogeneous linear differential equations with constant coefficients, it is useful to have a number whose square is -1. The complex number system was devised as a generalization of the real numbers to make up for the lack of a real number with the desired property. Similarly, mathematicians define *generalized functions* to include all true functions and also allow for objects such as our unit impulse function that satisfy the properties of both Equation (11) and Equation (12). The definition and study of generalized functions are far beyond the scope of a first course in differential equations; for our purposes, it is sufficient to think of the **Dirac delta function**[6] as a functionlike object that satisfies these two key properties. There is one additional property that is useful.

Consider a spring-mass system that is set into motion by a hammer blow of unit impulse delivered at time 0. Suppose we have an experimental apparatus to measure the force imparted by the blow as a function of time. We would observe that the hammer blow actually does take a finite amount of time and the imparted force is truly a function. The function would have a definite integral equal to 1, and it would be 0 for all but a very small amount of time. The nonzero portion of the graph would be very complicated. Intuitively, using the delta function to model this force ought to result in a simplification compared with the model that would result from the actual force function. If something happens very fast, it is reasonable to expect a simpler model to result from the approximation of "very fast" as "instantaneous." If this intuitive argument of why the delta function ought to be convenient is correct, there must be some mathematical property that serves to simplify results obtained with the delta function rather than an actual function that the delta function replaces. Indeed, definite integrals involving the delta function are very easy to evaluate.

Theorem 7.4.1

Sifting Property Let f be a function that is continuous on the whole real line, and let $0 \le t_0 < t_1$. Then

$$\int_{t_0}^{t_1} f(t)\delta(t-a)\,dt = \begin{cases} f(a) & t_0 \le a < t_1 \\ 0 & \text{otherwise} \end{cases}.$$

We can explain this property heuristically,[7] at least for $a \neq t_0$. Given a function f and a value $t = a$, we can define a constant function by $g(t) \equiv f(a)$. The two functions f and g are equal at the point $t = a$. The two functions $f(t)\delta(t-a)$ and $g(t)\delta(t-a)$ are therefore equal at $t = a$, and both of them are 0, by Equation (11), for all other values of t. Thus, $f(t)\delta(t-a) = f(a)\delta(t-a)$ and so

$$\int_{t_0}^{t_1} f(t)\delta(t-a)\,dt = \int_{t_0}^{t_1} f(a)\delta(t-a)\,dt = f(a)\int_{t_0}^{t_1} \delta(t-a)\,dt = f(a).$$

[6] The function is named in honor of the British mathematical physicist Paul A. M. Dirac. It is customary to define F_h to be nonzero on the interval $[-h, h]$ rather than $[0, h]$; the definition we are using allows impulses delivered at time 0 to be treated just as impulses delivered at times greater than 0.

[7] A more careful derivation is deferred to Exercise 15.

(The last result comes from the substitution $u = t - a$.) An immediate consequence of the sifting property is the formula for the Laplace transform of the delta function,

$$\mathcal{L}[\delta(t-a)] = e^{-as}, \qquad a \geq 0. \tag{13}$$

✦ INSTANT EXERCISE 5

Use Theorem 7.4.1 to derive Equation (13).

Differential Equations with Impulsive Forcing

Problems with impulsive forcing work out conveniently with the Laplace transform method because of the simple formula of Equation (13).

EXAMPLE 2

Consider a spring-mass system with unit mass, spring constant 4, and no damping. Suppose the mass is pulled down a distance 1 and released from rest. At some time $t = a$, an impulse of strength 4 is delivered to the mass. The initial-value problem is

$$y'' + 4y = 4\delta(t-a), \qquad y(0) = 1, \qquad y'(0) = 0.$$

Applying the Laplace transform to this problem gives the algebraic equation

$$s^2Y - s + 4Y = 4e^{-as};$$

thus,

$$Y(s) = \frac{s}{s^2+4} + 4\frac{e^{-as}}{s^2+4}.$$

Inverting the transform yields the result

$$y(t) = \cos 2t + 2\sin(2(t-a))H(t-a).$$

In particular, the motion after the impulse is

$$y(t) = \cos 2t + 2\sin(2t - 2a), \qquad t > a.$$

To understand the general characteristics of this motion, it helps to expand the second term, using the identity

$$\sin(x-y) = \sin x \cos y - \sin y \cos x.$$

This yields

$$y(t) = (1 - 2\sin 2a)\cos 2t + (2\cos 2a)\sin 2t, \qquad t > a.$$

Note that a is a parameter in the model, rather than a variable. We should think of the quantities in parentheses as the coefficients of the functions $\cos 2t$ and $\sin 2t$. Using Equation (7) from Section 3.1, we obtain a formula for the amplitude:

$$A^2 = (1 - 2\sin 2a)^2 + (2\cos 2a)^2 = 5 - 4\sin 2a.$$

Thus, the amplitude after the hammer blow could be as small as 1 or as large as 3. There is also a significant phase shift in most cases. Figure 7.4.4 shows the solution for the cases $a = \pi/4$, $a = \pi/2$, and $a = 3\pi/4$. Note that the solution follows the same curve prior to the hammer blow in each case.

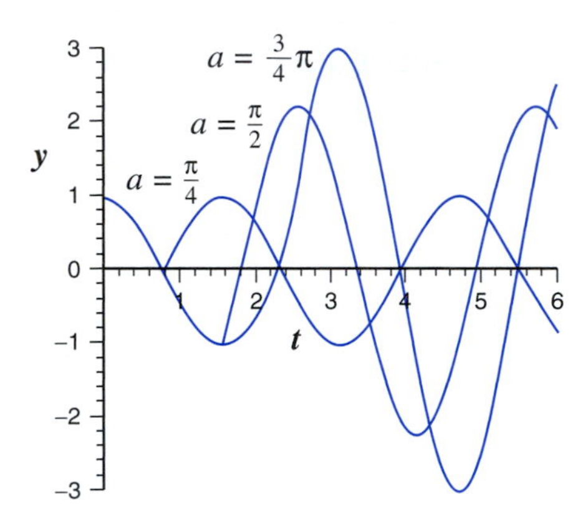

Figure 7.4.4
Some solutions of Example 2.

Model Problem 7.1 by the Laplace Transform Method

In Section 7.1, we used the method of undetermined coefficients to solve the problem

$$v'' + v = f_{\text{sq}}(t), \qquad v(0) = 0, \qquad v'(0) = 0,$$

where[8]

$$f_{\text{sq}}(t) = 1 + 2\sum_{n=1}^{\infty}(-1)^n H(t-n).$$

The procedure was complicated because of the piecewise character of the forcing function. Suppose we use the Laplace transform method instead. With $V = \mathcal{L}[v]$ and $F = \mathcal{L}\left[f_{\text{sq}}\right]$, the problem becomes

$$s^2 V + V = F = \frac{1}{s} + \frac{2}{s}\sum_{n=1}^{\infty}(-1)^n e^{-ns}.$$

This equation yields a solution for V:

$$V(s) = G(s) + 2\sum_{n=1}^{\infty}(-1)^n e^{-ns} G(s), \qquad G(s) = \frac{1}{s(s^2+1)}.$$

[8]See Exercise 15 of Section 7.1.

The inversion formula (4) yields the result

$$v(t) = g(t) + 2\sum_{n=1}^{\infty}(-1)^n g(t-n)H(t-n), \qquad g = \mathcal{L}^{-1}[G],$$

and formula 33, with $a = 0$ and $k = 1$, yields the result

$$g(t) = 1 - \cos t.$$

Combining these results leads to the solution of the voltage problem:

$$v(t) = 1 - \cos t + 2\sum_{n=1}^{\infty}(-1)^n[1 - \cos(t-n)]H(t-n). \tag{14}$$

This solution formula looks different from that obtained in Section 7.1, but they are in fact equivalent.

Laplace Transform or Undetermined Coefficients?

We've seen that any problem that can be solved by the method of undetermined coefficients can also be solved by the Laplace transform method. Which one should we prefer? For problems with forcing that can be written as a single formula without using step functions, the choice is a matter of personal preference. The student should work problems using both methods; this provides a way of checking results and experience in judging which method is more convenient for a given type of problem. For problems with piecewise-continuous forcing, a comparison of the two solutions of Model Problem 7.1 suggests that the solution formulas are more easily obtained by the Laplace transform method. The solution formula produced by the Laplace transform method also seems to have a small computational advantage. In an informal experiment, the time required to reproduce Figure 7.1.2 from the Laplace transform solution formula (14) was a little less than one-half that needed when using the formula obtained in Section 7.1 by the method of undetermined coefficients. Finally, problems with impulsive forcing cannot be solved by the method of undetermined coefficients at all.[9]

7.4 Exercises

In Exercises 1 through 4, find and sketch the inverse Laplace transform of the given function.

1. $G(s) = \dfrac{e^{-2s}}{s^2 + s - 2}$

2. $G(s) = \dfrac{2e^{-2s}}{s^2 - 4}$

3. $G(s) = \dfrac{(s-2)e^{-s}}{s^2 - 4s + 3}$

4. $G(s) = \dfrac{e^{-s} + e^{-2s} - e^{-3s}}{s}$

[9]They *can* be solved by the method of variation of parameters, which is reexamined in Section 7.5.

In Exercises 5 through 10, solve the given initial-value problems and sketch the graph of the solution.

5. $y'' + y = H(t - 3\pi), \qquad y(0) = 1, \qquad y'(0) = 0$

6. $y'' + 3y' + 2y = f(t), \qquad y(0) = 0, \qquad y'(0) = 0, \qquad f(t) = \begin{cases} 2 & t < 3 \\ 0 & t \geq 3 \end{cases}$

7. $y'' + y = g(t), \qquad y(0) = 0, \qquad y'(0) = 1, \qquad g(t) = \begin{cases} t & t < \pi \\ \pi & t \geq \pi \end{cases}$

8. $y'' + 3y' + 2y = \delta(t), \qquad y(0) = 0, \qquad y'(0) = 0$

9. $y'' + 2y' + 2y = \delta(t - \pi), \qquad y(0) = 1, \qquad y'(0) = 0$

10. $y'' + y = \delta(t - 2\pi), \qquad y(0) = 0, \qquad y'(0) = 1$

T 11. Let f_{tr} be the 2-periodic triangular wave illustrated in Figure 7.4.5. Determine and plot the current in a circuit with $L = C = 1$, $R = 0$, initially at rest and forced by the triangular wave.

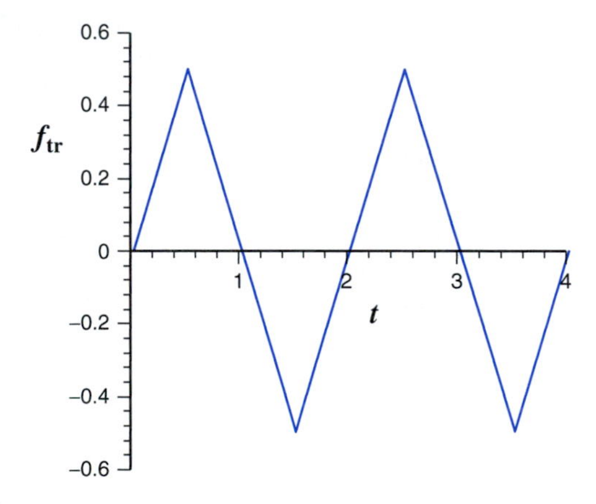

Figure 7.4.5
The triangular wave function of Exercise 11.

T 12. An LC circuit with unit inductance and capacitance is initially at rest. Beginning at time 0, the circuit is forced by an oscillating function that is alternately 1 and -1 over intervals of length 2 units. Determine the current in the circuit, and plot the results on the interval $0 < t < 20$.

13. The Dirac delta function can be thought of as the derivative of the Heaviside function. Assume that the symbol H' represents a generalized function that indicates the derivative of the Heaviside function H.

   a. Use the definition of H to explain why there can be no *function* H'.
   b. Show that properties (11) and (12) suggest that δ is the derivative of H.
   c. Use the definition of the delta function and the definition of the derivative to "derive" the result $\delta = H'$.

14. Consider the problem

$$y_h'' + y_h = F_h(t), \qquad y_h(0) = 0, \qquad y_h'(0) = 0,$$

where F_h is defined by Equation (9).

a. Solve the problem, using the method employed in this section to solve Model Problem 7.1.

b. Use the solution of part *a* to compute $\lim_{h \to 0} y_h$.

c. Use the method of variation of parameters to solve the problem

$$y'' + y = \delta(t), \qquad y(0) = 0, \qquad y'(0) = 0.$$

d. A solution to a spring-mass system with unit coefficients driven by a hammer blow of unit impulse could be obtained using either parts *a* and *b* or part *c*. How do these solutions compare? Which is easier? Which is the messier calculation?

15. Derive the sifting property of the Dirac delta function by computing

$$\lim_{h \to 0} \int_{-\infty}^{\infty} F_h(t-a) f(t)\, dt.$$

Hint: show that $\int_{-\infty}^{\infty} F_h(t-a) f(t)\, dt = \bar{f}(h)$, where$\bar{f}(h)$ is the average value of f on the interval $[a, a+h]$.

✦ 7.4 INSTANT EXERCISE SOLUTIONS

1. The inverse transform of F is determined by completing the square. We have

$$F(s) = \frac{2}{s(s^2 + 0.4s + 0.04 + 1.96)} = \frac{2}{s[(s+0.2)^2 + 1.4^2]}.$$

Now let $G(s+0.2) = F(s)$. Then the shifted-transform formula gives

$$f(t) = \mathcal{L}^{-1}[G(s+0.2)] = e^{-0.2t} g(t).$$

To determine g, we first must find $G(s)$ from

$$G(s+0.2) = \frac{2}{s[(s+0.2)^2 + 1.4^2]}.$$

We can replace s by $s - 0.2$; this gives us

$$G(s) = \frac{2}{(s-0.2)(s^2 + 1.4^2)}.$$

This matches formula 33 with $a = 0.2$ and $k = 1.4$. Thus,

$$g(t) = 2\,\frac{1.4e^{0.2t} - 1.4\cos 1.4t - 0.2\sin 1.4t}{1.4(0.2^2 + 1.4^2)} = e^{0.2t} - \cos 1.4t - \frac{1}{7}\sin 1.4t.$$

The final result (2) now follows.

2. Define $F(s)$ by $F(s) = 1/(s^2+4)$. This means that $f(t) = \mathcal{L}^{-1}[F(s)] = (\sin 2t)/2$. The inversion formula (4) then yields

$$\mathcal{L}^{-1}\left[\frac{e^{-3s}}{s^2+4}\right] = f(t-3)H(t-3) = \frac{1}{2}\sin(2t-6)H(t-3).$$

3. By definition, we have

$$f(t) = 1 - e^{-0.2t}\left(\cos 1.4t + \tfrac{1}{7}\sin 1.4t\right).$$

Straightforward differentiation yields

$$f'(t) = 0.2e^{-0.2t}\left(\cos 1.4t + \tfrac{1}{7}\sin 1.4t\right) - e^{-0.2t}\left(-1.4\sin 1.4t + 0.2\cos 1.4t\right) = \tfrac{10}{7}e^{-0.2t}\sin 1.4t.$$

4. $F_h(t)$ is 0 unless $0 < t < h$; hence, $\int_{-\infty}^{\infty} F_h(t)\,dt = \int_0^h F_h(t)\,dt = \int_0^h (1/h)\,dt = 1$.
5. By definition,

$$\mathcal{L}[\delta(t-a)] = \int_0^\infty e^{-st}\delta(t-a)\,dt.$$

Since e^{-st} is continuous, the sifting property yields the result

$$\mathcal{L}[\delta(t-a)] = e^{-st}|_{t=a} = e^{-as}, \qquad a \geq 0.$$

7.5 Convolution and the Impulse Response Function

Up to this point, the Laplace transform method has been used as an alternative to the method of undetermined coefficients. It can also be used as an alternative to the method of variation of parameters. The idea is to use the Laplace transform to obtain the solution to a nonhomogeneous problem with impulsive forcing and then to use that solution to obtain an integral solution formula.

Let p, r, y_0, and v_0 be constants and g a function. The problem

$$L[y] = y'' + py' + ry = g, \qquad y(0) = y_0, \qquad y'(0) = v_0$$

can be divided into two subproblems,

$$y_p'' + py_p' + ry_p = g, \qquad y_p(0) = 0, \qquad y_p'(0) = 0$$

and

$$y_h'' + py_h' + ry_h = 0, \qquad y_h(0) = y_0, \qquad y_h'(0) = v_0.$$

Once these subproblems have been solved, the solution of the original problem is $y = y_p + y_h$.

✦ INSTANT EXERCISE 1

Show that $y = y_p + y_h$ solves the problem for y given above.

This separation of a nonhomogeneous problem into subproblems is similar to the separation of such problems in Chapter 4. However, there is a subtle difference. In Chapter 4, we found *any* particular solution of the nonhomogeneous differential equation without specifying initial conditions, and therefore the correct coefficients in the complementary solution could not be determined independent from the particular solution. Here, we deliberately separate the nonhomogeneities in the problem so that one problem has homogeneous initial conditions and the other has a homogeneous differential equation; compared with the methods of Chapter 4, we are looking for a *specific* particular solution. The problem for y_h can be solved either by the method of Chapter 3 or by the Laplace transform method. Our focus here is on using the Laplace transform method to solve the problem for y_p, and it is for this reason that specific initial conditions must be assigned to the two subproblems. From here on, we drop the subscript p and consider the problem

$$y'' + py' + ry = g, \qquad y(0) = 0, \qquad y'(0) = 0. \tag{1}$$

MODEL PROBLEM 7.5

Solve the problem

$$y'' - y = 2\tan t, \qquad y(0) = 0, \qquad y'(0) = 0.$$

The Impulse Response Function

Think of the constants p and r in Equation (1) as fixed and the function g as a member, to be specified later, of the set of Laplace-transformable input functions. Of all possible choices for g, the most elementary in some sense is the delta function. The function q defined by

$$q'' + pq' + rq = \delta, \qquad q(0) = 0, \qquad q'(0) = 0 \tag{2}$$

is called the **impulse response function** for the operator L.[10] The Laplace transform $Q = \mathcal{L}[q]$ is called the **transfer function.**

We consider a two-stage method for constructing solutions for Equation (1); these stages consist of the distinct problems of determining the impulse response function and developing a formula to express the solution of Equation (1) in terms of the impulse response function.

Finding the Impulse Response Function

The impulse response function for Model Problem 7.5 is the solution of

$$q'' - q = \delta(t), \qquad q(0) = 0, \qquad q'(0) = 0.$$

The problem transforms to

$$s^2 Q - Q = 1,$$

which has solution $Q = 1/(s^2 - 1)$. The table of Laplace transforms yields the solution

$$q(t) = \sinh t.$$

[10]Some authors use the phrase *Green's function for the initial-value problem* rather than *impulse response function*. Here we choose the latter because it is descriptive of the properties of the function and to avoid confusion with Green's functions that arise in the context of partial differential equations. See, for example, Richard Haberman, *Applied Partial Differential Equations,* 4th ed., Prentice-Hall, Upper Saddle River, NJ, 2004.

It is interesting to note that this function is also the solution of the problem

$$q'' - q = 0, \qquad q(0) = 0, \qquad q'(0) = 1.$$

This correspondence is not a coincidence. In the language of the spring-mass model, we can think of the impulse provided by the delta function as adding a unit amount of momentum to the system. The mass coefficient in the differential equation is 1, so the effect of the delta function is to create an instantaneous unit increase in velocity. The impulse response problem for q applies to the situation the moment *before* the impulse is delivered, while the homogeneous problem for q with initial condition $q'(0) = 1$ applies to the situation the moment *after* the impulse. In other contexts, it is often possible to write an impulse response problem as an initial-value problem for a homogeneous equation by interpreting the unit impulse as a unit change of momentum. However, as we see here, the form with the impulsive nonhomogeneity in the differential equation is particularly convenient when the Laplace transform method of solution is used.

Convolutions and the Convolution Theorem

Returning to the general case, application of the Laplace transform to Equation (1) yields the equation

$$s^2Y + psY + rY = G,$$

and similarly, the Laplace transform of Equation (2) is

$$s^2Q + psQ + rQ = 1.$$

These algebraic equations have solutions

$$Y = \frac{G}{s^2 + ps + r}, \qquad Q = \frac{1}{s^2 + ps + r}.$$

What is interesting here is not the actual results for Y and Q, but their relationship:

$$Y = QG.$$

This last equation says that we can solve the problem by a three-step method:

1. Find $G = \mathcal{L}[g]$.
2. Transform the equation for the impulse response function and solve it for Q.
3. Compute y from $y = \mathcal{L}^{-1}[QG]$.

As always with the Laplace transform, the transform inversion step is the tricky one. We need something on the order of a "product rule" for Laplace transforms. It would be nice if the product of the transforms were the transform of the product, in other words, $\mathcal{L}[qg] = \mathcal{L}[q]\,\mathcal{L}[g]$. A simple example suffices to show that this guess is not correct.

EXAMPLE 1

Let $q = 1$ and $g = 1$. Then $\mathcal{L}[qg] = \mathcal{L}[1] = 1/s$, but $\mathcal{L}[q]\,\mathcal{L}[g] = 1/s^2$.

The situation is analogous to integration. One can separate sums from an integral, but one cannot factor an integral into two separate integrals. In integration, there is no rule for $\int f(t)g(t)\,dt$. Fortunately, there *is* a product rule for Laplace transform inversion. It is first necessary to understand an operation called the *convolution*.

Let f and g be two functions defined for all positive arguments. The **convolution** of f and g is defined by the formula

$$(f * g)(t) = \int_0^t f(\tau)g(t-\tau)\,d\tau.$$

Note that the convolution is an operation between two functions of a common variable. The convolution $f * g$ is a function of the same variable as each of f and g. The convolution operation satisfies some of the elementary properties that we might expect. It is commutative; that is,

$$f * g = g * f,$$

and it is linear, meaning

$$f * cg = c(f * g), \qquad f * (g + h) = (f * g) + (f * h)$$

for all functions f, g, and h for which the convolutions are defined and for all constants c (see Exercise 6).

EXAMPLE 2

Suppose we want to compute $t * e^{-t}$. We have

$$t * e^{-t} = \int_0^t \tau e^{\tau - t}\,d\tau = e^{-t}\int_0^t \tau e^{\tau}\,d\tau = e^{-t}[(\tau-1)e^{\tau}]|_{\tau=0}^{t} = e^{-t}[(t-1)e^{t}+1] = t - 1 + e^{-t}.$$

Alternatively, since the convolution is commutative, we could use

$$t * e^{-t} = \int_0^t (t-\tau)e^{-\tau}\,d\tau = (1 - t + \tau)e^{-\tau}|_{\tau=0}^{t} = e^{-t} - (1-t) = t - 1 + e^{-t}.$$

Notice that

$$\mathcal{L}[t] = \frac{1}{s^2}, \qquad \mathcal{L}\left[e^{-t}\right] = \frac{1}{s+1},$$

and

$$\mathcal{L}\left[t - 1 + e^{-t}\right] = \frac{1}{s^2} - \frac{1}{s} + \frac{1}{s+1} = \frac{s+1}{s^2(s+1)} - \frac{s(s+1)}{s^2(s+1)} + \frac{s^2}{s^2(s+1)} = \frac{1}{s^2(s+1)},$$

so

$$\mathcal{L}\left[t * e^{-t}\right] = \mathcal{L}[t]\,\mathcal{L}\left[e^{-t}\right].$$

Example 2 shows that $\mathcal{L}[f * g] = \mathcal{L}[f]\,\mathcal{L}[g]$ for the specific function pair t and e^{-t}. Could this "product rule" be true in general? The convolution theorem answers in the affirmative.

Theorem 7.5.1 **Convolution Theorem** If $\mathcal{L}[f] = F$ and $\mathcal{L}[g] = G$, then $\mathcal{L}[f * g] = FG$.

A proof of the convolution theorem is sketched out in Exercise 7.

Finding Particular Solutions Using the Convolution Theorem

We have already seen that the Laplace transform of the solution y of the problem

$$y'' + py' + ry = g, \qquad y(0) = 0, \qquad y'(0) = 0,$$

where p and r are constants, is given by $Y(s) = Q(s)G(s)$, with Q the transform of the impulse response function. The convolution theorem supplies the final piece of the puzzle.

Theorem 7.5.2 Let p and r be given real numbers. Let q be the impulse response function defined by

$$q'' + pq' + rq = \delta(t), \qquad q(0) = 0, \qquad q'(0) = 0.$$

If the convolution $q * g$ exists on an interval $0 < t < T$, then the problem

$$y'' + py' + ry = g(t), \qquad y(0) = 0, \qquad y'(0) = 0$$

has the solution

$$y = q * g$$

on the interval $0 < t < T$.

We derived Theorem 7.5.2 by using the Laplace transform, but the theorem is true even for cases where the function g is not Laplace-transformable. Consider Model Problem 7.5 as an example. We have already determined the result $q(t) = \sinh t$. The function given by $g(t) = 2\tan t$ is not Laplace-transformable. Theorem 7.5.2 asserts that the problem

$$y'' - y = 2\tan t, \qquad y(0) = 0, \qquad y'(0) = 0$$

has solution

$$y(t) = \int_0^t 2\sinh(t - \tau)\tan\tau \, d\tau \tag{3}$$

as long as the integral converges. For the interval $0 \le \tau < \pi/2$, the integrand is continuous; thus, the integral clearly converges over the interval $0 \le t < \pi/2$. The situation for $t \ge \pi/2$ is not so clear. In this case, $\pi/2$ is in the domain of the integral. Since the tangent function is not defined at $\pi/2$, the integral is improper. We then have

$$y(t) = \int_0^{\pi/2} 2\sinh(t - \tau)\tan(\tau)\, d\tau + \int_{\pi/2}^t 2\sinh(t - \tau)\tan(\tau)\, d\tau,$$

provided both improper integrals converge. The integrand is displayed in Figure 7.5.1 for the values $t = \pi/2$ and $t = 1.6 > \pi/2$. With $t = \pi/2$, the integrand is actually continuous, so the integral converges. For $t = 1.6$, the integrand approaches infinity as $\tau \to (\pi/2)^+$, making it unclear whether or not the integral converges. In fact, it does not (see Exercise 8).

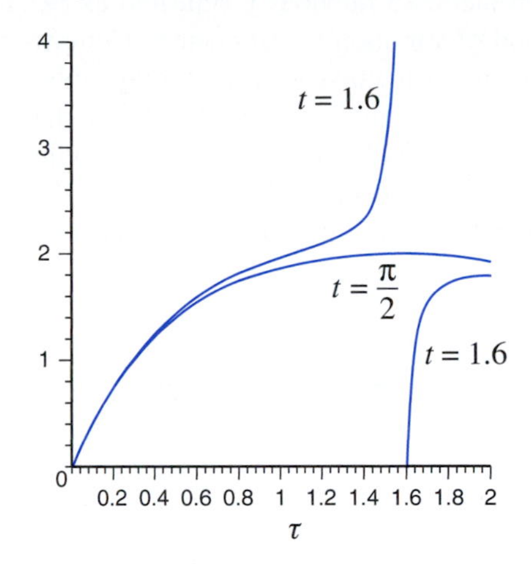

Figure 7.5.1
The integrand for the solution (3) for Model Problem 7.5.

The graph of the solution of Model Problem 7.5 appears in Figure 7.5.2. It is interesting to note that the solution does not become infinite as $t \to \pi/2$; rather, the solution exists and is finite at $t = \pi/2$ and does not exist for larger t. Instead of going off the top or bottom of the viewing window, the graph stops abruptly. One mathematical detail deserves special emphasis:

- Formulas that are derived under a given set of assumptions can sometimes be valid even when the derivation is not justified. We used the Laplace transform to discover Theorem 7.5.2, but the theorem can be proved without using the Laplace transform, simply by checking that the given formula satisfies the given problem. It is the proof that establishes that it is not necessary for g to be Laplace-transformable.

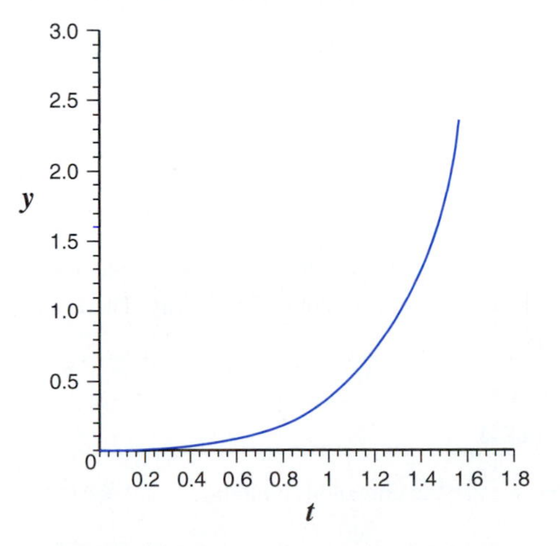

Figure 7.5.2
The solution (3) of Model Problem 7.5.

Convolution or Variation of Parameters?

Model Problem 7.5 previously appeared as Example 2 in Section 4.6, where it was solved by the method of variation of parameters. Comparison of the two solution methods is instructive. The convolution method is easier to remember because it requires only one formula, that of Theorem 7.5.2. When an integral form of the solution is all that is possible or desired, the convolution method has the advantage of giving a formula with a single integral, obtained with only the minimal effort needed to calculate the impulse response function. However, for cases where the integrals can be computed, it may be a little easier to obtain the final solution formula by using variation of parameters.

EXAMPLE 3

For the problem

$$y'' + y = 2\tan t, \qquad y(0) = 0, \qquad y'(0) = 0,$$

the impulse response function is

$$q(t) = \sin t,$$

so the solution, by Theorem 7.5.2, is

$$y(t) = \int_0^t 2\sin(t-\tau)\tan\tau\,d\tau.$$

This integral can be evaluated exactly. Using the trigonometric identity

$$\sin(a-b) = \sin a\cos b - \cos a\sin b,$$

we have

$$y(t) = 2\int_0^t (\sin t\cos\tau - \cos t\sin\tau)\tan\tau\,d\tau = 2\sin t\int_0^t \sin\tau\,d\tau - 2\cos t\int_0^t \sin\tau\tan\tau\,d\tau.$$

The second integral can be found in a table of integrals, so we obtain the result

$$\begin{aligned} y(t) &= 2\sin t\,(-\cos\tau)|_{\tau=0}^{t} - 2\cos t[\ln(\tan\tau + \sec\tau) - \sin\tau]|_{\tau=0}^{t} \\ &= -2\sin t\cos t + 2\sin t - 2\cos t\ln(\tan t + \sec t) + 2\cos t\sin t, \end{aligned}$$

or

$$y(t) = 2\sin t - 2\cos t\ln(\tan t + \sec t). \tag{4}$$

This result was previously obtained in Example 1 of Section 4.6; in that case, the integration did not require the use of the trigonometric identity. There were two integrals to compute, but one of them was very easy.

EXAMPLE 4

Suppose we use the convolution method to solve Model Problem 7.1. The impulse response function for

$$v'' + v = f_{sq}(t), \qquad v(0) = 0, \qquad v'(0) = 0, \qquad f_{sq}(t) = 1 + 2\sum_{n=1}^{\infty}(-1)^n H(t-n)$$

is $q(t) = \sin t$, so we have the solution formula

$$v(t) = q * f_{sq} = \int_0^t \left[1 + 2\sum_{n=1}^{\infty}(-1)^n H(t - \tau - n) \right] \sin\tau \, d\tau.$$

Interchanging the sum and the integral yields

$$v(t) = \int_0^t \sin\tau \, d\tau + 2\sum_{n=1}^{\infty}(-1)^n \int_0^t H(t - \tau - n)\sin\tau \, d\tau.$$

The integral has two possible values, depending on the sign of $t - n$. If $n \geq t$, then the argument of the Heaviside function is necessarily negative for $0 < \tau < t$, so the integral is 0. Otherwise, the Heaviside function is 1 for $\tau < t - n$. Then

$$\int_0^t H(t - \tau - n)\sin\tau \, d\tau = \int_0^{t-n} \sin\tau \, d\tau = -\cos\tau \,\Big|_{\tau=0}^{t-n} = 1 - \cos(t - n), \qquad n < t.$$

The integral can be written as a single formula using $H(t - n)$:

$$\int_0^t H(t - \tau - n)\sin\tau \, d\tau = [1 - \cos(t - n)]H(t - n).$$

Thus,

$$v(t) = 1 - \cos t + 2\sum_{n=1}^{\infty}(-1)^n[1 - \cos(t - n)]H(t - n).$$

It is arguable whether this method is easier or more difficult for this problem than the method used to solve it in Section 7.4.

It should also be remembered that the method of variation of parameters can be used for problems where the coefficients are not constant, provided the complementary solution is known. The convolution method also works for such problems, provided the impulse response function can somehow be determined. As so often happens, the conclusion is that the choice of method depends on the problem. If one method does not seem to work, try another. Using two different methods has a significant advantage over using either method alone because it improves the likelihood of a solution free of computational errors.

7.5 Exercises

In Exercises 1 through 4, find a particular solution by using Theorem 7.5.2. Compare the calculation with the corresponding problem (Exercises 1 through 4) from Section 4.6.

1. $y'' + 4y = \sec 2t, \qquad -\pi/4 < t < \pi/4$
2. $y'' + 4y = \sec t, \qquad -\pi/2 < t < \pi/2$
3. $y'' + 2y' + y = t^{-p}e^{-t}, \qquad t > 0,$ where p is a positive integer
4. $y'' + 2y' + y = e^{-t}/(1 + t^2)$

T 5. Let f_{tr} be the 2-periodic triangular wave illustrated in Figure 7.4.5. Determine and plot the current in a circuit with $L = C = 1$, $R = 0$, initially at rest, and forced by the triangular wave. Compare the calculations with Exercise 11 of Section 7.4.

6. Show that the convolution is linear; in other words, demonstrate that $f * cg = c(f * g)$ and $f * (g + h) = (f * g) + (f * h)$ for all functions f, g, and h for which the convolutions are defined and for all constants c.

7. Prove the convolution theorem.

   a. Use the definition of the Laplace transform to write down $\mathcal{L}[f * g]$. The result is an iterated double integral with t as the integration variable for the outside integral and τ for the inside.

   b. Notice that the integrand f does not depend on t. Since the integrand $f(\tau)g(t - \tau)$ is a simpler function of t than of τ, it makes sense to try to simplify the integral by changing the order of integration. Rewrite the integral from part *a* by reversing the order of integration.

   c. Use the substitution $u = t - \tau$ to replace the integration variable t. You should now have an iterated double integral with τ as the integration variable for the outside integral and u for the inside.

   d. The expressions $e^{s\tau}$ and $f(\tau)$ can be factored out of the inside integral because they are independent of u. Then observe that the inside integral is the Laplace transform of g. Complete the proof by factoring $G(s)$ out of the remaining integral.

8. Show that the solution (3) of Model Problem 7.5 does not exist for $t > \pi/2$ by demonstrating that the integral $I = -\int_{\pi/2}^{t} \sinh(t - \tau)\tan(\tau)\,d\tau$ diverges for $t = \pi/2 + x$, where x is small and positive.

   a. Explain why the divergence of the integral I for a particular value of $t > \pi/2$ means that the solution (3) does not exist for that value of t.

   b. Explain why it is only necessary to consider small values of x to show that the solution does not exist for $t > \pi/2$.

   c. Use the substitution $u = \tau - \pi/2$ to rewrite the integral.

   d. Use trigonometric identities to write $\tan(\pi/2 + u)$ as $-\cos u/\sin u$.

   e. Show that

$$I > \int_0^{x/2} \sinh(x - u)\frac{\cos u}{\sin u}\,du > \sinh\frac{x}{2}\int_0^{x/2}\frac{\cos u}{\sin u}\,du > 0.$$

   f. Compute

$$\int_a^{x/2}\frac{\cos u}{\sin u}\,du \qquad \text{for } 0 < a < \frac{x}{2}$$

   and use the result to show that the integral

$$\int_0^{x/2}\frac{\cos u}{\sin u}\,du$$

   diverges.

   g. Explain why the results of parts *e* and *f* prove that the integral I diverges.

✦ 7.5 INSTANT EXERCISE SOLUTION

1. Because of the linearity of the differential operator L, $L[y] = L[y_p] + L[y_h]$. It follows from $L[y_p] = g$ and $L[y_h] = 0$ that $L[y] = g$. Similarly, $y(0) = y_p(0) + y_h(0) = 0 + y_0 = y_0$, and the second initial condition is satisfied by the same argument.

Growth of a Structured Population

CASE STUDY 7

In recent years, demographic trends in the population of the United States have begun to influence public policy decisions. In particular, the average age of the population has risen, with the consequence that the ratio of workers who contribute to the nation's Social Security pool to retirees who draw funds from the pool is at a significantly lower level than at any time in the history of the Social Security program. Characteristics such as the average age of a population and fraction of a population over a certain age are quantitative characteristics. Most of the population models we have examined up to now have made no attempt to distinguish individuals in a population. The disease models of Chapter 6 distinguished individuals in a population by qualitative characteristics (susceptible, infective, and so on) only.

There is a fundamental distinction between a population divided into discrete classes such as healthy and sick and a population in which each individual is characterized by a continuous variable such as age. Biologists and social scientists often arbitrarily group members of a population into age classes in order to use a discrete model for the continuous process of aging. However, it is also possible to keep track of age structure with a model in which each individual ages continuously. In general, these models involve partial differential equations, but we can get a more manageable problem by limiting the scope of the investigation to keeping track of the birthrate in an age-structured population.

Given information about the numbers and ages of an initial population, the age-dependent birthrate, and an age-independent death rate, determine the overall birthrate and the population as a function of time.

Recall the general strategy of modeling for problems of mechanics. Each force F_i creates acceleration F_i/m. Since the whole is equal to the sum of the parts, we have the general rule that the total acceleration can be found by adding the sum of the accelerations due to each force. Thus,

$$a(t) = \sum \frac{F_i}{m}.$$

The same strategy can be followed to count the total birthrate in a population as a sum of the birthrates in the various segments of the population. The main difference is that there are infinitely many segments of a population in which each individual is marked by its continuously changing age.

Let $b(t)$ be the overall birthrate of the population, relative to some standard population size. Let μ be the age-independent relative mortality rate. Let x be a variable that denotes the age of individuals in a population. Suppose that individuals of age x give birth at a rate $f(x)$ per unit time. We should expect that f is zero for individuals too young or too old to give birth and reaches some peak value at some intermediate age. [For populations with two sexes, the model makes sense if it is restricted to the females in the population; thus, $f(x)$ is well defined, and only births of females are counted.] The function f is called the **fecundity** of the population.

In general, we have to count separately births to mothers who were part of the initial population and births to those who were born after time 0. In words, our model will be an equation that gives the total birthrate as a sum of births to original members (the first generation) plus births to members of subsequent generations. Let $b_0(t)$ be the birthrate at time t for the original population, and let $b_s(t)$ be the birthrate at time t for the subsequent generations. The equation for the birthrate will be based on the simple equation

$$b(t) = b_0(t) + b_s(t). \tag{1}$$

To keep the model simple, we assume that the initial population consists of a unit number of individuals, all of age 0, with more general models to be explored in Exercises 1 and 2. Since the relative mortality rate is μ, the size $p_0(t)$ of the initial population is governed by the initial-value problem

$$\frac{dp_0}{dt} = -\mu p_0, \qquad p_0(0) = 1.$$

This simple natural decay model was introduced in Section 1.1 and has appeared several times since. The solution is

$$p_0(t) = e^{-\mu t}.$$

To determine the rate of births to the original population, we must multiply the birthrate per individual by the number of individuals still alive. The individuals from the first generation are of age t at time t, so the birthrate is

$$b_0(t) = f(t)e^{-\mu t}.$$

Calculating the Birthrate for Subsequent Generations

For a fixed time t and age x, with $t > x$,[11] consider individuals of ages x to $x + dx$, where dx is some arbitrarily small value. The birthrate for such individuals is approximately $f(x)$, so the overall birthrate is $p(x, t)f(x)$, where p is the population of individuals in this group. To determine p, we have to consider that the individuals in this group were all born during the time interval $[t - x - dx,\ t - x]$. Since the overall birthrate at time $t - x$ was $b(t - x)$ and the group consists of individuals born in an interval of length dx, the initial number of individuals in this group must have been $b(t - x)\,dx$. This is not quite the number $p(x, t)$, because individuals in this group have been dying over the x time units between their birth at time $t - x$ and the current time t. Given a death rate of μ, the size of this group at time t is $p(x, t) = b(t - x)e^{-\mu x}\,dx$. Thus,

[11] If $t \le x$, then the cohort is part of the original generation, as the age is as large as the clock time.

the birthrate for these individuals at time t is

$$p(x,t)f(x) \approx f(x)b(t-x)e^{-\mu x}\,dx,$$

where the symbol $\approx$ indicates that we have used the constant birthrate $f(x)$ even though the group consists of individuals of ages x to $x+dx$. If we count all groups of different ages, we find the total birthrate at time t to be

$$b_s(t) \approx \sum_x f(x)b(t-x)e^{-\mu x}\,dx.$$

The approximation becomes exact in the limit $dx \to 0$, with the sum becoming the definite integral over the range of possible ages:

$$b_s(t) = \int_0^t f(x)b(t-x)e^{-\mu x}\,dx. \tag{2}$$

An Equation for the Birthrate

Substituting the results for b_0 and b_s into the birthrate equation (1), we have the integral equation

$$b(t) = f(t)e^{-\mu t} + \int_0^t f(x)b(t-x)e^{-\mu x}\,dx. \tag{3}$$

This equation is called the **renewal** equation because it indicates how the population is renewed by births.

Integral equations are very different from differential equations, and generally they are much more difficult. Observe first that the equation says that the current birthrate is dependent not just on the current situation, as in a differential equation model, but also on the entire history of the population. Mathematically, all values of b for times less than t combine to determine b at time t. The key to solving this integral equation is to observe that the integral can be written as a convolution. Let $k(t) = f(t)e^{-\mu t}$. Then the equation becomes

$$b(t) = k(t) + (k * b)(t).$$

While this equation is not a differential equation, it is an equation that can be simplified by using the Laplace transform, because of the convolution theorem. With $B = \mathcal{L}[b]$ and $K = \mathcal{L}[k]$, the Laplace transform of the birthrate equation is

$$B(s) = K(s) + K(s)B(s).$$

Thus,

$$B(s) = \frac{K(s)}{1-K(s)}. \tag{4}$$

It remains "only" to invert the transform, but as we have seen repeatedly, the inversion of the transform is problematic.

Given our success with the convolution theorem, there is an obvious plan of attack for recovering b from B. Simply let $Y = 1/(1-K)$, so that $B = YK$, whence $b = y * k$. However, this approach fails because of the nature of the Laplace transform. We have studied in some detail the requirements for a function f to have a Laplace transform, but we have not studied the requirements for a function in the s domain to be a Laplace transform.

Theorem C7.1 Let k be a Laplace-transformable function, and let $K = \mathcal{L}[k]$ and $Y = 1/(1 - K)$. Then $\mathcal{L}^{-1}[Y]$ does not exist.

Determining the Birthrate

Our first attempt to determine the birthrate by applying the convolution theorem to Equation (4) failed; instead, we compute B directly and then attempt to invert the transform. The problem now is to find b from

$$b = \mathcal{L}^{-1}[B], \qquad B(s) = \frac{K(s)}{1 - K(s)}, \qquad K(s) = \mathcal{L}\left[f(t)e^{-\mu t}\right](s). \tag{5}$$

The shifted-transform formula can be used to further simplify the result. Let $F = \mathcal{L}[f]$. Then $K(s) = F(s + \mu)$, so

$$B(s) = \frac{F(s + \mu)}{1 - F(s + \mu)}.$$

Now let $R(s + \mu) = B(s)$. Then $e^{-\mu t}r(t) = b(t)$ and

$$R(s) = \frac{F(s)}{1 - F(s)}.$$

The final result is as follows:

Suppose $f(x)$ is the age-dependent fecundity and μ the age-independent relative death rate. Then the birthrate is given by

$$b(t) = e^{-\mu t}r(t), \qquad r = \mathcal{L}^{-1}[R], \qquad R(s) = \frac{F(s)}{1 - F(s)}. \tag{6}$$

For any given set of data, we need only to calculate $F(s)$, substitute to get R, and invert the transform to obtain the final answer. It is also possible to solve the renewal equation (3) approximately (see Exercise 3).

Some Examples

Any reasonable choice for the age-dependent birthrate should satisfy $f(0) = 0$ and $\lim_{x\to\infty} f(x) = 0$. Also, the rate should increase to a maximum and then decrease, so there should be some value X such that $f' \geq 0$ for $x < X$ and $f' \leq 0$ for $x > X$. Absent any experimental data for f, we can still try to get a sense of how the birthrate behaves by choosing a function f that has these properties and also has a fairly simple Laplace transform. One idea would be to try choosing a constant value over some range of fertile ages $0 < a < x < b$:

$$f_1 = \begin{cases} \rho & a < x < b \\ 0 & \text{otherwise} \end{cases}.$$

Another idea would be to choose a function of the form

$$f_2 = \rho e^{-x/x_m},$$

which has its maximum at $x = x_m$. Figure C7.1 shows a typical member of each of these families.

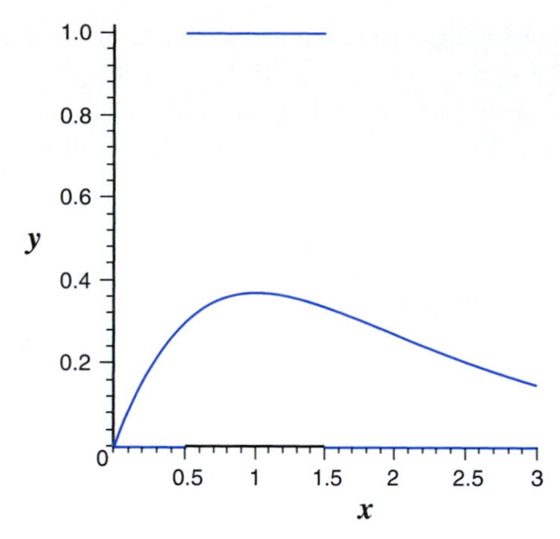

Figure C7.1
Two fecundity functions $f_1 = H\left(t - \frac{1}{2}\right) - H\left(t - \frac{3}{2}\right)$ and $f_2 = xe^{-x}$.

EXAMPLE 1

For $f_1 = \rho[H\left(t - \frac{1}{2}\right) - H\left(t - \frac{3}{2}\right)]$, with ρ a constant, we have

$$F(s) = \frac{\rho(e^{-s/2} - e^{-3s/2})}{s},$$

so

$$R(s) = \frac{\rho(e^{-s/2} - e^{-3s/2})}{s - \rho(e^{-s/2} + e^{-3s/2})}.$$

There is no simple formula to use to invert this transform, so instead the problem has to be solved numerically.

EXAMPLE 2

For $f_2 = \rho xe^{-x}$, with ρ a constant, the Laplace transform is

$$F(s) = \frac{\rho}{(s+1)^2},$$

so

$$R(s) = \frac{\rho/(s+1)^2}{1 - \rho/(s+1)^2} = \frac{\rho}{(s+1)^2 - \rho} = \frac{\rho}{s^2 + 2s + (1-\rho)}.$$

This transform fits formula 30 with a and b given by

$$a = \frac{-2 + \sqrt{4 - 4(1-\rho)}}{2} = \sqrt{\rho} - 1, \qquad b = -\sqrt{\rho} - 1.$$

Thus,

$$r(t) = \rho\frac{e^{(\sqrt{\rho}-1)t} - e^{(-\sqrt{\rho}-1)t}}{2\sqrt{\rho}}$$

and

$$b(t) = \frac{\sqrt{\rho}}{2}\left(e^{(\sqrt{\rho}-1-\mu)t} - e^{(-\sqrt{\rho}-1-\mu)t}\right).$$

Notice that the second term is a decaying exponential for any values of ρ and μ, so the long-term fate of the population depends on the sign of $\sqrt{\rho} - 1 - \mu$. If $\rho > (1+\mu)^2$, then the sign of this factor is positive and the birthrate shows exponential growth. If, however, $\rho < (1+\mu)^2$, then the birthrate decays to zero and the population dies out. The population stabilizes with the critical value $\rho = (1+\mu)^2$. For this latter case, the birthrate is

$$b(t) = \frac{1+\mu}{2}\left(1 - e^{-2(1+\mu)t}\right).$$

The birthrate is initially zero, because the population is all of age 0, but it gradually approaches its stable value as the population matures. See Figure C7.2.

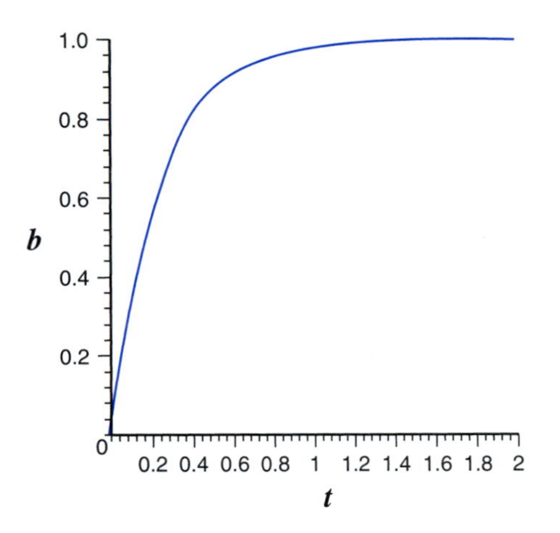

Figure C7.2
The birthrate for the case $f = 4xe^{-x}$ and $\mu = 1$.

The Total Population

To calculate the total population, we need to consider the populations of the original and subsequent generations separately. First, the population of the original generation has already been determined to be $p_0 = e^{-\mu t}$. The population of subsequent generations can be determined in a manner similar to the determination of the birthrate, with the result $p_s(t) = \int_0^t b(t-x)e^{-\mu x}\,dx = p_0(t) + (b * p_0)(t)$. Combining these, we have the total population as

$$P(t) = p_0(t) + (b * p_0)(t), \qquad p_0(t) = e^{-\mu t}. \tag{7}$$

EXAMPLE 3

Consider the case $f = 4xe^{-x}$, $\mu = 1$ from Example 2 and Figure C7.2. We have

$$p_0(t) = e^{-t}, \qquad b(t) = 1 - e^{-4t},$$

so

$$(b * p_0)(t) = \int_0^t e^{x-t}(1 - e^{-4x})\,dx = e^{-t}\int_0^t (e^x - e^{-3x})\,dx$$
$$= e^{-t}\left(e^x + \tfrac{1}{3}e^{-3x}\right)\Big|_{x=0}^t = 1 - \tfrac{4}{3}e^{-t} + \tfrac{1}{3}e^{-4t}.$$

Thus,

$$P(t) = 1 - \tfrac{1}{3}e^{-t} + \tfrac{1}{3}e^{-4t}.$$

The solution is shown in Figure C7.3.

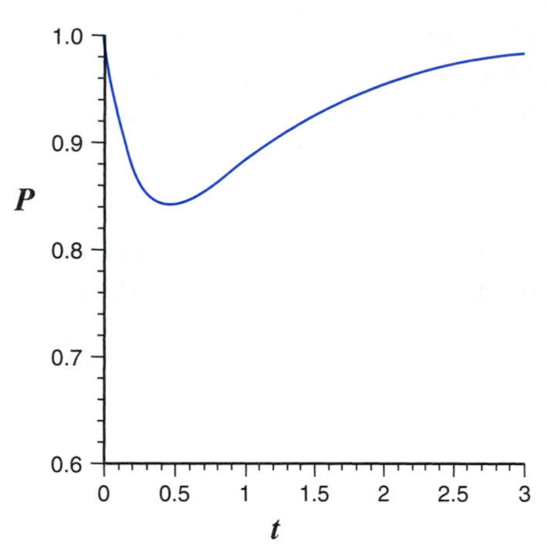

Figure C7.3
The birthrate for the case $f = 4xe^{-x}$ and $\mu = 1$.

Case Study 7 Exercises

T 1. Suppose the initial population is of age x_0 rather than age 0.

a. Derive the renewal equation for this case.

b. Determine a solution similar to (6) for the renewal equation. [*Hint:* you will have $b = b_0 + k * b$; you will also need to define the appropriate function g so that $B_0(s) = \mathcal{L}\left[g(t)e^{-\mu t}\right](s)$.]

c. Use the data for f and μ from Figure C7.2 to determine the birthrate and the total population.

d. Plot the results for the cases $x_0 = 0.2$ and $x_0 = 0.4$.

e. Describe how these results differ depending on the initial age (assuming $x_0 < 0.5$). Can you explain the reasons for these differences?

2. Suppose the population of individuals under the age of x in the initial population is

$$p_0(x, 0) = \int_0^x u(z)\,dz,$$

where $p_0(x, t)$ is the population at time t of those members of the original generation who are younger than age x at time t and u is some given function.

a. Determine $p_0(x, t)$.

b. Determine $b_0(t)$ and use it to derive the renewal equation for this case.

T 3. *a.* Suppose $\mu = 1$ and $f(x) = 4xe^{-x}$. Recall that $b_0(t) = f(t)e^{-\mu t}$. For all positive integers n, define b_n by

$$b_n(t) = f(t)e^{-\mu t} + \int_0^t f(x)b_{n-1}(t-x)e^{-\mu x}\,dx = b_0(t) + (b_0 * b_{n-1})(t).$$

Each b_n can be calculated in turn, beginning with b_1. Calculate b_1, b_2, and b_3.

b. Plot b_0, b_1, b_2, and b_3. Discuss the graphs. In particular, discuss how you can approximate the solution of the renewal equation by using a sequence of approximations. Solution procedures such as this are said to be **iterative.**

4. Suppose f' is Laplace-transformable. Use integration by parts to show that $\lim_{s\to\infty} F(s) = 0$. Explain why this proves that $1/(1-K)$ cannot be the Laplace transform of a function y whose derivative is Laplace-transformable. Also explain why this does not quite prove Theorem C7.1.

Vibrating Strings: A Focused Introduction to Partial Differential Equations

chapter

8

In Chapters 3 and 4, we studied the motion of a point mass in one direction, using the mathematical model of a linear oscillator. Now we turn to a related problem. Suppose we stretch a string horizontally between two fixed points. If we pluck the string with a vertical motion, the string will vibrate. We can think of the string as a set of infinitely many point masses, each connected to its neighbors and oscillating in the vertical direction. The displacement of the string depends on time, but it also depends on horizontal position, so displacement is a function of two independent variables, x and t. Instead of an ordinary differential equation with time derivatives, we have a partial differential equation with derivatives in both time and space. The resulting equation, called the *wave equation*, has interesting solutions.

As noted in the title, this chapter is intended to be a *focused* introduction to partial differential equations. There are three standard partial differential equations of mathematical physics, including the *heat equation* and *Laplace's equation* as well as the wave equation. Only the wave equation is studied in this chapter, with the other two equations making only a brief appearance. The solution methods for the three standard partial differential equations are often similar. By focusing on the wave equation alone, we are able to survey a broad selection of methods with less investment of time than would be necessary if our study included all three of the standard equations. At the conclusion of the chapter, the reader will have a lot of experience with the mathematics of waves. Those readers who want a more complete introduction to partial differential equations can follow their study of Chapter 8 with Section A.7 on the heat equation and Section A.8 on Laplace's equation.

We derive the wave equation in Section 8.1. Then we solve the *signaling problem* for the wave equation; these solutions introduce the idea of *traveling wave solutions*.

In Section 8.2 we find the general solution of the wave equation and use it to solve problems concerning the motion of long strings that are struck or plucked.

Sections 8.3 through 8.5 are concerned with the problem of solving the wave equation on a finite domain. We first determine *vibration modes* in Section 8.3. Vibration modes are used in Section 8.4 as the building blocks for *Fourier series* solutions of initial-boundary-value problems for the wave equation. In Section 8.5 we study the mathematics of Fourier series.

Case Study 8 uses the mathematics of Fourier series to study the sounds made by stringed instruments and percussion instruments.

8.1 Transverse Vibration of a String

Stretch an elastic band between the fingers of one hand and pluck it with the other hand. You will see it vibrate, with patterns that depend somewhat on the degree to which you have stretched the elastic band. If you have marked a dot on the elastic band, you will find that the dot moves only in a direction normal to the rubber band. This is **transverse** vibration.

Try repeating the above experiment, but with a different material. You can use a string, a telephone cord, a thin wire, or an elastic band of different weight. You will notice that the vibrations are different for different materials.

Suppose we want to develop the simplest reasonable mathematical model for a vibrating string and use it to examine wave propagation. We begin with the following assumptions:

- The cross-sectional area is small enough that the string can be thought of as occupying a curve in space. At rest, the string lies on the x axis.
- The string is made of uniform elastic material.[1] A good example is a guitar string.
- Each point on the string moves along an axis that is normal to the x axis. The displacement in this transverse direction is $u(x, t)$.
- The only force is that of tension, which holds the string taut. This force has magnitude T and acts tangentially to the string, in both directions.
- The string has a constant linear density[2] ρ.
- The displacement of the string from the x axis is small relative to the length of the string.

These assumptions are generally reasonable for a stretched guitar string. The most significant error lies in neglecting the small force of friction. A real guitar string eventually stops vibrating, but our idealized string will continue to vibrate forever.

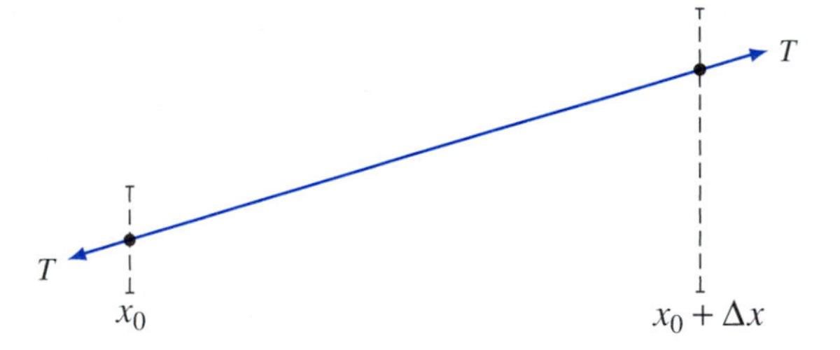

Figure 8.1.1
A small portion of a stretched string.

Consider a portion of the string, from x_0 to $x_0 + \Delta x$, as depicted in Figure 8.1.1. (The displacement of the string is exaggerated.) The crucial idea is that the motion of the string can be described by Newton's second law of motion, $\mathbf{F} = m\mathbf{a}$, with $\mathbf{F}$ the net force on the string.

[1]An **elastic** material is one that is deformed by application of a force but returns to its original geometry if the force is removed.

[2]Note that ρ is the mass of string per unit length, *not* the material density of the material from which the string is made. The mass per unit length is the product of the material density (mass/volume) and the cross-sectional area.

Since the motion of the string is strictly transverse, Newton's second law is applied only in the u direction. The transverse acceleration of the bit of string is

$$a \approx \frac{\partial^2 u}{\partial t^2}(x_0, t),$$

and the mass is

$$m \approx \rho\, \Delta x.$$

Let $F_P(x, t)$ and $F_T(x, t)$ be the parallel and transverse components, respectively, of the force exerted at a point by the portion of the string that is to the right of that point, as depicted in Figure 8.1.2. The net transverse force on the bit of string is then

$$\Delta F_T = F_T(x_0 + \Delta x, t) - F_T(x_0, t);$$

the sign of the second term is negative because the force acting at x_0 is caused by the portion of the string to the left of the bit rather than by the portion to the right.

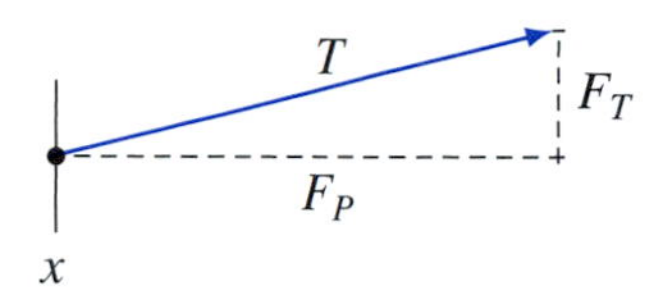

Figure 8.1.2
Components of the force exerted by the string to the right of x.

Substituting these approximations into $ma = \Delta F_T$ results in the equation

$$\rho\, \frac{\partial^2 u}{\partial t^2}(x_0, t)\, \Delta x \approx F_T(x_0 + \Delta x, t) - F_T(x_0, t),$$

which we can rewrite as

$$\rho\, \frac{\partial^2 u}{\partial t^2}(x_0, t) \approx \frac{F_T(x_0 + \Delta x, t) - F_T(x_0, t)}{\Delta x}.$$

This equation becomes exact in the limit $\Delta x \to 0$; thus,

$$\rho\, \frac{\partial^2 u}{\partial t^2}(x_0, t) = \lim_{\Delta x \to 0} \frac{F_T(x_0 + \Delta x, t) - F_T(x_0, t)}{\Delta x}.$$

The quantity on the right of this last equation is simply the partial derivative of F_T with respect to x, evaluated at (x_0, t). Thus, we have the equation

$$\rho\, \frac{\partial^2 u}{\partial t^2} = \frac{\partial F_T}{\partial x}. \tag{1}$$

To complete the model, it is necessary to compute F_T.

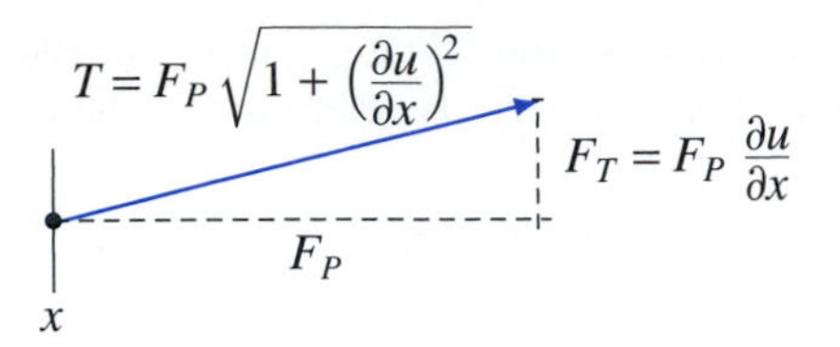

Figure 8.1.3
Components of the force exerted by the string to the right of x, in terms of the slope of the string.

Figure 8.1.3 shows the relationships among the tension force and its components in terms of the slope of the string. Eliminating F_P from the two equations in the figure yields the formula

$$F_T = T\frac{\partial u/\partial x}{\sqrt{1+(\partial u/\partial x)^2}}.$$

It is at this point that we use the simplifying approximation that the displacement is small compared with the length of the string. Accordingly, $\partial u/\partial x$ is small, and hence we have the approximation

$$\sqrt{1+\left(\frac{\partial u}{\partial x}\right)^2} \approx 1.$$

Taking this to be exact for our idealized model reduces the formula for the transverse force to

$$F_T = T\,\frac{\partial u}{\partial x}.$$

Substituting this formula into Equation (1) yields the final result:

$$\frac{\partial^2 u}{\partial t^2} = c^2\,\frac{\partial^2 u}{\partial x^2}, \qquad c^2 = \frac{T}{\rho}. \tag{2}$$

This is the **one-dimensional wave equation.** A large amount of interesting mathematics is associated with this deceptively simple equation. The remainder of this chapter is devoted largely to the study of this equation.

Partial Differential Equations

The wave equation is an example of a *linear partial differential equation.*

> A **partial differential equation** is an equation containing derivatives of an unknown function with respect to more than one independent variable. Suppose a partial differential equation is written as $L[u] = g$, where u is the unknown function and g depends only on the independent variables. The equation is **linear** if $L[c_1u_1 + c_2u_2] = c_1L[u_1] + c_2L[u_2]$.

The one-dimensional wave equation is one of three standard linear partial differential equations that are studied in any first course on the subject. The others are the one-dimensional heat equation and Laplace's equation in two dimensions. The **one-dimensional heat equation** is the equation

$$\frac{\partial u}{\partial t} = k\,\frac{\partial^2 u}{\partial x^2}, \tag{3}$$

where u is the temperature and k is a parameter that measures the rate of heat transport in the material. **Laplace's equation** in two dimensions is the equation

$$\frac{\partial^2 u}{\partial x^2} + \frac{\partial^2 u}{\partial y^2} = 0, \tag{4}$$

where u is the temperature in a flat plate or the displacement of a deformed elastic sheet.

At first glance, the differences between the three standard equations are minimal. The heat equation differs from the wave equation in that the time derivative is a first derivative rather than a second derivative; Laplace's equation differs from the wave equation in that the *sum* of the two second derivatives is 0 rather than the difference. These seemingly minor differences account for very large differences in the properties of the equations. The heat equation is the subject of Section A.7 and appears in some of the exercises in Chapter 8. Laplace's equation is the subject of Section A.8.

The Signaling Problem

The simplest problem that illustrates wave propagation is the **signaling problem,** in which a semi-infinite string is initially at rest and motion is introduced at the boundary of the string.

MODEL PROBLEM 8.1

Determine the vibration u for a very long string ($0 \le x < \infty$) initially at rest [$u(x, 0) = 0$] and motionless [$\partial u/\partial t(x, 0) = 0$], with the motion at the end $x = 0$ prescribed as

$$u(0, t) = f(t) = \begin{cases} \sin^3(\pi t) & 0 < t < 1 \\ 0 & \text{otherwise} \end{cases},$$

as shown in Figure 8.1.4. Assume the parameter value $c = 1$.

Your everyday experience of motion offers some intuition about Model Problem 8.1. The boundary motion is similar to what you would obtain by flicking the end of a long, taut rope. The subsequent motion ought to have each point on the rope execute the same motion as the boundary, but delayed as the disturbance travels down the rope. Thus, we anticipate that the function f that describes the boundary motion will strongly influence the solution for $u(x, t)$.

The displacement u must satisfy the nonhomogeneous **boundary condition**

$$u(0, t) = f(t) = \begin{cases} \sin^3(\pi t) & 0 < t < 1 \\ 0 & \text{otherwise} \end{cases},$$

as well as the differential equation

$$\frac{\partial^2 u}{\partial t^2} = \frac{\partial^2 u}{\partial x^2}, \qquad \text{for } x > 0,\ t > 0,$$

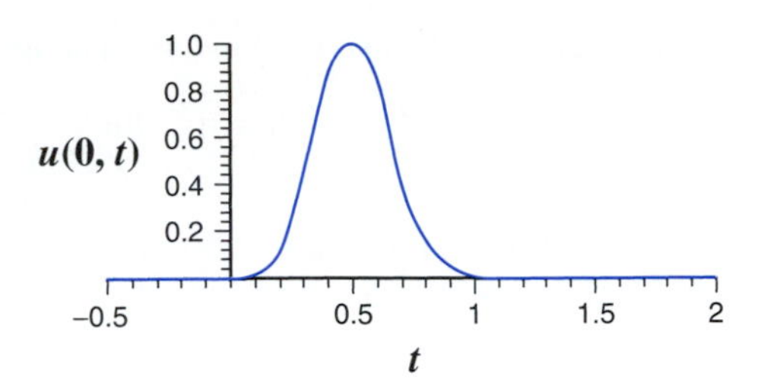

Figure 8.1.4
The boundary motion for Model Problem 8.1.

and the initial conditions

$$u(x, 0) = 0, \qquad \frac{\partial u}{\partial t}(x, 0) = 0.$$

Surprisingly, we can show that the function

$$u(x, t) = f(\tau), \qquad \tau = t - x,$$

satisfies all the requirements of the problem. The boundary condition is satisfied immediately because $\tau = t$ when $x = 0$. The initial conditions are also satisfied because $f(\tau) = 0$ whenever $\tau \le 0$, and $\tau \le 0$ is initially true for all x. To check the differential equation, note that

$$\frac{\partial u}{\partial t} = \frac{df}{d\tau}\frac{\partial \tau}{\partial t} = \frac{df}{d\tau},$$

while

$$\frac{\partial u}{\partial x} = \frac{df}{d\tau}\frac{\partial \tau}{\partial x} = -\frac{df}{d\tau}.$$

A second differentiation yields the results

$$\frac{\partial^2 u}{\partial t^2} = \frac{d^2 f}{d\tau^2}, \qquad \frac{\partial^2 u}{\partial x^2} = \frac{d^2 f}{d\tau^2},$$

as needed. Thus, the solution of Model Problem 8.1 is given by

$$u(x, t) = f(t - x) = \begin{cases} \sin^3(\pi[t - x]) & x < t < x + 1 \\ 0 & \text{otherwise} \end{cases}.$$

It is important to understand the qualitative behavior of the solutions of the model problem. The time history of the signal (Fig. 8.1.4) shows a single wave that begins at time 0, reaches a peak strength of 1 at time 1/2, and decreases to 0 at time 1. Now consider the time history of the solution at the point $x = 1$. This quantity is given by the formula

$$u(1, t) = f(t - 1) = \begin{cases} \sin^3(\pi[t - 1]) & 1 < t < 2 \\ 0 & \text{otherwise} \end{cases}.$$

The time history of the point $x = 1$ (Fig. 8.1.5) is identical to that of the boundary point $x = 0$, except that the disturbance begins at time $t = 1$ and ends at time $t = 2$. Similarly, the history at any point x_0 shows the same disturbance, beginning at time $t = x_0$ and ending at time $t = x_0 + 1$.

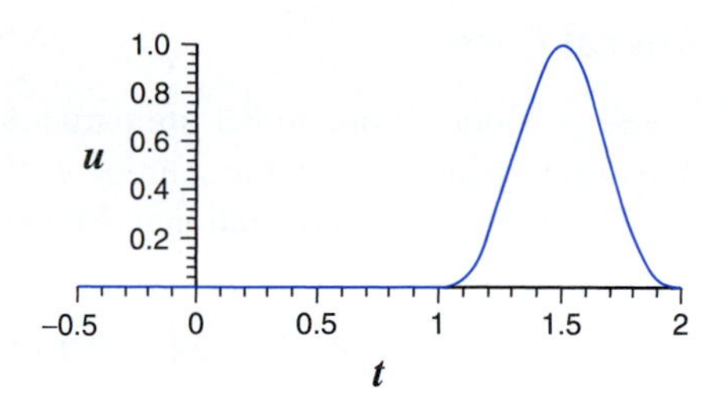

Figure 8.1.5
The time history at $x = 1$ for Model Problem 8.1.

✦ INSTANT EXERCISE 1

Sketch the time history at $x = 0.5$ for Model Problem 8.1.

It is also instructive to look at wave profiles. Figure 8.1.6 shows the wave at several different times. At $t = 0.5$, the boundary value has just reached the peak value of 1 and the wave extends to $x = 0.5$. At $t = 1$, the boundary value has dropped back down to 0 and the wave extends to $x = 1$. Thereafter, the wave moves to the right, advancing 1 unit in position for each unit of time. By $t = 3$, the wave extends from $x = 2$ to $x = 3$.

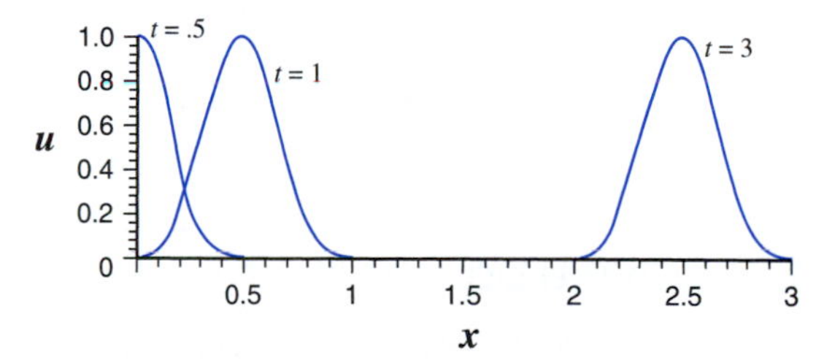

Figure 8.1.6
Wave profiles for Model Problem 8.1.

Figure 8.1.7 shows the solution in a three-dimensional plot. It is customary to orient the axes so that x increases to the right and t increases toward the top, rotated to obtain a good viewing perspective. The solid curves show the wave profiles at times 0.5, 1, 1.5, and so on.

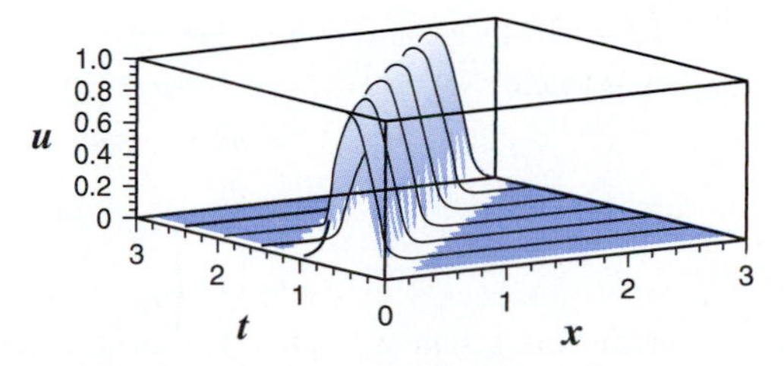

Figure 8.1.7
The solution of Model Problem 8.1.

The General Case

Two features of Model Problem 8.1 are particularly noteworthy. The solution *formula* depends significantly on the signal f, but the solution *method* does not use the function f in a fundamental way. We consider now a generalization of Model Problem 8.1.

Signaling Problem for the Wave Equation

Given a function f and a constant $c > 0$, find a function u that satisfies the differential equation

$$\frac{\partial^2 u}{\partial t^2} = c^2 \frac{\partial^2 u}{\partial x^2}, \qquad \text{for} \quad 0 < x < \infty, \quad t > 0$$

with boundary condition

$$u(0, t) = f(t)$$

and initial conditions

$$u(x, 0) = 0, \qquad \frac{\partial u}{\partial t}(x, 0) = 0.$$

The result we obtained in Model Problem 8.1 is applicable for any sufficiently smooth function f.

Theorem 8.1.1 Let f be a function with a continuous second derivative. Then the function $u = f(t - x/c)$ solves the signaling problem for the wave equation.

✦ INSTANT EXERCISE 2

Show that Model Problem 8.1 meets the requirements of Theorem 8.1.1.

If the signal is not smooth, then $f(t - x/c)$ fails to satisfy the wave equation at points where the second derivative does not exist. It is useful when working with the wave equation to admit solutions that are only piecewise-smooth. The wave equation has the property that function values propagate at speed c, and therefore discontinuities also propagate at speed c.

EXAMPLE 1

The signaling problem

$$\frac{\partial^2 u}{\partial t^2} = \frac{\partial^2 u}{\partial x^2}, \qquad x, t > 0,$$

$$u(0, t) = \begin{cases} |t - 1| & 0 \le t \le 2 \\ 0 & \text{otherwise} \end{cases},$$

$$u(x, 0) = 0, \qquad \frac{\partial u}{\partial t}(x, 0) = 0,$$

has the solution

$$u = \begin{cases} |t - x - 1| & x \le t \le x + 2 \\ 0 & \text{otherwise} \end{cases}.$$

Technically speaking, this solution does not exist at points $x = t - 2$, $x = t - 1$, and $x = t$ because the second partial derivatives do not exist at these points. However, it still makes sense to think of the solution formula as representing wave motion for all $x, t > 0$. Figure 8.1.8 shows some wave profiles.

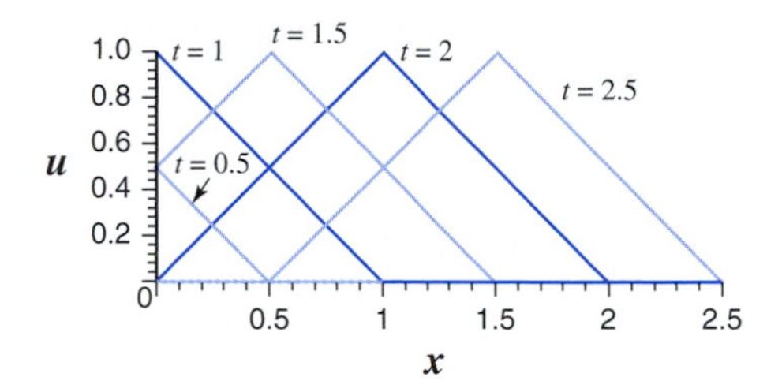

Figure 8.1.8
Wave profiles for Example 1.

Traveling Waves

A **traveling wave** is a function of the form $f(t - x/v)$ or $f(x - vt)$. The function f is the **waveform** or waveshape, and the constant v is the velocity of the wave. Negative values of v correspond to waves that move to the left. As a matter of convenience, it is common to use the form $f(t - x/v)$ for the specific case of waves caused by boundary motion, as in signaling problems, and the form $f(x - vt)$ for other cases.

EXAMPLE 2

Consider the possibility of traveling wave solutions for the one-dimensional heat equation,

$$\frac{\partial u}{\partial t} = k\frac{\partial^2 u}{\partial x^2}, \qquad k > 0.$$

Define functions f and g by

$$f(t) = e^t, \qquad g(t) = \sin t.$$

If $u = f(x - vt) = e^{x-vt}$, then $u_t = -ve^{x-vt}$ and $u_{xx} = e^{x-vt}$. Substituting these formulas into the heat equation yields $-ve^{x-vt} = ke^{x-vt}$; hence, the waveform f can propagate at speed $v = -k$. If, however, $u = g(x - vt) = \sin(x - vt)$, then $u_t = -v\cos(x - vt)$ and $u_{xx} = -\sin(x - vt)$. There is no speed v for which $u_t = ku_{xx}$. The heat equation has traveling wave solutions for only a very restricted set of waveforms.

Traveling wave solutions can occur for many partial differential equations, but generally such solutions can be found only for specific waveforms, as seen in Example 2. The wave equation, as we have seen, has solutions with *arbitrary* waveforms. This remarkable property is shared by the **advection** equation $u_t + cu_x = 0$ and the *spherically symmetric wave equation* in three dimensions.[3] In each of these cases, the waves can only propagate at a speed c that is characteristic of the medium. In the case of waves traveling along a string, this speed is proportional to the square root of the tension and inversely proportional to the square root of the linear density.

✦ INSTANT EXERCISE 3

Let f be an arbitrary function of one variable with continuous second derivative. Determine the wave velocities for traveling wave solutions of the wave equation (2) having waveform f.

A Comment about Sound Waves

Sound waves are not caused by transverse vibration, so the model of this section does not work for them. Nevertheless, models of acoustic vibration result in the same wave equation that we are studying in this chapter, and the same qualitative features apply: sound waves in one or three dimensions can have an arbitrary waveform, and all sound waves have a speed that is characteristic of the medium. Waves traveling through air at 1 atmosphere (atm) pressure and 20°C propagate at a speed of approximately 343 m/s, or 763 mi/h.

8.1 Exercises

In Exercises 1 through 6, (a) check the conditions of Theorem 8.1.1, (b) determine the solution of the indicated problem, (c) sketch a graph of the time history of the points $x = 0$ and $x = 6$, and (d) sketch the wave profile at time $t = 2$.

T 1.
$$\frac{\partial^2 u}{\partial t^2} = 4\frac{\partial^2 u}{\partial x^2}, \qquad \text{for } x > 0, \quad t > 0,$$
$$u(0,t) = \begin{cases} \sin^3 \dfrac{\pi t}{2} & 0 \le t \le 2 \\ 0 & \text{otherwise} \end{cases}, \qquad u(x,0) = 0, \qquad \frac{\partial u}{\partial t}(x,0) = 0.$$

T 2.
$$\frac{\partial^2 u}{\partial t^2} = \frac{\partial^2 u}{\partial x^2}, \qquad \text{for } x > 0, \quad t > 0,$$
$$u(0,t) = \begin{cases} \sin^3 \pi t & 0 \le t \le 2 \\ 0 & \text{otherwise} \end{cases}, \qquad u(x,0) = 0, \qquad \frac{\partial u}{\partial t}(x,0) = 0.$$

T 3.
$$\frac{\partial^2 u}{\partial t^2} = 4\frac{\partial^2 u}{\partial x^2}, \qquad \text{for } x > 0, \quad t > 0,$$
$$u(0,t) = \begin{cases} t - t^2 & 0 \le t \le 1 \\ 0 & \text{otherwise} \end{cases}, \qquad u(x,0) = 0, \qquad \frac{\partial u}{\partial t}(x,0) = 0.$$

[3] See Exercise 10.

T 4.
$$\frac{\partial^2 u}{\partial t^2} = 4\frac{\partial^2 u}{\partial x^2}, \qquad \text{for } x > 0, \quad t > 0,$$
$$u(0,t) = \begin{cases} (t-t^2)^3 & 0 \le t \le 1 \\ 0 & \text{otherwise} \end{cases}, \qquad u(x,0) = 0, \qquad \frac{\partial u}{\partial t}(x,0) = 0.$$

T 5.
$$\frac{\partial^2 u}{\partial t^2} = \frac{\partial^2 u}{\partial x^2}, \qquad \text{for } x > 0, \quad t > 0,$$
$$u(0,t) = \begin{cases} (1-\cos \pi t)^3 & 0 \le t \le 2 \\ 0 & \text{otherwise} \end{cases}, \qquad u(x,0) = 0, \qquad \frac{\partial u}{\partial t}(x,0) = 0.$$

T 6.
$$\frac{\partial^2 u}{\partial t^2} = \frac{\partial^2 u}{\partial x^2}, \qquad \text{for } x > 0, \quad t > 0,$$
$$u(0,t) = \begin{cases} t^3 - 4t^2 + 3t & 0 \le t \le 3 \\ 0 & \text{otherwise} \end{cases}, \qquad u(x,0) = 0, \qquad \frac{\partial u}{\partial t}(x,0) = 0.$$

7. Determine the functions f for which the one-dimensional heat equation (3) has traveling wave solutions, and find the wave velocities for those solutions.

8. Determine the functions f for which the **convection diffusion equation** $u_t + cu_x = ku_{xx}$ has traveling wave solutions, and find the wave velocities for those solutions.

The wave equation in one dimension has the property that arbitrary waveforms propagate with no loss in amplitude. This is the principle behind the "tin can telephone," which transmits sound accurately along a taut string or wire connecting two metal cans with only minimal loss of amplitude.[4]

9. Can oral communication occur in a two-dimensional medium?[5]

   a. The **radially symmetric wave equation** in two dimensions is
   $$\frac{\partial^2 u}{\partial t^2} = \frac{c^2}{r}\frac{\partial}{\partial r}\left(r\frac{\partial u}{\partial r}\right).$$
   Assume a solution of the form $u(r,t) = g(r)f(r - vt)$ for some function g and velocity v, with f an arbitrary waveform. Substitute into the wave equation and separately collect terms with factors of f, f', and f''.

   b. Since f must be arbitrary, the coefficients of f, f', and f'' should all be zero. Determine possible velocities v by setting the f'' coefficient to zero.

   c. Obtain two differential equations for the amplitude function g from the f and f' coefficients. Show that $g = 0$ is the only function that solves both equations. Conclude that

---

[4]The wave equation is based on a conceptual model that is not qualitatively accurate. There is a loss of amplitude in real one-dimensional wave motion, but not in the wave equation model. This qualitative inaccuracy is generally ignored in studies of wave motion, as it has practical relevance only for long times and distances.

[5]This exercise and Exercise 10 are adapted from T. Morley, "A simple proof that the world is three-dimensional," *SIAM Review*, vol. 27, pp. 69–71, 1985.

the assumption that waves of arbitrary waveform f can propagate in two dimensions is false.[6]

10. Can oral communication occur in a three-dimensional medium?

    a. The **spherically symmetric wave equation** is
$$\frac{\partial^2 u}{\partial t^2} = \frac{c^2}{r^2}\frac{\partial}{\partial r}\left(r^2 \frac{\partial u}{\partial r}\right).$$
    Assume a solution of the form $u(r, t) = g(r)f(r - vt)$ for some function g and velocity v, with f an arbitrary waveform. Substitute into the wave equation and separately collect terms with factors of f, f', and f''.

    b. Since f must be arbitrary, the coefficients of f, f', and f'' should all be zero. Determine possible velocities v by setting the f'' coefficient to zero.

    c. Obtain two differential equations for the amplitude function g from the f and f' coefficients. Find a one-parameter family of solutions for g. Conclude that waves of arbitrary waveform f can propagate in three dimensions with an amplitude proportional to the reciprocal of the distance to the source.

    d. Explain the significance of the result for the amplitude. In particular, what would happen if the amplitude in three-dimensional space behaved in a similar way to that in one-dimensional space?

11. The phenomenon known as the *soliton* (solitary wave) was first described by J. S. Russell in 1834:

    > I was observing the motion of a boat which was rapidly drawn along a narrow channel by a pair of horses, when the boat suddenly stopped—not so the mass of water in the channel which it had put in motion; it accumulated round the prow of the vessel in a state of violent agitation, then suddenly leaving it behind, rolled forward with great velocity, assuming the form of a large solitary elevation, a rounded, smooth, and well-defined heap of water, which continued its course along the channel apparently without change of form or diminution of speed. I followed it on horseback, and overtook it still rolling on at a rate of some 8 or 9 miles an hour, preserving its original figure some 30 feet long and a foot to a foot and a half in height.

    A mathematical model of water waves in a narrow channel eventually reduces to a nonlinear partial differential equation called the **Korteweg-deVries (KdV) equation** for the vertical displacement of the water surface as a function of time and distance:
$$\frac{\partial u}{\partial t} + \frac{\partial^3 u}{\partial x^3} + 6u\frac{\partial u}{\partial x} = 0, \qquad -\infty < x < \infty.$$

    a. Derive a third-order ordinary differential equation for a traveling wave solution $u(x, t) = vf(z)$, $z = x - vt$ of the KdV equation. (The extra factor v in this formula for u makes the calculations a little easier.)

    b. Obtain a first integral of the equation from part *a* by direct integration; then multiply the result by $2f'$ and integrate again to obtain a second integral. In each stage, the integration

[6] Sound transmitted in a two-dimensional medium, such as a "tin can telephone" with the cans connected by a metal sheet rather than a wire, propagates only with changes in the waveforms. The sound heard at the "receiver" is unintelligible. Many native peoples have discovered that it is possible to hear sounds propagated through the earth at a greater distance than the sounds can be heard through the air. Sound propagated through the earth is easier to *detect* than sound propagated through the air even though it cannot be *understood*.

constant can be determined by using Russell's observation that the wave is but a single heap of water (and therefore there is no disturbance far from $z = 0$).

c. At this point, you should have a first-order nonlinear differential equation for f. Assume that the peak of the wave is at $x - vt = 0$, so that $f'(0) = 0$. Use this fact and the differential equation to determine an initial condition for f.

d. Solve the initial-value problem for f to obtain a formula for a soliton. Are solitons limited to specific wave speeds?

e. Plot the soliton waveforms vf with $v = 1$ and $v = 4$.

T 12. The biologist R. A. Fisher (in 1937) proposed a model to describe the spatial spread of species. The model combines the logistic population growth equation with diffusion:

$$\frac{\partial u}{\partial t} - D\frac{\partial^2 u}{\partial x^2} = ru\left(1 - \frac{u}{K}\right).$$

Since new genetic traits have been seen to diffuse through populations at a constant speed, it seems reasonable to look for traveling wave solutions of Fisher's equation.

a. Determine an appropriate set of scales u_r, x_r, and t_r so that Fisher's equation can be written in terms of dimensionless variables $U = u/u_r$, $X = x/x_r$, and $T = t/t_r$ as

$$\frac{\partial U}{\partial T} - \frac{\partial^2 U}{\partial X^2} = U(1 - U).$$

b. Derive a second-order ordinary differential equation for a traveling wave solution $U(X, T) = F(Z)$, $Z = X - cT$ of Fisher's equation.

c. The ordinary differential equation from part *b* cannot be solved in closed form, but it can be studied by phase plane analysis. Begin by writing the equation as a system of equations using the variables F and $G = F'$.

d. Observe that the system has two critical points. Use the Jacobian (Section 6.7) to determine the stability of the critical points.

e. A traveling wave solution is possible if one of the critical points represents the situation behind the wave ($Z \to -\infty$) and the other represents the situation in front of the wave ($Z \to +\infty$). The waveform then corresponds to a trajectory that moves from one critical point to the other. Explain why $c < 2$ is unrealistic.

f. Use a computer algebra system to plot a trajectory connecting the critical points with $c = 4$. What are the values of u ahead of the wave? What is the value of u after the wave has passed? Explain why these values make biological sense.

✦ 8.1 INSTANT EXERCISE SOLUTIONS

1.

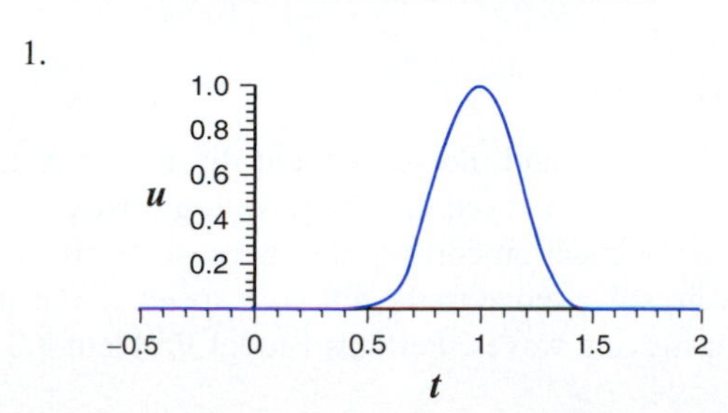

2. We have

$$f(t) = \begin{cases} \sin^3 \pi t & 0 < t < 1 \\ 0 & \text{otherwise} \end{cases};$$

hence,

$$f'(t) = \begin{cases} 3\pi \sin^2 \pi t \cos \pi t & 0 < t < 1 \\ 0 & \text{otherwise} \end{cases}$$

and

$$f''(t) = \begin{cases} 3\pi^2 \sin(\pi t)(2\cos^2 \pi t - \sin^2 \pi t) & 0 < t < 1 \\ 0 & \text{otherwise} \end{cases}.$$

The function f'' is clearly continuous on the intervals $(-\infty, 0)$, $(0, 1)$, and $(1, \infty)$. It is not immediately clear that f'' is continuous at the points 0 and 1. However, by calculation we have

$$\lim_{t\to 0^+} f(t) = \lim_{t\to 0^-} f(t) = 0, \qquad \lim_{t\to 1^+} f(t) = \lim_{t\to 1^-} f(t) = 0;$$

hence, f has a continuous second derivative.

3. Given $u = f(x - vt)$, we have $\partial^2 u/\partial t^2 = v^2 f'(x - vt)$ and $c^2(\partial^2 u/\partial x^2) = c^2 f'(x - vt)$. These quantities are equal if $v^2 = c^2$; thus, traveling waves can have wave velocities $\pm c$.

8.2 The General Solution of the Wave Equation

In Section 8.1, we considered the signaling problem for the wave equation. Another very general problem for the wave equation is the *initial-value problem* on an unbounded domain. We consider a simple example.

MODEL PROBLEM 8.2

Solve the problem consisting of the partial differential equation

$$\frac{\partial^2 u}{\partial t^2} = \frac{\partial^2 u}{\partial x^2}, \qquad \text{for } -\infty < x < \infty, \quad t > 0$$

with initial conditions

$$u(x, 0) = U(x) = 2e^{-x^2}, \qquad \frac{\partial u}{\partial t}(x, 0) = 0.$$

In the study of Model Problem 8.1 and the subsequent discussion of traveling wave solutions, especially Instant Exercise 3, we observed that the traveling waves $u = f(x - ct)$ and $u = f(x + ct)$ are solutions of the wave equation corresponding, respectively, to waves that move to the right with speed c and to waves that move to the left with speed c. The signaling problem of Section 8.1 requires only right-moving waves. Perhaps Model Problem 8.2 can be solved by a combination of waves moving in either direction. Since

$$u = f(x - t) + g(x + t)$$

solves the wave equation (with $c = 1$) for any functions f and g, all that needs to be done is to find functions f and g so that u satisfies the initial conditions. These requirements become

$$f(x) + g(x) = 2e^{-x^2}, \qquad -f'(x) + g'(x) = 0.$$

The second of these equations says that the derivatives of f and g are equal, which means that the functions can differ by a constant at most. Suppose $g(x) = f(x) + C_1$. Then the first equation becomes $2f(x) + C_1 = 2e^{-x^2}$, or $f(x) = e^{-x^2} - C$, where we have substituted $C = C_1/2$; then $g(x) = e^{-x^2} + C$. Thus, we have a one-parameter family of solutions for f and a corresponding one-parameter family of solutions for g. Substituting these results into the solution for u yields

$$u = e^{-(x-t)^2} + e^{-(x+t)^2}$$

regardless of the choice of C. Hence, there is no loss of generality in choosing $C = 0$, whereupon $f = g = U/2$. The initial displacement profile splits into two equal profiles, one that moves to the right and one that moves to the left. These waves overlap, so the appearance of the solution is not always similar to the initial profile. The solution of Model Problem 8.2, depicted in Figure 8.2.1, is a case in point. At $t = 1$, the overlap is large enough to make the profile look rather different. By $t = 2$, the amount of overlap is small enough that the profile looks very much like two halves of the initial profile receding from each other.

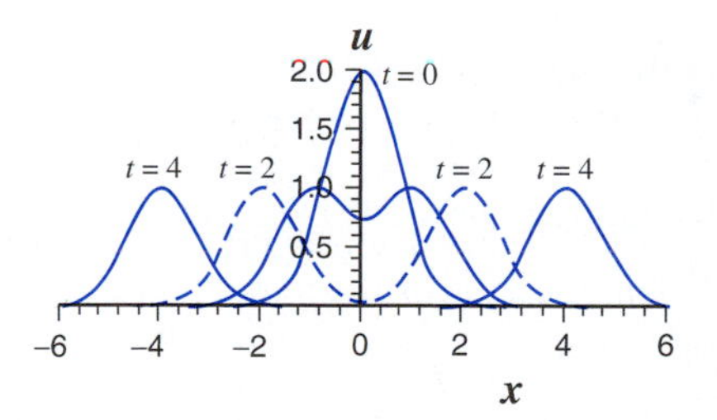

Figure 8.2.1
Wave profiles at $t = 0, 1, 2,$ and 4 for Model Problem 8.2.

Uniqueness of Solutions

In Section 8.1, we used the traveling wave $u = f(t - x/c)$ to solve the signaling problem for the wave equation. Now we have used a combination of traveling waves to solve an initial-value problem for the wave equation. At no point did we address the mathematical question of uniqueness. Might there be other solutions of the wave equation that are not traveling waves?

Define two new independent variables by

$$w = x + ct, \qquad z = x - ct.$$

By careful use of the chain rule, we can rewrite the wave equation in terms of w and z rather than x and t. Let U be a function of w and z that solves the wave equation, in the sense that we can calculate the displacement at position x and time t by first computing the appropriate point in the wz coordinate plane and then using the function U to determine the displacement. In mathematical notation, we have

$$u(x, t) = U(w(x, t), z(x, t)). \tag{1}$$

Using the chain rule, we can differentiate Equation (1) to obtain expressions for the derivatives of u in terms of the derivatives of U. For example,

$$\frac{\partial u}{\partial t} = \frac{\partial U}{\partial w}\frac{\partial w}{\partial t} + \frac{\partial U}{\partial z}\frac{\partial z}{\partial t} = c\,\frac{\partial U}{\partial w} - c\,\frac{\partial U}{\partial z},$$

where we have computed the derivatives of w and z by using the definitions of these quantities. A second differentiation with respect to time yields

$$\frac{\partial^2 u}{\partial t^2} = \frac{\partial}{\partial t}\left(\frac{\partial u}{\partial t}\right) = c^2\,\frac{\partial^2 U}{\partial w^2} - 2c^2\,\frac{\partial^2 U}{\partial w\,\partial z} + c^2\,\frac{\partial^2 U}{\partial z^2}. \tag{2}$$

The last step in the calculation assumes that the mixed partial derivatives are equal, which is true whenever U has continuous second derivatives. A similar calculation yields the result

$$c^2\,\frac{\partial^2 u}{\partial x^2} = c^2\,\frac{\partial^2 U}{\partial w^2} + 2c^2\,\frac{\partial^2 U}{\partial w\,\partial z} + c^2\,\frac{\partial^2 U}{\partial z^2}. \tag{3}$$

✦ INSTANT EXERCISE 1

Use the chain rule to derive Equation (2) by differentiating the expression for u_t with respect to t.

Substituting Equation (2) and Equation (3) into the wave equation yields, after simplification, the equation

$$\frac{\partial^2 U}{\partial w\,\partial z} = 0.$$

This version of the wave equation can be solved by straightforward integration. The w derivative of U_z is 0, so integrating with respect to w yields

$$\frac{\partial U}{\partial z} = F(z)$$

for any function F. This result can be integrated at once to yield

$$U = \int F(z)\,dz + g(w)$$

for any function g. Since F is arbitrary, its antiderivatives are also arbitrary, so we can rewrite the solution as

$$U = f(z) + g(w),$$

where both f and g are arbitrary. This calculation establishes the result of Theorem 8.2.1.

Theorem 8.2.1

If u is a function with continuous second derivatives that solves the wave equation, then u can be written as

$$u = f(x - ct) + g(x + ct) \tag{4}$$

for some functions f and g.

The Initial-Value Problem in General

There are two ways that Model Problem 8.2 can be generalized. We can make changes to the initial profile, and we can also introduce a nonzero initial velocity. We consider the general **initial-value problem** for the wave equation on an unbounded domain.

Initial-Value Problem for the Wave Equation

Given functions U and V for which $\lim_{|x|\to\infty} U(x) = 0$ and $\lim_{|x|\to\infty} V(x) = 0$, find the function u that satisfies the differential equation

$$\frac{\partial^2 u}{\partial t^2} = c^2 \frac{\partial^2 u}{\partial x^2}, \qquad \text{for} \quad -\infty < x < \infty, \quad t > 0, \tag{5}$$

with initial conditions

$$u(x, 0) = U(x), \qquad \frac{\partial u}{\partial t}(x, 0) = V(x). \tag{6}$$

The general initial-value problem can be solved in a manner similar to the way we solved Model Problem 8.2 (see Exercises 10 through 12). The result is summarized in Theorem 8.2.2.

Theorem 8.2.2

Let U be a function with a continuous second derivative, and let V be a function with a continuous first derivative. Then the unique solution of the initial-value problem

$$\frac{\partial^2 u}{\partial t^2} = c^2 \frac{\partial^2 u}{\partial x^2}, \qquad \text{for} \quad -\infty < x < \infty, \quad t > 0,$$

with

$$u(x, 0) = U(x), \qquad \frac{\partial u}{\partial t}(x, 0) = V(x),$$

is

$$u(x, t) = \frac{U(x - ct) + U(x + ct)}{2} + \frac{1}{2c} \int_{x-ct}^{x+ct} V(s)\, ds. \tag{7}$$

This solution can also be written in the form

$$u = f(x - ct) + g(x + ct),$$

with

$$f(x) = \frac{1}{2}U(x) - \frac{1}{2c}\int_0^x V(s)\,ds - C, \qquad g(x) = \frac{1}{2}U(x) + \frac{1}{2c}\int_0^x V(s)\,ds + C, \tag{8}$$

with C an arbitrary constant.

Equation (7) is known as **D'Alembert's[7] formula.**

✦ INSTANT EXERCISE 2

Verify that $\int_0^{x+ct} V(s)\,ds$ solves the wave equation for any function V that has a continuous derivative.

Now suppose an infinite string is set into motion by giving it an initial velocity with no initial displacement. This corresponds to the case $U = 0$. In this case, the waveforms are

$$g(x) = \frac{1}{2c}\int_0^x V(s)\,ds + C, \qquad f(x) = -g(x).$$

There is again a symmetry between the left-moving and right-moving waves, but this time one is the negative of the other.

EXAMPLE 1

Consider the initial-value problem for the infinite string with $c = 1$ and initial conditions

$$u(x,0) = 0, \qquad \frac{\partial u}{\partial t}(x,0) = \begin{cases} 12x(1-x^2)^2 & |x| < 1 \\ 0 & \text{otherwise} \end{cases}.$$

Integrating V, we have for $|x| < 1$,

$$g(x) = C + 1 - (1-x^2)^3;$$

we therefore choose $C = -1$ and obtain

$$g(x) = -(1-x^2)^3, \qquad |x| < 1.$$

Now suppose $x > 1$. Then

$$g(x) = \int_0^x V(s)\,ds - 1 = \int_0^1 V(s)\,ds - 1 = g(1) = 0.$$

The key point here is that the region $1 < s < x$ does not contribute anything to the integral because $V(s) = 0$ throughout that region. Similarly, we have $g(x) = 0$ for $x < -1$. Hence, we obtain the results

$$g(x) = \begin{cases} -(1-x^2)^3 & |x| < 1 \\ 0 & \text{otherwise} \end{cases}, \qquad f(x) = \begin{cases} (1-x^2)^3 & |x| < 1 \\ 0 & \text{otherwise} \end{cases}.$$

[7] The correct pronunciation is "Dah-lem-behrz."

The solution of the problem is then known in terms of these functions as

$$u = f(x - ct) + g(x + ct).$$

Profiles of the solution for times $t = 0, 0.1, 1$, and 4 are shown in Figure 8.2.2. The profile is initially flat, but it quickly emerges. By $t = 1$, we see the full waveforms, and subsequently we see the movement of the waves away from the origin.

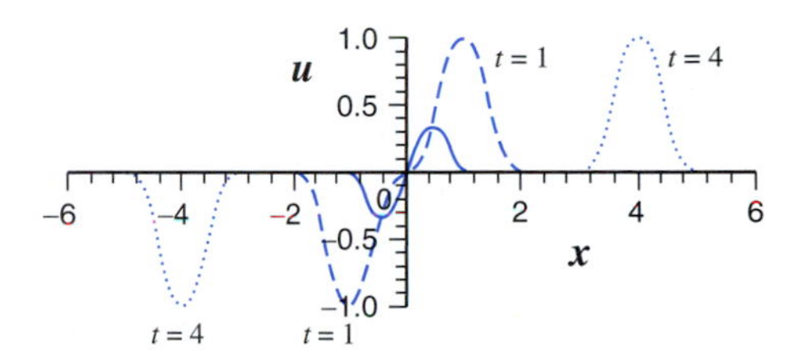

Figure 8.2.2
Wave profiles at $t = 0, 0.1, 1$, and 4 for Example 1.

✦ INSTANT EXERCISE 3

Derive the formula $g(x) = C + 1 - (1 - x^2)^3$ given in Example 1.

EXAMPLE 2

Consider the initial-value problem for the infinite string with $c = 1$ and initial conditions

$$u(x, 0) = 2e^{-x^2}, \qquad \frac{\partial u}{\partial t}(x, 0) = \begin{cases} 12x(1 - x^2)^2 & |x| < 1 \\ 0 & \text{otherwise} \end{cases}.$$

We have already computed solutions corresponding to the initial conditions for each of u and $\partial u/\partial t$ separately, and D'Alembert's formula shows that the solution of the complete problem is the sum of the two portions. The solution formula is then

$$u = f(x - t) + g(x + t),$$

where

$$f(x) = e^{-x^2} + (1 - x^2)^3, \qquad g(x) = e^{-x^2} - (1 - x^2)^3.$$

The easiest way to understand the behavior of the solution is to examine the functions f and g that constitute the traveling waves; these appear in Figure 8.2.3. From the figure, we see that the right-moving wave is larger than the left-moving wave, and that both functions are quite small outside the interval $[-2, 2]$. Since the waves travel at speed 1, it takes about 2 time units for the waves to "separate." At $t = 2$, the right-moving wave will be small outside of $[0, 4]$, and the left-moving wave will be small outside of $[-4, 0]$. Figure 8.2.4 shows the solution profiles at $t = 1$ and $t = 4$. As predicted, the solution profile at $t = 1$ shows continuing overlap of the two waves, while the solution profile at $t = 4$ shows the two waves receding from each other.

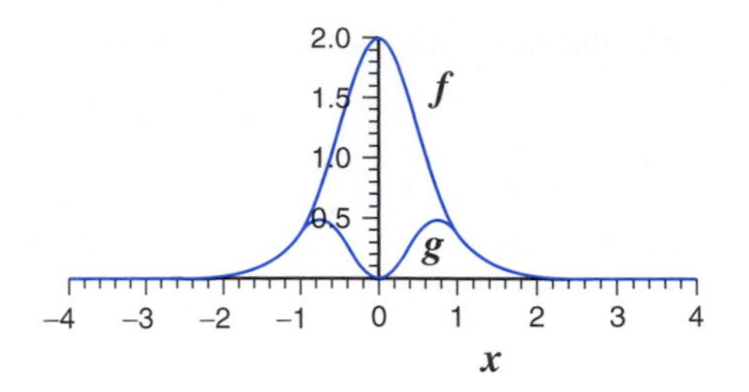

Figure 8.2.3
The functions f and g for Example 2.

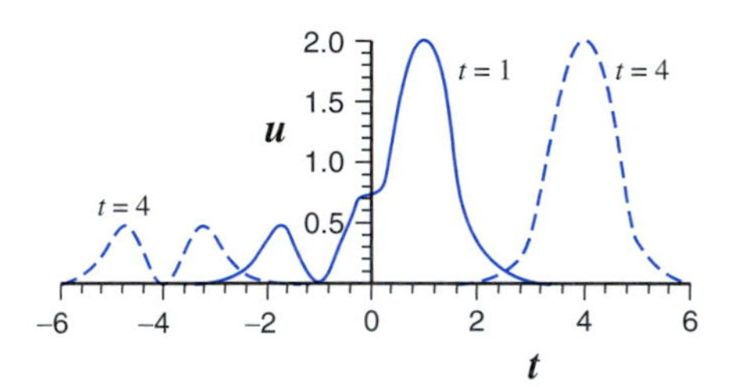

Figure 8.2.4
Wave profiles at $t = 1$ and 4 for Example 2.

Comparison of Ordinary and Partial Differential Equations

Our study of the wave equation has pointed out an important property of partial differential equations, namely, that the solutions depend in a fundamental way on the initial and/or boundary conditions. In the case of linear ordinary differential equations, the initial conditions serve only to determine the relative weight of the basic solutions provided by the differential equations; in contrast, the choice of boundary condition in a signaling problem and the initial conditions in an initial-value problem determine the shape of the traveling wave solutions of the wave equation. In general, the auxiliary conditions for a partial differential equation problem play a large role in determining the solution of the problem. Because of this crucial dependence of solutions on auxiliary conditions, general solutions *cannot* be obtained for most partial differential equations. The standard procedure for solving ordinary differential equation problems is to find a general solution and then employ the auxiliary conditions. This procedure is seldom possible with partial differential equations. We have been able to employ it with the wave equation because of the existence of a simple general solution formula. However, the use of a general solution with the wave equation becomes problematic for a finite spatial domain. We solve such problems in the remainder of Chapter 8 without using the general solution.

8.2 Exercises

In Exercises 1 through 6, (a) determine the solution of the indicated problem and (b) sketch the wave profiles at times 0, 3, and 6.

T 1.
$$\frac{\partial^2 u}{\partial t^2} = \frac{\partial^2 u}{\partial x^2}, \quad \text{for} \quad -\infty < x < \infty, \ t > 0,$$
$$u(x,0) = \frac{2}{1+x^2}, \qquad \frac{\partial u}{\partial t}(x,0) = 0.$$

T 2.
$$\frac{\partial^2 u}{\partial t^2} = \frac{\partial^2 u}{\partial x^2}, \quad \text{for} \quad -\infty < x < \infty, \ t > 0,$$
$$u(x,0) = \frac{x}{1+x^2}, \qquad \frac{\partial u}{\partial t}(x,0) = 0.$$

3.
$$\frac{\partial^2 u}{\partial t^2} = 4\frac{\partial^2 u}{\partial x^2}, \quad \text{for} \quad -\infty < x < \infty, \ t > 0,$$
$$u(x,0) = \begin{cases} 1-x^2 & -1 \le x \le 1 \\ 0 & \text{otherwise} \end{cases}, \qquad \frac{\partial u}{\partial t}(x,0) = 0.$$

4.
$$\frac{\partial^2 u}{\partial t^2} = 4\frac{\partial^2 u}{\partial x^2}, \quad \text{for} \quad -\infty < x < \infty, \ t > 0,$$
$$u(x,0) = \begin{cases} \sin \pi x & -2 \le x \le 2 \\ 0 & \text{otherwise} \end{cases}, \qquad \frac{\partial u}{\partial t}(x,0) = 0.$$

T 5.
$$\frac{\partial^2 u}{\partial t^2} = \frac{\partial^2 u}{\partial x^2}, \quad \text{for} \quad -\infty < x < \infty, \ t > 0,$$
$$u(x,0) = 0, \qquad \frac{\partial u}{\partial t}(x,0) = \frac{x}{1+x^2}.$$

T 6.
$$\frac{\partial^2 u}{\partial t^2} = 4\frac{\partial^2 u}{\partial x^2}, \quad \text{for} \quad -\infty < x < \infty, \ t > 0,$$
$$u(x,0) = 0, \qquad \frac{\partial u}{\partial t}(x,0) = \begin{cases} 1-x^2 & -1 \le x \le 1 \\ 0 & \text{otherwise} \end{cases}.$$

In Exercises 7 and 8, (a) plot f and g and (b) plot the wave profile at $t = 2$.

T 7.
$$\frac{\partial^2 u}{\partial t^2} = \frac{\partial^2 u}{\partial x^2}, \quad \text{for} \quad -\infty < x < \infty, \ t > 0,$$
$$u(x,0) = \frac{2}{1+x^2}, \qquad \frac{\partial u}{\partial t}(x,0) = \frac{x}{1+x^2}.$$

(See Exercises 1 and 5.)

T 8.
$$\frac{\partial^2 u}{\partial t^2} = 4\frac{\partial^2 u}{\partial x^2}, \quad \text{for} \quad -\infty < x < \infty, \ t > 0,$$
$$u(x,0) = \begin{cases} 1-x^2 & -1 \le x \le 1 \\ 0 & \text{otherwise} \end{cases}, \qquad \frac{\partial u}{\partial t}(x,0) = \begin{cases} 1-x^2 & -1 \le x \le 1 \\ 0 & \text{otherwise} \end{cases}.$$

(See Exercises 3 and 6.)

9. A solution can be found for the one-dimensional heat equation [Section 8.1, Equation (3)] under certain circumstances by reducing the equation to an ordinary differential equation.
   a. Assume that the heat equation has a solution of the form $u(x,t) = f(z)$, where $z = x/\sqrt{4kt}$. Substitute into the heat equation to obtain an ordinary differential equation for f.
   b. Solve the equation for f from part a to obtain a two-parameter family of solutions of the heat equation. (*Hint*: you can use the error functions defined in Section 2.4.)
   c. What restrictions on the initial and boundary conditions for the problem

$$\frac{\partial u}{\partial t} = k\frac{\partial^2 u}{\partial x^2}, \qquad -\infty < x < \infty, \qquad t > 0,$$

$$u(0,t) = g(t), \qquad u(x,0) = f(x), \qquad \lim_{x\to\infty} u = 0$$

   are necessary to have the family found in part b solve the initial and boundary conditions as well as the partial differential equation? The existence of restrictions means that the formula found in part b is *not* a general solution.

Exercises 10 through 12 develop the formula (7) for solutions of the initial-value problems for the wave equation.

10. Consider the initial-value problem for an infinite string initially motionless:

$$\frac{\partial^2 u}{\partial t^2} = c^2\frac{\partial^2 u}{\partial x^2}, \qquad \text{for} \quad -\infty < x < \infty, \quad t > 0,$$

   with

$$u(x,0) = U(x), \qquad \frac{\partial u}{\partial t}(x,0) = 0.$$

   a. Show that the functions f and g representing the right-moving waveform and the left-moving waveform can only differ by a constant. Without loss of generality, this means that we can assume $g = f$.
   b. Use the remaining initial condition to determine the function f in terms of the given function U.

11. Consider the initial-value problem for an infinite string initially at rest:

$$\frac{\partial^2 u}{\partial t^2} = c^2\frac{\partial^2 u}{\partial x^2}, \qquad \text{for} \quad -\infty < x < \infty, \quad t > 0,$$

   with

$$u(x,0) = 0, \qquad \frac{\partial u}{\partial t}(x,0) = V(x).$$

   a. Use the condition $u(x,0) = 0$ to determine the right-moving waveform f in terms of the left-moving waveform g.
   b. Use the result of part a to construct a formula for u in terms of a definite integral of g'.
   c. Use the condition $u_t(x,0) = V(x)$ to determine g' in terms of V, thereby obtaining a solution formula for u as a definite integral of V.

12. Let u_1 be the solution of

$$\frac{\partial^2 u}{\partial t^2} = c^2 \frac{\partial^2 u}{\partial x^2}, \qquad \text{for} \quad -\infty < x < \infty, \quad t > 0,$$

with

$$u(x, 0) = U(x), \qquad \frac{\partial u}{\partial t}(x, 0) = 0,$$

and let u_2 be the solution of

$$\frac{\partial^2 u}{\partial t^2} = c^2 \frac{\partial^2 u}{\partial x^2}, \qquad \text{for} \quad -\infty < x < \infty, \quad t > 0,$$

with

$$u(x, 0) = 0, \qquad \frac{\partial u}{\partial t}(x, 0) = V(x).$$

Show that

$$u = u_1 + u_2$$

is the solution of the problem

$$\frac{\partial^2 u}{\partial t^2} = c^2 \frac{\partial^2 u}{\partial x^2}, \qquad \text{for} \quad -\infty < x < \infty, \quad t > 0,$$

with

$$u(x, 0) = U(x), \qquad \frac{\partial u}{\partial t}(x, 0) = V(x).$$

This result is the **superposition principle** for linear partial differential equations and is analogous to the superposition principle for linear ordinary differential equations.

✦ 8.2 INSTANT EXERCISE SOLUTIONS

1. $$\frac{\partial^2 u}{\partial t^2} = \frac{\partial}{\partial t}\left(\frac{\partial u}{\partial t}\right) = \frac{\partial}{\partial w}\left(\frac{\partial u}{\partial t}\right)\frac{\partial w}{\partial t} + \frac{\partial}{\partial z}\left(\frac{\partial u}{\partial t}\right)\frac{\partial z}{\partial t} = c\,\frac{\partial}{\partial w}\left(\frac{\partial u}{\partial t}\right) - c\,\frac{\partial}{\partial z}\left(\frac{\partial u}{\partial t}\right)$$

$$= c\,\frac{\partial}{\partial w}\left(c\,\frac{\partial U}{\partial w} - c\,\frac{\partial U}{\partial z}\right) - c\,\frac{\partial}{\partial z}\left(c\,\frac{\partial U}{\partial w} - c\,\frac{\partial U}{\partial z}\right) = c^2\frac{\partial^2 U}{\partial w^2} - 2c^2\frac{\partial^2 U}{\partial w\,\partial z} + c^2\frac{\partial^2 U}{\partial z^2}.$$

2. Let $u(x, t) = \int_0^w V(s)\,ds$, with $w = x + ct$. Then

$$\frac{\partial u}{\partial t} = V(w)\frac{\partial w}{\partial t} = cV(x + ct).$$

Similarly,

$$\frac{\partial u}{\partial x} = V(x + ct).$$

Differentiating again, we have

$$\frac{\partial^2 u}{\partial t^2} = c^2 V(x+ct), \qquad \frac{\partial^2 u}{\partial x^2} = V(x+ct).$$

Hence, u satisfies the wave equation.

3. Let $w = 1 - s^2$. Then

$$g(x) = C + \int_0^x 6s(1-s^2)^2\,ds = C - 3\int_1^{1-x^2} w^2\,dw = C - (1+x^2)^3 + 1.$$

8.3 Vibration Modes of a Finite String

One of our goals in this chapter is to study the vibrations of a string of finite length L, such as a guitar string. Mathematically, this amounts to solving problems consisting of the wave equation on a finite domain together with auxiliary conditions. We consider the family of problems defined as follows:

Motion of a Plucked String

Given a function U for which $U(0) = U(L) = 0$, find the function u that satisfies the differential equation

$$u_{tt} = c^2 u_{xx}, \qquad \text{for} \quad 0 < x < L, \quad t > 0, \tag{1}$$

with boundary conditions

$$u(0,t) = 0, \qquad u(L,t) = 0 \tag{2}$$

and initial conditions

$$u(x,0) = U(x), \qquad u_t(x,0) = 0. \tag{3}$$

The conditions of Equations (2) are called **boundary conditions** because they prescribe properties of the solution at a specific point in the spatial domain of the problem. The boundary conditions in the plucked string problem specify no displacement at the ends of the string. Zero-displacement boundary conditions are called **Dirichlet conditions.**[8] Other kinds of boundary conditions are considered in the exercises. By choosing the initial velocity of the string to be zero, we are considering only the case of a *plucked* string. We could get a more general mathematics problem by allowing $u_t(x,0) = V(x)$. The case $U = 0$, $V \neq 0$ corresponds to a struck string, such as a string in a piano. Problems involving the wave equation on a finite domain are called **initial-boundary-value problems.**

[8]The correct pronunciation is "Di-ri-shlay."

Vibration Modes and Waveforms

Since we have a general solution for the wave equation, one idea for solving the plucked string problem is to find the functions f and g so that $f(x-ct)+g(x+ct)$ satisfy the auxiliary conditions. With boundary conditions at two points, we would have to allow for waves to reflect back and forth from $x=0$ to $x=L$. This does not turn out to be a satisfactory method. Instead, we have a method that is based on the observed phenomena of **standing waves;** these are waves whose shape is maintained in time and that vary only in magnitude. An example of a standing wave is the motion of a jump rope in the hands of a skilled jumper. Mathematically, a standing wave in one dimension corresponds to solutions u that can be written as

$$u = f(x)g(t). \tag{4}$$

The function f determines the **waveform.** Changes in the amplitude function g serve to increase or decrease the magnitude of the vibration without changing the shape. The solutions fg that satisfy the same boundary conditions as a given initial-boundary-value problem are the **vibration modes** for that combination of partial differential equation and boundary conditions.[9] Our long-term plan is to solve the plucked string problem by using linear combinations of vibration modes, but it is helpful first to focus on the vibration modes independent of the initial conditions. To find vibration modes, we look for standing wave solutions. How do we determine a standing wave solution when we don't yet know how to solve initial-boundary-value problems? The answer is that we look for functions f that give rise to a standing wave solution when used as the initial profile U.

MODEL PROBLEM 8.3

Find all functions f, and the corresponding functions g, for which the problem

$$u_{tt} = c^2 u_{xx}, \qquad \text{for} \quad 0 < x < L, \quad t > 0,$$

$$u(0,t) = 0, \qquad u(L,t) = 0,$$

$$u(x,0) = f(x), \qquad u_t(x,0) = 0,$$

has solutions given by

$$u = f(x)g(t).$$

The difference between the general plucked string problem and the vibration mode problem is subtle. The general plucked string problem uses a known, but arbitrary, initial profile. The vibration mode problem uses an unknown initial profile.

[9]Recall that the functions f and g in the general solution $f(x-ct)+g(x+ct)$ are called waveforms. The meaning of the term *waveform* here is slightly different, but similar. The waveform for a standing wave is the function that indicates the shape of the wave. Waves on an infinite domain consist of waveforms that move along the x axis, while standing waves consist of waveforms that change in amplitude without moving.

Ordinary Differential Equations for f and g

If $u = f(x)g(t)$, then

$$u_{xx} = f''(x)g(t), \qquad u_{tt} = f(x)g''(t),$$

where the prime symbol refers to the appropriate derivative. Thus, the wave equation becomes

$$f(x)g''(t) = c^2 f''(x)g(t),$$

which we can rearrange as

$$\frac{f''(x)}{f(x)} = \frac{1}{c^2}\frac{g''(t)}{g(t)}.$$

This last equation is interesting because the quantity on the left does not depend on time, while the quantity on the right does not depend on position. Since the two sides of the equation are equal for all x and t, it must be that neither quantity actually depends on either time or position. The individual functions f and f'' could still depend on x, and the functions g and g'' could still depend on t; however, the ratios f''/f and g''/g must be constant. We have deduced that both sides of the equation must have a common constant value, but we do not know the correct value(s). For the moment, we assume

$$\frac{f''(x)}{f(x)} = \frac{1}{c^2}\frac{g''(t)}{g(t)} = k,$$

where k is to be determined as part of the solution. Thus, we have two separate[10] linear ordinary differential equations for the functions f and g,

$$f''(x) = kf(x), \qquad g''(t) = kc^2\, g(t). \tag{5}$$

These equations need auxiliary conditions, which come from the auxiliary conditions of the original problem. Substituting $u = f(x)g(t)$ into the boundary conditions yields the conditions

$$f(0)g(t) = 0, \qquad f(L)g(t) = 0.$$

These equations are satisfied if $g \equiv 0$, but there can be no meaningful waveform if the amplitude is always 0. Given the requirement of nonzero amplitude, the boundary conditions for the partial differential equation reduce to boundary conditions on the function f:

$$f(0) = 0, \qquad f(L) = 0.$$

This simply says that the waveforms must have no displacement at the fixed ends, which is obvious anyway.

Similarly, the initial conditions become

$$f(x)g(0) = f(x), \qquad f(x)g'(0) = 0;$$

therefore

$$g(0) = 1, \qquad g'(0) = 0.$$

[10]The procedure we have discovered is often called **separation of variables** because it results in separate ordinary differential equations for the functions of x and t. There is a superficial resemblance to the method of separation of variables used to solve separable first-order ordinary differential equations.

We now have a **boundary value problem** for f,

$$f'' - kf = 0, \qquad f(0) = 0, \qquad f(L) = 0, \tag{6}$$

and an initial-value problem for g,

$$g'' - kc^2 g = 0, \qquad g(0) = 1, \qquad g'(0) = 0. \tag{7}$$

We have not yet studied boundary value problems for ordinary differential equations, so we do not know what to expect. The problem (6) is homogeneous, so we know that $f \equiv 0$ is a solution. However, our interest is in solutions that are not identically zero.

A Restriction on k

The substitution $f = e^{rx}$ into the differential equation for the waveforms (6) yields the characteristic equation

$$r^2 - k = 0.$$

Depending on the value of k, we could have two real roots, a single multiple root, or two nonreal roots for the characteristic equation. Consider first the case $k > 0$. Then there are real roots $r = \pm\sqrt{k}$. The general solution is then

$$f = Ae^{\sqrt{k}x} + Be^{-\sqrt{k}x}.$$

From $f(0) = 0$, we have $A + B = 0$; thus,

$$f = A\left(e^{\sqrt{k}x} - e^{-\sqrt{k}x}\right).$$

Alternatively, this formula can be written as

$$f = 2A \sinh(\sqrt{k}x).$$

The hyperbolic sine function is 0 only at $x = 0$; therefore, the boundary condition $f(L) = 0$ forces $A = 0$. Hence, there are *no* nonzero solutions with $k > 0$. Similarly, the case $k = 0$ yields no nonzero solutions (Exercise 3). Given the restriction $k < 0$, there is no loss of generality in replacing k by the substitution $k = -\lambda^2$.

The Waveforms

The substitution $k = -\lambda^2$ changes the boundary value problem for f to the form

$$f'' + \lambda^2 f = 0, \qquad f(0) = 0, \qquad f(L) = 0. \tag{8}$$

The boundary value problem of Equation (8) is the **eigenvalue problem** with fixed-end boundary conditions for the differential equation $f'' + \lambda^2 f = 0$. The terminology suggests a similarity with the eigenvalue problem of linear algebra. In both cases, we have a problem in which there is a zero solution for all values of a parameter λ, but the task is to find values of λ for which there are nonzero solutions.

The general solution of the differential equation $f'' + \lambda^2 f = 0$ is given immediately[11] by

$$f = A\cos\lambda x + B\sin\lambda x.$$

The boundary condition $f(0) = 0$ forces $A = 0$; then we have

$$f = B\sin\lambda x.$$

The boundary condition $f(L) = 0$ then yields the equation

$$B\sin\lambda L = 0.$$

Unlike in the previous cases, we are not forced to make $B = 0$. Nonzero solutions are possible provided

$$\sin\lambda L = 0,$$

and this occurs if

$$\lambda L = n\pi$$

for any integer n. In summary, we have eigenvalues

$$\lambda_n = \frac{n\pi}{L}, \qquad n = 1, 2, \ldots. \tag{9}$$

Each choice of n corresponds to a one-parameter family of waveforms

$$f_n(x) = B_n \sin\frac{n\pi x}{L}, \tag{10}$$

where we have introduced a subscript on the coefficient to indicate that each waveform has its own coefficient. The term *waveform* is based on the physical setting of the model; in mathematical terms, the solutions of an eigenvalue problem such as that in Equation (8) are called **eigenfunctions.** The first three waveforms (with $B_n = 1$ and $L = 1$) are illustrated in Figure 8.3.1.

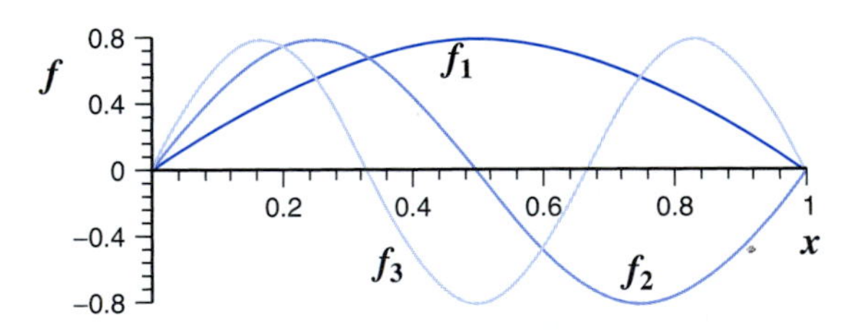

Figure 8.3.1
The waveforms $f_1(x) = \sin\pi x$, $f_2(x) = \sin 2\pi x$, $f_3(x) = \sin 3\pi x$.

✦ INSTANT EXERCISE 1

Explain why we do not need to consider negative values of n.

---

[11]Recall Section 3.1. Alternatively, the substitution $f = e^{rt}$ yields the equation $r^2 + \lambda^2 = 0$.

The Vibration Modes

To complete the determination of the vibration modes, we need to determine the amplitude function g_n corresponding to each of the waveforms f_n. By using the results for k, the solution for g_n is

$$g_n(t) = \cos \frac{n\pi ct}{L},$$

and the corresponding vibration modes for a plucked string with fixed ends are given by

$$u_n(x,t) = B_n \cos \frac{n\pi ct}{L} \sin \frac{n\pi x}{L}. \tag{11}$$

 **INSTANT EXERCISE 2**

Solve the problem (7) for the amplitude functions, with $k = -\lambda_n^2$, to obtain the formula for g_n.

It is interesting to note the relationship between the waveforms and the period of the vibration. The waveforms $f_1 = \sin(\pi x/L)$, $f_2 = \sin(2\pi x/L)$, $f_3 = \sin(3\pi x/L)$, depicted in Figure 8.3.1, have periods of $2L/c$, L/c, and $2L/(3c)$, respectively. As n increases, the waveform becomes more oscillatory and the period becomes shorter. In other words, the simpler waveforms correspond to the slower vibrations.

Guitars and Other Stringed Instruments

Of particular interest in the case of guitars and other musical instruments is the question of which factors determine the characteristics of the sound produced by the vibration. This question is addressed further in Case Study 8. For the moment, we consider only the pitch of the musical tone made by the vibration of the string. The pitch is related to the period of the $n = 1$ vibration mode, which is just the period of the amplitude function

$$g_1 = \cos \frac{\pi ct}{L};$$

i.e. $2L/c$

We define the **fundamental frequency** ν of the vibrating string to be the reciprocal of the first mode period. Thus,

$$\nu = \frac{\sqrt{T}}{2L\sqrt{\rho}}. \tag{12}$$

The fundamental frequency of the string is what determines the pitch. A vibration frequency of 440 cycles per second [hertz (Hz)] defines the pitch of A directly above middle C. An octave corresponds to a change in frequency of a factor of 2, so A below middle C has frequency 220. Half-steps are determined by a ratio equal to the 12th root of 2, which is approximately 1.05946. Thus, $A^{\#}$ below middle C has frequency 232; and middle C, three half-steps above A, has frequency 262. Because the factors in the formula for pitch are dependent on weather conditions, stringed instruments need to have a way of adjusting (tuning) each string to just the right frequency. The standard way of tuning a stringed instrument is to change the tension T. Only small changes in T are needed to tune a string that has been designed to give approximately the correct pitch.

The simplest way to design a stringed instrument is to keep the tension and density roughly fixed and use a different length for each pitch. This is the principle used to design a harp. It has the drawback of requiring a different string for each note.

For guitars and many other stringed instruments, notes of different pitch are obtained either by playing a different string *or* by changing the length of a given string by pressing the string down on a fingerboard. Each fret on a guitar corresponds to a half-step change in pitch, so the formula for the fundamental frequency indicates that the 12th fret on a guitar should correspond to a decrease in length of one-half. Examination of a guitar shows that this is indeed the case. However, changes in length on a fingerboard are not sufficient to give a large range to the instrument. To use a single guitar string to play a melodic line of three-octave range, it would be necessary to have 36 frets and decrease the string length by a factor of 8 on the fingerboard. This is not practical, so instruments such as a guitar or violin have several strings. The top and bottom string of a guitar correspond to a two-octave difference in pitch, but the strings are all the same basic length. The difference in pitch from the bottom string to the top is created primarily by changes in linear density.

Modes for Heat Flow and Changes in Boundary Conditions

There are heat flow modes that correspond to the vibration modes of the wave equation. Given boundary conditions that fix the temperature at both ends as zero, the eigenvalue problems are the same for the heat equation; however, the amplitude functions in heat flow modes are very different from those in vibration modes. Similarly, different boundary conditions lead to different eigenvalue problems and correspondingly different eigenfunctions. Several cases are considered in Exercises 5 through 9, and the subject is treated in detail in Section A.7.

8.3 Exercises

1. The strings of a standard guitar are 63 cm long. The A string is tuned so that its fundamental frequency is 440 vibrations per second. Determine the speed of sound c for this string.

2. Assume that guitar strings are strung with a specific amount of tension and are made of a material with a fixed mass density (mass per unit volume). If two such strings sound an octave apart, what is the ratio of their thicknesses?

3. Show that a plucked string has no waveforms for the case $k = 0$.

4. The mathematical model for a struck string, such as in a piano, is given by Equation (1) and Equation (2), along with the initial conditions
$$u(x, 0) = 0, \qquad u_t(x, 0) = V(x).$$
Determine the amplitude functions $g_n(t)$ for a struck string. Note that the waveforms $f_n(x)$ are the same as those for the plucked string of Model Problem 8.3.

5. The mathematical model for heat flow in a finite one-dimensional domain, with fixed temperature of 0 at both ends, is
$$u_t = ku_{xx}, \qquad \text{for} \quad 0 < x < L, \quad t > 0,$$
$$u(0, t) = 0, \qquad u(L, t) = 0,$$
$$u(x, 0) = U(x).$$
Determine the heat flow modes.

6. Consider the problem

$$u_t = ku_{xx}, \qquad \text{for} \quad 0 < x < L, \quad t > 0,$$
$$u_x(0,t) = 0, \quad u_x(L,t) = 0,$$
$$u(x,0) = U(x).$$

What do the boundary conditions imply about the ends of the rod? Determine the heat flow modes for this problem. Note that there is a mode with $k = 0$.

7. Consider the problem

$$u_t = ku_{xx}, \qquad \text{for} \quad 0 < x < L, \quad t > 0,$$
$$u(0,t) = 0, \qquad u_x(L,t) = 0,$$
$$u(x,0) = U(x).$$

What do the boundary conditions imply about the ends of the rod? Determine the heat flow modes for this problem.

T 8. Consider a vibrating circular membrane of radius a, such as the head of a snare drum. If we assume that the displacement of the head depends only on the radial coordinate r and the time t, the vibration is governed by the two-dimensional radially symmetric wave equation

$$u_{tt} = \frac{c^2}{r}\frac{\partial}{\partial r}(ru_r),$$

along with boundary conditions

$$u(a,t) = 0, \qquad |u(0,t)| < \infty,$$

and initial conditions

$$u(r,0) = U(r), \qquad u_t(r,0) = 0.$$

The boundary condition at $r = a$ indicates that the edge of the membrane is fixed in place, while the boundary condition at $r = 0$ says only that the displacement in the center of the membrane cannot be infinite.

a. Assume a solution of the form $u(r,t) = f(r)g(t)$. Derive the linear ordinary differential equation

$$rf'' + f' = krf,$$

where k is a constant to be determined as part of the solution. Also determine the appropriate boundary conditions for f.

b. Solve the differential equation for the waveforms, assuming $k = -\lambda^2$ for some $\lambda > 0$.[12]

c. Use the boundary condition at $r = 0$ to force one of the constants in the general solution to be zero. Then substitute the solution into the boundary condition at $r = a$ to obtain the algebraic equation

$$J_0(\lambda a) = 0.$$

d. The zeros of the function J_0 are tabulated in the libraries of several computer algebra systems and many reference books. In particular, the first two zeros are $z_1 \approx 2.405$ and

---

[12]The differential equation is a parametric Bessel equation. See Section 3.7.

$z_2 \approx 5.520$. Determine the waveforms corresponding to these two zeros. Plot these functions for the case where $a = 1$ and $f(0) = 1$.

9. Find the vibration modes for the spherically symmetric wave equation problem

$$u_{tt} = \frac{c^2}{r^2}\frac{\partial}{\partial r}(r^2 u_r),$$
$$u(a,t) = 0, \qquad |u(0,t)| < \infty,$$
$$u(r,0) = U(r), \qquad u_t(r,0) = 0.$$

[*Hint*: the differential equation for f can be simplified by the change of variables $w(r) = rf(r)$.]

✦ 8.3 INSTANT EXERCISE SOLUTIONS

1. The eigenfunctions are the same for n and $-n$ because $\sin(-nx) = -\sin(nx)$, and the extra factor of -1 is absorbed by the constant B_n.
2. The amplitude function problem (7) becomes

$$g_n'' + \frac{n^2\pi^2c^2}{L^2} g_n = 0, \qquad g_n(0) = 1, \qquad g_n'(0) = 0.$$

The differential equation has general solution

$$g_n(t) = c_1 \cos\frac{n\pi ct}{L} + c_2 \sin\frac{n\pi ct}{L}.$$

The initial condition on g_n' forces $c_2 = 0$, and the initial condition on g_n then forces $c_1 = 1$.

8.4 Motion of a Plucked String

In Section 8.3, we determined the vibration modes for a plucked string. The vibration modes allow us to solve a limited set of initial-value problems.

EXAMPLE 1

Consider the problem

$$u_{tt} = c^2 u_{xx}, \qquad \text{for} \quad 0 < x < L, \quad t > 0,$$
$$u(0,t) = 0, \qquad u(L,t) = 0,$$
$$u(x,0) = \sin\frac{\pi x}{L} - 2\sin\frac{3\pi x}{L}, \qquad u_t(x,0) = 0.$$

The functions

$$u_1(x,t) = \cos\frac{\pi ct}{L}\sin\frac{\pi x}{L}, \qquad u_3(x,t) = \cos\frac{3\pi ct}{L}\sin\frac{3\pi x}{L}$$

are solutions for the differential equation, boundary conditions, and initial velocity. Any function of the form

$$u = b_1 u_1 + b_3 u_3$$

also solves these equations, since they are all linear and homogeneous. Does one of these functions satisfy the initial-displacement condition? Substituting the solution family into the initial condition gives us

$$b_1 \sin \frac{\pi x}{L} + b_3 \sin \frac{3\pi x}{L} = \sin \frac{\pi x}{L} - 2 \sin \frac{3\pi x}{L}.$$

Obviously, $b_1 = 1$ and $b_3 = -2$. The initial-value problem has the solution

$$u(x,t) = \cos \frac{\pi c t}{L} \sin \frac{\pi x}{L} - 2 \cos \frac{3\pi c t}{L} \sin \frac{3\pi x}{L}.$$

We were able to solve the problem of Example 1 because the initial condition was chosen carefully to correspond to specific vibration modes. Other such problems can be solved in a similar manner.

✦ INSTANT EXERCISE 1

Solve the problem

$$u_{tt} = 4u_{xx}, \qquad \text{for} \quad 0 < x < 1, \quad t > 0,$$

$$u(0,t) = 0, \qquad u(1,t) = 0,$$

$$u(x,0) = \sin 2\pi x + 3 \sin 4\pi x, \qquad u_t(x,0) = 0.$$

We can now solve a few initial-value problems, but the method of Example 1 only works with a limited set of initial conditions. Is there a way to use the vibration modes to solve initial-value problems with arbitrary initial displacement?

MODEL PROBLEM 8.4

Solve the problem[13]

$$u_{tt} = u_{xx}, \qquad \text{for} \quad 0 < x < 1, \quad t > 0,$$

$$u(0,t) = 0, \qquad u(1,t) = 0,$$

$$u(x,0) = \phi(x) = \begin{cases} x & 0 < x < 0.5 \\ 1 - x & 0.5 < x < 1 \end{cases}, \qquad u_t(x,0) = 0.$$

---

[13] When we derived the wave equation in Section 8.1, we assumed that displacements were small relative to the length of the domain. Nevertheless, we are going to ignore that requirement when we choose initial conditions. Think of u and x as dimensionless lengths, in which case the assumption of small displacements means that the ratio of scales u_r/x_r, rather than the values of the variables u and x, is small. The calculations are messy enough without including an extra small constant factor in every term.

The initial displacement is shown in Figure 8.4.1.

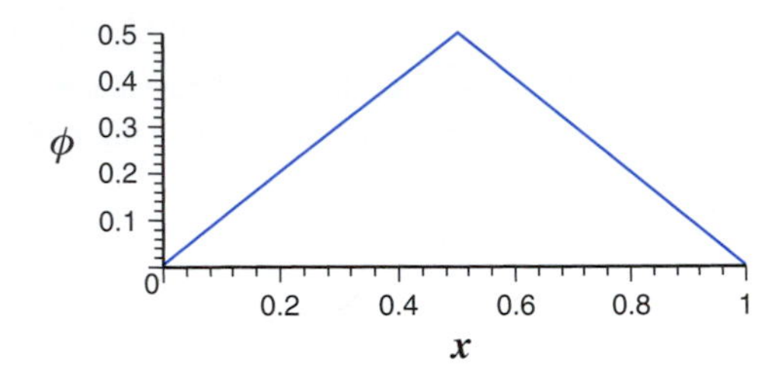

Figure 8.4.1
The initial displacement for Model Problem 8.4.

An Infinite-Parameter Family of Solutions

The initial displacement for Model Problem 8.4 is not a sine function or a finite combination of sine functions. Nevertheless, we have (from Sec. 8.3) an infinite collection of functions (with $L = 1$ and $c = 1$), given by the formula

$$u_n(x, t) = \cos n\pi t \sin n\pi x, \qquad n = 1, 2, 3, \ldots,$$

each of which solves the partial differential equation, both boundary conditions, and the initial condition for the velocity.

✦ INSTANT EXERCISE 2

Show that each function u_n satisfies the homogeneous equations

$$u_{tt} = u_{xx}, \qquad \text{for} \quad 0 < x < 1, \quad t > 0,$$

$$u(0, t) = 0, \qquad u(1, t) = 0, \qquad u_t(x, 0) = 0.$$

The partial differential equation, boundary conditions, and initial-velocity condition are all linear and homogeneous; therefore, finite linear combinations of the solutions, such as that which we used in Example 1, are also solutions of the homogeneous part of the problem. This suggests that the infinite-parameter family

$$u = \sum_{n=1}^{\infty} b_n \cos n\pi t \sin n\pi x \tag{1}$$

is a solution of the homogeneous part as well. Infinite sums do not necessarily have all the properties of finite sums, so some care is required in drawing conclusions about their behavior. All such technical issues are deferred to Section 8.5. The focus of this section is the question of whether the family defined by Equation (1) includes a function that represents the solution of Model Problem 8.4 in some sense. We need to consider only the nonhomogeneous initial condition. Model Problem 8.4 quickly reduces to the problem of finding the "correct" values of the coefficients $b_1, b_2, \ldots$ such that

$$\sum_{n=1}^{\infty} b_n \sin n\pi x = \phi(x) = \begin{cases} x & 0 < x < 0.5 \\ 1 - x & 0.5 < x < 1 \end{cases}. \tag{2}$$

We can think of Equation (2) as representing infinitely many equations because it must hold for all x values from 0 to 1. There are also infinitely many unknowns. It is not immediately clear whether the equation has unique solutions for all the coefficients, nor is it clear whether the formula obtained from such coefficients actually represents the function ϕ. The first of these issues is settled in this section, while the second is deferred to Section 8.5.

Evaluating the Coefficients

In practice, one uses a single formula to evaluate all the coefficients together. However, part of our interest lies in understanding where the formula comes from. For that reason, we begin by considering the problem of finding the first coefficient independent of the others.

Finding b_1 Suppose $n > 1$, and consider the integral $\int_0^1 \sin \pi x \sin n\pi x\, dx$. Using integration by parts or a table of integrals, we have the integration formula

$$\int \sin ax \sin bx\, dx = \frac{a \cos ax \sin bx - b \sin ax \cos bx}{b^2 - a^2}, \qquad \text{provided } a^2 \neq b^2. \tag{3}$$

With $a = \pi$ and $b = n\pi$, we get

$$\int \sin \pi x \sin n\pi x\, dx = \frac{\cos \pi x \sin n\pi x - n \sin \pi x \cos n\pi x}{(n^2 - 1)\pi}.$$

Thus,

$$\int_0^1 \sin \pi x \sin n\pi x\, dx = 0, \qquad n \neq 1. \tag{4}$$

This property allows us to determine the coefficient b_1 by a very clever method.

We first multiply both sides of Equation (2) by $\sin \pi x$ and integrate over the interval $0 < x < 1$:

$$\int_0^1 \sum_{n=1}^{\infty} b_n \sin \pi x \sin n\pi x\, dx = \int_0^1 \phi(x) \sin \pi x\, dx.$$

Assuming that the order of integration and summation can be reversed, we have

$$\sum_{n=1}^{\infty} b_n \int_0^1 \sin \pi x \sin n\pi x\, dx = \int_0^1 \phi(x) \sin \pi x\, dx.$$

So far, nothing seems to have been accomplished. However, the integrals inside the sum are the same integrals that appear in Equation (4). All the integrals are 0, *with the exception of the $n = 1$ integral*. The infinite sum reduces to a single term, and the equation becomes

$$b_1 \int_0^1 \sin^2 \pi x\, dx = \int_0^1 \phi(x) \sin \pi x\, dx.$$

The remaining integrals are easy to calculate, the one on the left by using the appropriate trigonometric identity and the one on the right by integrating by parts. The final result is

$$b_1 = \frac{4}{\pi^2}.$$

✦ INSTANT EXERCISE 3

Fill in the details in the calculation of b_1.

Let us be clear on what has been accomplished. We have not solved the model problem, nor have we demonstrated that it has a solution in the form of an infinite series. What we have demonstrated is this:

If the model problem has a solution of the form (1), then $b_1 = 4/\pi^2$.

Valuable insight is gained by comparing the actual initial condition $\phi(x)$ with the one-term approximation $(4/\pi^2)\sin \pi x$. These functions are illustrated in Figure 8.4.2. The one-term approximation is remarkably good, considering that the sine function does not have the right shape.

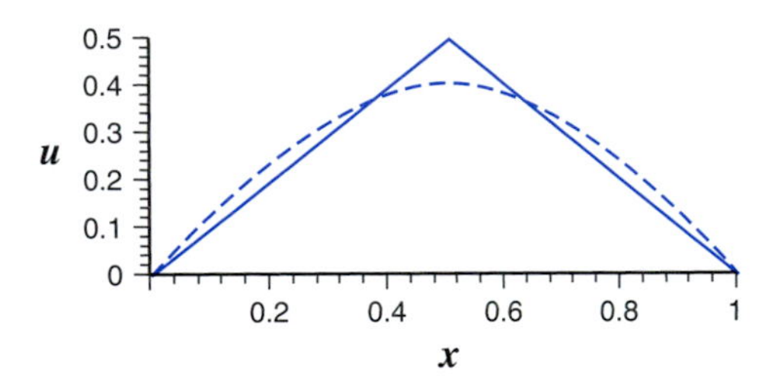

Figure 8.4.2
Graph of $b_1 \sin \pi x$ and $\phi(x)$ for Model Problem 8.4.

Finding the Other Coefficients The procedure for finding the other coefficients is similar to that of finding b_1. Using $a = 2\pi$ instead of $a = \pi$ in Equation (3) leads to the conclusion

$$\int_0^1 \sin 2\pi x \sin n\pi x \, dx = 0, \qquad n \neq 2.$$

Thus, we can find b_2 by multiplying Equation (2) by $\sin 2\pi x$ and integrating from 0 to 1. This ultimately yields the result $b_2 = 0$.

In similar fashion, we can compute all the other coefficients in the series. Each even-numbered coefficient is zero, and the odd-numbered coefficients are

$$b_1 = \frac{4}{\pi^2}, \qquad b_3 = -\frac{4}{9\pi^2}, \qquad b_5 = \frac{4}{25\pi^2}, \qquad b_7 = -\frac{4}{49\pi^2}, \quad \ldots. \tag{5}$$

Approximating the Solution

We now have a possible solution in the form of the infinite series (2), with coefficients given by Equation (5). Even if it is correct, we cannot actually compute the solution with infinitely many terms. Before we address the question of convergence,[14] it is interesting to examine finite-term approximations to the series. Specifically, we define the ***N*th partial sum** of the series to be the finite sum

$$S_N = \sum_{n=1}^{N} b_n u_n(x).$$

Thus, our one-term approximation is both S_1 and S_2. The next approximation is

$$S_4 = S_3 = \frac{4}{\pi^2}\left(\sin \pi x - \frac{1}{9}\sin 3\pi x\right).$$

Figure 8.4.3 shows that S_4 is a better approximation to the initial condition than is S_2.[15]

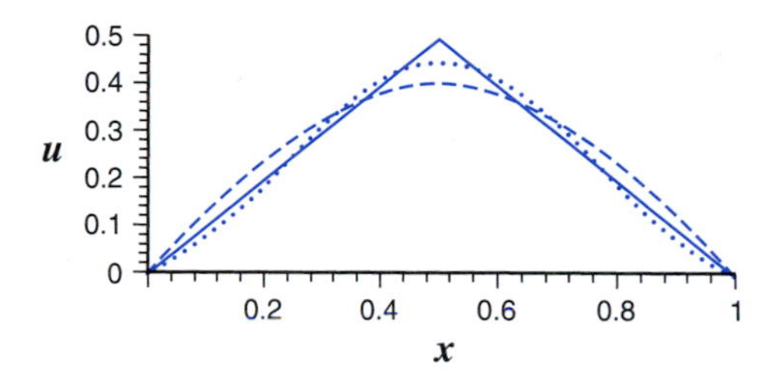

Figure 8.4.3
The partial sums S_4 and S_2 and the initial condition ϕ for Model Problem 8.4.

Adding more terms makes the approximation even better. Figure 8.4.4 shows that S_{10} is an excellent approximation. Without any formal mathematical argument, it certainly appears that the procedure works for this problem. Adding still more terms to the partial sum gives a better and better approximation.

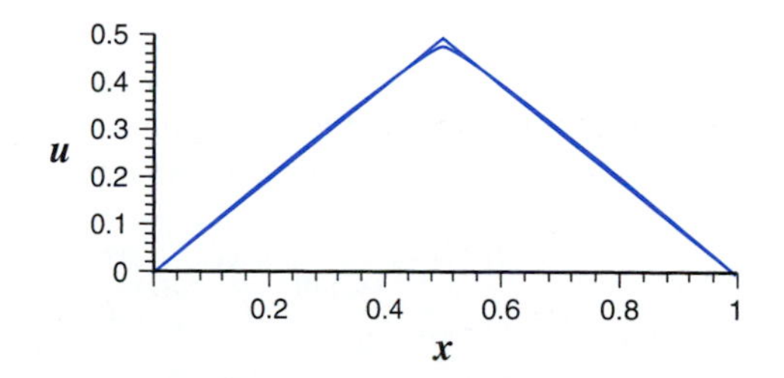

Figure 8.4.4
The partial sum S_{10} and the initial condition ϕ for Model Problem 8.4.

---

[14] Section 8.5.

[15] It is reasonable to expect higher-order approximations to be better than lower-order ones, so the first two approximations are best thought of as S_2 and S_4 rather than S_1 and S_3.

Visualization

We now have the solution of the model problem. We can visualize this solution by plotting solution profiles $u(x, t_i)$ with evenly spaced times t_i. The solution is periodic, with period 2. (The individual terms have periods of 2, $2\pi/3$, $2\pi/5$, ...) Thus, a good picture of the wave is obtained by using several times running from 0 to 1. Figure 8.4.5 shows the wave motion, using S_{10} to approximate the solution.

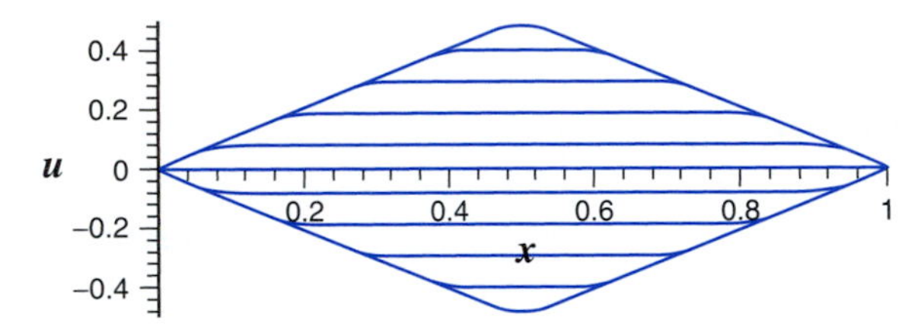

Figure 8.4.5
Wave profiles for Model Problem 8.4 at times 0, 0.1, 0.2, ..., 1.0 (top to bottom).

Fourier Sine Series

Model Problem 8.4 indicates the general plan for solving initial-value problems for the wave equation on a finite domain with $u = 0$ at the boundaries. The procedure does not depend on the initial conditions. Eventually, the problem reduces to one of finding a set of coefficients b_n so that

$$\sum_{n=1}^{\infty} b_n \sin \frac{n\pi x}{L} = \phi(x), \qquad 0 < x < L \tag{6}$$

for some function ϕ. The series in Equation (6) is called the **Fourier sine series,** and the coefficients are the **Fourier sine coefficients.** The key property that allows the coefficients b_n to be determined uniquely for nearly any function f is the integral formula

$$\int_0^L \sin \frac{m\pi x}{L} \sin \frac{n\pi x}{L}\, dx = 0, \qquad n \neq m. \tag{7}$$

It is this property that allows the series equation (6) to be decoupled into separate equations for each of the coefficients. Choose any positive integer m. Multiplying the series equation by $\sin(m\pi x/L)$ and integrating over the interval 0 to L yields

$$\int_0^L \sum_{n=1}^{\infty} b_n \sin \frac{m\pi x}{L} \sin \frac{n\pi x}{L}\, dx = \int_0^L f(x) \sin \frac{m\pi x}{L}\, dx.$$

Assuming that the integral can be distributed over the sum on the left side, the left side has infinitely many integrals, each with the same value of m and different values of n. The integral is 0 for those terms with $n \neq m$. The only integral that is not zero is the one in which $n = m$. Thus, the equation reduces to

$$b_m \int_0^L \sin^2 \frac{m\pi x}{L}\, dx = \int_0^L f(x) \sin \frac{m\pi x}{L}\, dx.$$

As before, we may simplify the integral on the left by using the trigonometric identity $\sin^2 x = (1 - \cos 2x)/2$, with the result

$$\frac{b_m L}{2} = \int_0^L f(x) \sin \frac{m\pi x}{L}\, dx.$$

The formula we have obtained for b_m works no matter which value we picked for m. It therefore also works in the general case; hence, we have the Fourier sine coefficient formula

$$b_n = \frac{2}{L} \int_0^L f(x) \sin \frac{n\pi x}{L}\, dx. \tag{8}$$

This coefficient formula allows the Fourier sine coefficients to be computed simultaneously, provided it is done carefully.

EXAMPLE 2

Suppose we want to use the general coefficient formula (8) to compute the Fourier sine coefficients for Model Problem 8.4. First we divide the integration interval into two portions:

$$b_n = 2\left[\int_0^{1/2} x \sin n\pi x\, dx + \int_{1/2}^1 (1 - x) \sin n\pi x\, dx\right].$$

Integrating by parts yields

$$b_n = \frac{2}{n\pi}\left(\int_0^{1/2} \cos n\pi x\, dx - \int_{1/2}^1 \cos n\pi x\, dx\right).$$

After computing the integrals, we have

$$b_n = \frac{4}{n^2\pi^2} \sin \frac{n\pi}{2}.$$

There are now three cases. If n is even, then $\sin(n\pi/2) = 0$. If n is $1, 5, 9$, and so on, then $\sin(n\pi/2) = 1$; and if n is 3, 7, and so on, then $\sin(n\pi/2) = -1$. We can combine the results for odd n into a single formula by noting that we get the positive sign if $(n - 1)/2$ is even and the negative sign if $(n - 1)/2$ is odd. In summary,

$$b_n = (-1)^{(n-1)/2} \frac{4}{n^2\pi^2} \quad \text{for } n \text{ odd}; \qquad b_n = 0 \quad \text{for } n \text{ even}.$$

✦ INSTANT EXERCISE 4

Supply the missing details in the calculation of the formula

$$b_n = \frac{2}{n\pi}\left(\int_0^{1/2} \cos n\pi x\, dx - \int_{1/2}^1 \cos n\pi x\, dx\right).$$

Even and Odd Functions It was not a coincidence that the even coefficients evaluated to 0; rather, it was a direct consequence of the symmetry of the function ϕ.

Theorem 8.4.1 Suppose f is defined on the interval $(0, L)$. If f is even about the midpoint $L/2$, then the even Fourier sine coefficients for f are zero; if f is odd about $L/2$, then the odd Fourier sine coefficients are zero.

This theorem is proved in Exercises 3 and 4.

8.4 Exercises

Recurrence relations are used in calculus to simplify the task of repeated integration. Similarly, we can construct recurrence relations to make it easy to find Fourier coefficients for polynomial functions. Let p be a polynomial, let n be a positive integer, let L be a positive number, and define a_n, b_n, a'_n, and b'_n by

$$a_n = \frac{2}{L}\int_0^L p(x)\cos\frac{n\pi x}{L}\,dx, \qquad a'_n = \frac{2}{L}\int_0^L p'(x)\cos\frac{n\pi x}{L}\,dx,$$

$$b_n = \frac{2}{L}\int_0^L p(x)\sin\frac{n\pi x}{L}\,dx, \qquad b'_n = \frac{2}{L}\int_0^L p'(x)\sin\frac{n\pi x}{L}\,dx.$$

Relationships between these quantities are explored in Exercises 1 and 2. These relationships can be used to advantage in many of the subsequent exercises.

1. Derive the formula

$$b_n = \frac{2}{n\pi}[p(0) - (-1)^n p(L)] + \frac{L}{n\pi}a'_n.$$

2. Derive a formula that gives a_n in terms of b'_n.

3. Show that the even Fourier sine coefficients of $f(x)$ are zero whenever $f(L-x) = f(x)$. Show that the odd coefficients for this case can be computed from

$$b_n = \frac{4}{L}\int_0^{L/2} f(x)\sin\frac{n\pi x}{L}\,dx.$$

Functions for which $f(L-x) = f(x)$ are **even** about $L/2$.

4. Show that the odd Fourier sine coefficients of $f(x)$ are zero whenever $f(L-x) = -f(x)$. Show that the even coefficients for this case can be computed from

$$b_n = \frac{4}{L}\int_0^{L/2} f(x)\sin\frac{n\pi x}{L}\,dx.$$

T 5. *a.* Solve

$$u_{tt} = u_{xx}, \qquad 0 < x < 1, \quad t > 0,$$

$$u(0,t) = 0, \qquad u(1,t) = 0,$$

$$u(x,0) = x(1-x), \qquad u_t(x,0) = 0.$$

b. Use a computer algebra system to plot the solution at times 0, 0.2, 0.4, 0.6, 0.8, and 1.

T 6. *a.* Solve

$$u_{tt} = u_{xx}, \qquad 0 < x < 1, \quad t > 0,$$
$$u(0, t) = 0, \qquad u(1, t) = 0,$$
$$u(x, 0) = 1 - \cos 2\pi x, \qquad u_t(x, 0) = 0.$$

b. Use a computer algebra system to plot the solution at times 0, 0.2, 0.4, 0.6, 0.8, and 1.

T 7. *a.* Solve

$$u_{tt} = 4u_{xx}, \qquad 0 < x < 1, \quad t > 0,$$
$$u(0, t) = 0, \qquad u(1, t) = 0,$$
$$u(x, 0) = 2x^3 - 3x^2 + x, \qquad u_t(x, 0) = 0.$$

b. Use a computer algebra system to plot the solution at times 0, 0.05, 0.1, 0.15, 0.2, and 0.25.

T 8. *a.* Solve

$$u_{tt} = 4u_{xx}, \qquad 0 < x < 1, \quad t > 0,$$
$$u(0, t) = 0, \qquad u(1, t) = 0,$$
$$u(x, 0) = 2x - 1 + \cos 3\pi x, \qquad u_t(x, 0) = 0.$$

b. Use a computer algebra system to plot the solution at times 0, 0.05, 0.1, 0.15, 0.2, and 0.25.

T 9. *a.* Solve

$$u_{tt} = u_{xx}, \qquad 0 < x < 1, \quad t > 0,$$
$$u(0, t) = 0, \qquad u(1, t) = 0,$$
$$u(x, 0) = \begin{cases} \dfrac{5x}{9} & 0 < x < 0.9 \\ 5 - 5x & 0.9 < x < 1 \end{cases}, \qquad u_t(x, 0) = 0.$$

b. Use a computer algebra system to plot the solution at times 0, 0.2, 0.4, 0.6, 0.8, and 1.
c. Compare the result with that of Model Problem 8.4. Note that the average initial displacement of both problems is the same.

T 10. *a.* Solve

$$u_{tt} = u_{xx}, \qquad 0 < x < 1, \quad t > 0,$$
$$u(0, t) = 0, \qquad u(1, t) = 0,$$
$$u(x, 0) = \begin{cases} x^2 & 0 < x < 0.5 \\ (1 - x)^2 & 0.5 < x < 1 \end{cases}, \qquad u_t(x, 0) = 0.$$

b. Use a computer algebra system to plot the solution at times 0, 0.2, 0.4, 0.6, 0.8, and 1.

11. If a string is struck rather than plucked, the initial displacement is zero and the initial velocity is nonzero. As with a plucked string, the solution can be found in the form

$$u = \sum_{n=1}^{\infty} b_n g_n(t) \sin \frac{n\pi x}{L}.$$

Determine the functions g_n for which this form satisfies the wave equation and the initial condition

$$u(x, 0) = 0.$$

Then determine the Fourier series coefficients b_n so that

$$u_t(x, 0) = \psi(x).$$

(See Exercise 4 of Section 8.3.)

T 12. *a.* Use the solution of Exercise 11 to solve the problem

$$u_{tt} = u_{xx}, \qquad 0 < x < 1, \quad t > 0,$$
$$u(0, t) = 0, \qquad u(1, t) = 0,$$
$$u(x, 0) = 0, \qquad u_t(x, 0) = x(1 - x).$$

b. Use a computer algebra system to plot the solution at times 0, 0.1, 0.2, 0.3, 0.4, and 0.5.

T 13. *a.* Use the solution of Exercise 11 to solve the problem

$$u_{tt} = u_{xx}, \qquad 0 < x < 1, \quad t > 0,$$
$$u(0, t) = 0, \qquad u(1, t) = 0,$$
$$u(x, 0) = 0, \qquad u_t(x, 0) = x^2(1 - x).$$

b. Use a computer algebra system to plot the solution at times 0, $\frac{1}{6}$, $\frac{1}{3}$, $\frac{1}{2}$, $\frac{2}{3}$, and $\frac{5}{6}$.

14. The flow of heat through a rod is governed by the partial differential equation

$$u_t = k u_{xx},$$

where k is the thermal diffusivity of the rod. Determine the solution formula for the initial-value problem

$$u_t = k u_{xx}, \qquad 0 < x < L, \quad t > 0,$$
$$u(0, t) = 0, \qquad u(L, t) = 0,$$
$$u(x, 0) = \phi(x).$$

(See Exercise 5 of Section 8.3.)

✦ 8.4 INSTANT EXERCISE SOLUTIONS

1. The initial conditions have terms corresponding to the second and fourth modes, so we begin with the two-parameter family

$$u(x, t) = b_2 \cos 4\pi t \sin 2\pi x + b_4 \cos 8\pi t \sin 4\pi x.$$

All functions in this family solve the differential equation, boundary conditions, and initial velocity. Substituting the solution family into the nonhomogeneous initial condition gives us

$$u(x, 0) = b_2 \sin 2\pi x + b_4 \sin 4\pi x.$$

Obviously, $b_2 = 1$ and $b_4 = 3$. The initial-value problem has the solution

$$u(x, t) = \cos 4\pi t \sin 2\pi x + 3 \cos 8\pi t \sin 4\pi x.$$

2. Each u_n clearly solves each of the algebraic equations $u(0, t) = 0$, $u(1, t) = 0$, and $\partial u/\partial t(x, 0) = 0$ by a simple substitution. Also,

$$\frac{\partial^2 u_n}{\partial t^2} = \frac{\partial^2 u_n}{\partial x^2} = -n^2\pi^2 u_n.$$

3. Using the definition of ϕ, we find

$$\int_0^1 \phi(x) \sin \pi x \, dx = \int_0^{0.5} x \sin \pi x \, dx + \int_{0.5}^1 (1 - x) \sin \pi x \, dx.$$

The remaining calculation is simplified by a substitution in the second integral. The idea is that both the function ϕ and the sine functions are symmetric about the point $x = 0.5$; hence the two integrals should be the same. Formally, the substitution $z = 1 - x$, applied to the second integral, yields

$$\int_{0.5}^1 (1 - x) \sin \pi x \, dx = \int_0^{0.5} z \sin(\pi - \pi z) \, dz.$$

Now, application of the trigonometric identity $\sin(\pi - \theta) = \sin\theta$ simplifies the integral to $\int_0^{0.5} z \sin \pi z \, dz$, which is equivalent to the first integral. Thus,

$$\int_0^1 \phi(x) \sin \pi x \, dx = 2 \int_0^{0.5} x \sin \pi x \, dx.$$

Integration by parts yields

$$2 \int_0^{0.5} x \sin \pi x \, dx = -2\frac{x}{\pi} \cos \pi x \Big|_0^{0.5} + \frac{1}{\pi} \int_0^{0.5} 2 \cos \pi x \, dx = 2\left(-\frac{x}{\pi} \cos \pi x + \frac{1}{\pi^2} \sin \pi x\right)\Big|_0^{0.5} = \frac{2}{\pi^2}.$$

4.

$$b_n = 2\left[\int_0^{1/2} x \sin n\pi x \, dx + \int_{1/2}^1 (1 - x) \sin n\pi x \, dx\right].$$

For the first integral, integrating by parts yields

$$2\left(-\frac{x}{n\pi} \cos n\pi x\Big|_0^{1/2} + \frac{1}{n\pi} \int_0^{1/2} \cos n\pi x \, dx\right) = -\frac{1}{n\pi} \cos \frac{n\pi}{2} + \frac{2}{n\pi} \int_0^{1/2} \cos n\pi x \, dx;$$

similarly, the second integral is

$$2\left(-\frac{1-x}{n\pi}\cos n\pi x\Big|_{1/2}^{1} - \frac{1}{n\pi}\int_{1/2}^{1}\cos n\pi x\,dx\right) = +\frac{1}{n\pi}\cos\frac{n\pi}{2} - \frac{2}{n\pi}\int_{1/2}^{1}\cos n\pi x\,dx.$$

Adding these two results yields the formula

$$b_n = \frac{2}{n\pi}\left(\int_0^{1/2}\cos n\pi x\,dx - \int_{1/2}^{1}\cos n\pi x\,dx\right).$$

8.5 Fourier Series

The work of Section 8.4 led to a "solution" of Model Problem 8.4. What we actually did was to look for a solution in the form of a Fourier sine series and compute the Fourier sine coefficients. This does not allow us to conclude that we have solved the problem. We assumed the existence of a Fourier series solution and then computed it, but the whole result is based on the untested assumption that there *is* a solution in the form of a Fourier series. We have no way, so far, of knowing whether the assumption of a Fourier series solution holds. Ultimately, this question is tied into the question of whether a Fourier series is actually equal to the function that was used to generate the coefficients. For the moment, we restrict our consideration to functions that are $2L$-periodic;[16] later we consider a more general class of functions.

MODEL PROBLEM 8.5

Explore the question of whether there is a set of functions $S = \{f_1, f_2, \ldots\}$ having the property that there is a unique set of coefficients a_n for each continuous $2L$-periodic function f such that $f = f_s$, where

$$f_s(x) = \sum_n^{\infty} a_n f_n(x). \tag{1}$$

The symbol f_s is used here merely as a label for the infinite series. Whether the sum actually means anything depends on the convergence of the series. Also, the values of n over which the sum in formula (1) is taken are omitted from the formula because it is not always convenient to start the sum at $n = 1$. Whatever the starting index, the sum must have infinitely many terms.[17]

The Inner Product and Orthogonal Functions

The discussion of vector spaces in Section 6.2 indicated that it is sometimes convenient to consider a set of functions, with the operations of addition and composition, as a vector space. It is surprising how successful this idea is. In the case of eigenfunctions for partial differential equations, we

[16] For our purposes, a function f is $2L$-*periodic* if $f(x + 2L) = f(x)$ for all x. For example, the function $\sin 2x$ is 2π-periodic. The fundamental period of this function is π rather than 2π; we consider the function to be both π-periodic and 2π-periodic.

[17] Of course there could be only finitely many nonzero terms for a specific f.

can go beyond the algebraic vector space properties discussed in Section 6.2 and add geometric properties to the analogy. The most important geometric property of the geometric vector spaces is that of *orthogonality*. For geometric vectors, the term *orthogonal* is synonymous with the term *perpendicular*; generalizations to nongeometric vectors require an algebraic notion of orthogonality. In the case of the space of three-dimensional geometric vectors, the **inner product** of two vectors $\mathbf{u} = u_1\mathbf{i} + u_2\mathbf{j} + u_3\mathbf{k}$ and $\mathbf{v} = v_1\mathbf{i} + v_2\mathbf{j} + v_3\mathbf{k}$ is defined by

$$\langle \mathbf{u}, \mathbf{v} \rangle = \mathbf{u} \cdot \mathbf{v} = u_1 v_1 + u_2 v_2 + u_3 v_3.$$

One can then show that two vectors are perpendicular if their inner product is 0.

It is also possible to define meaningful inner products[18] for spaces composed of functions defined on a closed interval; one such definition is given by

$$\langle f, g \rangle = \int_{-L}^{L} f(x) g(x)\, dx, \tag{2}$$

where f and g are two continuous $2L$-periodic functions. We can then say that two such functions are **orthogonal** on the interval $[-L, L]$ if their inner product is 0.

Orthogonal Expansions

In geometry, it is sometimes useful to be able to write vectors in three-dimensional space, using coordinate systems other than the Cartesian xyz system. Any system can be used to give a unique representation of a geometric vector, provided that the system is based on a set of three mutually perpendicular coordinate vectors. Similarly, the computation of unique coefficients in a series of the form (1) is possible only if the set of functions f_n is **mutually orthogonal,** meaning

$$\int_{-L}^{L} f_m(x) f_n(x)\, dx = 0, \qquad m \neq n. \tag{3}$$

As in the specific case of the Fourier sine series of Section 8.4, we can derive a formula for the coefficients a_n in Equation (1) by making use of orthogonality. Multiplying by $f_m(x)$ and integrating from $-L$ to L yields

$$\int_{-L}^{L} f(x) f_m(x)\, dx = \int_{-L}^{L} \sum_{n}^{\infty} a_n f_m(x) f_n(x)\, dx = \sum_{n}^{\infty} a_n \int_{-L}^{L} f_m(x) f_n(x)\, dx = a_m \int_{-L}^{L} f_m^2\, dx.$$

Thus, for *any* series of the form (1), where the functions in S are mutually orthogonal, the coefficients are given uniquely as

$$a_n = \frac{\int_{-L}^{L} f(x) f_n(x)\, dx}{\int_{-L}^{L} f_n^2(x)\, dx}, \tag{4}$$

provided that the integrals converge. This is guaranteed whenever the functions f and f_n are continuous and have at most finitely many maxima and minima on the interval $[-L, L]$, a criterion that is met for almost any function the student can think of. The series given by

$$f_s(x) = \sum_{n}^{\infty} a_n f_n(x)$$

[18] A "meaningful" inner product is one that satisfies certain required properties. See any elementary linear algebra book for details.

is called the **orthogonal expansion** of f with respect to the mutually orthogonal set $\{f_1, f_2, \ldots\}$.

Orthogonality of the functions f_n is sufficient to guarantee a unique orthogonal expansion for f, but not to guarantee that the Fourier series is actually equal to the function.

EXAMPLE 1

The functions $\sin n\pi x$ with $n = 1, 2, \ldots$ are mutually orthogonal on the interval $-\pi < x < \pi$. From the coefficient formula (4), we have

$$a_n = \frac{1}{\pi}\int_{-\pi}^{\pi} f(x)\sin nx\,dx.$$

Now consider $f(x) \equiv 1$ as a simple example. The Fourier sine coefficients all turn out to be zero. The orthogonality of the sine functions gives a unique orthogonal expansion, $f_s \equiv 0$, but the expansion is not equal to the function f.

✦ INSTANT EXERCISE 1

Fill in the details in Example 1; that is, compute $\int_{-\pi}^{\pi} \sin^2 nx\,dx$ and a_n.

A Complete Set of Orthogonal Functions

Example 1 shows that the function $f \equiv 1$ is orthogonal to all the functions in the set $S = \{\sin x, \sin 2x, \sin 3x, \ldots\}$. By definition, a function f that is orthogonal to the set S has all its Fourier coefficients equal to 0. Suppose we want a mutually orthogonal set with the property that any continuous $2L$-periodic function f is equal to its orthogonal expansion. We might start with S, but more functions are needed. Since $f \equiv 1$ is mutually orthogonal to S, we can add it to the set. The enlarged set is still mutually orthogonal, and the set of functions for which $f_s = f$ is larger. It turns out that many more functions can be added to the set S. In particular, the set

$$S = \{1, \cos t, \sin t, \cos 2t, \sin 2t, \cos 3t, \sin 3t, \ldots\}$$

is mutually orthogonal. Thus, every continuous 2π-periodic function having no more than a finite number of extrema in each period has a unique orthogonal expansion with respect to this set. The expansion can be written as

$$f_s(x) = \bar{f} + \sum_{n=1}^{\infty}(a_n \cos nx + b_n \sin nx), \tag{5}$$

where

$$a_n = \frac{1}{\pi}\int_{-\pi}^{\pi} f(x)\cos nx\,dx, \qquad b_n = \frac{1}{\pi}\int_{-\pi}^{\pi} f(x)\sin nx\,dx, \qquad \bar{f} = \frac{1}{2\pi}\int_{-\pi}^{\pi} f(x)\,dx. \tag{6}$$

We denote the coefficient of the function 1 by $\bar{f}$ rather than a_0 because of the observation that this coefficient represents the average value of the function f.[19] The constant term is therefore

[19] Most authors use the notation $a_0/2$ instead of our $\bar{f}$. The advantage of this notation is that then the formula for a_n is also correct for a_0. However, a_0 must still be calculated separately from the other a_n. An advantage of our notation is that for many functions it is easy to determine the average value $\bar{f}$ geometrically.

missing for any function with average value zero. Similarly, the sine terms are all zero if f is *even*, and the cosine terms are zero when f is *odd*, where f is **even** if $f(-x) = f(x)$ for all x and f is **odd** if $f(-x) = -f(x)$ for all x.

EXAMPLE 2

Consider the function (see Figure 8.5.1) defined by

$$f(x) = \begin{cases} \dfrac{\pi}{2} - |x| & |x| < \dfrac{\pi}{2} \\ 0 & \dfrac{\pi}{2} < |x| < \pi \\ f(|x| - 2\pi) & \text{otherwise} \end{cases} .$$

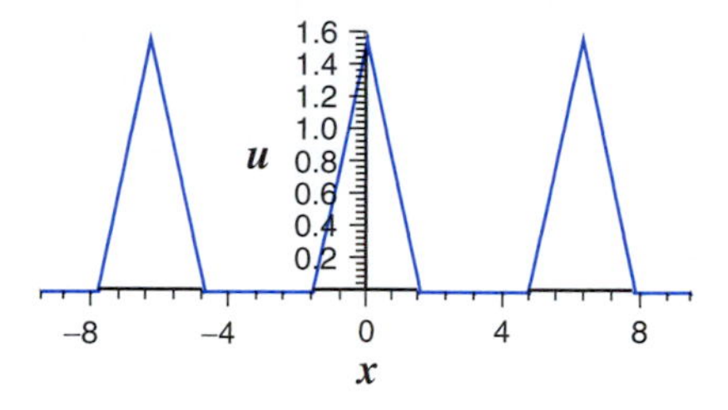

Figure 8.5.1
The function f of Example 2.

Note first that the function is even; hence, $b_n = 0$ for all n. The constant $\bar{f}$ can be determined geometrically by noting that the triangle in the center has height $\pi/2$ and average height $\pi/4$; since the triangle occupies one-half of the interval $|x| < \pi$, $\bar{f} = \pi/8$. The cosine coefficients are

$$a_n = \frac{1}{\pi}\int_{-\pi}^{\pi} f(x)\cos nx\,dx = \frac{2}{\pi}\int_0^{\pi/2}\left(\frac{\pi}{2} - x\right)\cos nx\,dx$$

$$= \left[\left(\frac{1}{n} - \frac{2x}{n\pi}\right)\sin nx - \frac{2}{n^2\pi}\cos nx\right]\Bigg|_0^{\pi/2} = \frac{2}{n^2\pi}\left(1 - \cos\frac{n\pi}{2}\right),$$

where the first equality comes from the principle that the integral of an even function is twice the integral of the portion to the right of $x = 0$ and the property $f = 0$ for $\pi/2 < x < \pi$. To finish, we note that $\cos(n\pi/2)$ is 0 for n odd, 1 for n divisible by 4, and -1 for all other n. Thus,

$$a_n = \frac{2}{n^2\pi}\begin{cases} 2 & n = 2, 6, 10, \ldots \\ 1 & n = 1, 3, 5, \ldots \\ 0 & n = 4, 8, 12, \ldots \end{cases} .$$

The series is then

$$f_s = \frac{\pi}{8} + \frac{2}{\pi}\left(\cos x + \frac{1}{9}\cos 3x + \frac{1}{25}\cos 5x + \cdots\right) + \frac{1}{\pi}\left(\cos 2x + \frac{1}{9}\cos 6x + \frac{1}{25}\cos 10x + \cdots\right).$$

The partial sum S_{10} is shown in Figure 8.5.2.[20] The series does indeed seem to be converging to the function.

---

[20]The partial sum S_{10} is the sum of the terms up to and including the $n = 10$ term. See Section 8.4.

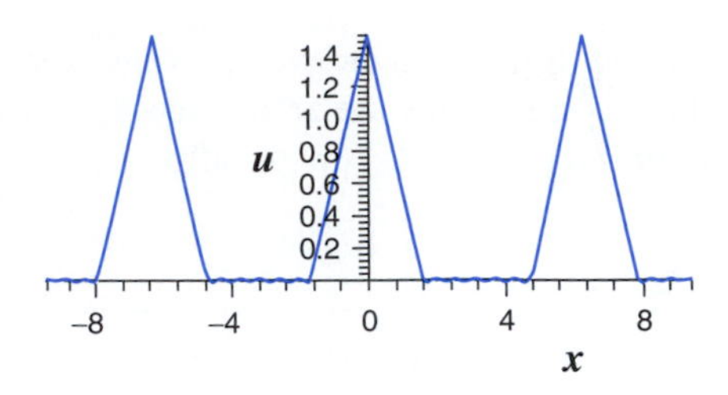

Figure 8.5.2
The partial sum S_{10} in the series of Example 2.

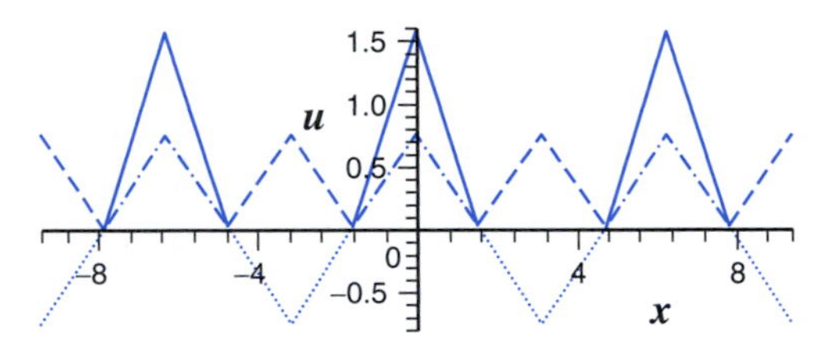

Figure 8.5.3
The sum of even-numbered terms (dashed), the sum of the odd-numbered terms (dotted), and the original function f (solid), all from Example 2.

It is also interesting to see how the different parts of the series work together to produce the function. Figure 8.5.3 shows the graph of f as a solid curve, the graph of the terms with even n (including the constant) as a dashed curve, and the terms with odd n as a dotted curve. The odd terms double the even terms on the intervals $(-\pi/2, \pi/2)$, $(3\pi/2, 5\pi/2)$, and so on and cancel the even terms on the other intervals. It is surprising how fast the series converges, given that the function f does not look very much like a sine or cosine function. In general, series defined by Equations (5) and (6) converge nicely for the well-behaved functions we have been considering.

Theorem 8.5.1 Let f be a continuous $2L$-periodic function that has only finitely many maxima and minima on the interval $[-L, L]$, and define Fourier coefficients by

$$a_n = \frac{1}{L}\int_{-L}^{L} f(x)\cos\frac{n\pi x}{L}\,dx, \qquad b_n = \frac{1}{L}\int_{-L}^{L} f(x)\sin\frac{n\pi x}{L}\,dx, \qquad n = 1, 2, \ldots. \tag{7}$$

Then

$$f(x) = \bar{f} + \sum_{n=1}^{\infty}\left(a_n \cos\frac{n\pi x}{L} + b_n \sin\frac{n\pi x}{L}\right), \tag{8}$$

where $\bar{f}$ is the average value of f. In other words, the series computed from the function f converges to the function f.

The set of functions given by

$$S_L = \left\{1, \cos\frac{\pi x}{L}, \sin\frac{\pi x}{L}, \cos\frac{2\pi x}{L}, \sin\frac{2\pi x}{L}, \cos\frac{3\pi x}{L}, \sin\frac{3\pi x}{L}, \ldots\right\}$$

is said to be a **complete** set of orthogonal functions on the interval $[-L, L]$.

Complete Sets of Orthogonal Functions and Eigenvalue Problems

Consider the eigenvalue problem

$$f'' - kf = 0, \qquad f(2L) = f(0), \qquad f'(2L) = f(0).$$

This problem consists of a differential equation along with two boundary conditions of a sort that we have not met before. Instead of prescribing a value of f or f' at a point, the conditions prescribe a relationship between values at two different points. These conditions serve to guarantee that the function is $2L$-periodic. Assuming $k = -\lambda^2 < 0$, the differential equation has general solution

$$f = a\cos\lambda x + b\sin\lambda x.$$

From the boundary conditions, we obtain the eigenvalues $\lambda_n = n\pi/L$, and the eigenfunctions are

$$\cos\frac{n\pi x}{L}, \qquad \sin\frac{n\pi x}{L}, \qquad n = 1, 2, \ldots.$$

✦ INSTANT EXERCISE 2

Fill in the details in the determination of the eigenvalues and eigenfunctions for the eigenvalue problem

$$f'' - \lambda^2 f = 0, \qquad f(2L) = f(0), \qquad f'(2L) = f(0).$$

We began by assuming $k < 0$, but all cases need to be considered. It turns out that there is an additional eigenfunction. With $k = 0$, the differential equation is $f'' = 0$, which has solutions $a + bx$. The boundary conditions require $a + 2Lb = a$ and $b = b$; thus, $b = 0$ and a is arbitrary. Including the eigenfunction $f \equiv 1$, we have exactly the set that we used to construct series for arbitrary 2π-periodic functions. In fact, the best way to find complete sets of orthogonal functions is to solve an appropriate differential equation eigenvalue problem.

Functions Defined on Finite Intervals and Piecewise-Continuous Functions

Suppose a function F is defined only on an interval $-L \le x \le L$, but satisfies the boundary conditions $F(-L) = F(L)$ and $F'(-L) = F'(L)$ and has only finitely many extrema. Let f be the function defined by

$$f(x) = \begin{cases} F(x) & -L \le x \le L \\ F(x \pm 2\pi) & \text{otherwise} \end{cases}.$$

The function f is said to be the **periodic extension** of F. Since the two functions are the same for $-L \le x \le L$, their Fourier series are the same. By Theorem 8.5.1, the Fourier series for f converges to the function f; hence, the Fourier series for F converges to f.

In the context of partial differential equations, boundary conditions are usually given at $x = 0$ and $x = L$, and the solution uses only the sine or cosine functions, depending on the boundary conditions. These Fourier sine and cosine series are closely related to the full Fourier series on the interval $-L \le x \le L$. To see this, suppose a function G is defined only on an interval $0 \le x \le L$ and satisfies the boundary conditions $G(L) = G(0) = 0$. Let F be defined on the interval $-L \le x \le L$ by

$$F(x) = \begin{cases} G(x) & 0 \le x \le L \\ -G(-x) & -L \le x \le 0 \end{cases}.$$

The function F is the **odd extension** of G. But now we have the same situation as in the previous discussion, with one difference. Since F is odd, its average is zero and its Fourier cosine coefficients are zero. Furthermore, its Fourier sine coefficients are

$$b_n = \frac{1}{L}\int_{-L}^{L} F(x)\sin\frac{n\pi x}{L}\,dx = \frac{2}{L}\int_0^L F(x)\sin\frac{n\pi x}{L}\,dx = \frac{2}{L}\int_0^L G(x)\sin\frac{n\pi x}{L}\,dx.$$

Thus, the Fourier sine series for a function defined on $[0, L]$ converges to the odd extension of the function. Similarly, if G is defined on $[0, L]$, then its Fourier cosine series converges to the **even extension** of G, defined by

$$F(x) = \begin{cases} G(x) & 0 \le x \le L \\ G(-x), & -L \le x \le 0 \end{cases}.$$

Note that the constant term must be included in any Fourier cosine series because the even extension of a function generally does not have average value zero.

Piecewise-Continuous Functions Theorem 8.5.1 does not indicate the full power of Fourier series, even with the extensions to functions that are defined on finite intervals. Fourier series are also useful for functions that are only piecewise-continuous, although the convergence of such series to the defining function is not quite as strong as that of Theorem 8.5.1.

EXAMPLE 3

Consider the function (see Figure 8.5.4)

$$f(x) = \begin{cases} 1 & |x| < \dfrac{\pi}{2} \\ 0 & \dfrac{\pi}{2} < |x| < \pi \\ f(|x| - 2\pi) & \text{otherwise} \end{cases}.$$

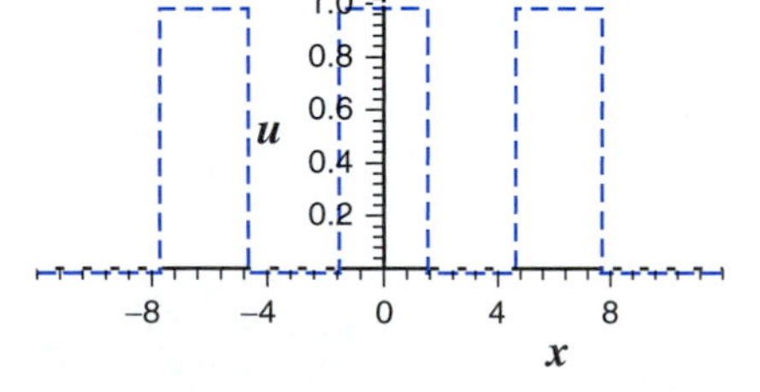

Figure 8.5.4
The square wave function of Example 3.

The Fourier series for this function is

$$f_s = \frac{1}{2} + \frac{2}{\pi}\left(\cos x - \frac{1}{3}\cos 3x + \frac{1}{5}\cos 5x - \cdots\right).$$

Sometimes it is convenient to rewrite the series in a form that includes only the nonzero terms. This requires a change in the summation index. The nonzero terms after a_0 are given by

$$\frac{2(-1)^{(n-1)/2}}{n\pi}\cos nx, \qquad n = 1, 3, 5, \ldots.$$

We need to define a new index k so that $n = 1, 3, 5, \ldots$ corresponds to $k = 1, 2, 3, \ldots$. This can be accomplished by defining

$$k = \frac{n+1}{2}.$$

The full series can then be written in summation notation as

$$f_s = \frac{1}{2} - \frac{2}{\pi}\sum_{k=1}^{\infty}\frac{(-1)^k}{2k-1}\cos(2k-1)x.$$

The series is illustrated in Figure 8.5.5.

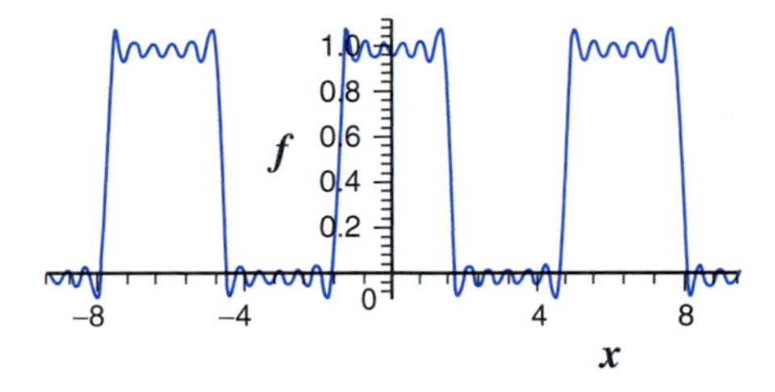

Figure 8.5.5
The partial sum S_6 of the full Fourier series for the square wave of Example 3.

✦ INSTANT EXERCISE 3

Derive the result for the coefficients a_n in Example 3.

The Fourier series in Example 3 *represents* the function from which it was constructed, but it does not appear to *be* that function. There are three features of the plots that are worth noting.

- At points where the original function is continuous, the series appears to be converging to the function, albeit slowly.
- At the points where the function changes between 0 and 1, the series appears to converge to the average value $\frac{1}{2}$.
- The series shows a lot of oscillation in the vicinity of the jumps and converges much more slowly near them as well.

These features are typical of Fourier series representations of functions that are only piecewise-continuous. The overshoot and oscillation at a jump is called the **Gibbs phenomenon.**[21]

[21] See, for example, Richard Haberman, *Applied Partial Differential Equations with Fourier Series and Boundary Value Problems*, 4th ed., Pearson, Upper Saddle River, NJ, 2004.

This phenomenon is explored further in Exercise 5. Theorem 8.5.2 summarizes the properties of Fourier series for functions that are only piecewise-smooth.

Theorem 8.5.2 Let F be a function defined on an interval $-L < x \leq L$, with piecewise-continuous derivative. Let f be the periodic extension of F, and define Fourier coefficients by Equations (7). Let f_s be the function defined by Equation (8). Then

1. At all points where f is continuous, $f_s(x) = f(x)$.
2. At any point a where f is not continuous,

$$f_s(a) = \frac{1}{2}\left[\lim_{x\to a^-} f(x) + \lim_{x\to a^+} f(x)\right].$$

8.5 Exercises

T 1. Let $F(x) = |x|$ for $-\pi < x < \pi$, and let f be its periodic extension. Find the Fourier coefficients for F, and plot the partial sum S_8 on the interval $-3\pi < x < 3\pi$.

T 2. Let $F(x) = x(x - \pi)$ for $0 \leq x \leq \pi$ and $F(x) = 0$ for $-\pi \leq x \leq 0$, and let f be the periodic extension of F. Find the Fourier coefficients for F, and plot the partial sum S_8 on the interval $-3\pi < x < 3\pi$. Also plot the partial sums of the sine series and the partial sums of the cosine series (including the constant) separately.

T 3. Let $F(x) = 1 - x^2$ for $-1 < x < 1$, and let f be its periodic extension. Find the Fourier coefficients for F and plot the partial sum S_8 on the interval $-3 < x < 3$.

T 4. Let $F(x) = 1/(1 + x^2)$ on the interval $-1 \leq x \leq 1$, and let f be its periodic extension. Find the Fourier coefficients for F, evaluate them numerically, and plot the partial sum S_8 on the interval $-3 < x < 3$.

T 5. Plot the approximations S_3 and S_9 for the function of Example 3. Compare the approximations with Figure 8.5.5. In particular, compare the number of oscillations in each portion of the graph and the maximum of the Gibbs phenomenon overshoot. Explain why your observations do not contradict Theorem 8.5.2.

T 6. Solve

$$\begin{aligned} u_{tt} &= u_{xx}, & 0 < x < 1, \quad t > 0,\\ u(0, t) &= 0, & u(1, t) = 0,\\ u(x, 0) &= 2x, & u_t(x, 0) = 0. \end{aligned}$$

Use a computer algebra system to plot the Fourier series solution at times 0, 0.25, 0.5, 0.75, and 1. Use the Gibbs phenomenon to discuss the results.

T 7. Solve

$$\begin{aligned} u_{tt} &= u_{xx}, & 0 < x < 1, \quad t > 0,\\ u(0, t) &= 0, & u(1, t) = 0,\\ u(x, 0) &= 1, & u_t(x, 0) = 0. \end{aligned}$$

Use a computer algebra system to plot the Fourier series solution at times 0, 0.25, 0.5, 0.75, and 1. Use the Gibbs phenomenon to discuss the results.

T 8. Solve

$$u_{tt} = u_{xx}, \qquad 0 < x < 1, \quad t > 0,$$
$$u(0,t) = 0, \qquad u(1,t) = 0,$$
$$u(x,0) = 1 - \cos \pi x, \qquad u_t(x,0) = 0.$$

Use a computer algebra system to plot the solution at times 0, 0.25, 0.5, 0.75, and 1.

T 9. *a.* Solve

$$u_{tt} = u_{xx}, \qquad 0 < x < 1, \quad t > 0,$$
$$u(0,t) = 0, \qquad u(1,t) = 0,$$
$$u(x,0) = 0, \qquad u_t(x,0) = 1.$$

b. Use a computer algebra system to plot the solution at times 0, 0.1, 0.2, 0.3, 0.4, and 0.5.

T 10. *a.* Solve

$$u_{tt} = u_{xx}, \qquad 0 < x < 1, \quad t > 0,$$
$$u(0,t) = 0, \qquad u(1,t) = 0,$$
$$u(x,0) = 0, \qquad u_t(x,0) = x.$$

b. Use a computer algebra system to plot the solution at times 0, 0.1, 0.2, 0.3, 0.4, and 0.5.

11. Consider the problem of determining the equilibrium displacement of a vibrating circular membrane of unit radius when the edge is given a fixed displacement. The problem is given by

$$\frac{1}{r}\frac{\partial}{\partial r}(ru_r) + \frac{1}{r^2}u_{\theta\theta} = 0,$$

along with the boundary conditions

$$u(1,\theta) = f(\theta), \qquad |u(0,\theta)| < \infty.$$

Derive a solution formula by using separation of variables. Note that boundary conditions in θ are implied by the geometry.

T 12. *a.* Use the solution formula of Exercise 11 to solve the problem

$$\frac{1}{r}\frac{\partial}{\partial r}(ru_r) + \frac{1}{r^2}u_{\theta\theta} = 0,$$
$$u(1,\theta) = f(\theta) = \begin{cases} \frac{1}{4} - \left|\frac{\theta}{\pi}\right| & -\frac{\pi}{4} < \theta < \frac{\pi}{4} \\ 0 & \text{otherwise} \end{cases}, \qquad |u(0,\theta)| < \infty.$$

b. Plot the cross sections $u(1,\theta)$, $u(0.75,\theta)$, $u(0.5,\theta)$, and $u(0.25,\theta)$, using the boundary condition for the first of these and the Fourier series solution for the remainder.

✦ 8.5 INSTANT EXERCISE SOLUTIONS

1.
$$\int_{-\pi}^{\pi} \sin^2 nx\,dx = \int_{-\pi}^{\pi} \left(\frac{1}{2} - \frac{\cos 2nx}{2}\right) dx = \left[\frac{x}{2} - \frac{\sin 2nx}{4n}\right]\Bigg|_{-\pi}^{\pi} = \pi,$$

so $a_n = \left[\int_{-\pi}^{\pi} f(x)\sin nx\,dx\right]/\pi$. For $f(x) = 1$,

$$a_n = \frac{1}{\pi}\int_{-\pi}^{\pi} \sin nx\,dx = \frac{-1}{n\pi}\cos nx\Big|_{-\pi}^{\pi} = \frac{-1}{n\pi}[\cos n\pi - \cos(-n\pi)] = \frac{-1}{n\pi}(\cos n\pi - \cos n\pi) = 0,$$

where we have used the trigonometric identity $\cos(-x) = \cos x$.

2. Given that $f(0) = a$ and $f'(0) = \lambda b$, the boundary conditions become

$$a\cos 2L\lambda + b\sin 2L\lambda = a, \qquad -\lambda a\sin 2L\lambda + \lambda b\cos 2L\lambda = \lambda b.$$

Rewriting these as a vector equation for the coefficients yields

$$\begin{pmatrix} \cos 2L\lambda - 1 & \sin 2L\lambda \\ -\sin 2L\lambda & \cos 2L\lambda - 1 \end{pmatrix} \begin{pmatrix} a_n \\ b_n \end{pmatrix} = \begin{pmatrix} 0 \\ 0 \end{pmatrix}.$$

Nontrivial solutions are possible when the determinant of the matrix is zero:

$$0 = \begin{vmatrix} \cos 2L\lambda - 1 & \sin 2L\lambda \\ -\sin 2L\lambda & \cos 2L\lambda - 1 \end{vmatrix} = \cos^2 2L\lambda - 2\cos 2L\lambda + 1 + \sin^2 2L\lambda = 2(1 - \cos 2L\lambda).$$

The eigenvalues must satisfy $\cos 2L\lambda = 1$, which means $2L\lambda = 2n\pi$ for any positive integer n. Thus, the eigenvalues are $\lambda_n = n\pi/L$, and the eigenfunctions are

$$\cos\frac{n\pi x}{L}, \text{ and } \sin\frac{n\pi x}{L}, \qquad n = 1, 2, \ldots.$$

3. As in Example 2, the sine coefficients are zero because the function f is even. Then

$$a_n = \frac{1}{\pi}\int_{-\pi}^{\pi} f(x)\cos nx\,dx = \frac{2}{\pi}\int_0^{\pi} f(x)\cos nx\,dx = \frac{2}{\pi}\int_0^{\pi/2} \cos nx\,dx = \frac{2}{n\pi}\sin nx\Big|_0^{\pi/2}$$
$$= \frac{2}{n\pi}\sin\frac{n\pi}{2}.$$

CASE STUDY 8

Stringed Instruments and Percussion

The vibration of a stretched string has been a recurring theme of this chapter, and some connections have been drawn between string vibration and music. We now try to make some stronger connections between the vibration caused by striking a stretched string and the tone produced by the string. It should be kept in mind that musical instruments are quite a bit more complicated than a mere vibrating string. When a guitar string is plucked, for example, the vibration of the string is communicated via wave motion to the sound box of the instrument. The music heard by the listener is really the vibration of the air in the sound box, as excited by the vibration of the string. Nevertheless, the string itself creates a musical tone, and the characteristics of that tone are determined by the characteristics of the vibration. The quality of any string or percussion instrument is a combination of the quality of the vibration transmitted from the original sound production mechanism and the characteristics of the sound box. With the simple vibration models of this chapter, we will not be able to examine questions about fine distinctions, such as why a plucked string on a violin sounds different from a plucked string on a guitar, but we will be able to obtain information about the general differences between classes of instruments. Our focus will be on the question of why it is that vibrating strings such as those of a guitar or piano produce musical instruments that are capable of melody and harmony, while a vibrating membrane, such as that of a bass drum or tom-tom, produces a sound that has only a vague sense of pitch. We focus on vibrations caused by striking rather than plucking because of the simplification we get by the use of impulsive forcing.

Characterize the tones made by a struck string and a struck membrane. In particular, determine why a struck string can be used to play a musical note and a struck membrane cannot.

Sound Quality for a Struck String

We consider a dimensionless version of the vibrating string problem with the assumption that the initial displacement is 0 and that the striking of the string at time 0 introduces an impulsive velocity. Thus, we have the differential equation

$$u_{tt} = 4u_{xx} \tag{1}$$

with boundary and initial conditions

$$u(0,t) = 0, \qquad u(1,t) = 0, \qquad u(x,0) = 0, \qquad u_t(x,0) = \delta(x-a), \tag{2}$$

where a marks the point that is struck and $c = 2$ is chosen as a matter of convenience.[22] Although it seems natural to strike a string in the center, a look inside a piano reveals that the hammers do not actually strike the string in the center; we therefore consider what effect the hammer placement might have on the tone.

The differential equation and boundary conditions are standard, so the details of the solution procedure can easily be omitted. The Fourier series solution for a struck string (given $L = 1$ and $c = 2$) has the form

$$u(x,t) = \sum_{n=1}^{\infty} b_n \sin 2n\pi t \sin n\pi x. \tag{3}$$

A quick check shows that this solution formula satisfies the differential equation, the boundary conditions, and the initial condition $u(x,0) = 0$. To satisfy the remaining initial condition, the constants b_n must be chosen so that

$$\delta(x-a) = \sum_{n=1}^{\infty} 2n\pi b_n \sin n\pi x.$$

Recall that the coefficients for the equation

$$f(x) = \sum_{n=1}^{\infty} b_n \sin n\pi x$$

[22]The δ "function" is discussed in Section 7.4. For our purposes, the only thing we need to know about this function is the property $\int_0^1 f(x)\delta(x-a)\,dx = f(a)$ for $0 < a < 1$.

are given by

$$b_n = 2\int_0^1 f(x)\sin n\pi x\,dx.$$

The problem at hand is a standard Fourier coefficient problem except with $2n\pi b_n$ instead of b_n. Thus,

$$2n\pi b_n = 2\int_0^1 \delta(x-a)\sin n\pi x\,dx = 2\sin n\pi a.$$

Hence, we obtain the solution

$$u(x,t) = \frac{1}{\pi}\sum_{n=1}^{\infty}\frac{\sin n\pi a}{n}\sin 2n\pi t\sin n\pi x. \tag{4}$$

It remains to interpret the results in terms of musical tone. This problem was studied in the 1860s in great detail by the German scientist Hermann Helmholtz, whose book *On the Sensations of Tone* is still read by students of the science of music. Helmholtz was able to show that perceived quality of sound is influenced by the relationships between the vibration mode frequencies and the relative magnitudes of the Fourier coefficients.

The nth vibration mode has period $1/n$, or frequency n. The various modes have frequencies that are integer multiples of the lowest, or **fundamental,** frequency. Since frequency corresponds to pitch, each mode contributes its own pitch to the overall sound. The strength of the contribution is determined by the Fourier coefficient $(\sin n\pi a)/n\pi$. These contributions can be displayed as vertical lines of length proportional to $|b_n|$ located at the position n on a number line. It is convenient to normalize these lengths by dividing them by a total length. In many cases, the arithmetic total does not exist, so instead we define a "geometric" total[23] by

$$T = \left(\sum_{n=1}^{\infty} b_n^2\right)^{1/2}.$$

We define the **relative amplitude** of the nth mode to be

$$A_n = \frac{|b_n|}{T}.$$

We consider two particular cases here, beginning with the symmetric case.

EXAMPLE 1

Suppose $a = \frac{1}{2}$. Then

$$b_n = \frac{\sin n\pi a}{n\pi} = \frac{1}{n\pi}\sin\frac{n\pi}{2} = \frac{1}{n\pi}\begin{cases} 1 & n = 1, 5, 9, \ldots \\ 0 & n = 2, 4, 6, \ldots \\ -1 & n = 3, 7, 11, \ldots \end{cases}.$$

Thus,

$$T^2 = \frac{1}{\pi^2}\left(1 + \frac{1}{9} + \frac{1}{25} + \frac{1}{49} + \cdots\right) = \frac{1}{8},$$

[23]The reader who has studied statistics should note the similarity to the standard deviation. The latter is a "geometric" total of the differences of the measurements from the mean.

where the infinite sum can be evaluated by a computer algebra system or a table of infinite sums. Thus, the relative amplitudes are

$$A_n = \frac{2\sqrt{2}}{n\pi}, \qquad n = 1, 3, 5, \ldots.$$

A plot of the amplitudes appears in Figure C8.1.

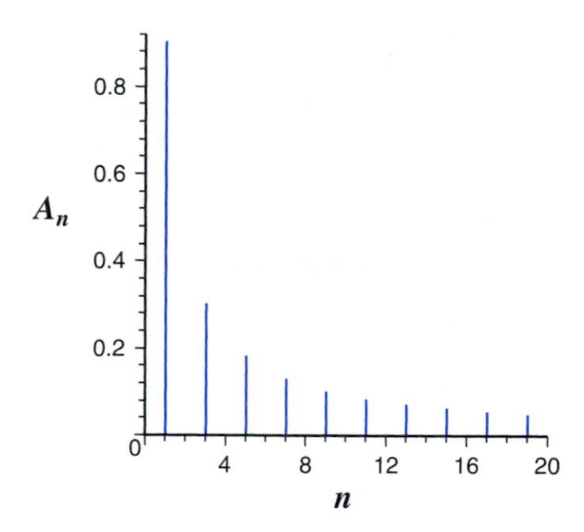

Figure C8.1
The relative amplitudes of the first 20 modes from Example 1.

Piano builders do not choose the midpoint of the string to be the spot struck by the hammers. Instead, pianos are generally designed so that the hammers strike the strings at a point about 12% of the way from one end of the string to the other. We consider this case in Example 2.

EXAMPLE 2

Suppose $a = 0.12$. Then

$$b_n = \frac{\sin 0.12n\pi}{n\pi}.$$

The magnitudes of the first 20 Fourier coefficients are given in Table C8.1, and the corresponding relative amplitudes are plotted in Figure C8.2.

Mode	1	2	3	4	5	6	7	8	9	10
$\lvert b_n \rvert$	0.117	0.1089	0.0960	0.0794	0.0606	0.0409	0.0219	0.0050	0.0088	0.0187
Mode	11	12	13	14	15	16	17	18	19	20
$\lvert b_n \rvert$	0.0244	0.0261	0.0241	0.0192	0.0125	0.0049	0.0023	0.0085	0.0129	0.0151

Table C8.1 The magnitudes of the first 20 Fourier coefficients from Example 2.

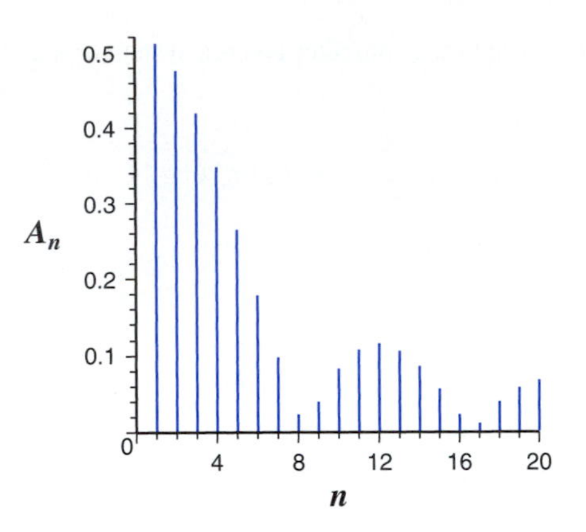

Figure C8.2
The relative amplitudes of the first 20 modes from Example 2.

Presumably, the choice of a point 12% from one end to the other resulted from experimentation. A concert pianist and a well-trained piano craftsman can hear the difference between pianos with hammers at 50% and 12%, although many people without special training cannot tell the difference. The amplitude plots of Figures C8.1 and C8.2 indicate that the sounds of the *strings* should be very different; however, the sound made by the strings is amplified by the soundboard.[24] The overall characteristics of the sound depend more on the soundboard design than on the string amplitudes; hence, the difference between the sound of a piano with hammers in the center and the same piano with hammers nearer to one end is somewhat less than the differences in Figures C8.1 and C8.2 suggest.

Sound Quality for a Struck Membrane

The partial differential equation for a vibrating membrane is

$$u_{tt} = c^2 \nabla^2 u, \tag{5}$$

where $\nabla^2 u$ is the **Laplacian** of u. The Laplacian is the two-dimensional analog of u_{xx}. The specific form of the Laplacian depends on the geometry of the vibrating region. Before specifying the geometry, we can separate the time variable from the spatial variables by assuming a solution of the form

$$u(\cdot, t) = g(t) f(\cdot),$$

where the symbol $\cdot$ simply refers to whatever spatial coordinates are appropriate for the problem at hand. As in the simple case of one spatial dimension, the assumption of a separated solution leads to the equation

$$\frac{g''}{c^2 g} = \frac{\nabla^2 f}{f} = -\lambda^2,$$

[24] The soundboard is a heavy metal plate inside the piano.

where the constant is taken to be negative in anticipation of the requirement that the solutions be oscillatory. The result is a pair of differential equations, one for the waveform f and the other for the amplitude function g:

$$g'' + c^2\lambda^2 g = 0, \qquad \nabla^2 f + \lambda^2 f = 0.$$

We consider here only objects that are set into motion by striking; hence, we are only interested in amplitude functions that are initially zero. The amplitude function for each mode is a solution of

$$g_n'' + c^2\lambda_n^2 = 0, \qquad g_n(0) = 0.$$

This problem has the family of solutions $g_n = C_n \sin c\lambda_n t$. Thus, the vibration modes are

$$u(\cdot, t) = \sin c\lambda_n t f(\cdot), \tag{6}$$

with f and λ_n determined by the eigenvalue problem consisting of the partial differential equation $\nabla^2 f + \lambda^2 f = 0$ with appropriate boundary conditions.[25]

We consider here a circular membrane of unit radius,[26] with spatial variation in the polar coordinates r and θ. In this case, the eigenvalue problem is

$$\frac{1}{r}\frac{\partial}{\partial r}(r f_r) + \frac{1}{r^2} f_{\theta\theta} + \lambda^2 f = 0, \qquad f(1, \theta) = 0. \tag{7}$$

To solve this problem, we might again try separation of variables. Suppose

$$f(r, \theta) = R(r)\Theta(\theta).$$

Substitution of this form into the partial differential equation yields

$$R''\Theta + \frac{1}{r}R'\Theta + \frac{1}{r^2}R\Theta'' + \lambda^2 R\Theta = 0,$$

which can be rewritten as

$$\frac{r^2 R'' + rR'}{R} + \frac{\Theta''}{\Theta} + r^2\lambda^2 = 0.$$

Collecting terms that are independent of r and terms that are independent of θ yields

$$\frac{r^2 R'' + rR'}{R} + r^2\lambda^2 = -\frac{\Theta''}{\Theta} = k,$$

where k is another separation constant. In addition to the explicit boundary condition at $r = 1$, there is an implicit boundary condition that the solution should be bounded at $r = 0$ as well as a pair of conditions requiring that Θ and Θ' be continuous. Thus, the separated problems for Θ and R are

$$\Theta'' + k\Theta = 0, \qquad \Theta(2\pi) = \Theta(0), \qquad \Theta'(2\pi) = \Theta'(0),$$

and

$$r^2R'' + rR' + (\lambda^2 r^2 - k)R = 0, \qquad |R(0)| < \infty, \qquad R(1) = 0.$$

---

[25]There is no loss of generality in taking $C_n = 1$, because each vibration mode will have an undetermined Fourier coefficient as a factor.

[26]There is no loss of generality in taking the radius of the membrane to be 1, as that only means that lengths are measured relative to the radius of the membrane rather than in arbitrary units such as centimeters or feet.

The boundary conditions on Θ make sense because $\theta = 0$ and $\theta = 2\pi$ represent the same point on the membrane. Hence, the boundary conditions for Θ are based on the requirement that the membrane be continuous. In effect, these conditions require that the eigenfunctions be 2π-periodic. The requirement of periodic solutions restricts the values of k to be nonnegative. Let $m = \sqrt{k}$. Then the Θ problem is

$$\Theta'' + m^2\Theta = 0, \qquad \Theta(2\pi) = \Theta(0), \qquad \Theta'(2\pi) = \Theta'(0).$$

The differential equation has solutions

$$\Theta = a\cos m\theta + b\sin m\theta,$$

and the periodicity requirement restricts m to integer values, including 0. Thus,

$$\Theta_0 = 1, \qquad \Theta_m = a_m\cos m\theta + b_m\sin m\theta, \qquad k_m = m^2, \quad m = 0, 1, 2, \ldots. \tag{8}$$

The remaining eigenvalue problem then becomes

$$r^2R'' + rR' + (\lambda^2r^2 - m^2)R = 0, \qquad |R(0)| < \infty, \qquad R(1) = 0, \qquad m = 0, 1, 2, \ldots.$$

The differential equation here is the **parametric Bessel equation**[27] of order m. There are no nonzero solutions for $\lambda = 0$ that satisfy the boundary conditions; the solutions with $\lambda > 0$ are

$$R = c_1J_m(\lambda r) + c_2Y_m(\lambda r),$$

where J_m and Y_m are the Bessel functions of the first and second kinds. The Y_m functions (Figure C8.3) are unbounded at the origin. Thus, we must have $c_2 = 0$. We can then take $c_1 = 1$ without loss of generality, so

$$R = J_m(\lambda r).$$

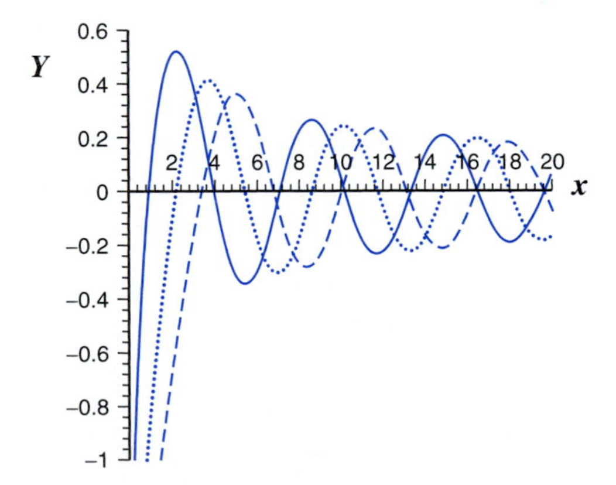

Figure C8.3
The Bessel functions Y_0 (solid), Y_1 (dotted), and Y_2 (dashed).

The boundary condition $R(1) = 0$ results in the equation

$$J_m(\lambda) = 0,$$

[27]See Exercise 9 of Section 3.7.

which determines the eigenvalues. Each J_m function (Figure C8.4) has the property that there is an infinite increasing sequence of positive numbers z_{mn} such that $J_m(z_{mn}) = 0$. These values, which can be found in computer algebra system libraries and reference books, form a two-parameter family of eigenvalues and a corresponding two-parameter family of eigenfunctions,

$$R_{mn} = J_m(z_{mn}r), \qquad m = 0, 1, 2, \ldots, \qquad n = 1, 2, 3, \ldots. \tag{9}$$

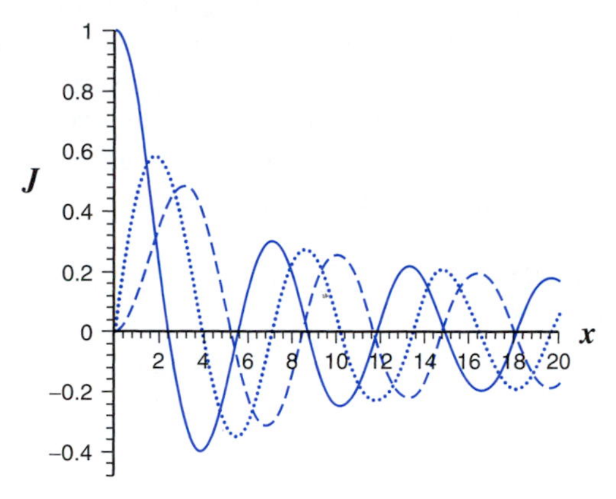

Figure C8.4
The Bessel functions J_0 (solid), J_1 (dotted), and J_2 (dashed).

The waveforms are the products of the R and Θ functions,

$$f_{mn} = (a_{mn} \cos m\theta + b_{mn} \sin m\theta)J_m(z_{mn}r), \qquad m = 0, 1, 2, \ldots, \qquad n = 1, 2, 3, \ldots, \tag{9}$$

and the corresponding vibration modes are

$$u_{mn} = (\sin cz_{mn}t)(a_{mn} \cos m\theta + b_{mn} \sin m\theta)J_m(z_{mn}r), \qquad m = 0, 1, 2, \ldots, \qquad n = 1, 2, 3, \ldots. \tag{10}$$

We could now solve the vibration for a given initial velocity, but to understand the tone quality of a drum, we need only to consider the vibration frequencies themselves. The vibration modes have freqüencies $cz_{mn}/(2\pi)$. The fundamental mode has $m = 0$ and $n = 1$, and the corresponding value $z_{01} = 2.4048$. The character of the tone is determined by the ratios of the other vibration mode frequencies to that of the fundamental. These are summarized in Table C8.2.

Mode	1	2	3	4	5	6	7	8	9	10
m	0	1	2	0	3	1	4	2	0	5
n	1	1	1	2	1	2	1	2	3	1
z_{mn}/z_{01}	1.000	1.593	2.136	2.295	2.653	2.917	3.155	3.500	3.598	3.647
Mode	11	12	13	14	15	16	17	18	19	20
m	3	6	1	4	7	2	0	8	5	3
n	2	1	3	2	1	3	4	1	2	3
z_{mn}/z_{01}	4.059	4.132	4.230	4.601	4.610	4.832	4.903	5.084	5.131	5.412

Table C8.2 The frequencies, relative to the fundamental frequency, of the first 20 modes for a vibrating circular membrane.

The information in Table C8.2 explains in striking fashion why the sound made by a stretched elastic membrane does not have any clear pitch. The higher vibration modes do not occur at frequencies closely related to the fundamental, so the sounds made by the various vibration modes conflict with one another. The result is a collection of unrelated tones that combine into a sound that has no discernible pitch. It is of course possible to make drums that have higher or lower pitches, but one cannot use a set of standard drums to play a tune. Tympani are drums with a rounded bottom. The vibration of the head is amplified by the vibration of the bottom so that some modes are enhanced and others weakened. The design gives the tympani a better sense of pitch than other drums. Tympani have enough of a sense of pitch to allow them to be tuned, but they do not have enough for them to be used as melodic instruments. It *is* possible to play tunes using the steel drums that are popular in some Caribbean cultures, but these drums have heads made of bent sheet metal rather than flat, elastic membranes.

Case Study 8 Exercises

1. Determine the maximum amplitude of the string in Example 1.
2. Show that the series that represents the sum of the amplitudes of the vibration modes in Example 1 does not converge.
3. Suppose the piano string of Example 1 is struck at a point one-seventh of the way from one end to the other. Determine the resulting vibration and plot the amplitudes.
4. Suppose a piano builder wants to maximize the amplitude of the third vibration mode. Determine the best place to have the hammer strike the string.
5. (T) A piano string of unit length and unit wave speed is struck by a flat hammer at time 0. The motion of the string is governed by the problem

$$u_{tt} = u_{xx}, \qquad 0 < x < 1, \quad t > 0,$$
$$u(0,t) = 0, \qquad u(1,t) = 0,$$
$$u(x,0) = 0, \qquad u_t(x,0) = \psi(x),$$

where

$$\psi(x) = \begin{cases} 0 & 0 < x < 0.45 \\ 1 & 0.45 < x < 0.55 \\ 0 & 0.55 < x < 1 \end{cases}.$$

a. Solve the problem.
b. Plot the amplitudes.
c. Plot the solution at times 0.125, 0.25, 0.375, 0.5, 0.625, 0.75, 0.875, and 1.0.

6. (T) A piano string of unit length and unit wave speed is struck by a rounded hammer at time 0. The motion of the string is governed by the problem

$$u_{tt} = u_{xx}, \qquad 0 < x < 1, \qquad t > 0,$$
$$u(0,t) = 0, \quad u(1,t) = 0,$$
$$u(x,0) = 0, \quad u_t(x,0) = \phi(x),$$

where

$$\phi(x) = \begin{cases} 0 & 0 < x < 0.45 \\ 1.5 - 600(x - 0.5)^2 & 0.45 < x < 0.55 \\ 0 & 0.55 < x < 1 \end{cases}.$$

a. Solve the problem.
b. Plot the amplitudes.
c. Plot the solution at times 0.125, 0.25, 0.375, 0.5, 0.625, 0.75, 0.875, and 1.0.

7. Compare the results of Exercises 5 and 6.
   a. Plot the functions ψ and ϕ.
   b. Determine the average values of ψ and ϕ over the width of the hammers. How do they compare?
   c. Is there a significant pattern in the amplitude plots that suggests a difference in sound quality?

8. Consider a stretched membrane that is initially at rest before being given a velocity $\psi(r)$.
   a. Explain why the solution does not depend on θ.
   b. Use the appropriate expression for the Laplacian to write the eigenvalue problem for the waveform $f(r)$.
   c. Solve the eigenvalue problem of part *b*, and use the results to construct a series solution for the problem $u_{tt} = c^2 \nabla^2 u$, with initial condition $u(r, 0) = 0$ and the usual boundary conditions.
   d. Use the facts

$$J_0'(r) = -J_1(r), \qquad [r J_1(r)]' = r J_0(r)$$

   to show that

$$\int_0^1 J_0(z_n r) J_0(z_m r)\, dr = 0, \qquad n \neq m.$$

   [*Hint*: show that there is a nonzero integral Q such that $\int_0^1 J_0(z_n r) J_0(z_m r)\, dr = z_m Q = z_n Q$.]
   e. Use the orthogonality property in part *d* and the formula

$$\int_0^1 J_0^2(z_m r) r\, dr = \frac{J_1^2(z_m)}{2}$$

   to obtain a formula for the series coefficients in terms of the specified initial velocity $\psi(r)$.

chapter

A

Some Additional Topics

The topics addressed here are optional in the sense that they can easily be omitted without loss of course continuity. They are included in the text to give instructors greater flexibility in the design of their course.

Section A.1 presents the standard method for solving first-order linear equations. The idea is to multiply the equation by a suitable function to render the equation integrable. The method is essentially equivalent to the method of variation of parameters that appears in Section 4.5, and the choice between the two is a matter of taste. The integrating factor method is included here for the benefit of those instructors who wish to present it either because they prefer the method over variation of parameters or because they prefer to study first-order linear equations in Chapter 2, where the topic traditionally appears. This section should be viewed as an alternative to Section 4.5, with no benefit to be gained from studying both of them except perhaps to develop a preference for one method or the other. This section is best studied after Section 2.4 or at the end of Chapter 2.

Section A.2 discusses the mathematical issues that arise in the proof of the standard existence and uniqueness theorem for initial-value problems for first-order equations. This topic is useful for those courses in which it is desired to incorporate some formal mathematical argumentation in addition to concepts, calculations, and mathematical models. This section is best studied along with Chapter 2.

Section A.3 discusses the mathematical issues that relate to the degree of success achieved in using numerical methods to obtain approximate solutions for initial-value problems. The section discusses numerical truncation error and the conditioning of a differential equation. This section is best studied at the end of Chapter 2.

Section A.4 briefly presents the method for obtaining a power series representation for the solution of a linear differential equation with variable coefficients. This is a topic whose usefulness has diminished in the era of modern computing. Nevertheless, some instructors find it of value in their treatment of differential equations. This section is best studied at the end of Chapter 3.

Section A.5 introduces some additional differential equations theory in the form of fundamental matrices and the matrix exponential. It is best studied after Section 6.6.

Section A.6 augments Chapter 6 on systems of differential equations by extending the solution methods for homogeneous linear systems to nonhomogeneous linear systems. It is best studied after Section A.5. The portion of Section A.5 that deals with the matrix exponential is not prerequisite to Section A.6, but the remainder of Section A.5 is necessary background for Section A.6.

Sections A.7 and A.8 augment the chapter on partial differential equations (Chap. 8) by extending the method of separation of variables to the one-dimensional heat equation and Laplace's equation. This material can be included at the end of Chapter 8.

A.1 Using Integrating Factors to Solve First-Order Linear Equations

Recall[1] that first-order linear equations are equations that can be written in the form

$$\frac{dy}{dt} + p(t)y = g(t). \tag{1}$$

These equations have a unique solution through any point where p and g are continuous.[2] This unique solution can always be found by a standard technique, which we consider here.

MODEL PROBLEM A.1

Solve the differential equation

$$y' + 2y = 2 + 4t.$$

The method is based on the idea of trying to integrate both sides of the differential equation with respect to the independent variable t. This is not possible for the equation in its given form. We can integrate y' and $2 + 4t$ immediately, but we cannot integrate $2y$ without first solving the equation.[3] However, suppose we multiply the differential equation by e^{2t}. The new form of the equation is then

$$e^{2t}y' + 2e^{2t}y = (2 + 4t)e^{2t}.$$

Now observe that the second coefficient on the left side, $2e^{2t}$, is the derivative of the first coefficient, e^{2t}. So the left side is e^{2t} times the derivative of y plus y times the derivative of e^{2t}. This looks like the result of a derivative calculation by the product rule. In fact, we can rewrite the equation as

$$\frac{d}{dt}(e^{2t}y) = e^{2t}y' + 2e^{2t}y = e^{2t}(y' + 2y) = (2 + 4t)e^{2t}.$$

The equation is integrable with respect to t in this form because the left side is the t derivative of $e^{2t}y(t)$. It does not matter that we don't know $y(t)$ yet, because we are integrating the *derivative* of $e^{2t}y$ rather than the *function* $e^{2t}y$. This integration gives us

$$e^{2t}y = \int (2 + 4t)e^{2t}\,dt = 2te^{2t} + C.$$

Now we can multiply by e^{-2t} to get the general solution

$$y = 2t + Ce^{-2t}.$$

---

[1] Section 2.4.

[2] Theorem 2.4.2.

[3] Keep in mind that when we talk about "integrating" an equation, we mean integrating it *with respect to* t. We can integrate $2y$ with respect to y, but that does not do any good.

✦ INSTANT EXERCISE 1

Compute $\int(2+4t)e^{2t}\,dt$.

This solution of Model Problem A.1 demonstrates the main idea of the method, but we have omitted a key detail. The success of the method depends on finding the right function (in this case e^{2t}) by which to multiply the differential equation. This function is called the **integrating factor,** and full development of the method requires a technique for computing it. At this point, it is at least clear that one needs to look only at the left side of the differential equation to choose the correct integrating factor. The integrating factor e^{2t} will work for any differential equation of the form $y' + 2y = g(t)$.

EXAMPLE 1

Consider the initial-value problem

$$y' + 2y = \tan t, \qquad y(0) = 1.$$

Multiplying the equation by the integrating factor e^{2t} yields

$$\frac{d}{dt}(e^{2t}y) = e^{2t}y' + 2e^{2t}y = e^{2t}(y' + 2y) = e^{2t}\tan t.$$

In principle, we can integrate both sides of this equation, but in practice we have a problem because the function $e^{2t}\tan t$ has no elementary antiderivative. This is less of a problem than it appears to be. Recall[4] that antiderivatives can be constructed with definite integrals. So in integrating the differential equation, we can *define* a function that has the correct derivative. For example, the derivative of

$$\int_0^t e^{2w}\tan w\,dw$$

is, by the fundamental theorem of calculus, $e^{2t}\tan t$. We can therefore integrate the differential equation to obtain

$$e^{2t}y = C + \int_0^t e^{2w}\tan w\,dw;$$

after applying the initial condition, we have

$$y = e^{-2t} + e^{-2t}\int_0^t e^{2w}\tan w\,dw.$$

Notice that in constructing an antiderivative, we could have used any value for the lower limit of integration except a value, such as $\pi/2$, for which the integrand is undefined. However, the lower limit of integration should always be taken to be the value of t corresponding to the initial condition, so that the integration constant can be evaluated immediately.

✦ INSTANT EXERCISE 2

Solve $y' + 2y = e^{-t}$.

---

[4]Theorem 1.2.1.

Overview of the Integrating Factor Method

The general procedure for solving any equation of the form

$$y' + py = g \tag{2}$$

consists of two steps:

1. Multiply the equation by a suitable integrating factor μ.
2. Integrate both sides of the equation with respect to t.

Finding an Integrating Factor for Model Problem A.1

Consider the differential equation family

$$y' + 2y = g.$$

Our task is to derive the integrating factor e^{2t}. The idea is to attempt to solve the equation by using an unknown integrating factor and then to insist that the integrating factor change the left side of the equation into a product rule derivative.

Assume an unknown integrating factor μ. Multiplying the differential equation by the integrating factor yields

$$\mu y' + 2\mu y = \mu g.$$

This is equivalent to the equation

$$e^{2t} y' + 2e^{2t} y = (2 + 4t)e^{2t}$$

in the solution of Model Problem A.1. The next step in the solution of Model Problem A.1 is

$$\frac{d}{dt}(e^{2t} y) = e^{2t} y' + 2e^{2t} y = e^{2t}(y' + 2y) = (2 + 4t)\, e^{2t};$$

equivalently, we have

$$\frac{d}{dt}(\mu y) = \mu y' + \mu' y = \mu(y' + 2y) = \mu g$$

for the case where g is unspecified and μ is not yet determined. Of the three equalities in this equation, the first follows directly from the product rule and the third follows directly from the differential equation; these two equalities hold regardless of the choice of μ. The second equality is different, for it requires

$$\mu' = 2\mu.$$

The solution process carries through *only if the integrating factor is chosen to satisfy* $\mu' = 2\mu$. The equation $\mu' = 2\mu$ is first-order and separable, so we can solve it by the method of Section 2.2; a result is $\mu = e^{2t}$. Note that we do *not* need the general solution of the differential equation for μ. Any one integrating factor works, so we do not need a formula for all possible integrating factors.

Finding an Integrating Factor in General

Suppose now that we have some other equation of the form (2). We again multiply the differential equation by the unknown integrating factor to obtain

$$\mu y' + p\mu y = \mu g.$$

We also have, by the product rule,

$$\frac{d}{dt}(\mu y) = \mu y' + \mu' y.$$

What is needed is a choice for μ that allows these equations to be combined. This choice must be a function that satisfies the differential equation

$$\mu' = p\mu.$$

This equation is separable, and it leads to

$$\int \frac{d\mu}{\mu} = \int p(t)\,dt.$$

The integral on the left is independent of p, so we can always obtain the result

$$\ln \mu = \int p(t)\,dt.$$

Note that we have omitted the absolute value from the left side. Since we need only one function μ, we can assume that μ is positive.

Theorem A.1.1 Any equation of the form

$$y' + py = g,$$

where p and g are continuous, can be made integrable by multiplying it by an integrating factor μ that satisfies

$$\mu' = p\mu.$$

EXAMPLE 2

Consider the problem

$$y' - 2ty = 1, \qquad y(0) = 1.$$

We want the equation

$$\mu y' - 2t\mu y = \mu$$

to have the form

$$\mu y' + \mu' y = \mu,$$

and this requires

$$\mu' = -2t\mu.$$

One solution of this equation is

$$\mu = e^{-t^2}.$$

Thus, the original differential equation is equivalent to

$$\frac{d}{dt}(e^{-t^2}y) = e^{-t^2}y' - 2te^{-t^2}y = e^{-t^2}.$$

The equation is now integrable. We can write a definite integral for the antiderivative of e^{-t^2}, or we can use the error function that was defined in Section 2.4. The result is

$$e^{-t^2}y = C + \frac{\sqrt{\pi}}{2}\,\text{erf}\,t.$$

The initial condition, along with the property $\text{erf}\,(0) = 0$, yields $C = 1$. The final result is

$$y = e^{t^2} + \frac{\sqrt{\pi}}{2}\,e^{t^2}\,\text{erf}\,t.$$

EXAMPLE 3

Consider the differential equation

$$x\,\frac{dy}{dx} + 2y = \frac{1}{x^2}, \qquad x > 0.$$

This equation is not in the standard form, so we first divide through by x:

$$y' + \frac{2}{x}y = \frac{1}{x^3}.$$

After multiplying by an unknown integrating factor, we have

$$\mu y' + \frac{2}{x}\mu y = \mu\frac{1}{x^3},$$

which we want to have the form

$$\mu y' + \mu' y = \mu\frac{1}{x^3}.$$

The integrating factor must therefore satisfy

$$\mu' = \frac{2}{x}\mu,$$

or

$$\frac{\mu'}{\mu} = \frac{2}{x}.$$

Integrating this equation yields (note that $x > 0$ from the problem statement and $\mu > 0$ by choice)

$$\ln\mu = 2\,\ln x = \ln x^2;$$

hence, $\mu = x^2$. With this integrating factor, we have

$$\frac{d}{dx}(x^2 y) = x^2 y' + 2xy = \frac{1}{x}.$$

This equation is integrable, with the result

$$x^2 y = C + \ln x.$$

(Again, no absolute value is needed because the problem statement specifies $x > 0$.) Hence, the solution is

$$y = (C + \ln x)x^{-2}.$$

A.1 Exercises

In Exercises 1 through 16, solve the given problem by using the integrating factor method.

1. $y' - 4y = te^t$
2. $y' + 2y = (t-1)e^{-t}$
3. $y' - 4y = te^{4t}$
4. $y' + y = (t-1)e^{-t}$
5. $y' - 3y = \cos t$
6. $y' + y = \sin 2t$
7. $y' + y = 1/(1+e^t)$
8. $y' + y = 1/(1+e^{2t})$
9. $y' + ty = 2t$
10. $y' - 4ty = t$
11. $y' + 2ty = e^{-t^2} \cos t$
12. $y' + 2ty = 2t^3$
13. $ty' + (1+t)y = 1$
14. $ty' + (1+t)y = e^t$
15. $y' - y \tan t = 1$
16. $y' + y \tan t = t \sin 2t$

In Exercises 17 through 24, solve the given initial-value problem and indicate the interval on which the solution is valid. You may find it convenient to use the error function, as in Example 2.

17. $ty' - 2y = 6t^5, \quad y(1) = 0$
18. $ty' + y = t \cos t, \quad y(\pi) = 0$
19. $(1+t^2)y' - 2ty = 1 + t^2, \quad y(0) = 0$
20. $(1+t^2)y' + 4ty = 2t/(1+t^2)^2, \quad y(0) = 1$
21. $y' + 2y \tan t = 1, \quad y(0) = 1$
22. $y' + y \tan t = \sin t, \quad y(0) = 0$
23. $y' - 8ty = 1, \quad y(0) = 1$
24. $y' - 2y = e^{-t^2}, \quad y(0) = 0$
25. Consider the differential equation

$$ty' + my = kt^n, \qquad t > 0,$$

where m and n are integers and k is any nonzero real number.

a. Find the general solution for the case where $n = -m$.
b. Find the general solution for the case where $n \neq -m$.
c. Suppose m and n are both positive. Explain why it is possible that the solution could be defined for all t. Does this contradict the existence and uniqueness theorem (Theorem 2.4.2)?
d. Suppose m and n are both negative. Is it possible that the solution could be defined for all t?

26. Let f be continuously differentiable, let g be continuous, and let y_0 be any real number. Derive a formula for the solution of the initial-value problem

$$fy' + f'y = g, \qquad y(0) = y_0.$$

What additional requirement must the function f satisfy to guarantee that this solution is valid?

27. Consider the family of initial-value problems

$$y' - y\tan t = \sin t, \qquad y(0) = k.$$

a. Find a formula for the solution and the interval on which the solution is valid.
b. Determine which values of k result in a solution that vanishes at some point in the interior of the interval of validity.
c. Determine which values of k result in a solution whose maximum occurs at $t = 0$. [*Hint*: It is best not to use the solution of the differential equation. Instead, differentiate the differential equation implicitly with respect to t to evaluate the second derivative at $y''(0)$.]

28. Consider the family of initial-value problems

$$ty' + (1+t)y = e^{-t}, \qquad y(0) = y_0.$$

a. What does the existence and uniqueness theorem (Theorem 2.4.2) say about this initial-value problem?
b. Find the general solution of the differential equation.
c. Find the unique value of y_0 for which the initial-value problem has a unique solution.

(*Note*: The answer to part c does not contradict the correct answer to part a.)

✦ A.1 INSTANT EXERCISE SOLUTIONS

1.

$$\int (2+4t)e^{2t}\,dt = \int (2+4t)\frac{d}{dt}\left(\frac{1}{2}e^{2t}\right)dt = (1+2t)e^{2t} - \int 2e^{2t}\,dt$$
$$= (1+2t)e^{2t} - e^{2t} + C = 2te^{2t} + C.$$

2. Multiplying $y' + 2y = e^{-t}$ by the integrating factor e^{2t} yields the equation

$$e^{2t}y' + 2e^{2t}y = e^{t},$$

which can be rewritten as

$$\frac{d}{dt}\left(e^{2t}y\right) = e^{t}.$$

Integration yields $e^{2t}y = e^{t} + C$, or $y = e^{-t} + Ce^{-2t}$.

A.2 Proof of the Existence and Uniqueness Theorem for First-Order Equations

Perhaps the single most important theoretical result in the subject of differential equations is the existence and uniqueness theorem, Theorem 2.4.1.

Theorem 2.4.1

Let R be a rectangle in the ty plane that contains the initial point (t_0, y_0) in its interior. If the functions f and $\partial f/\partial y$ are continuous throughout R, then the initial-value problem

$$\frac{dy}{dt} = f(t, y), \qquad y(t_0) = y_0,$$

has a unique solution on some interval $t_l < t < t_r$ that contains t_0.

The reader can get a sense of how mathematical theory is developed by considering the proof of Theorem 2.4.1. An interesting feature of the proof is that it actually deals with the subject of integral equations, a rich subject in its own right.[5] Before we begin the proof, it is convenient to restate the theorem in a slightly simpler, but fully general, formulation. Let new variables z and s be defined by

$$z = y - y_0, \qquad s = t - t_0.$$

The original problem in the ty plane can be recast in the sz plane as

$$\frac{dz}{ds} = g(s, z), \qquad z(0) = 0,$$

where $g(s, z) = f(s + t_0, z + y_0)$. The new problem is the same as the old one except that the initial condition is at the origin. There is therefore no loss in generality in simply assuming that the initial condition in the original problem is the specific point $(0, 0)$ rather than the arbitrary point (t_0, y_0). Similarly, there is no loss of generality in considering only intervals in t centered at the origin. These changes lead to Theorem A.2.1.

Theorem A.2.1

Let R be a rectangle in the ty plane that contains the origin in its interior. If the functions f and $\partial f/\partial y$ are continuous throughout R, then the initial-value problem

$$\frac{dy}{dt} = f(t, y), \qquad y(0) = 0, \tag{1}$$

has a unique solution on some interval $-T < t < T$.

Since there was no loss of generality in the restatement, a proof of Theorem A.2.1 establishes Theorem 2.4.1 as well.

An Integral Equation Formulation

Suppose the problem of Equation (1) has a solution. Then we can integrate the differential equation from time 0 to an arbitrary time t, provided $-T < t < T$. Example 1 serves to illustrate the procedure.

[5] An elementary treatment of linear integral equations can be found in Glenn Ledder, "A Simple Introduction to Integral Equations," *Mathematics Magazine*, 69(1996): 172–181.

EXAMPLE 1

Consider the initial-value problem

$$y' + y = 1, \qquad y(0) = 0.$$

The differential equation is linear and has constant coefficients; therefore, we can solve it by methods from Chapters 3 and 4 or by the method of Section A.1. The solution is

$$y = 1 - e^{-t}.$$

To recast the initial-value problem as an integral equation, we first rewrite the equation, using a dummy variable s in place of the independent variable t. Thus,

$$\frac{dy}{ds}(s) + y(s) = 1, \quad y(0) = 0.$$

Next we rearrange the equation and integrate from $s = 0$ to $s = t$, with the result

$$\int_0^t \frac{dy}{ds}(s)\,ds = \int_0^t [1 - y(s)]\,ds.$$

The left side is computed immediately from the fundamental theorem of calculus, yielding the integral equation

$$y(t) = \int_0^t [1 - y(s)]\,ds.$$

✦ INSTANT EXERCISE 1

Verify that $y = 1 - e^{-t}$ solves the integral equation $y(t) = \int_0^t [1 - y(s)]\,ds$.

The procedure of Example 1 works for any initial-value problem of the form given by Equation (1). Rewriting the problem with s in place of t, integrating with respect to s, and applying the fundamental theorem of calculus to the left side yield the equation

$$y(t) = \int_0^t f(s, y(s))\,ds. \tag{2}$$

Equation (2) is an **integral equation,** so named because the unknown function y appears inside a definite integral. Integral equations come in several varieties, this one being a *Volterra integral equation of the second kind.* A **Volterra integral equation** is one in which one of the limits of integration is the independent variable; Equation (2) is of the **second kind** because the unknown function appears outside the integral as well as inside. Fortunately, Volterra integral equations of the second kind are the integral equations with the most convenient properties.

The problems of Equations (1) and (2) are equivalent; therefore, any theorem about one corresponds to an equivalent theorem about the other. In stating the integral equation theorem that is equivalent to Theorem A.2.1, we make some additional simplifications to the hypotheses. There is no loss of generality in restricting t to be positive, as the proof for negative t would be identical to the proof for positive t. Nor is there any loss of generality in restricting the rectangle to be symmetric about the t axis. We therefore arrive at an integral equation theorem whose proof is sufficient to establish Theorem 2.4.1.

Theorem A.2.2

Let R be a rectangle defined by $0 \le s \le a$ and $-b \le y \le b$ such that the functions f and $\partial f/\partial y$ are continuous throughout R. Then the integral equation

$$y(t) = \int_0^t f(s, y(s))\,ds$$

has a unique solution on some interval $0 \le t < T$.

Even Theorem A.2.2 requires a lot of sophisticated mathematical analysis to prove. We will sketch the outline of the proof and complete some of the details. It helps to have an example problem in mind.

MODEL PROBLEM A.2

Investigate Theorem A.2.2 for the case

$$f(s, y) = \frac{2s - y}{1 - y}$$

corresponding to the initial-value problem

$$\frac{dy}{dt} = \frac{2t - y}{1 - y}, \qquad y(0) = 0.$$

In the context of Model Problem A.2, we have

$$f(s, y) = \frac{2s - y}{1 - y}, \qquad \frac{\partial f}{\partial y}(s, y) = \frac{2s - 1}{(1 - y)^2}.$$

Both f and $\partial f/\partial y$ are continuous everywhere except on the line $y = 1$. We can choose any $a > 0$ and $0 < b < 1$ for the rectangle R. Theorem A.2.2 claims that the integral equation has a solution on some interval $[0, T)$, but it gives no indication of the correct value of T. The rectangle could extend as far toward infinity as we like, and yet the solution may exist only for a small time beyond $t = 0$.

Picard Iteration

There are a variety of ways to prove existence theorems. If possible, it is best to find a **constructive proof,** which means that the proof includes a formula or algorithm that either computes the solution exactly or at least constructs a sequence that converges uniformly[6] to the solution. There is an algorithm called **Picard iteration**[7] that produces a sequence of functions that converge uniformly to the solution of the integral equation.

[6] *Uniform convergence* means that the convergence does not need to consider any portion of the function domain as being different from the rest. A formal definition can be found in books on advanced calculus or elementary real analysis.

[7] Picard iteration is named after the French mathematician Émile Picard, who used it to prove existence and uniqueness theorems in the 1890s.

Before working on the proof of Theorem A.2.2, we examine Picard iteration as a solution technique. Let $y_0 \equiv 0$. Then we can use the integral equation to define a new function y_1 by using the old function y_0 on the right side while substituting y_1 on the left side. Specifically,

$$y_1(t) = \int_0^t f(s, 0)\, ds.$$

Continuing in this manner, we have the general iteration formula

$$y_n(t) = \int_0^t f(s, y_{n-1}(s))\, ds, \qquad n = 1, 2, \ldots. \tag{3}$$

The terms $y_0, y_1, \ldots$ are called the **Picard iterates** for the integral equation. We refer to the sequence itself as the **Picard sequence** for a given function f.

Consider what might happen with a Picard sequence. It could be that some element y_N of the sequence is a solution of the integral equation. If so, then

$$y_{N+1}(t) = \int_0^t f(s, y_N(s))\, ds = y_N(t).$$

Thus, it is possible that the sequence converges to a solution in a finite number of steps. More commonly, the sequence converges to the solution of the integral equation, but only as $n \to \infty$.

EXAMPLE 2

Consider the integral equation

$$y(t) = \int_0^t [1 - y(s)]\, ds,$$

which we derived in Example 1. Given $y_0 = 0$, the next Picard iterates are

$$y_1(t) = \int_0^t 1\, ds = t, \quad y_2(t) = \int_0^t (1 - s)\, ds = t - \frac{t^2}{2}, \quad y_3(t) = \int_0^t \left(1 - s + \frac{s^2}{2}\right) ds = t - \frac{t^2}{2} + \frac{t^3}{6}.$$

These iterates are displayed together in Figure A.2.1, along with the solution. Certainly the iterates seem to be approaching the solution. In general, we have

$$y_n = -\sum_{i=1}^{n} \frac{(-t)^i}{i!}.$$

Example 2 illustrates what typically happens when Picard iteration works well. The sequence of iterates is a sequence of partial sums in the Taylor series approximation for the solution. As a theoretical matter, this is ideal because the Taylor series for a solution converges over some interval in t. As a practical matter, the Taylor series approximation for a solution is usually not as useful as a numerical solution. Observe that in Figure A.2.1, the improvement in the sequence of iterates is quite satisfactory near $t = 0$, but it is not very good at $t = 2$. Taylor series approximations are local approximations at the center, in this case $t = 0$. They are created solely from information at the center, so it should not be surprising that they are not very good far away from the center. This theme will recur in Section A.3 when differential equations are solved by power series

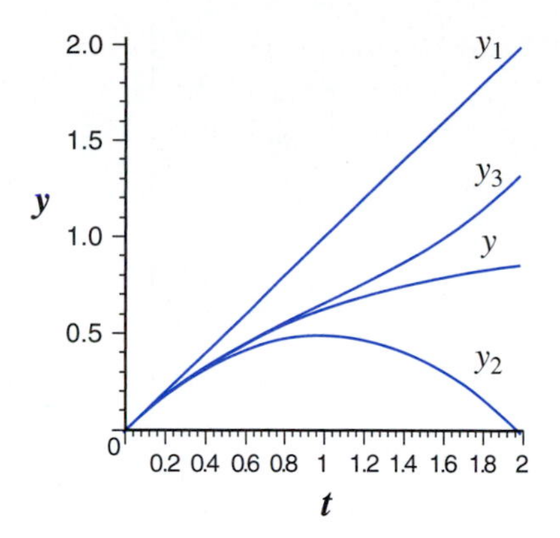

Figure A.2.1
The solution and first three Picard iterates for Example 2.

methods. Thus, the method of Picard iteration is not practical as a method of solving differential equations; however, our purpose here is to establish theory, and Picard iteration is excellent for that purpose.

Now consider the Picard iterates for Model Problem A.2. This problem is one that cannot be solved by any exact analytical method, so the solution can at best be approximated, using graphical (slope field), numerical (Euler, modified Euler, rk4), or analytical (Picard iteration, power series) methods. If we begin with $y_0 = 0$, the next iterate is

$$y_1 = \int_0^t 2s\, ds = t^2.$$

The next iterate,

$$y_2 = \int_0^t \frac{2s - s^2}{1 - s^2}\, ds = t - \frac{1}{2}\ln(1 - t) - \frac{3}{2}\ln(1 + t),$$

is somewhat more complicated. The next iterate,

$$y_3 = \int_0^t \frac{2s + \ln(1 - s) + 3\ln(1 + s)}{2 - 2s + \ln(1 - s) + 3\ln(1 + s)}\, ds,$$

is a definite integral that can only be evaluated numerically. As typically happens with nonlinear problems, the integrals become more complicated as n increases. A plot of these iterates appears in Figure A.2.2, along with the solution as determined by the rk4 method.

✦ INSTANT EXERCISE 2

Verify the result for y_2 for Model Problem A.2.

The Picard iterates are quite good for $t < 0.7$, but after that there are difficulties. The iterate y_2 is only defined for $t < 1$, and the iterate y_3 seems to have a smaller domain yet. This is not necessarily a problem, because the actual solution has an even smaller domain. But it does raise a

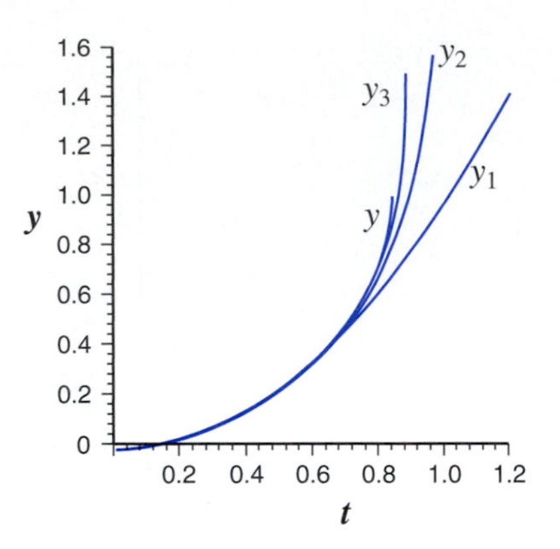

Figure A.2.2
The iterates y_1, y_2, and y_3 for Model Problem A.2, along with the solution of the problem.

question that needs to be answered: Is it possible that the domain of the Picard iterates just keeps getting smaller? Could it be that the Picard sequence eventually yields a term that does not exist at all?

Four Issues That Need to Be Settled

We turn now to the use of Picard iteration to establish Theorem A.2.2. Four statements must be demonstrated to prove the theorem:

1. There are infinitely many terms in the Picard sequence.
2. The Picard sequence converges uniformly to some function Y.
3. The limit function Y solves the integral equation.
4. The solution of the integral equation is unique.

We now discuss each of these issues in turn.

Is the Picard Sequence Infinite? Let R be a rectangular region $0 \leq t \leq a$, $-b \leq y \leq b$ that satisfies the requirements of Theorem A.2.2 for a given function f. The function f is continuous, and the region R is closed and bounded. These conditions guarantee that f has a global maximum and a global minimum in R, which in turn guarantees that $|f|$ has a global maximum. Define the number M by

$$M = \max_R |f|.$$

This guarantees that $|f(s, y)| \leq M$ at every point in R, which in turn places a bound on the magnitude of each Picard iterate:

$$|y_n(t)| \leq \left| \int_0^t f(s, y_{n-1}(s))\, ds \right| \leq \int_0^t |f(s, y_{n-1}(s))|\, ds \leq \int_0^t M\, ds = Mt \leq Ma.$$

The integral that defines the Picard iterate exists as long as the iterate stays inside R. If $Ma \le b$, then the iterate stays inside R for the whole interval $[0, a]$. Thus, for any choice of b, there is a positive value of a small enough that all iterates are defined for $0 \le t \le a$. Therefore, the possibility that the Picard sequence ends because each term has a smaller domain than the previous term is ruled out. For any given integral equation, it is possible to determine a guaranteed minimum interval of existence for the sequence of Picard iterates. This extra effort yields a guaranteed interval of existence for the solution of the problem, a conclusion that does not appear in the theorem. Exercise 9 directs you through the process for Model Problem A.2.

Does the Picard Sequence Converge Uniformly? This question is a difficult one. We begin with an example in which the details are not difficult.

EXAMPLE 3

The Picard iterates of Example 2 are all known explicitly as

$$y_n = -\sum_{i=1}^{n} \frac{(-t)^i}{i!}.$$

Each term is a partial sum of the series

$$Y = -\sum_{i=1}^{\infty} \frac{(-t)^i}{i!},$$

which is the Taylor series for the solution $y = 1 - e^{-t}$. Let E_n be the error in the approximation y_n; that is,

$$E_n(t) = |y_n(t) - y(t)|.$$

By Taylor's theorem,[8] for each t there exists a specific, but unknown, number z such that $0 \le z \le t$ and

$$E_n(t) = \left| \frac{d^{n+1}y}{dt^{n+1}}(z) \frac{t^{n+1}}{(n+1)!} \right|.$$

From the solution for y, we have

$$y' = e^{-t}, \qquad y'' = -e^{-t}, \qquad \ldots.$$

Thus, the error formula simplifies to

$$E_n(t) = \frac{e^{-z(t)} t^{n+1}}{(n+1)!}.$$

Let T be any positive number, and consider the interval $0 \le t \le T$. The largest value $e^{-z(t)}$ can take is 1, and the largest value t^{n+1} can take is T^{n+1}. Therefore,

$$E_n(t) \le \frac{T^{n+1}}{(n+1)!}.$$

The term on the right is larger than E_n and converges to 0 as $n \to \infty$; therefore, the sequence E_n converges to 0 and the sequence y_n converges to y. The convergence is *uniform* because the convergent upper bound is independent of t.

[8] See any calculus book.

We had a significant advantage in Example 3 that we do not have with Model Problem A.2 or in general: we knew in advance the function to which the sequence converges, whereas in Model Problem A.2 we do not. This allowed us to calculate formulas for E_n. In the model problem, we only have simple formulas up to y_2. See Exercises 6 and 8 for arguments that establish the uniform convergence of Picard sequences in general.

Does the Limit of the Picard Sequence Solve the Integral Equation? We would like to use the calculation

$$\begin{aligned} Y &= \lim_{n\to\infty} y_n \\ &= \lim_{n\to\infty} \int_0^t f[s, y_{n-1}(s)]\, ds \\ &= \int_0^t \lim_{n\to\infty} f[s, y_{n-1}(s)]\, ds \\ &= \int_0^t f\left[s, \lim_{n\to\infty} y_{n-1}(s)\right] ds \\ &= \int_0^t f[s, Y(s)]\, ds \end{aligned}$$

to establish that Y satisfies the integral equation. Each of the equalities needs to be justified. The first and last follow immediately from the convergence of the sequence, and the second follows from the definition of the Picard sequence. It is the third and fourth that require care. These two equalities are established by standard theorems in analysis, which prove that the limit can be moved inside the integral and then inside the function evaluation whenever the sequence converges uniformly and the function is continuous. A deep discussion of the mathematics involved in these arguments belongs in a course on analysis.[9]

Is the Solution of the Integral Equation Unique? Again, this question is easily resolved for a simple example.

EXAMPLE 4

Suppose we have found a solution Y of an integral equation. The standard plan for demonstrating uniqueness is to let Z be an arbitrary solution of the integral equation and show that $Y - Z = 0$, whence Z is the same function as Y. For the integral equation of Example 3, we have a formula for Y, but of course we have no formula for Z. What we do know about Z is that it, like Y, satisfies the integral equation. Thus, we have

$$Y(t) - Z(t) = \int_0^t [1 - Y(s)]\, ds - \int_0^t [1 - Z(s)]\, ds = \int_0^t [Y(s) - Z(s)]\, ds.$$

Now suppose $Y(t) - Z(t) \geq 0$. Then

$$|Y(t) - Z(t)| = \left| \int_0^t [Y(s) - Z(s)]\, ds \right| = \int_0^t |Y(s) - Z(s)|\, ds.$$

[9] See, for example, W. Rudin, *Principles of Mathematical Analysis*, 3rd ed., McGraw-Hill, New York, 1976.

The same result holds if $Y(t) - Z(t) \le 0$. If $Y - Z$ is sometimes positive and sometimes negative, then

$$|Y(t) - Z(t)| = \left| \int_0^t [Y(s) - Z(s)]\, ds \right| < \int_0^t |Y(s) - Z(s)|\, ds.$$

The result also follows intuitively from the observation that the first integral is a sum of positive and negative contributions, while the integral on the right is a sum of positive contributions. In general, we have

$$|Y(t) - Z(t)| \le \int_0^t |Y(s) - Z(s)|\, ds. \tag{4}$$

Now we define a function u by

$$u(t) = \int_0^t |Y(s) - Z(s)|\, ds \ge 0.$$

We can differentiate u by the fundamental theorem of calculus, with the result

$$u'(t) = |Y(s) - Z(s)|.$$

Thus, inequality (4) can be written as

$$u'(t) \le u(t), \tag{5}$$

where u, by its definition, must be a nonnegative function. At this point, it is possible to use inequality (5) to demonstrate that u is nonpositive. (See Exercise 5.) But if u is nonnegative *and* nonpositive, then $u \equiv 0$, which means that $Z = Y$. The arbitrary solution Z must be the same as the known solution Y; therefore the solution Y is unique.

The argument of Example 4 doesn't work for Model Problem A.2 or in general. The necessary arguments for the general case are developed in Exercises 6 and 7.

A.2 Exercises

In Exercises 1 through 4, compute the Picard iterates y_1, y_2, and y_3. Plot the iterates to see if they seem to be converging. Also compare with a plot of a numerical solution.

T 1. $y' = 1 - ty, \quad y(0) = 0$

T 2. $y' = t + t^2 y, \quad y(0) = 0$

T 3. $y' = t^2 + y^2, \quad y(0) = 0$

T 4. $y' = 1 - \cos y, \quad y(0) = 0$

5. This exercise completes Example 4.
   a. Use inequality (5) to prove $\left(e^{-t}u\right)' \le 0$.
   b. Use the result of part *a* to show that u must be nonpositive for $t \ge 0$. (*Hint*: you need to know the value of u at one point.)

6. A function f of one variable is said to satisfy a **Lipschitz condition** on an interval I if there exists a number K such that $|f(x_2) - f(x_1)| \le K|x_2 - x_1|$ for any $x_1, x_2 \in I$.
   a. Explain the meaning of the Lipschitz condition in terms of the graph of f. (*Hint*: divide both sides by $|x_2 - x_1|$ and think about the meaning of the quotient.)

b. Show that the function $f(x) = |x|$ satisfies a Lipschitz condition even though the function is not differentiable.

c. Consider two points on the graph of $f(x) = x^2$ for the interval $0 \le x \le 2$. Let $x_1 < x_2$ be arbitrary points in the interval. Define $g(x_1, x_2) = [f(x_2) - f(x_1)]/(x_2 - x_1)$. Determine the maximum value of g on the domain $0 \le x_1 \le 2$, $0 \le x_2 \le 2$. What connection does this calculation have with the Lipschitz condition?

d. Show that any differentiable function satisfies a Lipschitz condition on a closed interval. (*Hint*: use the mean value theorem from calculus.)

e. Suppose $f(t, y)$ and $\partial f/\partial y$ are continuous on a closed rectangle R. Use the result of part *d* to show that f satisfies a Lipschitz condition in y; that is, show that there is a number K such that $|f(t, y_2) - f(t, y_1)| \le K|y_2 - y_1|$ whenever t, y_1, and y_2 are chosen so that (t, y_1) and (t, y_2) are both in R.

Exercises 7 and 8 resolve some of the issues in the sketch of the proof of the existence and uniqueness theorem. Both exercises use the result of Exercise 6.

7. Show that solutions of the integral equation $y(t) = \int_0^t f(s, y(s))\, ds$ are unique whenever f is continuous on a rectangular region $0 \le s \le a$, $-b \le y \le b$ and satisfies a Lipschitz condition in y in the same region.

   a. Let Y and Z be solutions of the integral equation. Show that

$$|Z(t) - Y(t)| \le \int_0^t |f(s, Z(s)) - f(s, Y(s))|\, ds.$$

   b. Use the Lipschitz condition to show that

$$|Z(t) - Y(t)| \le K \int_0^t |Z(s) - Y(s)|\, ds,$$

   where K is the maximum value of $|\partial f/\partial y|$ on the region.

   c. Complete the demonstration by defining $u(t) = \int_0^t |Z(s) - Y(s)|\, ds$ as in Example 4.

8. Let $y_1, y_2, \ldots$ be the Picard sequence for a problem that meets the requirements for Theorem A.2.2. Prove that the sequence converges uniformly.

   a. Use the result of Exercise 6 to show that

$$|f(t, y_n(t)) - f(t, y_{n-1}(t))| \le K|y_n(t) - y_{n-1}(t)|.$$

   b. Let M be the maximum of $|f|$ on R. Use the result of part *a* to show that $|y_1(t) - y_0(t)| \le Mt$ on R.

   c. Use mathematical induction to show that $|y_n(t) - y_{n-1}(t)| \le MK^{n-1}t^n/n!$ on R for $n \ge 1$.

   d. Use the result of part *c* to show that

$$\left| \sum_{n=1}^{\infty} [y_n(t) - y_{n-1}(t)] \right| \le \frac{M}{K} e^{Ka}.$$

   Since the upper bound does not depend on t, the convergence of the series

$$\sum_{n=1}^{\infty} [y_n(t) - y_{n-1}(t)]$$

   is uniform.

   e. Use the result of part *d* to establish the uniform convergence of the Picard sequence.

9. In this exercise, we use the properties of the function f in Model Problem A.2 to determine a guaranteed interval of existence for the solution of the problem.
   a. Use slope field arguments to show that the solution of Model Problem A.2 (in the differential equation form) satisfies the restriction $0 \le y \le 2t$ for $t \le \frac{1}{2}$. (*Hint*: where in the ty plane is $y' = 2$?)
   b. Suppose $b < 1$ is given. Let D be the closed region bounded by $y = 0$, $y = b$, $y = 2t$, and $t = a$, with $a > \frac{1}{2}$. Show that f is always positive on D and that the maximum value of f on D occurs at the point (a, b).
   c. Use the result of part b to find a value for the maximum slope of the solution while it remains in D.
   d. Determine a condition on a and b so that the line of maximum slope passes through the point (a, b).
   e. Explain why the solution is guaranteed to exist for times up to the largest value of a that satisfies the condition of part d.
   f. Find the largest value of a that satisfies the condition of part d. [*Hint*: Think of the relation of part d as defining a function $a(b)$. A second equation comes from the condition that the desired point should be the maximum of this function. You now have a system of two algebraic equations to determine the largest value of a.]

✦ A.2 INSTANT EXERCISE SOLUTIONS

1. By using the solution of the initial-value problem,

$$\int_0^t [1 - y(s)]\, ds = \int_0^t e^{-s}\, ds = -e^{-s}\Big|_0^t = -e^{-t} + 1 = y(t).$$

2. By the method of partial fraction decomposition, there are constants A and B such that

$$\frac{s^2 - 2s}{s^2 - 1} = \frac{s^2 - 1 + 1 - 2s}{s^2 - 1} = 1 + \frac{1 - 2s}{s^2 - 1} = 1 + \frac{A}{s - 1} + \frac{B}{s + 1}.$$

To find these constants, we multiply by $s^2 - 1$, with the result

$$1 - 2s = A(s + 1) + B(s - 1).$$

Evaluating this equation at $s = 1$ yields $A = -\frac{1}{2}$; similarly, evaluation at $s = -1$ yields $B = -\frac{3}{2}$. Then

$$y_2 = \int_0^t \left(1 - \frac{1}{2}\frac{1}{s - 1} - \frac{3}{2}\frac{1}{s + 1}\right) ds = \left[t - \frac{\ln(|s - 1|)}{2} - 3\frac{\ln(|s + 1|)}{2}\right]\Bigg|_0^t$$

$$= t - \frac{\ln(1 - t)}{2} - 3\frac{\ln(1 + t)}{2}.$$

A.3 Error in Numerical Methods

Our experience in Sections 2.5 and 2.6 raises some interesting questions about numerical approximations for the solutions of differential equations.

1. What determines the amount of numerical error in an approximation?

2. Why does halving the step size tend to decrease the numerical error in Euler's method by one-half and the numerical error in the modified Euler method by one-quarter?
3. Are some differential equations more difficult to approximate numerically than others? If so, can this be predicted without doing numerical experiments?

None of these questions has a simple answer, but we will at least be able to offer a partial answer for each.

MODEL PROBLEM A.3

The initial-value problem

$$\frac{dy}{dt} = \frac{2y - 18t}{1+t}, \qquad y(0) = 4,$$

has solution $y = 4 + 8t - 5t^2$. Determine how many subdivisions are needed to ensure that the Euler approximation on the interval $[0, 2]$ has error no greater than 0.001.

✦ INSTANT EXERCISE 1

Verify the solution formula for Model Problem A.3.

The Meaning of *Error*

Any approximation of a function necessarily allows a possibility of deviation from the correct value of the function. **Error** is the term used to denote the amount by which an approximation fails to equal the exact solution.[10] Error occurs in an approximation for several reasons.

Truncation error in a numerical method is error that is caused by using simple approximations to represent exact mathematical formulas. The only way to completely avoid truncation error is to use exact calculations. However, truncation error can be reduced by applying the same approximation to a larger number of smaller intervals or by switching to a better approximation. Analysis of truncation error is the single most important source of information about the theoretical characteristics that distinguish good methods from poor ones. With a combination of theoretical analysis and numerical experiments, it is possible to estimate truncation error accurately.

Round-off error in a numerical method is error that is caused by using a discrete number of significant digits to represent real numbers on a computer. Since computers can retain a large number of digits in a computation, round-off error is problematic only when the approximation requires that the computer subtract two numbers that are nearly identical. This is exactly what happens if we apply an approximation to intervals that are too small. Thus, the effort to decrease truncation error can have the unintended consequence of introducing significant round-off error.

[10]In everyday vocabulary, we tend to use the term *error* to denote an avoidable human failing, such as when a baseball player mishandles the ball. In numerical analysis, error is a characteristic of an approximation correctly performed. Unacceptable error dictates a need for greater refinement or a better method.

Practitioners of numerical approximation are most concerned with truncation error, but they also try to restrict their efforts at decreasing truncation error to improvements that do not introduce significant round-off error. In our study of approximation error, we consider only truncation error. We seek information about error on both a local and global scale. *Local truncation error* is the amount of truncation error that occurs in one step of a numerical approximation. *Global truncation error* is the amount of truncation error that occurs in the use of a numerical approximation to solve a problem.

Taylor's Theorem and Approximations

The principal tool in the determination of truncation error is Taylor's theorem, a key result in calculus that provides a formula for the amount of error in using a truncated Taylor series to represent the corresponding function. Taylor's theorem in its most general form applies to approximations of any degree; here we present only the information specifically needed to analyze the error in Euler's method.

Theorem A.3.1

Let y be a function of one variable having a continuous second derivative on some interval $I = [0, t_f]$, and let f be a function of two variables having continuous first partial derivatives on the rectangular region $R = I \times [y_0 - r, y_0 + s]$, with $r, s > 0$.[11]

1. Given T and t in I, there exists some point $\tau \in I$ such that
$$y(t) = y(T) + y'(T)(t - T) + \frac{y''(\tau)}{2}(t - T)^2. \tag{1}$$
2. Given points (T, Y) and (T, y) in R, there exists some point $(T, \eta) \in R$ such that
$$f(T, y) = f(T, Y) + f_y(T, \eta)(y - Y), \tag{2}$$
where f_y is the partial derivative of f with respect to its second argument.

Theorem A.3.1 serves to quantify the idea that the difference in function values for a smooth function should vanish as the evaluation points become closer.

One can be a little more restrictive when specifying the range of possible values for τ and η; however, nothing is gained by doing so. We cannot use Theorem A.3.1 to *compute* the error in an approximation. The theorem provides formulas for the error, but the catch is that there is no way to determine τ and η without knowing the exact solution. It may seem that this catch makes the formulas useless, but this is not the case. We do know that τ and η are confined to a given closed interval, and therefore we can compute worst-case values for the quantities $y''(\tau)$ and $f_y(T, \eta)$.

EXAMPLE 1

Suppose we want to approximate the function $\ln(0.5 + t)$ near the point $t = 0.5$. We have

$$y(t) = \ln(0.5 + t), \qquad y(0.5) = 0, \qquad y'(t) = \frac{1}{0.5 + t}, \qquad y'(0.5) = 1, \qquad y''(t) = \frac{-1}{(0.5 + t)^2}.$$

[11] This notation means that the first variable is in the interval I and the second is in the interval $[y_0 - r, y_0 + s]$.

Let's assume that our goal[12] is to find an upper bound for the largest possible error in approximating $y(t)$ by $t - 0.5$ with $0 \leq t \leq 1$. The approximation formula of Equation (1) yields

$$y(t) = 0 + (t - 0.5) + \frac{y''(\tau)}{2}(t - 0.5)^2,$$

and the approximation error E is defined by

$$E = (t - 0.5) - y(t).$$

We therefore have

$$|E| = \frac{|y''(\tau)|}{2}(t - 0.5)^2,$$

with $0 \leq \tau \leq 1$. Now let

$$M = \max_{0 \leq t \leq 1} |y''(t)| = \max_{0 \leq t \leq 1} \left| \frac{-1}{(0.5 + t)^2} \right| = 4.$$

Given the range of possible τ values, the worst case is $|y''(\tau)| = M$; thus,

$$|E| \leq 2(t - 0.5)^2, \qquad t \in [0, 1].$$

Local Truncation Error for Euler's Method

Consider an initial-value problem

$$y' = f(t, y(t)), \qquad y(0) = y_0, \tag{3}$$

where f has continuous first partial derivatives on some region R defined by $0 \leq t \leq t_f$ and $y_0 - r \leq y \leq y_0 + s$. Euler's method is a scheme for obtaining an approximate value y_{n+1} for $y(t_{n+1})$ by using only the approximation y_n for $y(t_n)$ and the function f that calculates the slope of the solution curve through any point. Specifically, the method is defined[13] by the formula

$$y_{n+1} = y_n + hf(t_n, y_n), \qquad \text{where} \quad h = t_{n+1} - t_n. \tag{4}$$

We define the **global truncation error** at step n in any numerical approximation of (3) by

$$E_n = y_n - y(t_n). \tag{5}$$

Our aim is to find a worst-case estimate of the global truncation error in the numerical scheme for Euler's method, assuming that the correct solution and numerical approximation stay within R. This is a difficult task because we have so little to work with. We will start by trying to determine the relationship between the error at time t_{n+1} and the error at time t_n. Figure A.3.1 shows the relationships among the relevant quantities involved in one step of the approximation. The approximation and exact solution at each of the two time steps are related by the error definitions. The approximations at the two time steps are related by Euler's method. We still need to

[12]This problem is an artificial one because we know a formula for y and can therefore calculate the error exactly. However, the point is to illustrate how information about derivatives of y can be used to generate error estimates for cases where we *don't* have a formula for y.

[13]Section 2.5.

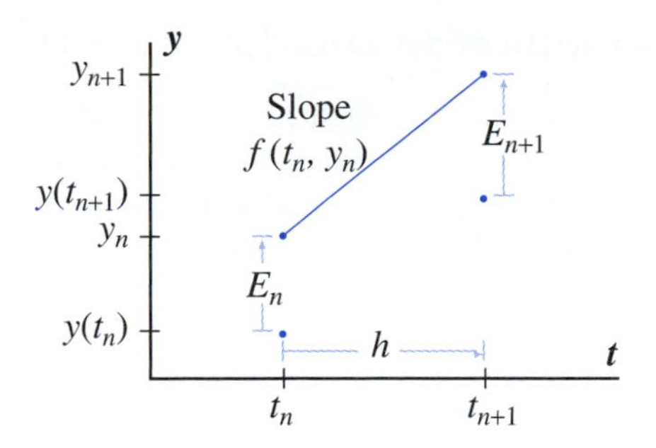

Figure A.3.1
The relationships between the solution values and Euler approximants.

have a relationship between the exact solution values at the two time steps, and this is where Theorem A.3.1 is needed.

The comparison of $y(t_{n+1})$ and $y(t_n)$ must begin with Equation (1). Combining Equation (1), with $t = t_{n+1}$ and $T = t_n$, and Equation (3) yields

$$y(t_{n+1}) = y(t_n) + hf(t_n, y(t_n)) + \frac{h^2}{2}y''(\tau). \tag{6}$$

This result is a step in the right direction, but it is not yet satisfactory. A useful comparison of $y(t_{n+1})$ with $y(t_n)$ can have terms consisting entirely of known quantities and error terms, but not quantities, such as $f(t_n, y(t_n))$, that need to be evaluated at points that are not known exactly. These quantities have to be approximated by known quantities. This is where Equation (2) comes into the picture. Substituting

$$f(t_n, y(t_n)) = f(t_n, y_n) + f_y(t_n, \eta)[y(t_n) - y_n] = f(t_n, y_n) - f_y(t_n, \eta)E_n$$

into Equation (6) gives us

$$y(t_{n+1}) = y(t_n) + hf(t_n, y_n) - hf_y(t_n, \eta)E_n + \frac{h^2}{2}y''(\tau). \tag{7}$$

Equation (7) meets our needs because every term is a quantity of interest, a quantity that can be evaluated, or an error term. Subtracting this equation from the Euler formula (4) yields

$$y_{n+1} - y(t_{n+1}) = y_n - y(t_n) + hf_y(t_n, \eta)E_n - \frac{h^2}{2}y''(\tau),$$

or

$$E_{n+1} = [1 + hf_y(t_n, \eta)]E_n - \frac{h^2}{2}y''(\tau). \tag{8}$$

This result indicates the relationship between the errors at successive steps. The **local truncation error** is defined to be the error in step $n + 1$ when there is no error in step n; hence, the local truncation error for Euler's method is $-h^2y''(\tau)/2$. The local truncation error has two factors of h, and we say that it is $\mathcal{O}(h^2)$.[14] The quantity $[1 + hf_y(t_n, \eta)]E_n$ represents the error at step $n + 1$ caused by the error at step n. This propagated error is larger than E_n if $f_y > 0$ and smaller than E_n if $f_y < 0$. Having $f_y < 0$ is generally a good thing because it causes truncation errors to diminish as they propagate.

---

[14]This is read as "big oh of h^2."

Global Truncation Error for Euler's Method

As we see by Equation (8), the error at a given step consists of error propagated from the previous steps along with error created in the current step. These errors might be of opposite signs, but the quality of the method is defined by the worst case rather than the best case. The worst case occurs when y'' and E_n have opposite algebraic signs and $f_y > 0$. Thus, we can write

$$|E_{n+1}| \leq \left[1 + h|f_y(t_n, \eta)|\right] |E_n| + \frac{|y''(\tau)|}{2}h^2. \tag{9}$$

The quantity $f_y(t_n, \eta)$ cannot be evaluated, but we do at least know that f_y is continuous. We can also show that y'' is continuous: by using the chain rule, we can differentiate the differential equation (3) to obtain the formula

$$y'' = f_t + f_y y' = f_t + f f_y. \tag{10}$$

The problem statement requires f to have continuous first derivatives, so y'' must also be continuous. Equation (10) also gives us a way to calculate $y''(\tau)$ at any point $(\tau, y(\tau))$ in R. Given that the region R is closed and bounded, we can apply the theorem from multivariable calculus that says that a continuous function on a closed and bounded region has a maximum value and a minimum value. Hence, we can define positive numbers K and M by

$$K = \max_{(t,y)\in R} |f_y(t, y)| < \infty, \qquad M = \max_{(t,y)\in R} |(f_t + f f_y)(t, y)| < \infty. \tag{11}$$

For the worst case, we have to assume $f_y(t_n, \eta) = K$ and $|y''(\tau)| = |(f_t + f f_y)(\tau, y(\tau))| = M$, even though neither of these is likely to be true. Substituting these bounds into the error estimate of Equation (9) puts an upper bound on the size of the truncation error at step $n + 1$ in terms of the size of the truncation error at step n and the global properties of f:

$$|E_{n+1}| \leq (1 + Kh)|E_n| + \frac{Mh^2}{2}. \tag{12}$$

In all probability, there is a noticeable gap between actual values of $|E_{n+1}|$ and the upper bound in Equation (12). The upper bound represents not the worst case for the error, but the error for the unlikely case where each quantity has its worst possible value.

Some messy calculations (see Exercise 5) are needed to obtain an upper bound for the global truncation error from the error bound of Equation (12). Theorem A.3.2 summarizes the main result.

Theorem A.3.2 Let I and R be defined as in Theorem A.3.1, and let K and M be defined as in Equation (11). Let $y_1, y_2, \ldots, y_N$ be the Euler approximations at the points $t_n = nh$. If the points (t_n, y_n) and $(t_n, y(t_n))$ are in R, then the error in y_n is bounded by

$$|E_n| \leq \frac{M}{2K}(e^{Kt_n} - 1)h.$$

There are two important conclusions to draw from Theorem A.3.2. The first is that the error vanishes as $h \to 0$. Thus, the truncation error can be made arbitrarily small by reducing h, although doing so eventually causes large round-off errors. The second conclusion is that the error in the method is approximately proportional to h. The **order** of a numerical method is the number of factors of h in the global truncation error estimate for the method. Euler's method is therefore a **first-order** method. Halving the step size in a first-order method reduces the error by approximately a factor of 2. As with Euler's method, the order of most methods is 1 less than the power of h in the local truncation error.

Returning to Model Problem A.3, we can compute the global error bound by finding the values of K and M. Given the function f from Model Problem A.3, we have

$$f_y(t, y) = \frac{2}{1+t}, \qquad (f_t + ff_y)(t, y) = \frac{2y - 36t - 18}{(1+t)^2}.$$

The interval I is $[0, 2]$, as prescribed by the problem statement. The largest and smallest values of y on I can be determined by graphing the exact solution, finding the vertex algebraically, or applying the standard techniques of calculus. By any of these methods,[15] we see that $0 \le y \le 7.2$. Clearly, f_y is always positive and achieves its largest value in R on the boundary $t = 0$; hence, $K = 2$. It can be shown that $f_t + ff_y$ is negative throughout R. Since $f_t + ff_y$ is increasing in y, its minimum (maximum in absolute value) must occur on the line $y = 0$. In fact, the minimum occurs at the origin, so $M = 18$.

✦ INSTANT EXERCISE 2

Derive the formula for $f_t + ff_y$ for Model Problem A.3. Also verify that $f_t + ff_y$ is negative throughout R and that the minimum of $f_t + ff_y$ on R occurs at the origin.

With these values of K and M for Model Problem A.3, Theorem A.3.2 yields the estimate

$$|E_n| \le 4.5\left(e^{2t_n} - 1\right)h.$$

In particular, at $t_N = 2$ the error estimate is approximately $241h$. Thus, setting $241h < 0.001$ is sufficient to achieve the desired error tolerance. Since $Nh = 2$, the conclusion is that $N = 482{,}000$ is large enough to guarantee the desired accuracy. This number is larger than what would be needed in actual practice, owing to our always assuming the worst in obtaining the error bound. The important part of Theorem A.3.2 is the order of the method, not the error estimate. Knowing the order of the method allows us to obtain an accurate estimate of the error from numerical experiments. Figure A.3.2 shows the error in the Euler approximations at $t = 2$ using various step sizes. The points fall very close to a straight line, indicating the typical first-order behavior. From the slope of the line, we estimate the actual numerical error to be approximately $E_N = 30h$, which is roughly one-eighth of that computed using the error bound of Theorem A.3.2. This still indicates that 60,000 steps are required to obtain an error within the specified tolerance of 0.001. The desired accuracy ought better be attempted by using a better method.

Conditioning

Why does Euler's method do so poorly with Model Problem A.3? Recall the interpretation of the local truncation error formula (8). Propagated error increases with n if $f_y > 0$ and decreases

[15] We are admittedly cheating by using the solution formula, which we do not generally know. In practice, one can use a numerical solution to estimate the range of possible y values.

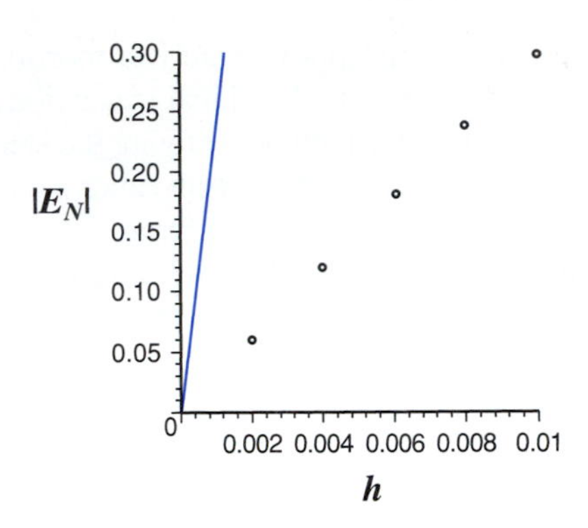

Figure A.3.2
The Euler method error in $y(2)$ for Model Problem A.3, using several values of h, along with the theoretical error bound $241h$.

with n if $f_y < 0$. A differential equation is said to be **well-conditioned** if $\partial f/\partial y < 0$ and **ill-conditioned** if $\partial f/\partial y > 0$. Ill-conditioned equations are ones for which the truncation errors are magnified as they propagate. The reason that ill-conditioned problems magnify errors can be seen by an examination of the slope field for an ill-conditioned problem.

Figure A.3.3 shows the slope field for Model Problem A.3 along with the numerical solution using just four steps. The initial point (0, 4) is on the correct solution curve, but the next approximation point (0.5, 8) is not. This point is, however, on a different solution curve. If we could avoid making any further error beyond $t = 0.5$, the numerical approximation would follow *a* solution curve from then on, but it would not be the *correct* solution curve. The amount of error ultimately caused by the error in the first approximation step can become larger or smaller at later times, depending on the relationship between the different solution curves. Figure A.3.4 shows our approximation again, along with the solution curves that pass through all the points used in

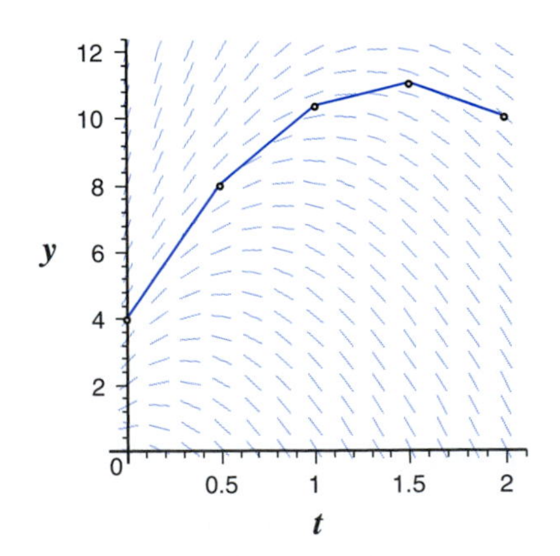

Figure A.3.3
The Euler method approximation for Model Problem A.3 using $h = 0.5$, along with the slope field.

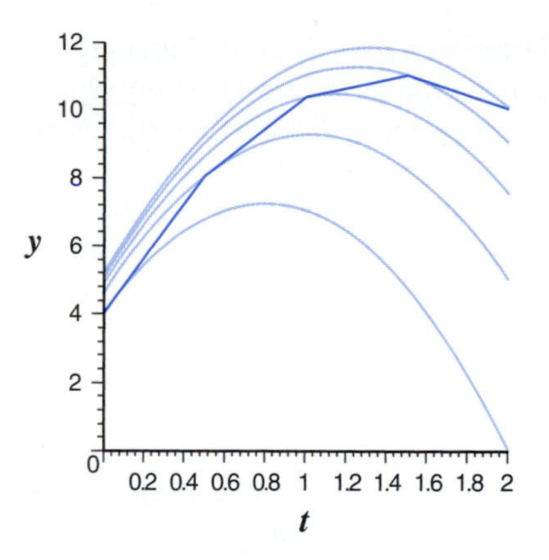

Figure A.3.4
The Euler method approximation for Model Problem A.3 using $h = 0.5$, along with the solution curves that pass through the points used in the approximation.

the approximation. The solution curves are spreading apart with time, carrying the approximation farther away from the correct solution. The error introduced at each step moves the approximation onto a solution curve that is farther yet from the correct one.

Recall that for Model Problem A.3, we have

$$\frac{\partial f}{\partial y} = \frac{2}{1+t}.$$

The magnitude of f_y ranges from 2 at the beginning to $\frac{2}{3}$ at $t = 2$, and yet this degree of ill-conditioning is sufficient to cause the large truncation error. The error propagation is worse for a differential equation that is more ill-conditioned than that in our model problem. In general, Euler's method is inadequate for any ill-conditioned problem. Well-conditioned problems are "forgiving" in the sense that numerical error moves the solution onto a different path that converges toward the correct path. Nevertheless, Euler's method is inadequate even for well-conditioned problems if a high degree of accuracy is required, owing to the slow first-order convergence. Commercial software generally uses fourth-order methods similar to the rk4 method introduced in Section 2.6. For most purposes, the second-order modified Euler method is adequate.

A.3 Exercises

1. Consider the problem

$$\frac{dy}{dt} = \frac{8e^{-t}}{3+y}, \qquad y(0) = 0,$$

on the interval $0 \le t \le 1$. (This was Model Problem 2.5.)

a. Determine the values of K and M. You may assume that the solution and approximations stay within the interval $0 \leq y \leq 2$.

b. Use Theorem A.3.2 to predict the error in the Euler approximations at $t = 0.2, 0.4, \ldots, 1.0$ for $h = 0.025$.

c. Compare the predicted errors with the actual error data reported in Table 2.5.2. In particular, determine the ratio of actual error to predicted error for each t. Explain the pattern of the results in terms of the properties of the function $8e^{-t}/(3+y)$.

T 2. In the text, we showed that approximately 60,000 steps are required to approximate $y(2)$ with error less than 0.001 by using Euler's method.

a. Determine by experiment the number of steps required to achieve the desired tolerance by using the modified Euler method. How many evaluations of the function f are computed in the approximation?

b. Repeat part *a* with the rk4 method.

c. In general, what is the recommended plan for numerical solution of ill-conditioned problems?

3. In this exercise, we demonstrate that the local truncation error for the trapezoidal rule is $\mathcal{O}(h^3)$. It is sufficient to calculate the local truncation error in the first step of the approximation (and easier than calculating the error in the nth step because there is no propagated error in the first step). Since we anticipate error at $\mathcal{O}(h^3)$ rather than $\mathcal{O}(h^2)$, as in Euler's method, we need to use a form of Taylor's theorem that includes an additional term. Assume that the function f in the initial-value problem

$$y' = f(t, y), \qquad y(0) = y_0,$$

has continuous second derivatives for all t and y, which is sufficient to guarantee that y has continuous third derivatives and allows us to ignore consideration of the region R. Thus, Taylor's theorem implies the formula

$$y(t) = y(T) + y'(T)(t-T) + y''(T)\frac{(t-T)^2}{2} + y'''(\tau)\frac{(t-T)^3}{6}$$

where T and t are any two times.

a. Use the given formula to write the Taylor theorem approximation for $y(t_1)$ in terms of y and its derivatives at t_0. Use it again to write the approximation for $y(t_0)$ in terms of y and its derivatives at t_1. Solve this last equation for $y(t_1)$ and average the result with the first equation.

b. Use the differential equation to replace the y' terms in the result of part *a* with the function f evaluated at the appropriate points. [Keep in mind that $y(t_0) = y_0$.]

c. Subtract the result of part *b* from the trapezoidal rule formula for y_1 (see Sec. 2.6).

d. Use Taylor's theorem to reduce the pair of $\mathcal{O}(h)$ terms in the result of part *c* to a term that includes the error E_1. Also use Taylor's theorem to reduce the pair of $\mathcal{O}(h^2)$ terms to an $\mathcal{O}(h^3)$ term. Solve for E_1.

4. *a.* Let y_1 be the trapezoidal method approximation of $y(t_1)$ for the problem

$$y' = f(t, y), \qquad y(0) = y_0,$$

and let Y_1 be the modified Euler method approximation of the same quantity. Use Taylor's theorem to determine the order of $Y_1 - y_1$.

b. Given that the trapezoidal method is second-order, use the result of part *a* to determine the order of the modified Euler method.

5. In this exercise, we guide you through the process of obtaining Theorem A.3.2 from the local truncation error bound of Equation (12).

   a. Compute the error bounds for $|E_1|$, $|E_2|$, and $|E_3|$, and use them to develop a bound for $|E_n|$ of the form

$$|E_n| \le \frac{Mh^2}{2} S,$$

   where S is a finite sum of n terms.

   b. Write down the sum for $(1 + Kh)S$ and subtract the sum for S from it to obtain a simple formula for KhS. Use this result to replace the sum in the error bound.

   c. Show that $1 + Kh < e^{Kh}$ when $Kh < 1$. (*Hint*: what is the Taylor series for e^x?)

   d. Combine the results of parts *b* and *c* to obtain the error bound in Theorem A.3.2.

T 6. In a realistic numerical problem, the correct solution is not available for use in determining the error in the approximation. In this exercise, we consider a simple method for including an error estimate along with the numerical approximation.

   a. Let $y(t)$ be the unknown value of the function y at time t. Let $y(t; h)$ be the approximation of $y(t)$ obtained by a first-order method using step size h. Use the approximation values $y(t; h)$ and $y(t; 2h)$ to obtain an estimate for $y(t)$ that is better than $y(t; h)$. Compare this "exact value" with $y(t; h)$ to estimate the error in $y(t; h)$.

   b. Use Euler's method with $h = 0.01$ to approximate $y(1)$ and the error in the approximation for the problem

$$y' = t^2 + y^2, \qquad y(0) = 0.$$

   c. Use the rk4 method with $h = 0.01$ or a commercial numerical solver to determine the value of $y(1)$. Assuming that this is the correct value, how does the actual error in the approximation of part *b* compare with the error approximation computed in part *b*?[16]

7. A common application of Newton's law of cooling (Section 1.1) in textbooks is the problem of determining the time of death of a homicide victim. Suppose the corpse is discovered at 6 A.M. in a room where the ambient temperature is 68°F. The temperature of the corpse is 77°F. The detective waits until 7 A.M. and then measures the temperature to be 74°F.

   a. Estimate the time of death, assuming that normal body temperature is 98.6°F.

   b. Suppose the victim had a fever at the time of death, so that the actual starting body temperature was 100°F. How much error does the assumption of normal body temperature at death cause in this case?

   c. Suppose the temperature at 6 A.M. was actually 76.8°F and the temperature at 7 A.M. was actually 74.2°F. How much error does the imprecision of temperature measurement cause?

[16]It can be shown that the "method" of part *a* is second-order. In general, the technique of combining two or more approximations to obtain a better approximation is called the **Richardson extrapolation.**

d. Explain why the small errors in measurement in part *c* lead to relatively large errors in the estimated time. (*Hint*: Determine whether the equation for cooling is well-conditioned or ill-conditioned. Note that we know the final value and an intermediate value and we are trying to calculate the initial value. This is like running the cooling process backward.)

e. When you watch a crime show on TV, you see the investigator estimate the time of death from just one measurement. For this to work, what additional assumption must the investigator make?

✦ A.3 INSTANT EXERCISE SOLUTIONS

1. From $y = 4 + 8t - 5t^2$, we have $y' = 8 - 10t$. Thus, $(1+t)y' = 8 - 2t - 10t^2$ and $2y - 18t = 8 - 2t - 10t^2$.

2. We have

$$\frac{\partial f}{\partial t} = \frac{-18(1+t) + (18t - 2y)}{(1+t)^2} = \frac{-2y - 18}{(1+t)^2}$$

and

$$f\frac{\partial f}{\partial y} = \frac{2y - 18t}{1+t}\,\frac{2}{1+t} = \frac{4y - 36t}{(1+t)^2}.$$

Thus,

$$y'' = \frac{-2y - 18}{(1+t)^2} + \frac{4y - 36t}{(1+t)^2} = \frac{2y - 36t - 18}{(1+t)^2}.$$

Given $y = 0$, $y'' = g(t) = -18(1+2t)/(1+t)^2$. The derivative of this function is

$$g' = -18\frac{2(1+t)^2 - 2(1+2t)(1+t)}{(1+t)^4} = -18\frac{2(1+t) - 2(1+2t)}{(1+t)^3} = \frac{36t}{(1+t)^3} \geq 0;$$

hence, the minimum of g occurs at the smallest t, namely, $t = 0$.

A.4 Power Series Solutions

Many important and interesting differential equations cannot be solved exactly. Instead, one must settle for qualitative information and/or an approximate solution. Qualitative information can be obtained by graphical methods (Chapter 5). Approximate solutions can be obtained by numerical methods, such as the rk4 (Section 2.6) method. It is also possible to use analytical methods to obtain formulas that approximate solutions. One of these methods attempts to find an approximate solution in the form of a power series centered at the point of the initial condition. A full treatment of this subject is far beyond the scope of an elementary course in differential equations.[17] Our presentation here is intended to serve as a quick introduction to the subject.

We assume that initial conditions are given at $x = 0$ and that the coefficient functions are polynomials.

[17] See, for example, Carl M. Bender and Steven A. Orszag, *Advanced Mathematical Methods for Scientists and Engineers*, McGraw-Hill, New York, 1978.

MODEL PROBLEM A.4*a*

The Airy function Ai is the solution of the initial-value problem

$$y'' = xy, \qquad y(0) = c_1, \qquad y'(0) = -c_2,$$

where

$$c_1 = \frac{1}{\pi}\int_0^\infty \cos\frac{t^3}{3}\,dt \approx 0.3550280539,$$

$$c_2 = \frac{1}{\pi}\int_0^\infty t\sin\frac{t^3}{3}\,dt \approx 0.2588194038.$$

Find the solution of this initial-value problem in the form

$$\mathrm{Ai}\,(x) = \sum_{n=0}^{\infty} a_n x^n = a_0 + a_1 x + a_2 x^2 + \cdots$$

or show that no such solution exists.

Simply put, the power series method consists of replacing all occurrences of the unknown function in the initial-value problem with the assumed power series. It is instructive to attempt to solve Model Problem A.4*a* using the power series in their expanded forms. Thus, we assume a solution

$$y = a_0 + a_1 x + a_2 x^2 + a_3 x^3 + \cdots,$$

with the coefficients to be determined. Unlike most other methods of solution, the power series method does not require solution of the differential equation before application of the initial conditions; indeed, there are advantages to applying the initial conditions first. Substituting the power series into the condition $y(0) = c_1$, we immediately see that $a_0 = c_1$. Similarly, the condition $y'(0) = -c_2$ requires $a_1 = -c_2$. With these values, the power series becomes

$$y = c_1 - c_2 x + a_2 x^2 + a_3 x^3 + \cdots.$$

From this formula, we have

$$y'' = 2a_2 + 6a_3 x + 12a_4 x^2 + 20a_5 x^3 + 120a_6 x^4 + \cdots$$

and

$$xy = c_1 x - c_2 x^2 + a_2 x^3 + a_3 x^4 + \cdots.$$

Substituting these into the Airy equation yields the equation

$$2a_2 + 6a_3 x + 12a_4 x^2 + 20a_5 x^3 + 120a_6 x^4 + \cdots = c_1 x - c_2 x^2 + a_2 x^3 + a_3 x^4 + \cdots.$$

Two power series are equal only when all the coefficients are equal. Equating the coefficients in these series determines the values of the previously unknown coefficients. Thus, we must have

$$2a_2 = 0, \qquad 6a_3 = c_1, \qquad 12a_4 = -c_2, \qquad 20a_5 = a_2, \qquad 120a_6 = a_3,$$

and so on. These can be solved consecutively, with the results

$$a_2 = 0, \qquad a_3 = \frac{c_1}{6}, \qquad a_4 = -\frac{c_2}{12}, \qquad a_5 = 0, \qquad a_6 = \frac{a_3}{120} = \frac{c_1}{720}.$$

A General Formula for the Coefficients

Insofar as power series methods involve only simple manipulations of polynomials, they are not at all difficult. The difficulty comes when we try to obtain a general formula for the coefficients so as to decrease the extent of tedious calculations. In summation notation, the important quantities are

$$y' = \sum_{n=1}^{\infty} n a_n x^{n-1}, \qquad y'' = \sum_{n=2}^{\infty} (n-1)n a_n x^{n-2}, \qquad xy = \sum_{n=0}^{\infty} a_n x^{n+1}.$$

Thus, the differential equation in power series form is

$$\sum_{n=2}^{\infty} (n-1)n a_n x^{n-2} = \sum_{n=0}^{\infty} a_n x^{n+1}.$$

The series on the left begins with the constant term $2a_2$, at $n = 2$, and the one on the right begins with $a_0 x$ at $n = 0$. These terms have different powers of x, so they are not equal to each other. If instead we compare the $n = 2$ terms, we get $2a_2$ and $a_2 x^3$, and these are not to be set equal either. How do we arrange to compare the corresponding terms in the two series?

The answer is that we have to write the series so that the corresponding terms in each series have the same power of x. It doesn't do to have one series use x^{n-2} and the other x^{n+1}. Look again at the series for xy in expanded form: $xy = a_0 x + a_1 x^2 + \cdots$. We want the first term to be the $n = 1$ term, not the $n = 0$ term, so that each term has the factor x^n. The coefficients are no longer a_n, however; instead, the coefficients are a_{n-1}. Thus,

$$xy = \sum_{n=0}^{\infty} a_n x^{n+1} = \sum_{n=1}^{\infty} a_{n-1} x^n.$$

(The reader should expand these two series to confirm that they are the same.) Similarly, the second-derivative series is

$$y'' = \sum_{n=2}^{\infty} (n-1)n a_n x^{n-2} = \sum_{n=0}^{\infty} (n+1)(n+2) a_{n+2} x^n.$$

✦ INSTANT EXERCISE 1

Use the expanded version of y'' to derive the sums $\sum_{n=2}^{\infty} (n-1)n a_n x^{n-2}$ and $\sum_{n=0}^{\infty} (n+1)(n+2) a_{n+2} x^n$.

By using the series forms where n is the power rather than the coefficient, we have the equation

$$\sum_{n=0}^{\infty} (n+1)(n+2) a_{n+2} x^n = \sum_{n=1}^{\infty} a_{n-1} x^n.$$

The series are now "lined up" so that terms that ought to be equated are actually equated. For $n = 0$ we have $2a_2 = 0$, and for $n > 0$ we have $(n+1)(n+2)a_{n+2}x^n = a_{n-1}x^n$. Solving this last equation for a_{n+2} (the coefficient with the larger subscript), we have

$$a_{n+2} = \frac{a_{n-1}}{(n+1)(n+2)}, \qquad n \geq 1. \tag{1}$$

Equation (1) is called the **recurrence relation** for the power series solution. From it, we can calculate $a_3 = a_0/6 = c_1/6$, $a_4 = a_1/12 = -c_2/12$, and all other remaining coefficients.

Increasing the Efficiency of the Method

The key to the power series method is careful bookkeeping. It is possible to do all the manipulations easily without ever expanding any of the power series. Given

$$y = \sum_{n=0}^{\infty} a_n x^n,$$

the derivative is

$$y' = \sum_{n=0}^{\infty} n a_n x^{n-1}.$$

However, the first term is zero, so the series really starts at $n = 1$. Hence,

$$y' = \sum_{n=1}^{\infty} n a_n x^{n-1}.$$

Note that it is not wrong to start the series at $n = 0$; starting it with a nonzero term is a matter of taste. Continuing, we have

$$y'' = \sum_{n=2}^{\infty} (n-1) n a_n x^n.$$

This form for y'' is correct, but it is not convenient because the term number is the coefficient subscript rather than the power of x. To fix this, let $m = n - 2$. Then the series is

$$y'' = \sum_{m=0}^{\infty} (m+1)(m+2) a_{m+2} x^m.$$

Similarly, the series $xy = \sum_{n=0}^{\infty} a_n x^{n+1}$ is fixed with the replacement index $m = n + 1$, resulting in

$$xy = \sum_{n=0}^{\infty} a_n x^{n+1} = \sum_{m=1}^{\infty} a_{m-1} x^m.$$

After getting the recurrence relation

$$a_{m+2} = \frac{a_{m-1}}{(m+1)(m+2)}, \qquad m \geq 1,$$

we could change back to the previous index (by $n = m + 2$) as

$$a_n = \frac{a_{n-3}}{(n-1)(n)}, \qquad n \geq 3.$$

EXAMPLE 1

The Legendre equation of order 3 is

$$(1-x^2)y'' - 2xy' + 12y = 0.$$

To find solutions for this equation, we assume a series solution

$$y = \sum_{n=0}^{\infty} a_n x^n.$$

Then

$$y' = \sum_{n=1}^{\infty} n a_n x^{n-1}, \qquad xy' = \sum_{n=1}^{\infty} n a_n x^n, \qquad y'' = \sum_{n=2}^{\infty} (n-1) n a_n x^{n-2}, \qquad x^2 y'' = \sum_{n=2}^{\infty} (n-1) n a_n x^n.$$

Note that the term $(1-x^2)y''$ needs to be considered as two separate terms, y'' and x^2y''. Of the four terms in the differential equation, only one, y'', does not currently have n as the power of x in the general term. For that one term, let $m = n-2$, but let $m = n$ for all the others. The differential equation then becomes

$$\sum_{m=0}^{\infty} (m+1)(m+2)\, a_{m+2} x^m - \sum_{m=2}^{\infty} (m-1) m a_m x^m - \sum_{m=1}^{\infty} 2 m a_m x^m + \sum_{m=0}^{\infty} 12 a_m x^m = 0.$$

The series can be added term by term for the terms from $m = 2$ and up. Sometimes the starting index of a series can be decreased 1 because the additional term(s) has (have) the value 0. In the second summation, extra terms corresponding to $m = 1$ and $m = 0$ have value 0, as does an extra term in the third summation corresponding to $m = 0$; thus, the series can all be started at $m = 0$:

$$\sum_{m=0}^{\infty} (m+1)(m+2)\, a_{m+2} x^m - \sum_{m=0}^{\infty} (m-1) m a_m x^m - \sum_{m=0}^{\infty} 2 m a_m x^m + \sum_{m=0}^{\infty} 12 a_m x^m = 0.$$

Summing the coefficients of x_m for $m \geq 0$ yields the recurrence relation

$$(m+1)(m+2)a_{m+2} - (m-1)m a_m - 2m a_m + 12 a_m = 0,$$

which we solve for the coefficient with the largest subscript to obtain

$$a_{m+2} = \frac{m^2 + m - 12}{(m+1)(m+2)} a_m = \frac{(m-3)(m+4)}{(m+1)(m+2)} a_m, \qquad m \geq 0.$$

Finally, we can change from m back to n with $n = m+2$ to get the final version,

$$a_n = \frac{(n-5)(n+2)}{(n-1)(n)} a_{n-2}, \qquad n \geq 2.$$

The recurrence relation determines the coefficients beginning with the x^2 term. If initial conditions are given, then they determine the coefficients a_0 and a_1; otherwise, these can be taken as arbitrary integration constants. From the recurrence relation,

$$a_2 = -6a_0, \qquad a_3 = -\tfrac{5}{3}a_1, \qquad a_4 = -\tfrac{1}{2}a_2 = 3a_0,$$
$$a_5 = 0, \qquad a_6 = \tfrac{4}{15}a_4 = \tfrac{4}{5}a_0, \qquad a_7 = 0, \qquad \ldots.$$

Thus, the solution can be written in expanded form as

$$y = a_0 + a_1 x - 6a_0 x^2 - \tfrac{5}{3}a_1 x^3 + 3a_0 x^4 + \tfrac{4}{5}a_0 x^6 + \tfrac{3}{7}a_0 x^8 + \cdots.$$

Two linearly independent solutions can be isolated by choosing $a_0 = 1$ and $a_1 = 0$, followed by $a_0 = 0$ and $a_1 = 1$:

$$y_1 = 1 - 6x^2 + 3x^4 + \tfrac{4}{5}x^6 + \tfrac{3}{7}x^8 + \cdots, \qquad y_2 = x - \tfrac{5}{3}x^3.$$

In this case, one of the solutions is a finite polynomial rather than a power series. This solution is a multiple of the Legendre polynomial of order 3 (see Section 2.4, Exercise 11).

Does the Method Always Work?

Example 2 illustrates one of the difficulties that can arise with linear differential equations with variable coefficients.

EXAMPLE 2

Consider the differential equation

$$xy'' + y' + xy = 0,$$

which is the Bessel equation of order 0. The substitution $y = \sum_{n=0}^{\infty} a_n x^n$ results in the equation

$$a_1 + \sum_{m=1}^{\infty}\left[(m+1)^2 a_{m+1} + a_{m-1}\right]x^m = 0.$$

Assuming no initial conditions are supplied, we take a_0 to be arbitrary, but the coefficient a_1 must be 0 for the constant term to cancel. After that, the remaining coefficients come from the recurrence relation $(m+1)^2 a_{m+1} + a_{m-1} = 0$ for $m \geq 1$, or

$$a_n = -\frac{1}{n^2}a_{n-2}, \qquad n \geq 2.$$

The solution is

$$y = a_0 - \frac{1}{4}a_0x^2 + \frac{1}{4 \cdot 16}a_0x^4 - \frac{1}{4 \cdot 16 \cdot 36}a_0x^6 + \cdots.$$

By taking $a_0 = 1$, we get the Bessel function of the first kind of order 0 (see Section 3.7), which we see has the power series representation

$$J_0(x) = 1 - \frac{1}{4}x^2 + \frac{1}{4 \cdot 16}x^4 - \frac{1}{4 \cdot 16 \cdot 36}x^6 + \cdots.$$

If we label the terms using $k = n/2$, we can rewrite the series as

$$J_0(x) = \sum_{k=0}^{\infty} \frac{(-1)^k x^{2k}}{4^k (k!)^2}.$$

This is the power series definition of the function J_0.

There are some new features in Example 2 that merit attention. We expect the coefficients a_0 and a_1 for a second-order differential equation to be prescribed by the initial conditions. If no initial conditions are prescribed, we expect to get the general solution, a two-parameter family. For the Bessel equation of order 0, the differential equation determines the value of a_1; thus, the series method yields a one-parameter family of solutions rather than a two-parameter family. The second solution, as we saw in Section 3.7, is the function Y_0, which has the property $\lim_{x\to 0} Y_0(x) = -\infty$. This solution does not have a power series centered at $x = 0$ because it is not even defined at $x = 0$.

In the standard form, the Bessel equation of order 0 is

$$y'' + \frac{1}{x}y' + y = 0.$$

The y' coefficient is not continuous at $x = 0$, so the conditions of the existence and uniqueness theorem (Theorem 2.4.2) are not met. The theorem does not guarantee a unique solution for all sets of initial conditions at the origin; the results of Example 2 show that there is a unique solution if and only if the initial condition on y' is $y'(0) = 0$. Theorem 2.4.2 can be modified slightly to incorporate the power series method.

Theorem A.4.1 Let x_0 be a real number, and let I be an open interval $(x_0 - R,\ x_0 + R)$ for some $R > 0$. If the functions $p_i (i = 1, \ldots, n)$ and g have power series representations of the form $\sum_{n=0}^{\infty} a_n(x - x_0)$ that converge on I, then the initial-value problem

$$\frac{d^n y}{dx^n} + p_1(x)\frac{d^{n-1}y}{dx^{n-1}} + p_2(x)\frac{d^{n-2}y}{dx^{n-2}} + \cdots + p_{n-1}(x)\frac{dy}{dx} + p_n(x)y = g(x),$$

$$y(x_0) = y_0,\ \frac{dy}{dx}(x_0) = y_0',\ \ldots,\ \frac{d^{n-1}y}{dx^{n-1}}(x_0) = y_0^{n-1}$$

has a unique solution throughout the interval I that can be written in the form

$$y = \sum_{n=0}^{\infty} a_n(x - x_0)^n.$$

A point x_0 for which the coefficient functions in a linear differential equation satisfy the requirements for Theorem A.4.1 is called an **ordinary point.** A point where the conditions are not met, such as $x = 0$ for the Bessel equation of order 0, is called a **singular point.** Power series solutions do not always exist at singular points.

Characteristics of Power Series Solutions

When we say that a power series converges, we mean that a partial sum can be made arbitrarily accurate *on a particular interval* by using more terms. However, a power series approximation that is very accurate on one interval is frequently inaccurate on a larger interval. Figure A.4.1 shows the approximation made by using the first 16 nonzero terms of the series for $\mathrm{Ai}(x)$, along with the actual Airy function. The approximation is quite good up to a point, but shows catastrophic error beyond that point.

It is instructive to compare the performance of the power series approximation with that of a numerical approximation. Figure A.4.2 compares the 24-term power series approximation over the interval $[0, 5]$ with a modified Euler approximation over the same interval. The error in the

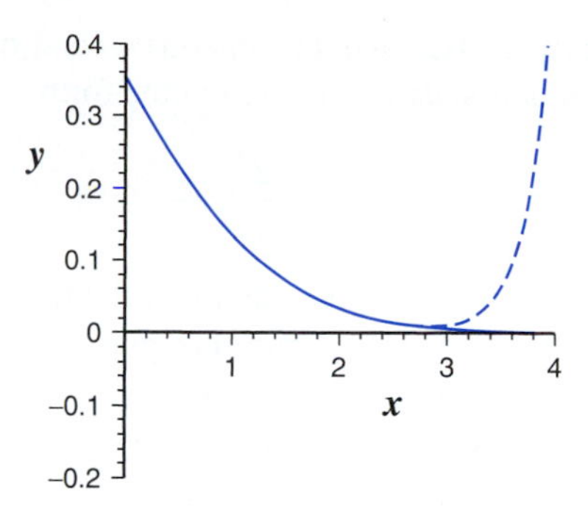

Figure A.4.1
The first 16 terms of the series for Ai (dashed) along with the function Ai.

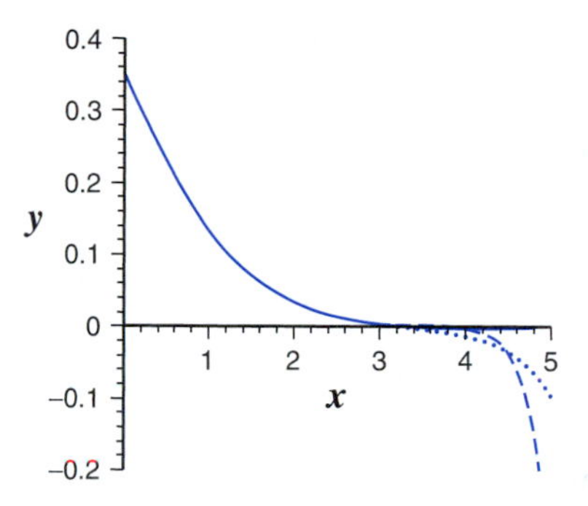

Figure A.4.2
The first 24 terms of the series for Ai (dashed) and the modified Euler solution (dotted) with $\Delta t = 0.025$ along with the function Ai.

numerical approximation grows slightly as x increases;[18] in contrast, the error in the power series approximation grows quite rapidly as x increases. Which is preferred between the power series and modified Euler methods is perhaps a matter of taste. However, commercial numerical solvers, such as those provided by Maple, Matlab, and Mathematica, are far superior to the modified Euler method. In particular, the best numerical solvers have subroutines that monitor the approximate errors and choose a smaller step size as needed. An additional problem with the use of power series approximations is that the power series solution may have a convergence interval that is smaller than the interval of existence of the solution.

Power Series at Regular Singular Points

Example 2 demonstrates that it is sometimes possible to find a power series solution even at a singular point. In exploring this issue further, we restrict consideration to a limited case. Our aim is to indicate the variety of possibilities, but not to provide an exhaustive treatment.[19]

---

[18]The differential equation is ill-conditioned (see Sec. A.3).

[19]See Bender and Orszag, op. cit., for a complete treatment of series approximation.

Suppose we have a homogeneous second-order linear equation for which 0 is a singular point. Such an equation can be written in the form

$$cx^2 \frac{d^2y}{dx^2} + q(x)x\frac{dy}{dx} + r(x)y = 0, \qquad c > 0, \tag{2}$$

for some pair of functions q and r. If the functions q and r are both analytic,[20] then the singular point $x = 0$ is a **regular singular point.** To keep matters from being overly complicated, we assume that q and r are polynomials.

Recall that Cauchy–Euler equations,[21] which are of the form

$$cx^2 \frac{d^2y}{dx^2} + qx\frac{dy}{dx} + ry = 0, \qquad c > 0,$$

often have solutions of the form x^α. If q and r are polynomials rather than constants, it seems reasonable that the local approximation to the solution should still be of the form x^α.

MODEL PROBLEM A.4*b*

Look for a solution of the form

$$y = x^\alpha \sum_{n=0}^{\infty} a_n x^n \tag{3}$$

for the Bessel equation of order 1

$$x^2y'' + xy' + \left(x^2 - 1\right)y = 0.$$

There are a number of ways to approach Model Problem A.4*b*. One way is to rewrite the differential equation by using the substitution

$$y(x) = x^\alpha z(x). \tag{4}$$

Differentiating Equation (4), we have

$$y' = x^\alpha z' + \alpha x^{\alpha-1}z, \qquad y'' = x^\alpha z'' + 2\alpha x^{\alpha-1}z' + \alpha(\alpha-1)x^{\alpha-2}z,$$

and therefore

$$xy' = x^{\alpha+1}z' + \alpha x^\alpha z, \qquad x^2y'' = x^{\alpha+2}z'' + 2\alpha x^{\alpha+1}z' + \alpha(\alpha-1)x^\alpha z. \tag{5}$$

Substituting the results of Equation (5) into the Bessel equation of order 1 results in a differential equation for z:

$$\left[x^{\alpha+2}z'' + 2\alpha x^{\alpha+1}z' + \alpha(\alpha-1)x^\alpha z\right] + \left(x^{\alpha+1}z' + \alpha x^\alpha z\right) + (x^2-1)x^\alpha z = 0;$$

[20] A function is *analytic* if it has derivatives of all orders.

[21] Section 3.7.

removing the common factor of x^α and simplifying yields

$$x^2 z'' + (2\alpha + 1)xz' + (\alpha^2 - 1)z + x^2 z = 0. \tag{6}$$

The assumption $y = x^\alpha \sum_{n=0}^{\infty} a_n x^n$ implies $a_0 \neq 0$ (else we should have picked a different value of α), so we now substitute the series approximation

$$z = \sum_{n=0}^{\infty} a_n x^n, \qquad a_0 \neq 0, \tag{7}$$

into Equation (6). This yields the equation

$$\sum_{n=2}^{\infty}(n-1)na_n x^n + (2\alpha+1)\sum_{n=1}^{\infty} na_n x^n + (\alpha^2 - 1)\sum_{n=0}^{\infty} a_n x^n + \sum_{n=0}^{\infty} a_n x^{n+2} = 0.$$

The last term needs to be fixed because the factor x^2 changes the powers of x inside the series. We have, with the substitution $m = n + 2$, $\sum_{n=0}^{\infty} a_n x^{n+2} = \sum_{m=2}^{\infty} a_{m-2} x^m$. Leaving the starting indices unchanged, we therefore have the equation

$$\sum_{m=2}^{\infty}(m-1)ma_m x^m + (2\alpha+1)\sum_{m=1}^{\infty} ma_m x^m + (\alpha^2 - 1)\sum_{m=0}^{\infty} a_m x^m + \sum_{m=2}^{\infty} a_{m-2} x^m = 0. \tag{8}$$

Only one term on the left side of Equation (8) makes a contribution for $m = 0$. Therefore, equating the $m = 0$ coefficients yields

$$(\alpha^2 - 1)a_0 = 0.$$

Since $a_0 \neq 0$, we arrive at the equation

$$\alpha^2 - 1 = 0. \tag{9}$$

This equation is called the **indicial equation,** and it determines whether or not there are any suitable values of α.

The Case $\alpha = 1$ With $\alpha = 1$, Equation (8) becomes

$$\sum_{m=2}^{\infty}(m-1)ma_m x^m + 3\sum_{m=1}^{\infty} ma_m x^m + \sum_{m=2}^{\infty} a_{m-2} x^m = 0,$$

or

$$3a_1 x + \sum_{m=2}^{\infty}\left[(m^2 + 2m)a_m + a_{m-2}\right]x^m = 0.$$

We therefore obtain the result $a_1 = 0$ and the recurrence relation

$$(m^2 + 2m)\,a_m + a_{m-2} = 0, \qquad m \geq 2.$$

Finally, we can solve for a_m and obtain (by substituting n back in for m) the result

$$a_n = -\frac{1}{n(n+2)}a_{n-2}, \qquad n \geq 2. \tag{10}$$

Thus, all odd coefficients are 0. We have found the one-parameter family of solutions

$$y = a_0 x \left[1 - \frac{x^2}{(2)(4)} + \frac{x^4}{(2)(4)(4)(6)} - \frac{x^6}{(2)(4)(4)(6)(6)(8)} + \cdots \right].$$

The Case $\alpha = -1$ Equation (8) with $\alpha = -1$ is

$$-a_1 x + \sum_{m=2}^{\infty} \left[(m^2 - 2m) a_m + a_{m-2} \right] x^m = 0.$$

Thus, $a_1 = 0$ and we obtain the recurrence relation

$$a_n = -\frac{1}{n(n-2)} a_{n-2}, \qquad n \geq 2.$$

This looks almost identical to what we obtained for $\alpha = 1$, but this time there is a problem: the factor $n - 2$ in the denominator means that a_2 does not exist. The conclusion is that the assumption of a solution of the form $y = x^{-1} \sum_{n=0}^{\infty} a_n x^n$ is wrong.[22]

A General Result

Is there any way to determine in advance how many power series solutions Equation (2) has for given coefficient functions q and r? The answer is summarized in Theorem A.4.2, which we state without proof.

Theorem A.4.2 Let q and r be polynomials. If the degree of q is at least 1 and the degree of r is at least 2, then $x = 0$ is an ordinary point of the differential equation

$$cx^2 \frac{d^2 y}{dx^2} + q(x) x \frac{dy}{dx} + r(x) y = 0.$$

Otherwise, the differential equation has at least one solution of the form

$$y = x^\alpha \sum_{n=0}^{\infty} a_n x^n$$

if and only if

$$4cr(0) < [q(0) - c]^2.$$

The value(s) of α can be found from the indicial equation

$$c\alpha^2 + [q(0) - c]\alpha + r(0) = 0. \tag{11}$$

If there are two real values of α that differ by an integer, there may not be a power series solution for the smaller of the two values. If there *are* two power series solutions, those solutions comprise a linearly independent set.

[22] It is possible to find a second solution of the Bessel equation of order 1 in a form that combines a power series with a logarithm function.

A.4 Exercises

In Exercises 1 through 4, use the power series method to solve the given initial-value problem. (Determine the recurrence relation and the first six nonzero terms.)

1. $2y'' - xy' + xy = 0, \quad y(0) = -1, \quad y'(0) = 2$
2. $y'' + 2xy' - y = 0, \quad y(0) = 1, \quad y'(0) = 3$
3. $(1+x)y'' = 3y, \quad y(0) = 2, \quad y'(0) = -1$
4. $(1-x)y'' + xy' - 2y = 0, \quad y(0) = -1, \quad y'(0) = 1$

In Exercises 5 through 8, use the power series method to find two linearly independent solutions for the given differential equation. (Determine the recurrence relation and the first four nonzero terms.)

5. $(1+x^2)y'' + 2xy' - 2y = 0$
6. $y'' + y' - xy = 0$
7. $y'' - xy' + y = 0$
8. $(x^2+1)y'' = 2y$
9. The parabolic cylinder functions of order ν are solutions of the differential equation
$$y'' + \left(\nu + \tfrac{1}{2} - \tfrac{1}{4}x^2\right)y = 0.$$
Use the power series method to determine the general solution of the differential equation. (Determine the recurrence relation and the first three nonzero terms of each series.)
10. Chebyshev's equation is
$$(1-x^2)y'' - xy' + p^2y = 0,$$
where p is a nonnegative constant.
   a. Find two linearly independent series solutions.
   b. Use the ratio test to determine the convergence interval of the solutions.
   c. Show that there is a polynomial solution of degree p if p is an integer.
11. The binomial theorem can be derived by a power series argument.
   a. Find a power series solution for $(1+x)y' = my$ with initial condition $y(0) = 1$, where m is any real number.
   b. Solve the problem by separation of variables.
   c. Obtain the binomial theorem by comparing the solutions of parts *a* and *b*.
12. Use the recurrence relation for the Airy function Ai to construct a full power series representation for the function. [*Hint*: it works out best if you try to massage it into the form
$$\text{Ai}(x) = c_1 - c_2x + \sum_{n=1}^{\infty} c_1 \frac{\Pi_1}{m_1!}x^{p_1} - c_2\frac{\Pi_2}{m_2!}x^{p_2},$$
where m_1, m_2, p_1, and p_2 are integers that depend on n, and Π_1 and Π_2 are products in which the number of factors depends on n.]
13. The Bessel equation of order ν is
$$x^2y'' + xy' + (x^2 - \nu^2)y = 0.$$
For what values of ν does Theorem A.4.2 guarantee two linearly independent solutions?

14. Apply the power series method to the Bessel equation of order 2

$$x^2y'' + xy' + (x^2 - 4)y = 0.$$

Discuss the results in light of Theorem A.4.2.

15. *a.* Use the power series method to find series representations for two linearly independent solutions of

$$x^2y'' + xy' + \left(x^2 - \tfrac{1}{4}\right)y = 0.$$

b. Explain why the results do not contradict Theorem A.4.2.
c. Use the power series solutions from part *a* to write the two linearly independent solutions in terms of elementary functions.

16. Apply the power series method to the equation $4xy'' + 2y' + y = 0$.

17. Apply the power series method to the equation $2x^2y'' + xy' - (x + 1)y = 0$.

18. Apply the power series method to the equation $x^2y'' - 3xy' + (4x + 4)y = 0$.

19. Consider the differential equation

$$x^2y'' + (3x - 1)y' + y = 0.$$

a. Assume a solution of the form

$$y = \sum_{n=0}^{\infty} a_n x^{\alpha+n}, \qquad a_0 \neq 0.$$

Substitute this form into the differential equation,[23] and then discard the common factor of x^α.
b. Determine the one suitable value of α.
c. Determine the recurrence relation.
d. Use the recurrence relation to get a series representation for the solution.
e. Explain why the "solution" obtained in part *d* is not correct.
f. Does this contradict Theorem A.4.2?
g. What happens if you use the method in the text for this problem?

20. Consider the case of the regular singular point with q constant.

a. Show that Equation (6) reduces to

$$c\sum_{n=2}^{\infty}(n-1)na_n x^n + (2c\alpha + q)\sum_{n=1}^{\infty} na_n x^n + [r(x) - r(0)]\sum_{n=0}^{\infty} a_n x^n = 0.$$

b. Show that $a_1 = -r'(0)a_0/(2c\alpha + q)$ provided $r(0) \neq q(q - 2c)/(4c)$.
c. Determine what happens if $r(0) = q(q - 2c)/(4c)$.

---

[23]This is the standard way of deriving the indicial equation.

✦ A.4 INSTANT EXERCISE SOLUTION

1. We have $y'' = 2a_2 + 6a_3x + (3)(4)a_4x^2 + (4)(5)a_5x^3 + \cdots$. With n representing the subscript of the coefficient, the first term is $n = 2$, there are $n - 2$ factors of x, and the algebraic factor is $(n - 1)n$. Thus, $y'' = \sum_{n=2}^{\infty}(n-1)na_nx^{n-2}$. Alternatively, with n representing the power of x, the first term is $n = 0$, the coefficient is a_{n+2}, and the algebraic factor is $(n + 1)(n + 2)$; hence, $y'' = \sum_{n=0}^{\infty}(n+1)(n+2)a_{n+2}x^n$.

A.5 Matrix Functions

A **matrix function** is a function whose output is a matrix. The rules of calculus generally carry over to matrix functions, just as they do for vector functions. Consider, for example, the simple fact that the scalar equation $x' = ax$ has a solution $x = e^{at}$. By analogy, it seems reasonable to think that the vector equation $\mathbf{x}' = \mathbf{A}\mathbf{x}$, where $\mathbf{A}$ is constant, has a solution $\mathbf{x} = e^{\mathbf{A}t}$.

MODEL PROBLEM A.5

Can we reasonably define the *matrix exponential* $e^{\mathbf{A}t}$ so that it solves the differential equation $\mathbf{x}' = \mathbf{A}\mathbf{x}$ in some sense and satisfies properties analogous to those of the scalar exponential e^{at}?

We begin with some preliminary ideas in the algebra and calculus of matrix functions.

Matrix Multiplication

Let $\mathbf{A}$ be an $m \times p$ matrix and let $\mathbf{B}$ be a $p \times n$ matrix.We define **matrix multiplication** by specifying a formula for each entry of the product $\mathbf{AB}$:

$$(\mathbf{AB})_{ij} = \sum_{k=1}^{p} a_{ik}b_{kj}. \tag{1}$$

Note that the product $\mathbf{AB}$ has the same number of rows as $\mathbf{A}$ and the same number of columns as $\mathbf{B}$ and that the ij entry of $\mathbf{AB}$ is the dot product of the ith row of $\mathbf{A}$ with the jth column of $\mathbf{B}$.

EXAMPLE 1

Let

$$\mathbf{A} = \begin{pmatrix} 0 & 1 & 2 \\ 3 & 4 & 0 \\ 0 & 1 & 1 \end{pmatrix}, \qquad \mathbf{B} = \begin{pmatrix} 1 & 2 & 0 \\ 2 & 0 & 1 \\ 0 & 2 & 1 \end{pmatrix}.$$

Then $(\mathbf{AB})_{11}$ is the dot product of the first row of **A** with the first column of **B:** $(\mathbf{AB})_{11} = (0)(1) + (1)(2) + (2)(0) = 2$. Proceeding in this fashion, we obtain the full product

$$\mathbf{AB} = \begin{pmatrix} 2 & 4 & 3 \\ 11 & 6 & 4 \\ 2 & 2 & 2 \end{pmatrix}.$$

✦ INSTANT EXERCISE 1

Compute the product **BA**, using **A** and **B** from Example 1.

The result of Instant Exercise 1 illustrates one of the surprising properties of matrix multiplication—it is not commutative. The product **AB** is defined only when the number of columns of **A** and the number of rows of **B** are the same. (In the context of systems of differential equations, the matrices are square, and matrix multiplication is always defined for square matrices of the same dimension.)

If it happens that $\mathbf{AB} = \mathbf{I}$ for two square matrices **A** and **B,** then **B** is the **inverse** of **A** and vice versa. The inverse of **A** is denoted $\mathbf{A}^{-1}$. Calculation of inverses requires more computation than most other matrix calculations. The most convenient technique is the row reduction technique we used to solve matrix algebraic equations in Section 3.2. Recall that there are three row reduction operations: (*a*) changing the order of the rows, (*b*) multiplying rows by nonzero scalars, and (*c*) adding multiples of a row to other rows. Row reduction of the augmented matrix $[\mathbf{A}|\mathbf{I}]$ for an equation $\mathbf{AA}^{-1} = \mathbf{I}$ does not change the solution of the equation. To solve the equation for $\mathbf{A}^{-1}$, we row-reduce the augmented matrix $[\mathbf{A}|\mathbf{I}]$ until the left side is **I.** Suppose the right side is then a matrix **B.** The augmented matrix $[\mathbf{I}|\mathbf{B}]$ represents the equation $\mathbf{IA}^{-1} = \mathbf{B}$; hence, $\mathbf{A}^{-1} = \mathbf{B}$. This procedure works whenever **A** is nonsingular;[24] hence, nonsingular matrices have inverses. It can also be shown (see Exercise 13) that nonsingular matrices are the only matrices that have inverses.

EXAMPLE 2

Let

$$\mathbf{A} = \begin{pmatrix} 0 & 1 & -3 \\ 1 & -1 & 0 \\ -1 & -2 & 3 \end{pmatrix}.$$

Then

$$[\mathbf{A}|\mathbf{I}] = \left[\begin{array}{ccc|ccc} 0 & 1 & -3 & 1 & 0 & 0 \\ 1 & -1 & 0 & 0 & 1 & 0 \\ -1 & -2 & 3 & 0 & 0 & 1 \end{array}\right].$$

[24] See the definition of *nonsingular* in Section 3.2.

We proceed by columns from left to right, each column proceeding by the three manipulations, as needed, in the order *abc*. We begin with column 1. The first step is to choose which of the rows in the original augmented matrix is most conveniently placed at the top. The current top row cannot be scaled so that its entry in the first column is a 1, so it cannot be chosen. Either of the other two can be used, but the second row seems to be simpler. Hence, we switch the first and second rows:

$$[\mathbf{A}|\mathbf{I}] \cong \left[\begin{array}{rrr|rrr} 1 & -1 & 0 & 0 & 1 & 0 \\ 0 & 1 & -3 & 1 & 0 & 0 \\ -1 & -2 & 3 & 0 & 0 & 1 \end{array}\right].$$

Manipulation b is not necessary because the 1–1 entry in the augmented matrix is what is needed for the identity matrix. We complete column 1 by adding row 1 to row 3 to obtain 0 for the 1–3 entry:

$$[\mathbf{A}|\mathbf{I}] \cong \left[\begin{array}{rrr|rrr} 1 & -1 & 0 & 0 & 1 & 0 \\ 0 & 1 & -3 & 1 & 0 & 0 \\ 0 & -3 & 3 & 0 & 1 & 1 \end{array}\right].$$

We continue with column 2. The first step is to choose which row to place as row 2. Note that row 1 cannot be moved because column 1 is now set. Of the remaining rows, the current row 2 is better left in place to avoid the need for manipulation b. All that remains for column 2 is to add the correct multiples of row 2 to rows 1 and 3 in order to get the 1–2 and 3–2 entries to be 0. We obtain

$$[\mathbf{A}|\mathbf{I}] \cong \left[\begin{array}{rrr|rrr} 1 & 0 & -3 & 1 & 1 & 0 \\ 0 & 1 & -3 & 1 & 0 & 0 \\ 0 & 0 & -6 & 3 & 1 & 1 \end{array}\right].$$

We continue with the third column. There is no longer any choice for the row to place third. If the 3–3 entry is 0, it means that the matrix is singular and there is no inverse. Here, we simply divide the third row by -6 to change the 3–3 entry to 1:

$$[\mathbf{A}|\mathbf{I}] \cong \left[\begin{array}{rrr|rrr} 1 & 0 & -3 & 1 & 1 & 0 \\ 0 & 1 & -3 & 1 & 0 & 0 \\ 0 & 0 & 1 & -\frac{1}{2} & -\frac{1}{6} & -\frac{1}{6} \end{array}\right].$$

Finally, the third column is completed by adding multiples of the third row to the others:

$$[\mathbf{A}|\mathbf{I}] \cong \left[\begin{array}{rrr|rrr} 1 & 0 & 0 & -\frac{1}{2} & \frac{1}{2} & -\frac{1}{2} \\ 0 & 1 & 0 & -\frac{1}{2} & -\frac{1}{2} & -\frac{1}{2} \\ 0 & 0 & 1 & -\frac{1}{2} & -\frac{1}{6} & -\frac{1}{6} \end{array}\right] = [\mathbf{I}|\mathbf{B}].$$

Thus,

$$\mathbf{A}^{-1} = \frac{1}{6}\begin{pmatrix} -3 & 3 & -3 \\ -3 & -3 & -3 \\ -3 & -1 & -1 \end{pmatrix}.$$

Inverses for 2 × 2 Matrices

There is a simple formula for the inverse of any nonsingular 2×2 matrix. If $\mathbf{A}$ is defined by

$$\mathbf{A} = \begin{pmatrix} a & b \\ c & d \end{pmatrix},$$

then

$$\mathbf{A}^{-1} = \frac{1}{\det A} \begin{pmatrix} d & -b \\ -c & a \end{pmatrix}. \tag{2}$$

EXAMPLE 3

Let $\mathbf{A} = \begin{pmatrix} e^t & 2e^t \\ 3e^t & 4e^t \end{pmatrix}$. From Equation (2), the inverse is

$$\mathbf{A}^{-1} = \frac{1}{-2e^{2t}} \begin{pmatrix} 4e^t & -2e^t \\ -3e^t & e^t \end{pmatrix} = \frac{1}{2} \begin{pmatrix} -4e^{-t} & 2e^{-t} \\ 3e^{-t} & -e^{-t} \end{pmatrix}.$$

The techniques for finding inverses are summarized in Algorithm A.5.1.

Algorithm A.5.1

Finding the Inverse of a Nonsingular Matrix To find the inverse of a nonsingular matrix $\mathbf{A}$, row-reduce the augmented matrix $[\mathbf{A}|\mathbf{I}]$ to the form $[\mathbf{I}|\mathbf{B}]$. Then $\mathbf{A}^{-1} = \mathbf{B}$. Alternatively, if $\mathbf{A}$ is a 2×2 nonsingular matrix, its inverse can be computed by Equation (2).

As in Example 3, Equation (2) is particularly convenient when $\mathbf{A}$ is a matrix function.

Calculus of Matrix Functions

The derivative of a matrix $\mathbf{F}(t)$ is found simply by differentiating each entry in $\mathbf{F}$. Thus,

$$(\mathbf{F}')_{ij} = (\mathbf{F}_{ij})'. \tag{3}$$

Many of the rules of ordinary calculus carry over to matrix functions; in particular, we need the product rule, which holds for matrix–vector products as well as matrix–matrix products:

$$(\mathbf{FG})' = \mathbf{FG}' + \mathbf{F}'\mathbf{G}. \tag{4}$$

Because matrix multiplication is not commutative, it is necessary to retain the original order of the factors.

Fundamental Matrices

Recall (Sec. 6.4) that the general solution of a vector differential equation $\mathbf{x}' = \mathbf{A}(t)\mathbf{x}$, with $\mathbf{A}$ an $n \times n$ matrix, can be written as

$$\mathbf{x} = c_1\mathbf{x}^{(1)} + c_2\mathbf{x}^{(2)} + \cdots + c_n\mathbf{x}^{(n)}.$$

The solutions $\mathbf{x}^{(k)}$ together constitute a **fundamental set** of solutions for the differential equation. The matrix

$$\mathbf{\Psi}(t) = \begin{pmatrix} x_1^{(1)} & x_1^{(2)} & \cdots & x_1^{(n)} \\ x_2^{(1)} & x_2^{(2)} & \cdots & x_2^{(n)} \\ \vdots & \vdots & \vdots & \vdots \\ x_n^{(1)} & x_n^{(2)} & \cdots & x_n^{(n)} \end{pmatrix}, \tag{5}$$

whose columns are the vector solutions in the given fundamental set, is said to be a **fundamental matrix** for the differential equation. Fundamental matrices are always nonsingular (see Exercise 15).

The general solution of a homogeneous linear system can be expressed concisely in terms of a fundamental matrix. By the definition of matrix multiplication, we have

$$\mathbf{\Psi c} = \begin{pmatrix} x_1^{(1)} & x_1^{(2)} & \cdots & x_1^{(n)} \\ x_2^{(1)} & x_2^{(2)} & \cdots & x_2^{(n)} \\ \vdots & \vdots & \vdots & \vdots \\ x_n^{(1)} & x_n^{(2)} & \cdots & x_n^{(n)} \end{pmatrix} \begin{pmatrix} c_1 \\ c_2 \\ \vdots \\ c_n \end{pmatrix} = \begin{pmatrix} x_1^{(1)}c_1 + x_1^{(2)}c_2 + \cdots + x_1^{(n)}c_n \\ x_2^{(1)}c_1 + x_2^{(2)}c_2 + \cdots + x_2^{(n)}c_n \\ \vdots \quad \vdots \quad \vdots \quad \vdots \\ x_n^{(1)}c_1 + x_n^{(2)}c_2 + \cdots + x_n^{(n)}c_n \end{pmatrix}$$

$$= c_1\mathbf{x}^{(1)} + c_2\mathbf{x}^{(2)} + \cdots + c_n\mathbf{x}^{(n)} = \mathbf{x}.$$

Hence, the solution of the system

$$\mathbf{x}' = \mathbf{A}(t)\mathbf{x}$$

is

$$\mathbf{x} = \mathbf{\Psi}(t)\,\mathbf{c}.$$

From this solution, we can derive an important property of fundamental matrices. Substituting the solution $\mathbf{\Psi c}$ into the differential equation yields the result

$$\mathbf{\Psi}'(t)\mathbf{c} = \mathbf{A}(t)\mathbf{\Psi}(t)\,\mathbf{c},$$

or

$$(\mathbf{\Psi}' - \mathbf{A\Psi})\,\mathbf{c} = 0.$$

Since this result holds for *all* vectors **c**, it follows that the matrix $\mathbf{\Psi}' - \mathbf{A\Psi}$ is the zero matrix.[25] Therefore,

$$\mathbf{\Psi}' = \mathbf{A\Psi}. \tag{6}$$

Equation (6) is a **matrix differential equation.**

EXAMPLE 4

The differential equation $\mathbf{x}' = \mathbf{Ax}$, with $\mathbf{A} = \begin{pmatrix} -3 & 1 \\ -1 & -1 \end{pmatrix}$, served as Model Problem 6.6. The solution of this problem was found to be

$$\mathbf{x} = e^{-2t}\left[c_1\begin{pmatrix} 1 \\ 1 \end{pmatrix} + c_2\begin{pmatrix} t \\ 1+t \end{pmatrix}\right].$$

[25] The rule that $\mathbf{Ac} = \mathbf{0}$ for all **c** implies **A** is the zero matrix serves as a *cancelation law* for matrix multiplication. Technically, we ought to prove it before we use it. See Exercise 16 for a proof of Equation (6) that is based on direct calculation.

From this solution, we obtain the fundamental matrix

$$\Psi = \begin{pmatrix} e^{-2t} & te^{-2t} \\ e^{-2t} & (1+t)e^{-2t} \end{pmatrix}.$$

The Matrix Exponential

We turn now to the problem of defining the matrix exponential. Since the scalar exponential function satisfies the equation

$$e^{at} = \sum_{n=0}^{\infty} \frac{a^n t^n}{n!} = 1 + at + \frac{a^2t^2}{2} + \frac{a^3t^3}{6} + \cdots + \frac{a^n t^n}{n!} + \cdots,$$

it makes sense to try to define the **matrix exponential** by

$$\Theta(t) = e^{\mathbf{A}t} = \sum_{n=0}^{\infty} \frac{\mathbf{A}^n t^n}{n!} = \mathbf{I} + t\mathbf{A} + \frac{t^2}{2}\mathbf{A}^2 + \frac{t^3}{6}\mathbf{A}^3 + \cdots + \frac{t^n}{n!}\mathbf{A}^n + \cdots. \tag{7}$$

Note, first, that the infinite sum converges because of the factor $n!$ in the denominator (Exercise 17). Second, each of the quantities $\mathbf{A}^n$ is a matrix product and therefore also a matrix. These are then multiplied by scalar functions and summed, so the result is a matrix that is a function of t. We have assigned to the matrix exponential the name $\Theta(t)$ to emphasize that it is a matrix function. In the examples that follow, we use the notation $\mathbf{O}$ to denote the zero matrix, thereby distinguishing it from the zero vector $\mathbf{0}$.

EXAMPLE 5

Let $\mathbf{A} = \begin{pmatrix} 2 & 4 \\ -1 & -2 \end{pmatrix}$. Then $\mathbf{A}^2 = \mathbf{O}$. In this case, the matrix exponential is easy to calculate because it has only a finite number of terms. We have

$$\Theta = \mathbf{I} + t\mathbf{A} = \begin{pmatrix} 1+2t & 4t \\ -t & 1-2t \end{pmatrix}.$$

EXAMPLE 6

Let $\mathbf{A} = \begin{pmatrix} 2 & 4 \\ 1 & 2 \end{pmatrix}$. Then $\mathbf{A}^2 = \begin{pmatrix} 8 & 16 \\ 4 & 8 \end{pmatrix} = 4\mathbf{A}$. This last result can be used to determine the remaining terms; for example, $\mathbf{A}^3 = \mathbf{A}^2\mathbf{A} = (4\mathbf{A})(\mathbf{A}) = 4\mathbf{A}^2 = 16\mathbf{A}$. In general, we have $\mathbf{A}^n = 4^{n-1}\mathbf{A}$ for $n > 0$. Thus,

$$\Theta = \mathbf{I} + t\mathbf{A} + \frac{4t^2}{2}\mathbf{A} + \cdots + \frac{4^{n-1}t^n}{n!}\mathbf{A} + \cdots,$$

which we can further manipulate into the form

$$\Theta = \mathbf{I} + \frac{1}{4}\left[4t + \frac{(4t)^2}{2} + \cdots + \frac{(4t)^n}{n!} + \cdots\right]\mathbf{A} = \mathbf{I} + \frac{1}{4}\left[\left(\sum_{n=0}^{\infty} \frac{(4t)^n}{n!}\right) - 1\right]\mathbf{A}.$$

The infinite sum in this last result is e^{4t}; hence,

$$\Theta = \mathbf{I} + \tfrac{1}{4}(e^{4t} - 1)\mathbf{A} = \mathbf{I} - \tfrac{1}{4}\mathbf{A} + \tfrac{1}{4}e^{4t}\mathbf{A} = \begin{pmatrix} (e^{4t}+1)/2 & e^{4t}-1 \\ (e^{4t}-1)/4 & (e^{4t}+1)/2 \end{pmatrix}.$$

The Matrix Exponential and Differential Equations

As we've seen, the matrix exponential is a *matrix* function. Thus, it certainly cannot be the solution of the *vector* differential equation $\mathbf{x}' = \mathbf{A}\mathbf{x}$. One can show, however, that the fundamental matrix is a solution of the matrix differential equation $\mathbf{X}' = \mathbf{A}\mathbf{X}$ (Exercise 18).

EXAMPLE 7

Let

$$\mathbf{A} = \begin{pmatrix} 2 & 4 \\ 1 & 2 \end{pmatrix}, \qquad \Theta(t) = e^{\mathbf{A}t} = \begin{pmatrix} (e^{4t}+1)/2 & e^{4t}-1 \\ (e^{4t}-1)/4 & (e^{4t}+1)/2 \end{pmatrix},$$

as in Example 6. Then

$$\Theta' = \begin{pmatrix} 2e^{4t} & 4e^{4t} \\ e^{4t} & 2e^{4t} \end{pmatrix} = \begin{pmatrix} 2 & 4 \\ 1 & 2 \end{pmatrix} e^{4t}$$

and

$$\mathbf{A}\Theta = \begin{pmatrix} 2 & 4 \\ 1 & 2 \end{pmatrix} \begin{pmatrix} (e^{4t}+1)/2 & (e^{4t}-1) \\ (e^{4t}-1)/4 & (e^{4t}+1)/2 \end{pmatrix} = \begin{pmatrix} 2e^{4t} & e^{4t} \\ 4e^{4t} & 2e^{4t} \end{pmatrix} = \begin{pmatrix} 2 & 4 \\ 1 & 2 \end{pmatrix} e^{4t}.$$

Thus, Θ satisfies the equation $\mathbf{X}' = \mathbf{A}\mathbf{X}$.

✦ INSTANT EXERCISE 2

Show that the matrix exponential in Example 5 satisfies the differential equation $\mathbf{X}' = \mathbf{A}\mathbf{X}$ with the matrix $\mathbf{A}$ of Example 5.

Next, consider the initial-value problems for scalar and vector differential equations. The initial-value problem

$$x' = ax, \qquad x(0) = x_0,$$

has the solution $x = x_0 e^{at}$, so perhaps the initial-value problem

$$\mathbf{x}' = \mathbf{A}\mathbf{x}, \qquad \mathbf{x}(0) = \mathbf{x}_0, \tag{8}$$

has the solution

$$\mathbf{x} = e^{\mathbf{A}t}\mathbf{x}_0. \tag{9}$$

Equation (9) is at least consistent, because the quantities on both sides are vectors. This conjecture is correct (Exercise 19), and it has the consequence of connecting the matrix exponential for $\mathbf{A}$ with a fundamental matrix for $\mathbf{x}' = \mathbf{A}\mathbf{x}$ (Exercise 20). Theorem A.5.1 summarizes the important properties of the matrix exponential function.

Theorem A.5.1 Let $\mathbf{A}$ be a constant square matrix. Then

1. The solution of the initial-value problem
$$\mathbf{X}' = \mathbf{AX}, \qquad \mathbf{X}(0) = \mathbf{I}$$
is
$$\mathbf{\Theta} = e^{\mathbf{A}t}.$$
2. The solution of the initial-value problem
$$\mathbf{x}' = \mathbf{Ax}, \qquad \mathbf{x}(0) = \mathbf{x}_0,$$
is
$$\mathbf{x} = e^{\mathbf{A}t}\mathbf{x}_0.$$
3. The matrix exponential function is a fundamental matrix for $\mathbf{x}' = \mathbf{Ax}$ and can be computed from any fundamental matrix $\mathbf{\Psi}$ by the formula
$$\mathbf{\Theta}(t) = \mathbf{\Psi}(t)\mathbf{\Psi}^{-1}(0). \tag{10}$$

EXAMPLE 8

Consider again the problem of Example 4. From the fundamental matrix $\mathbf{\Psi}$ given there and formula (10), we can calculate the matrix exponential. We have

$$\mathbf{\Psi}(t) = \begin{pmatrix} e^{-2t} & te^{-2t} \\ e^{-2t} & (1+t)e^{-2t} \end{pmatrix}, \qquad \mathbf{\Psi}(0) = \begin{pmatrix} 1 & 0 \\ 1 & 1 \end{pmatrix}, \qquad \mathbf{\Psi}^{-1}(0) = \begin{pmatrix} 1 & 0 \\ -1 & 1 \end{pmatrix}.$$

Thus,

$$\mathbf{\Theta} = \begin{pmatrix} e^{-2t} & te^{-2t} \\ e^{-2t} & (1+t)e^{-2t} \end{pmatrix}\begin{pmatrix} 1 & 0 \\ -1 & 1 \end{pmatrix} = \begin{pmatrix} (1-t)e^{-2t} & te^{-2t} \\ -te^{-2t} & (1+t)e^{-2t} \end{pmatrix}.$$

A.5 Exercises

In Exercises 1 through 8, find the inverse of the given matrix.

1. $\mathbf{\Theta} = \begin{pmatrix} 1+2t & 4t \\ -t & 1-2t \end{pmatrix}$

2. $\mathbf{\Psi} = \begin{pmatrix} e^{2t} & -e^{-t} \\ 3e^{2t} & 2e^{-t} \end{pmatrix}$

3. $\mathbf{\Psi} = \begin{pmatrix} e^{-2t} & te^{-2t} \\ e^{-2t} & (1+t)e^{-2t} \end{pmatrix}$

4. $\mathbf{\Theta} = \begin{pmatrix} (e^{4t}+1)/2 & e^{4t}-1 \\ (e^{4t}-1)/4 & (e^{4t}+1)/2 \end{pmatrix}$

5. $\mathbf{A} = \begin{pmatrix} 0 & 1 & -1 \\ 1 & 1 & 0 \\ -1 & 0 & 1 \end{pmatrix}$

6. $\mathbf{A} = \begin{pmatrix} 3 & 2 & 0 \\ 2 & 0 & 0 \\ 1 & 1 & 3 \end{pmatrix}$

7. $\mathbf{A} = \begin{pmatrix} -3 & 1 & -1 \\ -2 & 0 & -1 \\ -1 & 1 & -2 \end{pmatrix}$

8. $\mathbf{A} = \begin{pmatrix} -5 & -3 & -3 \\ 3 & 1 & 3 \\ 6 & 6 & 4 \end{pmatrix}$

In Exercises 9 through 12, determine the matrix $\mathbf{\Theta}$ for $\mathbf{x}' = \mathbf{A}\mathbf{x}$ by using (*a*) the series definition of the matrix exponential $e^{\mathbf{A}t}$ and (*b*) a fundamental matrix $\mathbf{\Psi}$, as in Example 8.

9. $\mathbf{A} = \begin{pmatrix} 2 & 1 \\ 0 & 2 \end{pmatrix}$

10. $\mathbf{A} = \begin{pmatrix} 4 & -3 \\ 8 & -6 \end{pmatrix}$

11. $\mathbf{A} = \begin{pmatrix} -2 & 1 \\ -5 & 2 \end{pmatrix}$

12. $\mathbf{A} = \begin{pmatrix} -1 & 1 \\ -4 & 3 \end{pmatrix}$

13. Show that only nonsingular matrices have inverses. (*Hint*: Assume that a square matrix $\mathbf{A}$ has an inverse. Can you solve any equation $\mathbf{A}\mathbf{x} = \mathbf{b}$?)

14. Derive formula (2) for the inverse of a 2×2 matrix by applying Algorithm A.5.1 to the matrix $\begin{pmatrix} a & b \\ c & d \end{pmatrix}$.

15. Show that all fundamental matrices are nonsingular.

16. Derive Equation (6) by using the definition of $\mathbf{\Psi}$.

17. Use the ratio test (see any calculus text) to show that the series of Equation (7) converges for any matrix $\mathbf{A}$.

18. Prove statement 1 of Theorem A.5.1. In other words, show that the fundamental matrix is a solution of the matrix differential equation $\mathbf{X}' = \mathbf{A}\mathbf{X}$ by differentiating Equation (7); also verify the initial condition.

19. Prove statement 2 of Theorem A.5.1. In other words, show that Equation (9) identifies the solution of the initial-value problem (8).

20. Prove statement 3 of Theorem A.5.1. For the second part, you may assume that initial-value problems for constant-coefficient matrix differential equations, such as $\mathbf{X}' = \mathbf{A}\mathbf{X}$, have unique solutions. [*Hint*: For the first part, consider the set of n initial value problems $\mathbf{x}' = \mathbf{A}\mathbf{x}$, $\mathbf{x}(0) = \mathbf{x}_0$, where $\mathbf{x}_0$ is a vector consisting of a single 1 and all other entries 0. Then use statement 2.]

✦ A.5 INSTANT EXERCISE SOLUTIONS

1.

$$\mathbf{BA} = \begin{pmatrix} 6 & 9 & 2 \\ 0 & 3 & 5 \\ 6 & 9 & 1 \end{pmatrix}.$$

2. We have

$$\mathbf{A} = \begin{pmatrix} 2 & 4 \\ -1 & -2 \end{pmatrix}, \qquad \mathbf{\Theta} = \begin{pmatrix} 1+2t & 4t \\ -t & 1-2t \end{pmatrix}.$$

Thus,

$$\Theta' = \begin{pmatrix} 2 & 4 \\ -1 & -2 \end{pmatrix}, \qquad \mathbf{A}\Theta = \begin{pmatrix} 2 & 4 \\ -1 & -2 \end{pmatrix}\begin{pmatrix} 1+2t & 4t \\ -t & 1-2t \end{pmatrix} = \begin{pmatrix} 2 & 4 \\ -1 & -2 \end{pmatrix}.$$

A.6 Nonhomogeneous Linear Systems

The methods for solving nonhomogeneous linear equations that appeared in Sections 4.3, 4.5, and 4.6 can be adapted to nonhomogeneous linear systems. The details are somewhat messier, and a greater mastery of linear algebra is required. In general, we consider the system

$$\mathbf{x}' = \mathbf{A}(t)\mathbf{x} + \mathbf{g}(t). \tag{1}$$

This system is linear, so it can be thought of as having the form

$$\mathbf{L}[\mathbf{x}] = \mathbf{g},$$

where the linear operator $\mathbf{L}$ is defined by

$$\mathbf{L}[\mathbf{x}] = \mathbf{x}' - \mathbf{A}\mathbf{x}.$$

As in the case of nonhomogeneous scalar differential equations, the problem can be partitioned into separate problems of finding the complementary solution, which is the general solution of the associated homogeneous system

$$\mathbf{x}' = \mathbf{A}(t)\mathbf{x},$$

and finding a particular solution. We assume that the complementary solution is known. It is also possible to split the forcing function $\mathbf{g}$ into a sum of terms and then determine the full particular solution by adding the particular solutions for each.[26]

We consider three methods here: undetermined coefficients, variation of parameters, and diagonalization. Each of these methods has its own set of advantages and disadvantages, so each is worth study.

1. The method of undetermined coefficients[27] can sometimes be very efficient, but it can be used only if $\mathbf{A}$ is a constant matrix and g is limited to generalized exponential functions and their sums. The method of undetermined coefficients is particularly inconvenient when the characteristic value of g is an eigenvalue of $\mathbf{A}$.
2. Variation of parameters can be used for any nonhomogeneous linear system, provided the general solution of $\mathbf{x}' = \mathbf{A}\mathbf{x}$ is known. As with variation of parameters for scalar equations,[28] the drawback of the method is that it can require a substantial amount of integration.

[26] See Sections 4.2 and 4.3 for the scalar case.

[27] See Section 4.3 for the scalar case.

[28] Sections 4.5 and 4.6.

3. The method of diagonalization is generally easier to use than the method of variation of parameters, but the price for that advantage is that its use is restricted to systems in which $\mathbf{A}$ is constant and nondeficient.[29]

We consider these methods in turn, applying each to the same model problem.

MODEL PROBLEM A.6

Find the general solution of the system

$$\mathbf{x}' = \begin{pmatrix} 1 & -2 \\ 1 & 4 \end{pmatrix} \mathbf{x} + \begin{pmatrix} e^{2t} \\ 36t \end{pmatrix}.$$

Note that the general solution of the associated homogeneous problem[30] is

$$\mathbf{x} = c_1 \begin{pmatrix} -2 \\ 1 \end{pmatrix} e^{2t} + c_2 \begin{pmatrix} -1 \\ 1 \end{pmatrix} e^{3t}.$$

To solve Model Problem A.6, either we can determine a full particular solution with one calculation, or we can find separate particular solutions for the problems

$$\mathbf{x}' = \begin{pmatrix} 1 & -2 \\ 1 & 4 \end{pmatrix} \mathbf{x} + \begin{pmatrix} 0 \\ 36t \end{pmatrix}, \qquad \mathbf{x}' = \begin{pmatrix} 1 & -2 \\ 1 & 4 \end{pmatrix} \mathbf{x} + \begin{pmatrix} 1 \\ 0 \end{pmatrix} e^{2t},$$

and sum the results.

Undetermined Coefficients

The method of undetermined coefficients for systems is identical to the method used for scalar equations, provided the characteristic value of the generalized exponential forcing function is not a characteristic value of the homogeneous equation. When the characteristic value of the forcing function *is* a characteristic value of the homogeneous equation, then the method must be applied carefully.

EXAMPLE 1

Think of Model Problem A.6 as

$$\mathbf{x}' - \mathbf{A}\mathbf{x} = (\mathbf{u} + \mathbf{w}t) + \mathbf{z}e^{2t},$$

where

$$\mathbf{u} = \begin{pmatrix} 0 \\ 0 \end{pmatrix}, \qquad \mathbf{w} = \begin{pmatrix} 0 \\ 36 \end{pmatrix}, \qquad \mathbf{z} = \begin{pmatrix} 1 \\ 0 \end{pmatrix}.$$

This form is analogous to the scalar case. The forcing term $\mathbf{u} + \mathbf{w}t$ is a generalized exponential function with characteristic value 0 and degree 1, and the term $\mathbf{z}e^{2t}$ is a generalized exponential function with characteristic value 2 and degree 0. We consider the two particular solutions separately.

---

[29] See Section 6.2 for the definition of deficient matrices.

[30] Instant Exercise 3 of Section 6.4.

The characteristic values of the differential operator $\mathbf{L} = d/dt - \mathbf{A}$ are the eigenvalues of the matrix $\mathbf{A}$, which were determined to be 2 and 3 in the process of finding the complementary solution. Hence, a trial solution of characteristic value 0 will yield an image with the same degree. The trial solution has to be general enough to have all first-degree polynomials with vector coefficients as possible images. Hence, we take the trial solution

$$\mathbf{y} = \mathbf{b} + \mathbf{c}t.$$

Substituting this into $\mathbf{x}' - \mathbf{A}\mathbf{x} = \mathbf{u} + \mathbf{w}t$ yields

$$\mathbf{c} - \mathbf{A}\mathbf{b} - \mathbf{A}\mathbf{c}t = \mathbf{u} + \mathbf{w}t.$$

The t coefficients result in the equation

$$\mathbf{A}\mathbf{c} = -\mathbf{w},$$

which we solve to obtain

$$\mathbf{c} = \begin{pmatrix} -12 \\ -6 \end{pmatrix}.$$

Then the constant terms are $\mathbf{A}\mathbf{b} = \mathbf{c} - \mathbf{u}$; thus,

$$\mathbf{b} = \begin{pmatrix} -10 \\ 1 \end{pmatrix},$$

and the particular solution is

$$\mathbf{x}_{p1} = \begin{pmatrix} -10 - 12t \\ 1 - 6t \end{pmatrix}.$$

For the other forcing function, the natural trial solution $\mathbf{y} = \mathbf{b}e^{2t}$ does not work, but the situation is a little different from the scalar case. If $\mathbf{b}$ is *not* an eigenvector, then the image of the trial solution under $\mathbf{L}$ is nonzero. But if it *is* an eigenvector, then the image will be $\mathbf{0}$. Of course we don't know $\mathbf{b}$ until after we have solved the problem, so we have to allow for *both* possibilities. A term of degree 1 needs to be added, but only if the vector constant is an eigenvector. The correct trial solution is thus

$$\mathbf{y} = (\mathbf{b} + c\mathbf{v}t)e^{2t}, \qquad \mathbf{v} = \begin{pmatrix} -2 \\ 1 \end{pmatrix};$$

note that $\mathbf{v}$ is the eigenvector that appears in the complementary solution corresponding to the eigenvalue 2. The first term has a nonzero image under $\mathbf{L}$ if $\mathbf{b}$ is not an eigenvector, and the second term has an image of degree 0 because $\mathbf{v}$ is an eigenvector. Both terms are likely to contribute to the particular solution. Unlike the scalar case, we cannot omit the degree 0 term from the trial solution. In the vector case, having the characteristic value of the forcing function match that of the operator adds an extra unknown (the scalar c) to the trial solution as well as raises the degree. With the correct trial solution, we have

$$\begin{aligned} \mathbf{z}e^{2t} &= \left[(\mathbf{b} + c\mathbf{v}t)\, e^{2t}\right]' - \mathbf{A}(\mathbf{b} + c\mathbf{v}t)\, e^{2t} = 2(\mathbf{b} + c\mathbf{v}t)\, e^{2t} + c\mathbf{v}e^{2t} - (\mathbf{A}\mathbf{b} + c\mathbf{A}\mathbf{v}t)\, e^{2t} \\ &= [c(2\mathbf{v} - \mathbf{A}\mathbf{v})t + (2\mathbf{b} + c\mathbf{v} - \mathbf{A}\mathbf{b})]\, e^{2t}. \end{aligned}$$

Now, $\mathbf{Av} = 2\mathbf{v}$ because $\mathbf{v}$ is an eigenvector with eigenvalue 2; hence, the t term disappears. From the remaining term, we have (using $2\mathbf{b} = 2\mathbf{Ib}$)

$$(\mathbf{A} - 2\mathbf{I})\mathbf{b} = c\mathbf{v} - \mathbf{z},$$

or

$$\begin{pmatrix} -1 & -2 \\ 1 & 2 \end{pmatrix}\begin{pmatrix} b_1 \\ b_2 \end{pmatrix} = c\begin{pmatrix} -2 \\ 1 \end{pmatrix} - \begin{pmatrix} 1 \\ 0 \end{pmatrix} = \begin{pmatrix} -2c-1 \\ c \end{pmatrix}.$$

The augmented matrix for this linear algebraic system is

$$\left(\begin{array}{cc|c} -1 & -2 & -2c-1 \\ 1 & 2 & c \end{array}\right) \cong \left(\begin{array}{cc|c} 1 & 2 & c \\ 0 & 0 & -c-1 \end{array}\right).$$

The system has a solution only if $c = -1$, in which case the vector $\mathbf{b}$ must satisfy the equation $b_1 + 2b_2 = -1$. The simplest choice is $b_1 = -1$, $b_2 = 0$, yielding the particular solution

$$\mathbf{x}_{p2} = \left[\begin{pmatrix} -1 \\ 0 \end{pmatrix} - \begin{pmatrix} -2 \\ 1 \end{pmatrix} t\right] e^{2t} = \begin{pmatrix} -1+2t \\ -t \end{pmatrix} e^{2t}.$$

The full particular solution is

$$\mathbf{x}_p = \begin{pmatrix} (-1+2t)e^{2t} - 10 - 12t \\ -te^{2t} + 1 - 6t \end{pmatrix}. \tag{2}$$

Note that there are infinitely many other formulas that could have been obtained for $\mathbf{x}_p$ in Example 1 by making different choices of c_1 and c_2 in the calculation of $\mathbf{x}_{p2}$.

The method of undetermined coefficients, where the characteristic value of $\mathbf{g}$ is not an eigenvalue of $\mathbf{A}$ or is at most an eigenvalue of multiplicity 1, is summarized in Algorithm A.6.1. The case where the characteristic value of $\mathbf{g}$ is an eigenvalue of $\mathbf{A}$ of multiplicity 2 is considered in Exercise 15.

Algorithm A.6.1

Solving $\mathbf{x}' = \mathbf{Ax} + \mathbf{g}(t)$ by Undetermined Coefficients The matrix $\mathbf{A}$ must be constant and $\mathbf{g}$ must be a generalized exponential function.

1. Solve the associated homogeneous problem $\mathbf{x}' = \mathbf{Ax}$ and use the solutions to construct a fundamental matrix $\mathbf{\Psi}(t)$. Let $\gamma + i\omega$ and k be the characteristic value and degree, respectively, of $\mathbf{g}$.
2. Determine the appropriate trial solution.
   - If $\gamma + i\omega$ is not an eigenvalue of $\mathbf{A}$, assume a particular solution of the form

$$\mathbf{y} = e^{\gamma t}[\mathbf{b}(t)\cos\omega t + \mathbf{d}(t)\sin\omega t],$$

where $\mathbf{b}$ and $\mathbf{d}$ are vector polynomials of degree k.

- If $\gamma + i\omega$ is an eigenvalue of $\mathbf{A}$ of multiplicity 1, assume a particular solution of the form

$$\mathbf{y} = e^{\gamma t}\{[c_1\mathbf{v}t^{k+1} + \mathbf{b}(t)]\cos\omega t + [c_2\mathbf{v}t^{k+1}\mathbf{d}(t)]\sin\omega t\},$$

where $\mathbf{v}$ is an eigenvector corresponding to $\gamma + i\omega$ and $\mathbf{b}$ and $\mathbf{d}$ are vector polynomials of degree k.

3. Determine the coefficients in the particular solution by substituting the trial solution into the nonhomogeneous differential equation.
4. The general solution is $\mathbf{\Psi}(t)\,\mathbf{c} + \mathbf{x}_p(t)$, where $\mathbf{c}$ is an arbitrary vector constant and $\mathbf{x}_p$ is the particular solution from step 3.

Variation of Parameters

Observe the analogy between the system and the corresponding scalar equation. The scalar equation $x' = a(t)x$ has a general solution that can be written as $\psi(t)c$, where ψ is any one solution and c is an arbitrary scalar constant. The vector equation $\mathbf{x}' = \mathbf{A}(t)\mathbf{x}$ has a general solution that can be written as $\mathbf{\Psi}(t)\,\mathbf{c}$, where $\mathbf{\Psi}$ is any matrix that solves the matrix equation $\mathbf{\Psi}' = \mathbf{A\Psi}$ and $\mathbf{c}$ is an arbitrary vector constant. This analogy suggests that the method of variation of parameters that we used in Section 4.5 for scalar equations can be recast to be suitable for vector equations. The idea of variation of parameters for a nonhomogeneous scalar (first-order) linear equation is to look for a solution of the form $\psi(t)u(t)$, where $c\psi(t)$ is the general solution of the corresponding homogeneous equation. Similarly, variation of parameters for a vector equation consists of the search for a vector $\mathbf{u}$ such that the general solution of the nonhomogeneous system is

$$\mathbf{x} = \mathbf{\Psi}(t)\mathbf{u}(t) + \mathbf{\Psi}(t)\,\mathbf{c}, \tag{3}$$

where $\mathbf{\Psi}$ is a fundamental matrix for the system. Substituting the first term of this expression into the original nonhomogeneous differential equation (1) yields

$$\mathbf{\Psi}'(\mathbf{u}+\mathbf{c}) + \mathbf{\Psi}\mathbf{u}' = \mathbf{A\Psi}(\mathbf{u}+\mathbf{c}) + \mathbf{g}.$$

The first term on each side cancels because $\mathbf{\Psi}' = \mathbf{A\Psi}$; hence, the vector function $\mathbf{u}$ must be the general solution of the equation

$$\mathbf{\Psi}\mathbf{u}' = \mathbf{g}. \tag{4}$$

All that remains is to solve Equation (4) *algebraically* for $\mathbf{u}'$ and then to integrate the result to obtain $\mathbf{u}$.

EXAMPLE 2

From the general solution given in the problem statement of Model Problem A.6, we have a fundamental matrix

$$\mathbf{\Psi}(t) = \begin{pmatrix} -2e^{2t} & -e^{3t} \\ e^{2t} & e^{3t} \end{pmatrix}.$$

Hence,

$$\begin{pmatrix} -2e^{2t} & -e^{3t} \\ e^{2t} & e^{3t} \end{pmatrix} \begin{pmatrix} u_1' \\ u_2' \end{pmatrix} = \begin{pmatrix} e^{2t} \\ 36t \end{pmatrix}.$$

This system has the solution

$$u_1' = -1 - 36te^{-2t}, \qquad u_2' = e^{-t} + 72te^{-3t}.$$

Integration yields

$$u_1 = -t + (18t + 9)e^{-2t}, \qquad u_2 = -e^{-t} - (24t + 8)e^{-3t}.$$

Finally, the particular solution is

$$\mathbf{x}(t) = \boldsymbol{\Psi}(t)\mathbf{u}(t) = \begin{pmatrix} (2t+1)e^{2t} - 12t - 10 \\ -(t+1)e^{2t} - 6t + 1 \end{pmatrix}.$$

✦ INSTANT EXERCISE 1

Why is the solution of Example 2 different from the solution (2) of Example 1?

In general, we have the following algorithm:

Algorithm A.6.2

Solving $\mathbf{x}' = \mathbf{A}(t)\mathbf{x} + \mathbf{g}(t)$ by Variation of Parameters

1. Solve the associated homogeneous problem $\mathbf{x}' = \mathbf{A}(t)\mathbf{x}$ and use the solutions to construct a fundamental matrix $\boldsymbol{\Psi}(t)$.
2. Solve the algebraic system $\boldsymbol{\Psi}\mathbf{u}' = \mathbf{g}$ for $\mathbf{u}'$.
3. Find a function $\mathbf{u}$ that is an antiderivative of $\mathbf{u}'$.
4. The particular solution is $\boldsymbol{\Psi}(t)\mathbf{u}(t)$, and the general solution is $\boldsymbol{\Psi}(t)\mathbf{c} + \boldsymbol{\Psi}(t)\mathbf{u}(t)$, where $\mathbf{c}$ is an arbitrary vector constant.

In practice, step 1 is generally possible only if $\mathbf{A}$ is constant or if $\mathbf{A}(t) = \mathbf{B}/t$, where $\mathbf{B}$ is constant (see Exercise 13). The remaining steps can always be done in principle, although the antiderivatives may need to be given as definite integrals and the algebra can get messy.

The method of variation of parameters can be used to derive a formula for the solution of any nonhomogeneous linear system. The idea is to solve the system in step 2 formally. The fundamental matrix is nonsingular by construction; hence it has an inverse. Multiplying $\boldsymbol{\Psi}\mathbf{u}' = \mathbf{g}$ by $\boldsymbol{\Psi}^{-1}$ produces the formula $\mathbf{u}' = \boldsymbol{\Psi}^{-1}\mathbf{g}$. Then, assuming initial conditions $\mathbf{x}(t_0) = \mathbf{x}_0$, we can construct an antiderivative, $\mathbf{u}(t) = \int_{t_0}^{t} \boldsymbol{\Psi}^{-1}(s)\mathbf{g}(s)\,ds$. We now have a general solution

$$\mathbf{x}(t) = \boldsymbol{\Psi}(t)\mathbf{c} + \boldsymbol{\Psi}(t)\int_{t_0}^{t} \boldsymbol{\Psi}^{-1}(s)\mathbf{g}(s)\,ds. \tag{5}$$

Diagonalization

Let **A** be a constant nondeficient matrix. Choose a linearly independent set of n eigenvectors $\mathbf{v}^{(1)}, \mathbf{v}^{(2)}, \ldots, \mathbf{v}^{(n)}$ and form a matrix **T,** using the eigenvectors as the columns. Let **D** be the matrix whose elements are $d_{kk} = \lambda_k$, where λ_k is the eigenvalue that corresponds to the eigenvector $\mathbf{v}^{(k)}$, and $d_{ij} = 0$ for $i \neq j$. (A matrix with zeros for all the off-diagonal entries is a **diagonal** matrix.) It follows from the linear independence of the vectors in **T** that **T** is nonsingular; it can be further shown that

$$\mathbf{T}^{-1}\mathbf{A}\mathbf{T} = \mathbf{D} \tag{6}$$

and that the eigenvalues of **D** are the same as those of **A** (see Exercise 14).

Now consider the system (1) with **A** constant and nondeficient. Define a new dependent variable **y** by

$$\mathbf{y} = \mathbf{T}^{-1}\mathbf{x}. \tag{7}$$

Multiplying both sides of this equation on the left by **T** yields the formula

$$\mathbf{x} = \mathbf{T}\mathbf{y};$$

using this formula to replace **x** in the system (1) yields

$$\mathbf{T}\mathbf{y}' = \mathbf{A}\mathbf{T}\mathbf{y} + \mathbf{g}(t).$$

A differential equation for **y** follows from multiplying on the left by $\mathbf{T}^{-1}$ and applying Equation (6):

$$\mathbf{y}' = \mathbf{D}\mathbf{y} + \mathbf{T}^{-1}\mathbf{g}(t). \tag{8}$$

The system for **y** is much more convenient than the original system for **x** because the matrix **D** is diagonal. This means that the system is decoupled. All the methods for solving scalar equations are available to use for each of the components of **y.** The process of decoupling a system using a matrix of eigenvectors is called **diagonalization.** It is an excellent example of the power of linear algebra to simplify problems in other areas of mathematics.

EXAMPLE 3

The transformation matrix **T** and diagonalized matrix **D** for Model Problem A.6 are

$$\mathbf{T} = \begin{pmatrix} -2 & -1 \\ 1 & 1 \end{pmatrix}, \qquad \mathbf{D} = \begin{pmatrix} 2 & 0 \\ 0 & 3 \end{pmatrix}.$$

Then

$$\mathbf{T}^{-1} = \begin{pmatrix} -1 & -1 \\ 1 & 2 \end{pmatrix}, \qquad \mathbf{T}^{-1}\mathbf{g} = \begin{pmatrix} -e^{2t} - 36t \\ e^{2t} + 72t \end{pmatrix}.$$

The decoupled system is

$$\mathbf{y}' = \begin{pmatrix} 2 & 0 \\ 0 & 3 \end{pmatrix}\mathbf{y} + \begin{pmatrix} -e^{2t} - 36t \\ e^{2t} + 72t \end{pmatrix},$$

or

$$y_1' - 2y_1 = -e^{2t} - 36t, \qquad y_2' - 3y_2 = e^{2t} + 72t.$$

These scalar problems can be solved by either undetermined coefficients or variation of parameters; here, the former seems the better choice. For y_{1p}, 2 is a characteristic value of multiplicity 1 for the homogeneous part of the differential equation; hence, the trial solution needs to be $y_{1p} = Bte^{2t} + Ct + D$ rather than the usual $y_{1p} = Be^{2t} + Ct + D$. Substituting this trial solution into the differential equation yields

$$-e^{2t} - 36t = (2Bte^{2t} + Be^{2t} + C) - 2(Bte^{2t} + Ct + D) = Be^{2t} - 2Ct + (C - 2D).$$

Thus, $B = -1$, $C = 18$, and $D = 9$, and the particular solution is

$$y_{1p} = -te^{2t} + 18t + 9.$$

Similarly,

$$y_{2p} = -e^{2t} - 24t - 8.$$

To get the desired particular solution for the original problem, we need only compute $\mathbf{x}$ from

$$\mathbf{x} = \mathbf{T}\mathbf{y} = \begin{pmatrix} -2 & -1 \\ 1 & 1 \end{pmatrix} \begin{pmatrix} -te^{2t} + 18t + 9 \\ -e^{2t} - 24t - 8 \end{pmatrix} = \begin{pmatrix} (2t+1)e^{2t} - 12t - 10 \\ -(t+1)e^{2t} - 6t + 1 \end{pmatrix}.$$

The result is the same as that of Example 2.

✦ INSTANT EXERCISE 2

Compute $\mathbf{T}^{-1}\mathbf{A}\mathbf{T}$ for Model Problem A.6, and make sure that you get $\mathbf{D}$.

✦ INSTANT EXERCISE 3

Solve the y_2 problem in Example 3 using the method of undetermined coefficients.

The diagonalization method is summarized in Algorithm A.6.3.

Algorithm A.6.3

Solving $\mathbf{x}' = \mathbf{A}\mathbf{x} + \mathbf{g}(t)$ by Diagonalization

1. Solve the associated homogeneous problem $\mathbf{x}' = \mathbf{A}\mathbf{x}$, and use the eigenvectors and eigenvalues to construct the matrices $\mathbf{T}$ and $\mathbf{D}$.
2. Compute $\mathbf{T}^{-1}$ and $\mathbf{T}^{-1}\mathbf{g}(t)$.
3. Find a solution of $\mathbf{y}' = \mathbf{D}\mathbf{y} + \mathbf{T}^{-1}\mathbf{g}(t)$.
4. The particular solution is $\mathbf{T}\mathbf{y}(t)$, and the general solution is $\mathbf{\Psi}(t)\mathbf{c} + \mathbf{T}\mathbf{y}(t)$, where $\mathbf{c}$ is an arbitrary vector constant and $\mathbf{\Psi}$ a fundamental matrix constructed from the results of step 1.

Note that the algorithm is inconvenient if **A** has any nonreal eigenvalues. The matrices **D** and **T** are nonreal in this case, although the particular solutions are real (see Exercises 11 and 12).

Choice of Methods

There are subtle distinctions between problems that seem to work out better with one method and problems that seem to work out better with another method. For example, the rules for choosing the trial solution and the calculations for determining the coefficients for the method of undetermined coefficients can be quite complicated for those cases where the forcing function has a characteristic value that is also a characteristic value of the operator. While the method of undetermined coefficients is generally the easiest of the three for the simple case, one might well prefer one of the other methods when the characteristic values match. The method of diagonalization is particularly convenient when the eigenvalues of **A** are all real; however, it is messy if any of the eigenvalues are nonreal (and cannot be used at all if the matrix is deficient).

A.6 Exercises

In Exercises 1 through 8, find a particular solution of the system $\mathbf{x}' = \mathbf{A}\mathbf{x} + \mathbf{g}$, using (*a*) undetermined coefficients, (*b*) variation of parameters, and (*c*) diagonalization; or state why the method cannot be used.

1. $\mathbf{A} = \begin{pmatrix} 1 & 0 \\ 2 & 3 \end{pmatrix}, \quad \mathbf{g}(t) = \begin{pmatrix} 3 \\ e^{3t} \end{pmatrix}$ (see Section 6.4, Exercise 3)

2. $\mathbf{A} = \begin{pmatrix} 1 & 3 \\ 2 & 6 \end{pmatrix}, \quad \mathbf{g}(t) = \begin{pmatrix} t \\ 0 \end{pmatrix}$ (see Section 6.4, Exercise 4)

3. $\mathbf{A} = \begin{pmatrix} -1 & -2 \\ 1 & -4 \end{pmatrix}, \quad \mathbf{g}(t) = \begin{pmatrix} e^{t} \\ e^{3t} \end{pmatrix}$ (see Section 6.4, Exercise 5)

4. $\mathbf{A} = \begin{pmatrix} 3 & -1 \\ 5 & -3 \end{pmatrix}, \quad \mathbf{g}(t) = \begin{pmatrix} e^{-t} \\ e^{t} + e^{-t} \end{pmatrix}$ (see Section 6.4, Exercise 6)

5. $\mathbf{A} = \begin{pmatrix} 1 & 0 \\ 2 & 1 \end{pmatrix}, \quad \mathbf{g}(t) = \begin{pmatrix} t \\ 3 \end{pmatrix}$

6. $\mathbf{A} = \begin{pmatrix} 2 & 1 \\ 0 & 1 \end{pmatrix}, \quad \mathbf{g}(t) = \begin{pmatrix} 1 \\ 2 \end{pmatrix} e^{2t}$ (see Section 6.4, Exercise 8)

7. $\mathbf{A} = \begin{pmatrix} 2 & 0 & -1 \\ 0 & 2 & 1 \\ 0 & -1 & 0 \end{pmatrix}, \quad \mathbf{g}(t) = \begin{pmatrix} 0 \\ 1 \\ 0 \end{pmatrix} e^{2t}$ (see Section 6.6, Exercise 7)

8. $\mathbf{A} = \begin{pmatrix} -2 & 0 & 1 \\ 2 & -4 & 1 \\ -1 & 0 & -4 \end{pmatrix}, \quad \mathbf{g}(t) = \begin{pmatrix} 2 \\ 1 \\ 0 \end{pmatrix} e^{-4t}$ (see Section 6.6, Exercise 9)

9. *a.* Use the method of variation of parameters to find a particular solution for $\mathbf{x}' = \mathbf{A}\mathbf{x} + \mathbf{g}$ with

$$\mathbf{A} = \begin{pmatrix} 1 & 0 \\ 2 & 1 \end{pmatrix}, \qquad \mathbf{g}(t) = \begin{pmatrix} e^{-t^2} \\ 0 \end{pmatrix}$$

b. Explain why variation of parameters is the method of choice for this problem.

10. *a.* Use the method of diagonalization to find a particular solution for $\mathbf{x}' = \mathbf{A}\mathbf{x} + \mathbf{g}$ with

$$\mathbf{A} = \begin{pmatrix} 3 & -1 & 0 \\ 2 & 0 & 0 \\ -1 & 1 & 2 \end{pmatrix}, \qquad \mathbf{g}(t) = \begin{pmatrix} e^{2t} \\ 0 \\ 2te^{2t} \end{pmatrix}$$

(see Section 6.4, Exercise 12).

b. Offer an argument to justify the view that diagonalization is the method of choice for this problem.

11. Consider the problem $\mathbf{x}' = \mathbf{A}\mathbf{x} + \mathbf{g}$, where

$$\mathbf{A} = \begin{pmatrix} 0 & -1 \\ 4 & 0 \end{pmatrix}, \qquad \mathbf{g} = \begin{pmatrix} 4 \\ 8 \end{pmatrix} e^{2t}.$$

a. Determine the eigenvalues and a set of eigenvectors for the matrix.
b. Use the method of diagonalization to find a particular solution and the general solution.
c. Use the method of undetermined coefficients to find a particular solution and the general solution.

12. Consider the problem $\mathbf{x}' = \mathbf{A}\mathbf{x} + \mathbf{g}$, where

$$\mathbf{A} = \begin{pmatrix} 3 & 1 \\ -1 & 3 \end{pmatrix}, \qquad \mathbf{g} = \begin{pmatrix} 0 \\ 4 \end{pmatrix} e^{2t}.$$

a. Determine the eigenvalues and a set of eigenvectors for the matrix.
b. Use the method of diagonalization to find a particular solution and the general solution.
c. Use the method of undetermined coefficients to find a particular solution and the general solution.

13. *a.* Use a change of independent variable to convert the equation

$$t\frac{d\mathbf{x}}{dt} = \mathbf{B}\mathbf{x} + \mathbf{g}(t), \qquad t > 0,$$

into a nonhomogeneous linear differential equation with constant coefficients.

b. Solve the problem

$$t\frac{d\mathbf{x}}{dt} = \mathbf{B}\mathbf{x} + \mathbf{g}(t), \qquad t > 0, \qquad \mathbf{B} = \begin{pmatrix} 1 & 3 \\ 1 & -1 \end{pmatrix}, \qquad \mathbf{g}(t) = \begin{pmatrix} 0 \\ 5t^3 \end{pmatrix}.$$

14. Let $\mathbf{A}$ be a constant nondeficient matrix. Choose a linearly independent set of n eigenvectors $\mathbf{v}^{(1)}, \mathbf{v}^{(2)}, \ldots, \mathbf{v}^{(n)}$ and form a matrix $\mathbf{T}$, using the eigenvectors as the columns. Let $\mathbf{D}$ be the diagonal matrix whose elements are $d_{kk} = \lambda_k$, where λ_k is the eigenvalue that corresponds to the eigenvector $\mathbf{v}^{(k)}$.

a. Show that $\mathbf{AT} = \mathbf{TD}$. Then use the nonsingularity of $\mathbf{T}$ to derive Equation (6).

b. Show that the eigenvalues of **D** are the same as those of **A.**

c. Find the eigenvectors of **D.**

15. *a.* Solve

$$\mathbf{x}' = \begin{pmatrix} 3 & -2 \\ 2 & -1 \end{pmatrix} \mathbf{x} + \begin{pmatrix} 1 \\ -1 \end{pmatrix} e^t$$

by the method of variation of parameters.

b. Determine the appropriate trial solution for the particular solution for any problem of the form

$$\mathbf{x}' = \mathbf{A}\mathbf{x} + \mathbf{w}e^{at},$$

where **A** is a 2×2 matrix with an eigenvalue $\lambda = a$ of algebraic multiplicity 2 and geometric multiplicity 1 and **w** is any constant vector.

c. Use the trial solution of part *b* to solve

$$\mathbf{x}' = \begin{pmatrix} -4 & 1 \\ -4 & 0 \end{pmatrix} \mathbf{x} + \begin{pmatrix} 1 \\ 0 \end{pmatrix} e^{-2t}.$$

16. Suppose the spring-mass system of Figure A.6.1 (also see Exercise 11 of Sec. 6.5) is connected to a motor that exerts a force of cos $2t$ on the mass that is on the left. The system is initially at rest. Determine the subsequent motion. In particular, plot the deviation from equilibrium of the distance between the masses as a function of time.

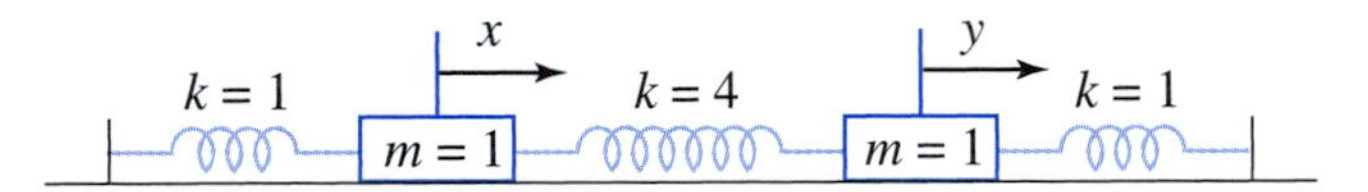

Figure A.6.1
The mechanical system for Exercise 16.

✦ A.6 INSTANT EXERCISE SOLUTIONS

1. Any solution of the nonhomogeneous equation can be the particular solution. Had we chosen $c_1 = 1$ and $c_2 = -1$ in Example 1, we would have gotten the same particular solution as in Example 2.

2.

$$\mathbf{T}^{-1}\mathbf{A}\mathbf{T} = \begin{pmatrix} -1 & -1 \\ 1 & 2 \end{pmatrix} \begin{pmatrix} 1 & -2 \\ 1 & 4 \end{pmatrix} \begin{pmatrix} -2 & -1 \\ 1 & 1 \end{pmatrix} = \begin{pmatrix} -1 & -1 \\ 1 & 2 \end{pmatrix} \begin{pmatrix} -4 & -3 \\ 2 & 3 \end{pmatrix} = \begin{pmatrix} 2 & 0 \\ 0 & 3 \end{pmatrix}.$$

3. The forcing functions have characteristic values 2 and 0, which are not eigenvalues of the homogeneous equation. Hence, a particular solution can be found from the family $y_{2p} = Be^{2t} + Ct + D$. We have $e^{2t} + 72t = y_{2p}' - 3y_{2p} = (2Be^{2t} + C) - 3(Be^{2t} + Ct + D) = -Be^{2t} - 3Ct + (C - 3D)$. Hence, $B = -1$, $C = -24$, and $D = -8$; the particular solution is $y_{2p} = -e^{2t} - 24t - 8$.

A.7 The One-Dimensional Heat Equation

Consider a thin solid rod made of a uniform material and wrapped in insulation. Suppose you hold the rod by one end and plunge the other end of the rod into a pot of boiling water. If you continue to hold the rod in this manner, you will soon feel the temperature increase in the end of the rod that is in your hand. The temperature at your end is increasing because the material of the rod conducts heat energy from points of higher temperature to points of lower temperature. Suppose your rod is equipped with a device that measures the temperature at the point halfway between the two ends. The device will show that the temperature at the midpoint is also increasing, at a faster rate than the temperature at your end. Thus, the temperature in the rod depends on both time and position. A mathematical model for this heat flow process will yield a partial differential equation with two independent variables, the time t and position x. Note that in our conceptual model there is no heat flow from the rod to the insulation, or else we would need more than one spatial coordinate in the mathematical model.

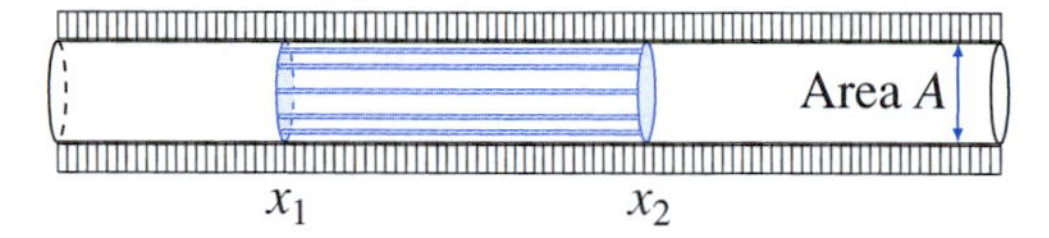

Figure A.7.1
An arbitrary portion of a solid insulated rod.

Energy is a *conserved* quantity, which means that the total change is the net sum of energy increases. Heat flow models are based on this idea—one can write expressions for heat gain by each mechanism and equate the sum to the overall rate of heat gain. We consider an arbitrary portion of the rod, with endpoints x_1 and x_2 (Figure A.7.1). The overall rate of change is the time derivative of the total amount of heat energy. The number of heat flow mechanisms is quite limited. There is the mechanism of heat conduction, which causes heat to flow across the boundaries at x_1 and x_2. There is also a possible mechanism of heat creation within the region, as is the case if there is an electric current traveling through the rod. These are the only mechanisms permitted by the conceptual model.[31] Figure A.7.2 indicates the form of the equation based on conservation of heat energy.

$$\frac{d}{dt}\begin{pmatrix}\text{total amount}\\ \text{of heat energy}\end{pmatrix} = \begin{matrix}\text{rate of flow}\\ \text{coming in at } x_2\end{matrix} + \begin{matrix}\text{rate of flow}\\ \text{coming in at } x_1\end{matrix} + \begin{matrix}\text{rate of production}\\ \text{within the section}\end{matrix}$$

Figure A.7.2
A schematic diagram of the conservation of heat in a portion of a solid rod.

[31] In some circumstances, heat flows by being carried along with moving material. An example of this occurs when you turn on the hot water faucet. The water coming out of the faucet gets hotter, not because something is heating the water molecules themselves, but because the cold water molecules are coming out of the faucet and being replaced by hot water molecules from the hot water heater at the other end of the pipe. Our rod, however, is made of solid material, so this type of heat transfer does not occur.

The total amount of heat energy in the region depends on the temperature, the size of the region, and the material out of which the rod is made. The material is characterized by its *heat capacity* c, which is the amount of energy required to raise the temperature of a unit mass by 1 degree. Let u be the temperature, which we suppose for the moment to be constant. Then cu has the dimension of energy per unit mass. The mass of the portion of rod is $\rho A(x_2 - x_1)$, where ρ is the material density (mass per unit volume) and A is the cross-sectional area of the region. We conclude that the total heat energy is $\rho c A u(x_2 - x_1)$. The only problem with this calculation is that the temperature is *not* constant because the portion of rod from x_1 to x_2 is not arbitrarily small. We could, however, divide up the region from x_1 to x_2 into a large number of very small lengths, add the total amounts of energy ($\rho c A u \,\Delta x$) in the subregions, and take a limit as the subregion width approaches 0. The result is a definite integral:

$$\text{Total amount of heat energy} = \int_{x_1}^{x_2} \rho c A u(x,t)\,dx.$$

Next, we consider the rate of heat flow across the boundaries of the region. At any point x, the amount of heat flow is proportional to the area A of the boundary. Since the flow of heat is caused by temperature differences, it seems reasonable that the rate of flow should be proportional to the spatial derivative of the temperature as well.[32] A positive temperature gradient at x_2 corresponds to an increase in the heat in the region, while a positive temperature gradient at x_1 corresponds to a decrease. Thus,

$$\text{Net rate of flow coming in at } x_1 \text{ and } x_2 = \kappa A[u_x(x_2,t) - u_x(x_1,t)],$$

where u_x is the partial derivative of u with respect to x and κ is a physical property of the material of the rod. Since the net rate of flow has the dimension of energy per unit time, the property κ, called the *thermal conductivity,* has the dimension of energy per unit time per unit area.

We can write down an expression for the rate of heat production without postulating a specific mechanism. Simply let Q be the rate of energy production per unit mass at the point x. If this quantity is constant, then the total rate of production is $\rho A Q(x_2 - x_1)$. In general, we should expect the production rate to be variable; as in the case of the total heat energy calculation, we subdivide the region and pass to the integral. The overall production rate is then

$$\text{Rate of production between } x_1 \text{ and } x_2 = \int_{x_1}^{x_2} \rho A Q(x,t)\,dx.$$

Assembling all the pieces yields the equation

$$\frac{d}{dt}\int_{x_1}^{x_2} \rho c u(x,t)\,dx = \kappa[u_x(x_2,t) - u_x(x_1,t)] + \int_{x_1}^{x_2} \rho Q(x,t)\,dx,$$

where we have removed the common factor A.

The net flow across the boundaries can be written as a definite integral by using the fundamental theorem of calculus:

$$u_x(x_2,t) - u_x(x_1,t) = \int_{x_1}^{x_2} u_{xx}(x,t)\,dx,$$

[32]This is **Fourier's law of heat conduction.**

where we have used u_{xx} for the second partial derivative of u with respect to x. It is also possible to interchange the order of differentiation and integration in the term on the left side of the heat balance equation, but we must note that the time derivative on the inside of the integral is a partial derivative. Using u_t for the partial derivative of u with respect to time, and making these two changes, we have

$$\int_{x_1}^{x_2} \rho c u_t(x,t)\,dx = \kappa \int_{x_1}^{x_2} u_{xx}(x,t)\,dx + \int_{x_1}^{x_2} \rho Q(x,t)\,dx.$$

Now all the terms can be combined into a single integral to yield the equation

$$\int_{x_1}^{x_2} [\rho c u_t(x,t) - \kappa u_{xx}(x,t) - \rho Q(x,t)]\,dx = 0.$$

This last equation makes a remarkable statement. The given integral is 0, *no matter what points are chosen for x_1 and x_2*. This says that if we graph the integrand with respect to x and then compute the area under an arbitrary portion of the x axis, the area under the curve is always 0. The only way this can happen is if the integrand itself is 0.[33] We therefore arrive at the partial differential equation

$$u_t = k u_{xx} + q, \tag{1}$$

where $k = \kappa/(\rho c)$ and $q = Q/c$. The parameter k is called the *thermal diffusivity*. The complete heat flow model is a nonhomogeneous partial differential equation, but our interest is in the special case in which there is no production mechanism. We have therefore arrived at the **one-dimensional heat equation**

$$u_t = k u_{xx}. \tag{2}$$

As in the case of the wave equation, there is a lot of interesting mathematics in the deceptively simple heat equation.

Boundary Conditions for Heat Flow

As with the wave equation,[34] a complete problem for the heat equation on a finite domain requires specification of auxiliary conditions at each boundary. Several types are common.

1. **Dirichlet[35] conditions** prescribe the temperature at a point:

$$u(0,t) = A(t), \qquad u(L,t) = B(t).$$

2. **Neumann[36] conditions** prescribe the flow per unit area at a point. From our derivation of the heat equation, the flow per unit area is κu_x; thus, the Neumann conditions are

$$\kappa u_x(0,t) = A(t), \qquad \kappa u_x(L,t) = B(t).$$

---

[33]This fact is known as the *Dubois–Raymond lemma* (pronounced "doo-bwah ray-mohn").

[34]Sections 8.3 through 8.5.

[35]"Di-ri-shlay."

[36]"Noy-mahn."

3. **Robin**[37] **conditions** are used in situations where the ends of the rod are subjected to environmental heating or cooling. This is similar to Newton's law of cooling for situations where the temperature is spatially independent. The specific form of the Robin condition depends on whether it is being applied to the left end of the rod or the right end:

$$\kappa u_x(0,t) = h[u(0,t) - A(t)], \qquad \kappa u_x(L,t) = -h[u(L,t) - B(t)],$$

where $h > 0$ is a proportionality constant and $A(t)$ and $B(t)$ are the temperatures of the environment at $x = 0$ and $x = L$, respectively. Note that these conditions reduce to Neumann conditions as $h \to 0$ and to Dirichlet conditions as $h \to \infty$.

4. It is also possible to have one kind of boundary condition at $x = 0$ and another kind at $x = L$. Such problems are said to have *mixed* boundary conditions.

✦ INSTANT EXERCISE 1

Explain the choice of signs in the Robin conditions. To simplify the explanation, assume that $A = B = 0$ and $u > 0$.

Initial-Boundary-Value Problems for the Heat Equation

Initial-boundary-value problems for the heat equation consist of the heat equation, two boundary conditions, and an initial condition. The solution procedure is the same as that used for the wave equation in Section 8.4. We focus on one example.

MODEL PROBLEM A.7

An aluminum rod of length 1 ft is initially at room temperature (20°C). One end is placed in boiling water at time 0, and the other end is insulated. Determine the subsequent temperature in the rod.

Let $T(x,t)$ be the temperature in degrees Celsius at a distance x feet from the hot end, t minutes after the end at $x = 0$ is immersed. Let k be the thermal diffusivity. The temperature is governed by the problem

$$T_t = kT_{xx}, \qquad 0 \le x \le 1, \tag{3}$$

$$T(0,t) = 100, \qquad T_x(1,t) = 0, \tag{4}$$

$$T(x,0) = 20. \tag{5}$$

This problem is not correctly formulated for the separation of variables technique because the boundary condition at $x = 0$ is nonhomogeneous. Observe that the temperature profile as $t \to \infty$ is easily determined by setting T_t equal to 0 and ignoring the initial condition. The result is the

[37] "Roh-ban."

steady-state temperature problem

$$T_s'' = 0, \qquad T_s(0) = 100, \qquad T_s'(1) = 0,$$

where $T_s(x)$ is the steady-state temperature. The solution of this simple problem is the constant $T_s \equiv 100$. Suppose we subtract T_s from T to obtain a new temperature variable, say,

$$w = T - T_s = T - 100.$$

With w in place of T, the problem becomes

$$w_t = kw_{xx}, \qquad 0 \le x \le 1,$$

$$w(0,t) = 0, \qquad w_x(1,t) = 0,$$

$$w(x,0) = -80.$$

The new problem has homogeneous boundary conditions.

We can further simplify the problem by scaling. Define a new temperature and time by

$$u = \frac{w}{80} = \frac{T - 100}{80}, \qquad \tau = kt.$$

With these changes, we arrive at the problem

$$u_\tau = u_{xx}, \qquad 0 \le x \le 1, \tag{6}$$

$$u(0,\tau) = 0, \qquad u_x(1,\tau) = 0, \tag{7}$$

$$u(x,0) = -1. \tag{8}$$

Once this problem has been solved, we can recover the solution of the original problem as

$$T(x,t) = 100 + 80\,u(x,kt), \tag{9}$$

where we must be careful to report k in square feet per minute.

Construction of a Series Solution The search for solutions of the form

$$u = f(x)g(t)$$

results in the eigenvalue problem

$$f'' - kf = 0, \qquad f(0) = 0, \qquad f'(1) = 0,$$

along with the differential equation

$$g' - kg = 0$$

for the amplitude function. There are no nontrivial solutions with $k \ge 0$, so we may set $k = -\lambda^2$. The resulting problem,

$$f'' + \lambda_n^2 f = 0, \qquad f(0) = 0,$$

has solutions of the form

$$f_n = b_n \sin \lambda_n x.$$

The remaining boundary condition leads to the trigonometric equation

$$\cos \lambda_n = 0;$$

hence, the eigenvalues are

$$\lambda_n = \frac{(2n-1)\pi}{2}, \qquad n = 1, 2, \ldots.$$

The amplitude functions must satisfy

$$g_n' + \lambda_n g_n = 0;$$

hence, we have

$$g_n(t) = e^{-(2n-1)^2\pi^2 t/4}.$$

We therefore obtain a Fourier series solution for the homogeneous problem of Equations (6) and (7):

$$u = \sum_{n=1}^{\infty} b_n e^{-(2n-1)^2\pi^2 t/4} \sin \frac{(2n-1)\pi x}{2}. \tag{10}$$

Determination of Fourier Coefficients Substituting the solution (10) into the nonhomogeneous initial condition (8) yields the algebraic equation

$$-1 = \sum_{n=1}^{\infty} b_n \sin \frac{(2n-1)\pi x}{2}.$$

The eigenfunctions can be shown to be orthogonal with respect to the inner product[38]

$$\langle f, g \rangle = \int_0^1 f(x) g(x)\, dx.$$

Hence, we can isolate a coefficient b_m by multiplying the algebraic equation by the factor

$$\sin \frac{(2m-1)\pi x}{2}$$

and integrating. The result is

$$-\int_0^1 \sin \frac{(2m-1)\pi x}{2}\, dx = b_m \int_0^1 \sin^2 \frac{(2m-1)\pi x}{2}\, dx,$$

or

$$\frac{b_m}{2} = -\frac{2}{(2m-1)\pi} \cos \frac{(2m-1)\pi x}{2}\bigg|_0^1 = -\frac{2}{(2m-1)\pi}.$$

Thus,

$$b_m = -\frac{4}{(2m-1)\pi}.$$

[38] Section 8.5.

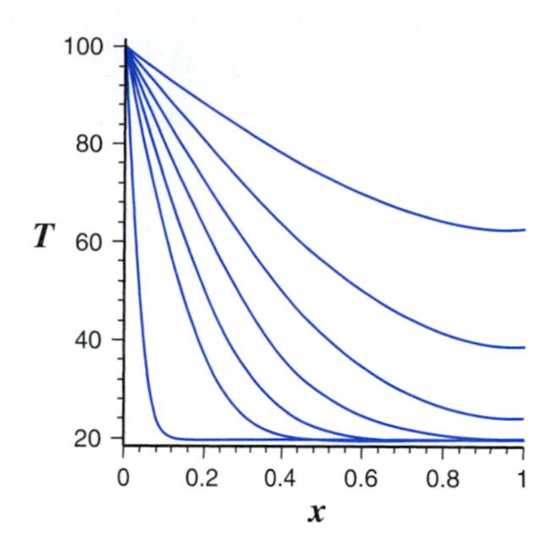

Figure A.7.3
Temperature profiles for Model Problem A.7 at times 1 s, 15 s, 30 s, 1 min, 2 min, 4 min, and 8 min, from bottom to top.

Substituting this result and that of Equation (10) into Equation (9) yields the solution of Model Problem A.7 as

$$T = 100 - \frac{320}{\pi}\sum_{n=1}^{\infty}\frac{1}{2n-1}e^{-(2n-1)^2\pi^2 kt/4}\sin\frac{(2n-1)\pi x}{2}. \tag{11}$$

The thermal diffusivity of aluminum is approximately 0.79 cm^2/s, which corresponds to 0.051 ft^2/min. With this value, we obtain the approximate solution

$$T = 100 - \frac{320}{\pi}\sum_{n=1}^{\infty}\frac{1}{2n-1}e^{-0.126(2n-1)^2 t}\sin\frac{(2n-1)\pi x}{2}.$$

The solution is displayed in Figures A.7.3 and A.7.4. Figure A.7.3 shows the evolution of the temperature profile over the first 8 min. Figure A.7.4 shows the time history of the temperatures at the insulated end and at the midpoint.

A.7 Exercises

Exercises 1 through 5 consider heat flow with Dirichlet conditions at both ends. Separation of variables yields, via the substitution $u = f(x)g(t)$, the eigenvalue problem

$$f'' + \lambda^2 f = 0, \qquad f(0) = 0, \qquad f(1) = 0,$$

given homogeneous Dirichlet conditions. This eigenvalue problem is identical to the eigenvalue problem obtained for the wave equation with homogeneous Dirichlet conditions. The solutions are

$$f(x) = b_n \sin n\pi x, \qquad \lambda = n\pi.$$

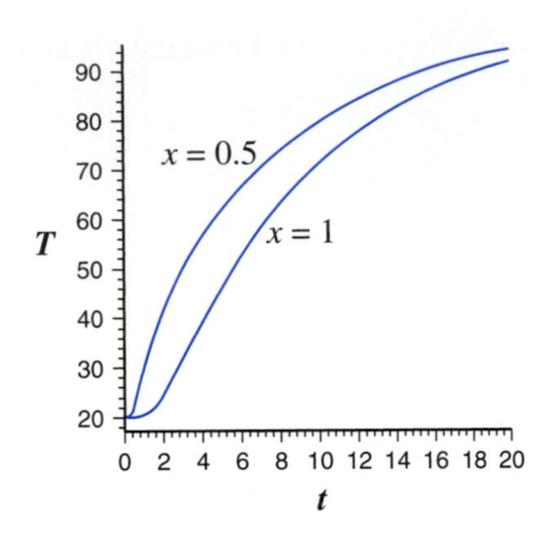

Figure A.7.4
The temperature at the insulated end ($x = 1$) and midpoint ($x = 0.5$) for Model Problem A.7, with time in minutes.

1. Derive a solution formula for the Dirichlet problem

$$u_t = ku_{xx}, \qquad 0 < x < 1, \quad t > 0,$$
$$u(0, t) = 0, \qquad u(1, t) = 0,$$
$$u(x, 0) = \phi(x).$$

T 2. *a.* Use the result of Exercise 1 to solve

$$u_t = u_{xx}, \qquad 0 < x < 1, \quad t > 0,$$
$$u(0, t) = 0, \qquad u(1, t) = 0,$$
$$u(x, 0) = x(1 - x).$$

b. Use a computer algebra system to plot the solution at times 0, 0.05, 0.1, 0.15, and 0.2.

T 3. *a.* Use the result of Exercise 1 to solve

$$u_t = u_{xx}, \qquad 0 < x < 1, \quad t > 0,$$
$$u(0, t) = 0, \qquad u(1, t) = 0,$$
$$u(x, 0) = \frac{1}{2h}\begin{cases} 1 & 0.5 - h < x < 0.5 + h \\ 0 & \text{otherwise} \end{cases}, \qquad 0 < h < 0.5.$$

(*Hint*: The definite integral for b_n leads to a formula with two terms. These terms can be combined into one term by a trigonometric identity.)

b. Observe that the total amount of heat in the problem of part *a* is independent of h. Obtain a formula for the temperature due to a unit amount of heat concentrated at the point $x = 0.5$ by taking a limit as $h \to 0$ of the result of part *a*.

c. The heat flow caused by a unit input of heat in the center of the rod can also be modeled by the problem

$$u_t = u_{xx}, \qquad 0 < x < 1, \quad t > 0,$$
$$u(0,t) = 0, \qquad u(1,t) = 0,$$
$$u(x,0) = \delta(x - \tfrac{1}{2}).$$

Solve this problem and compare the result with that of part *b*. (See Section 7.4 for information about the delta function.)

d. Plot the temperature $u(0.5, t)$ for the $h = 0.1$ and $h = 0$ cases for the time interval $0.001 \le t \le 0.1$. Why can't you plot $u(0.5, t)$ on the interval $0 \le t \le 0.1$?

e. Discuss the use of the delta function to model concentrated heat input. In particular, describe the advantages and disadvantages of using the problem of part *c* as compared to the problem of part *a* with a small value of h.

T 4. *a.* Use the result of Exercise 1 to solve

$$u_t = u_{xx}, \qquad 0 < x < 1, \quad t > 0,$$
$$u(0,t) = 0, \qquad u(1,t) = 0,$$
$$u(x,0) = \begin{cases} x & 0 < x < 0.5 \\ 1 - x & 0.5 < x < 1 \end{cases}.$$

b. Use a computer algebra system to plot the solution at times 0, 0.05, 0.1, 0.15, and 0.2.

5. Determine the temperature in the rod of Model Problem A.7 for the case where the end at $x = 1$ is kept at room temperature instead of being insulated. Plot temperature profiles and compare with Figure A.7.3.

T 6. *a.* Solve

$$u_t = u_{xx}, \qquad 0 < x < 1, \quad t > 0,$$
$$u(0,t) = 0, \qquad u_x(1,t) = 0,$$
$$u(x,0) = \begin{cases} x & 0 < x < 0.5 \\ 1 - x & 0.5 < x < 1 \end{cases}.$$

b. Use a computer algebra system to plot the solution at times 0, 0.05, 0.1, 0.15, and 0.2.

c. Compare the result with that of Exercise 4.

7. Derive a solution formula for the Neumann problem

$$u_t = u_{xx}, \qquad 0 < x < 1, \quad t > 0,$$
$$u_x(0,t) = 0, \qquad u_x(1,t) = 0,$$
$$u(x,0) = \phi(x).$$

(*Hint*: check to see if 0 is an eigenvalue.)

T 8. *a.* Use the result of Exercise 7 to solve the problem

$$u_t = u_{xx}, \qquad 0 < x < 1, \quad t > 0,$$
$$u_x(0,t) = 0, \qquad u_x(1,t) = 0,$$
$$u(x,0) = \begin{cases} x & 0 < x < 0.5 \\ 1 - x & 0.5 < x < 1 \end{cases}.$$

b. Use a computer algebra system to plot the solution at times 0, 0.05, 0.1, 0.15, and 0.2.
c. Compare the result with that of Exercise 4.

9. Consider the problem

$$u_t = u_{xx}, \qquad 0 < x < 1, \quad t > 0,$$
$$u(0,t) = 0, \qquad u_x(1,t) = -u(1,t),$$
$$u(x,0) = \phi(x).$$

a. Use the assumption $u = f(x)g(t)$ to derive an eigenvalue problem for f and an ordinary differential equation for g.
b. Show that the eigenvalues are the positive solutions of the equation

$$\lambda + \tan\lambda = 0.$$

c. Without calculating values for λ_n, determine the form of the series solution for the differential equation and boundary conditions.
d. Derive the formula

$$\int_0^1 \sin^2 \lambda_n x \, dx = \frac{1 + \cos^2 \lambda_n}{2}.$$

e. Complete the solution of the problem by deriving an integral formula for the Fourier coefficients.

T 10. *a.* A rod of length 1 and thermal diffusivity 1 is initially at a uniform temperature $u = 1$. Determine the subsequent temperature $u(x,t)$, for the case where the ends are kept at temperature 0. Use a computer algebra system to plot the solution at times 0.02, 0.04, ..., 0.1. (The first six terms in the series are sufficient.)
b. A rod of length 1 and thermal diffusivity 1 has initial temperature $u = 10$ for $0 < x < 0.1$ and $u = 0$ for $0.1 < x < 1$. Determine the subsequent temperature $u(x,t)$, for the case where the ends are kept at temperature 0. Use a computer algebra system to plot the solution at times 0.02, 0.04, ..., 0.1. (The first six terms in the series are sufficient.)
c. Discuss the results of parts *a* and *b*. (*Hint*: What is the total amount of heat in the initial situation in each problem? In what ways are the temperature profiles similar and in what ways are they different? Do the differences make sense?)

11. Consider the problem of determining the temperature in a thin insulated ring of unit radius (think of a wire of length 2π bent into the shape of a unit circle) with unit thermal diffusivity. The problem is given by

$$u_t = u_{\theta\theta}, \qquad 0 \le \theta < \pi,$$

along with the initial condition

$$u(\theta, t) = \phi(\theta).$$

Derive a solution formula, using separation of variables. Note that boundary conditions in θ are implied by the geometry.

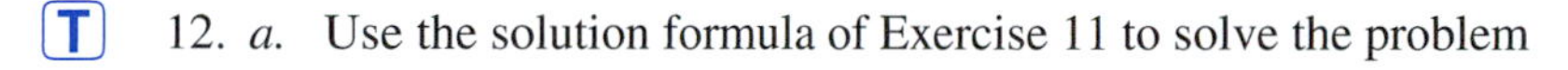

12. *a.* Use the solution formula of Exercise 11 to solve the problem

$$u_t = u_{\theta\theta},$$

$$u(\theta, t) = \begin{cases} \frac{1}{4} - \left|\frac{\theta}{\pi}\right| & -\frac{\pi}{4} < \theta < \frac{\pi}{4} \\ 0 & \text{otherwise} \end{cases}.$$

b. Use a computer algebra system to plot the solution at times 0, 0.05, 0.1, 0.15, and 0.2.

13. Derive the three-dimensional heat equation

$$u_t = k\,\nabla^2 u = k\nabla\cdot(\nabla u).$$

(*Hint*: The derivation follows the same general plan as that in the text for the one-dimensional heat equation. You may assume there are no internal heat sources. The total amount of heat is a volume integral, and the net heat flux is a surface integral. Use the divergence theorem to rewrite the flux as a volume integral.)

✦ A.7 INSTANT EXERCISE SOLUTION

1. The assumptions $A = B = 0$ and $u > 0$ mean that the temperature in the rod is higher than the temperature of the surroundings. Since the rod is cooling at the boundaries, the temperature at points near the boundaries must be a little higher than the temperature at the boundaries themselves. This means that u_x is positive at $x = 0$ and negative at $x = L$. The algebraic signs in the conditions are chosen to agree with these conclusions.

A.8 Laplace's Equation

The heat equation in two or three dimensions is

$$u_t = k\,\nabla^2 u,$$

where $\nabla^2 u = \nabla \cdot (\nabla u)$ is the **Laplacian** of u. Suppose the temperature u is prescribed along a simple closed curve in the xy plane, and let R be the region inside the curve. Given enough time, the temperature in R approaches a steady distribution that is determined by the problem

$$\nabla^2 u = 0 \text{ on } R, \qquad u = h \text{ on } \partial R, \tag{1}$$

where ∂R is the boundary of R and h is a continuous function[39] defined on that boundary. This is the **Dirichlet problem** for **Laplace's equation.**

In general, problems with Laplace's equation are amenable to analytical solution only when the geometry is very simple. Problems on disks or rectangles can be solved by separation of variables, but problems defined on more complicated regions must in general be solved numerically. We begin with an example of a Dirichlet problem on a rectangle.

MODEL PROBLEM A.8*a*

Solve the problem

$$\begin{aligned} u_{xx} + u_{yy} = 0, &\qquad 0 < x, y < 2, \\ u(x, 0) = 0, &\qquad u(x, 2) = 0, \\ u(0, y) = 0, &\qquad u(2, y) = y(2 - y). \end{aligned}$$

Model Problem A.8*a* can be solved by separation of variables in rectangular coordinates. The general plan is to construct a series solution of the homogeneous portion of the problem and then to determine the coefficients that allow the solution to satisfy the nonhomogeneous boundary condition. Problems with more than one nonhomogeneous condition are considered later.

Solving Model Problem A.8*a*

When we solve the wave equation (Chapter 8) and the heat equation (Section A.7), we make the substitution $u = f(x)g(t)$ and obtain an eigenvalue problem for f. The eigenvalue problem consists of an ordinary differential equation with two boundary conditions. Model Problem A.8*a* is a partial differential equation with two spatial variables and no time variable. The question arises: which variable in the separation of variables method gives us the eigenvalue problem? The question is answered by examining the boundary conditions for the partial differential equation. Only homogeneous boundary conditions are used in the search for separated solutions. The boundary condition $u(2, y) = y(2 - y)$ can be satisfied only by choosing the correct Fourier coefficient values. The problem for the factor that depends on x is therefore going to have only one boundary condition. The eigenvalue problem needed to solve Model Problem A.8*a* must be the problem for the factor that depends on y. To that end, we assume a solution of the form

$$u = f(y)g(x). \tag{2}$$

This substitution changes the partial differential equation to the form

$$-\frac{f''}{f} = \frac{g''}{g} = k.$$

[39]The problem statement can be modified slightly to allow the function h to be only piecewise-continuous. In this case, the problem has to be stated more carefully, since a function smooth enough to satisfy the differential equation cannot be discontinuous on the boundary. The situation is similar to the cases in Section 8.5 where the initial data are not continuous or are not consistent with the boundary data.

(We could just as easily have written this equation as $f''/f = -g''/g = k$, as the ultimate outcome is the same in both cases.) The eigenvalue problem, using the boundary conditions at constant y boundaries, is

$$f'' + kf = 0, \qquad f(0) = 0, \qquad f(2) = 0.$$

This is the same eigenvalue problem (with $L = 2$) that we found for the Dirichlet problems for both the wave equation and the heat equation. We know from experience that the solutions have to be trigonometric, so we conclude that $k > 0$. Hence, the eigenvalue problem is

$$f'' + \lambda^2 f = 0, \qquad f(0) = 0, \qquad f(2) = 0. \tag{3}$$

Its solutions are the eigenvalues and eigenfunctions

$$\lambda_n = \frac{n\pi}{2}, \qquad f_n = b_n \sin\frac{n\pi y}{2}. \tag{4}$$

The remaining homogeneous problem is

$$g_n'' - \frac{n^2\pi^2}{4} g_n = 0, \qquad g_n(0) = 0. \tag{5}$$

The solution of the differential equation for g is normally written as

$$g_n = c_1 e^{n\pi x/2} + c_2 e^{-n\pi x/2};$$

however, the boundary condition $g_n(0) = 0$ suggests that it is better to use the hyperbolic functions. Therefore we write the solution of the differential equation as

$$g_n = c_1 \cosh\frac{n\pi x}{2} + c_2 \sinh\frac{n\pi x}{2}.$$

The boundary condition determines $c_1 = 0$, and c_2 is arbitrary. Since g_n is to be multiplied by f_n, which already has an arbitrary constant, there is no harm in taking $c_2 = 1$. Thus, the x-dependent factor is

$$g_n = \sinh\frac{n\pi x}{2},$$

and the full separated solution is

$$u_n = b_n \sinh\frac{n\pi x}{2} \sin\frac{n\pi y}{2}.$$

We arrive at the series solution

$$u = \sum_{n=0}^{\infty} b_n \sinh\frac{n\pi x}{2} \sin\frac{n\pi y}{2} \tag{6}$$

for the homogeneous portion of Model Problem A.8a.

To complete the solution, we need to determine the correct values for the coefficients b_n so that the nonhomogeneous boundary condition is satisfied. The problem to be solved is

$$y(2 - y) = \sum_{n=0}^{\infty} b_n \sinh n\pi \sin\frac{n\pi y}{2}.$$

This is not quite the usual Fourier series coefficient problem, but we can recast it into the usual form as

$$y(2 - y) = \sum_{n=0}^{\infty} c_n \sin\frac{n\pi y}{2}, \qquad c_n = b_n \sinh n\pi. \tag{7}$$

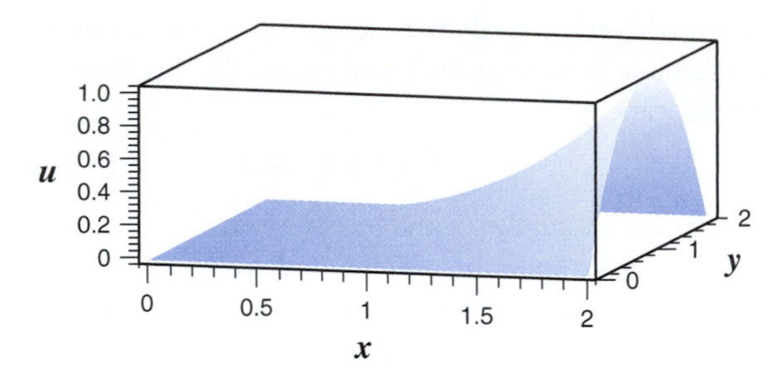

Figure A.8.1
The solution of Model Problem A.8a.

This is now the standard Fourier coefficient problem with $L = 2$, and it has the solution

$$c_n = \int_0^2 y(2-y)\sin\frac{n\pi y}{2}\,dy.$$

After all calculations are completed, the solution of Model Problem A.8a is found to be

$$u = \frac{16}{\pi^3}\sum_{n=1}^{\infty}\frac{1-(-1)^n}{n^3}\frac{\sinh(n\pi x/2)}{\sinh n\pi}\sin\frac{n\pi y}{2}. \tag{8}$$

The solution is plotted in Figure A.8.1.

Using Geometry, Symmetry, and Superposition

The limitation of the method of separation of variables for Laplace's equation lies in the requirement that the boundary conditions be applicable to just one of the factors in the separated solutions. This means that the boundary conditions must be applied at boundaries of the form variable = constant, and this places a severe restriction on the spatial domains of suitable problems. Rectangles, circles, and sectors of circles are always suitable, but it is seldom possible to solve any problems on other shapes, the exception being when the geometry can be extended to one of these suitable shapes.

MODEL PROBLEM A.8b

Solve the problem

$$u_{xx} + u_{yy} = 0, \qquad (x, y) \in R,$$

$$u(x, 0) = 0, \qquad u(0, y) = h(y) = y(2-y), \qquad u(x, 2-x) = 0,$$

where R is the region bounded by $y = 0$, $x = 0$, and $y = 2 - x$.

It is often advantageous to use a visual depiction of the domain and boundary conditions for problems with Laplace's equation. The idea is to sketch the region and label each portion of the boundary with information indicating the appropriate boundary condition. Figure A.8.2 depicts the data for Model Problem A.8b and a related problem on a square domain that we call Problem S.

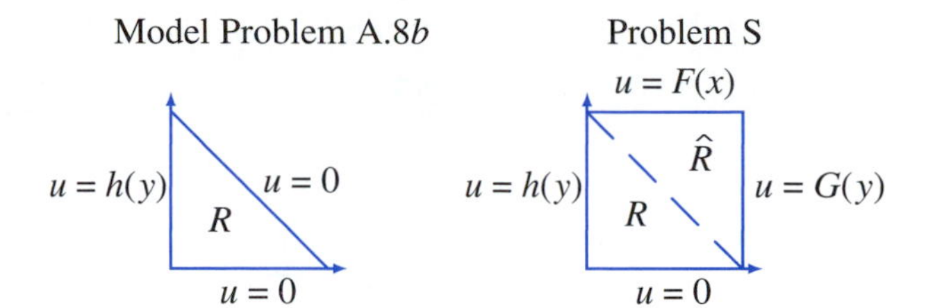

Figure A.8.2
The domain and boundary conditions for Model Problem A.8b and Problem S.

The triangular domain of Model Problem A.8b does not work for separation of variables, but the square domain of Problem S does work. The two problems share common boundaries $x = 0$ and $y = 0$ and common boundary conditions on those boundaries. The boundary conditions on the new boundaries at $x = 2$ and $y = 2$ in Problem S are given as functions F and G, which are as yet unspecified. The key idea is this: *If we can choose functions F and G so that the solution of Problem S satisfies the boundary condition $u(x, 2 - x) = 0$ of Model Problem A.8b, then the problems will have the same solution at all points within R.*

In most cases where this "domain extension" method can be used, it is not too difficult to determine the correct boundary conditions by inspection. Consider the point (1.99, 0.01). This point is very close to the lower right-hand corner of the domain of Problem S. It seems apparent that the solution at that point is going to be largely independent of the boundary functions h and F. Furthermore, it seems clear that the sign of $u(1.99, 0.01)$ is going to be determined by the sign of G near $y = 0$. We want to have $u(1.99, 0.01) = 0$, because this value is prescribed for Model Problem A.8b, suggesting that h near $y = 0$ should be neither positive nor negative. Having come this far, it is not too much of a stretch to suggest that $G \equiv 0$ is the correct boundary function. It is a bit more of a stretch to guess the correct choice for F. Since we are motivated by the thinking about G, our first thought is that F should be negative, since h is positive and the effects of the two functions must cancel on the line $x + y = 2$. Moreover, the magnitudes of the two functions must be comparable if the cancelation is to occur. The logical conclusion of this thought experiment is $F(x) = -h(x)$. It is important to emphasize that this is only a sophisticated guess at this point.

We have a second difficulty to overcome: even with $G \equiv 0$, Problem S has two nonhomogeneous boundary conditions. The problem of more than one nonhomogeneous boundary condition is not actually a difficult one to address, given that Laplace's equation is linear. Our studies of linear ordinary differential equations have frequently benefited from the linearity property

$$L[y_1 + y_2] = L[y_1] + L[y_2],$$

where L is a linear operator in one variable. This property holds for linear partial differential equations also. Problem S can be broken up into two separate problems, each with just one nonhomogeneous boundary condition; the solution of Problem S, and therefore also of Model Problem A.8b, is the sum of the two solutions. The boundary conditions for the two subproblems are illustrated in Figure A.8.3.

Problems 1 and 2 could be solved by the method of separation of variables, but instead we can take advantage of the solution of Model Problem A.8a. Let $u_a(x, y)$ be the solution of Model

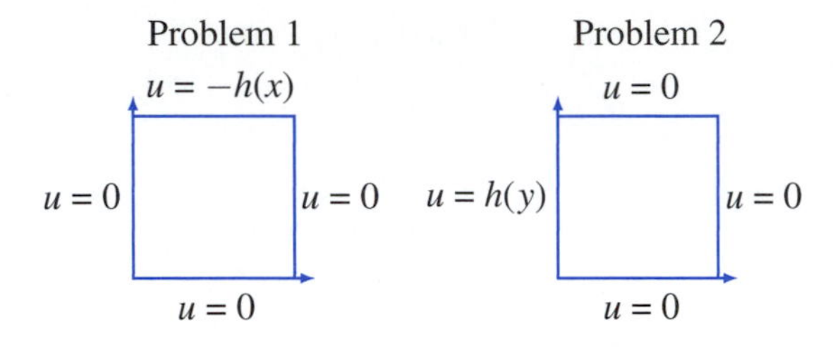

Figure A.8.3
The domain and boundary conditions for Problems 1 and 2.

Problem A.8a and consider Problem 1. If we switch the variables x and y, the resulting problem is

$$\begin{gathered} u_{xx} + u_{yy} = 0, \\ u(0, y) = 0, \qquad u(2, y) = 0, \\ u(x, 0) = 0, \qquad u(x, 2) = -h(y), \end{gathered}$$

which is the same as Model Problem A.8a except for the minus sign on the boundary condition at $y = 2$. The solution of Problem 1 is therefore

$$u_1(x, y) = -u_a(y, x).$$

We need check only the boundary conditions. We have

$$\begin{gathered} u_1(0, y) = -u_a(y, 0) = 0, \qquad u_1(2, y) = -u_a(y, 2) = 0, \\ u_1(x, 0) = -u_a(0, x) = 0, \qquad u_1(x, 2) = -u_a(2, x) = -x(2 - x) = -h(x), \end{gathered}$$

all of which agree with the data for Problem 1. Problem 2 is also similar to Model Problem A.8a. In this case, the boundary data appear at $x = 0$ instead of $x = 2$. We can rewrite Problem 2 by replacing x by $X = 2 - x$, with the resulting problem

$$\begin{gathered} u_{XX} + u_{yy} = 0, \\ u(2, y) = h(y), \qquad u(0, y) = 0, \\ u(X, 0) = 0, \qquad u(X, 2) = 0. \end{gathered}$$

This is exactly Model Problem A.8a, except with X instead of x, so the solution is $u_a(X, y)$, or

$$u_2(x, y) = u_a(2 - x, y).$$

Finally, the solution of Model Problem A.8b is the sum of the solutions of Problems 1 and 2; assuming that everything has been done correctly, we have $u = u_a(2 - x, y) - u_a(y, x)$, or

$$u = \frac{16}{\pi^3} \sum_{n=1}^{\infty} \frac{1 - (-1)^n}{n^3 \sinh n\pi} \left[\sinh \frac{n\pi(2 - x)}{2} \sin \frac{n\pi y}{2} - \sinh \frac{n\pi y}{2} \sin \frac{n\pi x}{2} \right]. \tag{9}$$

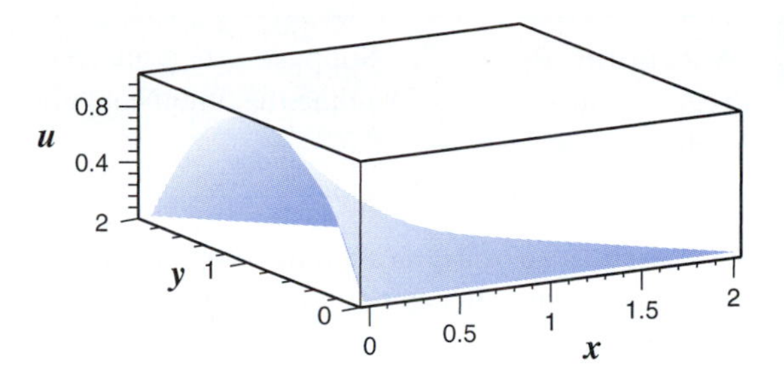

Figure A.8.4
The solution of Model Problem A.8*b*.

✦ **INSTANT EXERCISE 1**

Show that u as defined by Equation (9) satisfies the homogeneous boundary conditions of Model Problem A.8*b*.

The solution of Model Problem A.8*b* appears in Figure A.8.4.

General Properties of Laplace's Equation and Its Solutions

As noted earlier, Laplace's equation can also be solved on circular domains. In particular, consider the Dirichlet problem for Laplace's equation on a unit disk,

$$u_{rr} + \frac{1}{r}u_r + \frac{1}{r^2}u_{\theta\theta} = 0, \tag{10}$$

$$u(1, \theta) = h(\theta), \qquad h(2\pi) = h(0), \tag{11}$$

where we assume h is continuous. The solution of this problem can be shown (Exercise 7) to be

$$u = \bar{h} + \sum_{n=1}^{\infty} r^n (a_n \cos n\theta + b_n \sin n\theta), \tag{12}$$

where

$$\bar{h} = \frac{1}{2\pi}\int_0^{2\pi} h(\theta)\,d\theta, \qquad \frac{1}{\pi}\int_0^{2\pi} h(\theta)\cos n\theta\,d\theta, \qquad \frac{1}{\pi}\int_0^{2\pi} h(\theta)\sin n\theta\,d\theta. \tag{13}$$

Solutions of Laplace's equations (in any geometry) are called **harmonic functions.** These functions have two very useful properties, the *mean value property* and the *maximum principle*. For simplicity, we state these properties for Laplace's equation in two dimensions. The proofs are deferred to Exercises 8 and 9.

Theorem A.8.1

Mean Value Property Suppose u is a continuous solution of Laplace's equation on a region R in the plane. Let P be a point in the interior of R, and let C be a circle centered at P and contained in R. Then the value of u at P is equal to the average of the values of u on C.

Theorem A.8.2

Maximum Principle Suppose u is a nonconstant solution of Laplace's equation on a region R in the plane. Then the maximum and minimum values of u in R can occur only at points on the boundary of R.

As a consequence of the mean value property and the maximum principle, the Dirichlet problem for Laplace's equation can be shown to be *well posed.*

Theorem A.8.3

Well-Posedness Let h be a continuous function on a simple closed curve ∂R. Then the Dirichlet problem for Laplace's equation on the region R is **well posed;** that is,

1. It has a solution.
2. The solution is unique.
3. The solution varies continuously with the problem data h.

Requirements 2 and 3 of Theorem A.8.3 are established with the aid of the maximum principle (see Exercise 10).

A.8 Exercises

1. *a.* Solve the problem

$$\begin{aligned} u_{xx} + u_{yy} &= 0, \qquad 0 < x, y < 1, \\ u(x, 0) &= 0, \qquad u(x, 1) = x, \\ u(0, y) &= 0, \qquad u(1, y) = y. \end{aligned}$$

   b. Use a computer algebra system to graph the solution. What happens if you do not use a lot of terms? What is it about this problem that is responsible for this behavior?

   c. Plot $u(x, x)$.

T

2. *a.* Solve the problem

$$\begin{aligned} u_{xx} + u_{yy} &= 0, \qquad 0 < x, y < 2, \\ u_y(x, 0) &= 0, \qquad u_y(x, 2) = 0, \\ u(0, y) &= 0, \qquad u(2, y) = y(2 - y). \end{aligned}$$

   (*Hint*: make sure you use *all* the eigenvalues.)

   b. Use a computer algebra system to graph the solution. Check your solution by making sure that the graph shows the right behavior at $x = 2$.

3. Solve the problem

$$\begin{aligned} u_{rr} + \frac{1}{r}u_r + \frac{1}{r^2}u_{\theta\theta} &= 0, \\ u(1, \theta) &= \sin\theta, \end{aligned}$$

and use a computer algebra system to graph the solution.

4. Solve the problem

$$u_{rr} + \frac{1}{r}u_r + \frac{1}{r^2}u_{\theta\theta} = 0,$$
$$u(1, \theta) = \theta(2\pi - \theta),$$

and use a computer algebra system to graph the solution.

T 5. Consider the problem

$$u_{xx} + u_{yy} = 0, \qquad 0 < x < 2, \qquad 0 < y < 1,$$
$$u(x, 0) = x(2 - x), \qquad u_y(x, 1) = 0,$$
$$u(0, y) = 0, \qquad u(2, y) = 0.$$

a. Solve the problem by separation of variables. Use a computer algebra system to graph the solution.
b. Construct a Dirichlet problem on the region $0 \le x, y \le 2$ that is equivalent to the given problem. Solve this problem. Verify that the solution satisfies the condition $u_y(x, 1) = 0$.
c. Compare the solutions of parts a and b. Are the individual modes u_n the same? If not, is there any reason to prefer one of the solutions over the other?

T 6. Solve the problem

$$u_{xx} + u_{yy} = 0, \qquad (x, y) \in R,$$
$$u(x, 0) = x(2 - x), \qquad u(x, x) = 0, \qquad u(x, 2 - x) = 0,$$

where R is the region bounded by $y = 0$, $y = x$, and $y = 2 - x$. Use a computer algebra system to graph the solution.

7. Derive the solution

$$u(r, \theta) = \bar{h} + \sum_{n=1}^{\infty} r^n(a_n \cos n\theta + b_n \sin n\theta),$$

$$\bar{h} = \frac{1}{2\pi}\int_0^{2\pi} h(\theta)\, d\theta, \qquad \frac{1}{\pi}\int_0^{2\pi} h(\theta)\cos n\theta\, d\theta, \qquad \frac{1}{\pi}\int_0^{2\pi} h(\theta)\sin n\theta\, d\theta$$

for Laplace's equation on the unit disk,

$$u_{rr} + \frac{1}{r}u_r + \frac{1}{r^2}u_{\theta\theta} = 0,$$
$$u(1, \theta) = h(\theta).$$

8. Use Equation (12) to prove Theorem A.8.1. [*Hint*: use polar coordinates centered at the point P, and let $h(\theta) = u(a, \theta)$.]

9. Use Theorem A.8.1 to prove Theorem A.8.2.

10. Use Theorem A.8.2 to prove the uniqueness of solutions of Laplace's equation. (*Hint*: Let $w = u - v$, where u and v are solutions of Laplace's equation on some region R. Show that $w \equiv 0$.)

T 11. Use the method of **eigenfunction expansion** to solve the problem

$$u_{rr} + \frac{1}{r}u_r + \frac{1}{r^2}u_{\theta\theta} = 0, \qquad 0 < r < 1, \qquad 0 < \theta < \pi,$$

$$u(1, \theta) = 0, \qquad u(r, 0) = 1 - r, \qquad u(r, \pi) = 1 - r.$$

a. The boundary conditions at $\theta = 0$ and $\theta = \pi$ are nonhomogeneous. To make a problem with homogeneous boundary conditions, replace u with

$$u = 1 - r + w(r, \theta).$$

Determine the problem for w. Note that the differential equation for w is nonhomogeneous.

b. Determine the eigenfunctions f_n for the homogeneous problem

$$w_{rr} + \frac{1}{r}w_r + \frac{1}{r^2}w_{\theta\theta} = 0, \qquad 0 < r < 1, \qquad 0 < \theta < \pi,$$

$$w(r, 0) = 0, \qquad w(r, \pi) = 0.$$

(You do not need to include arbitrary coefficients, as these will arise later.)

c. Assume a solution of the w problem of the form

$$w = \sum_{n=1}^{\infty} g_n(r) f_n(\theta).$$

Substitute this form into the differential equation to obtain an equation of the form

$$\sum_{n=1}^{\infty} L[g_n] f_n(\theta) = r,$$

where L is a linear differential operator in the independent variable r.

d. Use the Fourier coefficient formula to obtain a differential equation

$$L[g_n] = c_n r$$

for some coefficient sequence c_n. What boundary conditions must each g_n satisfy at $r = 0$ and at $r = 1$?

e. Solve the g_n problems to complete the solution. Note that g_1 must be determined separately, but all other g_n can be determined with n unspecified.

f. Use a computer algebra system to graph the solution.

✦ A.8 INSTANT EXERCISE SOLUTION

1. From Equation (9),

$$u_n = \frac{16}{\pi^3} \frac{1 - (-1)^n}{n^3 \sinh n\pi} \left[\sinh \frac{n\pi(2 - x)}{2} \sin \frac{n\pi y}{2} - \sinh \frac{n\pi y}{2} \sin \frac{n\pi x}{2} \right].$$

We have $u_n(x, 0) = 0$ because $\sinh 0 = 0$. At $x + y = 2$, we have

$$u_n(2 - y, y) = \frac{16}{\pi^3} \frac{1 - (-1)^n}{n^3 \sinh n\pi} \sinh \frac{n\pi y}{2} \left[\sin \frac{n\pi y}{2} - \sin \frac{n\pi(2 - y)}{2} \right].$$

Using the trigonometric identity

$$\sin(a - b) = \sin a \cos b - \sin b \cos a,$$

we have

$$\sin \frac{n\pi y}{2} - \sin \frac{n\pi(2 - y)}{2} = \sin \frac{n\pi y}{2} - \left[\sin n\pi \cos \frac{n\pi y}{2} - \sin \frac{n\pi y}{2} \cos n\pi \right]$$
$$= \left(\sin \frac{n\pi y}{2} \right) (1 + \cos n\pi) = [1 + (-1)^n] \sin \frac{n\pi y}{2}.$$

Thus,

$$u_n(2 - y, y) = \frac{16}{\pi^3} \frac{[1 - (-1)^n][1 + (-1)^n]}{n^3 \sinh n\pi} \sinh \frac{n\pi y}{2} \sin \frac{n\pi y}{2} = 0,$$

the last result following from

$$[1 - (-1)^n][1 + (-1)^n] = 1 - (-1)^{2n} = 0.$$

Answers to Odd-Numbered Problems

Answers to Section 1.1

1. (a) The surface would cool more quickly, leading to a decreased cooling rate. It will take longer for the coffee to cool. **(b)** Styrofoam is a good insulator, so the cooling would take place almost exclusively at the top surface. It will take longer for the coffee to cool. **(c)** The metal spoon would provide an additional pathway for the heat to escape. It will take less time for the coffee to cool.

3. (a) $\dfrac{dT}{dt} = -k(T - 10), \quad T(0) = 68$ **(b)** 48°F, so there should be no problem **(c)** We assumed a constant outdoor temperature. Most likely the actual outdoor temperature would be cooler, so the estimate needs to be revised downward.
5. 50% **7. (a)** 37% **(b)** $kt_t = \ln 10$
9. $z = 17e^{-kt}/(9 + 8e^{-kt})$
11. (a) $y' = -ky, \ \ y(0) = Q_0 + 10$; the solution is $y = (Q_0 + 10)e^{-kt}$ **(b)** The problem must be solved separately on the time intervals (0, 6), (6, 12), (12, 18), and (18, 48). The values of Q_0 are 0, 3.35, 4.47, and 4.84, respectively.

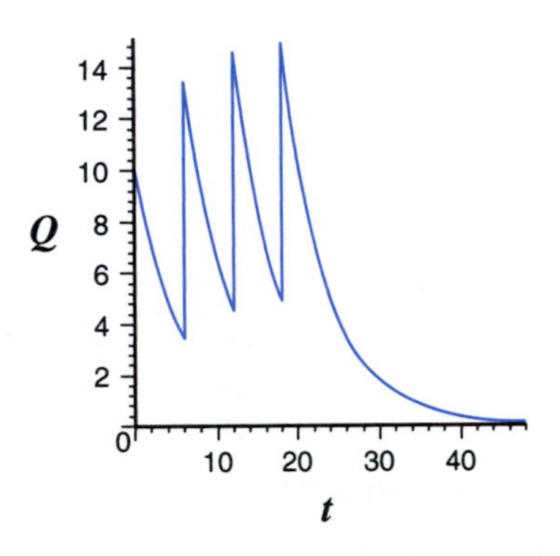

13. (a) The continuous rate of payment is $226.04 per month. **(b)** The actual monthly payment is $226.45. The use of a model assuming continuous payments makes very little error. **15.** 483 million—the actual population will probably be less because the birthrate is much less in 2000 than in 1970 and will likely continue to decrease.

Answers to Section 1.2

1. (a) Ordinary, second-order **(b)** Partial, second-order **(c)** Ordinary, second-order
3. $\left(-\frac{4}{5}e^{-2t} + 12e^{3t}\right) - 3\left(\frac{2}{5}e^{-2t} + 4e^{3t}\right) = -2e^{-2t}$
5. $\left(-3\cos t + \frac{9}{10}e^{3t}\right) + \left(3\cos t + \frac{1}{10}e^{3t}\right) = e^{3t}$
7. $x^2(6x) - 6(x^3 - 2) = 12$
9. $x(1 + x^2)^{-1/2} = x/y$
11. $\left(\dfrac{1}{2} - 2Ce^{-2t}\right) + 2\left(\dfrac{t}{2} - \dfrac{1}{4} + Ce^{-2t}\right) = t$
13. (a) $r = 3$ **(b)** $r = -2$ and $r = -1$ **(c)** $r = -2$
15. (a) $A = 1$ **(b)** $A = -\frac{1}{4}$ **(c)** No possible A
17. $y \equiv 0$ **19.** $y = 3e^{4t}$
21. $y = -\frac{3}{4}\ln(1 + 2t) + \frac{3}{2}t + 1$

(The absolute value is not needed because the function is not continuous at $t = -\frac{1}{2}$.)

23. $y = -\ln(\cos t) + y_0$ on $(-\pi/2, \pi/2)$
25. $y = \displaystyle\int_1^x \frac{e^{-s}}{s}\,ds + 3$
27. (a) $y' = \dfrac{2}{\sqrt{\pi}}e^{-x^2}, \quad y(0) = 0$
(b) $\sqrt{\pi}\,te^{t^2}\operatorname{erf} t + 1 = 2t\left(\dfrac{\sqrt{\pi}}{2}e^{t^2}\operatorname{erf} t\right) + 1 = 2ty + 1$
29. $y = \dfrac{2}{1 - 4t}, \quad \left(-\infty, \dfrac{1}{4}\right)$
31. $y = \dfrac{2}{2 - e^{2t}}, \quad \left(-\infty, \dfrac{\ln 2}{2}\right)$
33. $y = -\dfrac{16}{\sqrt{1 - 8t}}, \quad \left(-\infty, \dfrac{1}{8}\right)$

Answers to Section 1.3

1. $m \dfrac{d^2y}{dt^2} = -\alpha y - \beta \dfrac{dy}{dt}, \quad m, \alpha, \beta > 0$
The equation is second-order.

3. (a) $R + y$ is the distance from the center of the earth.
(b) $A = -mgR^2$ **(c)** $m \dfrac{d^2y}{dt^2} = -\dfrac{A}{(R+y)^2}$, no

5. $y^2 = -2e^t + y_0^2 + 2, \quad \left(0, \ln\left(1 + y_0^2/2\right)\right)$

Answers to Section 2.1

3. (a) $t_r = \sqrt{h_0}/k$ **(b)** $dt/dh = -1/(k\sqrt{h})$
(c) $t_e = 2\sqrt{h_0}/k$ **(d)** One-half of the total drain time
(e) $dH/d\tau = -\sqrt{H}, \quad H(0) = 0$

5. (a) $s'' + bs' + g\sin\theta = 0$
(b) $\theta'' + b\theta' + (g/L)\sin\theta = 0$
(c) $\theta'' + c\theta' + \sin\theta = 0, \quad c = b\sqrt{L/g}$

7. (a) $\dfrac{d^2Z}{d\tau^2} = -\dfrac{\alpha}{(1+Z)^2}$,

$$Z(0) = 0, \quad \frac{dZ}{d\tau}(0) = 1, \quad \alpha = \frac{gR^2}{V^2}$$

(b) $V^2/(gR)$
(c) $\dfrac{d^2Z}{d\tau^2} = -\dfrac{1}{(1+\alpha^{-1}Z)^2}, \quad Z(0) = 0, \quad \dfrac{dZ}{d\tau}(0) = 1$
(d) α must be large
(e) V must be considerably less than $\sqrt{gR}$.

9. $\tau_Z = \displaystyle\int_0^Z \frac{1}{1-z} e^{-\rho z/(1+\beta z)}\, dz$

11. $\tau_1 > e^{-\rho/(1+\beta)} \displaystyle\int_0^Z \frac{1}{1-z}\, dz$
The integral on the right diverges; therefore τ_1 does also.

13. (a) $\dfrac{dQ}{dt} = -\dfrac{rQ}{V}, \quad Q(0) = Q_0$
(b) $Q(t) = Q_0 e^{-rt/V}$ **(c)** $t = (V \ln 10)/r$, which is 5.9 years for Lake Erie and 425 years for Lake Superior

15. (a) $\dfrac{dQ}{dt} = rC_0 - \dfrac{r}{V}Q, \quad Q(0) = 0$
(b) $Q = C_0 V\left(1 - e^{-rt/V}\right)$
(c) For $0 \le t \le 20$, $C/C_0 = 1 - e^{-rt/V}$. For $t > 20$, $C/C_0 = \left(e^{20r/V} - 1\right)e^{-rt/V}$. **(d)** The flatter curve is for Lake Superior.

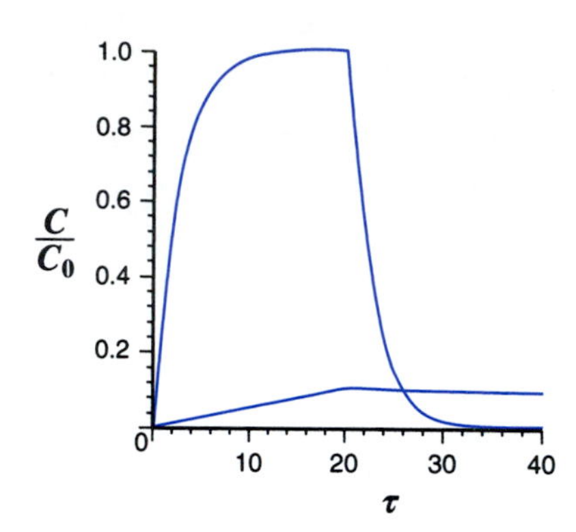

17. (a) $\dfrac{dQ}{dt} = -kQ + \dfrac{Q_0}{T}, \quad Q(0) = 0$
(b) $Q = \dfrac{Q_0}{kT}\left(1 - e^{-kt}\right)$
(c) $T = Q_0/(kQ_T)$. This value does not give a maximum of exactly Q_T, but it is close and definitely less than Q_T.
(d) $Q = Q_T\left(e^{Q_0/Q_T} - 1\right)e^{-kt}, \quad \text{for} \quad t > \dfrac{Q_0}{kQ_T}$
(e) For $0 \le \tau \le S$, $y = S^{-1}(1 - e^{-\tau})$. For $\tau > S$, $y = S^{-1}(e^S - 1)e^{-\tau}$. **(f)** The curve with the earlier peak is $S = 1$. A larger value of S means a greater total dose relative to the maximum safe dose. Thus, the drug must be administered over a longer time.

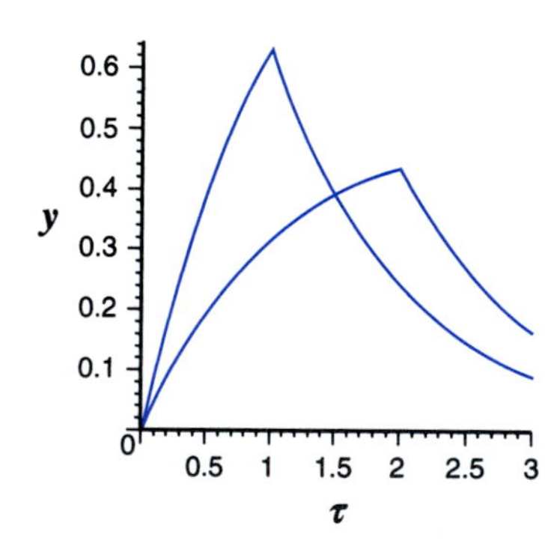

Answers to Section 2.2

1. Separable, $y = t^2/2 - t + 3$
3. Separable, $y = 1/\sqrt{1-2t}$
5. Separable, $y = 1/(e^t - te^t + C)$

7. Separable, $y = \ln[(\ln x)^2/2 + C]$

9. Separable, $y = \pm\sqrt{C - 2x^2}$

11. Separable, $y = \left[A \exp\left(\frac{3}{2}e^{x^2}\right) - 1\right]^{1/3}$

13. $y = \dfrac{5}{1 - 5t^2}$, $\left(-\dfrac{1}{\sqrt{5}}, \dfrac{1}{\sqrt{5}}\right)$

15. $y = \dfrac{1}{1 - \sin t}$, $\left(-\dfrac{3\pi}{2}, \dfrac{\pi}{2}\right)$

17. $y = -\sqrt{5x^2 - 1}$, $\left[\dfrac{1}{\sqrt{5}}, \infty\right]$

19.

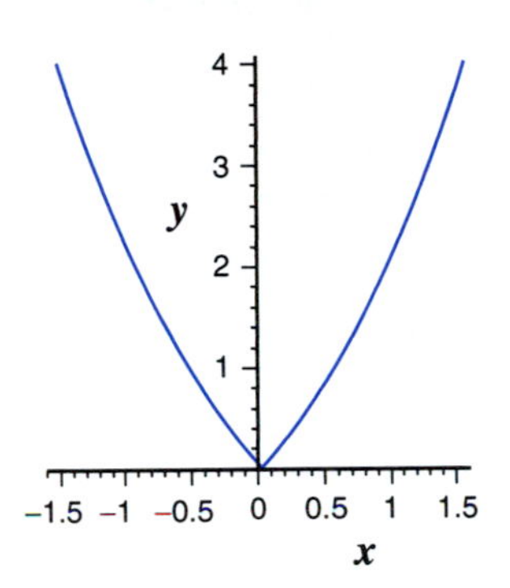

21.

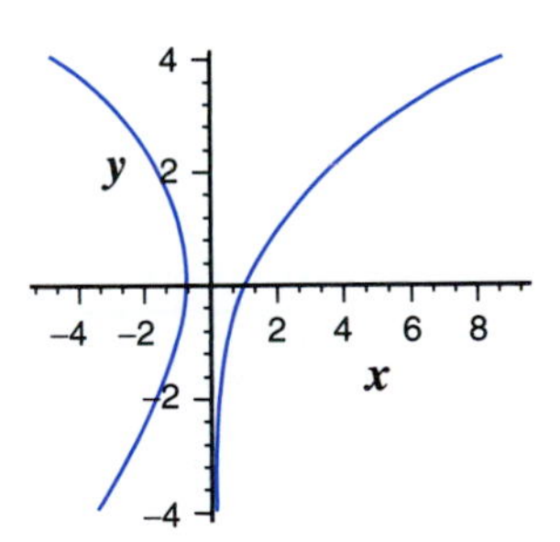

23. **(a)** $y = 2 \pm \sqrt{2t + C}$

(b) $y = \begin{cases} 2 + \sqrt{2t + (a-2)^2} & a > 2 \\ 2 - \sqrt{2t + (a-2)^2} & a < 2 \end{cases}$

Note that the solution does not exist if $a = 2$.

(c) $[-(a-2)^2/2, \infty)$.

25. $y = \sqrt{2(C - x)}$

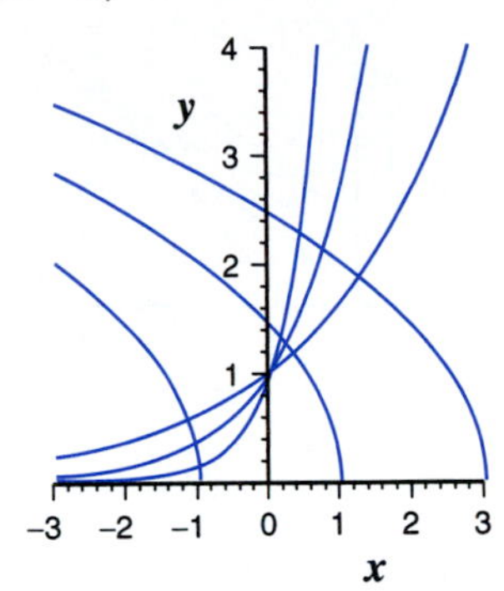

27. $y^2 + \dfrac{x^2}{2} = C^2$

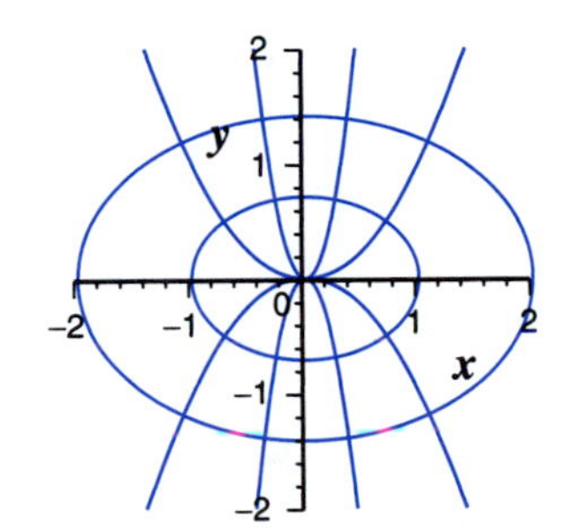

29. $y = 2t + Ce^{-t}$

31. $y = t^2 + Ce^{-t}$

33. $y = x \ln|x| + Cx$

35. $y = \pm x\sqrt{C - 2\ln|x|}$

37. $y = x - 2 + Ce^{-x/2}$

39. $y^2 = x^2 - 1 + Ae^{-x^2}$

41. Any equation of the form

$$y\frac{dy}{dx} = xf(y^2 - x^2)$$

Answers to Section 2.3

1.

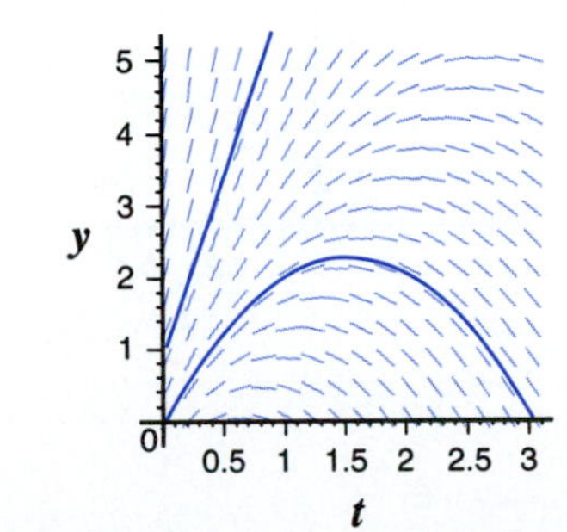

3. Solutions approach $y = 2t$.

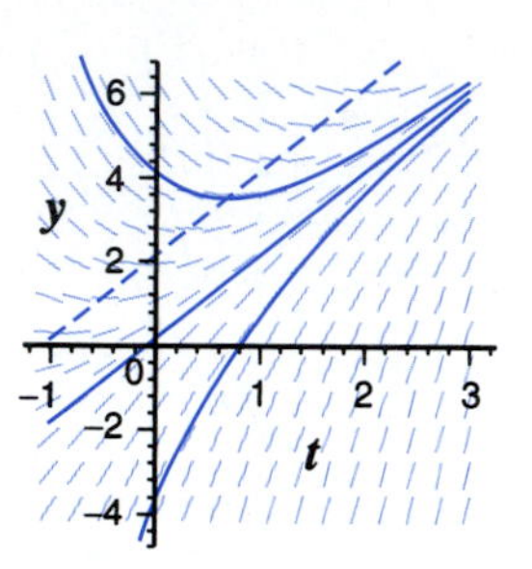

5. Solutions approach $y = t^2$.

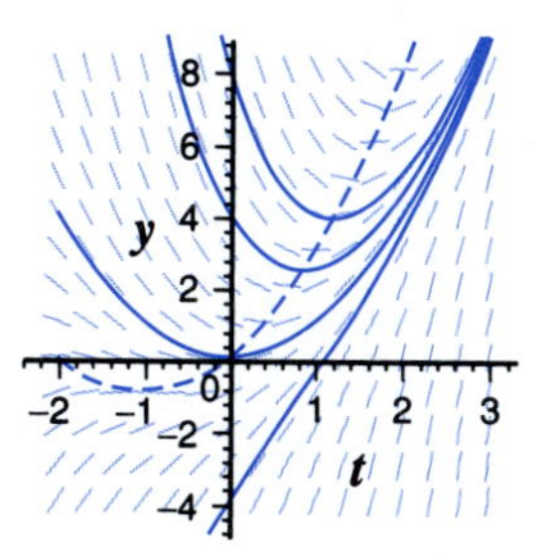

7.

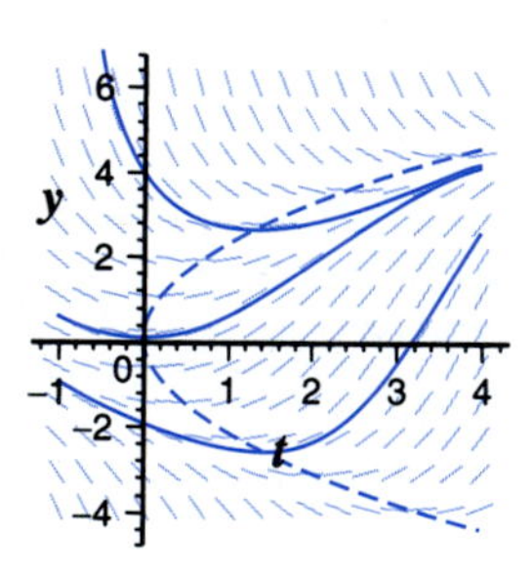

9.

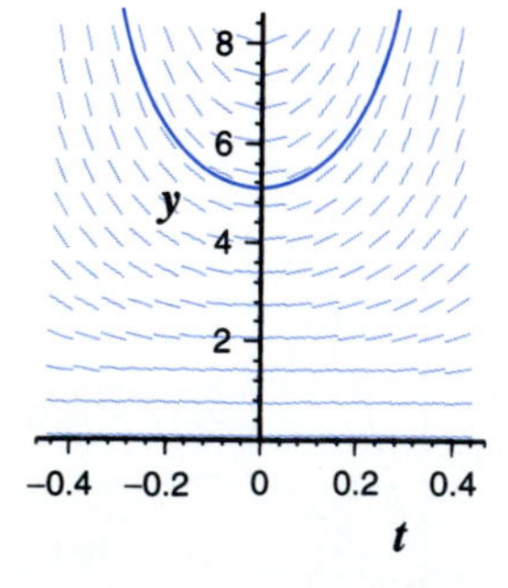

11.

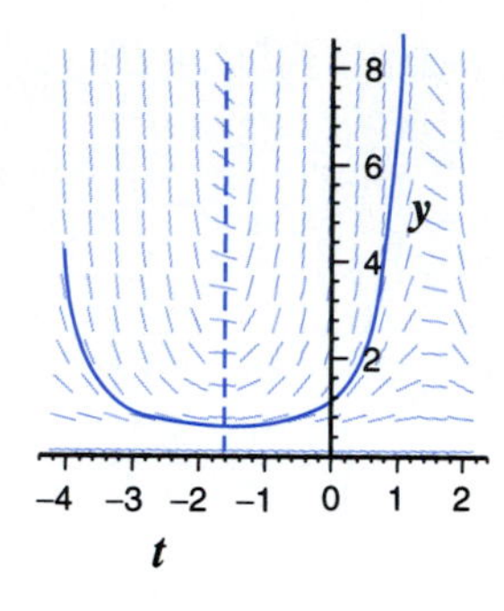

13. The solution approaches $y = -\sqrt{5}\,x$.

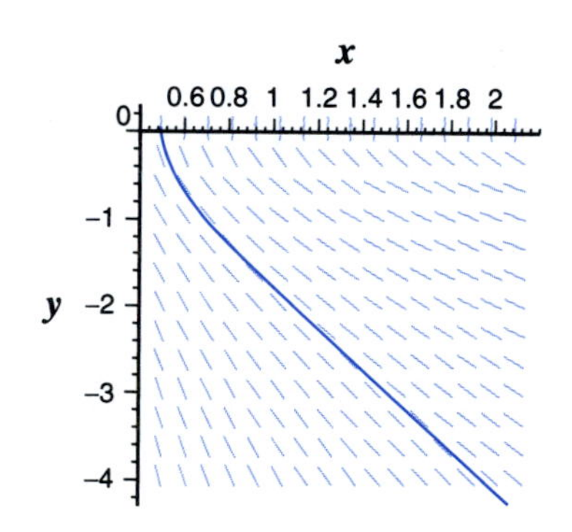

15.

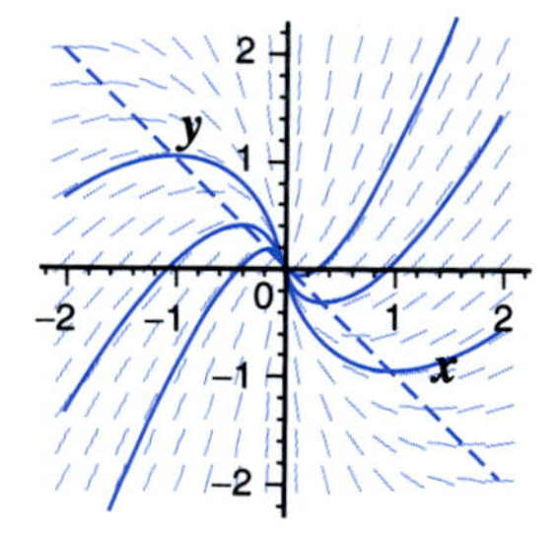

17.

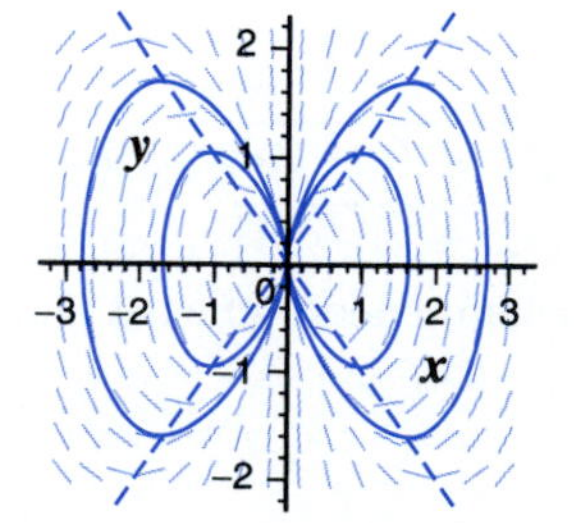

19.

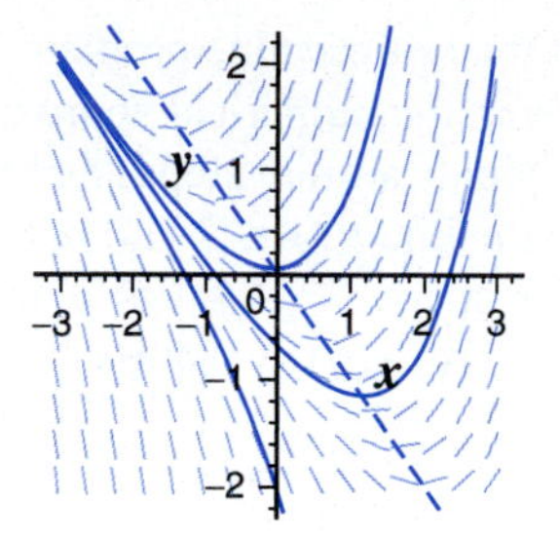

21.

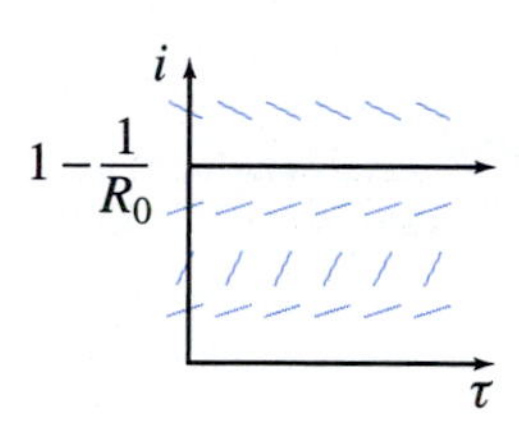

If $i(0) > 0$, then the solutions all approach $i = 1 - 1/R_0$ as $t \to \infty$.

Answers to Section 2.4

1. Any rectangular region not containing the line $x = -1$ or $y = 0$ **3.** Any rectangular region lying wholly inside the circle $t^2 + y^2 = 25$ and not containing any part of the line $y + t = -1$

5. $(-1, 2)$ **7.** $\left(-\frac{\pi}{2}, \frac{\pi}{2}\right)$ **9.** $(-\infty, 1)$

11. **(a)** $(-1, 1)$ **(b)** Yes, initial conditions given at $x = 1$ or $x = -1$ **(c)** $n = 0,\ y = 1;\quad n = 1,\ y = x;$ $n = 2,\ y = (3x^2 - 1)/2;\quad n = 3,\ y = (5x^3 - 3x)/2$ **(d)** No **13.** **(a)** $h = (2 - t)^2$ **(b)** The bucket is empty.

(c) $h = \begin{cases} (2-t)^2 & 0 < t \le 2 \\ 0 & t > 2 \end{cases}$

(d) You must show that the function is continuous and satisfies the differential equation, even at $t = 2$. **(e)** No

(f)

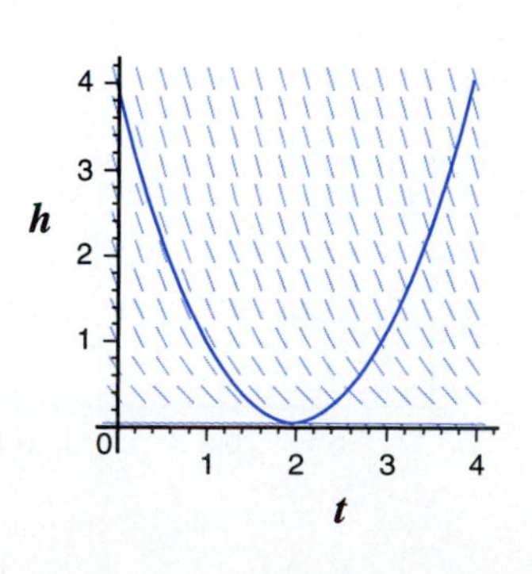

15. **(a)** Nothing, because the hypotheses of the theorem are not met **(b)** $x = 1 + t$ and $x = 1 - t$ are both solutions.

(c)

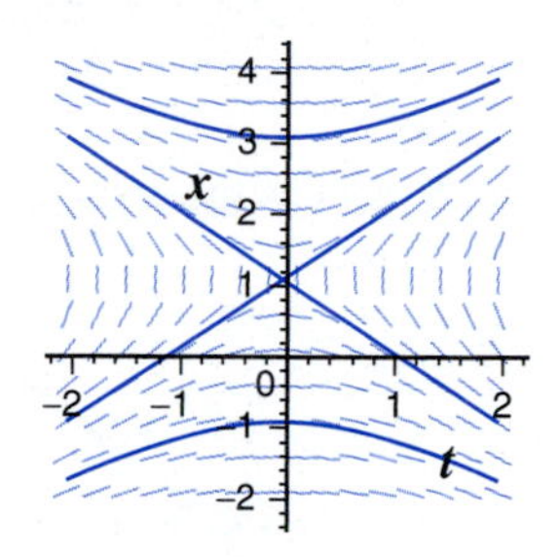

(d) No

19. $\dfrac{\partial y}{\partial t} = \dfrac{-kx \exp\left[-x^2/(4kt)\right]}{2\sqrt{\pi}(kt)^{3/2}} = k\,\dfrac{\partial^2 y}{\partial x^2}$

Answers to Section 2.5

1.

t	**Approximation** $\Delta t = 0.1$	**Solution** $y(t)$	**|Error|** $\Delta t = 0.1$
0.1	2.000	2.010	0.010
0.2	2.020	2.037	0.017
0.3	2.058	2.082	0.024
0.4	2.112	2.141	0.029

3.

	Approximation Δt			**Solution**	**|Error|** Δt		
t	**0.02**	**0.01**	**0.005**	$y(t)$	**0.02**	**0.01**	**0.005**
0.1	5.206	5.234	5.248	5.263	0.057	0.029	0.015
0.4	16.45	19.36	21.61	25.00	8.55	5.64	3.39

Halving the step size reduces the error by about one-half at $t = 0.1$; the improvement at $t = 0.4$ is less.

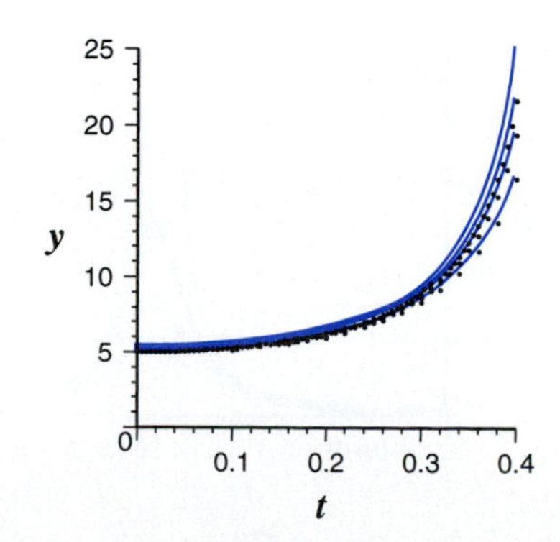

5.

	Approximation Δt			Solution	|Error| Δt		
t	**0.2**	**0.1**	**0.05**	$y(t)$	**0.2**	**0.1**	**0.05**
2	3.571	3.514	3.486	3.459	0.112	0.055	0.027

Halving the step size reduces the error by about one-half. |Error| $\approx 0.54\,\Delta t$

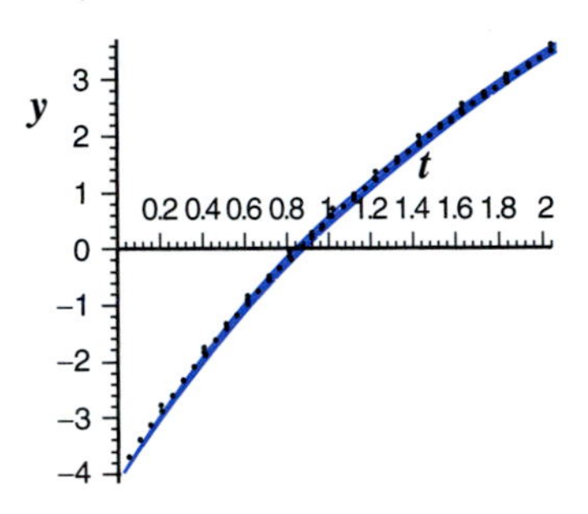

7.

	Approximation Δt			Solution	|Error| Δt		
t	**0.2**	**0.1**	**0.05**	$y(t)$	**0.2**	**0.1**	**0.05**
2	4.232	4.618	4.822	5.033	0.801	0.415	0.211

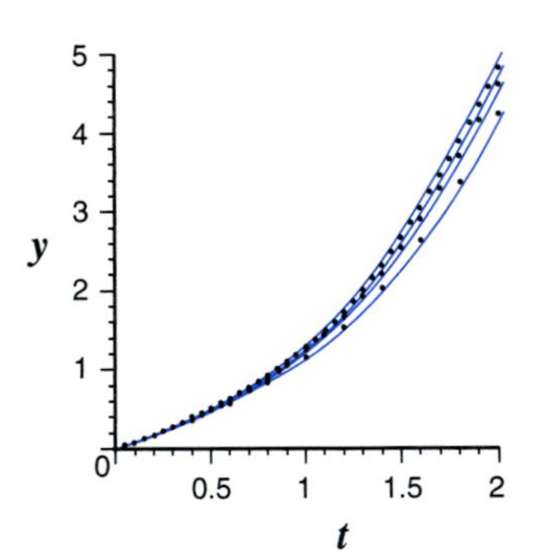

9.

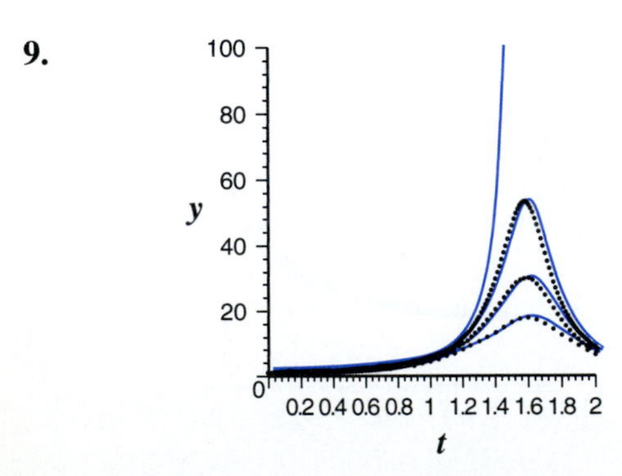

The solution of the initial-value problem exists only for $t < \pi/2$. However, the numerical approximation scheme always yields a finite slope; therefore, it "jumps" across the line $t = \pi/2$ and begins to follow a different solution curve. **11.** **(a)** $y(1) \approx 0.4789$ **(b)** $y(1) \approx 0.4846$ **(c)** $y(1) \approx 0.4903$

Answers to Section 2.6

1.

	Approximation			Solution	|Error|		
t	**Euler**	**M. Euler**	**rk4**	$y(t)$	**Euler**	**M. Euler**	**rk4**
0.2	2.0200	2.0400		2.0374	0.0174	0.0026	
0.4	2.1122	2.1448	2.1408	2.1406	0.0286	0.0040	0.0002

Each approximation for $t = 0.2$ uses two evaluations of the derivative function, and each approximation for $t = 0.4$ uses four evaluations of the derivative function. The very large differences in accuracy are due to the differences in the methods.

3.

	Approximation Δt			Solution	|Error| Δt		
t	**0.05**	**0.025**	**0.0125**	$y(t)$	**0.05**	**0.025**	**0.0125**
0.1	5.2613	5.2629	5.2631	5.2632	0.00187	0.00027	0.00004
0.4	21.70	23.74	24.61	25.00	3.30	1.26	0.39

Halving the step size reduces the error by about one-sixth at $t = 0.1$; the error at $t = 0.4$ is reduced by roughly one-third.

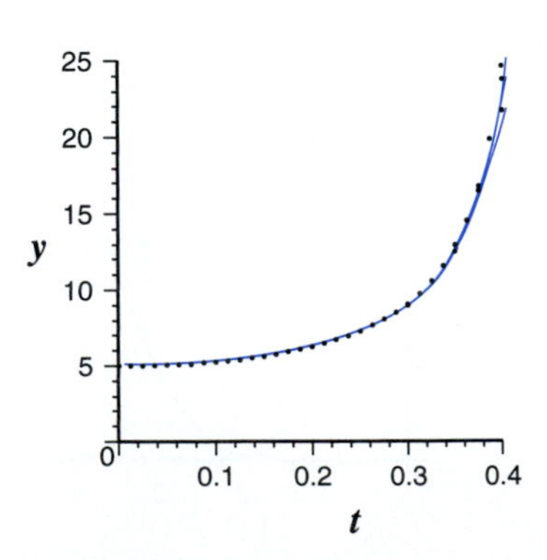

	Approximation Δt		Solution	$\lvert\text{Error}\rvert$ Δt	
t	0.1	0.05	$y(t)$	0.1	0.05
0.1	5.26320	5.26316	5.26316	4.6×10^{-5}	3.4×10^{-6}
0.4	24.13	24.87	25.00	0.87	0.13

Halving the step size reduces the error by about one-fourteenth at $t = 0.1$; the error at $t = 0.4$ is reduced by roughly one-sixth.

5. Modified Euler

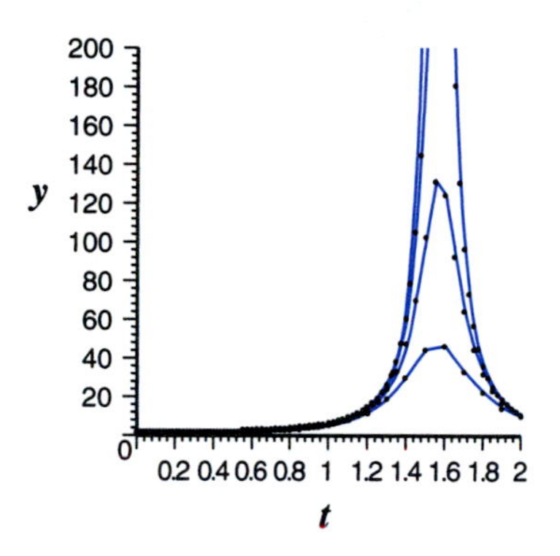

Runge–Kutta

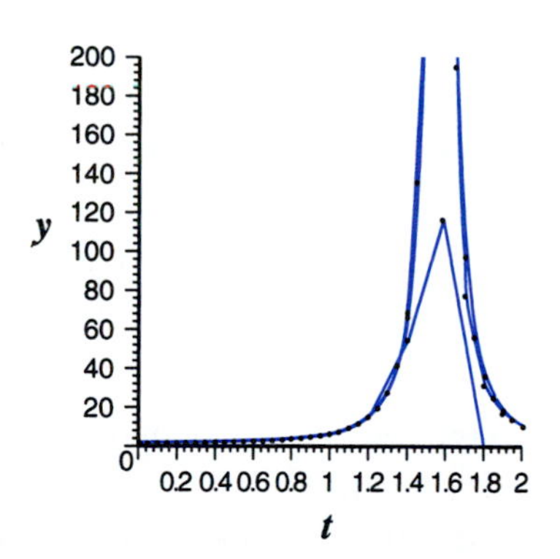

Both the modified Euler and rk4 approximations are better at tracking the sudden increase in slope as $t \to \pi/2$. Neither is able to identify that the solution does not exist for $t \geq \pi/2$.

7.

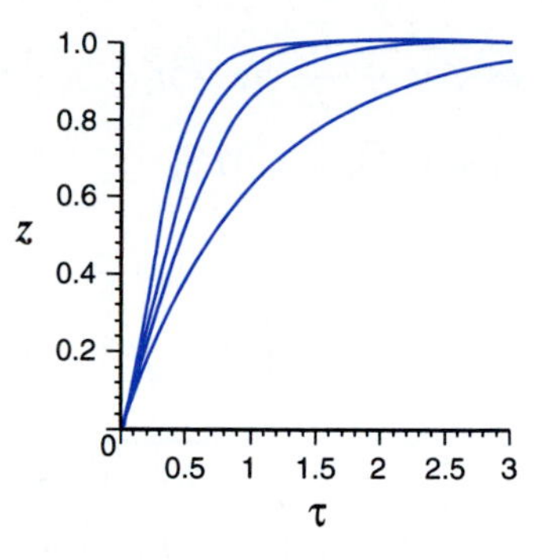

The curves are for β values of 0, 1, 2, and 4, from bottom to top. Increasing the heat liberated by the reaction makes the reaction accelerate faster and approach completion earlier.

Answers to Section 3.1

1. $y = -\cos t - \sqrt{3}\sin t = 2\cos\left(t - \dfrac{4\pi}{3}\right)$

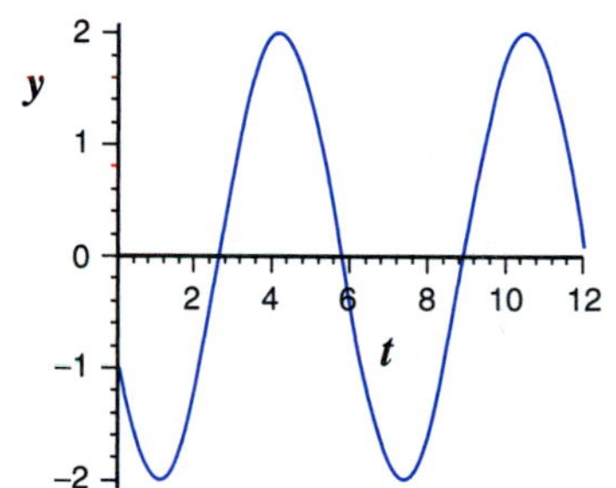

3. $y = \cos\sqrt{3}t - \dfrac{1}{\sqrt{3}}\sin\sqrt{3}t$

$= \dfrac{2}{\sqrt{3}}\cos\left(\sqrt{3}t + \dfrac{\pi}{6}\right)$

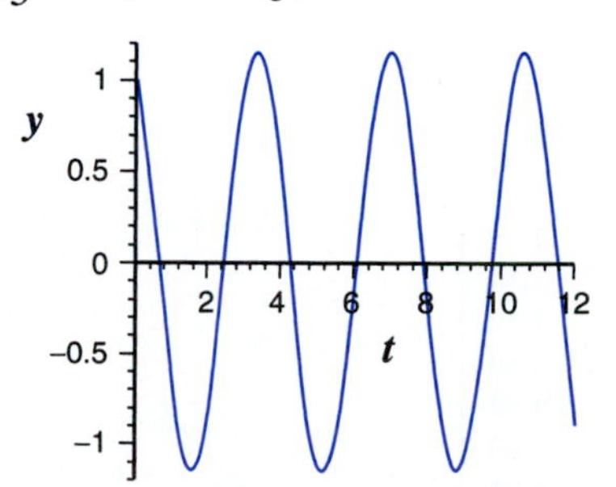

5. $y = -2\cos\sqrt{5}t - \dfrac{6}{\sqrt{5}}\sin\sqrt{5}t$

$\approx 3.347\cos\left(\sqrt{5}t - 4.072\right)$

7. $k = 40\pi^2$ N/m. The period is 1 s. **9.** $m = 0.065$ slug **11.** $y'' + 196y = 0$, $y(0) = -2$, $y'(0) = -1$ The period is $\pi/7$ s, the amplitude is 2.001 cm, and the shift is 3.177.

13. $y'' + 64y = 0$, $y(0) = 0.25$, $y'(0) = 1$ The period is $\pi/4$ s, the amplitude is $\sqrt{5}/8 \approx$ 0.280 cm, and the shift is 0.464.

15. **(a)** $\dfrac{d^2\theta}{dt^2} + \dfrac{g}{L}\theta = 0$ and $T_0 = \dfrac{2\pi\sqrt{L}}{\sqrt{g}}$

(b) $\left(\dfrac{d\theta}{dt}\right)^2 - \dfrac{2g}{L}\cos\theta = C = -\dfrac{2g}{L}\cos A$

(c) $\dfrac{dt}{d\theta} = -\sqrt{\dfrac{L}{2g}}\dfrac{1}{\sqrt{\cos\theta - \cos A}}$, $t(A) = 0$, $t(0) = \dfrac{T}{4}$

(d) $F = \dfrac{\sqrt{2}}{\pi}\displaystyle\int_0^A \dfrac{d\theta}{\sqrt{\cos\theta - \cos A}}$

(e) $F = \dfrac{\sqrt{2}}{\pi}\displaystyle\int_{\cos A}^1 \dfrac{dx}{\sqrt{x - \cos A}\sqrt{1 - x^2}}$

(f)

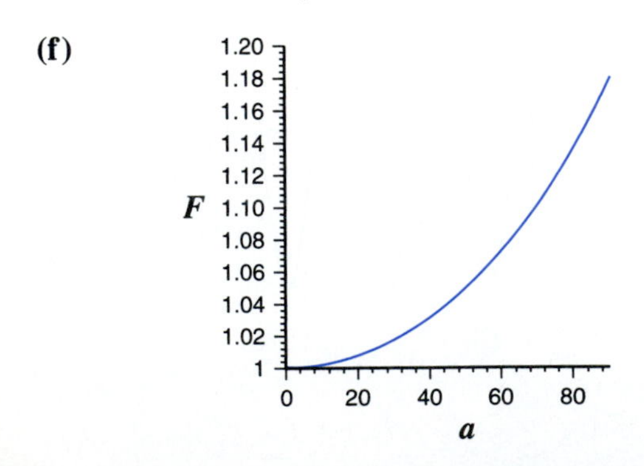

29.6° for 1 min/h, 6.04° for 1 min/day, and 2.28° for 1 min/week

Answers to Section 3.2

1. $x = 2$, $y = 0$, $z = 3$
3. $x = 1$, $y = 2$, $z = -3$
5. $x = 10$, $y = -3$, $z = 6$
7. The system has many solutions.
9. The system has many solutions.
11. -3 **13.** -62 **15.** 320

17. No solutions if $c = 2$; otherwise $x = \dfrac{3c - 7}{c - 2}$ and $y = \dfrac{1}{c - 2}$

19. No solutions if $c = 3.2$; otherwise $x = \dfrac{48 - 18c}{16 - 5c}$, $y = \dfrac{8}{16 - 5c}$, and $z = \dfrac{-2c}{16 - 5c}$

21. $x = 3c$ and $y = c$ for any real number c

Answers to Section 3.3

1. $L[y](t) = 2t^3 + 4t$
3. $L[y](t) = 2e^{-t}(\sin t + \cos t)$
5. $L[y](t) = 6c_1e^{-2t}$ **7.** $L[y](t) = -6c_1\sin 2t$
9. $y'' + 2y = 2t^3 + 4t$ **11.** $y = c_1 + c_2t + c_3e^{2t}$
13. $y = c_1e^t + c_2te^t$
15. **(a)** $(-\infty, \infty)$ **(e)** $y = c_1e^{-2t} + c_2e^{3t}$ **(f)** $y = (e^{3t} - e^{-2t})/5$ **(g)** $(-\infty, \infty)$
17. **(a)** $(0, \infty)$ **(e)** $y = c_1x^2 + c_2x^2\ln x$ **(f)** $y = x^2 - 3x^2\ln x$ **(g)** $(0, \infty)$
19. **(a)** $(-\infty, \infty)$ **(e)** $y = c_1 + c_2t + c_3e^{-2t}$ **(f)** $y = 1$ **(g)** $(-\infty, \infty)$

Answers to Section 3.4

1. $y = c_1e^{-3t} + c_2e^{-t}$ **3.** $y = c_1e^{2t} + c_2e^{-2t}$
5. $y = c_1 + c_2e^{-4t} + c_3e^{-t}$ **7.** $y = 6e^{-2t} - 4e^{-3t}$
Long-time behavior $y \approx 6e^{-2t}$

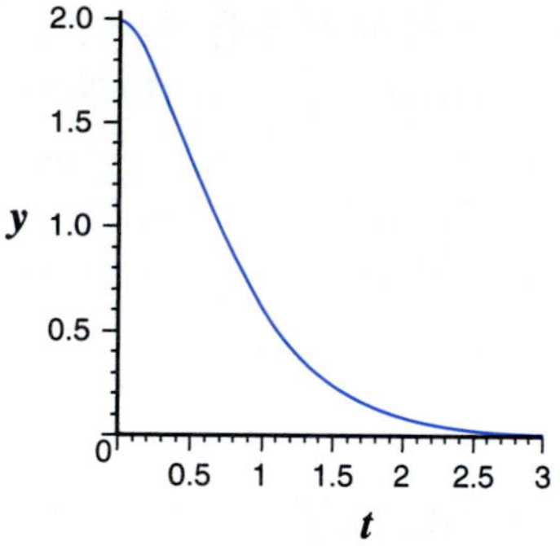

9. $y = (4e^{-4t} + 8e^{2t})/3$
Long-time behavior $y = 8e^{2t}/3$

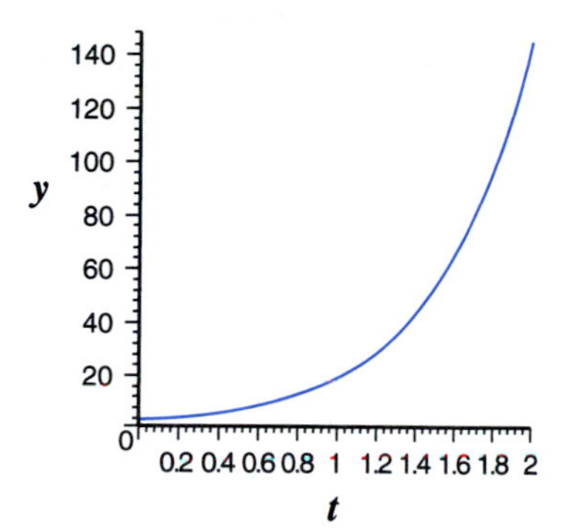

11. **(a)** $b = 4$ **(b)** $y = 1.077e^{-0.536t} - 0.077e^{-7.46t}$
(c) $s > 7.46$

Answers to Section 3.5

1. $y = c_1 \cos \sqrt{5}t + c_2 \sin \sqrt{5}t$
3. $y = e^{2t}(c_1 \cos 2t + c_2 \sin 2t)$
5. $y = c_1 + c_2e^t + c_3e^{-t}$
7. The solution is $y = e^{-2t}(2 \cos t + 4 \sin t)$; the envelope is $y_e = \pm 2\sqrt{5}e^{-2t}$.

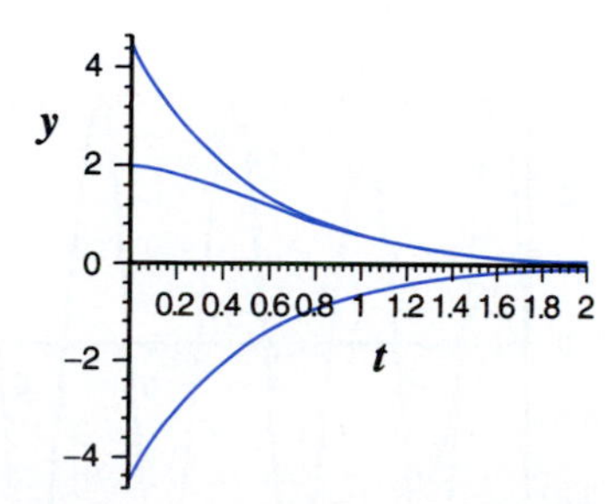

9. The solution is $y = (2/\sqrt{5})e^{-t} \sin \sqrt{5}t$; the envelope is $y_e = \pm(2/\sqrt{5})e^{-t}$.

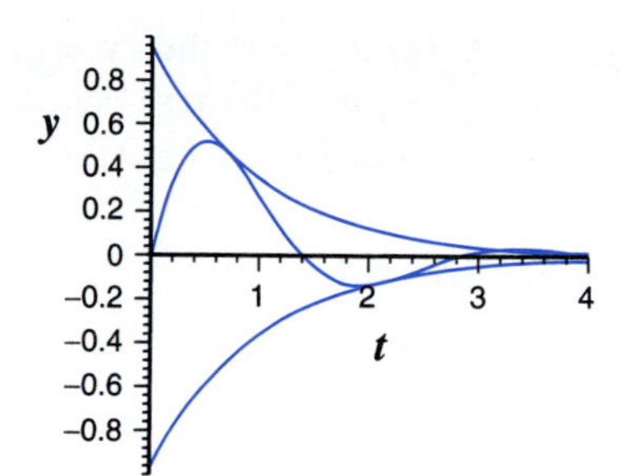

11. $\beta e^{2\alpha t}$

13. The solution is $y = e^{-t}\left(\cos \sqrt{3}t + \dfrac{1}{\sqrt{3}} \sin \sqrt{3}t\right)$;

the envelope is $y_e = \pm 2e^{-t}/\sqrt{3}$.

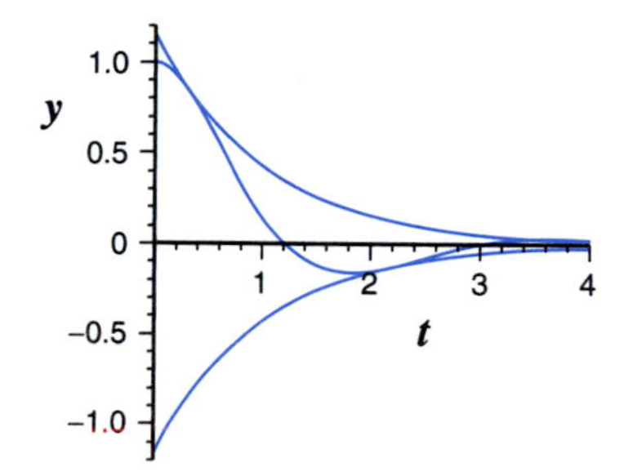

Answers to Section 3.6

1. $y = (2 - 3t)e^{2t}$ **3.** $y = 1 + 2e^{3t} \cos 2t$
5. $y = (c_1 + c_2x^3)e^x$
7. $y = e^x\left(c_1 + c_2 \displaystyle\int_0^x \exp e^s \, ds\right)$
9. $x^{-1/2} \sin x \left(c_1 + c_2 \displaystyle\int_0^x \sqrt{\sin s} \, ds\right)$
11. $y = e^{x^2}(c_1 + c_2 \operatorname{erf} x)$
13. $y = (c_1 + c_2t + c_3t^2)e^{-t}$

Answers to Section 3.7

1. $y = -x^{3/2} + 2x^{-1}$, $y \approx 2x^{-1}$ as $x \to 0$
3. $y = 2x^2 + 7x^2 \ln(-x)$ [Note that the solution exists only for $x < 0$.] Also $\lim_{x\to 0} y = 0$.
5. $0 < \beta < \frac{1}{4}$
7. $\dfrac{d^2y}{d\tau^2} + 2\beta \dfrac{dy}{d\tau} + y = 0$
The parameter β measures the relative amount of damping, with $\beta < 1$ underdamped and $\beta > 1$

overdamped. **9.** **(a)** If $\nu = 0$, then $y = c_1 + c_2 \ln r$. Otherwise, $y = c_1 r^{\nu} + c_2 r^{-\nu}$ **(b)** $y(r) = c_1 J_\nu(\lambda r) + c_2 Y_\nu(\lambda r)$ **11.** $y = c_1 x J_1(x) + c_2 x Y_1(x)$

Answers to Section 4.1

1. $i = 1.5e^{-1000t}$ **3.** $i = 2.5te^{-5t}$

5. $i = e^{-0.1t}\left(\cos\beta t - \dfrac{0.1}{\beta}\sin\beta t\right)$ where $\beta = \sqrt{0.99}$

7. **(a)** $Li' + Ri = v_E, \quad i(0) = i_0$ **(b)** $y = i - v_E/R$ yields the problem $Ly' + Ry = 0, \quad y(0) = i_0 - v_E/R$
(c) $y = (i_0 - v_E/R)e^{-Rt/L}$
(d) $i = v_E/R + (i_0 - v_E/R)e^{-Rt/L}$ **(e)** v_E/R
(f) Oscillation; nonzero steady current
9. **(a)** The voltage in the primary winding when the switch is closed is $v_L = 12e^{-448t}$. **(b)** The voltage in the primary winding when the switch is opened is $v_L = -e^{-224t}(497\sin 18{,}568t + 12\cos 18{,}568t)$.
(c) The spark plug ignites when the switch is opened.
11. **(a)** $L_0 = 1$
(b) $i = 0.5e^{-0.5t}$ for $L = 0$

$i = e^{-0.5t}\sin 0.5t$ for $L = 2$

$i = 0.5e^{-0.2t}\sin 0.4t$ for $L = 5$

$i = \frac{1}{3}e^{-0.1t}\sin 0.3t$ for $L = 10$

(c)

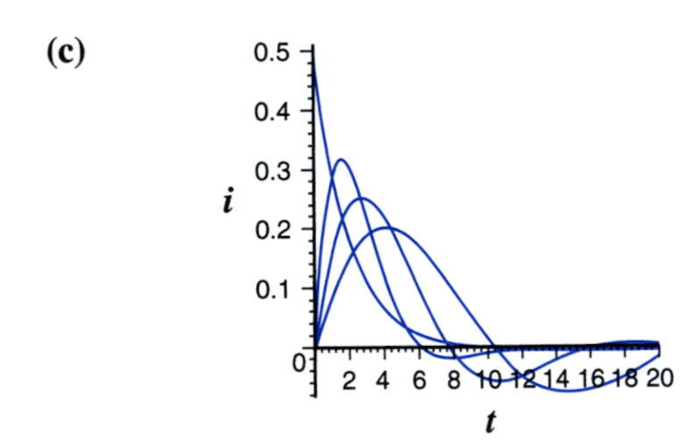

Answers to Section 4.2

1. $y = c_1 e^{-2t} + c_2 e^{2t} - \frac{1}{3}e^t$
3. $y = (c_1 + c_2 t)e^{-3t} + \frac{1}{25}e^{2t}$
5. $y = c_1 e^{-4t} + \frac{4}{25}\cos 3t + \frac{3}{25}\sin 3t$
7. $y = c_1\cos 2t + c_2\sin 2t - \frac{1}{5}\sin 3t$
9. $y = 2e^{-2t} + e^t$
11. $y = \frac{93}{29}e^{-5t} + \frac{15}{29}\sin 2t - \frac{6}{29}\cos 2t$
13. **(a)** $y_p = \frac{5}{13}\cos 2t - \frac{1}{13}\sin 2t$ **(b)** Not possible **(c)** $y + p = \frac{3}{13}\cos 2t + \frac{2}{13}\sin 2t + \frac{4}{25}\cos 4t - \frac{3}{25}\sin 4t$ **15.** **(a)** $L[1] = 2$ **(b)** $L[t] = 1 + 2t$
(c) $L[t^2] = 2t + 2t^2$ **(d)** $y_p = \frac{3}{4} - \frac{1}{2}t + \frac{1}{2}t^2$
(e) $y_p = A_0 + A_1 t + \cdots + A_k t^k$

Answers to Section 4.3

1. **(a)** Yes, characteristic value 3, degree 1 **(b)** Yes, characteristic value 0, degree 3 **(c)** No **(d)** Yes, characteristic value $1 \pm 3i$, degree 1
3. $L[te^{3t}] = (7 + 10t)e^{3t}$ **5.** $L[te^t] = 3e^t$
7. $y_p = (2t - t^2)e^{-t}$ **9.** $y_p = (-0.07 + 0.1t)e^{3t}$
11. $y_p = \frac{4}{3}te^t$ **13.** $y_p = \frac{1}{17}e^t(\cos 2t + 4\sin 2t)$
15. $y_p = \frac{1}{36}e^t(19 - 30t + 18t^2)$
17. $y_p = \frac{1}{4}te^{2t} - (\frac{2}{9} + \frac{1}{3}t)e^t$
19. $y_p = \frac{1}{3}e^{3t} - \frac{1}{3}\cos 3t$
21. $y = c_1 x + c_2 x^2 + \frac{1}{6}x^4$
23. $c_1 x + c_2 x^2 + (\frac{1}{2}\ln x - \frac{3}{4})x^3$
25. **(a)** $L[Y] = At^2 + B(1 + t^3) + C(2t + t^4)$
(b) The first has solution $y_p = 5 + t^2$; the second does not work.

Answers to Section 4.4

3. $y = -\dfrac{2}{\sqrt{35}}e^{-t/2}\sin\left(\dfrac{\sqrt{35}}{2}t\right) + \dfrac{1}{3}\sin 3t$

The steady-state solution is $y_p = \frac{1}{3}\sin 3t$.

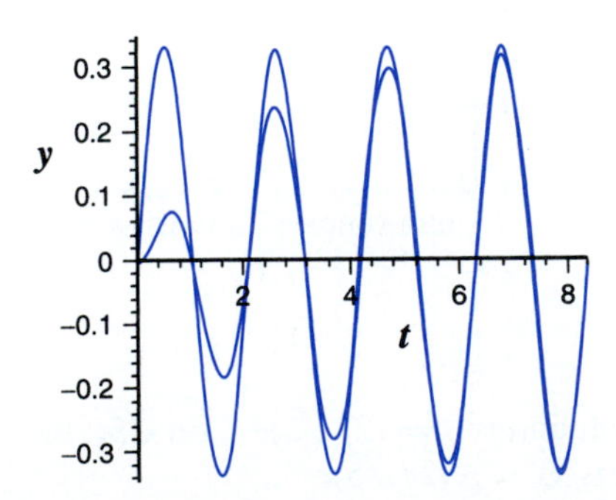

5. $y = c_1 e^{r_1 t} + c_2 e^{r_2 t} - \frac{1}{267}(5\cos 5t + 8\sin 5t)$, with $r_1 = \frac{-3+\sqrt{5}}{2}$, $r_2 = \frac{-3-\sqrt{5}}{2}$, $c_1 = \frac{5+19\sqrt{5}}{534}$, $c_2 = \frac{5-19\sqrt{5}}{534}$
The steady-state solution is
$y_p = 0.0353\cos(5t + 2.13)$.

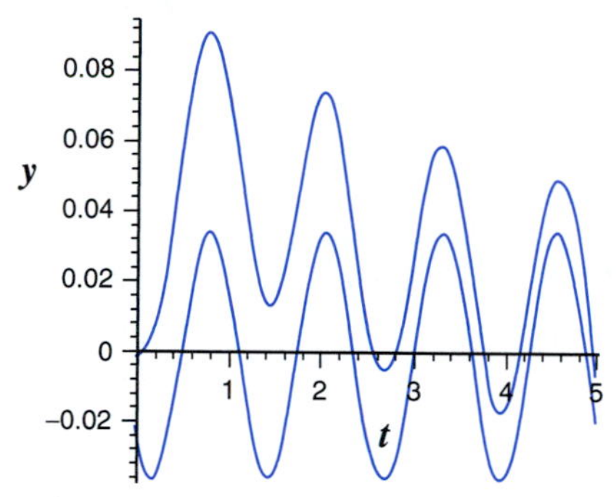

7. $y_p = -\dfrac{i}{3}e^{3it}$. The amplitude is $\frac{1}{3}$, which is the same as in Exercise 3.

9. **(a)** $A = [(k^2 - 16)^2 + 16]^{-1/2}$ **(b)** The maximum amplitude occurs when $k = 4$.

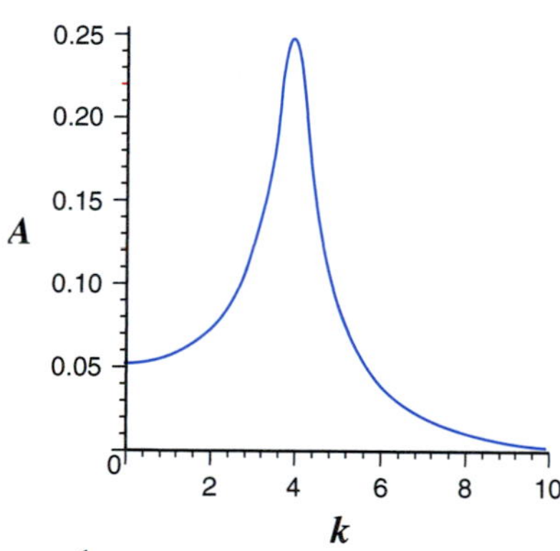

11. **(a)** $y = \dfrac{1}{2.04}(\cos 1.4t - \cos 2t)$

(b)

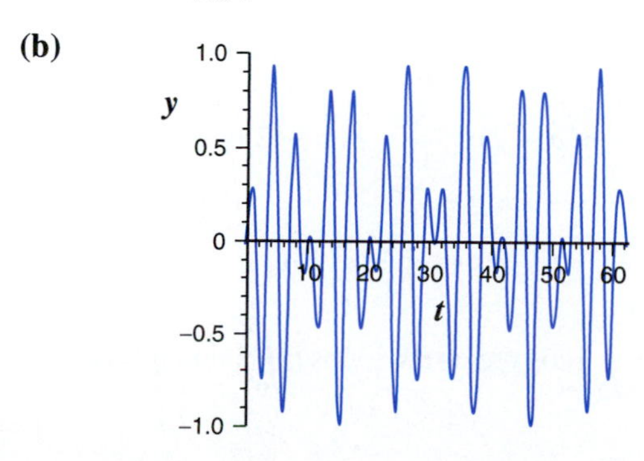

(d) $y = \frac{1}{2}(\cos\sqrt{2}\,t - \cos 2t)$

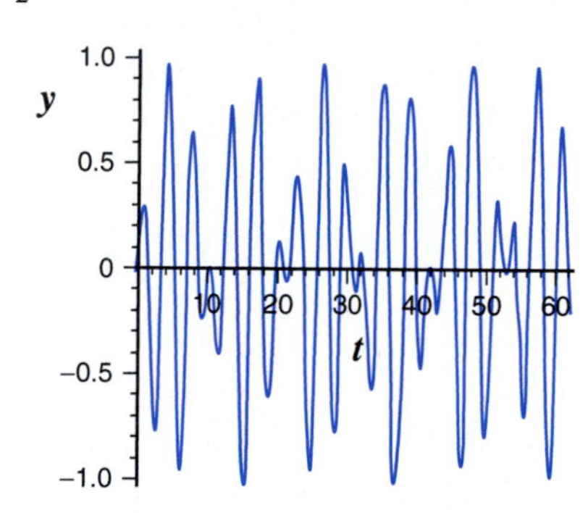

(f) β must be a rational number. Suppose $\beta = p/q$, where p and q are a relatively prime pair of integers. If p is even, then the period of $\cos 2t - \cos\beta t$ is q; if p is odd, then the period of $\cos 2t - \cos\beta t$ is $2q$.

13. **(a)** $T' + kT = 20k - 5k\cos\dfrac{\pi t}{12}$

(b) $T_s = 20 - \dfrac{60k}{\sqrt{\pi^2 + 144k^2}}$

$$\cos\left(\frac{\pi t}{12} - \arcsin\frac{\pi}{\sqrt{\pi^2 + 144k^2}}\right)$$

(c)

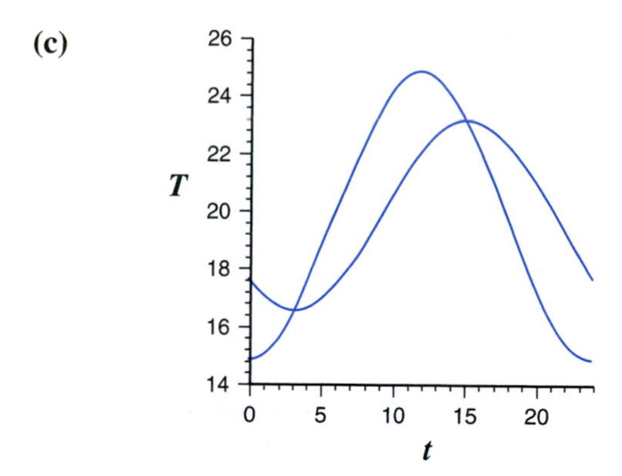

Answers to Section 4.5

1. $y = -\left(\dfrac{1}{9} + \dfrac{t}{3}\right)e^t + Ce^{4t}$

3. $y = \frac{1}{2}t^2 e^{4t} + Ce^{4t}$

5. $y = \frac{1}{10}\sin t - \frac{3}{10}\cos t + Ce^{3t}$

7. $y = [C + \ln(1 + e^t)]e^{-t}$

9. $y = 2 + Ce^{-t^2/2}$

11. $y = (C + \sin t)e^{-t^2}$

13. $y = \frac{1}{t}(1 + Ce^{-t})$

15. $y = \tan t + C \sec t$

17. $y = 2t^5 - 2t^2, \quad (-\infty, \infty)$

19. $y = (1 + t^2) \arctan t, \quad (-\infty, \infty)$

21. $y = (\cos^2 t)(1 + \tan t), \quad \left(-\frac{\pi}{2}, \frac{\pi}{2}\right)$

23. $y = e^{4t^2} + \frac{\sqrt{\pi}}{4} e^{4t^2} \operatorname{erf}(2t), \quad (-\infty, \infty)$

25. $y = \left(\frac{2}{5}t^{-1} + Ct^4\right)^{-1/2}$

27. $P = \frac{K}{1 + CKe^{-rt}}$

29. **(a)** $y = (k \ln t)t^{-m} + Ct^{-m}$
(b) $y = \frac{k}{m+n} t^n + Ct^{-m}$ **(c)** $c = 0$, no **(d)** No

31. **(a)** $y = (\cos t)[k - \ln(\cos t)], \quad \left(-\frac{\pi}{2}, \frac{\pi}{2}\right)$
(b) $y \to 0$ **(c)** $k \le 0$ **(d)** $k \ge 1$ (Showing that 0 is a local maximum for $k = 1$ is difficult. The idea is to write y' as a power series in t and then examine the asymptotic behavior of y' as $t \to 0$.)

33. $y = \begin{cases} 1 & t \le T \\ e^{T-t} & t > T \end{cases}$

35. **(b)** $P_1 = \exp\left[\frac{\ln 2}{12}\left(t - \frac{18}{\pi} \sin \frac{\pi t}{6}\right)\right]$

(c) $P_p = -R \exp\left[\frac{\ln 2}{12}\left(t - \frac{18}{\pi} \sin \frac{\pi t}{6}\right)\right]$
$$\int_0^t \exp\left[-\frac{\ln 2}{12}\left(\tau - \frac{18}{\pi} \sin \frac{\pi \tau}{6}\right)\right] d\tau$$

(d)
$$P = \left\{1 - R \int_0^t \exp\left[-\frac{\ln 2}{12}\left(\tau - \frac{18}{\pi} \sin \frac{\pi \tau}{6}\right)\right] d\tau\right\}$$
$$\exp\left[\frac{\ln 2}{12}\left(t - \frac{18}{\pi} \sin \frac{\pi t}{6}\right)\right]$$

(e) $R = 0.0543$

(f)

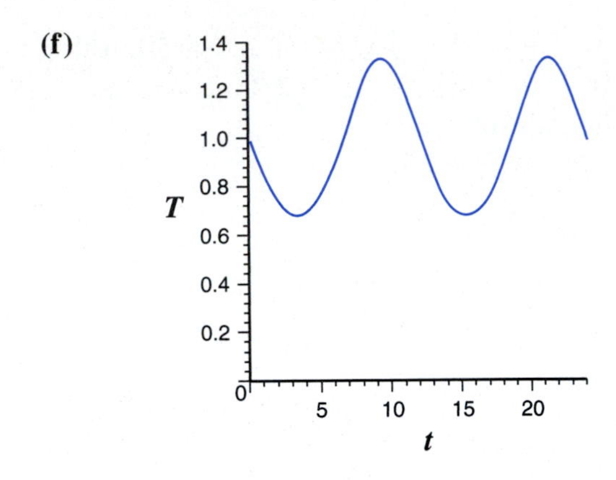

Answers to Section 4.6

1. $y_p = (1 + 2t - t^2)e^{-t}$

3. $y_p = \left(-\frac{7}{100} + \frac{1}{10}t\right)e^{3t}$

5.
$y = c_1 \cos 2t + \left(c_2 + \frac{1}{2}t\right) \sin 2t + \frac{1}{4} \cos(2t) \ln(\cos 2t)$

7.
$$y = e^{-t}(c_1 + c_2 t + t \ln t) \quad \text{for } p = 1$$
$$y = e^{-t}(c_1 + c_2 t - \ln t) \quad \text{for } p = 2$$
$$y = e^{-t}\left[c_1 + c_2 t + \frac{t^{2-p}}{(2-p)(1-p)}\right] \quad \text{for } p = 3, 4, \ldots$$

9.
$$y = x\left[c_1 + \int_0^x \frac{se^{-s}}{(1-s)^2} ds\right] + e^x\left[c_2 - \int_0^x \frac{s^2 e^{-2s}}{(1-s)^2} ds\right]$$

11. $y = c_1 x^{-1/2} \cos x + c_2 x^{-1/2} \sin x + \frac{1}{8} x^{1/2} \sin x$

13. **(b)** $y_p = 1$ for $g(x) \equiv 1$, $y_p = -x^2$ for $g(x) = x^2$, and $y_p = a - bx^2$ for $g(x) = a + bx^2$
(c)
$$y_p = -\frac{1}{p-1}x^p - \frac{p}{p-1}x^{p-1} - px^{p-2} - p(p-2)x^{p-3} - p(p-2)(p-3)x^{p-4} - \cdots - p(p-2)!$$
(d) $y_p = -1 - (1 + x) \ln x - e^x E_1(x)$

15. **(a)**
$$y_p = \sin t \int_0^t (\cos s) g(s)\, ds - \cos t \int_0^t (\sin s) g(s)\, ds$$

(b) $u_1 = \frac{1}{4}\sin 2t - \frac{t}{2}, \quad u_2 = \frac{1}{4} - \frac{1}{4}\cos 2t,$

$y_p = \frac{1}{2}\sin t - \frac{t}{2}\cos t$

(c)

$$y_p = \left(\frac{1}{4} - \frac{1}{4}\cos 2T\right)\sin t + \left(\frac{1}{4}\sin 2T - \frac{T}{2}\right)\cos t$$

(d) $A = \sqrt{\frac{1}{8} - \frac{1}{8}\cos 2T - \frac{T}{4}\sin 2T + \frac{T^2}{4}}$

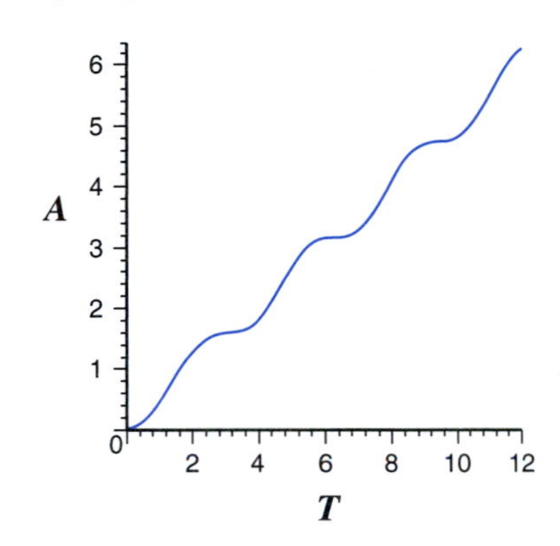

(e) $T = 0.5\pi$, $T = 0.75\pi$, and $T = \pi$

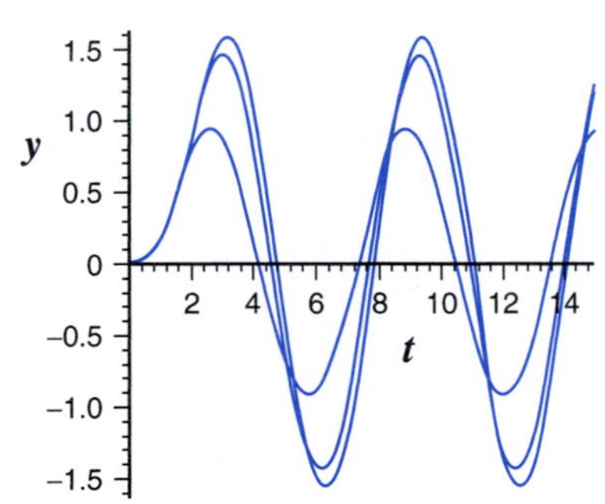

$T = \pi$, $T = 1.25\pi$, and $T = 1.5\pi$

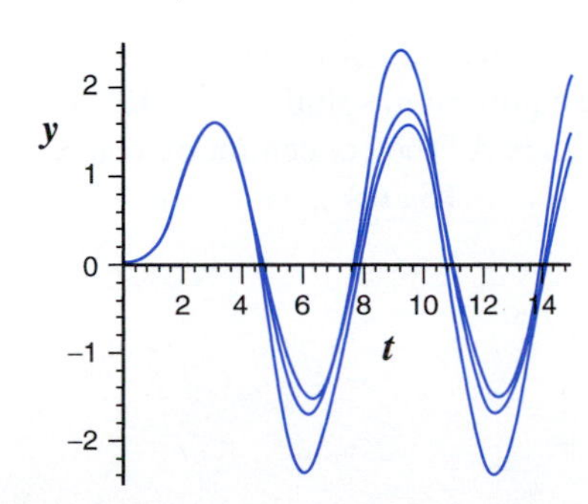

17. (a) $y_p = u_1 y_1 + u_2 y_2 + u_3 y_3$

(b) $\begin{pmatrix} y_1 & y_2 & y_3 \\ y_1' & y_2' & y_3' \\ y_1'' & y_2'' & y_3'' \end{pmatrix} \begin{pmatrix} u_1' \\ u_2' \\ u_3' \end{pmatrix} = \begin{pmatrix} 0 \\ 0 \\ g \end{pmatrix}$

(c)

$$u_1' = \frac{(y_2 y_3' - y_2' y_3)g}{W[y_1, y_2, y_3]}, \quad u_2' = \frac{-(y_1 y_3' - y_1' y_3)g}{W[y_1, y_2, y_3]},$$

$$u_3' = \frac{(y_1 y_2' - y_1' y_2)g}{W[y_1, y_2, y_3]}$$

19. $y_p = \left(\frac{1}{4} - \frac{1}{2}t\right)e^t$

21. (a) $y_c = c_1 + c_2 x + c_3 \cosh x + c_4 \sinh x$

(c) $y = c_3(\cosh x - 1) + c_4(\sinh x - x) + y_p$, where

$$y_p = -\int_0^x s w(s)\,ds + x\int_0^x w(s)\,ds + \int_0^x \sinh(s - x)\,w(s)\,ds$$

(d) $y_p = 0$ for $x \le 0.9$, $\quad y_p = 10 - 10\cosh(x - 0.9) + 5(x - 0.9)^2$ for $x > 0.9$

(e) $y \approx 1.487(\sinh x - x) - 1.100(\cosh x - 1) + y_p$

(f)

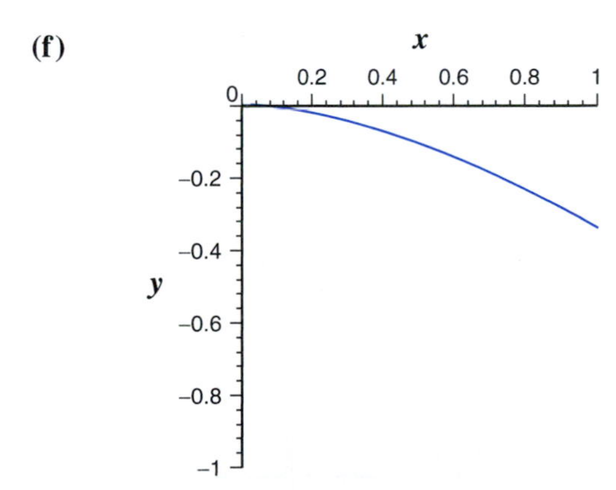

Answers to Section 5.1

1. The population increases if p is less than 10 and decreases if p is greater than 10.

3. Only system b is a predator-prey model, with y as the predator.

5. Model c is a competing species model. An increase of either population is bad for both of the populations.

7. $X = bx/a$, $Y = sy/r$, $\tau = rt$, and $k = a/r$

9. $\dfrac{dx}{dt} = rx\left(1 - \dfrac{x}{K}\right) - sxy, \quad \dfrac{dy}{dt} = csxy - my$

11. (a) $\dfrac{dS}{dt} = -rSI + \gamma I, \quad \dfrac{dI}{dt} = rSI - \gamma I$

(b) Assumption 2 says that the rate is proportional to the number of contacts between a susceptible person and an infected person. Assumption 3 says that the relative rate of recovery is constant. Both of these are reasonable guesses in the absence of more detailed information. **(c)** Recovery from the flu makes a person immune to that strain of flu. Models for the flu need to include a recovered class.

13. (a) $\dfrac{dw}{dt} = -kwx, \quad \dfrac{dx}{dt} = ckwx - mx$

(b) $\dfrac{1}{t_r}\dfrac{dW}{d\tau} = -kx_rWX, \quad \dfrac{1}{t_r}\dfrac{dX}{d\tau} = ckw_rWX - mX$

(c) $w_r = m/(ck), \quad t_r = 1/m, \quad x_r = m/k$

Answers to Section 5.2

1.

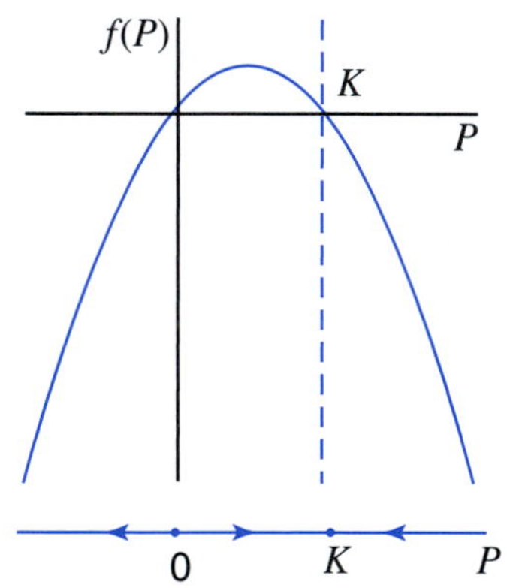

0 K P

3. (a) 0 and 2

(b)

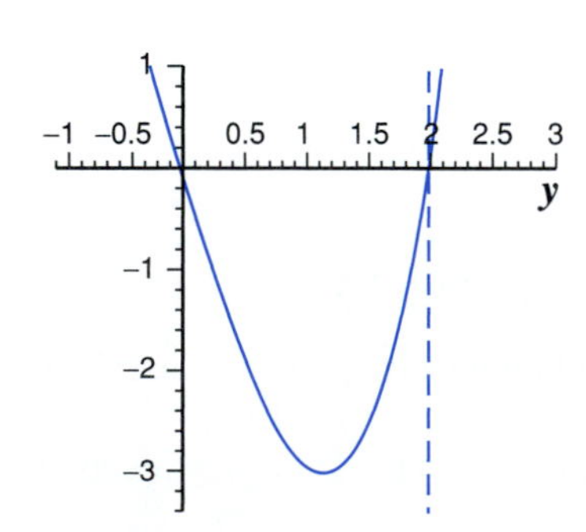

(c)

0 2 y

(d) 0 is stable and 2 is unstable

5. (a) 0

(b)

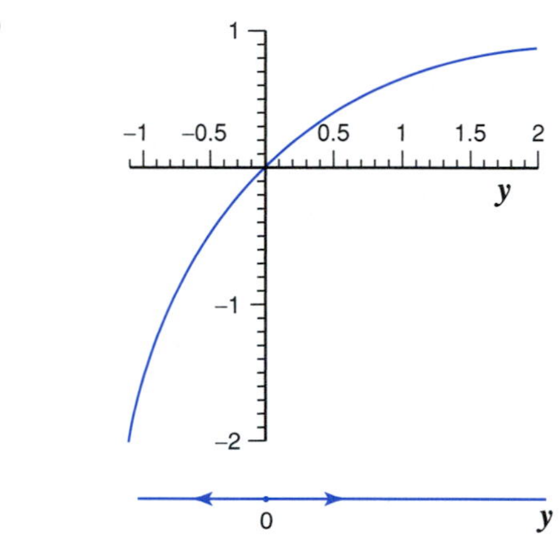

(c)

0 y

(d) 0 is unstable

7. (a) 0

(b)

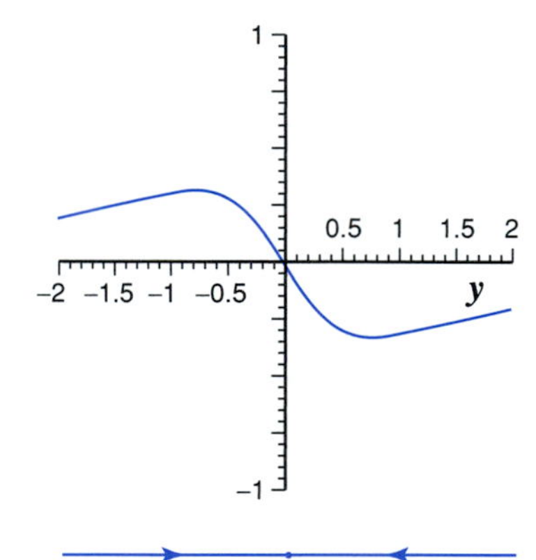

(c)

0 y

(d) 0 is stable

9. Exercise 4. The critical value, $y = 1$, corresponds to an unstable equilibrium solution. **13.** The stable concentration is A. The concentration reaches $A/2$ at time $(\ln 2)/B$. **15. (c)** $y_e = 1 - h$

(e)

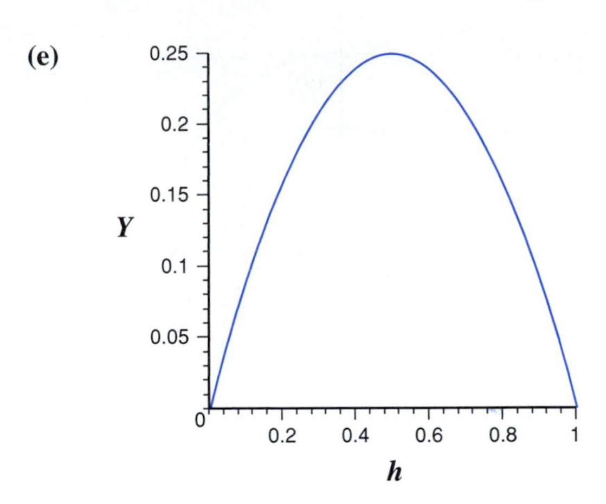

(f) $h = \frac{1}{2}$

Answers to Section 5.3

1. $y\,\dfrac{dy}{dx} = 1, \quad 2x = y^2 + C$

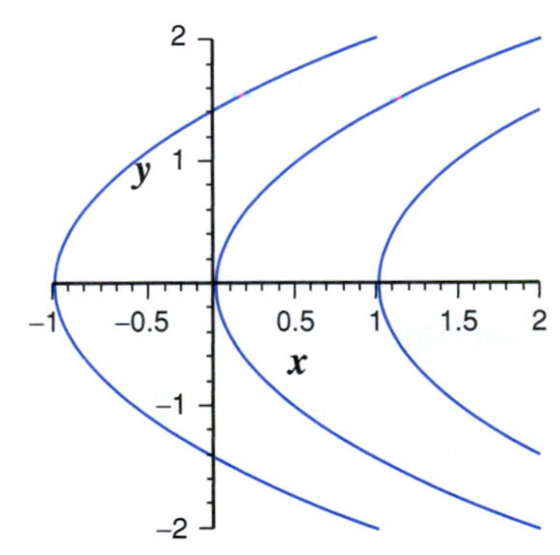

3. $3y^2\,\dfrac{dy}{dx} = e^x, \quad y^3 = e^x + C$

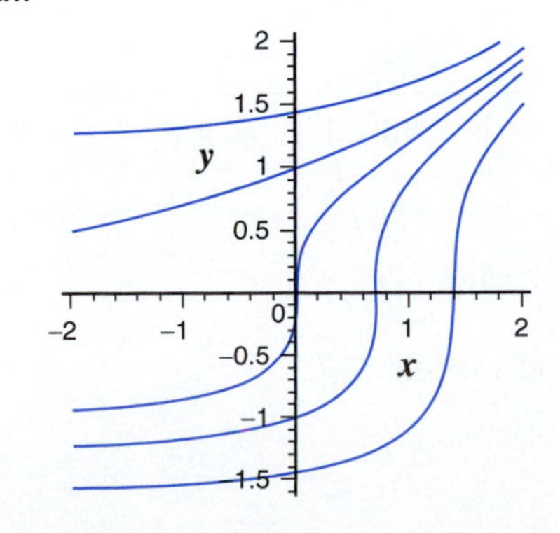

5. $v\,\dfrac{dv}{dy} = -y^2, \quad 3v^2 + 2y^3 = C$

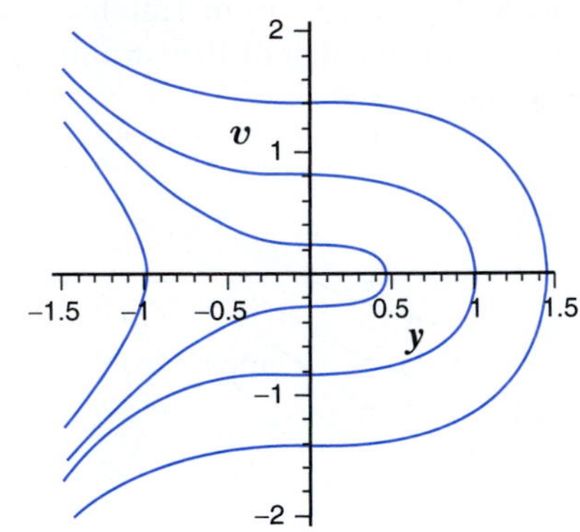

7. $\dfrac{dY}{dX} = \dfrac{Y(1-X)}{X(1-Y)}, \quad Y - \ln Y = X - \ln X + C$

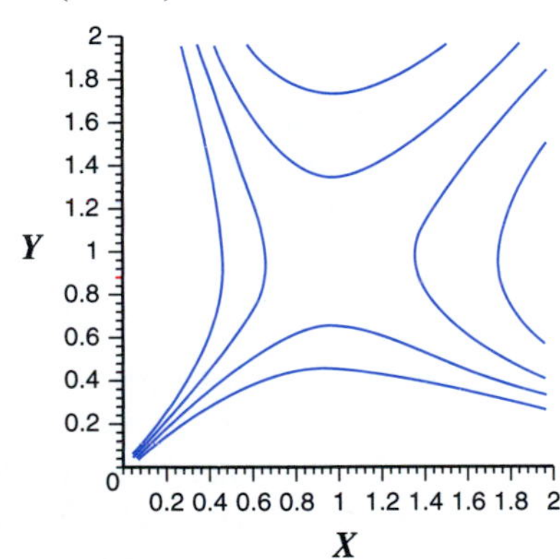

9.

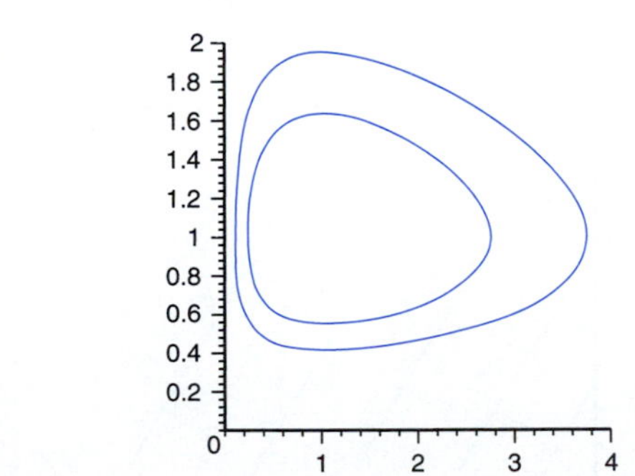

Smaller k makes the population of prey vary more and the population of predators vary less.

11. (a) The loss rate of each force is proportional to the size of the other force. This is consistent with a battle consisting of forces firing at each other from a distance. **(b)** $y^2 - x^2 = y_0^2 - x_0^2$ **(c)** $\sqrt{3}/2$, which is approximately 87% **(d)** The French would have won, with 19 surviving ships. **(e)** The trajectory corresponding to the numbers from Trafalgar is the uppermost, with x the number of British ships and y the number of French ships.

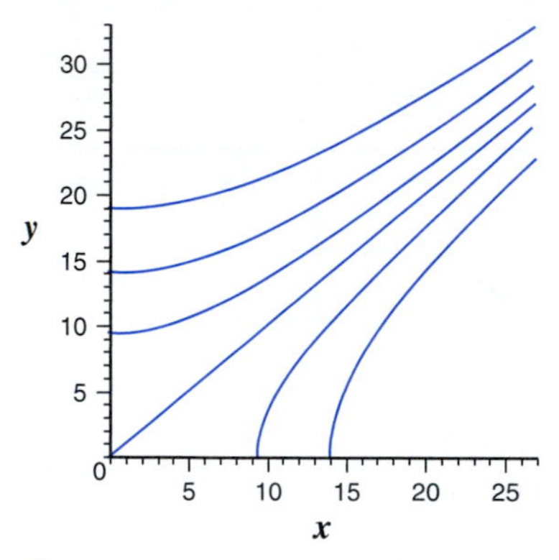

13. **(a)** $x'' - x = 0$ **(b)** $x = x_0 \cosh t - y_0 \sinh t$
(d) The shorter battle ends at $t = 0.26$ and the larger at $t = 0.95$. The 15 French survivors from the smaller battle could enter the larger battle before it had progressed very far.

15. $\dfrac{dI}{dS} = \dfrac{\gamma - rS}{rS}, \quad I = \dfrac{\gamma}{r} \ln S - S + C$
The condition is necessary so that $I'(0) > 0$.

17. **(a)** $v_I = \sqrt{6g} \approx 7.67$ m/s **(b)** $v_{cc} = -49$ m/s
(c) $v_{oc} = -7$ m/s **(d)** The terminal velocity is just less in magnitude than the safe impact speed. **(e)** The thin horizontal line indicates the terminal velocity.

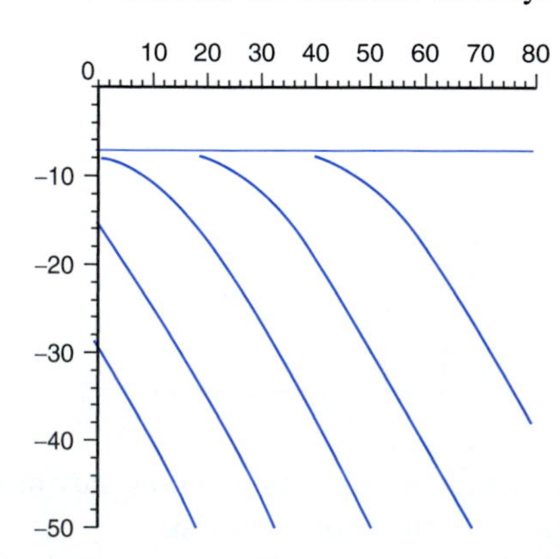

(f) 50.2 m

19. **(a)** $\omega \dfrac{d\omega}{d\theta} = (a^2 \cos\theta - 1) \sin\theta$
(b) $\omega^2 + a^2 \cos^2\theta - 2\cos\theta = C$

(c)

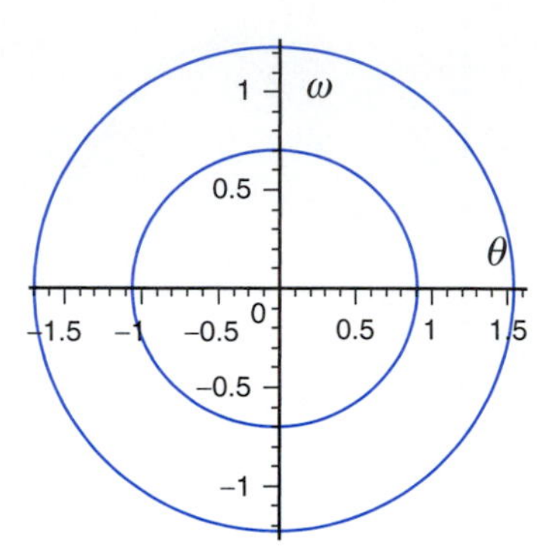

(d)

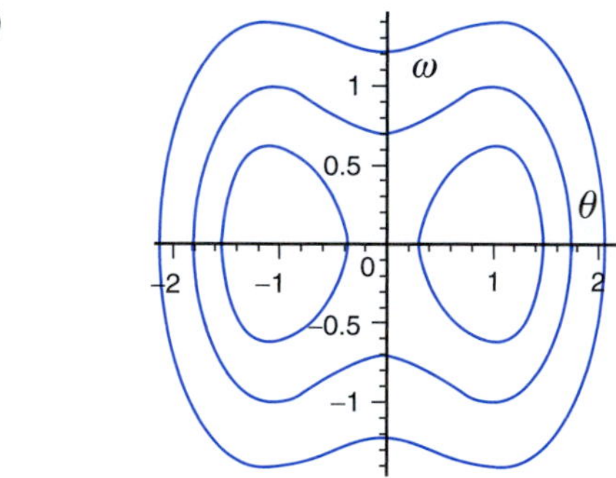

Answers to Section 5.4

1. $(r, c) = (0, 0)$ and $(r, c) = (4, 3/2)$
3. $(n\pi, n\pi)$ for all integers n

5.

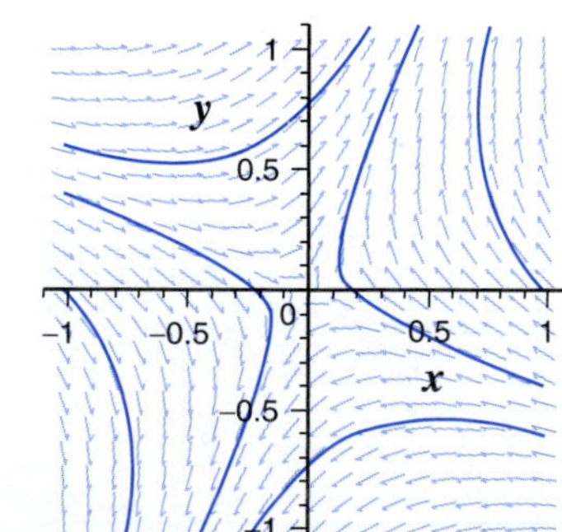

Unstable and a saddle

7.

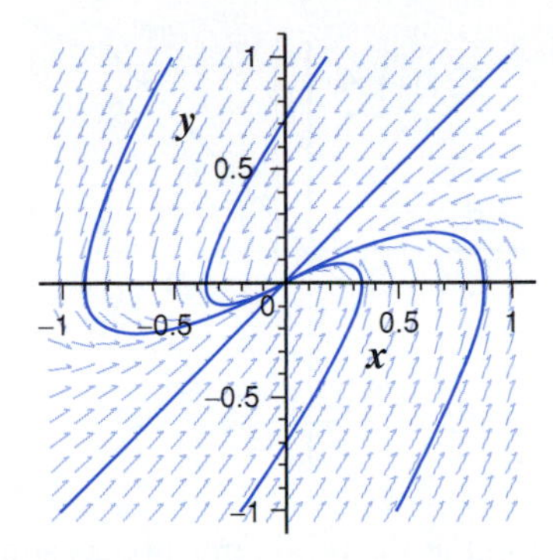

Asymptotically stable

9.

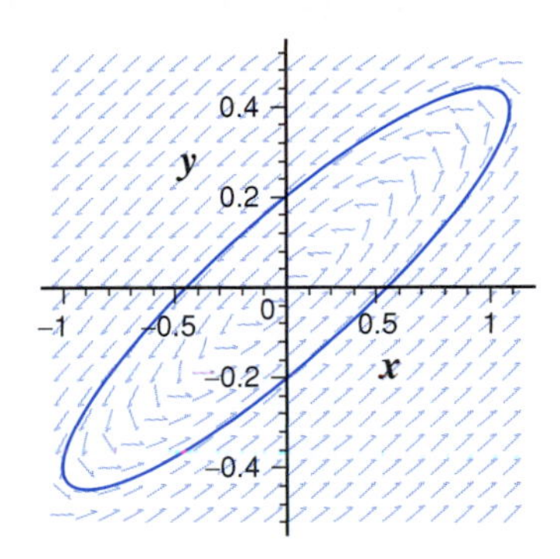

Stable and a center

11. (a) (0, 0) and (1, 1)

(b)

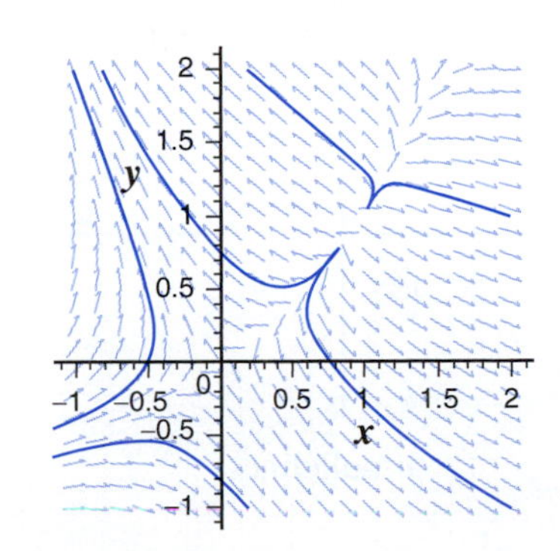

(c) Both are unstable **(d)** (0, 0)

(e)

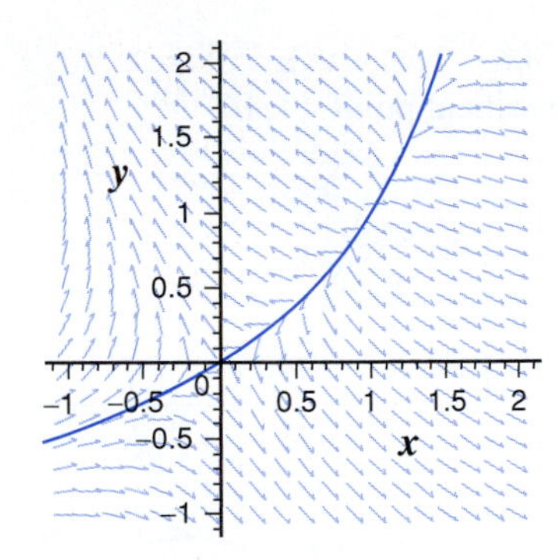

13. (a) (0, 0) and (2, 1)

(b)

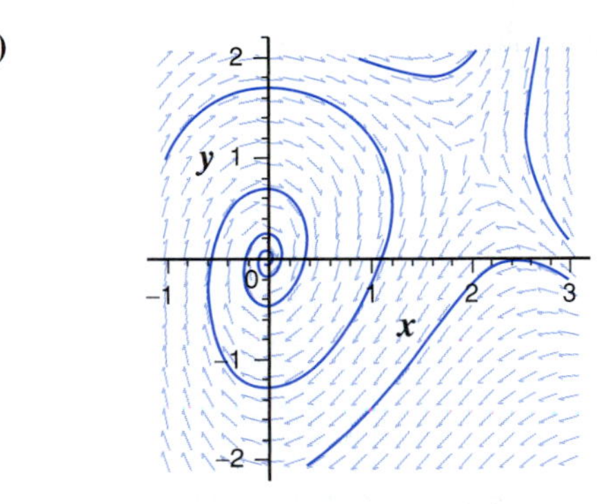

(c) (0, 0) is asymptotically stable and (2, 1) is unstable
(d) (2, 1)

(e)

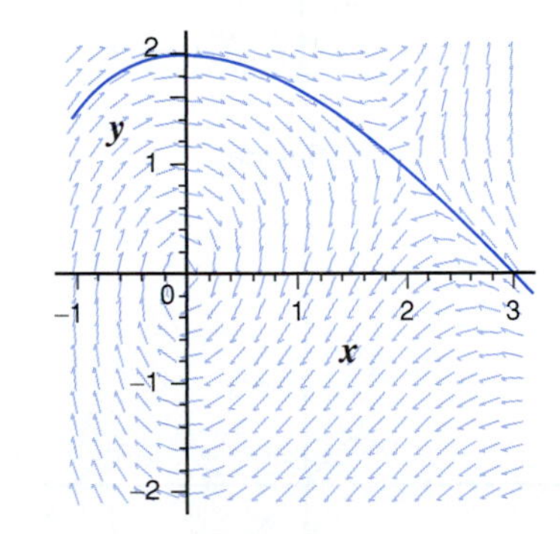

15. $2rv' = 3(1 - 1/r - v^2), \quad r' = v$
There is one critical point, (1, 0).

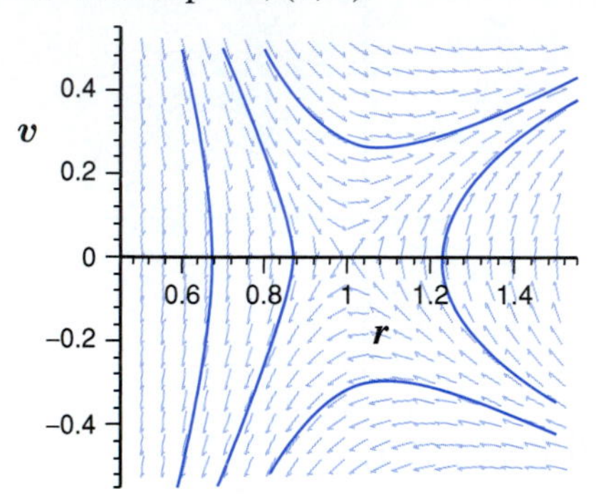

Unstable and a saddle

Answers to Section 5.5

1.

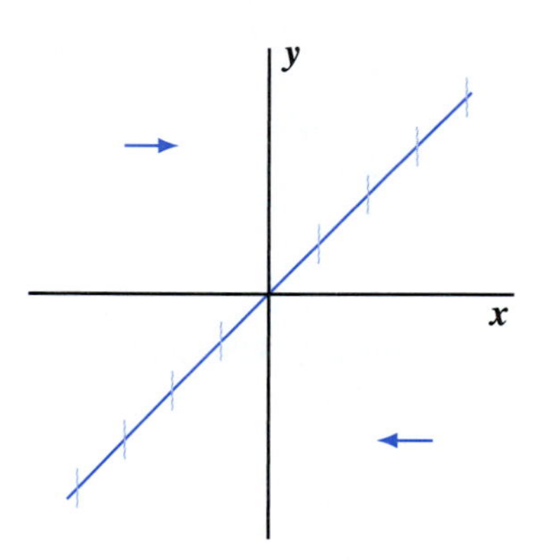

x-nullclines

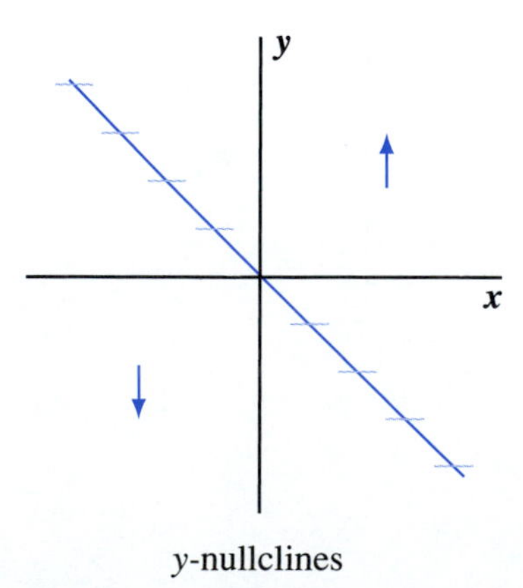

y-nullclines

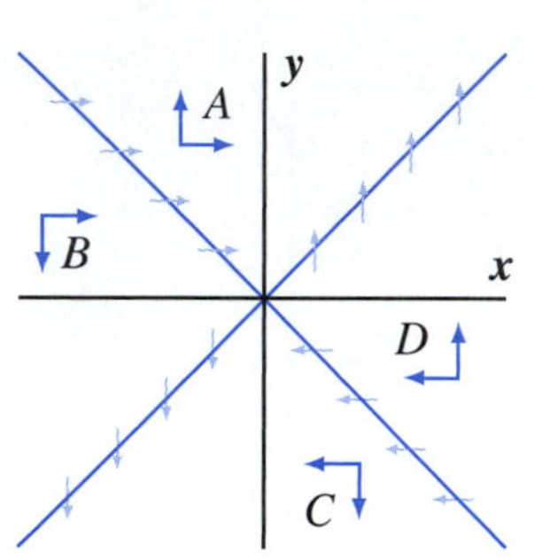

The critical point at the origin must be a saddle point. There must be a separatrix that passes through region B (above the x axis) and region D (below the x axis).

3.

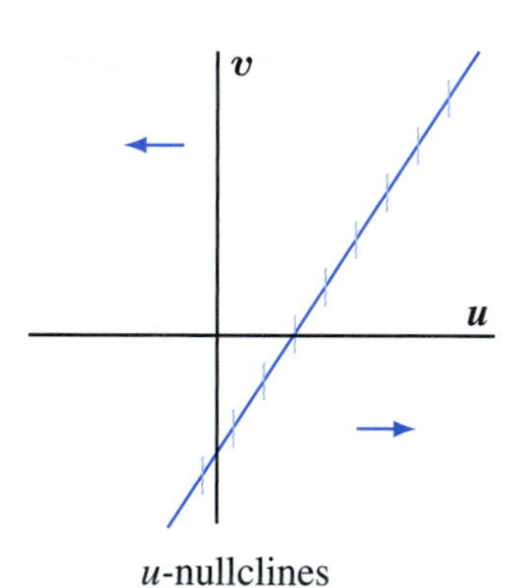

u-nullclines

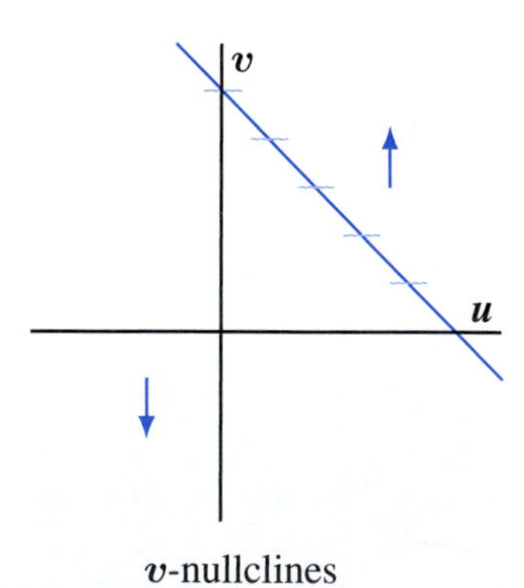

v-nullclines

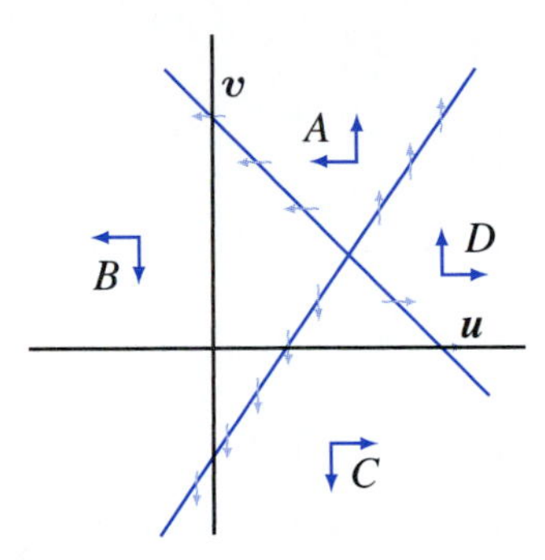

The critical point at (3, 2) is unstable. There are no separatrices. The phase portrait (below) confirms this conclusion.

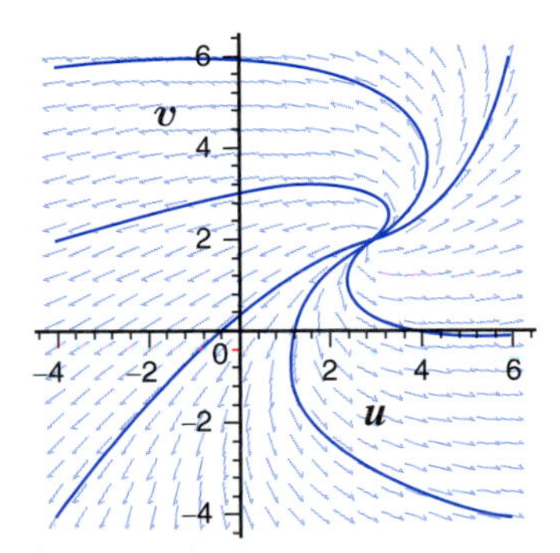

5.

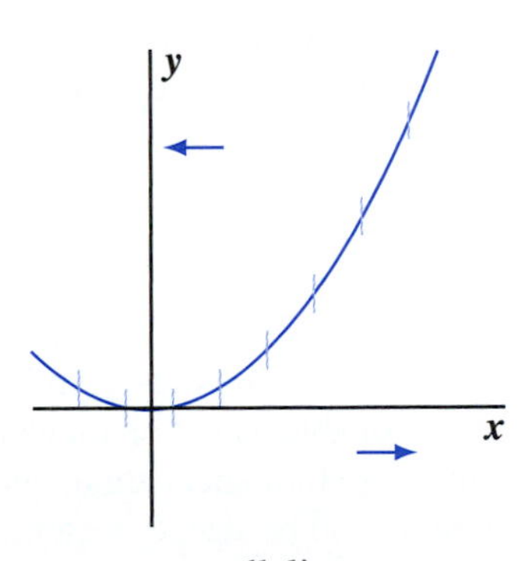

x-nullclines

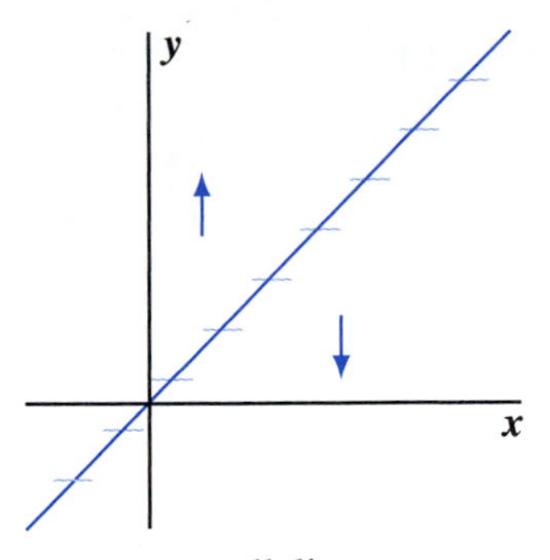

y-nullclines

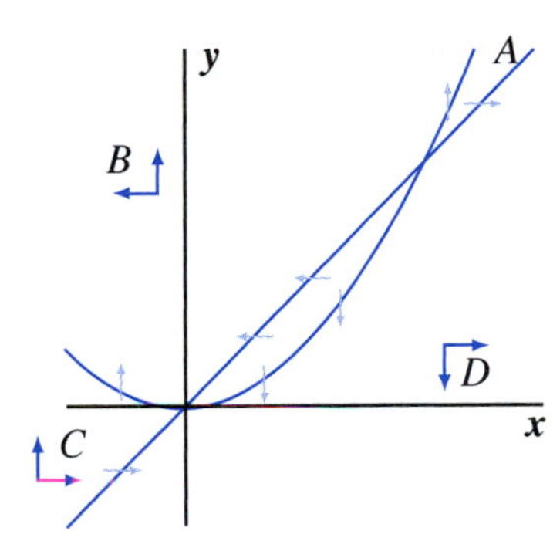

The origin must be a saddle point, and the point (1, 1) must be unstable. There must be separatrices that pass from region C to the origin and from the point (1, 1) into region A and through the central region to the origin.

7.

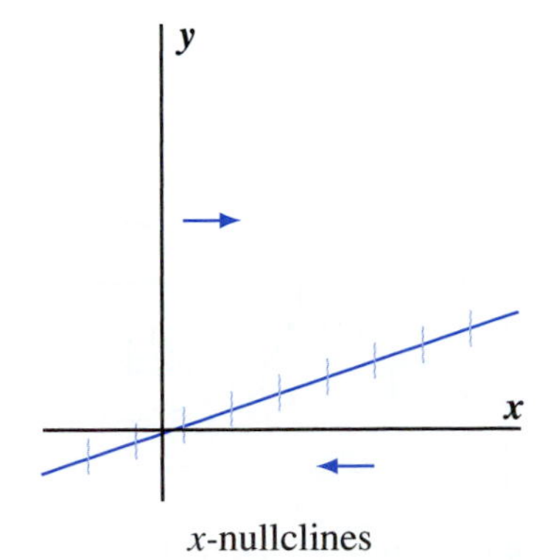

x-nullclines

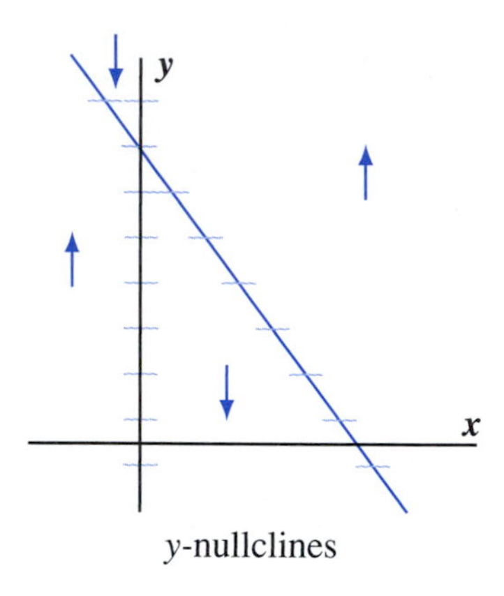

y-nullclines

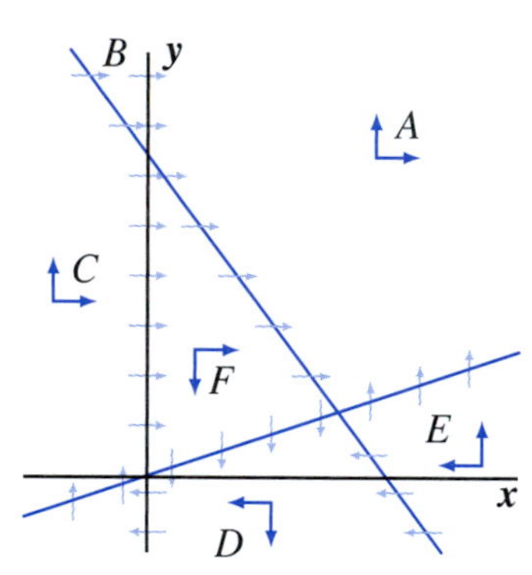

The critical point at (2, 1) must be a saddle point. The critical point at the origin cannot be classified with information from the nullcline diagram alone. There must be a separatrix that passes through the regions C, F, and E.

9.

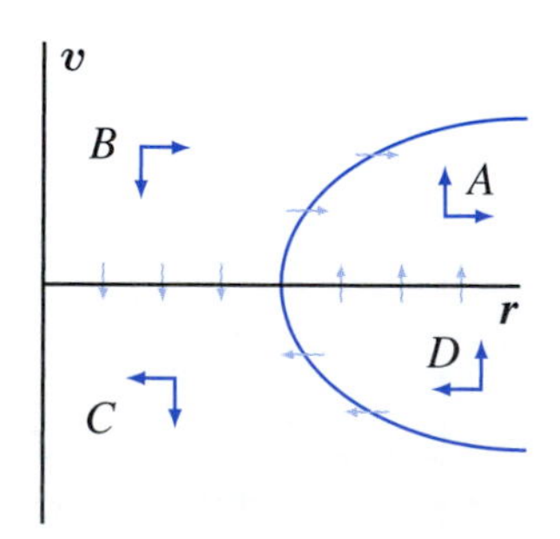

A separatrix passes through the regions B and D. Bubbles that are to the upper right of the separatrix continue to grow, and bubbles to the lower left eventually disappear.

11. (a)

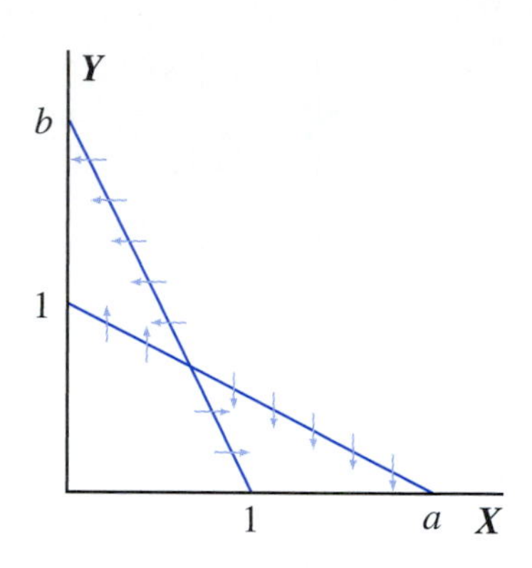

(b)

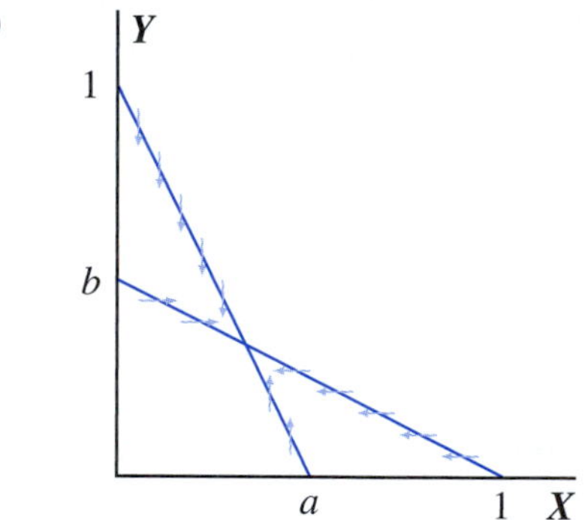

(a) A separatrix runs from the origin to the saddle point at $\left(\frac{ab-a}{ab-1}, \frac{ab-b}{ab-1}\right)$ and beyond. Depending on the initial point, trajectories go to either $(a, 0)$ or $(0, b)$. The model predicts that one species will drive the other one to local extinction. This does seem to be a realistic possibility. **(b)** All trajectories approach the critical point at $\left(\frac{a-ab}{1-ab}, \frac{b-ab}{1-ab}\right)$. The model predicts that the species will coexist. This does seem to be a realistic possibility. (It turns out that this case, while mathematically reasonable, is not biologically reasonable, as there are biological arguments that require the Y nullcline to be steeper than the X nullcline. The result that case a is the only realistic one is called the **law of competitive exclusion**.)

Answers to Section 6.1

1. $x = Ae^{3t}, \quad y = \frac{A}{5}e^{3t} + Be^{-2t}$

3. $x = -c_1e^{-2t} + c_2(1-t)e^{-2t}, \quad y = c_1e^{-2t} + c_2te^{-2t}$

5. $x = c_1 - c_2e^{-3t} - \frac{1}{3}(1+t),$
$y = 2c_1 + c_2e^{-3t} - \frac{2}{3}t$

7. $x = 2e^{-2t}\sin 3t, \quad y = -2e^{-2t}\cos 3t$

9. $x = c_1e^{2t} + 2c_2e^{t} - \frac{3}{65}\cos 3t + \frac{48}{130}\sin 3t,$

$y = c_1e^{2t} + c_2e^{t} + \frac{7}{130}\cos 3t + \frac{9}{130}\sin 3t$

11. $x = c_1e^{2t} - 3c_2e^{-2t} + \frac{1}{4}(1-3t)e^{-2t},$

$y = c_1e^{2t} + c_2e^{-2t} + \frac{1}{4}te^{-2t}$

13. $x = (c_1 + c_2t)e^{t} + c_3e^{-t}, \quad y = 1 - c_2e^{t} - 2c_3e^{-t}$

15. $Y = \dfrac{1}{\sqrt{2}} \times$

$$\begin{cases} 4\sqrt{2} + \lambda_2 e^{\lambda_1\tau} - \lambda_1 e^{\lambda_2\tau} & \tau < T, \\ \lambda_2\left(1 - e^{-\lambda_1 T}\right)e^{\lambda_1\tau} - \lambda_1\left(1 - e^{-\lambda_2 T}\right)e^{\lambda_2\tau} & \tau > T \end{cases}$$

$X = \dfrac{1}{4\sqrt{2}} \times$

$$\begin{cases} 4\sqrt{2} + (1+\lambda_2)e^{\lambda_1\tau} - (1+\lambda_1)e^{\lambda_2\tau} \\ (1+\lambda_2)\left(1 - e^{-\lambda_1 T}\right)e^{\lambda_1\tau} - (1+\lambda_1)\left(1 - e^{-\lambda_2 T}\right)e^{\lambda_2\tau} \end{cases}$$

where $\lambda_1 = -3 + 2\sqrt{2}$ and $\lambda_2 = -3 - 2\sqrt{2}$

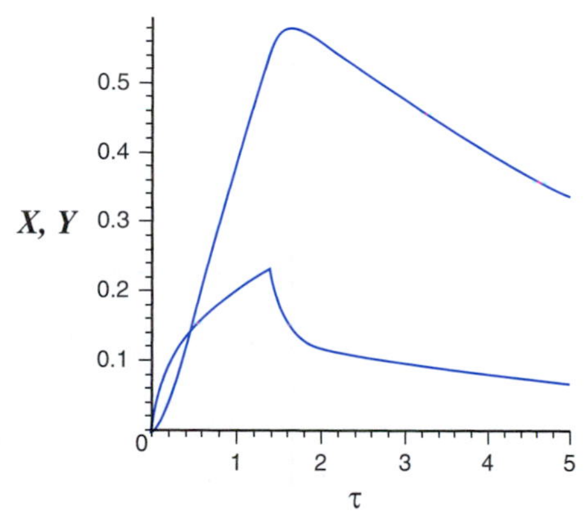

17. **(a)** $x' = 49.3 - 0.0361x + 0.0124y + 0.000035z,$

$y' = 0.0111x - 0.0286y,$

$z' = 0.0039x - 0.000035z$

(b) $x_{eq} \approx 1800, \quad y_{eq} \approx 700, \quad z_{eq} \approx 200{,}000$

(c)

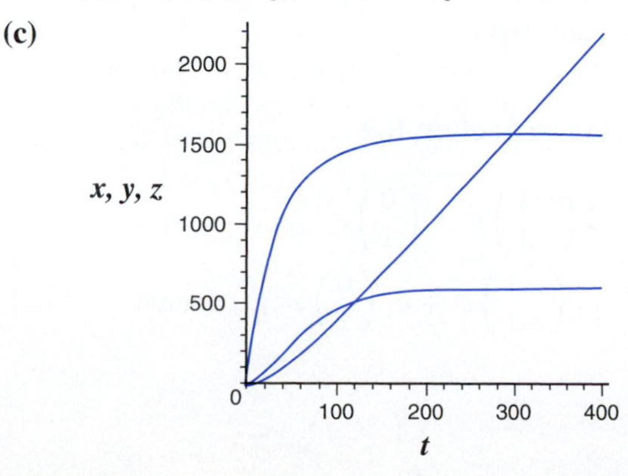

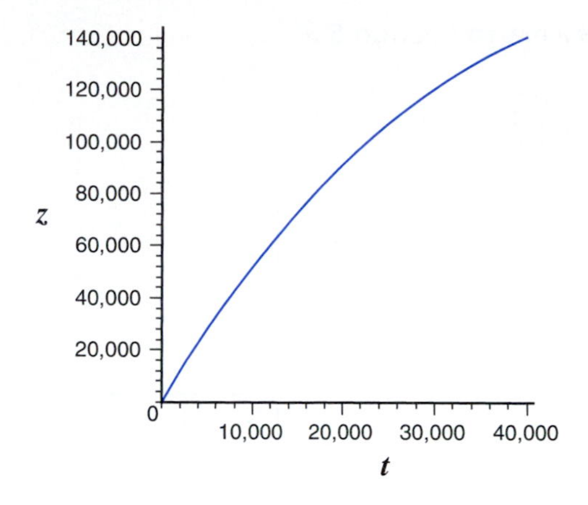

Answers to Section 6.2

1. **0** **3.** $c\begin{pmatrix}1\\1\\-3\end{pmatrix}$ **5.** $c\begin{pmatrix}0\\1\end{pmatrix}$ for $\lambda = 3$, $c\begin{pmatrix}-1\\1\end{pmatrix}$ for $\lambda = 1$ **7.** $c\begin{pmatrix}2\\1\end{pmatrix}$ for $\lambda = -2$, $c\begin{pmatrix}1\\1\end{pmatrix}$ for $\lambda = -3$ **9.** $c\begin{pmatrix}1\\-3\\2\end{pmatrix}$ for $\lambda = 1$, $c\begin{pmatrix}1\\-1\\2\end{pmatrix}$ for $\lambda = -1$, deficient **11.** $c\begin{pmatrix}2\\1\\3\end{pmatrix}$ for $\lambda = 4$, $c\begin{pmatrix}4\\-8\\1\end{pmatrix}$ for $\lambda = -1$, $c\begin{pmatrix}0\\0\\1\end{pmatrix}$ for $\lambda = 3$

13. $c\begin{pmatrix}-1\\0\\1\end{pmatrix}$ for $\lambda = 1$, $c\begin{pmatrix}-2\\1\\0\end{pmatrix}$ for $\lambda = 2$, $c\begin{pmatrix}0\\-1\\1\end{pmatrix}$ for $\lambda = 3$ **15.** $c\begin{pmatrix}-3\\1\\-3\end{pmatrix}$ for $\lambda = 1$, $c_1\begin{pmatrix}2\\0\\1\end{pmatrix} + c_2\begin{pmatrix}2\\1\\0\end{pmatrix}$ for $\lambda = 2$ **17.** $c\begin{pmatrix}1\\1\\-3\end{pmatrix}$ for $\lambda = 0$, $c\begin{pmatrix}0\\0\\1\end{pmatrix}$ for $\lambda = 1$, deficient

Answers to Section 6.3

1. $c\begin{pmatrix} 0 \\ 1 \end{pmatrix}$ and $c\begin{pmatrix} -1 \\ 1 \end{pmatrix}$

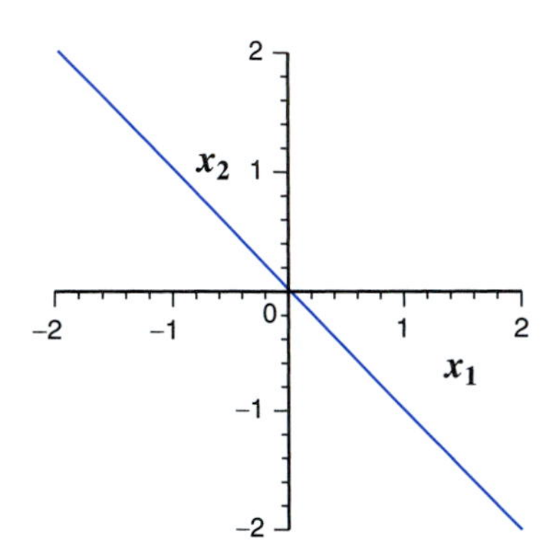

3. $c\begin{pmatrix} 2 \\ 1 \end{pmatrix}$ and $c\begin{pmatrix} 1 \\ 1 \end{pmatrix}$

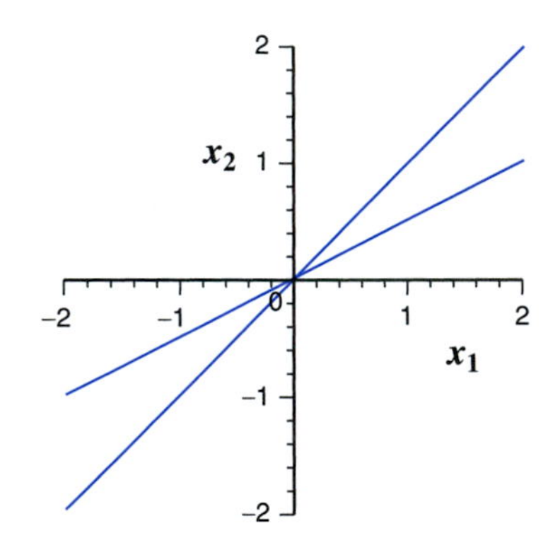

5. $c\begin{pmatrix} 1 \\ 0 \end{pmatrix}$ and $c\begin{pmatrix} 1 \\ -3 \end{pmatrix}$

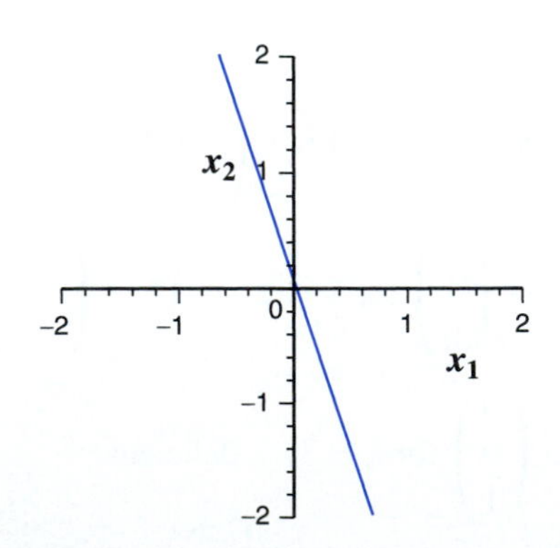

7. No linear trajectories

9. $c\begin{pmatrix} 5 \\ -2 \end{pmatrix}$ and $c\begin{pmatrix} 1 \\ 1 \end{pmatrix}$

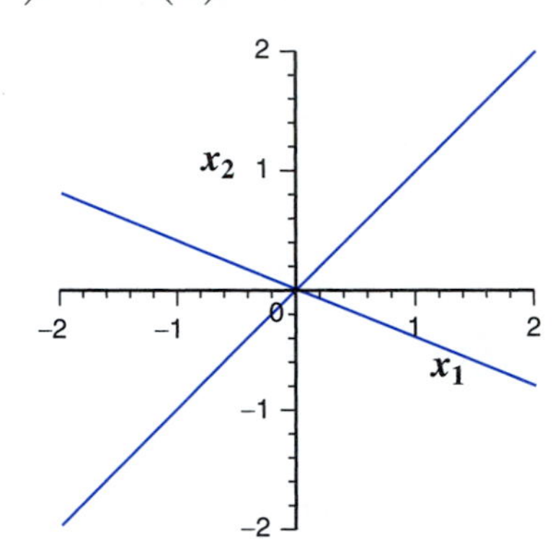

11. $c\begin{pmatrix} 1 \\ 1 \end{pmatrix}$ and $c\begin{pmatrix} 1 \\ -2 \end{pmatrix}$

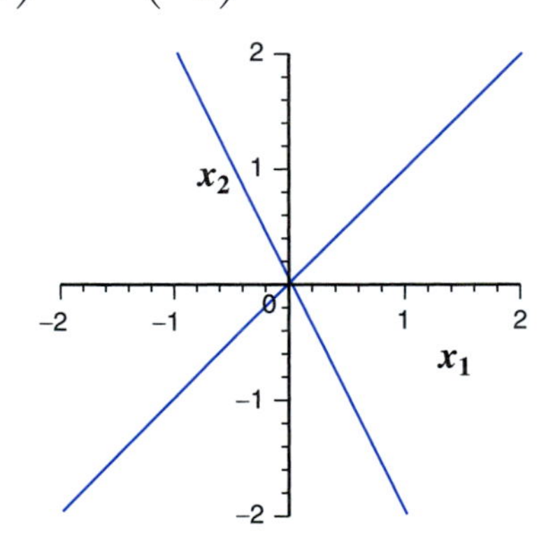

13. $c\begin{pmatrix} 1 \\ -3 \\ 2 \end{pmatrix}$ and $c\begin{pmatrix} 1 \\ -1 \\ 2 \end{pmatrix}$ **15.** $c\begin{pmatrix} -3 \\ 1 \\ -3 \end{pmatrix}$ and all lines in the plane $-x_1 + 2x_2 + 2x_3 = 0$ that pass through the origin

Answers to Section 6.4

1. $\mathbf{x} = 3\begin{pmatrix} 1 \\ -1 \end{pmatrix}e^{t} + \begin{pmatrix} 0 \\ 1 \end{pmatrix}e^{3t}$

3. $\mathbf{x} = c_1\begin{pmatrix} 1 \\ -1 \end{pmatrix}e^{t} + c_2\begin{pmatrix} 0 \\ 1 \end{pmatrix}e^{3t}$, source

5. $\mathbf{x} = c_1 \begin{pmatrix} 1 \\ 1 \end{pmatrix} e^{-3t} + c_2 \begin{pmatrix} 2 \\ 1 \end{pmatrix} e^{-2t}$, sink

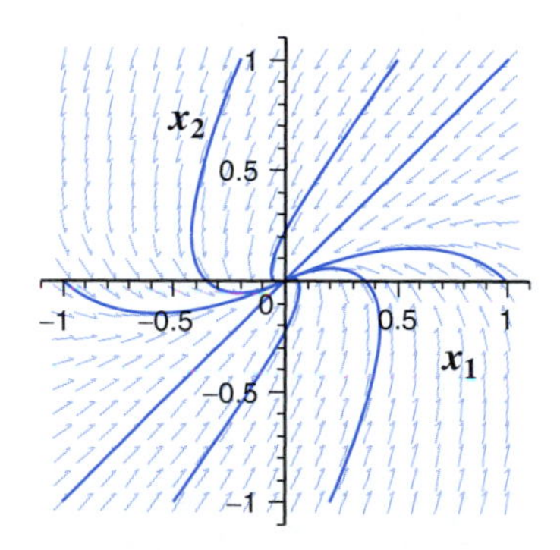

7. $\mathbf{x} = c_1 \begin{pmatrix} 1 \\ 0 \end{pmatrix} e^{2t} + c_2 \begin{pmatrix} 1 \\ -3 \end{pmatrix} e^{-t}$, saddle point

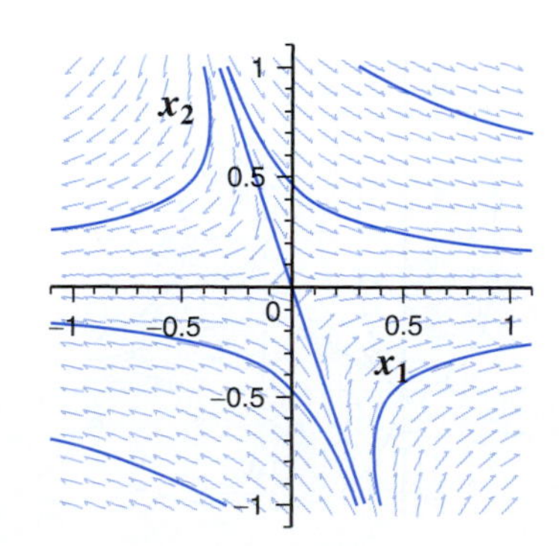

9. Saddle point

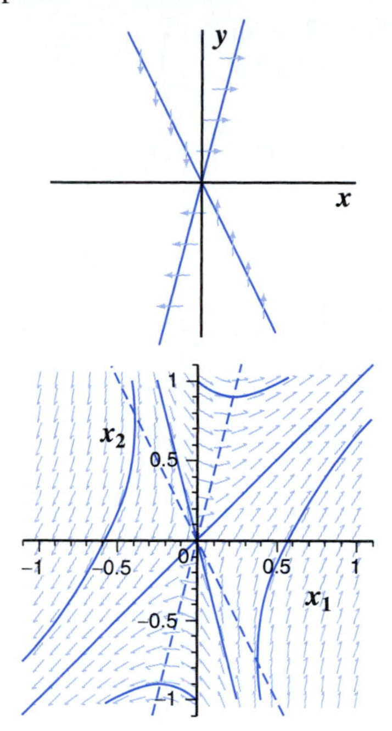

11. There aren't enough linear trajectories.

13. $\mathbf{x} = c_1 \begin{pmatrix} -3 \\ 1 \\ -3 \end{pmatrix} e^{t} + c_2 \begin{pmatrix} 2 \\ 0 \\ 1 \end{pmatrix} e^{2t} + c_3 \begin{pmatrix} 2 \\ 1 \\ 0 \end{pmatrix} e^{2t}$

15. $\mathbf{x} = c_1 \begin{pmatrix} 2 \\ 1 \\ 3 \end{pmatrix} e^{4t} + c_2 \begin{pmatrix} 0 \\ 0 \\ 1 \end{pmatrix} e^{3t} + c_3 \begin{pmatrix} 4 \\ -8 \\ 1 \end{pmatrix} e^{-t}$

Answers to Section 6.5

1. **(a)** $\lambda = \pm 2i$, stable

(b) $\mathbf{x} = \begin{pmatrix} \cos 2t - \sin 2t \\ -2\cos 2t - 2\sin 2t \end{pmatrix}$

3. **(a)** $\lambda = 3 \pm i$, unstable spiral

(b) $\mathbf{x} = e^{3t} \begin{pmatrix} 2\cos t + \sin t \\ \cos t - 2\sin t \end{pmatrix}$

5. **(a)** $\lambda = -2 \pm 3i$, asymptotically stable

(b) $\mathbf{x} = e^{-2t} \begin{pmatrix} -3\sin 3t \\ \cos 3t + \sin 3t \end{pmatrix}$

7. **(a)** $\lambda = 4 \pm 2i$ and $\lambda = -1$, unstable

(b) $\mathbf{x} = \begin{pmatrix} e^{4t}(\cos 2t - \sin 2t) \\ -2e^{4t} \sin 2t \\ e^{-t} \end{pmatrix}$

11. **(a)** $x'' = -(k_1 + k_2)x + k_2 y$, $y'' = k_2 x - (k_2 + k_3)y$

(b)

$$\mathbf{x}' = \mathbf{A}\mathbf{x}, \quad \mathbf{A} = \begin{pmatrix} 0 & 0 & 1 & 0 \\ 0 & 0 & 0 & 1 \\ -(k_1 + k_2) & k_2 & 0 & 0 \\ k_2 & -(k_2 + k_3) & 0 & 0 \end{pmatrix}$$

(c) $\lambda^2 = \dfrac{-(k_1 + 2k_2 + k_3) \pm \sqrt{(k_1 - k_3)^2 + 4k_2^2}}{2}$, so λ^2 is real. Since $\lambda^4 + (k_1 + 2k_2 + k_3)\lambda^2 + (k_1k_2 + k_1k_3 + k_2k_3) = 0$, both values of λ^2 are negative. Thus, all values of λ are purely imaginary. The origin is stable, but not asymptotically stable. **(d)** $x = c_1 \cos t + c_2 \sin t - c_3 \cos 3t - c_4 \sin 3t$, $y = c_1 \cos t + c_2 \sin t + c_3 \cos 3t + c_4 \sin 3t$ **(e)** If b is the equilibrium distance between the masses, then the distance between the masses is $b + y - x = b + 2c_3 \cos 3t + 2c_4 \sin 3t$, which is periodic with period $2\pi/3$. The midpoint is at a distance $(L + x + y)/2 = L/2 + c_1 \cos t + c_2 \sin t$ from the left end; this is periodic with period 2π.
(f) $x = -0.9 \cos t - 0.1 \cos 3t$, $x = -0.9 \cos t + 0.1 \cos 3t$

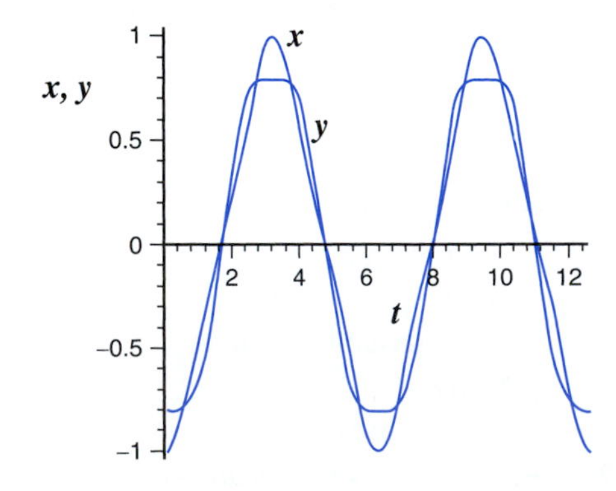

Answers to Section 6.6

1. $\mathbf{x} = e^t \begin{pmatrix} 1 + 2t \\ -2t \end{pmatrix}$ **3.** $\mathbf{x} = e^t \begin{pmatrix} 2 + 21t \\ 5 - 21t \end{pmatrix}$

5. $\mathbf{x} = e^{3t} \begin{pmatrix} 2 + t \\ 3 + t \end{pmatrix}$ **7.** $\mathbf{x} = \begin{pmatrix} e^{2t} - (1 + 2t)e^t \\ (1 + 2t)e^t \\ (1 - 2t)e^t \end{pmatrix}$

9. $\mathbf{x} = \begin{pmatrix} (1 + 2t)e^{-3t} \\ (1 + 2t)e^{-3t} - e^{-4t} \\ (1 - 2t)e^{-3t} \end{pmatrix}$

11. **(b)** Eigenvectors must satisfy $v_1 + v_2 + 2v_3 = 0$; by setting one of the coordinates to b, setting another to c, and computing the third, we obtain eigenvectors $\mathbf{u}$ and $\mathbf{v}$. One possibility is $\mathbf{u} = \begin{pmatrix} -1 \\ 1 \\ 0 \end{pmatrix}$ and $\mathbf{v} = \begin{pmatrix} -2 \\ 0 \\ 1 \end{pmatrix}$.

(c) Any function of the form

$\mathbf{x}^{(3)} = bte^{-t} \begin{pmatrix} 1 \\ 1 \\ -1 \end{pmatrix} + e^{-t}\mathbf{w}$ works, provided $b \neq 0$ and $w_1 + w_2 + 2w_3 = b$. **(e)** $\mathbf{x} = \begin{pmatrix} 1 + t \\ t \\ -t \end{pmatrix} e^{-t}$

Answers to Section 6.7

1. **(a)** (0, 0) and (4/3, 3/2)

(b) $J = \begin{pmatrix} -3 + 2y & 2x \\ 3y & -4 + 3x \end{pmatrix}$ **(c)** The origin is asymptotically stable (sink), and (4/3, 3/2) is unstable (saddle).

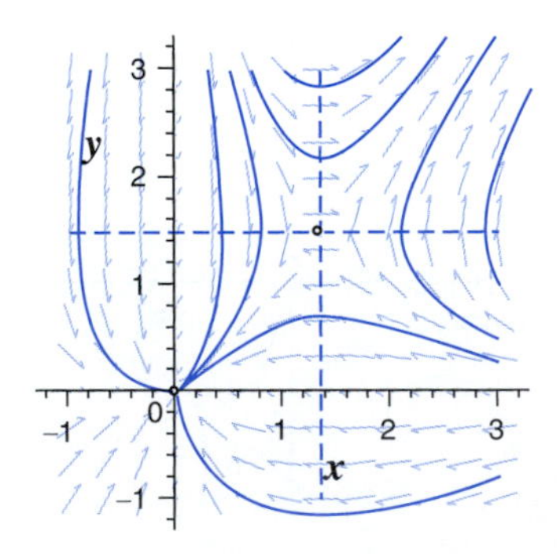

3. **(a)** (0, 0), (0, 3), and (2, 0)

(b) $J = \begin{pmatrix} 2 - 2x - y & -x \\ -y & 3 - x - 2y \end{pmatrix}$ **(c)** The origin is unstable (source), (0, 3) is asymptotically stable (sink), and (2, 0) is unstable (saddle).

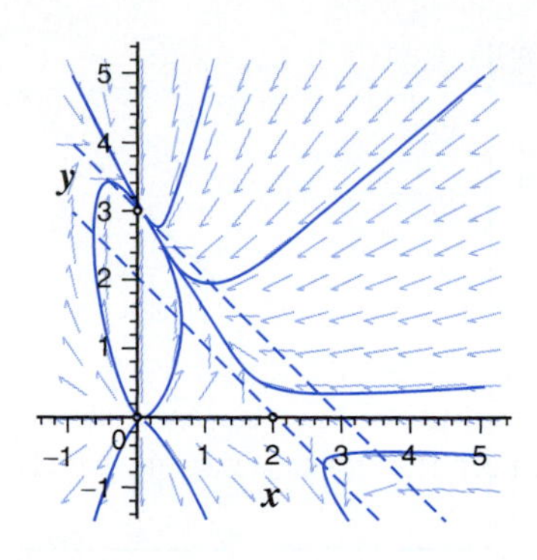

5. **(a)** $(0, 0)$, $(1, 0)$, and $(-1, 0)$

(b) $J = \begin{pmatrix} 0 & 1 \\ 1 - 3x^2 & -1 \end{pmatrix}$

(c) The origin is unstable (saddle), $(1, 0)$ and $(-1, 0)$ are asymptotically stable (spiral).

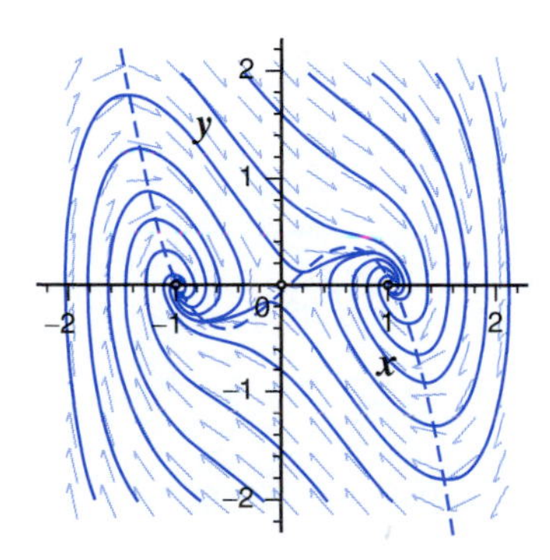

7. Asymptotically stable

11. **(a)** $(0, 0, 0)$, $(1, 0, 0)$, $(0, 1, 0)$, $(1, 1, 0)$, $(0.5, 0, 0.5)$, $(0, 0.5, 0.5r)$, and $\left(\frac{2-r}{2(1+r)}, \frac{2r-1}{2(1+r)}, \frac{3r}{2(1+r)}\right)$.
The last of these is valid only for $0.5 < r < 2$.

(b) $J = \begin{pmatrix} 1 - 2x - z & 0 & -x \\ 0 & r - 2ry - z & -y \\ 2z & 2z & 2x + 2y - 1 \end{pmatrix}$

(d) XZ is asymptotically stable if $r < 0.5$ and YZ is asymptotically stable if $r > 2$.

(e) XYZ is asymptotically stable for $0.5 < r < 2$, its full range of existence.

Answers to Section 7.1

1.

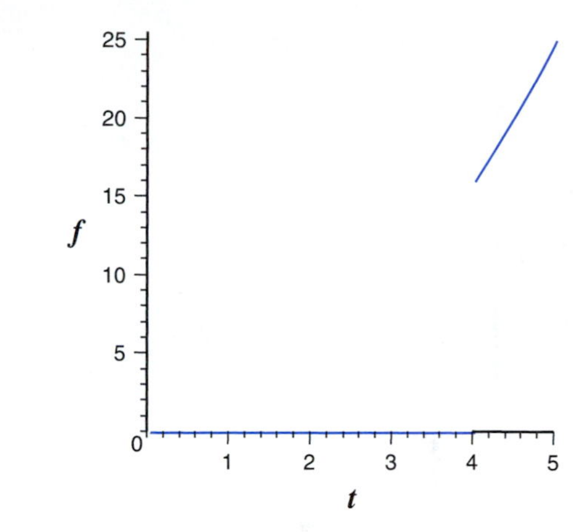

$f(t) = H(t - 4)t^2$

3.

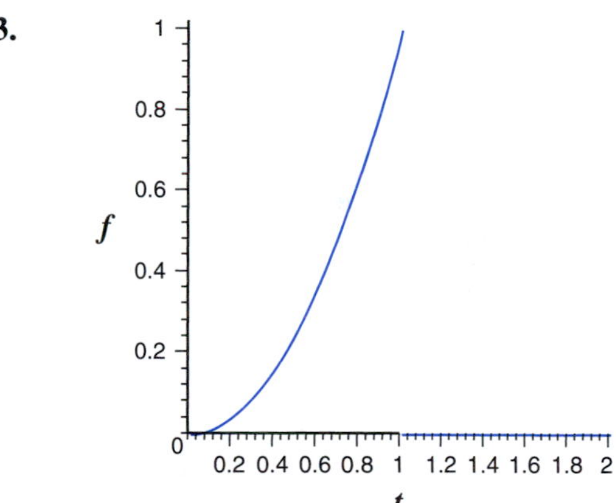

$f(t) = [1 - H(t - 1)]t^2$

5.

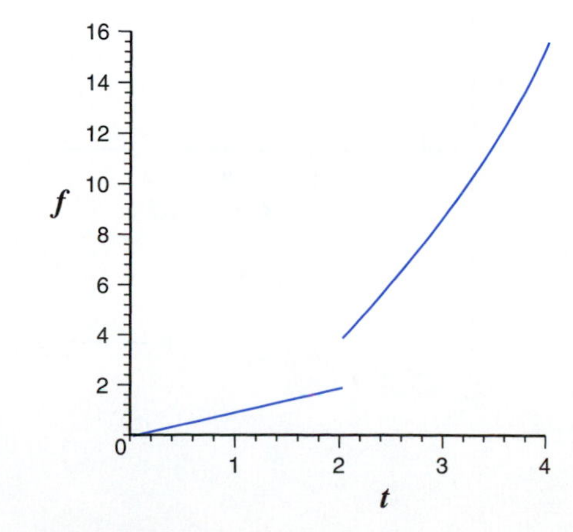

$f(t) = [1 - H(t-2)]t + H(t-2)t^2$

7.

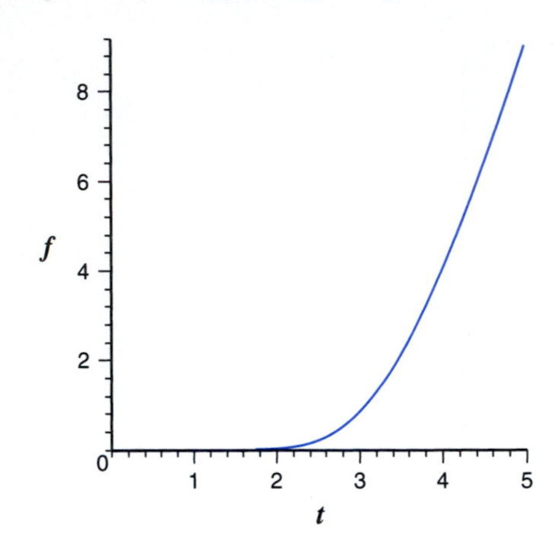

$f(t) = H(t-2)(t-2)^2$

9.

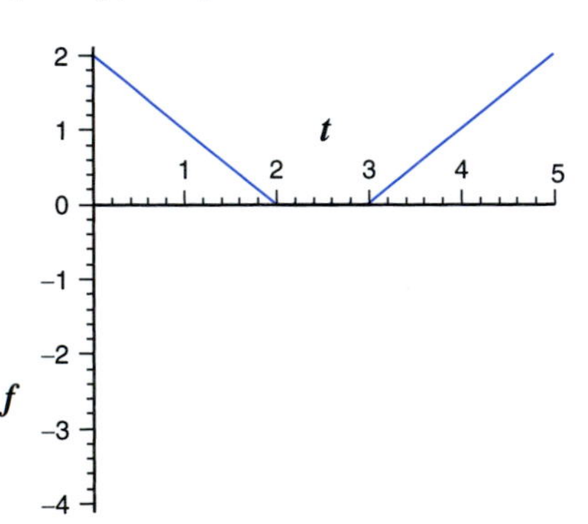

$f(t) = [1 - H(t-2)](2-t) + H(t-3)(t-3)$

11.

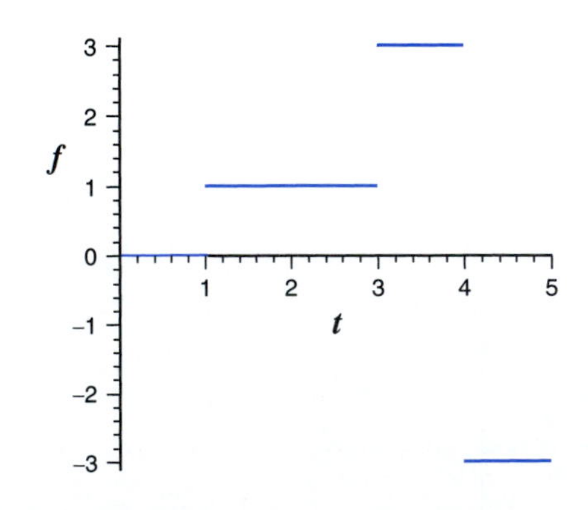

13.

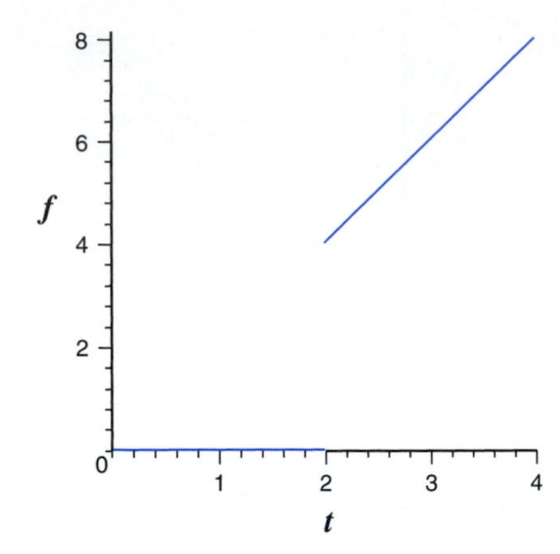

15. $f_{sq}(t) = 1 + 2\sum_{n=1}^{\infty}(-1)^n H(t-n)$

Answers to Section 7.2

1. $\mathcal{L}[f] = \dfrac{24}{s^5} - \dfrac{18}{s^4} + \dfrac{1}{s^2}$

3. $\mathcal{L}[f] = \dfrac{3}{s^2+9} + 2\sqrt{\dfrac{\pi}{s^3}}$

9. $\mathcal{L}[f] = \dfrac{1}{s}\left(e^{-s} + 2e^{-3s} - 6e^{-4s}\right)$

11. $\mathcal{L}[f] = \dfrac{2e^{-s}}{s^2}$

13. $\mathcal{L}[f] = 2e^{-4s}\left(\dfrac{1}{s^3} + \dfrac{4}{s^2} + \dfrac{8}{s}\right)$

15. $\mathcal{L}[f] = \dfrac{2}{s^3} - e^{-s}\left(\dfrac{2}{s^3} + \dfrac{2}{s^2} + \dfrac{1}{s}\right)$

17. $\mathcal{L}[f] = \dfrac{1}{s^2} + e^{-2s}\left(\dfrac{2}{s^3} + \dfrac{3}{s^2} + \dfrac{2}{s}\right)$

19. $\mathcal{L}[f] = \dfrac{2e^{-2s}}{s^3}$

21. $\mathcal{L}[f] = \dfrac{2}{s} - \dfrac{1}{s^2}\left(1 - e^{-2s} - e^{-3s}\right)$

23. $\mathcal{L}[f] = \dfrac{1}{s^2} - \dfrac{1}{s(e^s - 1)}$

25. $\mathcal{L}\left[f''\right] = s^2F(s) - sf(0) - f'(0)$

Answers to Section 7.3

1. $\mathcal{L}^{-1}[F] = \frac{3}{2}\sin 2t$ **3.** $\mathcal{L}^{-1}[F] = \frac{4}{5}e^{t} - \frac{4}{5}e^{-4t}$
5. $\mathcal{L}^{-1}[F] = 2e^{-t}\cos 2t$
7. $\mathcal{L}^{-1}[F] = e^{t}(2\cos t + 3\sin t)$
9. $\mathcal{L}^{-1}[F] = e^{-2t}(-2\cos t + 5\sin t)$
11. $y = 2e^{-t} - e^{-2t}$ **13.** $y = \cos(\sqrt{2}\,t)$

Answers to Section 7.4

1. $\mathcal{L}^{-1}[G] = \frac{1}{3}\left(e^{t-2} - e^{-2(t-2)}\right)H(t-2)$

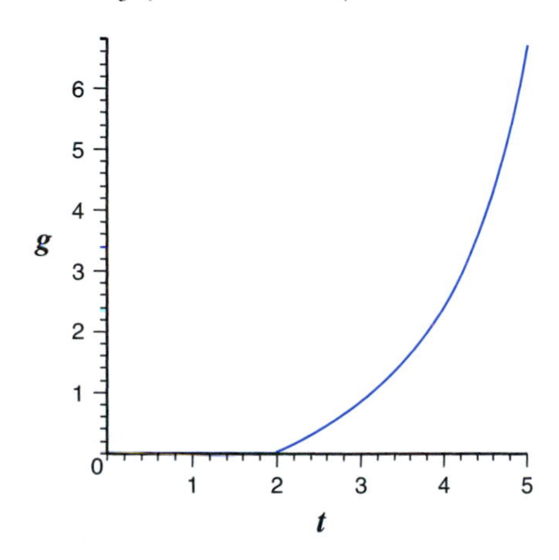

3. $\mathcal{L}^{-1}[G] = \frac{1}{2}\left(e^{3(t-1)} + e^{t-1}\right)H(t-1)$

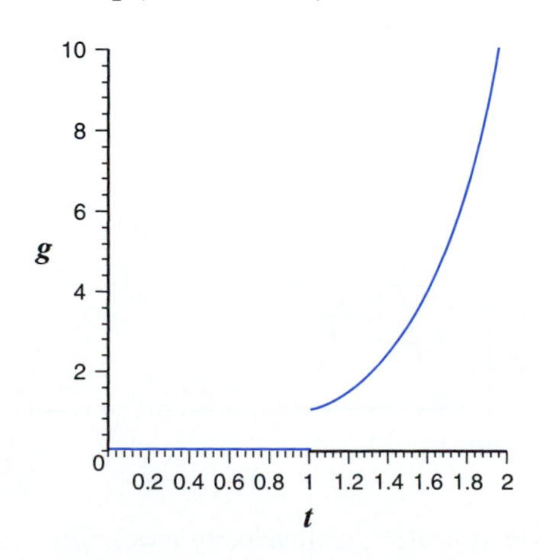

5. $y = \cos t + (1 + \cos t)H(t - 3\pi)$

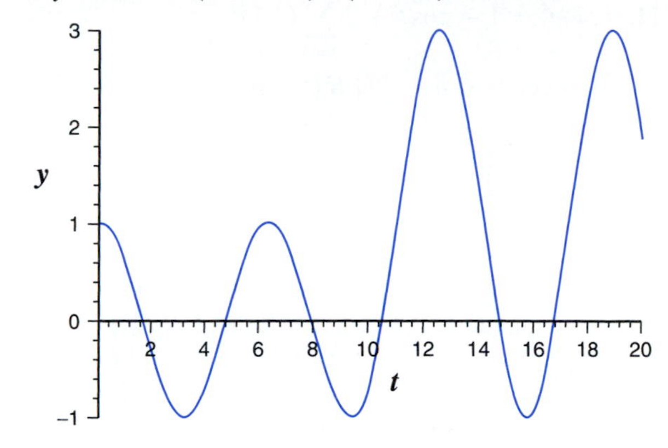

7. $y = t + (\pi - t - \sin t)H(t - \pi)$

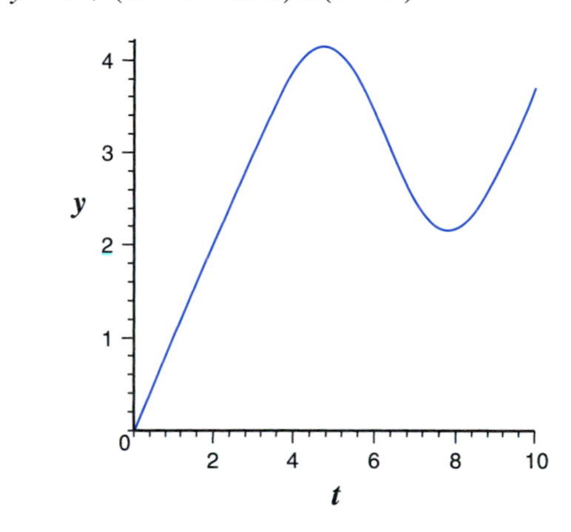

9. $y = e^{-t}(\cos t + \sin t) - e^{\pi - t}(\sin t)H(t - \pi)$

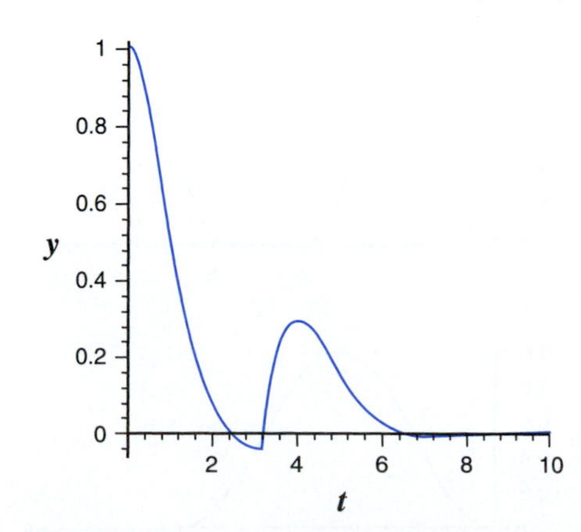

11. $i = v' = 1 - \cos t + 2\sum_{n=1}^{\infty}(-1)^n$
$\left[1 - \cos\left(t - n + \frac{1}{2}\right)\right] H\left(t - n + \frac{1}{2}\right)$

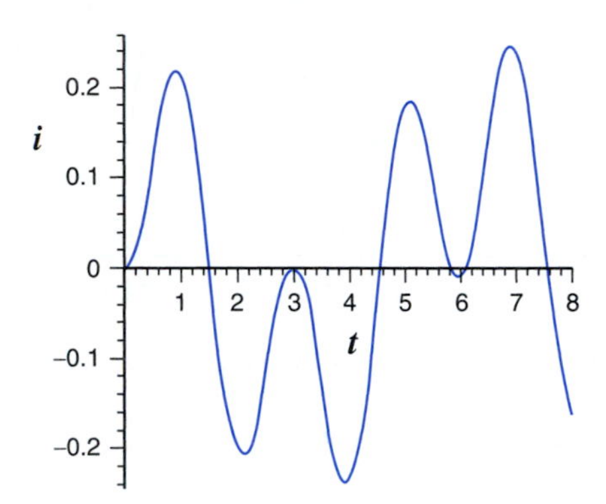

Answers to Section 7.5

1. $y = \frac{t}{2}\sin 2t + \frac{1}{4}\cos 2t\,\ln(\cos 2t)$

3. $y = \frac{1}{p^2 - 3p + 2}t^{2-p}e^{-t}$

5. The solution and graph are identical to those of Exercise 11 of Section 7.4.

Answers to Section 8.1

1. **(a)** Yes

(b) $u = \begin{cases} \sin^3\left[\pi\left(\frac{t}{2} - \frac{x}{4}\right)\right] & \frac{x}{2} \le t \le \frac{x}{2} + 2 \\ 0 & \text{otherwise} \end{cases}$

(c)

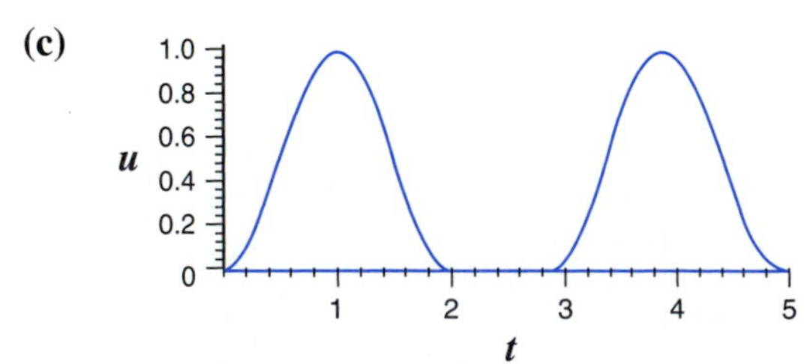

(d)

3. **(a)** No

(b) $u = \begin{cases} t - \frac{x}{2} - \left(t - \frac{x}{2}\right)^2 & \frac{x}{2} \le t \le \frac{x}{2} + 1 \\ 0 & \text{otherwise} \end{cases}$

(c)

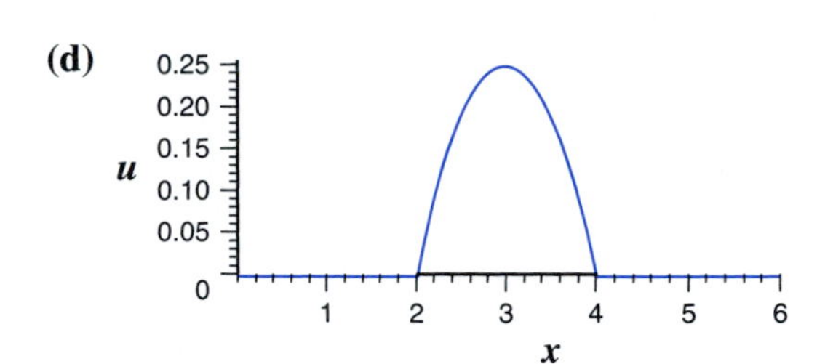

(d)

5. **(a)** Yes

(b) $u = \begin{cases} [1 - \cos(\pi t - \pi x)]^3 & x \le t \le x + 2 \\ 0 & \text{otherwise} \end{cases}$

(c)

(d)

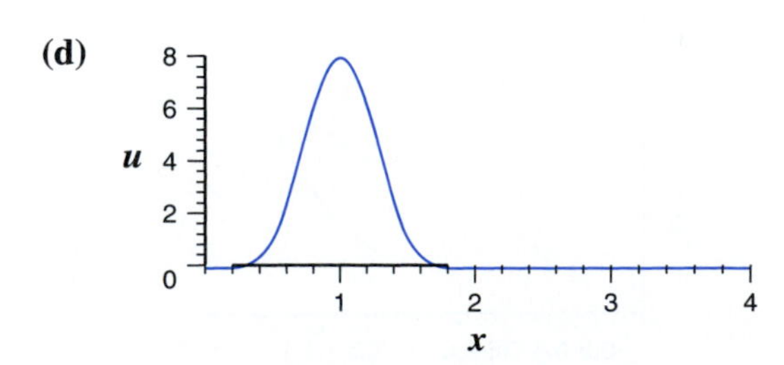

7. $f(t) = A + Be^{at}$, with velocity $v = -ak$

9. **(a)** $(c^2 - v^2)gf'' + c^2\left(2g' + \frac{g}{r}\right)f'$ $+ c^2\left(g'' + \frac{g'}{r}\right)f = 0$ **(b)** $v = \pm c$ **(c)** $2rg' + g = 0$ and $rg'' + g' = 0$ **11.** **(a)** $f''' + 6vff' - vf' = 0$ **(b)** $f'^2 + 2vf^3 - vf^2 = 2c_1 f + c_2$ **(c)** $f(0) = \frac{1}{2}$ **(d)** $f(z) = \frac{1}{2}\operatorname{sech}^2\frac{\sqrt{v}z}{2}$ The wave speed can be any positive value.

(e)

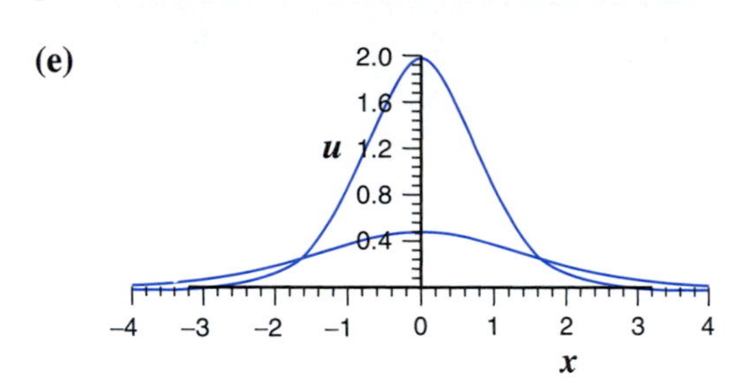

Answers to Section 8.2

1. **(a)** $u = \dfrac{1}{1+(x-t)^2} + \dfrac{1}{1+(x+t)^2}$

(b)

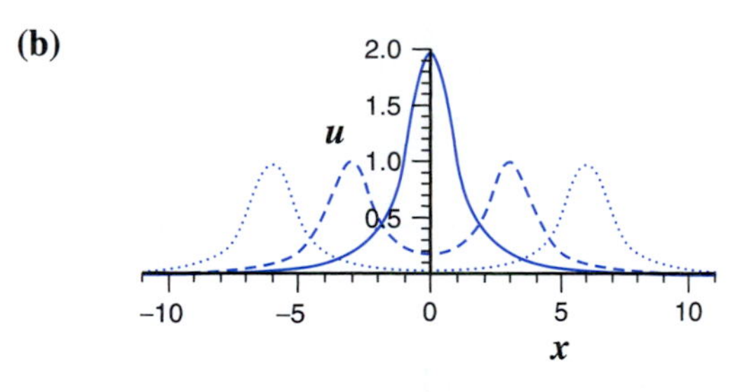

3. **(a)** $u = \dfrac{1}{2}\begin{cases} 1-(x-2t)^2 & 2t-1 \le x \le 2t+1 \\ 0 & \text{otherwise} \end{cases}$ $+ \dfrac{1}{2}\begin{cases} 1-(x+2t)^2 & -2t-1 \le x \le -2t+1 \\ 0 & \text{otherwise} \end{cases}$

(b)

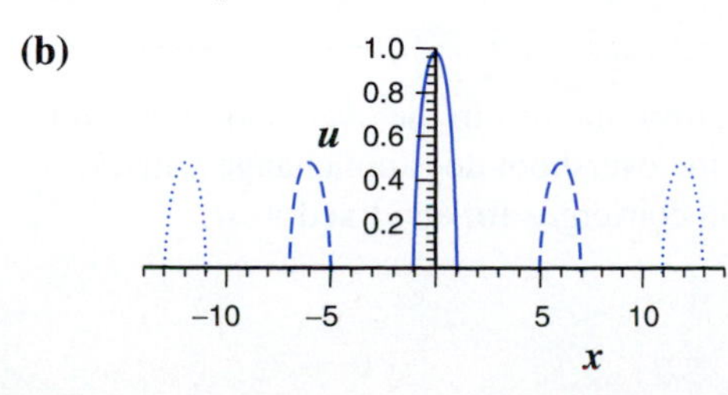

5. **(a)** $u = \frac{1}{4}\ln[1+(x+t)^2] - \frac{1}{4}\ln[1+(x-t)^2]$

(b)

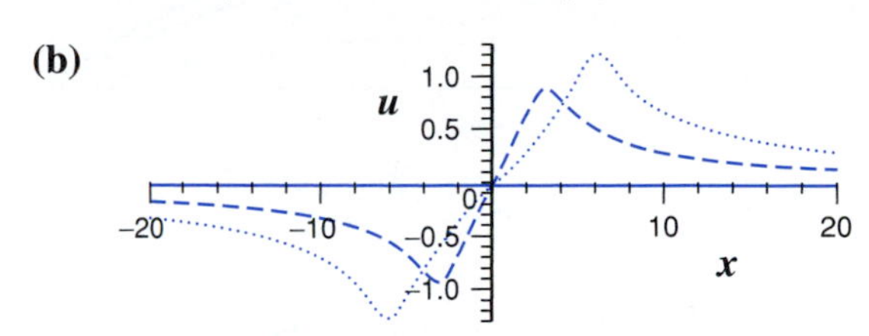

7. **(a)**

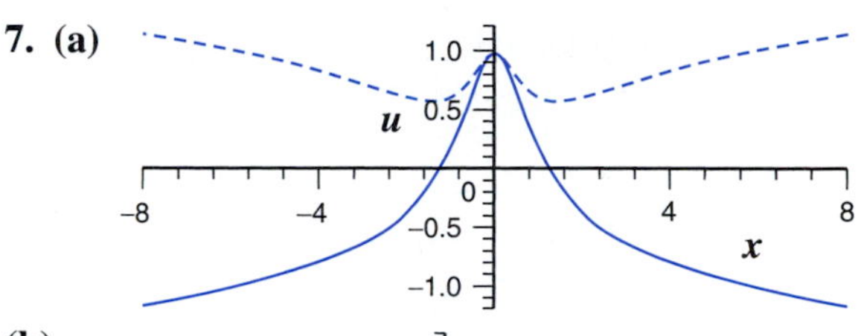

(b)

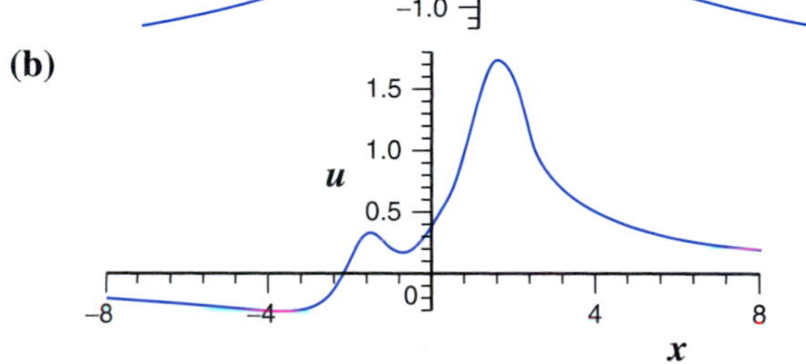

9. **(a)** $f'' + 2zf' = 0$
(b) $f = A + B\operatorname{erf} z$ or $f = A\operatorname{erfc} z + B\operatorname{erf} z$
(c) g must be constant and f must be identically 0

Answers to Section 8.3

1. 554.4 m/s

5. $u_n = B_n e^{-n^2\pi^2 k/L^2} \sin\dfrac{n\pi x}{L}, \quad n = 1, 2, \ldots$

7. $u_n = B_n e^{-(2n-1)^2\pi^2 k/(4L^2)} \sin\dfrac{(2n-1)\pi x}{2L},$ $n = 1, 2, \ldots$

9. $u_n = \dfrac{B_n}{r}\cos\dfrac{n\pi ct}{a}\sin\dfrac{n\pi r}{a}, \quad n = 1, 2, \ldots$

Answers to Section 8.4

5. **(a)** $u = \sum_{n=1}^{\infty} b_n \cos n\pi t \sin n\pi x,$

$$b_n = \frac{8}{n^3\pi^3} \text{ for } n \text{ odd}, \quad b_n = 0 \text{ for } n \text{ even}$$

(b)

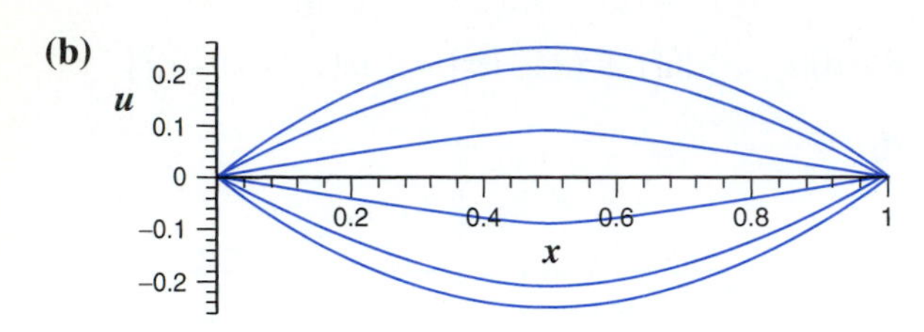

7. (a) $u = \sum_{n=2}^{\infty} b_n \cos 2n\pi t \sin n\pi x, \quad b_n = \dfrac{24}{n^3\pi^3}$ for n even, $\quad b_n = 0$ for n odd

(b)

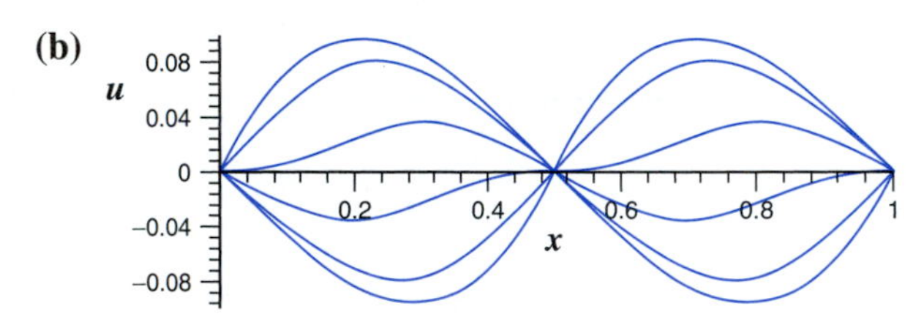

9. (a) $u = \dfrac{100}{9\pi^2} \sum_{n=1}^{\infty} \dfrac{\sin(0.9n\pi)}{n^2} \cos n\pi t \sin n\pi x$

(b)

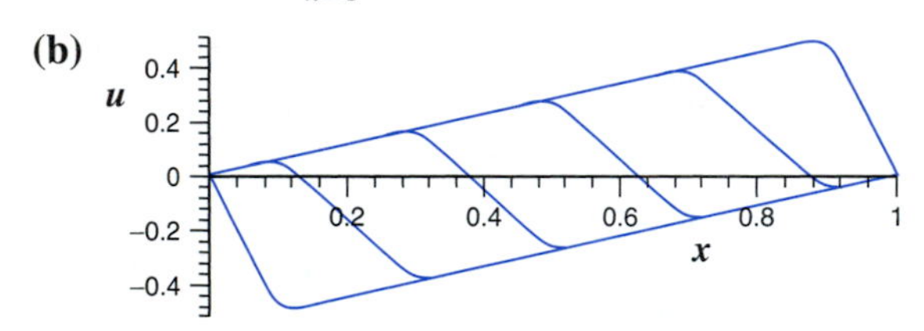

11.

$$g_n(t) = \sin \frac{n\pi ct}{L}, \quad b_n = \frac{2}{n\pi c} \int_0^L \psi(x) \sin \frac{n\pi x}{L}\, dx$$

or

$$g_n(t) = \frac{L}{n\pi c} \sin \frac{n\pi ct}{L}, \quad b_n = \frac{2}{L} \int_0^L \psi(x) \sin \frac{n\pi x}{L}\, dx$$

13. (a) $u = -\dfrac{4}{\pi^4} \sum_{n=1}^{\infty} \dfrac{1 + 2(-1)^n}{n^4} \sin n\pi t \sin n\pi x$

(b)

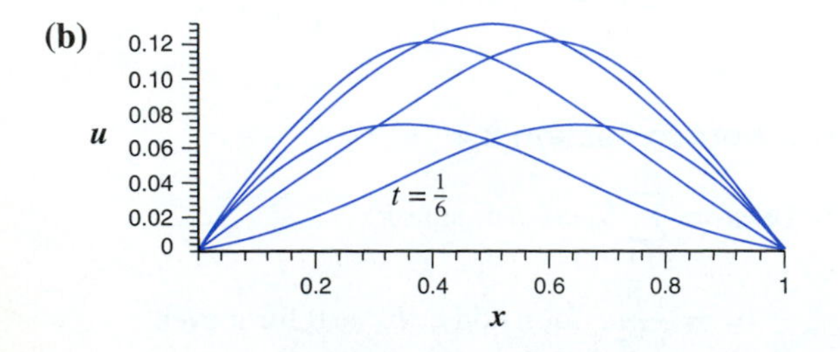

Answers to Section 8.5

1. $\bar{f} = \dfrac{\pi}{2}, \qquad a_n = -\dfrac{4}{n^2\pi}, \qquad n = 1, 3, \ldots$

All other coefficients are 0.

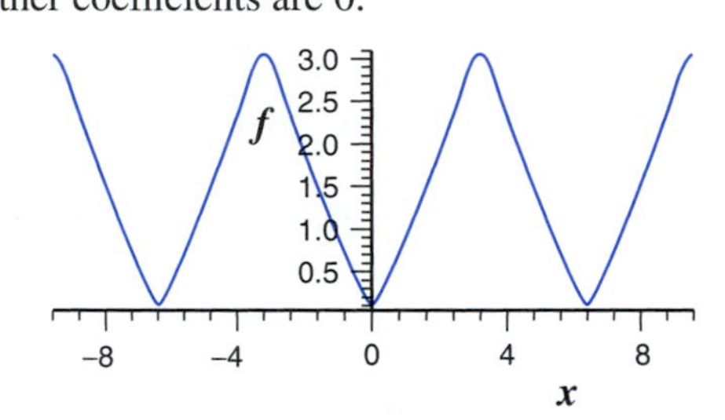

3. $\bar{f} = \dfrac{2}{3}, \quad a_n = \dfrac{4(-1)^{n+1}}{n^2\pi^2}, \quad b_n = 0$

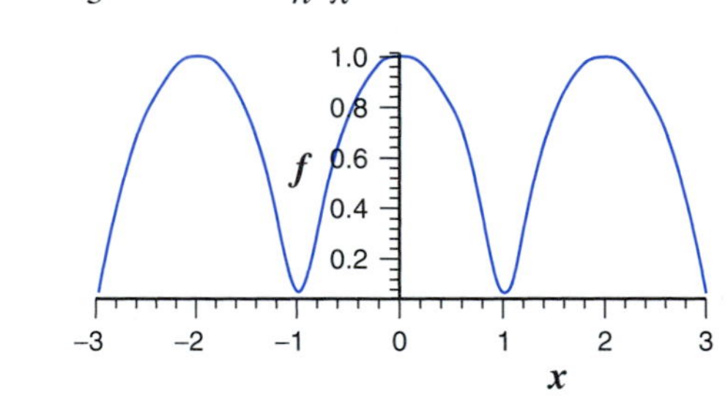

5.

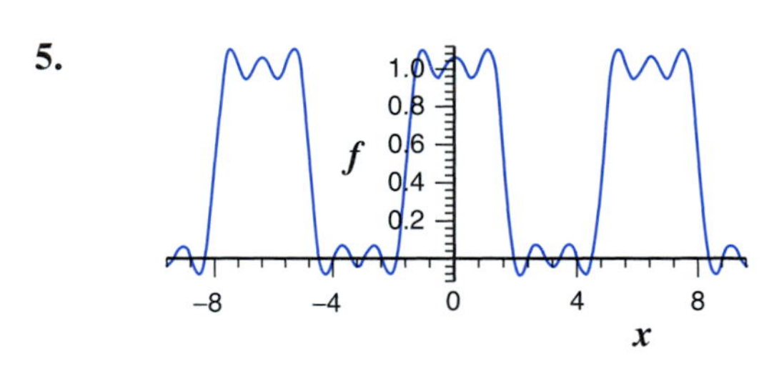

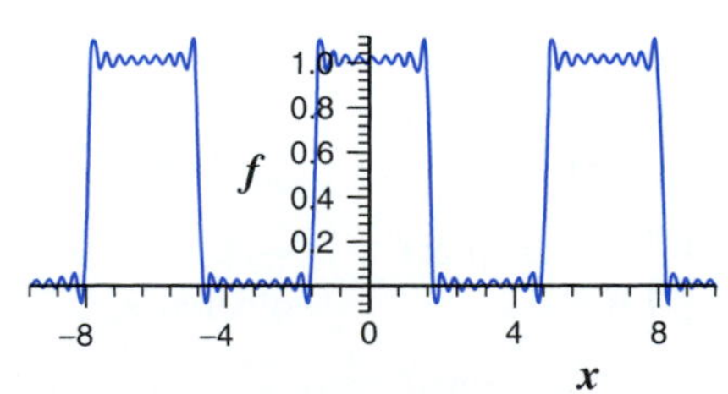

The number of oscillations in the overshoot is N, and the height of the overshoot does not change with N. The series still converges for any fixed x.

7. (a) $u = \sum_{n=1}^{\infty} b_n \cos n\pi t \sin n\pi x, \quad b_n = \dfrac{4}{n\pi}$ for n odd, $\quad b_n = 0$ for n even

(b)

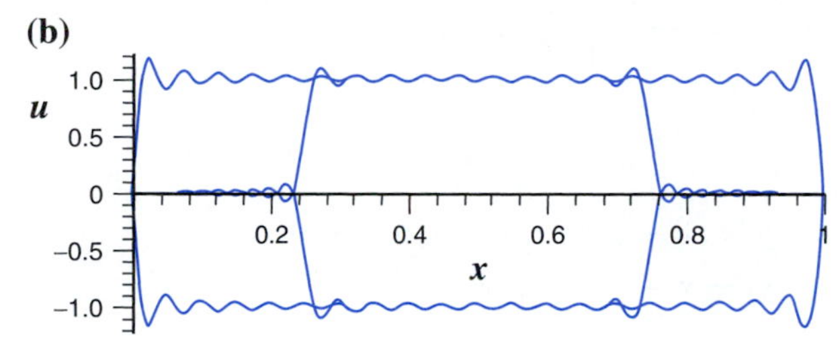

9. (a) $u = \sum_{n=1}^{\infty} b_n \cos(n\pi t)\sin(n\pi x), \quad b_n = \dfrac{4}{n^2\pi^2}$ for n odd, $\quad b_n = 0$ for n even

(b)

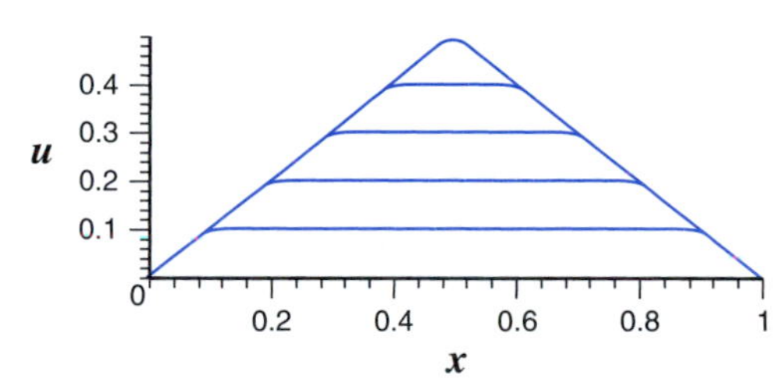

11. $u = \bar{f} + \sum_{n=1}^{\infty} r^n (a_n \cos n\theta + b_n \sin n\theta)$, with

$$\bar{f} = \frac{1}{2\pi}\int_0^{2\pi} f(\theta)\,d\theta, \quad a_n = \frac{1}{\pi}\int_0^{2\pi} f(\theta)\cos n\theta\,d\theta,$$

$b_n = \dfrac{1}{\pi}\displaystyle\int_0^{2\pi} f(\theta)\sin n\theta\,d\theta$ The integrals can be taken over any interval of length 2π.

Answers to Section A.1

1. $y = Ce^{4t} - \left(\frac{1}{9} + \frac{1}{3}t\right)e^t$ **3.** $y = Ce^{4t} + \frac{1}{2}t^2 e^{4t}$

5. $y = Ce^{3t} + \frac{1}{10}\sin t - \frac{3}{10}\cos t$

7. $y = Ce^{-t} + e^{-t}\ln(1 + e^t)$ **9.** $y = 2 + Ce^{-t^2/2}$

11. $y = Ce^{-t^2} + e^{-t^2}\sin t$

13. $y = Ct^{-1}e^{-t} + t^{-1}$ **15.** $y = \tan t + C\sec t$

17. $y = 2t^5 - 2t^2$ **19.** $y = (1 + t^2)\arctan t$

21. $y = \cos^2 t + \cos t \sin t$

23. $y = e^{4t^2}\left(1 + \dfrac{\sqrt{\pi}}{4}\operatorname{erf} 2t\right)$

25. (a) $y = Ct^{-m} + kt^{-m}\ln t$
(b) $y = Ct^{-m} + \dfrac{k}{m+n}t^n$ **(c)** The solution is defined for all t if $C = 0$. This does not contradict Theorem 2.4.1 because the theorem never asserts that there is no solution. **(d)** No

27. (a) $y = k\sec t + \dfrac{1}{2}\sin t \tan t$ on $-\dfrac{\pi}{2} < t < \dfrac{\pi}{2}$
(b) $-\frac{1}{2} \le k \le 0$ **(c)** $k \le -1$

Answers to Section A.2

1. $y_1 = t, \quad y_2 = t - \dfrac{t^3}{3}, \quad y_3 = t - \dfrac{t^3}{3} + \dfrac{t^5}{15}$

3. $y_1 = \dfrac{t^3}{3}, \quad y_2 = \dfrac{t^3}{3} + \dfrac{t^7}{63},$

$$y_3 = \frac{t^3}{3} + \frac{t^7}{63} + \frac{2t^{11}}{2079} + \frac{t^{15}}{59{,}535}$$

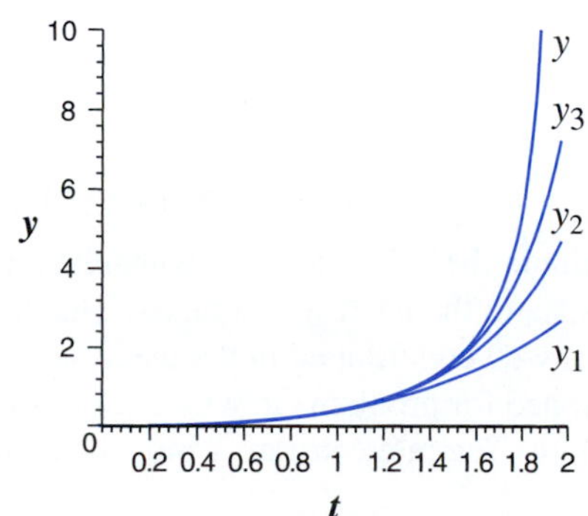

9. (c) $\dfrac{2a - b}{1 - b}$ **(d)** $a(2a - b) = b(1 - b)$

(f) $\dfrac{1 + 2\sqrt{2}}{7}$

Answers to Section A.3

1. (a) $K = 8/9$ and $M = 136/27$
(b), (c)

t	Actual error	Predicted error	Ratio
0.2	0.009	0.014	0.64
0.4	0.014	0.030	0.47
0.6	0.018	0.050	0.36
0.8	0.020	0.073	0.27
1.0	0.021	0.101	0.21

The problem is well conditioned, so propagated errors decrease. Also, $f_t + ff_y$ decreases in magnitude with t, so the actual local truncation errors are smaller for larger t.

3. (a) $y(t_1) = y(t_0) + \dfrac{h}{2}[y'(t_0) + y'(t_1)] + \dfrac{h^2}{4}[y''(t_0) - y''(t_1)] + \dfrac{h^3}{12}[y'''(\tau_1) + y'''(\tau_2)]$, where τ_1 and τ_2 are unknown points.
(b) $y(t_1) = y_0 + \dfrac{h}{2}[f(t_0, y_0) + f(t_1, y(t_1))] + \dfrac{h^2}{4}[y''(t_0) - y''(t_1)] + \dfrac{h^3}{12}[y'''(\tau_1) + y'''(\tau_2)]$
(c) $E_1 = \dfrac{h}{2}[f(t_0, y_0) - f(t_1, y(t_1))] + \dfrac{h^2}{4}[y''(t_1) - y''(t_0)] - \dfrac{h^3}{12}[y'''(\tau_1) + y'''(\tau_2)]$
(d) $E_1 = \dfrac{3y'''(\tau_3) - y'''(\tau_1) - y'''(\tau_2)}{6[2 - hf_y(t_1, \eta)]}h^3$, where τ_3 and η are also unknown points.

5. (a) $|E_n| \le \dfrac{Mh^2}{2}\sum_{i=0}^{n-1}(1 + Kh)^i$
(b) $|E_n| \le \dfrac{Mh}{2K}[(1 + Kh)^n - 1]$
(d) $|E_n| \le \dfrac{M}{2K}\left(e^{Kt_n} - 1\right)h$ **7. (a)** 2:59 A.M., to the nearest minute **(b)** 2:52 A.M., to the nearest minute **(c)** 2:26 A.M., to the nearest minute **(d)** The differential equation is well conditioned in the usual sense, so it is ill conditioned for problems in which time is run backward. **(e)** The rate constant k must be assumed.

Answers to Section A.4

1. $a_n = \dfrac{(n-2)a_{n-2} - a_{n-3}}{2n(n-1)}, \quad n \ge 3$
$y = -1 + 2x + \frac{1}{4}x^3 - \frac{1}{12}x^4 + \frac{3}{160}x^5 - \frac{7}{720}x^6 + \cdots$

3. $a_n = \dfrac{-(n-1)(n-2)a_{n-1} + 3a_{n-2}}{n(n-1)}, \quad n \ge 2$
$y = 2 - x + 3x^2 - \frac{3}{2}x^3 + \frac{3}{2}x^4 - \frac{9}{8}x^5 + \cdots$

5. $a_n = -\dfrac{n-3}{n-1}a_{n-2}, \quad n \ge 2$
$y_1 = 1 + x^2 - \frac{1}{3}x^4 + \frac{1}{5}x^6 + \cdots, \quad y_2 = x$

7. $a_n = -\dfrac{1}{n}a_{n-2}, \quad n \ge 2$
$y_1 = 1 - \frac{1}{2}x^2 + \frac{1}{8}x^4 - \frac{1}{48}x^6 + \cdots,$
$y_2 = x - \frac{1}{3}x^3 + \frac{1}{15}x^5 - \frac{1}{105}x^7 + \cdots$

9.
$$y = c_1\left\{1 - \frac{\nu + \frac{1}{2}}{2}x^2 + \left[\frac{1}{48} + \frac{(\nu + \frac{1}{2})^2}{24}\right]x^4 + \cdots\right\} + c_2\left\{x - \frac{\nu + \frac{1}{2}}{6}x^3 + \left[\frac{1}{80} + \frac{(\nu + \frac{1}{2})^2}{120}\right]x^5 + \cdots\right\}$$

11. (a) $y = 1 + mx + \dfrac{(m-1)m}{2}x^2 + \cdots + \dfrac{(m-n+1)(m-n+2)\cdots m}{n!}x^n + \cdots$
(b) $y = (1 + x)^m$ **13.** Any positive value other than $n/2$ with n an integer

15. $y_1 = \sqrt{x}\left(1 - \dfrac{x^2}{3!} + \dfrac{x^4}{5!} + \cdots\right) = \dfrac{1}{\sqrt{x}}\sin x,$
$y_2 = \dfrac{1}{\sqrt{x}}\left(1 - \dfrac{x^2}{2!} + \dfrac{x^4}{4!} + \cdots\right) = \dfrac{1}{\sqrt{x}}\cos x$

17. The indicial equation has roots 1 and $-\frac{1}{2}$. These yield two solutions,
$$y_1 = x\left(1 + \frac{x}{1\cdot 5} + \frac{x^2}{1\cdot 5\cdot 2\cdot 7} + \frac{x^3}{1\cdot 5\cdot 2\cdot 7\cdot 3\cdot 9} + \cdots\right)$$
and
$$y_2 = \frac{1}{\sqrt{x}}\left(1 - x - \frac{x^2}{2} - \frac{x^3}{2\cdot 3\cdot 3} - \frac{x^4}{2\cdot 3\cdot 3\cdot 4\cdot 5} - \cdots\right).$$

19. (a) $\sum_{n=0}^{\infty}\left[(\alpha + n)^2 + 2(\alpha + n) + 1\right]a_n x^n = \sum_{n=-1}^{\infty}(\alpha + n + 1)a_{n+1}x^n$ **(b)** $\alpha = 0$ **(c)** $a_n = na_{n-1}$
(d) $y = \sum_{n=0}^{\infty} n!x^n$

(e) The series does not converge. **(f)** No
(g) The indicial equation is not defined because $q(0)$ is not defined.

Answers to Section A.5

1. $\boldsymbol{\Theta}^{-1} = \begin{pmatrix} 1-2t & -4t \\ t & 1+2t \end{pmatrix}$

3. $\boldsymbol{\Psi}^{-1} = \begin{pmatrix} (1+t)e^{2t} & -te^{2t} \\ -e^{2t} & e^{2t} \end{pmatrix}$

5. $\mathbf{A}^{-1} = \dfrac{1}{2}\begin{pmatrix} -1 & 1 & -1 \\ 1 & 1 & 1 \\ -1 & 1 & 1 \end{pmatrix}$

7. $\mathbf{A}^{-1} = \dfrac{1}{4}\begin{pmatrix} -1 & -1 & 1 \\ 3 & -5 & 1 \\ 2 & -2 & -2 \end{pmatrix}$

9. $\boldsymbol{\Theta} = \begin{pmatrix} e^{2t} & te^{2t} \\ 0 & e^{2t} \end{pmatrix}$

11. $\boldsymbol{\Theta} = \begin{pmatrix} \cos t - 2\sin t & \sin t \\ -5\sin t & \cos t + 2\sin t \end{pmatrix}$

Answers to Section A.6

1. $\mathbf{x}_p = \begin{pmatrix} -3 \\ te^{3t} + 2 \end{pmatrix}$ **3.** $\mathbf{x}_p = \begin{pmatrix} -\frac{1}{15}e^{3t} + \frac{5}{12}e^{t} \\ \frac{2}{15}e^{3t} + \frac{1}{12}e^{t} \end{pmatrix}$

5. $\mathbf{x}_p = \begin{pmatrix} -1-t \\ 1+2t \end{pmatrix}$ **7.** $\mathbf{x}_p = \begin{pmatrix} t \\ 2 \\ -1 \end{pmatrix} e^{2t}$

9. $\mathbf{x}_p = \begin{pmatrix} e^{t}\int_0^t e^{-s-s^2}\,ds \\ 2e^{t}\int_0^t (t-s)e^{-s-s^2}\,ds \end{pmatrix}$

11. **(a)** $\lambda = 2i$, $\mathbf{v} = \begin{pmatrix} 1 \\ -2i \end{pmatrix}$ and $\lambda = -2i$, $\mathbf{v} = \begin{pmatrix} 1 \\ 2i \end{pmatrix}$

(b) $\mathbf{x}_p = \begin{pmatrix} 0 \\ 4 \end{pmatrix} e^{2t}$ **13.** **(a)** $z = \ln t$ changes the problem to $\dfrac{d\mathbf{x}}{dz} = \mathbf{Bx} + \mathbf{g}(e^{z})$

(b) $\mathbf{x} = \begin{pmatrix} 3t^3 + 3c_1t^2 + c_2t^{-2} \\ 2t^3 + c_1t^2 - c_2t^{-2} \end{pmatrix}$

15. **(a)** $\mathbf{x} = c_1\begin{pmatrix} 1 \\ 1 \end{pmatrix} e^{t} + c_2\begin{pmatrix} -2t \\ 1-2t \end{pmatrix} e^{t}$
$$+ \begin{pmatrix} t + 2t^2 \\ -t + 2t^2 \end{pmatrix} e^{t}$$

(b) $\mathbf{y} = (\mathbf{b}t + c\mathbf{v}t^2)e^{at}$, where $\mathbf{v}$ is an eigenvector

(c) $\mathbf{x} = c_1\begin{pmatrix} 1 \\ 2 \end{pmatrix} e^{-2t} + c_2\begin{pmatrix} t \\ 1+2t \end{pmatrix} e^{-2t}$
$$+ \begin{pmatrix} t - t^2 \\ -2t^2 \end{pmatrix} e^{-2t}$$

Answers to Section A.7

1. $u = \displaystyle\sum_{n=1}^{\infty} b_n e^{-n^2\pi^2 kt} \sin n\pi x,$

$b_n = 2\displaystyle\int_0^1 \phi(x) \sin n\pi x\, dx$

3. **(a)** $u = 2\displaystyle\sum_{n=1}^{\infty} \frac{\sin hn\pi}{hn\pi} \sin\frac{n\pi}{2}\, e^{-n^2\pi^2 kt} \sin n\pi x$

(b) $u = 2\displaystyle\sum_{n=1}^{\infty} \sin\frac{n\pi}{2}\, e^{-n^2\pi^2 kt} \sin n\pi x$

(d)

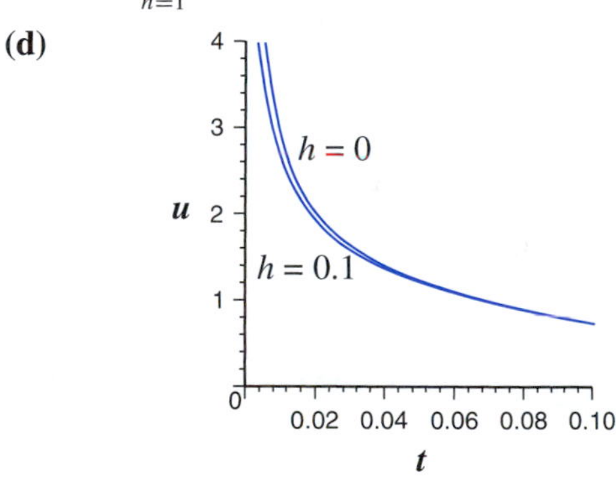

The $h = 0$ solution is unbounded as $t \to 0$.

5. $T = 100 - 80x - \dfrac{160}{\pi}\displaystyle\sum_{n=1}^{\infty} \frac{1}{n} e^{-n^2\pi^2 kt} \sin n\pi x$

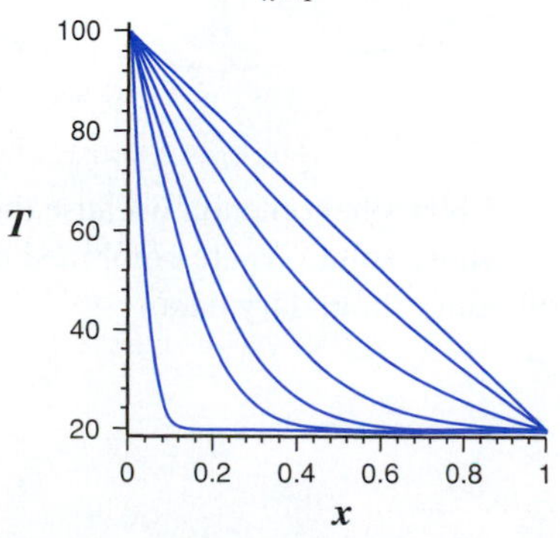

7. $u = \bar{f} + \sum_{n=1}^{\infty} a_n e^{-n^2\pi^2 t} \cos n\pi x,$

$\bar{f} = \int_0^1 \phi(x)\,dx, \quad a_n = 2\int_0^1 \phi(x)\cos n\pi x\,dx$

9. (a) $f'' + \lambda^2 f = 0, \quad f(0) = 0,$

$f'(1) + f(1) = 0,$ and $g' + \lambda^2 g = 0$

(c) $u = \sum_{n=1}^{\infty} b_n e^{-\lambda_n^2 t} \sin \lambda_n x$

(e) $b_n = \dfrac{2}{1+\cos^2\lambda_n} \int_0^1 \phi(x) \sin \lambda_n x\,dx$

11. $u = \bar{f} + \sum_{n=1}^{\infty} a_n e^{-n^2 t} \cos nx + \sum_{n=1}^{\infty} b_n e^{-n^2 t} \sin nx,$

where

$\bar{f} = \dfrac{1}{2\pi}\int_0^{2\pi} \phi(x)\,dx, \quad a_n = \dfrac{1}{\pi}\int_0^{2\pi}\phi(x)\cos nx\,dx,$

$b_n = \dfrac{1}{\pi}\int_0^{2\pi}\phi(x)\sin nx\,dx$

Answers to Section A.8

1. (a)

$$\frac{2}{\pi}\sum_{n=1}^{\infty}\frac{(-1)^{n+1}}{n\sinh n\pi}(\sinh n\pi y \sin n\pi x + \sinh n\pi x \sin n\pi y)$$

(b)

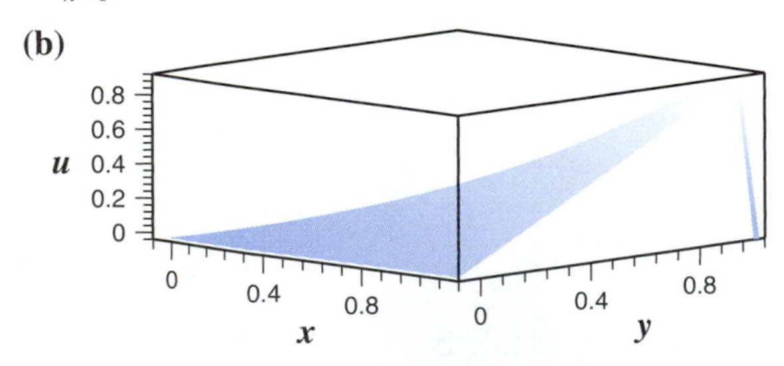

One sees the Gibbs' phenomenon because the two subproblems whose solutions are combined in part *a* do not have continuous boundary data.

(c)

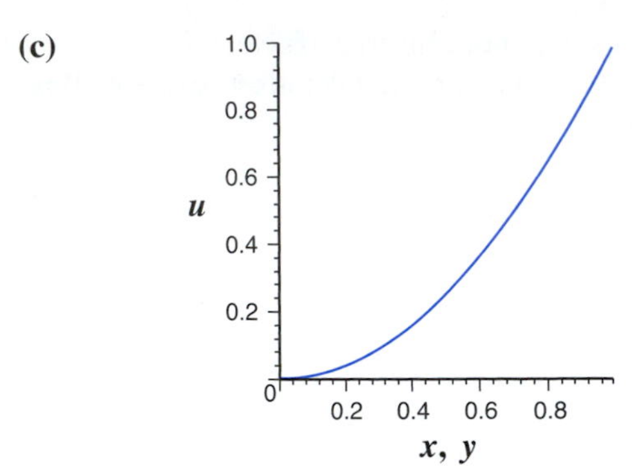

3. $u = r\sin\theta$

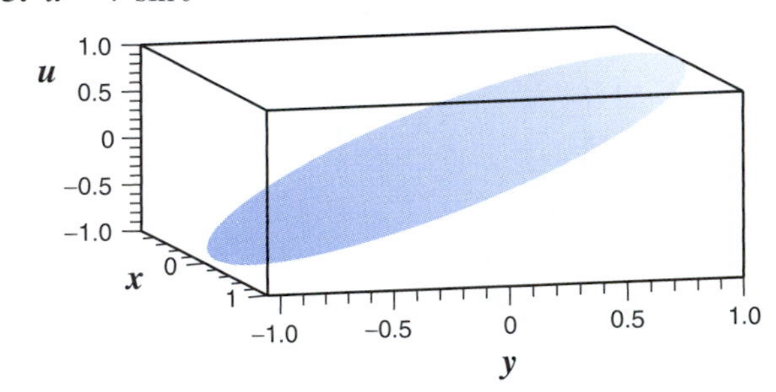

5. (a)

$$u = \frac{16}{\pi^3}\sum_{n=1}^{\infty}\frac{1-(-1)^n}{n^3\cosh(n\pi/2)}\cosh\frac{n\pi(1-y)}{2}\sin\frac{n\pi x}{2}$$

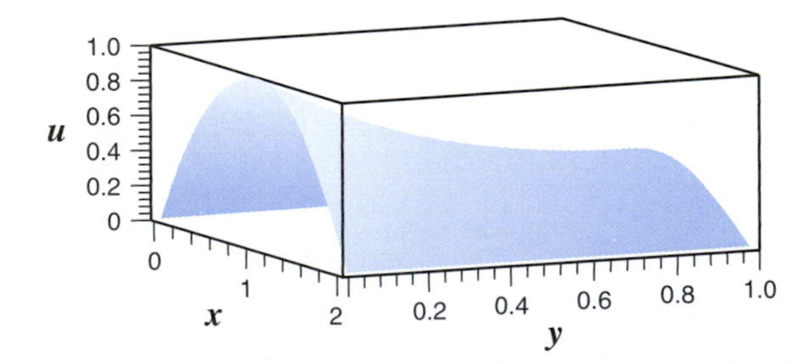

(b)

$$u = \frac{16}{\pi^3}\sum_{n=1}^{\infty}\frac{1-(-1)^n}{n^3\sinh n\pi}\left[\sinh\frac{n\pi y}{2} + \sinh\frac{n\pi(2-y)}{2}\right]$$

$\sin\dfrac{n\pi x}{2}$ **(c)** The solutions are different. The solution of part *b* is slightly preferable; it converges a little faster because of the factor $\sinh n\pi$ rather than $\cosh n\pi/2$ in the denominator. This is a minor point for this problem, since the first three terms of the solution of part *a* give an approximation accurate to within 1%.

11. **(a)** $r^2 w_{rr} + r w_r + w_{\theta\theta} = r, \quad w(1, \theta) = 0,$ $w(r, 0) = w(r, \pi) = 0$ **(b)** $f_n(\theta) = \sin n\theta$

(c) $\sum_{n=1}^{\infty} \left(r^2 g_n'' + r g_n' - n^2 g_n\right) \sin n\theta = r$

(d) $r^2 g_n'' + r g_n' - n^2 g_n = \dfrac{2[1 - (-1)^n]}{n\pi} r,$
$g_n(0) < \infty, \quad g_n(1) = 0$

(e) $u = 1 - r + \dfrac{2}{\pi} r \ln r \sin\theta - \dfrac{2r}{\pi} \sum_{n=3}^{\infty} \dfrac{1 - (-1)^n}{n^3 - n}$
$(1 - r^{n-1}) \sin n\theta$

(f)

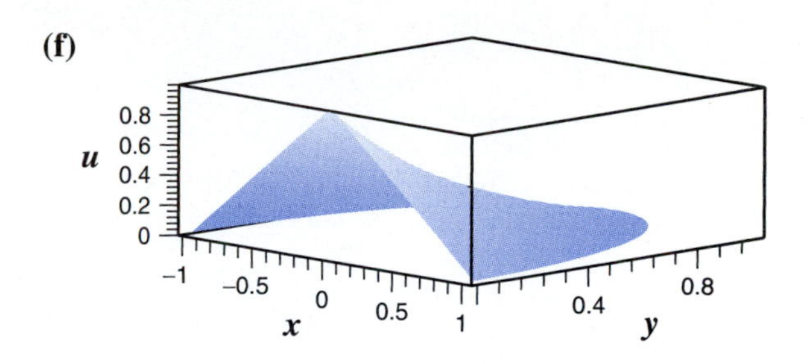

Index

A

Abel's theorem, 157
Absolute rate of change, 2
Advection, 486
Aging spring, 195–196
Air resistance, 55, 121
Airy equation, 92, 271, 571, 581
Airy functions, 92–93, 571, 581
Algebraic multiplicity, 354, 393
Amplitude, 131, 244
Angular velocity, 311
Asymptotic approximation, 60
Asymptotically stable, 293, 315, 383, 385, 404
Augmented matrix, 140, 584
Automobile ignition, 217
Autonomous equations, 16, 80, 81, 281
Average decay time, 418

B

Bacteria, 311
Basic reproductive number, 51
Basketball, 39, 124
Bead on a hoop, 312
Beam deflection, 273
Beats, 251–252
Becquerel, Henri, 9
Bernoulli, Daniel, 288
Bernoulli equation, 258
Bernoulli's law, 52
Bessel equations, 94, 193, 198, 575, 581
Bessel functions, 193, 537
Binomial theorem, 581
Boundary conditions, 273, 481, 500
Boundary value problem, 503
Brachistochrone problem, 28
Braun, M., 40
Bubble growth, 319, 327

C

Cannonballs, 124
Capital growth, 10, 13
Carbon-14, 45–46
Carbon monoxide, 62–63
Cauchy-Euler equations, 192–193, 238
Center, 383
Characteristic equation, 164, 167, 351
Characteristic polynomial, 164, 167, 351
Characteristic value, 164, 167, 184, 230
Chebyshev's equation, 581
Chemical kinetics, 55–57, 85, 115
Chemical mixing, 61–63
Chemical reactor, 296
Cigarette smoke, 62–63
Circular frequency, 130
Cofactor expansion, 145
Combat models, 309
Competing species models, 286, 317, 324, 325, 328, 408
Complementary error function, 91
Complementary solution, 215
Complete elliptic integral, 137
Complete set (of orthogonal functions), 522, 525
Complex conjugate, 379
Complex eigenvalues, 362, 379, 383
Complex exponential, 170–171, 243–244
Components of mathematical models, 37
Conceptual model, 32, 35, 56, 411
Constructive proof, 551
Convection-diffusion equation, 487
Convolution, 272, 463
Convolution theorem, 464
Cooling, 2, 7–8
Cooperating species models, 286, 324, 326
Coupled systems
 of algebraic equations, 140
 of ordinary differential equations, 122, 283
Critical points. *See* Equilibrium solutions
Critically-damped linear oscillators, 182–183

D

D'Alembert's formula, 494
D'Alembert's method, 179
Damping forces, 53, 311
Dating old paint, 40–44
Decay, 4, 8, 9, 80, 470
Decoupled systems
 of algebraic equations, 140
 of ordinary differential equations, 283
Deficient matrices, 354, 389, 390
Degree (of generalized exponential functions), 230
Depreciation, 10
Determinant, 143–146
Diagonalization, 598–599
Differential equation for trajectories, 304, 331
Differential operator, 150, 152
Digestion, 200–204
Dimension, 16, 346
Dimensionless variable, 49
Dirac delta function, 454
Direction field, 313
Dirichlet conditions, 500, 605
Dirichlet problem, 614
Disease, 47, 63, 287, 310, 415
Drag forces, 54, 55, 121
Drugs, 63, 236, 296, 337
Dubois-Reymond lemma, 605

E

Eigenfunction, 504
Eigenfunction expansion, 622
Eigenspace, 345, 348, 350
Eigenvalue, 345, 348, 350
 algebraic multiplicity, 354
 complex, 379
 geometric multiplicity, 354
Eigenvalue problem, 348, 503
Eigenvector, 348, 379
Electric circuit, 210, 212, 216, 241, 274, 388, 424, 448
Elementary functions, 86
Elementary row operations, 141
Endemic disease, 50, 58, 63, 85, 415
Envelope (of a solution), 174–175

Equilibrium solutions, 7, 17, 292, 315, 383, 385, 404
- asymptotically stable, 293, 315, 385, 404
- asymptotically stable spirals, 383
- centers, 383
- saddle points, 316
- semistable, 296
- sinks, 373
- sources, 373
- stable, 17, 316
- unstable, 17, 293, 316
- unstable spirals, 383

Equivalent systems of algebraic equations, 140
erf, 29, 91, 546
erfc, 91
Error, 102, 106, 560
Error functions, 29, 91, 546
Escape velocity, 305
Euler approximation, 101
Euler, Leonhard, 97
Euler's formula, 171
Euler's method, 97–98, 103, 111, 562
Evaporating raindrops, 23
Even extension, 526
Even function, 515, 523
Existence and uniqueness, 85, 88, 90, 367, 548
Explicit solution, 67, 70
Exponential forcing, 221
Exponential order, 432

F

Family of functions, 18
Family of solutions, 18, 19, 27, 154
Fecundity, 470
First integral, 132, 136
First-order differential equations
- existence of unique solutions, 88
- linear, 74, 90, 253, 545
- separable, 64, 68

Fisher's equation, 489
Forced oscillation, 213, 241, 245
Forces, 53–54
Fort Sumter, 124
Fourier coefficients, 524, 608
Fourier series, 514, 524
Fourier sine coefficients, 514, 524
Frequency, 131
Fundamental frequency, 505, 532
Fundamental matrix, 367, 376, 587, 590
Fundamental set, 367, 587
Fundamental theorem of calculus, 19

G

Gain, 275
General solution, 69, 150
Generalized eigenvector, 393
Generalized exponential functions, 229–231
Generalized functions, 454
Geometric multiplicity, 354, 371, 393
Gibbs' phenomenon, 527
Global truncation error, 562, 564
Gompertz model, 296
Gravitational force, 39, 53
Green's function for the initial-value problem, 461
Guitars, 505

H

Half-life, 6, 10
Harmonic functions, 619
Hawking, Stephen, 31
Heat equation, 96, 481, 485, 498, 506, 518, 603–605, 613
Heat flow, 189, 506, 603
Heaviside function, 425
Holling functional response, 409
Homogeneous
- linear differential equations, 154, 159, 167, 185–186
- linear systems, 366, 368

Hooke's law, 54
Hunting, 297
Hydrocodone bitartrate, 13
Hydrogen peroxide decomposition, 13

I

Identity matrix, 139
Ill-conditioned, 106, 566
Image (of a differential operator), 150, 231
Implicit solution, 67, 71
Impulse, 452
Impulse response function, 461
Impulsive forcing, 448, 451
Indicial equation, 579
Initial condition, 22, 155
Initial-boundary-value problem, 500, 606
Initial-value problem, 5, 22, 23, 27, 367, 439–440, 490, 493
Inner product, 521
Installment loans, 14
Insulated ring, 612
Integral curve, 76
Integral equation, 549–550
Integral transform, 430
Integrating factor, 543–545
Integration by substitution, 66
Interacting populations, 283, 324–326, 408–409
Interval of existence, 26, 27, 93
Inverse, 584, 586
Inverse Laplace transform, 441
Inverse square law, 39
Investment, 10

J

Jacobian, 403
Jellyfish digestion, 200

K

Keisch, Bernard, 40
Korteweg-de Vries equation, 488

L

Lakes Erie and Superior, 61, 62
Lanchester model, 309
Laplace transform, 440
- periodic functions, 436
- piecewise-defined functions, 434

Laplace's equation, 481, 614
Laplacian, 534, 613
Lead-210, 41
Lead poisoning, 342
Leaking bucket model, 51–52, 58, 94
Legendre equation, 94, 180, 187, 574
Legendre polynomials, 94, 187
Leslie matrix, 355
Libby, Willard, 40
Limit cycle, 405
Linear approximation, 400
Linear combination, 151, 154
Linear dependence, 159
Linear differential equations, 74, 90, 153, 186, 545
- Airy equation, 92, 271, 571
- Bessel equation, 94, 193–197, 198, 575
- Cauchy-Euler equations, 192–193, 238
- Chebyshev equation, 581
- existence of unique solutions, 90, 99
- homogeneous, 154, 159–160, 167, 186
- nonhomogeneous, 238, 448, 460

Linear differential operator, 152, 225, 228
Linear independence, 159
Linear oscillator, 126, 129, 165, 174, 182, 208–216, 387, 422
- critically damped, 182–183
- damped, 129, 241–245
- forced, 129, 213–216, 240–250
- overdamped, 165–167
- underdamped, 174–177, 245–250

Linear span, 151
Linear systems, 403
- of algebraic equations, 139
- of ordinary differential equations, 366, 368, 592

Linear trajectory, 357
Linearization, 403
Lipschitz condition, 557
Loans, 14
Local truncation error, 561, 563
Logistic growth, 280, 295, 297
Longtime behavior, 82, 290, 317, 372
Lorenz system, 409
Lotka-Volterra model, 281–282, 309, 316, 324, 325, 327
Lottery model, 14

M

Malthus, Thomas, 280
Mathematical modeling, 31, 38
Mathematical models
- aging spring, 195–196
- automobile ignition, 217
- bead on a hoop, 312
- beam deflection, 273
- bubble growth, 319, 327
- chemical kinetics, 55–57, 85, 115
- chemical mixing, 61–63
- chemical reactor, 296
- combat, 309
- competing species, 286, 317, 324, 325, 328, 408
- cooling, 2, 7–8
- cooperating species, 286, 324, 326
- dating old paint, 40–44
- decay, 4, 8, 9, 80, 470
- digestion, 200–204
- disease, 47, 63, 287, 310, 415
- drugs, 63, 236, 296, 337
- electric circuit, 210, 212, 216, 241, 274, 388, 424, 448
- endemic disease, 50, 58, 63, 85, 415
- Fort Sumter, 124
- Gompertz, 296
- heat flow, 189, 506, 603
- Holling functional response, 409
- hunting, 297
- hydrocodone bitartrate, 13
- hydrogen peroxide decomposition, 13
- installment loans, 14
- insulated ring, 612
- interacting populations, 283, 324–326, 408–409
- investment and depreciation, 10
- jellyfish digestion, 200
- Lanchester combat model, 309
- lead poisoning, 342
- leaking bucket, 51–52, 58, 94
- logistic growth, 280, 295, 297
- Lotka-Volterra, 281–282, 309, 316, 324, 325, 327
- lottery, 14
- motion of a projectile, 54, 121, 301
- natural decay, 2, 4, 7–9, 470
- natural growth, 2, 11, 280
- Newton's law of cooling, 7, 11, 569
- nonlinear spring, 195–196, 398
- one-dimensional heat equation, 96, 481, 485, 498, 506, 603–605
- one-dimensional wave equation, 480
- overdamped oscillator, 165–167
- pendulum, 58–59, 136 , 310
- phantom rain, 23
- pharmacokinetics, 63, 263, 336
- photographic film, 13
- pianos, 251–252, 533
- plucked string, 500, 508
- pollution, 61–63
- population growth, 14
- predator-prey, 283, 286, 316, 325, 327, 328, 408–409
- projectile motion, 54, 121, 301
- radiative cooling, 297
- radioactive decay, 9, 46
- radiocarbon dating, 45–46
- RC series circuit, 210
- raindrops, 23
- RL series circuit, 216
- RLC series circuit, 212, 241, 274, 424, 448
- rocket flight, 60
- rumors, 58, 287
- Schaefer model, 297
- Shroud of Turin, 46
- silver bromide, 13
- skydiving, 311
- smallpox, 288
- spring-mass system, 126, 169, 178, 195–196, 198, 387, 455
  - critically damped, 182–183
  - damped, 129
  - forced, 129, 240, 602
  - overdamped, 165–167
  - underdamped, 174–177
- state lotteries, 14
- Stonehenge, 46
- stringed instruments, 505, 531
- struck membrane, 534
- struck string, 506, 518, 531
- tennis, 124
- thrown ball, 32, 35, 40
- Tomb of Sneferu, 46
- Trafalgar, 309
- tuning a piano, 251–252
- tuning circuit for a radio, 274
- undamped oscillator, 129–135, 272, 422
- underdamped oscillator, 174–177
- United States population, 14
- uranium decay, 9, 10, 41
- vapor bubbles, 319, 327
- vibrating circular membrane, 507, 529, 534
- vibrating string, 478–480
- volleyball serve, 118
- waste treatment, 288, 311
- water drainage, 51
- water flow, 208–209

Mathematical resonance, 247
Matrix
- augmented, 140, 584
- deficient, 350, 389, 390
- identity, 139
- inverse, 584, 586
- nonsingular, 142, 584, 586
- singular, 143
- square, 139
- symmetric, 365
- upper triangular, 148

Matrix differential equation, 587
Matrix exponential, 588
Matrix function, 583, 586
Matrix multiplication, 583
Matrix product, 139
Maximum principle, 620
Mean decay time, 418
Mean value property, 619
Minitangent, 77
Modified Euler method, 112–113, 568
Motion of a projectile, 54, 121, 301
Multiplicity, 184
Mutually orthogonal, 73, 521

N

n-parameter family, 18
Nth partial sum, 513
Natural decay, 2, 4, 7–9, 470
Natural growth, 2, 11, 280
Neumann conditions, 605
Newton, Isaac, 7
Newton's law of cooling, 7, 11, 569
Newton's second law of motion, 34, 53
Nondimensionalization, 49, 284–286, 302–303, 414
Nonlinear spring, 195–196, 398
Nonlinear system of ordinary differential equations, 398
Nonsingular matrix, 142, 584, 586
Nuclear weapon tests, 46
Null isocline, 80
Nullcline, 80, 320
Nullspace, 348

O

Odd, 515, 523
Odd extension, 526
One-dimensional heat equation, 96, 481, 485, 498, 506, 603–605
One-dimensional wave equation, 480
Order
  of a differential equation, 16
  of a numerical method, 113, 565
Ordinary point, 576
Orthogonal, 521
Orthogonal expansion, 522
Orthogonal trajectories, 73–74
Overdamped oscillator, 165–167

P

Parameters, 37, 49
Parametric Bessel equation, 198, 536
Partial differential equation, 16, 480
Partial sum, 513
Particular solution, 215, 228
Pendulum, 58–59, 136 , 310
Period, 130
Periodic extension, 525
Periodic forcing, 448
Periodic function, 436, 524
Periodicity condition, 202
Phantom rain, 23
Pharmacokinetics, 63, 263, 336
Phase line, 294
Phase plane, 299
Phase portraits, 303
Phase shift, 132–134
Photographic film, 13
Physical laws, 33
Physical resonance, 247
Pianos, 251–252, 533
Picard iterate, 552
Picard iteration, 551
Picard sequence, 552
Piecewise-continuous forcing, 262, 448
Piecewise-continuous function, 262, 424–425, 526
Piecewise-defined function, 434, 450
Plucked string, 500, 508
Pollution, 61–63
Population growth, 14
Power series, 552, 576
Predator-prey models, 283, 286, 316, 325, 327, 328, 408–409
Principle of superposition, 154, 367, 499
Projectile motion, 54, 121, 301

Q

Qualitative analysis, 319
Quasi-periodic, 175–176

R

Radially-symmetric wave equation, 487
Radiative cooling, 297
Radioactive decay, 9, 46
Radiocarbon dating, 45–46
Radium-226, 41
RC series circuit, 210
Recurrence relation, 573
Reduction of order, 179
Regular singular point, 578
Relative amplitude, 532
Relative rate of change, 2
Resonance, 246–250
Restoring force, 54
Raindrops, 23
Repeated characteristic values, 179–181, 183–184
Richardson extrapolation, 569
RL series circuit, 216
RLC series circuit, 212, 241, 274, 424, 448
Robin conditions, 606
Rocket flight, 60
Round-off error, 560
Routh-Hurwitz criteria, 407
Row reduction, 141–142
Rumors, 58, 287
Runge-Kutta methods, 114, 508

S

s domain, 431
Saddle point, 316, 372
Scalar, 345
Scale, 49
Schaefer model, 297
Semistable equilibrium solution, 296
Separable equation, 64, 68
Separated form, 64, 68
Separation of variables, 502, 614
Separatrix, 317, 319
Shifted-transform formula, 445
Shroud of Turin, 46
Sifting property, 454
Signaling problem, 481, 484
Signum function, 54
Silver bromide, 13
Simple harmonic motion, 130
Singular matrix, 143
Singular point, 576
Singular solution, 69
Singular systems, 374
Sink, 373
SIR model, 288, 310, 411
SIS model, 287
Skydiving, 311
Slope field, 76–78, 97
Smallpox, 288
Soliton, 488
Solution, 4, 5, 17, 69, 150
Solution curve, 76, 78, 88
Solution space, 346
Sound wave, 486, 488
Source, 373
Span, 151
Special functions, 86, 91
  Airy functions, 92–93, 571, 581
  Bessel functions, 94, 193
  complementary error function, 91
  complete elliptic integral, 137
  error function, 29, 91, 96, 546
  Legendre polynomials, 94
  parabolic cylinder function, 581
Special substitution techniques, 74–75
Spherically-symmetric wave equation, 486, 488

Spiral equilibrium solution, 383
Spring-mass system, 126, 169, 178, 195–196, 198, 387, 455
critically damped, 182–183
damped, 129
forced, 129, 240, 602
overdamped, 165–167
underdamped, 174–177
Square matrix, 139
Square wave, 422, 456
Stability, 292–293
Stable equilibrium solution, 17, 316, 385
Standing wave, 501
State lotteries, 14
State (of a system), 281
Steady-state, 189, 216
Stefan's law, 297
Stonehenge, 46
Stringed instruments, 505, 531
Strontium-90, 12
Struck membrane, 534
Struck string, 506, 518, 531
Substitution, 66
Subspace, 355
Superposition principle, 154, 367, 499
Systems
of linear algebraic equations, 138
coupled, 140
decoupled, 140
equivalent, 140
of ordinary differential equations, 16
coupled, 283
decoupled, 283
linear, 368, 592
nonhomogeneous, 592
nonlinear, 398
singular, 374

T

Taylor series, 552
Taylor's theorem, 561
Tennis, 124
Three-dimensional heat equation, 613
Thrown ball, 32, 35, 40
Time series graph, 305
Tomb of Sneferu, 46
Trafalgar, 309
Trace, 396
Trajectory, 299, 303, 304
Transfer function, 461
Transient, 216
Transverse vibration, 478
Trapezoidal method, 111, 568
Traveling wave, 485
Trial solution, 221, 227, 233
Triangular matrix, 148
Triangular wave, 458, 468
Truncation error, 561–563
Tuning a piano, 251–252
Tuning circuit for a radio, 274

U

Uncoupled system
of algebraic equations, 140
of differential equations, 283
Undamped oscillator, 129–135, 272, 422
Underdamped oscillator, 174–177
Undetermined coefficients, 221, 227, 423, 593, 595
Uniqueness, 17, 25, 85, 88, 90, 367, 491
Unit impulse function, 453
Unit step function, 425
United States population, 14
Unstable equilibrium solution, 17, 293, 316, 383, 385, 404
Upper triangular matrix, 148
Uranium decay, 9, 10, 41

V

Van der Pol equation, 404
Van Meegeren art forgery, 41
Vapor bubbles, 319, 327
Variation of parameters, 253, 265, 596
Vector, 139, 346
Vector field, 314
Vector space, 345–346
Verhulst, P. F., 280
Vibrating circular membrane, 507, 529, 534
Vibrating string, 478–480
Vibration mode, 501, 505
Volleyball serve, 118
Volterra integral equation, 550

W

Waste treatment, 288, 311
Water drainage, 51
Water flow, 208–209
Wave equation, 480, 484, 490, 493
Wave profile, 483
Waveform, 485, 501, 503
Well posed, 620
Well-conditioned, 566
Wronskian, 157, 169, 268

Z

Zero isocline, 80